T0212882

Lecture Notes in Artificial Intelligence   9457

Subseries of Lecture Notes in Computer Science

LNAI Series Editors

Randy Goebel
  University of Alberta, Edmonton, Canada
Yuzuru Tanaka
  Hokkaido University, Sapporo, Japan
Wolfgang Wahlster
  DFKI and Saarland University, Saarbrücken, Germany

LNAI Founding Series Editor

Joerg Siekmann
  DFKI and Saarland University, Saarbrücken, Germany

More information about this series at http://www.springer.com/series/1244

Bernhard Pfahringer · Jochen Renz (Eds.)

AI 2015: Advances in Artificial Intelligence

28th Australasian Joint Conference
Canberra, ACT, Australia,
November 30 – December 4, 2015
Proceedings

 Springer

Editors
Bernhard Pfahringer
The University of Waikato
Hamilton
New Zealand

Jochen Renz
The Australian National University
Canberra, ACT
Australia

ISSN 0302-9743   ISSN 1611-3349 (electronic)
Lecture Notes in Artificial Intelligence
ISBN 978-3-319-26349-6   ISBN 978-3-319-26350-2 (eBook)
DOI 10.1007/978-3-319-26350-2

Library of Congress Control Number: 2015953780

LNCS Sublibrary: SL7 – Artificial Intelligence

Springer Cham Heidelberg New York Dordrecht London
© Springer International Publishing Switzerland 2015
This work is subject to copyright. All rights are reserved by the Publisher, whether the whole or part of the material is concerned, specifically the rights of translation, reprinting, reuse of illustrations, recitation, broadcasting, reproduction on microfilms or in any other physical way, and transmission or information storage and retrieval, electronic adaptation, computer software, or by similar or dissimilar methodology now known or hereafter developed.
The use of general descriptive names, registered names, trademarks, service marks, etc. in this publication does not imply, even in the absence of a specific statement, that such names are exempt from the relevant protective laws and regulations and therefore free for general use.
The publisher, the authors and the editors are safe to assume that the advice and information in this book are believed to be true and accurate at the date of publication. Neither the publisher nor the authors or the editors give a warranty, express or implied, with respect to the material contained herein or for any errors or omissions that may have been made.

Printed on acid-free paper

Springer International Publishing AG Switzerland is part of Springer Science+Business Media
(www.springer.com)

Preface

This volume contains the papers presented at the 28th Australasian Joint Conference on Artificial Intelligence 2015 (AI 2015), which was held in Canberra, Australia, November 30 to December 4, 2015. The conference is the premier event for artificial intelligence in Australasia and provides a forum for researchers and practitioners across all subfields of artificial intelligence to meet and discuss recent advances. AI 2015 took place at the QT Canberra hotel in the heart of Canberra, Australia's capital. The venue provided a stimulating environment for the discussion of a broad range of aspects of artificial intelligence. This year we were co-located with the Logic Summer School and also with ACRA 2015, the Australasian Conference on Robotics and Automation, with which we shared a joint session, a joint workshop, and a social event.

The technical program of AI 2015 comprised a number of high-quality papers that were selected in a thorough, double-blind reviewing process with at least three expert reviews per paper. Out of 102 submissions, our senior Program Committee with the help of an experienced international Program Committee selected 39 long papers and 18 short papers for presentation at the conference and inclusion in these proceedings. Papers were submitted by authors from 21 countries from five continents, demonstrating the broad international appeal of our conference. In addition to the 57 paper presentations, we had three keynote talks by high-profile speakers:

- Wolfram Burgard, Albert Ludwigs University Freiburg, Germany
- Kate Smith-Miles, Monash University, Australia
- Toby Walsh, NICTA and UNSW Sydney, Australia

AI 2015 also featured an exciting selection of workshops and tutorials that were free for all conference participants to attend. The two workshops were:

- AI-15 Workshop on Deep Learning and its Applications in Vision and Robotics, organised by Juxi Leitner (QUT), Anoop Cherian (ANU), and Sareh Shirazi (QUT)
- KR Conventicle 2015 in the memory of Norman Foo (1943 – 2015)

The four tutorials were on:

- Deep Learning by Lizhen Qu
- Fundamentals of Computational Social Choice, by Haris Aziz and Nicholas Mattei
- Mathematical Optimization in Supply Chain Logistics, by Thomas Kalinowski
- A Data Analytics View of Genomics, by Cheng Soon Ong

The conference was complemented by a student symposium as well as a social program that included a conference dinner at the National Museum of Australia.

A large number of people and organizations helped make AI 2015 a success. First and foremost, we would like to thank the authors for contributing and presenting their latest work at the conference. Without their contribution this conference would not have been possible. The same is true for the members of the conference organization.

Our special thanks go to our general chairs, Michael Maher and Sylvie Thiebaux, the local organization chairs, Kathryn Merrick and Sameer Alam, the other members of the conference organization, George Leu, Stephen Gould, Nina Narodytska, Carleton Coffrin, and Jiangjun Tang, as well as our student volunteers. We also thank the 18 members of our senior Program Committee, the members of our Program Committee, as well as additional reviewers who were all very dedicated and timely in their contributions to selecting the best papers for presentation at AI 2015.

We are grateful for support and sponsorship by the *Artificial Intelligence* journal, the College of Engineering and Computer Science of the Australian National University, UNSW Canberra, NICTA, AAAI, the Australian Computer Society, to Appazure (who built a free conference app for us to be used during the conference), and also the free conference management system EasyChair, which was used for putting together this volume. Last but not the least, we thank Springer for their sponsorship and their support in preparing and publishing this volume in the *Lecture Notes in Computer Science Series*.

September 2015   Bernhard Pfahringer
                                                                                          Jochen Renz

Organization

Conference Chairs

Michael Maher   UNSW Canberra, Australia
Sylvie Thiebaux   The Australian National University and NICTA, Australia

Program Chairs

Bernhard Pfahringer   University of Waikato, New Zealand
Jochen Renz   The Australian National University, Australia

Local Organizing Chairs

Kathryn Merrick   UNSW Canberra, Australia
Sameer Alam   UNSW Canberra, Australia

Finance and Sponsorship Chair

George Leu   UNSW Canberra, Australia

Workshop Chair

Stephen Gould   The Australian National University, Australia

Tutorial Chair

Nina Narodytska   University of Toronto, Canada

Student Symposium Chair

Carleton Coffrin   NICTA, Australia

Publicity Chair and Webmaster

Jiangjun Tang   UNSW Canberra, Australia

Senior Program Committee

Abdul Sattar   Griffith University, Australia
Abhaya Nayak   Macquarie University, Australia
Ajit Narayanan   Auckland University of Technology, New Zealand

Cecile Paris	CSIRO, Australia
Chengqi Zhang	University of Technology Sydney, Australia
Claude Sammut	The University of New South Wales, Australia
Fatih Porikli	The Australian National University and NICTA, Australia
Hussein Abbass	UNSW Canberra, Australia
Ian Watson	The University of Auckland, New Zealand
Marcus Hutter	The Australian National University, Australia
Mengjie Zhang	Victoria University of Wellington, New Zealand
Michael Thielscher	The University of New South Wales, Australia
Patrik Haslum	The Australian National University, Australia
Stephen Cranefield	University of Otago, New Zealand
Svetha Venkatesh	Deakin University, Australia
Tanja Mitrovic	University of Canterbury, New Zealand
Timothy Baldwin	The University of Melbourne, Australia
Wai Yeap	Auckland University of Technology, New Zealand

Program Committee

Harith Al-Sahaf	Victoria University of Wellington, New Zealand
David Albrecht	Monash University, Australia
Sagaya Amalathas	UNITAR International University, Malaysia
Quan Bai	Auckland University of Technology, New Zealand
Michael Bain	UNSW, Australia
Mike Barley	University of Auckland, New Zealand
Edwin Bonilla	The University of New South Wales, Australia
Richard Booth	University of Luxembourg, Luxembourg
Adi Botea	IBM Research, Ireland
Wray Buntine	Monash University, Australia
Mark Carman	Monash University, Australia
Gang Chen	Victoria University of Wellington, New Zealand
Ling Chen	University of Technology, Sydney, Australia
Vic Ciesielski	RMIT University, Australia
Gokberg Cinbis	Milsoft, Turkey
Nathalie Colineau	DSTO, Australia
Amélie Cordier	LIRIS, France
Mayank Daswani	Australian National University, Australia
James Delgrande	Simon Fraser University, Canada
Jeremiah D. Deng	University of Otago, New Zealand
Grant Dick	University of Otago, New Zealand
Minh Do	NASA Ames Research Center, USA
Alan Dorin	Monash University, Australia
David Dowe	Monash University, Australia
Lan Du	Macquarie University, Australia
Masoud Faraki	ANU-NICTA, Australia
Xiaoying Gao	Victoria University of Wellington, New Zealand
Xiaoyu Ge	Australian National University, Australia

Tom Gedeon	Australian National University, Australia
Stephen Gould	Australian National University, Australia
Garrison Greenwood	Portland State University, USA
Peter Gregory	Teesside University, UK
Hans W. Guesgen	Massey University, New Zealand
Christian Guttmann	IVBAR, Sweden
Ben Hachey	University of Sydney, Australia
James Harland	RMIT University, Australia
Bernhard Hengst	UNSW, Australia
Jose Hernandez-Orallo	Universitat Politecnica de Valencia, Spain
Geoffrey Holmes	University of Waikato, New Zealand
Hisao Ishibuchi	Osaka Prefecture University, Japan
Yaochu Jin	University of Surrey, UK
Bourhane Kadmiry	C.I., New Zealand
Sarvnaz Karimi	CSIRO, Australia
Sankalp Khanna	Griffith University, Australia
Michael Kirley	The University of Melbourne, Australia
Frank Klawonn	Ostfalia University of Applied Sciences, Germany
Alistair Knott	University of Otago, New Zealand
Yun Sing Koh	University of Auckland, New Zealand
Willem Labuschagne	University of Otago, New Zealand
Gerhard Lakemeyer	RWTH Aachen University, Germany
Jérôme Lang	LAMSADE, France
Tor Lattimore	University of Alberta, Canada
Jae-Hee Lee	The Australian National University, Australia
Gang Li	Deakin University, Australia
Jason Li	The Australian National University, Australia
Sanjiang Li	University of Technology, Sydney, Australia
C.P. Lim	Deakin University, Australia
Carlos Linares Lopez	Universidad Carlos III de Madrid, Spain
Nir Lipovetzky	University of Melbourne, Australia
Jiamou Liu	Auckland University of Technology, New Zealand
Jing Liu	Xidian University, China
Guodong Long	UTS, Australia
Hui Ma	Victoria University of Wellington, New Zealand
Daniele Magazzeni	King's College London, UK
Stephen Marsland	Massey University, New Zealand
Moffat Mathews	University of Canterbury, New Zealand
Robert Mattmüller	University of Freiburg, Germany
Michael Mayo	University of Waikato, New Zealand
Brendan Mccane	University of Otago, New Zealand
Yi Mei	RMIT University, Australia
Thomas Meyer	UKZN and CSIR Meraka, New Zealand
Eva Millan	Universidad de Málaga, Spain
Rei Miyata	University of Tokyo, Japan

Diego Molla	Macquarie University, Australia
Masud Moshtaghi	The University of Melbourne, Australia
Parma Nand	Auckland University of Technology, New Zealand
Nina Narodytska	Carnegie Mellon University, USA
Abhaya Nayak	Macquarie University, Australia
M.A. Hakim Newton	Griffith University, Australia
Hien Nguyen	Siemens, USA
Scott Nowson	Xerox Research Centre Europe, France
Oliver Obst	CSIRO, Australia
Yew-Soon Ong	Nanyang Technological University, Singapore
Nir Oren	University of Aberdeen, UK
Mehmet Orgun	Macquarie University, Australia
Huseyin Ozkan	Koc University, Turkey
Russel Pears	Auckland University of Technology, New Zealand
Duc Pham	Griffith University, Australia
Dinh Phung	Deakin University, Australia
Will Radford	Xerox Research Centre Europe, France
David Rajaratnam	University of New South Wales, Australia
Miguel Ramírez	RMIT University, Australia
Santu Rana	Deakin University, Australia
Patricia Riddle	University of Auckland, New Zealand
Goce Ristanoski	NICTA, Australia
Ji Ruan	Auckland University of Technology, New Zealand
Jonathan Rubin	University of Auckland, New Zealand
Abdallah Saffidine	The University of New South Wales, Australia
Bahar Salehi	Shiraz University, Iran
Philip Sallis	Auckland University of Technology, New Zealand
Mathieu Salzmann	NICTA, Australia
Paulo E. Santos	FEI, Brazil
Sebastian Sardina	RMIT University, Australia
Abeed Sarker	Arizona State University, USA
Ken Satoh	National Institute of Informatics and Sokendai, Japan
Torsten Schaub	University of Potsdam, Germany
Steven Schockaert	Cardiff University, UK
Rolf Schwitter	Macquarie University, Australia
Amir Shareghi Najar	University of Canterbury, New Zealand
Andy Song	RMIT University, Australia
Akshay Soni	StumbleUpon, USA
Hannes Strass	Leipzig University, Germany
Hanna Suominen	NICTA, Australia
Lech Szymanski	University of Otago, New Zealand
Yusuf Tas	NICTA, Australia
Truyen Tran	Deakin University, Australia
Ivor Tsang	UTS, Australia
Takehito Utsuro	University of Tsukuba, Japan
Keith Vander Linden	Calvin College, USA

Brijesh Verma	Central Queensland University, Australia
Karin Verspoor	The University of Melbourne, Australia
Bao Vo	Swinburne University of Technology, Australia
Kewen Wang	Griffith University, Australia
Lipo Wang	NTU, Singapore
Renata Wassermann	University of São Paulo, Brazil
Amali Weerasinghe	The University of Adelaide, Australia
Martin Wehrle	University of Basel, Switzerland
Peter Whigham	University of Otago, New Zealand
Mark Whitty	UNSW, Australia
Stefan Woelfl	University of Freiburg, Germany
Diedrich Wolter	University of Bamberg, Germany
Frank Wolter	University of Liverpool, UK
Kit Wong	Callaghan Innovation, New Zealand
Wilson Wong	RMIT University, Australia
Brendon J. Woodford	University of Otago, New Zealand
Bing Xue	Victoria University of Wellington, New Zealand
Nitin Yadav	RMIT University, Australia
Yi Yang	University of Technology, Sydney, Australia
John Yearwood	Deakin University, Australia
Gary Yen	Oklahoma State University, USA
Nayyar Zaidi	Monash University, Australia
Dongmo Zhang	University of Western Sydney, Australia
Peng Zhang	University of Technology, Sydney, Australia
Dengji Zhao	University of Southampton, UK
Tianqing Zhu	Deakin University, Australia
Zhiqiang Zhuang	Griffith University, Australia

Additional Reviewers

Wu Chen
Sarah Erfani
Tom Everitt
Alexander Feldman
Michael E. Houle
Ryutaro Ichise
Guifei Jiang
Farhan Khan
Trung Le

Shaowu Liu
Tu Dinh Nguyen
Christian Schulz-Hanke
Ihsan Utlu
Zhe Wang
Amail Weerasinghe
Stefan Woelfl
Ping Xiong

Contents

Exploiting the Beta Distribution-Based Reputation Model in Recommender System

Ahmad Abdel-Hafez[✉] and Yue Xu

Queensland University of Technology, Brisbane, Australia
{a.abdelhafez,yue.xu}@qut.edu.au

Abstract. Reputation systems are employed to measure the quality of items on the Web. Incorporating accurate reputation scores in recommender systems is useful to provide more accurate recommendations as recommenders are agnostic to reputation. The ratings aggregation process is a vital component of a reputation system. Reputation models available do not consider statistical data in the rating aggregation process. This limitation can reduce the accuracy of generated reputation scores. In this paper, we propose a new reputation model that considers previously ignored statistical data. We compare our proposed model against state-of the-art models using top-N recommender system experiment.

Keywords: Reputation system · Ratings aggregation · Beta distribution · Recommender system

1 Introduction

Reputation systems are acquiring increasing credibility among web users because these systems provide a metric with which product quality can be evaluated. They are currently considered essential components of e-commerce or product review websites, where they provide methods for collecting and aggregating users' ratings to enable the calculation of the overall reputation scores of products, users, or services (Shapiro 1982). Generated reputation scores influence customer decisions regarding items, since they are typically used to compare the quality of different available items.

In this paper, we focus on using ratings feedback in building item reputation scores. The simple mean method is the most straightforward approach to aggregate user ratings for the purpose of generating item reputations (Garcin et al. 2009). The mean provides a magnitude value of all ratings with reasonable accuracy. The median, which is also used to represent a reputation score, is more stable than the mean (Garcin et al. 2009). Reputation scores are critical components of feedback systems because of their increased influence on online users. Any minor improvement in the accuracy of reputation scores can noticeably affect website performance. An increasing number of aggregators have therefore been developed to enhance the accuracy of reputation scores (Abdel-Hafez et al. 2015, Bharadwaj and Al-Shamri 2009, Lauw et al. 2012).

Many reputation systems have recently been put forward, with the majority embedding one or more factors in the rating aggregation process to enhance the accuracy of reputation scores. These factors include the time at which a rating was provided, the

© Springer International Publishing Switzerland 2015
B. Pfahringer and J. Renz (Eds.): AI 2015, LNAI 9457, pp. 1–13, 2015.
DOI: 10.1007/978-3-319-26350-2_1

reputation of the user who provided this rating, and trust among users (Leberknight et al. 2012, Resnick et al. 2000, Wang et al. 2008). These factors are usually regarded as weights assigned to ratings during the aggregation process. The weighted mean method is a typical approach (Sabater and Sierra 2002). User- and time-related factors are independent of rating aggregation methods and can be incorporated into any aggregation technique, such as the simple mean method and the Dirichlet (Jøsang and Haller 2007), fuzzy (Bharadwaj and Al-Shamri 2009), and NDR (Abdel-Hafez et al. 2015) models.

Some of the proposed reputation models include other factors, such as the uncertainty of available ratings. These methods can produce more accurate reputation scores than those generated by the simple mean method (Jøsang and Haller 2007, Bharadwaj and Al-Shamri 2009, Abdel-Hafez et al. 2015) and are considered state-of-the-art models in reputation research. Despite these advantages, however, most existing reputation models do not explicitly consider the number (count) of ratings and the frequency of rating levels in the rating aggregation process. Rating count refers to the total number of ratings assigned to an item. Rating level pertains to a rating value, and the frequency of a rating level refers to the number of users who have rated an item with the rating value. In general, we believe that rating weights should relate to the frequency of rating levels and rating count. The frequency of rating levels for an item reflects how users view an item. For example, more instances of rating level 5 than rating level 2 indicate that the item is favored by a larger number of customers. The rating count of an item reflects the reliability of rating usage in building reputation scores; the higher the number of ratings assigned to an item, the larger the number of opinions that the ratings can reflect, and thus, the more accurate the item's reputation derived on the basis of these ratings.

In this paper, we propose a novel reputation method called the beta distribution-based reputation (BetaDR) model, which takes both rating level frequency and rating count into consideration in deriving item reputations.

2   Related Work

Reputation systems can be used to assess many objects, such as webpages, products, services, users, and peer-to-peer networks; these systems reflect what is generally said or believed about a target object (Abdel-Hafez et al. 2014b). An item's reputation is calculated on the basis of the ratings provided by many users, and a specific aggregation method is used for the calculation. Many methods use the weighted mean as an aggregator of ratings, wherein weight can represent a rater's reputation, the time at which a rating was provided, or the distance between the current reputation score and a recently received rating. Shapiro (1982) confirmed that time is important in calculating reputation scores; hence, the time decay factor has been widely used in reputation systems (Jøsang and Haller 2007, Leberknight et al. 2012, Wang et al. 2008). Leberknight et al. (2012) discussed the volatility of online ratings in an effort to reflect the current trend of users' ratings. The authors used the weighted mean, in which previous ratings have less weight than do current ones. Riggs and Wilensky (2001) performed collaborative quality filtering based on the principle of identifying the most reliable users. Lauw et al. (2012) classified users into lenient and strict users in their proposed leniency-aware quality model.

Jøsang and Haller (2007) introduced a multinomial Bayesian probability distribution reputation system based on Dirichlet probability distribution. The authors indicated that Bayesian reputation systems provide a statistically sound basis for computing reputation scores. A major contribution of their proposed model is its introduction of uncertainty to the reputation calculation process. The smaller the rating count involved, the higher the impact of the uncertainty addition. This model therefore provides more accurate reputation values when only a few ratings are assigned to an item.

Using fuzzy models is an equally popular approach in calculating reputation scores because fuzzy logic provides rules for reasoning with fuzzy measures, such as trustworthiness. These measures are typically used to describe reputation. Sabater and Sierra (2002) proposed the REGRET reputation system, which defines a reputation measure that considers the individual, social, and ontological dimensions. Bharadwaj and Al-Shamri (2009) put forward a fuzzy computational model for trust and reputation. The authors define the reputation of a user as the accuracy of his/her prediction regarding other users' ratings for different items. The authors also introduced the reliability metric, which represents the degree of reliability of a computed score.

Most recently, Abdel-Hafez et al. (2015) proposed a normal distribution-based reputation model (NDR), which is described as a weighted mean reputation system, wherein weights are generated using a normal distribution curve. In their work, the median rating and the ratings close to it acquire higher weights than do other ratings. The authors also put forward a modified NDR model with uncertainty (NDRU). Both models perform well on sparse and dense datasets. However, neither model explicitly considers rating count because under a small number of ratings, the median rating is unstable and uninformative. This shortcoming can negatively affect the accuracy of the reputations generated by NDR or NDRU.

3   The Beta Distribution-Based Reputation Model

In this section, we introduce a new rating aggregation method that generates item reputation scores. First, we use the arithmetic mean method as the naïve method. Second, the term "rating level" is used to represent the number of possible rating values that can be assigned to a specific item by a user. Let us consider, for example, a five-star rating system with possible rating values of $\{1, 2, 3, 4, 5\}$. Under this system, we say that we have five rating levels—one for each possible rating value.

As previously stated, the weighted mean method is the most frequently used approach in rating aggregation, and weights usually represent time decay or reviewer reputation. In the naïve method, the weight of each rating is $1/n$, where n is the number of ratings for an item. Regardless of whether we use the simplest mean method or the weighted mean methods that consider time or other user-related factors, the frequency of each rating level and the rating count of an item are not explicitly taken into consideration. For example, we assume that an item receives a set of ratings $\langle 2, 2, 2, 2, 3, 5, 5 \rangle$, under the simplest mean method, the weight assigned to each of the ratings is $1/7$. Although rating level 2 has a frequency higher than those of all other rating levels, the ratings in this level are assigned the same weight as those ascribed to other ratings, i.e., $1/7$. This example shows how rating level frequencies are disregarded in the weight calculation process.

Most ratings aggregators, such as the naïve, weighted mean and NDR (Abdel-Hafez et al. 2015) methods, disregard rating count as a measurement of the reliability of available ratings in reflecting item reputation. A situation that may occur in some cases and by chance is when a new item is introduced, the first few raters have similar opinions (either positive or negative) about an item. In such cases, available item ratings are insufficient for producing reliable reputation scores. Generally, the fewer the number of ratings assigned to an item, the less accurate the aggregated rating for this item.

We propose the use of the weighted mean to aggregate ratings. A more important feature of our approach is that weights are generated following two principles. First, the more frequent a rating level, the higher the weights assigned to the ratings at that level. Second, different weighting strategies should be used to calculate the rating weights of an item with few ratings and an item with many ratings. The beta distribution is suitable for use in the proposed reputation model given that it enables the flexibility necessary to satisfy the two principles.

3.1   Normal Distribution-Based Reputation Model

Abdel-Hafez et al. (2015) proposed the use of the probability density function (PDF) of the normal distribution to generate weights for the ratings of an item and then produce the item's reputation score by aggregating the ratings through the weighted mean method. This method (denoted as NDR) considers the frequency of ratings in the rating aggregation process. Assuming that the ratings fall under normal distribution (bell shape) the middle ratings are assigned higher weights than the ratings falling at the two ends of the distribution curve. Figure 1 shows the weights assigned to a list of ratings $\langle 2, 2, 2, 2, 3, 5, 5 \rangle$.

Garcin et al. (2009) studied and compared several reputation aggregators, including the mean, weighted mean, mode, and median. The authors demonstrated that the median is a more accurate representative of reputation because it is more informative and stable. The use of a bell-shaped normal distribution guarantees that middle ratings will be assigned weights higher than those allocated at the curve edges (extreme ratings). The middle ratings represent the median rating and the ratings close to it. Assigning higher weights to these ratings therefore enables a more accurate estimation of reputation score, as indicated in (Abdel-Hafez et al. 2015).

In the experiment discussed in (Abdel-Hafez et al. 2015), the NDR method exhibits higher accuracy when used on dense datasets than on sparse datasets. This result is attributed to the method's disregard of rating count in the weighting process. Under a small number of available ratings, therefore, the frequencies of rating levels are insufficient to produce accurate aggregation. The NDRU method is an attempt to overcome the unreliability problem. To this end, uncertainty is incorporated into the original NDR. This modification enhances accuracy over sparse datasets. Nevertheless, both NDR and NDRU assign higher weights to middle ratings by using a bell-shaped distribution to generate rating weights, regardless of the rating count of an item.

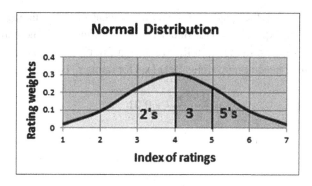

Fig. 1. Example of NDR normalized weights for 7 ratings

3.2 Weighting Based on the Standard Beta Distribution

As mentioned earlier, the main problem with the NDR and NDRU models (Abdel-Hafez et al. 2015) is the constancy of the bell distribution shape. Although the methods provide more accurate aggregations over dense datasets, their performance on sparse datasets is unimpressive. This failure establishes that assigning higher weights to middle ratings works well only on dense datasets. Over sparse datasets, this approach becomes insufficient when only the frequency of rating levels is considered; ensuring accuracy necessitates that the total number of ratings for an item (i.e., rating count) be taken into account as well. In the cases where the ratings count is relatively low, we propose to assign higher weights to extreme ratings to reduce the middle ratings contributions in reputation scores.

The beta distribution shows potential for generating different shapes, thus emphasizing middle or extreme ratings on the basis of shape parameters that can be related to dataset statistics. The standard beta distribution is generally a continuous probability distribution that is defined on the interval of $(0, 1)$, $0 < x_i < 1$. Its probability density function is presented in Eq. (1).

$$\text{Beta}\left(x_i\right) = \frac{\Gamma\left(\alpha + \beta\right)}{\Gamma\left(\alpha\right)\Gamma\left(\beta\right)} x_i^{\alpha-1} \left(1 - x_i\right)^{\beta-1} \tag{1}$$

where Γ represents the Gamma function, and α and β are two parameters that can determine distribution shape. Different values of shape parameters provide a variety of shapes that can flexibly model various datasets. Our proposed method is thus described as a weighted mean method, wherein weights are generated by the beta distribution. The crucial issue here is to determine shape parameters α and β to produce the desired distribution shape, which is used to generate ratings weights for every single item.

Suppose that we have n ratings for a specific item P, represented as a vector $R_P = r_0, r_1, r_2, \ldots, r_{n-1}$, where r_0 is the smallest rating, and r_{n-1} is the largest rating, i.e., $r_0 \leq r_1 \leq r_2 \leq \ldots \leq r_{n-1}$. To aggregate the ratings, we need to compute the weight associated with each rating, which is also represented as a vector $W_P = w_0, w_1, w_2, \ldots, w_{n-1}$.

As previously discussed, the weights of the ratings are calculated using the beta distribution PDF given in Eq. (1), where $\text{Beta}(x_i)$ is the weight of the rating at index $i = 0, \ldots, n - 1$. For the n ratings r_i in R_p, we design Eq. (2) to evenly select n values x_i within $[0, 1]$, thereby generating weights $\text{Beta}(x_i)$ for rating r_i, $i = 0, \ldots, n - 1$.

$$x_i = \frac{0.98 \times i}{n - 1} + 0.01 \tag{2}$$

By using Eq. (2), we derive $x_0 = 0.01, \cdots, x_{n-1} = 0.99$. The generated weights $\text{Beta}(x_i)$ are then normalized, so that the summation of all the weights is equal to 1. We generate a unified weight for every rating level and then use it to calculate the final reputation score. Normalized weights $\langle w_0, w_1, w_2, \ldots, w_{n-1} \rangle$ are calculated in Eq. (3), where $\sum_{i=0}^{n-1} w_i = 1$.

$$w_i = \frac{\text{Beta}(x_i)}{\sum_{j=0}^{n-1} \text{Beta}(x_j)} \tag{3}$$

3.3 Reputation Score Generation

We separate ratings into groups on the basis of rating levels, with each group containing ratings of the same level. $R^l = \left\langle r_0^l, r_1^l, r_2^l, \ldots, r_{|R^l|-1}^l \right\rangle$, $l = 1, 2, \ldots, k$, for each rating $r \in R^l$, $r = 1$. The set of all the ratings for item P is $R_p = \bigcup_{l=1}^{k} R^l$. The corresponding weights of the ratings in R^l are represented as $W^l = \left\langle w_0^l, w_1^l, w_2^l, \ldots, w_{|R^l|-1}^l \right\rangle$. The final reputation score is calculated as the weighted mean for each rating level by using Eq. (4), where level weight LW^l is the summation of the weights of every rating that belongs to level l.

$$\text{BetaDR}_p = \sum_{l=1}^{k} \left(1 \times LW^l\right), \qquad LW^l = \sum_{j=0}^{|R^l|-1} w_j^l \tag{4}$$

3.4 The Beta Distribution Shapes

Figure 2 shows three beta distribution shapes (and thus, three weighting distributions) for the simple rating example in Fig. 1. Shapes 1, 2, and 3 are generated for $\alpha = 2$ and $\beta = 5$, $\alpha = \beta = 5$, and $\alpha = 5$ and $\beta = 2$, respectively. The median rating is considered the centroid of the ratings, and it separates all the other ratings into two groups: the lower group, which contains all the ratings less than the median, and the upper group, which comprises all the ratings larger than the median. The median rating in the example illustrated in Fig. 2 is in index 4. The figure shows that for Shape 1 with $\alpha = 2$ and $\beta = 5$

(i.e., α < β), the lower group is assigned weights higher than those obtained by the upper group; for Shape 3 with α = 5 and β = 2 (i.e., α > β) the upper group acquires weights higher than those assigned to the lower group. These results indicate that in the two cases, the ratings in the two groups contribute differently to the reputation calculation. Generally, no evidence justifies the allocation of higher weights to either group. We propose to equally consider the weights for the two groups in reputation calculation; that is, in the proposed method, the weights assigned to both groups are equal, as in the case illustrated by Shape 2 in Fig. 2. In this case, the shape of the weight distribution is symmetric.

Symmetry is an important feature of the generated shape, which occurs when the two shape parameters are equal, α = β. A symmetric shape indicates that a line can split the shape into two pieces that are each other's mirror (Bury 1999). We use the symmetric shape for the beta distribution at all times to ensure fairness and the equal contribution of low and high ratings. In general, constantly using symmetric shapes in the proposed method is considered crucial for it to fulfil its purpose.

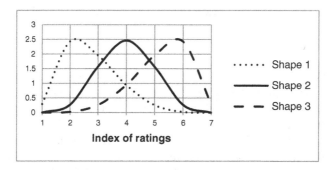

Fig. 2. The effect of using different values of α and β on PDF shape of the beta distribution using Table 1 example.

Figure 3 shows an example of the three different symmetric shapes of the beta distribution PDF. The U shape of the beta distribution is generated when shape parameters α = β < 1. The figure indicates that the extreme ratings—the first indexed rating (lowest rating value) and the last indexed rating (highest rating value)—are assigned the highest weights. The weights of the extreme ratings depend on the depth of the U shape. The lower the values of α and β, the deeper the curve will be, indicating higher weights for the extreme ratings. When α and β approach 1, the curve takes on a more flattened shape, thereby increasing the weights assigned to the middle ratings and decreasing those allocated to the extreme ratings.

In the case wherein shape parameters α = β = 1, the beta distribution PDF produces a uniform distribution [0, 1]. All the ratings have the same weights $w_i = \frac{1}{n}$. Figure 3 depicts the uniform distribution as a straight line. This case illustrates the naïve method, wherein the weights of all the ratings are unified.

The last shape illustrated in Fig. 3 is the bell shape, which is generated when the values of shape parameters $\alpha = \beta > 1$. In the bell shape case, the median rating and the ratings close to it are assigned weights higher than those provided for the ratings far from the median. Under larger shape parameters, the bell shape becomes sharper, thus increasing the weight given to the median rating.

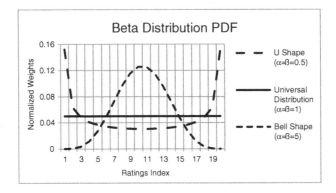

Fig. 3. Different symmetric PDF shapes of the beta distribution using 20 ratings

The reputation score of an item is derived from the ratings assigned to this item. As previously stated, the number of ratings for an item is important to generate an accurate item reputation score. This requirement indicates that if an item has a small number of ratings, then the reputation score generated by these ratings may be less reliable than those generated by the use of a high number of ratings. The rating count for an item should therefore be taken into consideration in deriving the reputation score for this item. Conversely, the distribution of rating count over items can differ across various application domains. For example, on average, the movies featured in a movie review website may receive hundreds or thousands of ratings, whereas the cars in a car selling website may receive only a few ratings. Directly using absolute rating counts in deriving reputations would therefore generate bias from one domain to another. To address this problem, we propose adopting the ratio between the rating count of an item and the average rating count for all the items in a domain. This ratio, called the item rating relative count (IRRC), is used to measure the rating count of an item, as calculated in Eq. (5):

$$\text{IRRC} = \frac{n_i}{\bar{n}}, \quad \bar{n} = \frac{\sum_{p_i \in M} n_i}{|M|} \tag{5}$$

where n_i is the rating count of an item p_i, and $\bar{n}$ denotes the average rating count of the items in a domain, assuming that M is the set of items in the domain.

The most important issue in this study is our proposal to control the shape of the beta distribution for an item's ratings. We suggest using IRRC as key factor to determine the distribution shape. Since the beta distribution shape is determined by the values of α and β. The proposed method for calculating the two shape parameters is

$$\alpha = \beta = \text{IRRC} \tag{6}$$

4 Incorporating Reputation in Recommender System

In this section we employ a method proposed by Abdel-Hafez et al. (2014a), to merge the recommender system generated ranked list of items with the reputation generated one in order to produce the Top N recommendations. This method was adopted because of its generality, as it separates the implementation of the recommender system, the reputation system, and the merging process. We use the top N recommender system in order to evaluate our proposed beta distribution-based reputation model. We implement the reputation-aware recommender system with the baseline reputation models and compare the results when we use our proposed reputation model. In this section we will describe the weighted Borda count method (Abdel-Hafez et al. 2014a) briefly. First we describe some definitions.

- Users: $U = \{u_1, u_2, \ldots, u_{|U|}\}$ is a set of users who have rated at least one item.
- Items: $P = \{p_1, p_2, \ldots, p_{|P|}\}$ is a set of items that are rated at least one time by a user in U.
- Users-Ratings: This is a user-rating matrix defined as a mapping $ur:U \times P \rightarrow [0, r]$. If the user u_i has rated the item p_j with rating a, then $ur(u_i, p_j) = a$; otherwise, $ur(u_i, p_j) = 0$ such that $0 < a <= r$, and r is the maximum rating.
- Item-Reputation Score: $S = \{s_1, s_2, \ldots, s_{|P|}\}$, where s_i is the reputation score for item p_i.
- Item Recommendation Score: $T = \{t_1, t_2, \ldots, t_{|P|}\}$ where t_i is the recommendation score for item p_i. This value is used to generate the candidate list of top-M recommendation using Eq. (7).

$$\text{TopM}_{u_i} = \underset{1 \rightarrow M}{\text{argmax}} \, T_{u_i}, u_i \in U \tag{7}$$

4.1 The Weighted Borda-Count Method

Using the Borda-count (BC) (Dummett 1998) method the first ranked candidates given the score N and the next one is $N - 1$, and so on. Every item that is outside the Top-N list will receive a score of zero. For an item $p \in P$, the sum of the BCs for this item is denoted $SBC(p)$. The items with the highest SBC will appear at the top of the list. The BC method was adopted to merge a recommendation list and a reputation list. For a user u and an item $p \in P$, let $BC_{rec}(p)$ be the BC of p in the recommendation list and $BC_{rep}(p)$ the BC of p in the reputation list. Then, $SBC(p) = BC_{rec}(p) + BC_{rep}(p)$. The Top-N recommendations for the user u are defined in Eq. (8).

$$\text{TopN}_u^{BC} = \underset{p \in P}{\text{argmax}^N} \, SBC(p) \tag{8}$$

The weighted Borda-count (WBC) method introduces a weight in the BC method. The weighted sum of BC (*WSBC*) and the top-N recommendations are defined in Eqs. (9), and (10), where $0 < \alpha < 1$:

$$\text{WSBC}(p) = \alpha \times \text{BC}_{\text{rec}}(p) + (1 - \alpha) \times \text{BC}_{\text{rep}}(p) \tag{9}$$

$$\text{TopN}_u^{\text{WBC}} = \text{argmax}_{p \in P}^N \text{WSBC}(p) \tag{10}$$

4.2   Using Personalized Item Reputation

An item's reputation is the global community opinion about it. At a specific time, the ranking of items based on item reputation is the same for all users. Therefore, Abdel-Hafez et al. (2014a) proposed to use personalized reputation for items to tackle this problem. The idea is to build a user-preference profile based on previous user ratings, and then to use this profile to filter the items that were outside the preference scope.

- Implicit Item Categories $C = \{C_1, C_2, \ldots, C_{|C|}\}$ is the set of categories wherein items in P belong to $C_i = \{p | p \in P\}$ and $C_i \cap C_j = \emptyset$.
- User Item Preference $P_u = \left\{ p | p \in P, ur(u, p) \geq \frac{(r+1)}{2} \right\}$, r is the maximum ring and contains all the user's preferred items.
- User Category Preference $F_u = \{C_i | C_i \in C, (C_i \cap P_u) \neq \emptyset\}$ contains item categories in which the user's preferred or positively rated items belong. A user category preference F_u is a set of categories that are preferred by the user u.

The personalized reputation was defined in Eq. (11), where S_p is the reputation for the item p.

$$\text{PIR}_p = \begin{cases} S_p, & p \in C_i, C_i \in F_u \\ 0, & \text{Otherwise} \end{cases} \tag{11}$$

5   Experiment

We conducted the top-N recommender system experiment. We aimed to demonstrate that combining item reputation with user-based CF could enhance the accuracy of the top-N recommendations.

5.1   Dataset

We used the MovieLens movie ratings dataset extracted from Grouplens.org. We used this dataset in three different ways: (1) using all (2) using only 10 %, and (3) using only 5 % of the ratings (Abdel-Hafez et al. 2014c). The purpose of the three tests was to observe whether the reputation method enhances recommendation accuracy over dense and sparse datasets. Table 1 presents some of the statistics for each dataset. We split each dataset into training and testing sets by randomly selecting 80 % of each user's ratings into a training dataset and the rest into a testing dataset. We performed a 5-fold experiment, each time a different 20 % of the dataset was selected for testing. We calculated the average of the results at the end. Sparsity for the datasets was calculated using Eq. (12).

$$\text{Sparsity} = 1 - \frac{\#\text{ of Ratings}}{\#\text{ of Users} \times \#\text{ of Items}} \tag{12}$$

Table 1. Datasets statistics

	MovieLens 5 % (ML5)	MovieLens 10 % (ML10)	MovieLens Complete (MLC)
Number of ratings	6,515	13,077	100,000
Sparsity	0.99589	0.99175	0.93695
Min ratings per user	5	10	20
Max ratings per user	36	73	737
Average ratings per user	6.849	13.867	106.044
Min ratings per movie	0	0	1
Max ratings per movie	59	114	583
Average ratings per movie	3.840	7.774	59.453

5.2 Experiment Settings

We conducted the experiment in three runs for each dataset using the values of the recommendation list $top-N = 20$, the candidate list $top-M = 60$, and the nearest neighbors $K = 20$. The recommended item was considered a hit if it appeared in the user-testing dataset and the user has granted the item a **rating** $>= 3$. The evaluation metrics used are precision, recall, and F1-score.

We implemented the user-based recommender system introduced in (Sarwar et al. 2000). We also implemented two baselines, the Dirichlet reputation model (DIR) (Jøsang and Haller 2007), and the normal distribution based reputation model with uncertainty (NDRU) (Abdel-Hafez et al. 2015). We compare the baseline models with the proposed BetaDR model.

5.3 Results and Discussion

Table 2 shows the precision, recall, and F1-scores for each of the implemented methods over the three tested datasets. We compare the proposed BetaDR model with the two state-of-the-art models in two different settings, firstly, using only the general reputation scores, and secondly, using the personalized version of each of the reputation models. We notice that the proposed BetaDR produces better results using both settings. The uncertainty in the ML5 and ML10 datasets is high; hence, DIR and NDRU methods provide similar results as both methods add uncertainty to their aggregation equations. In contrast, the BetaDR uses different method of mixing U and Bell shapes distributions to consider the count of ratings per item. This method uplifts the popular items in the ranked list. In the complete dataset, MLC, we notice an improvement of the NDRU model over the DIR one, as it emphasizes the ratings distribution. The BetaDR still performs better than the NDRU as it emphasizes both the rating distribution and the ratings count per item.

Table 2. Results of top-N recommendation accuracy using three datasets

Used Reputation Method with CF	ML5			ML10			MLC		
	Precision	Recall	F1-score	Precision	Recall	F1-score	Precision	Recall	F1-score
N/A	0.0061	0.0684	0.0112	0.0079	0.0723	0.0142	0.0283	0.0229	0.0253
Mean	0.0067	0.0651	0.0122	0.0079	0.0725	0.0142	0.0289	0.0237	0.0259
DIR	0.0075	0.0665	0.0135	0.0079	0.0728	0.0143	0.0301	0.0259	0.0278
NDRU	0.0075	0.0665	0.0135	0.0079	0.0729	0.0143	0.0336	0.0283	0.0307
BetaDR	0.0091	0.0742	0.0162	0.0097	0.0812	0.0173	0.0362	0.0394	0.0377
P-Mean	0.0112	0.1134	0.0204	0.0129	0.0793	0.0222	0.0363	0.0351	0.0357
P-DIR	0.0130	0.1239	0.0235	0.0145	0.0849	0.0248	0.0398	0.0402	0.0400
P-NDRU	0.0131	0.1249	0.0237	0.0146	0.0858	0.0250	0.0465	0.0448	0.0456
P-BetaDR	0.0178	0.1337	0.0314	0.0192	0.1015	0.0323	0.0519	0.0489	0.0504

6   Conclusions

In this paper, we have proposed a new aggregation method for generating reputation scores for items on the basis of customers' ratings. The proposed method is described as a weighted mean method that generates weights using the beta distribution. The essential question we targeted is how to determine the appropriate beta distribution shape for different datasets. In order to calculate the shape parameters we use the ratio between the rating count of an item and the average rating count for all the items in a domain. We provided an experiment using recommender-aware reputation model and compared the results with two of the state-of-the-art reputation models. The results show improvement for the proposed BetaDR model over the DIR and NDRU models.

References

Abdel-Hafez, A., Tang, X., Tian, N., Xu, Y.: A reputation-enhanced recommender system. In: Luo, X., Yu, J.X., Li, Z. (eds.) ADMA 2014. LNCS, vol. 8933, pp. 185–198. Springer, Heidelberg (2014)

Abdel-Hafez, A., Xu, Y., Jøsang, A.: A rating aggregation method for generating product reputations. In: Proceedings of the 25th ACM conference on Hypertext and Social Media, ACM, pp. 291–293 (2014b)

Abdel-Hafez, A., Xu, Y., Jøsang, A.: A normal-distribution based rating aggregation method for generating product reputations. Web Intell. 13(1), 43–51 (2015)

Abdel-Hafez, A., Xu, Y., Tian N.: Item reputation-aware recommender systems. In: Proceedings of the 16th International Conference on Information Integration and Web-based Applications & Services. ACM, pp. 79–86 (2014c)

Bharadwaj, K.K., Al-Shamri, M.Y.H.: Fuzzy computational models for trust and reputation systems. Electron. Commer. Res. Appl. **8**(1), 37–47 (2009)

Bury, K.: Statistical Distributions in Engineering. Cambridge University Press, Cambridge (1999)

Dummett, M.: The Borda count and agenda manipulation. Soc. Choice Welfare **15**(2), 289–296 (1998)

Garcin, F., Faltings, B., Jurca, R.: Aggregating reputation feedback. In: Proceedings of the First International Conference on Reputation: Theory and Technology, Italian National Research Council, pp. 62–74 (2009)

Jøsang, A., Haller, J.: Dirichlet reputation systems. In: Proceedings of the Second International Conference on Availability, Reliability and Security, IEEE, pp. 112–119 (2007)

Lauw, H.W., Lim, E.P., Wang, K.: Quality and leniency in online collaborative rating systems. ACM Trans. Web (TWEB) **6**(1), 4 (2012)

Leberknight, C.S., Sen, S., Chiang, M.: On the volatility of online ratings: an empirical study. In: Shaw, M.J., Zhang, D., Yue, W.T. (eds.) WEB 2011. LNBIP, vol. 108, pp. 77–86. Springer, Heidelberg (2012)

Resnick, P., Kuwabara, K., Zeckhauser, R., Friedman, E.: Reputation systems. Commun. ACM **43**(12), 45–48 (2000)

Riggs, T., Wilensky, R.: An algorithm for automated rating of reviewers. In: Proceedings of the First ACM/IEEE-CS Joint Conference on Digital Libraries, ACM, pp. 381–387 (2001)

Sabater, J., Sierra, C.: Reputation and social network analysis in multi-agent systems. In: Proceedings of the First International Joint Conference on Autonomous Agents and Multiagent Systems, Springer, Berlin Heidelberg, pp. 475–482

Sarwar, B., Karypis, G., Konstan, J., Riedl, J.: Analysis of recommendation algorithms for e-commerce. In: Proceedings of the 2nd ACM Conference on Electronic Commerce, ACM, pp. 158–167

Shapiro, C.: Consumer information, product quality, and seller reputation. Bell J. Econ. **13**(1), 20–35 (1982)

Wang, B.C., Zhu, W.Y., Chen, L.J.: Improving the Amazon review system by exploiting the credibility and time-decay of public reviews. In: Proceedings of the International Conference on Web Intelligence and Intelligent Agent Technology, IEEE/WIC/ACM, pp. 123–126 (2008)

A Heuristic Search Approach to Find Contrail Avoidance Flight Routes

Rubai Amin and Sameer Alam[✉]

School of Engineering and IT, University of New South Wales,
Canberra, Australia
{r.amin,s.alam}@adfa.edu.au

Abstract. Contrails are line-shaped clouds or "condensation trails," composed of ice particles that are visible behind jet aircraft engines. Contrails can affect the formation of clouds effecting climate change. This paper proposes an integrated model of atmosphere, airspace and flight routing with a Gradient Descent based heuristic search algorithm to find Contrail avoidance trajectories with climb, descent and vector maneuvers. Trade off analysis of Contrails avoidance with fuel burn/CO2 and distance flown is also presented.

Keywords: Heuristic search · Gradient descent · Environmental impact · Air traffic

1 Introduction

The Intergovernmental Panel on Climate Change (IPCC) special report on Aviation and the Global Atmosphere has accepted that secondary aviation emissions such as Contrails can have a climate impact comparable to the CO2 from the combustion process and may add to greenhouse gas effect [1].

Recent advances in avionics and onboard computing power on an aircraft may help pilots plan their routes that may mitigate the environmental impact of aviation. In this paper we propose a simple yet effective heuristic search i.e. Gradient descent algorithm to identify contrail avoidance trajectories with route choices for pilots and air traffic planners. The research contribution of this paper is in integrating a heuristic search algorithm with atmospheric, airspace and air traffic route models to generate realistic aircraft trajectories that can be implemented in real time.

2 Background

Contrails (also known as condensation trails) consist of tiny ice particles and are formed by water vapors in the exhaust of the aircraft engine, given the right weather conditions. Contrail formation is well understood by the Schmidt–Appleman criterion [2]. Persistent contrail increases the cloud cover over the Earth and in turn reflects solar radiation. This contributes significantly to greenhouse gas effect [1].

© Springer International Publishing Switzerland 2015
B. Pfahringer and J. Renz (Eds.): AI 2015, LNAI 9457, pp. 14–20, 2015.
DOI: 10.1007/978-3-319-26350-2_2

There are several suggested technological and operational modifications in an aircraft operation that can manage contrail formations. In this paper, we focus on Flight path management as means of avoiding contrail formations as it can be easily achievable yet cost effective. A flight path management option for contrail avoidance includes:

- Alteration of existing flight routes
- Climb/decent and vector based on daily conditions

An integrated approach with a combination of contrail model, atmosphere, airspace and flight route and an effective search strategy on three axes (Climb, Descend and Turn) can be more suitable and adoptable. Tradeoffs between increase in distance flown and additional fuel burn have to be taken into account. The key research problem this paper looks into is how can contrails formation be avoided or minimized by integrating a heuristic search algorithm with atmospheric, airspace and flight route models.

We used Gradient Descent method [3] as heuristic search algorithm to identify contrail avoidance trajectories. Gradient descent is an iterative method that is given an initial point, and follows the negative of the gradient in order to move the point toward a critical point, which is hopefully the desired local minimum. The key research question this paper looks into are:

1. How to integrate atmospheric data with air traffic management to identify alternative flight paths?
2. What are the tradeoffs between avoiding contrail formation and fuel burn/CO2 emissions and distance flown/time?

3   Methodology

Since the problem is two folds i.e. integrating different models and a search algorithm to work over them, proposed methodology consisting of following models:

3.1   Airspace Model

The airspace model consists of geo-spatial data points of the Australian airspace. The data is derived from Aeronautical Information Publication (AIP) for the Australian region and contains airspace specific information such as lateral dimensions, special use airspace, restricted zones, etc.

3.2   Atmospheric Model

The atmospheric model is developed by using data from the Bureau of Meteorology. It provides wind and temperature data for $5° \times 5°$ grids and 12 elevations. Further 3D interpolation is done to obtain data for $1° \times 1°$ grid. Humidity data is then obtained from nearest weather station.

3.3 Persistent Contrail Model

Persistent contrails are of interest because they increase the cloudiness of the atmosphere. Persistent contrails often evolve and spread into extensive cirrus cloud cover that is indistinguishable from naturally occurring cloudiness. Changes in cloudiness are important because clouds help control the temperature of the Earth's atmosphere. Persistent contrail model is developed as follows [4]: A region is identified as contrail formation region if the relative humidity $>r_{crit}$

$$where \ r_{crit} = \frac{G\left(T - T_{contr}\right) + e_{sat}^{liq}\left(T_{contr}\right)}{e_{sat}^{liq}(T)}$$

A region is identified as persistent contrail formation region if relative humidity $>r_{crit}$ and relative humidity with respect to ice $>100\,\%$

$$this \ is \ given \ by \ RHi = (RHw)\frac{6.0612e^{18.102T/(249.52+T)}}{6.1162e^{22.577T/(273.78+T)}}$$

Where:

$$G = \frac{EI_{H_2O}C_pP}{\varepsilon Q(1 - \eta)} \ and$$

- T = ambient temperature
- $T_{contr} = -46.46 + 9.43\ln(G - 0.053) + 0.72\ln^2(G - 0.053)$
- $G = \frac{EI_{H_2O}C_pP}{\varepsilon Q(1-\eta)}$
- e_{sat}^{liq} = Saturation vapor pressure
- EI_{H_2O} = Emission index of water vapor
- C_p = Isobaric heat capacity of air
- P = Pressure
- ε = Ratio of molecular masses of water and dry air
- Q = Specific combustion heat
- η = Average propulsion efficiency of the jet engine

3.4 Flight Route Model

Upper airspace routes in Australian airspace are obtained from Air Traffic Service Manual and integrated into the airspace model. Only those flight routes are considered that are in the upper atmosphere where contrail formation is possible.

3.5 Integration of Contrail, Flight Route, Atmosphere and Airspace

All four models were integrated and as illustrated in Fig. 1, regions of contrail formation can be seen in the Australian airspace with routes identified.

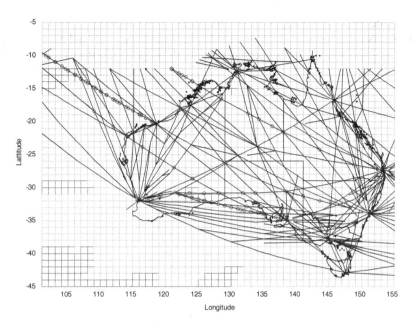

Fig. 1. An integrated view of flight routes, contrail regions, airspace and flights.

3.6 Algorithm Design

We then developed an algorithm for contrail avoidance. There are three maneuvers possible, climb, vector and descent as illustrated in Fig. 2, for a given flight.

The cost function J is developed as follows:

$$\text{MIN } J = \int_{t_0}^{t_f} [(SX(t))^T \dot{X}(t) + C_t + C_f f + C_r r(x,y)]dt$$

Where Ct is Cost coefficient of time

Cf is Cost coefficient of fuel

Cr is Cost coefficient of penalty areas

r(x,y) which is the Penalty function (i.e. Contrails formation regions) is defined as

$r(x,y) = \sum_i \frac{1}{d_i^2}$ where di is the distance between the aircraft and centre of the ith region

that potentially form persistent contrails.

Gradient Descent algorithm is adopted as follows to minimize the cost function:

(a) Checks if a given route between two successive waypoints crosses or enters a contrail formation region

(b) If route segment crosses contrail formation region then either the latitude or the altitude of the second point is randomly increased or decreased by a small delta.

(c)   Latitude is chosen based on the shortest distance to the next waypoint.
(d)   This process is repeated until the distance between the aircraft and centre of the ith region that potentially forms persistent contrails becomes less than an epsilon value.

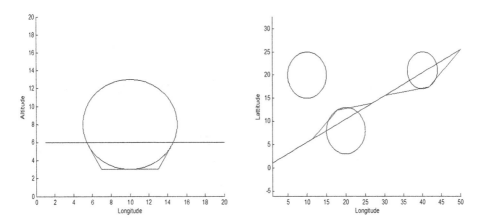

Fig. 2. Climb/descent and vector maneuvers on sample data by Gradient Descent. Straight lines (original trajectory), circumventing line (contrail avoidance trajectory), contrail regions (circles).

4   Experiment Design

In experiments we first identified routes which were crossing the contrail formation regions. Routes passing through regions of expected persistent contrail formation at the flight level altitude of 33,000 ft are highlighted with blue colored waypoints in Fig. 1. Three of these identified Air Traffic Service routes were chosen, J141, T97 and B340; for application of our algorithm.

5   Results

Simulation results show that the contrail formation regions were successfully avoided by employing different strategies such as climb, descent, vector and combination of them. Figure 3 shows different maneuvers generated by the Gradient Descent algorithm for contrail avoidance for some selected routes identified in Fig. 2.

The top section of Fig. 3 shows a Descend maneuver for contrail avoidance. However, it leads to 2.48 % increase in distance flown and fuel burn.

The middle section of Fig. 3 shows a vector (Turn) maneuver for contrail avoidance. However, it leads to 2.04 % increase in distance flown and fuel burn. The bottom section of Fig. 3 shows a Climb maneuver for contrail avoidance. However, it leads to 5.56 % increase in distance flown and fuel burn.

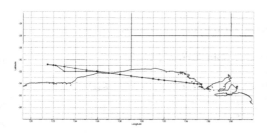

Route Name	Original Route				Modified Route				Additional distance, time, fuel, CO_2 (%)
	Distance (km)	Time (hours)	Fuel (kg)	CO_2 emitted (kg)	Distance (km)	Time (hours)	Fuel (kg)	CO_2 emitted (kg)	
ATS ROUTE T97	1357	1.56	7267	22978	1390	1.60	7444	23538	2.48

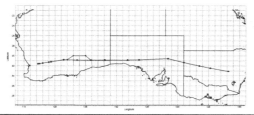

Route Name	Original Route				Modified Route				Additional distance, time, fuel, CO_2 (%)
	Distance (km)	Time (hours)	Fuel (kg)	CO_2 emitted (kg)	Distance (km)	Time (hours)	Fuel (kg)	CO_2 emitted (kg)	
ATS ROUTE J141	2985	3.44	15983	50539	3046	3.51	16309	51570	2.04

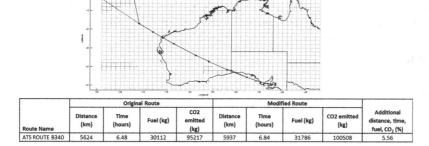

Route Name	Original Route				Modified Route				Additional distance, time, fuel, CO_2 (%)
	Distance (km)	Time (hours)	Fuel (kg)	CO_2 emitted (kg)	Distance (km)	Time (hours)	Fuel (kg)	CO_2 emitted (kg)	
ATS ROUTE B340	5624	6.48	30112	95217	5937	6.84	31786	100508	5.56

Fig. 3. Top (Descend maneuver), middle (Vector maneuver) and bottom (climb maneuver)

6 Analysis and Future Work

Gradient Descent approach efficiently minimized the cost function and successfully managed to avoid contrail formation regions. However, the flight path modification also leads to increased fuel burn/CO2 emissions (2.0–5.0 %) and distance flown.

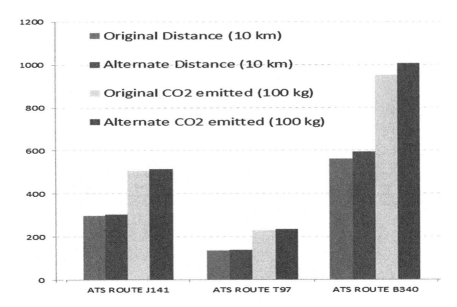

Fig. 4. Distance flown and CO2 emissions in contrail avoidance routes as compared to original routes in the three respective routes with three different routing strategies.

Results indicate that turn maneuvers were cost effective strategy in contrail avoidance. As illustrated in Fig. 4, contrail avoidance on longer routes (ATS Route 8340) has shown increase in distance flown and CO2 emission as compared to smaller routes (T97). For future work we will combine contrail avoidance trajectories with other traffic flow management strategies and employ multi-objective optimization approach.

References

1. Intergovernmental Panel on Climate Change (IPCC), Penner, J.E., Lister, D.H., Griggs, D.J., Dokken, D.J., McFarland, M. (eds.): Aviation and the Global Atmosphere. Cambridge University Press, Cambridge (1999)
2. Mannstein, H., Meyer, R., Wendling, P.: Operational detection of contrails from NOAA-AVHRR data. Int. J. Remote Sensing **20**, 1641–1660 (1999)
3. Mitchell, T.: Machine Learning. McGraw Hill, New York (1997)
4. Mannstein, H., Spichtinger, P., Gierens, K.: A note on how to avoid contrail cirrus. Transp. Res. Part D Transport Environ. **10**(5), 421–426 (2005)

Temporal Conjunctive Queries in Expressive Description Logics with Transitive Roles

Franz Baader, Stefan Borgwardt$^{(\boxtimes)}$, and Marcel Lippmann

Theoretical Computer Science, TU Dresden, Dresden, Germany
{franz.baader,stefan.borgwardt,marcel.lippmann}@tu-dresden.de

Abstract. In Ontology-Based Data Access (OBDA), user queries are evaluated over a set of facts under the open world assumption, while taking into account background knowledge given in the form of a Description Logic (DL) ontology. In order to deal with dynamically changing data sources, temporal conjunctive queries (TCQs) have recently been proposed as a useful extension of OBDA to support the processing of temporal information. We extend the existing complexity analysis of TCQ entailment to very expressive DLs underlying the OWL 2 standard, and in contrast to previous work also allow for queries containing transitive roles.

1 Introduction

Given a (man-made or natural) dynamical system that changes its states over time, it is sometimes useful to monitor the behavior of the system in order to detect and then react to critical situations [2]. To achieve this, one can monitor the running system using sensors (e.g., heart rate and blood pressure sensors for a patient) and store the (possibly aggregated and preprocessed) values in a database. Critical situations (such as "blood pressure too high") can then be described by database queries, and detecting them can be realized through query answering. However, such a pure database solution is unsatisfactory for several reasons. First, one cannot assume that the sensors provide a complete description of the current state of the system, which clashes with the closed world assumption used by database systems. Second, though one usually does not have a complete specification of the system's behavior, one may have some background knowledge restricting the possible states of the system, which can help to detect more situations.

These two problems are addressed by so-called ontology-based data access (OBDA) [14,27], where (i) the preprocessed and aggregated data are stored in a Description Logic (DL) ABox, which is interpreted with open world assumption, and (ii) the background knowledge is represented in a TBox (ontology) expressed in an appropriate DL. DLs [5] can be used to formalize knowledge using concepts, which represent sets of elements of an application domain, and roles, which describe binary relations between elements. For example, the concept Patient can be used to model the set of all patients in a hospital, while isTreatedWith represents

---

This work was partially supported by the DFG in the CRC 912 (HAEC).

© Springer International Publishing Switzerland 2015
B. Pfahringer and J. Renz (Eds.): AI 2015, LNAI 9457, pp. 21–33, 2015.
DOI: 10.1007/978-3-319-26350-2_3

a relationship between patients and treatments. Concept constructors can then be used to build complex concepts out of atomic concepts and roles. For example, Patient $\sqcap$ $\exists$isTreatedWith.Antibiotics describes patients treated with antibiotics. In the TBox, one can state subconcept-superconcept relationships, such a $\exists$isTreatedWith.Antibiotics $\sqsubseteq$ $\exists$finding.BacterialInfection, which says that antibiotics treatment is given only if there is a bacterial infection. In the ABox, one can state specific facts about individuals, such as isTreatedWith(BOB, PENICILLIN).

When monitoring a dynamical system, the situation to be recognized may also depend on states of the system at different points in time (such as "fluctuating heart rate"). For this reason, OBDA was extended to the temporal case in [1,4]. In [4] the complexity of answering temporal conjunctive queries (TCQs) w.r.t. TBoxes was investigated for TBoxes expressed in DLs between $\mathcal{ALC}$ and $\mathcal{SHQ}$. The results are concerned both with data complexity (which is measured only in the size of the data) and with combined complexity (which additionally takes the size of the query and the TBox into account). In addition, the paper considers rigid concepts and roles, whose interpretations must not change over time.

We extend the results of [4] in two directions. First, while being quite expressive, $\mathcal{SHQ}$ does not contain the constructors nominals and inverse roles, which are quite useful in many applications. Here, we also consider logics that have these two constructors. However, the main difference is that, though $\mathcal{SHQ}$ can express transitivity of roles and sub-role relationships, transitive roles and roles with transitive subroles must not occur in queries in [4]. In the present paper, we dispense with this restriction, which unfortunately leads to a dramatic increase in complexity that reflects the results for standard (atemporal) queries (see [15,21] and Table 2).

As an example that illustrates the benefit of transitive roles in queries, assume that we want to recognize patients who have previously had myocarditis, i.e., an inflammation of the heart muscle. This can be expressed using the TCQ

Patient(x) $\wedge$
$\bigcirc^{-} \diamondsuit^{-} \big(\exists y, z.\mathsf{partOf}(y, x) \wedge \mathsf{Heart}(y) \wedge \mathsf{partOf}(z, y) \wedge \mathsf{Muscle}(z) \wedge \mathsf{Inflamed}(z) \big).$

This query is looking for a patient that, at some past time point, had (as part) a heart that itself had as part a muscle that was inflamed. In this example, we assume that the role partOf is transitive and rigid. Transitivity implies that the inflamed muscle was also part of the patient and rigidity ensures that the heart is not part of different patients at different points in time. In addition, we assume that Heart and Muscle are rigid (hearts and muscles stay hearts and muscles over time), but Patient and Inflamed are non-rigid (the muscle may, e.g., cease to be inflamed and the patient may be discharged).

In the next section, we introduce the DLs investigated in this paper, as well as TCQs and their semantics. We also give an overview over the already known and the new complexity results (see Table 2). Section 3 investigates the complexity of answering certain atemporal queries in a fine-grained way. The reason is that, similar to [4], we split the task of answering TCQs into propositional temporal reasoning on the one hand, and answering atemporal queries on the other hand. In Sect. 4, we then determine the combined complexity of answering TCQs whereas in Sect. 5 we deal with the data complexity. Full proofs of our results can be found in an accompanying technical report [3].

2   Preliminaries

In this section, we recall the basic notions of DLs and TCQs. Throughout the paper, let N_C, N_R, and N_I be non-empty, pairwise disjoint sets of *concept names*, *role names*, and *individual names*, respectively.

Definition 1 (Syntax of DLs). *A* role *is either a role name* $r \in N_R$ *or an inverse role* r^-. *The set of* concepts *is inductively defined starting from concept names* $A \in N_C$ *using the constructors in the second part of Table 1, where* r, s *are roles,* $a, b \in N_I$, $n \in \mathbb{N}$, *and* C, D *are concepts. The third part of Table 1 shows how* axioms *are defined. A* TBox *is a finite set of general concept inclusions (GCIs), an* RBox *is a finite set of role inclusions and transitivity axioms, and an* ABox *is a finite set of concept and role assertions. A* knowledge base (KB) $\mathcal{K} = (\mathcal{A}, \mathcal{T}, \mathcal{R})$ *consists of an ABox* $\mathcal{A}$, *a TBox* $\mathcal{T}$, *and an RBox* $\mathcal{R}$.

In the DL $\mathcal{ALC}$, negation, conjunction, and existential restriction are the only allowed constructors. Also, no inverse roles, role inclusions and transitivity axioms are allowed in $\mathcal{ALC}$. Additional letters denote different concept constructors or types of axioms: $\mathcal{I}$ means inverse roles, $\mathcal{O}$ means nominals, $\mathcal{Q}$ means at-most restrictions, and $\mathcal{H}$ means role inclusions. For example, the DL $\mathcal{ALCHI}$ extends $\mathcal{ALC}$ by role inclusions and inverse roles. The extension of $\mathcal{ALC}$ with transitivity axioms is denoted by $\mathcal{S}$. Hence, the DL allowing for all the constructors and types of axioms introduced here is called $\mathcal{SHOIQ}$. We sometimes write $\mathcal{L}$-concept ($\mathcal{L}$-KB, ...) for some DL $\mathcal{L}$ to make clear which DL is used.

Table 1. Syntax and semantics of DLs

	Syntax	Semantics
Inverse role	r^-	$\{(e, d) \mid (d, e) \in r^{\mathcal{I}}\}$
Negation	$\neg C$	$\Delta^{\mathcal{I}} \setminus C^{\mathcal{I}}$
Conjunction	$C \sqcap D$	$C^{\mathcal{I}} \cap D^{\mathcal{I}}$
Existential restriction	$\exists r.C$	$\{d \in \Delta^{\mathcal{I}} \mid$ there is $e \in C^{\mathcal{I}}$ with $(d, e) \in r^{\mathcal{I}}\}$
Nominal	$\{a\}$	$\{a^{\mathcal{I}}\}$
At-most restriction	$\leqslant n\, r.C$	$\{d \in \Delta^{\mathcal{I}} \mid \sharp\{e \in C^{\mathcal{I}} \mid (d, e) \in r^{\mathcal{I}}\} \leq n\}$
General concept inclusion	$C \sqsubseteq D$	$C^{\mathcal{I}} \subseteq D^{\mathcal{I}}$
Concept assertion	$C(a)$	$a^{\mathcal{I}} \in C^{\mathcal{I}}$
Role assertion	$r(a, b)$	$(a^{\mathcal{I}}, b^{\mathcal{I}}) \in r^{\mathcal{I}}$
Role inclusion	$r \sqsubseteq s$	$r^{\mathcal{I}} \subseteq s^{\mathcal{I}}$
Transitivity axiom	$\mathsf{trans}(r)$	$r^{\mathcal{I}} = (r^{\mathcal{I}})^+$

Definition 2 (Semantics of DLs). *An interpretation is a pair $\mathcal{I} = (\Delta^{\mathcal{I}}, \cdot^{\mathcal{I}})$, where $\Delta^{\mathcal{I}}$ is a non-empty domain, and $\cdot^{\mathcal{I}}$ is a mapping assigning a set $A^{\mathcal{I}} \subseteq \Delta^{\mathcal{I}}$ to every $A \in N_C$, a binary relation $r^{\mathcal{I}} \subseteq \Delta^{\mathcal{I}} \times \Delta^{\mathcal{I}}$ to every $r \in N_R$, and a domain element $a^{\mathcal{I}} \in \Delta^{\mathcal{I}}$ to every $a \in N_I$, such that $a^{\mathcal{I}} \neq b^{\mathcal{I}}$ for all $a, b \in N_I$ with $a \neq b$* (unique name assumption (UNA)). *This function is extended to roles and concepts as shown in Table 1, where $\sharp S$ denotes the cardinality of the set S.*

Moreover, $\mathcal{I}$ is a model *of the axiom α (written $\mathcal{I} \models \alpha$) if the condition in the third part of Table 1 is satisfied, where $\cdot^+$ denotes the transitive closure. Furthermore, $\mathcal{I}$ is a model of a set of axioms $\mathcal{X}$ (written $\mathcal{I} \models \mathcal{X}$) if it is a model of all axioms $\alpha \in \mathcal{X}$, and $\mathcal{I}$ is a model of a KB $\mathcal{K} = (\mathcal{A}, \mathcal{T}, \mathcal{R})$ (written $\mathcal{I} \models \mathcal{K}$) if is a model of $\mathcal{A}$, $\mathcal{T}$, and $\mathcal{R}$. We call $\mathcal{K}$* consistent *if it has a model.*

For an RBox $\mathcal{R}$, we call a role name $r \in N_R$ *transitive* (w.r.t. $\mathcal{R}$) if every model of $\mathcal{R}$ is a model of trans(r). Moreover, r is a *subrole* of a role name $s \in N_R$ (w.r.t. $\mathcal{R}$) if every model of $\mathcal{R}$ is a model of $r \sqsubseteq s$. Finally, r is *simple w.r.t. $\mathcal{R}$* if it has no transitive subrole. Deciding whether $r \in N_R$ is simple can be done in time polynomial in the size of $\mathcal{R}$ by a simple reachability test. Unfortunately, the problem of deciding whether a given $\mathcal{SHQ}$-KB $\mathcal{K} = (\mathcal{A}, \mathcal{T}, \mathcal{R})$ is consistent is undecidable in general [19]. To regain decidability, we need to make the following syntactic restriction: if $\leqslant n\, r.C$ occurs in $\mathcal{K}$, then r must be simple w.r.t. $\mathcal{R}$.

To better separate the influence the ABox has on the complexity of reasoning, we assume in the following that assertions use only *names* that must also occur in the TBox or the RBox. One can still simulate a complex concept assertion $C(a)$ using $A(a)$ and $A \equiv C$, where the latter stands for $A \sqsubseteq C$ and $C \sqsubseteq A$.

Before we can define temporal queries, we need to lift the notions of knowledge bases and interpretations to a temporal setting. We assume that there are designated sets $N_{RC} \subseteq N_C$ of *rigid concept names* and $N_{RR} \subseteq N_R$ of *rigid role names*, whose interpretation does not change over time. All individual names are implicitly rigid. A concept or role name that is not rigid is called *flexible*.

Definition 3 (TKB). *A tuple $\mathcal{K} = ((\mathcal{A}_i)_{0 \leq i \leq n}, \mathcal{T}, \mathcal{R})$, consisting of a finite sequence of ABoxes $\mathcal{A}_i$, a TBox $\mathcal{T}$, and an RBox $\mathcal{R}$, is called a* temporal knowledge base *(TKB). Let $\mathfrak{J} = (\mathcal{I}_i)_{i \geq 0}$ be an infinite sequence of interpretations $\mathcal{I}_i = (\Delta, \cdot_i^{\mathcal{I}})$ over a fixed domain Δ. Then $\mathfrak{J}$ is a model of $\mathcal{K}$ (written $\mathfrak{J} \models \mathcal{K}$) if*

- $\mathcal{I}_i \models \mathcal{A}_i$ *for all $i, 0 \leq i \leq n$,*
- $\mathcal{I}_i \models \mathcal{T}$ *and $\mathcal{I}_i \models \mathcal{R}$ for all $i \geq 0$, and*
- $\mathfrak{J}$ *respects* rigid names, *i.e., $x^{\mathcal{I}_i} = x^{\mathcal{I}_j}$ for all $x \in N_I \cup N_{RC} \cup N_{RR}$ and $i, j \geq 0$.*

We denote the set of all individual names occurring in a TKB $\mathcal{K}$ by $\mathsf{Ind}(\mathcal{K})$. TCQs are defined by combining conjunctive queries via the operators of LTL [4,26].

Definition 4 (Syntax of TCQs). *Let N_V be a set of variables. A conjunctive query (CQ) is of the form $\exists y_1, \ldots, y_m.\psi$, where $y_1, \ldots, y_m \in N_V$ and ψ is a finite conjunction of atoms of the form $A(z_1)$ (concept atom), $r(z_1, z_2)$ (role atom), or $z_1 \approx z_2$ (equality atom), where $A \in N_C$, $r \in N_R$, and $z_1, z_2 \in N_V \cup N_I$.*

Temporal conjunctive queries (TCQs) *are built inductively from CQs, using the constructors* $\neg\phi_1$ *(negation),* $\phi_1 \wedge \phi_2$ *(conjunction),* $\bigcirc\phi_1$ *(next),* $\bigcirc^-\phi_1$ *(previous),* $\phi_1 \cup \phi_2$ *(until), and* $\phi_1 S \phi_2$ *(since), where* ϕ_1 *and* ϕ_2 *are TCQs.*

In contrast to [4], we allow non-simple roles to occur in CQs. A *union of conjunctive queries (UCQ)* is a disjunction of CQs, defined as $\phi_1 \vee \phi_2 := \neg(\neg\phi_1 \wedge \neg\phi_2)$. A *CQ-literal* is either a CQ or a negated CQ. We denote the set of individual names occurring in a TCQ ϕ by $\mathsf{Ind}(\phi)$, the set of variables occurring in ϕ by $\mathsf{Var}(\phi)$, and the set of free variables of ϕ by $\mathsf{FVar}(\phi)$. If $\mathsf{FVar}(\phi) = \emptyset$, we call ϕ *Boolean*. As in [4], we assume without loss of generality that all CQs are connected, i.e., all variables and individual names are related (transitively) by roles.

Definition 5 (Semantics of TCQs). *An interpretation* $\mathcal{I} = (\Delta, \cdot^{\mathcal{I}})$ *is a model of a Boolean CQ* ϕ *(written* $\mathcal{I} \models \phi$*) if there is a* homomorphism *of* ϕ *into* $\mathcal{I}$*, which is a mapping* $\pi\colon \mathsf{Var}(\phi) \cup \mathsf{Ind}(\phi) \to \Delta$ *with* $\pi(a) = a^{\mathcal{I}}$ *for all* $a \in \mathsf{Ind}(\phi)$*;* $\pi(z) \in A^{\mathcal{I}}$ *for all concept atoms* $A(z)$ *in* ϕ*;* $(\pi(z_1), \pi(z_2)) \in r^{\mathcal{I}}$ *for all role atoms* $r(z_1, z_2)$ *in* ϕ*; and* $\pi(z_1) = \pi(z_2)$ *for all equality atoms* $z_1 \approx z_2$ *in* ϕ*.*

An infinite sequence of interpretations $\mathfrak{I} = (\mathcal{I}_i)_{i \geq 0}$ *over a common domain* Δ *is a* model *of a Boolean TCQ* ϕ *at time point* $i \geq 0$ *iff* $\mathfrak{I}, i \models \phi$ *holds, where*

$$
\begin{array}{lll}
\mathfrak{I}, i \models \exists y_1, \ldots, y_m.\psi & \text{iff} & \mathcal{I}_i \models \exists y_1, \ldots, y_m.\psi \\
\mathfrak{I}, i \models \neg\phi_1 & \text{iff} & \mathfrak{I}, i \not\models \phi_1 \\
\mathfrak{I}, i \models \phi_1 \wedge \phi_2 & \text{iff} & \mathfrak{I}, i \models \phi_1 \text{ and } \mathfrak{I}, i \models \phi_2 \\
\mathfrak{I}, i \models \bigcirc\phi_1 & \text{iff} & \mathfrak{I}, i+1 \models \phi_1 \\
\mathfrak{I}, i \models \bigcirc^-\phi_1 & \text{iff} & i > 0 \text{ and } \mathfrak{I}, i-1 \models \phi_1 \\
\mathfrak{I}, i \models \phi_1 \cup \phi_2 & \text{iff} & \text{there is } k \geq i \text{ with } \mathfrak{I}, k \models \phi_2 \\
& & \text{and } \mathfrak{I}, j \models \phi_1 \text{ for all } j, i \leq j < k \\
\mathfrak{I}, i \models \phi_1 S \phi_2 & \text{iff} & \text{there is } k, 0 \leq k \leq i \text{ with } \mathfrak{I}, k \models \phi_2 \\
& & \text{and } \mathfrak{I}, j \models \phi_1 \text{ for all } j, k < j \leq i
\end{array}
$$

Given a TKB $\mathcal{K} = ((\mathcal{A}_i)_{0 \leq i \leq n}, \mathcal{T}, \mathcal{R})$*, we say that* $\mathfrak{I}$ *is a* model *of* ϕ *w.r.t.* $\mathcal{K}$ *if* $\mathfrak{I} \models \mathcal{K}$ *and* $\mathfrak{I}, n \models \phi$*. We call* ϕ satisfiable *w.r.t.* $\mathcal{K}$ *if it has a model w.r.t.* $\mathcal{K}$*, and it is* entailed *by* $\mathcal{K}$ *(written* $\mathcal{K} \models \phi$*) if every model* $\mathfrak{I}$ *of* $\mathcal{K}$ *satisfies* $\mathfrak{I}, n \models \phi$*.*

For a TCQ ϕ*,* $\mathfrak{a}\colon \mathsf{FVar}(\phi) \to \mathsf{Ind}(\mathcal{K})$ *is a* certain answer *to* ϕ *w.r.t.* $\mathcal{K}$ *if* $\mathcal{K} \models \mathfrak{a}(\phi)$*, where* $\mathfrak{a}(\phi)$ *is obtained from* ϕ *by replacing the free variables using* $\mathfrak{a}$*.*

As usual [4], in the following we consider only the TCQ entailment problem, which can be used to compute all certain answers. For this purpose, we analyze the satisfiability problem, which has the same complexity as non-entailment. We examine both the *combined complexity* where the whole TKB and the TCQ are considered as the input, and the *data complexity*, where TBox, RBox, and TCQ are fixed, i.e., the complexity is measured only w.r.t. the sequence of ABoxes.

In the remainder of this section, we recall the basic approach from [4] to decide satisfiability by splitting it into two separate satisfiability problems, one for the temporal component and one for the DL component. In the following, let $\mathcal{K} = ((\mathcal{A}_i)_{0 \leq i \leq n}, \mathcal{T}, \mathcal{R})$ be a TKB and ϕ be a TCQ to be checked for satisfiability. The *propositional abstraction* ϕ^p of ϕ is the propositional

LTL-formula obtained from ϕ by replacing all CQs by propositional variables. We assume that $\alpha_1, \ldots, \alpha_m$ are the CQs occurring in ϕ, and that each α_i is replaced by the propositional variable p_i, $1 \leq i \leq m$. We now consider a set $\mathcal{S} = \{X_1, \ldots, X_k\} \subseteq 2^{\{p_1, \ldots, p_m\}}$, which intuitively specifies the worlds that are allowed to occur in an LTL-structure satisfying ϕ^{p} at time point n, and a mapping $\iota \colon \{0, \ldots, n\} \to \{1, \ldots, k\}$, which assigns a world $X_{\iota(i)}$ to each input ABox $\mathcal{A}_i$.

Definition 6 (t-satisfiability). *The LTL-formula ϕ^{p} is t-satisfiable w.r.t. $\mathcal{S}$ and ι if there exists an LTL-structure $\mathfrak{J} = (w_i)_{i \geq 0}$ such that $\mathfrak{J}, n \models \phi^{\mathsf{p}}$, $w_i \in \mathcal{S}$ for all $i \geq 0$, and $w_i = X_{\iota(i)}$ for all i, $0 \leq i \leq n$.*

However, finding $\mathcal{S}$ and ι and then testing t-satisfiability is not sufficient for checking whether ϕ has a model w.r.t. $\mathcal{K}$. We must also check whether $\mathcal{S}$ can indeed be induced by some sequence of interpretations that is a model of $\mathcal{K}$, in the following sense.

Definition 7 (r-satisfiability). *The set $\mathcal{S}$ is r-satisfiable w.r.t. ι and $\mathcal{K}$ if there exist interpretations $\mathcal{J}_1, \ldots, \mathcal{J}_k, \mathcal{I}_0, \ldots, \mathcal{I}_n$ that share the same domain, respect rigid names, are models of $\mathcal{T}$ and $\mathcal{R}$, and additionally each $\mathcal{J}_i$, $1 \leq i \leq k$, is a model of $\chi_i := \bigwedge_{p_j \in X_i} \alpha_j \wedge \bigwedge_{p_j \notin X_i} \neg \alpha_j$, and each $\mathcal{I}_i$, $0 \leq i \leq n$, is a model of $\mathcal{A}_i$ and $\chi_{\iota(i)}$.*

The following was shown in [4] for $\mathcal{SHQ}$, but is actually independent of any specific DL.

Proposition 8. *ϕ is satisfiable w.r.t. $\mathcal{K}$ iff there are a set $\mathcal{S}$ and a mapping ι such that $\mathcal{S}$ is r-satisfiable w.r.t. ι and $\mathcal{K}$, and ϕ^{p} is t-satisfiable w.r.t. $\mathcal{S}$ and ι.*

The complexity of the t-satisfiability problem is obviously also DL-agnostic, and hence we can reuse another result from [4].

Proposition 9. *Deciding t-satisfiability of ϕ^{p} w.r.t. $\mathcal{S}$ and ι can be done in EXP w.r.t. combined complexity, and in P w.r.t. data complexity.*

Table 2 gives an overview over all known complexity results for TCQ entailment. We distinguish the cases that (i) no rigid names are allowed ($\mathsf{N_{RC}} = \mathsf{N_{RR}} = \emptyset$); (ii) only rigid concept names are allowed, but no rigid role names ($\mathsf{N_{RR}} = \emptyset$); and (iii) arbitrary rigid names are allowed. The first row of the table contains the known results for $\mathcal{ALC}/\mathcal{ALCHQ}$ [4][1], and in this paper we derive the upper bounds for cases (ii) and (iii) marked in bold font. Unfortunately, we leave open the precise data complexity for case (iii), as was the case in [4]. A question mark indicates that the precise complexity is unknown even for the atemporal CQ entailment problem. For $\mathcal{SHOIQ}$, it is not even known whether this problem is decidable, while for $\mathcal{ALCHOIQ}$ it is only known to be decidable, but no better upper bound has been found so far [25,28]. The shown lower bounds follow from the complexity of satisfiability of $\mathcal{ALC}$-LTL formulae [4,6] and the complexity of atemporal CQ entailment. More precisely, the latter problem is CO-NP-hard

---

[1] Actually, that paper considers $\mathcal{SHQ}$, but restricts the roles in CQs to be simple.

Table 2. Summary of known and new complexity results for TCQ entailment, where contributions of this paper are highlighted in boldface. Settings: (i) no rigid names are allowed, (ii) only rigid concept names are allowed, and (iii) arbitrary rigid names are allowed.

	Data complexity			Combined complexity		
	(i)	(ii)	(iii)	(i)	(ii)	(iii)
$\mathcal{ALC} - \mathcal{ALCHQ}$ [4]	co-NP	co-NP	$\leq$ Exp	Exp	co-NExp	2-Exp
$\mathcal{ALCO} - \mathcal{ALCHOQ}/\mathcal{ALCHOI}$	co-NP	**co-NP**	$\leq$ **Exp**	$\geq$ co-NExp	?	**2-Exp**
$\mathcal{S} - \mathcal{SQ}$	co-NP	**co-NP**	$\leq$ **Exp**	$\geq$ co-NExp	?	**2-Exp**
$\mathcal{SO} - \mathcal{SOQ}$	$\geq$co-NP	?	$\leq$ **Exp**	$\geq$ co-NExp	?	**2-Exp**
$\mathcal{SH}/\mathcal{ALCI} - \mathcal{SHIQ}$	co-NP	**co-NP**	$\leq$ **Exp**	2-Exp	2-Exp	**2-Exp**
$\mathcal{SHO} - \mathcal{SHOQ}/\mathcal{SHOI}$	$\geq$co-NP	?	$\leq$ **Exp**	2-Exp	2-Exp	**2-Exp**
$\mathcal{ALCOIQ} - \mathcal{ALCHOIQ}$	$\geq$co-NP	?	**decidable**	$\geq$co-2-NExp	?	**decidable**
$\mathcal{SOIQ} - \mathcal{SHOIQ}$	$\geq$co-NP	?	?	$\geq$co-2-NExp	?	?

in data complexity already for $\mathcal{ALE}$ [29]. Under combined complexity, it is co-NExp-hard for $\mathcal{ALCO}$ [23] and $\mathcal{S}$ [15], 2-Exp-hard for $\mathcal{SH}$ [15] and $\mathcal{ALCI}$ [21], and co-2-NExp-hard for $\mathcal{ALCOIQ}$ [18].

3   Atemporal Queries in $\mathcal{SHIQ}$, $\mathcal{SHOQ}$, and $\mathcal{SHOI}$

Since our results about TCQ entailment are based on reductions to conjunctions of CQ-literals, we first analyze in more detail the case of such atemporal queries. In a nutshell, we reduce the satisfiability of such a conjunction to UCQ non-entailment and exploit existing algorithms for this problem. We consider only the logics $\mathcal{SHIQ}$, $\mathcal{SHOQ}$, and $\mathcal{SHOI}$ that have the *quasi-forest model property* [12], which means that every consistent KB formulated in one of these logics has a model that basically consists of several tree-shaped structures whose roots are arbitrarily interconnected by roles (disregarding role connections due to nominals or transitive roles).

To show the results in the following sections, however, we need to conduct a more fine-grained analysis of the complexity of the atemporal query entailment algorithms. The main insight is that, while UCQ entailment in $\mathcal{SHIQ}$, $\mathcal{SHOQ}$, and $\mathcal{SHOI}$ is in 2-Exp w.r.t. combined complexity, the *number* of CQs in the UCQ only has an exponential influence on the complexity of this decision problem. Likewise, for data complexity, assuming that the number of CQs in the UCQ is linear instead of constant usually has no influence on the complexity. Unfortunately, to the best of our knowledge, the precise data complexity of UCQ entailment is known only for $\mathcal{SHIQ}$, $\mathcal{ALCHOQ}$, and $\mathcal{ALCHOI}$, while for $\mathcal{SHOQ}$ and $\mathcal{SHOI}$ it is still open [25].

In the following, we consider the *size* of a CQ ψ (written $|\psi|$) to be the number of symbols in ψ, ignoring constant expressions like '(' and '$\wedge$', considering each name and variable to be of size 1, and further ignoring the prefix $\exists y_1, \ldots, y_m$ since these variables also occur in the atoms of ψ. For example, $\exists x, y.r(x,y) \wedge A(x)$ has size 5. We could also assume that each name or variable is represented by a binary string denoting its name, and hence of size logarithmic in the size of ψ, but this would not affect our complexity results. Similarly, the size of a knowledge base is computed by ignoring the concept constructors, and hence considers only the number of occurrences of names in the axioms.

Lemma 10. *Let* $\psi = \rho_1 \wedge \cdots \wedge \rho_\ell \wedge \neg\sigma_1 \cdots \wedge \neg\sigma_o$ *be a Boolean conjunction of CQ-literals,* $\mathcal{K} = (\mathcal{A}, \mathcal{T}, \mathcal{R})$ *be a KB formulated in* $\mathcal{SHIQ}$, $\mathcal{SHOQ}$, *or* $\mathcal{SHOI}$, *and* $\|\psi\| := \max\{|\rho_1|, \ldots, |\rho_\ell|, |\sigma_1|, \ldots, |\sigma_o|\}$. *Then the satisfiability of* ψ *w.r.t.* $\mathcal{K}$ *can be decided by a deterministic algorithm in time bounded by* $2^{p(\ell, o, |\mathcal{K}|)^{p'(\|\psi\|)}}$, *for two polynomials* p *and* p'.

In the case of $\mathcal{SHIQ}$, $\mathcal{ALCHOQ}$, *or* $\mathcal{ALCHOI}$, *if* $\mathcal{T}$, $\mathcal{R}$, *and* $\|\psi\|$ *are fixed, then satisfiability of* ψ *w.r.t.* $\mathcal{K}$ *can be decided by a nondeterministic algorithm in time bounded by* $p(\ell, o, |\mathcal{A}|)$ *for some polynomial* p.

Proof. As in [4], we reduce the decision whether ψ has a model w.r.t. $\mathcal{K}$ to a UCQ non-entailment problem. We instantiate the positive CQs $\rho_1, \ldots, \rho_\ell$ by omitting the existential quantifiers and replacing all variables by fresh individual names. The set $\mathcal{A}_\rho$ of all resulting assertions can be viewed as an additional ABox. To ensure that the UNA is satisfied, we additionally consider equivalence relations $\approx$ on $\mathsf{Ind}(\mathcal{A} \cup \mathcal{A}_\rho)$ with the additional restriction that no two names from $\mathsf{Ind}(\mathcal{A})$ may be equivalent. We denote by $\mathcal{A}_\approx$ the ABox resulting from $\mathcal{A}_\rho$ by replacing each new individual name by a fixed representative of its equivalence class, where this representative is an element of $\mathsf{Ind}(\mathcal{A})$ whenever possible. It can be shown as in [4] that ψ is satisfiable w.r.t. $\mathcal{K}$ iff there is an equivalence relation $\approx$ with

$$(\mathcal{A} \cup \mathcal{A}_\approx, \mathcal{T}, \mathcal{R}) \not\models \sigma_1 \vee \cdots \vee \sigma_o. \tag{1}$$

Note that the number of equivalence relations $\approx$ is exponential in the total number of variables in $\rho_1, \ldots, \rho_\ell$, which is bounded by $\ell \cdot \|\psi\|$, but each is of size polynomial in $\ell \cdot \|\psi\|$. Hence, one can either enumerate all such equivalences in time exponential in $\ell \cdot \|\psi\|$, or guess one of them in time polynomial in $\ell \cdot \|\psi\|$.

We now consider the case that $\mathcal{K}$ is formulated in $\mathcal{SHIQ}$. By [16, Lemma 23], for the non-entailment test (1), it suffices to find a so-called *extended knowledge base* $\mathcal{K}' = (\mathcal{A} \cup \mathcal{A}_\approx \cup \mathcal{A}', \mathcal{T} \cup \mathcal{T}', \mathcal{R})$, where $\mathcal{A}'$ and $\mathcal{T}'$ are formulated in $\mathcal{SHIQ}^\sqcap$, i.e., $\mathcal{SHIQ}$ extended by role conjunctions, such that $\mathcal{K}'$ is consistent. By [16, Lemma 20 and Definition 21], the size of each $(\mathcal{A}', \mathcal{T}')$ is bounded by $p(o \cdot (|\mathcal{K}| + \ell \cdot \|\psi\|))^{p(\|\psi\|)}$ for some polynomial p, where the term $\ell \cdot \|\psi\|$ represents the size of the additional ABox $\mathcal{A}_\approx$. The bound given in [16] is exponential in the total size of the UCQ, i.e., $o \cdot \|\psi\|$, but the exponential blowup comes only from the rewriting of each individual CQ σ_i. Moreover, all pairs $(\mathcal{A}', \mathcal{T}')$ can be enumerated in time bounded by $2^{p(o \cdot (|\mathcal{K}| + \ell \cdot \|\psi\|))^{p(\|\psi\|)}}$. It is important to note that the size of the longest role conjunction occurring in $(\mathcal{A}', \mathcal{T}')$ is bounded by a polynomial in $\|\psi\|$. Hence, by [16, Lemma 28], one can check the consistency of $\mathcal{K}'$ in time $2^{p'(o \cdot (|\mathcal{K}| + \ell \cdot \|\psi\|))^{p'(\|\psi\|)}}$ for some polynomial p'. Thus, we can decide satisfiability of ψ w.r.t. $\mathcal{K}$ by enumerating all relations $\approx$ and extended KBs as above and testing each of them for consistency within the claimed time bound.

If $\mathcal{T}$, $\mathcal{R}$, and $\|\psi\|$ are fixed, then one can guess $\approx$ in time polynomial in ℓ. Following the proof of [16, Theorem 35], one can also guess $\mathcal{K}'$ in time $p(o \cdot (|\mathcal{A}| + \ell))$, and the following consistency test can be done in (deterministic) polynomial time in the size of the ABox $\mathcal{A} \cup \mathcal{A}_{\approx} \cup \mathcal{A}'$, which is polynomial in $o \cdot (|\mathcal{A}| + \ell)$. This establishes the second bound for the case of $\mathcal{SHIQ}$.

The proof of the remaining cases can be found in the technical report. For $\mathcal{SHOQ}$, we use algorithms developed in [17,20]. For $\mathcal{SHOI}$, we analyze the automata-based construction from [12,13] based on fully enriched automata [9]. For $\mathcal{ALCHOQ}$ and $\mathcal{ALCHOI}$ under the assumption that $\mathcal{T}$, $\mathcal{R}$, and $\|\psi\|$ are fixed, we obtain the claimed results using a tableaux algorithm introduced in [24]. $\qquad\square$

4 Combined Complexity of TCQ Entailment

Let $\mathcal{K} = ((\mathcal{A}_i)_{0 \le i \le n}, \mathcal{T}, \mathcal{R})$ be a TKB, ϕ be a TCQ, and assume for now that a set $\mathcal{S} = \{X_1, \dots, X_k\} \subseteq 2^{\{p_1, \dots, p_m\}}$ and a mapping $\iota \colon \{0, \dots, n\} \to \{1, \dots, k\}$ are given. For our complexity results, we employ the *copying* technique from [4,6] for deciding whether $\mathcal{S}$ is r-satisfiable w.r.t. ι and $\mathcal{K}$. The idea is to introduce enough copies of all flexible names in order to combine the separate satisfiability tests of Definition 7 into one big atemporal satisfiability test.

Formally, for all i, $1 \le i \le k+n+1$, and every *flexible* concept name A (every *flexible* role name r) occurring in $\mathcal{T}$ or $\mathcal{R}$, we introduce a copy $A^{(i)}$ $(r^{(i)})$. We call $A^{(i)}$ $(r^{(i)})$ the i-th copy of A (r). The conjunctive query $\alpha^{(i)}$ (the axiom $\beta^{(i)}$) is obtained from a CQ α (an axiom β) by replacing every flexible name by its i-th copy. Similarly, for $1 \le \ell \le k$, the conjunction of CQ-literals $\chi_\ell^{(i)}$ is obtained from χ_ℓ (see Definition 7) by replacing each CQ α_j by $\alpha_j^{(i)}$. Finally, we define

$$\chi_{\mathcal{S},\iota} := \bigwedge_{1 \le i \le k} \chi_i^{(i)} \wedge \bigwedge_{0 \le i \le n} \left(\chi_{\iota(i)}^{(k+i+1)} \wedge \bigwedge_{\alpha \in \mathcal{A}_i} \alpha^{(k+i+1)} \right),$$

$$\mathcal{T}_{\mathcal{S},\iota} := \{\beta^{(i)} \mid \beta \in \mathcal{T} \text{ and } 1 \le i \le k+n+1\},$$

$$\mathcal{R}_{\mathcal{S},\iota} := \{\gamma^{(i)} \mid \gamma \in \mathcal{R} \text{ and } 1 \le i \le k+n+1\}.$$

The following result, which reduces r-satisfiability to an atemporal satisfiability problem, was shown in [4] for $\mathcal{SHQ}$ with simple roles in queries, but it remains valid in our setting since it does not depend on the DL under consideration.

Proposition 11. *The set $\mathcal{S}$ is r-satisfiable w.r.t. ι and $\mathcal{K}$ iff $\chi_{\mathcal{S},\iota}$ is satisfiable w.r.t. $(\mathcal{T}_{\mathcal{S},\iota}, \mathcal{R}_{\mathcal{S},\iota})$.*

Together with Lemma 10, this allows us to show our first complexity results.

Theorem 12. *Let $\mathcal{L}$ be a DL that contains $\mathcal{ALCI}$ or $\mathcal{SH}$ and is contained in $\mathcal{SHIQ}$, $\mathcal{SHOQ}$, or $\mathcal{SHOI}$. Then TCQ entailment in $\mathcal{L}$ is 2-EXP-complete w.r.t. combined complexity, and in EXP w.r.t. data complexity.*

Proof. The lower bound directly follows from 2-EXP-hardness of CQ entailment in $\mathcal{SH}$ [15] and $\mathcal{ALCI}$ [21]. To check a TCQ ϕ for satisfiability w.r.t. a TKB $\mathcal{K}$, we first enumerate all possible sets $\mathcal{S}$ and mappings ι, which can be done in 2-EXP. For each of these double exponentially many pairs $(\mathcal{S}, \iota)$, we then check t-satisfiability of ϕ^p w.r.t. $\mathcal{S}$ and ι in exponential time (see Proposition 9) and test $\mathcal{S}$ for r-satisfiability w.r.t. ι and $\mathcal{K}$ (using Proposition 11). By Proposition 8, ϕ has a model w.r.t. $\mathcal{K}$ iff at least one pair passes both tests.

For the r-satisfiability test, observe that the conjunction of CQ-literals $\chi_{\mathcal{S},\iota}$ contains exponentially many (negated) CQs, each of size polynomial in the size of ϕ, and that $\mathcal{T}_{\mathcal{S},\iota}$ and $\mathcal{R}_{\mathcal{S},\iota}$ are of exponential size in the size of $\mathcal{K}$. By Lemma 10, the satisfiability of $\chi_{\mathcal{S},\iota}$ w.r.t. $(\mathcal{T}_{\mathcal{S},\iota}, \mathcal{R}_{\mathcal{S},\iota})$ can thus be checked in double exponential time in the size of ϕ and $\mathcal{K}$. For the data complexity, observe that the number of CQs in $\chi_{\mathcal{S},\iota}$ is linear in the size of the input ABoxes, and their size only depends on ϕ (the size of a single assertion is constant). Moreover, $\mathcal{T}_{\mathcal{S},\iota}$ and $\mathcal{R}_{\mathcal{S},\iota}$ are of size linear in n. Lemma 10 thus yields the claimed upper bound.   □

By the same arguments, it is easy to see that TCQ entailment in $\mathcal{ALCHOIQ}$ is decidable since this is the case for UCQ (non-)entailment [28].

5   Data Complexity Without Rigid Roles

To obtain a tight bound on the data complexity if we disallow rigid role names, we follow a different approach from [4]. Similarly to the previous section, we decide r-satisfiability of $\mathcal{S}$ w.r.t. ι and $\mathcal{K}$ by constructing conjunctions of CQ-literals of which we want to check satisfiability. However, we do not compile the whole r-satisfiability check into just one conjunction. More precisely, we define the conjunctions of CQ-literals $\gamma_i \wedge \chi_{\mathcal{S}}$, $0 \le i \le n$, w.r.t. $(\mathcal{T}_{\mathcal{S}}, \mathcal{R}_{\mathcal{S}})$, where

$$\gamma_i := \bigwedge_{\alpha \in \mathcal{A}_i} \alpha^{(\iota(i))}, \quad \chi_{\mathcal{S}} := \bigwedge_{1 \le i \le k} \chi_i^{(i)},$$

$$\mathcal{T}_{\mathcal{S}} := \{\beta^{(i)} \mid \beta \in \mathcal{T} \text{ and } 1 \le i \le k\},$$

$$\mathcal{R}_{\mathcal{S}} := \{\gamma^{(i)} \mid \gamma \in \mathcal{R} \text{ and } 1 \le i \le k\}.$$

This separates the consistency checks for the individual ABoxes $\mathcal{A}_i$, $1 \le i \le n$, from each other. For r-satisfiability, we additionally have to make sure that rigid consequences of the form $A(a)$ for a rigid concept name $A \in \mathsf{N_{RC}}$ and an individual name $a \in \mathsf{N_I}$ are shared between all the conjunctions $\gamma_i \wedge \chi_{\mathcal{S}}$. It suffices to do this for the set $\mathsf{RCon}(\mathcal{T})$ of rigid concept names occurring in $\mathcal{T}$ since those that occur only in ABox assertions cannot affect the entailment of the TCQ ϕ.

For this purpose, we guess a set $\mathcal{D} \subseteq 2^{\mathsf{RCon}(\mathcal{T})}$ that fixes the combinations of rigid concept names that are allowed to occur in the models of $\gamma_i \wedge \chi_{\mathcal{S}}$, and a function $\tau \colon \mathsf{Ind}(\phi) \cup \mathsf{Ind}(\mathcal{K}) \to \mathcal{D}$ that assigns to each individual name one such combination. To express this formally, we extend the TBox by the axioms in

$$\mathcal{T}_{\mathcal{D}} := \{A_Y \equiv C_Y \mid Y \in \mathcal{D}\},$$

where A_Y are fresh rigid concept names and, for every $Y \subseteq \mathsf{RCon}(\mathcal{T})$,

$$C_Y := \bigsqcap_{A \in Y} A \sqcap \bigsqcap_{A \in \mathsf{RCon}(\mathcal{T}) \setminus Y} \neg A.$$

The size of $\mathcal{T}_\tau$ is bounded polynomially in the sizes of $\mathcal{D}$ and $\mathsf{RCon}(\mathcal{T})$, which are constant w.r.t. data complexity. We now extend the conjunctions $\gamma_i \wedge \chi_{\mathcal{S}}$ by

$$\rho_\tau := \bigwedge_{a \in \mathsf{Ind}(\phi) \cup \mathsf{Ind}(\mathcal{K})} A_{\tau(a)}(a)$$

in order to fix the behavior of the rigid concept names on the named individuals.

We need one more definition to formulate the main lemma of this section. We say that an interpretation $\mathcal{I}$ *respects* $\mathcal{D}$ if

$$\mathcal{D} = \{Y \subseteq \mathsf{RCon}(\mathcal{T}) \mid \text{there is a } d \in \Delta^{\mathcal{I}} \text{ with } d \in (C_Y)^{\mathcal{I}}\},$$

which means that every combination of rigid concept names in $\mathcal{D}$ is realized by a domain element of $\mathcal{I}$, and conversely, the domain elements of $\mathcal{I}$ may only realize those combinations that occur in $\mathcal{D}$.

Lemma 13. *Let the DL $\mathcal{L}$ be contained in $\mathcal{SHIQ}$, $\mathcal{ALCHOQ}$, or $\mathcal{ALCHOI}$. If $\mathsf{N_{RR}} = \emptyset$, then $\mathcal{S}$ is r-satisfiable w.r.t. ι and $\mathcal{K}$ iff there exist $\mathcal{D} \subseteq 2^{\mathsf{RCon}(\mathcal{T})}$ and $\tau \colon \mathsf{Ind}(\phi) \cup \mathsf{Ind}(\mathcal{K}) \to \mathcal{D}$ such that each $\gamma_i \wedge \chi_{\mathcal{S}} \wedge \rho_\tau$, $0 \le i \le n$, has a model w.r.t. $(\mathcal{T}_{\mathcal{S}} \cup \mathcal{T}_{\mathcal{D}}, \mathcal{R}_{\mathcal{S}})$ that respects $\mathcal{D}$.* □

The restriction imposed by $\mathcal{D}$ can be expressed as the conjunction of CQ-literals

$$\sigma_{\mathcal{D}} := (\neg \exists x. A_{\mathcal{D}}(x)) \wedge \bigwedge_{Y \in \mathcal{D}} \exists x. A_Y(x),$$

where $A_{\mathcal{D}}$ is a fresh concept names that is restricted by adding the axiom $A_{\mathcal{D}} \equiv \bigsqcap_{Y \in \mathcal{D}} \neg A_Y$ to the TBox. We denote by $\mathcal{T}'_{\mathcal{S}}$ the resulting extension of $\mathcal{T}_{\mathcal{S}} \cup \mathcal{T}_{\mathcal{D}}$, and have now reduced the r-satisfiability of $\mathcal{S}$ w.r.t. ι and $\mathcal{K}$ to the consistency of $\gamma_i \wedge \chi_{\mathcal{S}} \wedge \rho_\tau \wedge \sigma_{\mathcal{D}}$ w.r.t. $(\mathcal{T}'_{\mathcal{S}}, \mathcal{R}_{\mathcal{S}})$.

Theorem 14. *Let $\mathcal{L}$ be a DL that contains $\mathcal{ALE}$ and is contained in $\mathcal{SHIQ}$, $\mathcal{ALCHOQ}$, or $\mathcal{ALCHOI}$. Then TCQ entailment in $\mathcal{L}$ is CO-NP-complete w.r.t. data complexity.*

Proof. The lower bound follows from CO-NP-hardness of instance checking in $\mathcal{ALE}$ [29]. To test satisfiability of a TCQ ϕ w.r.t. a TKB $\mathcal{K}$, we employ the same approach as before, but instead guess $\mathcal{S}$ and ι. Since $\mathcal{S}$ is of constant size in the size of the ABoxes and ι is of linear size, this can be done in nondeterministic polynomial time. The t-satisfiability test for Proposition 8 can be done in polynomial time by Proposition 9, and for the r-satisfiability test, we use Lemma 13.

Following the reduction described above, we guess a set $\mathcal{D} \subseteq 2^{\mathsf{RCon}(\mathcal{T})}$ and a function $\tau \colon \mathsf{Ind}(\phi) \cup \mathsf{Ind}(\mathcal{K}) \to \mathcal{D}$, which can be done in nondeterministic polynomial time since $\mathcal{D}$ only depends on $\mathcal{T}$ and τ is of size linear in the size of the input ABoxes. Next, we check the satisfiability of the polynomially many conjunctions $\gamma_i \wedge \chi_{\mathcal{S}} \wedge \rho_\tau \wedge \sigma_{\mathcal{D}}$ w.r.t. $(\mathcal{T}'_{\mathcal{S}}, \mathcal{R}_{\mathcal{S}})$. Note that $\chi_{\mathcal{S}}$, $\sigma_{\mathcal{D}}$, $\mathcal{T}'_{\mathcal{S}}$, and $\mathcal{R}_{\mathcal{S}}$ do not depend on the input ABoxes, while γ_i and ρ_τ are of polynomial size. Furthermore, the size of the CQs in γ_i and ρ_τ is constant. Hence, Lemma 10 yields the desired NP upper bound for these satisfiability tests. □

6   Conclusions

Query answering w.r.t. DL ontologies is currently a very active research area. We have extended complexity results for very expressive DLs underlying the web ontology language OWL 2 to the case of temporal queries. Our results show that, w.r.t. worst-case complexity, adding a temporal dimension often comes for free. In fact, in all sublogics of $\mathcal{SHOIQ}$, the upper bounds for the combined complexity of TCQ entailment obtained in this paper for the temporal case coincide with the best known upper bounds for atemporal query entailment (even in the presence of rigid roles). From the application point of view, data complexity is more important since the amount of data is often very large, and in comparison the size of the background knowledge and the user query is small. We have shown that, in many cases, the atemporal data complexity of CO-NP does not increase if we consider TCQs with rigid concepts (specifically, in $\mathcal{ALCHOQ}$, $\mathcal{ALCHOI}$, $\mathcal{SHIQ}$, and sublogics). For the remaining logics of Table 2, it is an open problem to find a CO-NP algorithm even in the atemporal case.

As part of future work, we will try to obtain CO-NP upper bounds even in the presence of rigid roles, and study extensions of TCQs with concrete domains and inconsistency-tolerant semantics. Since CO-NP is already a rather negative result for data complexity, we could also try to find restricted formalisms with lower data complexity. On the one hand, one could take a less expressive DL to formulate the background ontology, which has already been investigated for $\mathcal{EL}$ [11] and $DL\text{-}Lite_{horn}^{\mathcal{H}}$ [10], but only the latter choice reduces the data complexity (to ALOGTIME). On the other hand, one could investigate whether the data complexity can be reduced by imposing additional restrictions on the TBox or CQs, as has been done in the atemporal case [7,8,22].

References

1. Artale, A., Kontchakov, R., Wolter, F., Zakharyaschev, M.: Temporal description logic for ontology-based data access. In: Proceedings of IJCAI 2013, pp. 711–717 (2013)
2. Baader, F.: Ontology-based monitoring of dynamic systems. In: Proceedings of KR 2014, pp. 678–681 (2014)
3. Baader, F., Borgwardt, S., Lippmann, M.: Temporal conjunctive queries in expressive DLs with non-simple roles. LTCS-Report 15–17, Chair of Automata Theory, TU Dresden (2015). http://lat.inf.tu-dresden.de/research/reports.html
4. Baader, F., Borgwardt, S., Lippmann, M.: Temporal query entailment in the description logic $\mathcal{SHQ}$. J. Web Semant. 33, 71–93 (2015)
5. Baader, F., Calvanese, D., McGuinness, D.L., Nardi, D., Patel-Schneider, P.F. (eds.): The Description Logic Handbook: Theory, Implementation, and Applications, 2nd edn. Cambridge University Press, Cambridge (2007)
6. Baader, F., Ghilardi, S., Lutz, C.: LTL over description logic axioms. ACM T. Comput. Log. 13(3), 21:1–21:32 (2012)
7. Bienvenu, M., Ortiz, M., Šimkus, M., Xiao, G.: Tractable queries for lightweight description logics. In: Rossi, F. (ed.) Proceedings of IJCAI 2013, pp. 768–774 (2013)

8. Bienvenu, M., ten Cate, B., Lutz, C., Wolter, F.: Ontology-based data access: a study through disjunction datalog, CSP, and MMSNP. ACM T. Database Syst. **39**(4), 33:1–33:44 (2014)
9. Bonatti, P.A., Lutz, C., Murano, A., Vardi, M.Y.: The complexity of enriched μ-calculi. Log. Meth. Comput. Sci. **4**(3:11), 1–27 (2008)
10. Borgwardt, S., Thost, V.: Temporal query answering in DL-Lite with negation. In: Proceedings of GCAI 2015 (2015, to appear)
11. Borgwardt, S., Thost, V.: Temporal query answering in the description logic $\mathcal{EL}$. In: Proceedings of IJCAI 2015, pp. 2819–2825 (2015)
12. Calvanese, D., Eiter, T., Ortiz, M.: Regular path queries in expressive description logics with nominals. In: Proceedings of IJCAI 2009, pp. 714–720 (2009)
13. Calvanese, D., Eiter, T., Ortiz, M.: Answering regular path queries in expressive description logics via alternating tree-automata. Inf. Comput. **237**, 12–55 (2014)
14. Decker, S., Erdmann, M., Fensel, D., Studer, R.: ONTOBROKER: ontology based access to distributed and semi-structured information. In: Meersman, R., Tari, Z., Stevens, S. (eds.) Database Semantics. IFIP, vol. 11, pp. 351–369. Springer, New York (1999)
15. Eiter, T., Lutz, C., Ortiz, M., Šimkus, M.: Query answering in description logics with transitive roles. In: Proceedings of IJCAI 2009, pp. 759–764 (2009)
16. Glimm, B., Horrocks, I., Lutz, C., Sattler, U.: Conjunctive query answering for the description logic $\mathcal{SHIQ}$. J. Artif. Intell. Res. **31**(1), 157–204 (2008)
17. Glimm, B., Horrocks, I., Sattler, U.: Unions of conjunctive queries in $\mathcal{SHOQ}$. In: Proceedings of KR 2008, pp. 252–262 (2008)
18. Glimm, B., Kazakov, Y., Lutz, C.: Status $\mathcal{QIO}$: an update. In: Proceedings of DL 2011, pp. 136–146 (2011)
19. Horrocks, I., Sattler, U., Tobies, S.: Practical reasoning for very expressive description logics. L. J. IGPL **8**(3), 239–263 (2000)
20. Lippmann, M.: Temporalised description logics for monitoring partially observable events. Ph.D. thesis, TU Dresden, Germany (2014)
21. Lutz, C.: The complexity of conjunctive query answering in expressive description logics. In: Armando, A., Baumgartner, P., Dowek, G. (eds.) Automated Reasoning. LNCS, vol. 5195, pp. 179–193. Springer, Heidelberg (2008)
22. Lutz, C., Wolter, F.: Non-uniform data complexity of query answering in description logics. In: Brewka, G., Eiter, T., McIlraith, S.A. (eds.) Proceedings of KR 2012, pp. 297–307 (2012)
23. Ngo, N., Ortiz, M., Šimkus, M.: The combined complexity of reasoning with closed predicates in description logics. In: Proceedings of DL 2015, pp. 249–261 (2015)
24. Ortiz, M., Calvanese, D., Eiter, T.: Data complexity of query answering in expressive description logics via tableaux. J. Autom. Reasoning **41**(1), 61–98 (2008)
25. Ortiz, M., Šimkus, M.: Reasoning and query answering in description logics. In: Reasoning Web. 8th International Summer School, Chap. 1, pp. 1–53 (2012)
26. Pnueli, A.: The temporal logic of programs. In: Proceedings of SFCS 1977, pp. 46–57 (1977)
27. Poggi, A., Calvanese, D., De Giacomo, G., Lembo, D., Lenzerini, M., Rosati, R.: Linking data to ontologies. J. Data Semant. **10**, 133–173 (2008)
28. Rudolph, S., Glimm, B.: Nominals, inverses, counting, and conjunctive queries or: why infinity is your friend!. J. Artif. Intell. Res. **39**(1), 429–481 (2010)
29. Schaerf, A.: On the complexity of the instance checking problem in concept languages with existential quantification. J. Intell. Inf. Syst. **2**(3), 265–278 (1993)

Scaling up Multi-island Competitive Cooperative Coevolution for Real Parameter Global Optimisation

Kavitesh K. Bali[1,2][(✉)] and Rohitash Chandra[1,2]

[1] School of Computing Information and Mathematical Sciences,
University of South Pacific, Suva, Fiji
[2] Artificial Intelligence and Cybernetics Research Group,
Software Foundation, Nausori, Fiji
{bali.kavitesh,c.rohitash}@gmail.com

Abstract. A major challenge in using cooperative coevolution (CC) for global optimisation is the decomposition of a given problem into subcomponents. Variable interaction is a major constraint that determines the decomposition strategy of a problem. Hence, finding an optimal decomposition strategy becomes a burdensome task as interdependencies between decision variables are unknown for these problems. In recent related work, a multi-island competitive cooperative coevolution (MICCC) algorithm was introduced which featured competition and collaboration of several different decomposition strategies. MICCC used five different *uniform* problem decomposition strategies that were implemented as independent islands. This paper presents an analysis of the MICCC algorithm and also extends it to more than five islands. We incorporate arbitrary (*non-uniform*) problem decomposition strategies as additional islands in MICCC and monitor how each different problem decomposition strategy contributes towards the global fitness over different stages of optimisation.

1 Introduction

Cooperative coevolution (CC) [1] is an evolutionary algorithm that implements divide and conquer paradigm to decompose complex problems into subcomponents [2]. Cooperative coevolution (CC) [1] is an explicit means of problem decomposition in the context of evolutionary algorithms (EAs) [3]. A major challenge in using CC for large-scale optimisation is problem decomposition [4]. Without prior knowledge of the internal structure in terms of variable interactions or inter-dependencies [5], it is quite difficult to group interacting variables into an effective decomposition in order to take full advantage of cooperative coevolution. It has been shown that placement of interacting variables into separate subcomponents degrades the optimisation performance significantly [1,6]. To remedy this limitation, several problem decomposition methods have recently been proposed which automatically detect variable interaction and group them

© Springer International Publishing Switzerland 2015
B. Pfahringer and J. Renz (Eds.): AI 2015, LNAI 9457, pp. 34–48, 2015.
DOI: 10.1007/978-3-319-26350-2_4

accordingly in order to minimise inter-dependencies and help converge to better quality solutions [4, 7–9].

In the context of CC for global optimisation, it is quite clear that there are numerous different ways of subdividing different classes of problems. There is no unique decomposition strategy for problems such as fully-separable, fully non-separable or overlapping functions [10]. In a fully-separable problem, all of the decision variables can be optimised independently. In principle, a complete decomposition in which each variable is placed in a separate subcomponent is the most efficient decomposition strategy. However, a recent study showed that the performance of CC is very sensitive to the decomposition strategy for fully-separable problems [11]. Some partially separable problems may also contain a relatively high dimensional fully-separable subcomponent. Poor decomposition strategies of such subcomponents also affects the optimisation process [11].

It is quite challenging to find an effective decomposition strategy for the different classes of problems given the hurdle of extensive empirical studies [11]. One way to eliminate the need for finding optimal decomposition strategy is through adaptation. MLSoft is a relevant example which utilized a very simple reinforcement learning approach to dynamically adapt the decomposition strategy for fully-separable problems [11]. Recently, an alternative method to adaptation known as multi-island competitive cooperative coevolution (MICCC) was proposed which alleviated the need to find an optimal decomposition strategy [12]. MICCC is the successor to competitive island-based cooperative coevolution (CICC) which utilizes only two islands for solving different classes of global optimisation problems efficiently [13, 14]. CICC algorithm was originally designed for training recurrent neural networks on chaotic time series problems [15, 16].

In MICCC, a maximum of five different *uniform* problem decomposition strategies were implemented as islands that competed and collaborated with each other during evolution. The MICCC algorithm ensures that different problem decomposition strategies are given an opportunity during the course of evolution. In a uniform problem decomposition strategy, a problem is divided into equal sized subcomponents. Conversely, non-uniform problem decomposition strategies contain a range of different sized subcomponents.

It has yet not been established on to what extent MICCC can be effective if it is employed with a larger pool of islands. There has not been any investigation on the impact of having a hybrid pool of uniform and non-uniform problem decomposition strategies. This paper attempts to address this research gap through the following goals:

- To scale up MICCC algorithm and observe the behaviour as the number of islands increases.
- To find out if competition and collaboration of uniform and non-uniform problem decomposition strategies can improve the performance during the course of optimisation.
- To analyze the contributions of each of the problem decomposition strategies and identify the stronger islands during evolution.

The organization of the rest of this paper is as follows. Section 2 describes the proposed method and its application to the different classes of problems. Experimental results and their analysis are provided in Sect. 3. Section 4 concludes the paper with discussion of future work.

2   Multi-island Competitive Cooperative Coevolution

In this section, we provide details of the Multi-Island Competitive Cooperative Coevolution (MICCC) algorithm that enforces competition and collaboration between various different problem decomposition strategies that are implemented as islands.

Algorithm 1. Multi-Island Competitive Cooperative Coevolution

---

Stage 1: Initialization:
while *Island-n* $\leq$ *MaxNumIslands* **do**
 | Cooperatively evaluate Island-n
end

Stage 2: Evolution:
while *FuncEval* $\leq$ *GlobalEvolutionTime* **do**
  while *Island-n* $\leq$ *MaxNumIslands* **do**
    while *FuncEval* $\leq$ *Island-Evolution-Time* **do**
      foreach *Sub-population at Island-n* **do**
        foreach *Cycle in Max-Cycles* **do**
          foreach *Generation in Max-Generations* **do**
            | Create new individuals using genetic operators
            | Cooperative Evaluation of Island-n
          end
        end
      end
    end
  end

  Stage 3: Competition: Compare and mark the island with the best fitness.

  Stage 4: Collaboration: Inject the best individual from *Winner* island into all the other islands.
end

In MICCC [12], five different *uniform* problem decomposition strategies (same-sized subcomponents) are constructed as islands that compete and collaborate. These islands are evolved in isolation by independent G3-PCX [17] algorithm. The islands enforce competition by comparing their solutions after a fixed time (implemented as fitness evaluations) and exchange the best solution between the islands. Interaction and migration occurs between the different islands when evolutionary processes carry on for defined number of fitness evaluations or generations. During interaction, solutions of the winner island is migrated to those who lose the competition. The key aspects of the MICCC algorithm are initialization, evolution, competition and collaboration.

2.1 Initialization

In MICCC, a problem decomposition strategy is implemented as an island. To enforce an unbiased competition, all the islands begin search with the same genetic materials in the population. At the beginning, all the sub-populations of Island One are initialized with random-real number values from a domain specified in Table 1. These real values (from Island One) are copied into the sub-populations of the rest of the islands each of which are constructed with unique problem decomposition strategies.

In MICCC, the number of fitness evaluation depends on the number of sub-populations used in the respective island. An island with higher number of subcomponents will acquire more fitness evaluations for each cycle. A cycle is complete when all the sub-populations of an island have been cooperatively evolved for n number of generations. Therefore, each of the islands evolve by different island evolution time (in terms of fitness evaluations) until they have all reached maximum evolution time.

2.2 Cooperative Coevolution

Once the islands have been initialised, they are evolved in isolation simultaneously for a predefined time in the usual round robin fashion through cooperative coevolution. According to Algorithm 1, this predefined time is termed as *island-evolution time*. The island evolution time is established by the number of cycles that makes the required number of fitness evaluations for each of the different islands. Once evolved, cooperative evaluation of individuals in the respective sub-populations is done by concatenating the chosen individual from a given sub-population with the best individuals from the rest of the sub-populations [1].

2.3 Competition

In the competition phase of MICCC algorithm, fitness comparison of all the islands take place through a ranking mechanism whereby the islands with higher fitness are ranked higher while the low performing islands are ranked lower. The island with the best fitness is marked as the winner island. In the case when two islands have the same fitness (fitness tie), the winner island is randomly selected.

2.4 Collaboration: Inter-island Interaction and Solution Migration

In the collaboration stage of the MICCC algorithm, the actual interactions and migrations between different islands occur. Here, the best solution of the winner is copied and injected into to the runner-up islands. This migration of the best feasible solution is able assist and motivate the other islands to compete fairly in the next round.

The transfer of best solutions from one island to the rest is done via a context vector [18]. As an island wins, the best individuals from each of the subcomponents need to be carefully concatenated into a temporary context vector.

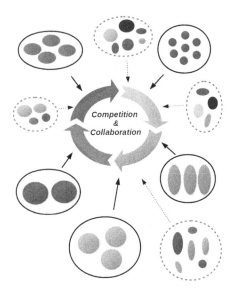

Fig. 1. Scaled up MICCC algorithm employing a hybrid pool size of nine islands which constitutes of five islands having uniform decomposition strategies while the remaining four islands have non-uniform problem decomposition strategies. Uneven sizes and colors of circular shapes represent *non-uniform* problem decomposition strategies (varied-size subcomponents) of the four additional islands that have dashed-line boundaries (Color figure online).

The best solutions are then split from the context vector and are then injected into the respective subcomponents of each of the runner-up islands. The runner-up islands which receive the best (injected) solution are cooperatively evaluated to ensure that the newly injected solution has a fitness. The best fitness of the winning island is also migrated alongside the best solution, to the rest of the other islands. Moreover, since the fitness of the best solution from the last sub-population carries a stronger solution, this fitness value is transferred and is used to override the fitness of the best solutions of all the sub-populations of the runnerup islands.

3   Simulation and Analyses

In this section, we perform a scale up study and evaluate the performances of the multi-island algorithms with a max pool size of nine islands. These additional islands are constructed with arbitrary problem decomposition strategies. For this study, we construct a hybrid pool size of nine islands which constitutes of five uniform decomposition strategies and four non-uniform decomposition strategies as shown in Fig. 1. These extended multi-island instances are first compared briefly against the standalone CC implementations. Next, we attempt to compare these extended multi-island instances with the original MICCC that featured fives islands [12], and observe trends or correlations (if any) of introducing additional arbitrary problem decomposition strategies for competition.

Furthermore, in-depth analyses are provided about which islands have been most dominant during the course of evolution by observing the islands that win the different phases of competition.

The generalized generation gap with parent-centric crossover evolutionary algorithm (G3-PCX) [17] is used as the subcomponent optimizer. We use a pool size of 2 parents and 2 offspring as presented in [17].

3.1   Benchmark Problems and Parameter Settings

The experimental results in this paper are based on eight benchmark problems used in [13] and are selected considering the level of difficulty, the scope of separability and the nature of problem,i.e. unimodal or multimodal listed by Table 1. These different classes of problems enable us to examine if the proposed method is suitable in a wide range of problems and if we could also highlight the limitations of MICCC. Furthermore, we introduce different problem decomposition strategies of 100 dimensions as inputs for competition in the multi-island algorithms. Since we are not sure about the right decomposition strategy before evolution, arbitrary problem decomposition strategies are included from the sixth island onwards. In each case, the mean and standard deviation of fitness errors $(f(x)-f(*x))$ of 25 runs are has been reported in the next subsection. The maximum number of fitness evaluations was set to 1500000.

Table 1. Problem definitions [19–21]

Problem	Name	Optimum	Range	Multi-modal	Fully separable
f_1	Ellipsoid	0	$[-5, 5]$	No	Yes
f_2	Shifted Sphere	-450	$[-100, 100]$	No	Yes
f_3	Schwefel's Problem 1.2	0	$[-5, 5]$	No	Yes
f_4	Rosenbrock	0	$[-5, 5]$	Yes	No
f_5	Shifted Rosenbrock	390	$[-100, 100]$	Yes	No
f_6	Rastrigin	0	$[-5, 5]$	Yes	Yes
f_7	Shifted Rastrigin	-330	$[-5, 5]$	Yes	Yes
f_8	Shifted Griewank	-180	$[-600, 600]$	Yes	No

Table 2. Island implementations of scaled up MICCC-9*. Island-1 to Island-5 possess uniform problem decomposition while Island-6 through Island-9 are constructed with non-uniform subcomponent sizes.

Island	1	2	3	4	5	6	7	8	9
PD	20×5	10×10	4×25	5×20	50×2	[22-26-25-27*]	[8-12-8-9-8-15-7-11-9-11*]	[12-10-8-9-8-15-7-11-9-11]	[20-29-18-33]

Table 3. Competition with max seven different problem decomposition strategies (MICCC-6* and MICCC-7*) results compared to individual decomposition strategies used in isolation

Func.	Stats.	Standard CC							MICCC-6*	MICCC-7*
		20×5	10×10	4×25	5×20	50×2	[22-26-25-27*]	[8-12-8-9-8-15-7-11-9-11*]		
f_1	Mean	6.83e+00	4.80e-98	3.76e-98	5.62e-98	2.01e-98	5.09e-98	5.69e-98	4.25e-89	0.00e+00
	StDev	2.03e+00	2.36e-98	1.45e-98	1.63e-98	1.70e-98	7.28e-99	7.93e-99	1.95e-89	0.00e+00
f_2	Mean	3.43e+05	9.03e-03	1.08e-12	8.98e-13	0.27e+01	8.55e-13	9.39e-13	2.39e-13	1.70e-13
	StDev	1.45e+04	1.03e-03	2.03e-13	1.21e-13	2.34e-01	5.62e-14	9.05e-14	1.40e-14	1.40e-14
f_3	Mean	5.00e-03	7.98e-51	1.13e-50	1.18e-50	1.54e-50	1.27e-50	7.53e-51	0.00e+00	0.00e+00
	StDev	1.00e-03	6.98e-51	1.03e-50	1.07e-50	2.01e-50	5.81e-51	3.51e-51	0.00e+00	0.00e+00
f_4	Mean	2.59e+02	5.16e+01	1.11e+02	6.31e-01	1.57e+02	7.97e-01	5.81e+01	3.47e+01	8.12e+01
	StDev	2.30e+01	0.10e+01	1.30e+01	2.03e-01	1.34e+01	6.25e-01	0.87e+01	0.64e+01	9.25e-01
f_5	Mean	7.95e+10	3.80e+01	9.01e+01	3.71e+01	5.67e+04	1.92e+02	6.05e+01	0.00e+00	0.00e+00
	StDev	1.96e+10	1.02e+01	1.20e+01	0.34e+01	3.49e+03	1.15e+02	0.69e+01	0.00e+00	0.00e+00
f_6	Mean	1.87e+01	2.70e+02	4.86e+02	4.73e+02	1.10e+02	4.83e+02	4.01e+02	1.13e-13	0.00e+00
	StDev	0.43e+01	1.20e+02	2.21e+02	1.03e+02	0.70e+02	2.15e+01	1.27e+01	7.50e-01	0.00e+00
f_7	Mean	8.83e+02	5.02e+02	7.56e+02	6.91e+02	1.22e+02	7.53e+02	5.97e+04	7.76e+01	8.70e+01
	StDev	2.23e+02	1.34e+02	2.78e+02	0.45e+02	0.94e+02	3.17e+01	2.35e+01	0.57e+01	3.57e+01
f_8	Mean	2.81e+03	5.11e-13	3.63e-03	5.12e-13	0.43e+01	4.83e-13	7.67e-13	8.52e-14	8.52e-14
	StDev	1.02e+03	2.09e-13	1.67e-03	2.87e-13	0.23e+01	1.82e-13	1.31e-13	1.13e-14	1.13e-14

Table 4. Comparison between CICC, MICCC-5, MICCC-6*, MICCC-7*, MICCC-8* and MICCC-9* algorithms

Func.	Stats.	CICC/MICCC instances					
		CICC [13]	MICCC-5 [12]	MICCC-6*	MICCC-7*	MICCC-8*	MICCC-9*
f_1	Mean	3.76e-99	5.59e-99	4.25e-89	0.00e+00	8.30e-317	2.12e-317
	StDev	2.06e-99	1.02e-99	1.95e-89	0.00e+00	1.81e-317	4.26e-317
f_2	Mean	7.78e-13	1.73e-13	2.39e-13	1.70e-13	2.09e-13	2.05e-13
	StDev	1.34e-13	1.03e-14	1.40e-13	2.27e-13	1.51e-14	1.41e-14
f_3	Mean	0.00e+00	0.00e+00	0.00e+00	0.00e+00	0.00e+00	0.00e+00
	tDev	0.00e+00	0.00e+00	0.00e+00	0.00e+00	0.00e+00	0.00e+00
f_4	Mean	7.93e+01	7.43e+01	3.47e+01	8.12e+01	6.29e+01	7.82e+01
	StDev	1.78e+01	0.78e+01	0.64e+01	9.25e-01	0.24e-01	0.41e-01
f_5	Mean	0.00e+00	0.00e+00	0.00e+00	0.00e+00	0.00e+00	0.00e+00
	StDev	0.00e+00	0.00e+00	0.00e+00	0.00e+00	0.00e+00	0.00e+00
f_6	Mean	1.40e+01	9.95e-01	1.13e-13	0.00e+00	9.95e-01	9.95e-01
	StDev	1.34e+00	1.45e-01	7.50e-01	0.00e+00	0.11e+01	0.11e+01
f_7	Mean	3.92e+02	9.75e+01	7.76e+01	8.70e+01	8.66e+01	7.96e+01
	StDev	2.09e+01	1.45e+01	0.57e+01	3.57e+01	0.51e+01	0.65e+01
f_8	Mean	1.99e-13	8.53e-14	8.52e-14	8.52e-14	8.52e-14	8.52e-14
	StDev	1.98e-13	1.04e-14	1.13e-14	1.13e-14	1.13e-14	1.13e-14

3.2 Results and Analysis

This section provides an analysis of increasing the number of islands to the original MICCC [12] that has a max pool size of five islands. We evaluate the performances of the extended instances of MICCC with pool size of six, seven, eight and nine islands. As mentioned earlier, the first five islands are constructed with uniform problem decomposition strategies [12]. The rest of the additional islands proposed for this extended study of MICCC possess arbitrary (non-uniform) problem decomposition strategies. The island implementations of each of nine suboptimal decompositions is given by Table 2. Experimental results have been summarised in Tables 3 and 4. The scaled up instances of MICCC are marked with an asterisk(*) to distinguish them from the existing MICCC and CICC algorithms [12–14].

3.3 Performance Analysis of MICCC as the Number of Islands Increase

Table 3 shows the experimental results of MICCC-6* Island and MICCC-7* Island algorithms alongside each of their standalone decomposition strategies. The results in Table 3 have shown that competition and collaboration of several different problem decomposition strategies can generate better quality solutions than each of the standalone CC implementations. MICCC-6* Island algorithm had just one arbitrary decomposition strategy in its competition pool and managed to perform generally than almost all its counterparts on problems identified (Function f_2–f_8). It recorded similar solutions as the standalone CC for Function f_1. MICCC-7* Island algorithm had two additional arbitrary decomposition strategies and managed to outperform all the rest of the standalone problem decomposition strategies for all the problems. This superior performances of the island algorithms over the standard CC have been common observations in CICC and MICCC [12–14]. Since we are focusing on the behaviour of MICCC as the number of islands increase, we omit the comparisons of MICCC with standalone CC for the cases of eight and nine islands. We are interested to discover any trends or observations (if any) gathered by increasing the original size of the pool of competition for MICCC which was limited to five islands. Table 4 provides a set of comparative data for CICC, MICCC and extended versions of MICCC* which were tested on the eight benchmark problems.

According to Table 4, it can be seen that CICC and MICCC with pool size of 2, 5 and 6 islands performed equally well on uni-modal and fully separable problems (Function f_1–f_3) as they recorded similar solution accuracy. However, on a closer inspection, one can see that the extended versions MICCC-7*, MICCC-8* and MICCC-9* performed considerably better for f_1 than the rest of MICCC with lower number of pools. These extended versions performed equally well as MICCC-5 and CICC on the multi-modal and non-separable Shifted Rosenbrock problem (Function f_5). MICCC-6* recorded the best quality solutions for f_4. Another observation is that MICCC-7* performed outstandingly better than the rest of the algorithms for f_6. On that note, the extended versions MICCC-8* and MICCC-9* performed equally well as MICCC-5, but outperformed the existing CICC-2 island algorithm for f_6. For multi-modal problems (Function f_7 and Function f_8), the extended instances of MICCC with pool size of six, seven, eight and nine islands managed to improve the accuracy and quality of the solutions when compared to the existing MICCC-5 island algorithm. From these observations, we can generalise that the scaled up MICCC* with a wider pool of nine islands is superior to the existing MICCC which is limited to five islands. Experimental results show that quality and precision of the solutions improve while utilising more islands in the competition. Arbitrary decomposition strategies have shown to be highly beneficial. It is advantageous to have a hybrid pool of uniform and non-uniform problem decomposition strategies competing and collaborating together to converge to a high quality solution. In this manner, the MICCC algorithm preserves diversity and combats premature convergence for complex multi-modal problems.

3.4 Island Competition Analysis over the Evolutionary Process

In this section, we study the behaviour of the islands during the evolutionary process. We check which islands are dominant and how each of the different islands contribute towards the global fitness. Figures 2, 3 and 4 show the competition between the different islands (problem decomposition strategies) monitored at different stages of the optimisation phase for each of the eight problems.

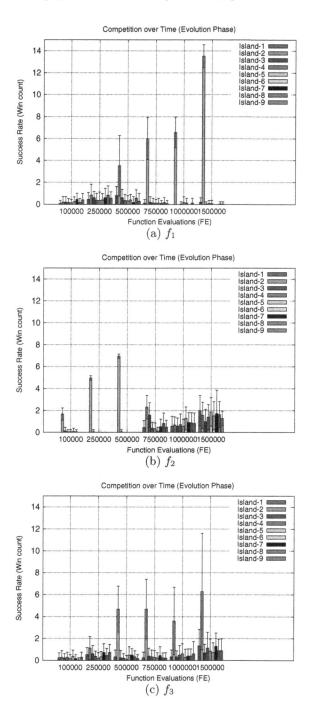

Fig. 2. Competition over time for the success rate (win count) of Functions f_1–f_3 for MICCC-9* Island algorithm.

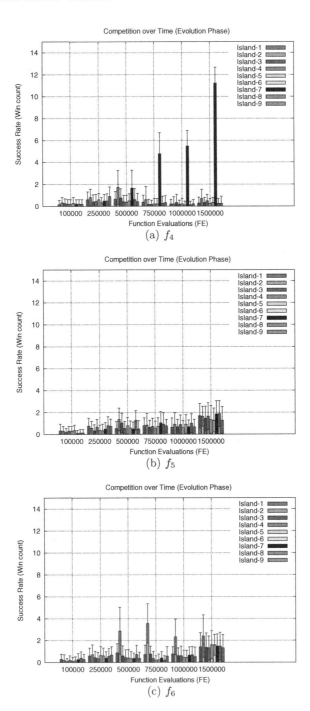

Fig. 3. Competition over time for the success rate (win count) of Functions f_4–f_6 for MICCC-9* Island algorithm.

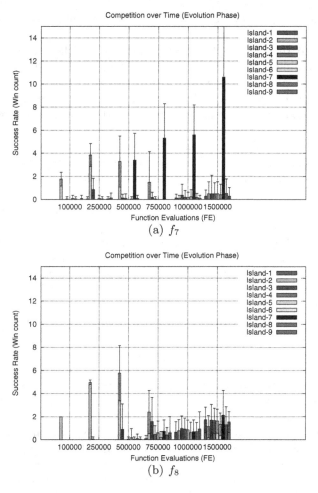

Fig. 4. Competition over time for the success rate (win count) of Functions f_7–f_8 for MICCC-9* Island algorithm.

For the purpose of this study, we monitor the competition of MICCC-9* algorithm which was implemented with the max pool size of nine different islands. Analyses of the islands that win the competition at different stages of the optimisation phase are provided by measuring their success rate (win counts). Each of the bars of the respective graphs show the average number of times an island wins the competition over a time interval of 250 000 function evaluations. The islands compete and collaborate until a fixed termination criteria of 1 500 000 fitness evaluations is reached. The competition pool of MICCC-9* consists of five uniform and four arbitrary problem decomposition strategies as previously defined in Table 2. Since different types of problems feature unique search landscapes, we can observe that the competition success rate, that is, how often each island wins the competition vary across all the different problems throughout the course of evolution.

According to the bar graphs of Figs. 2, 3 and 4, Island 2 has been quite dominant for uni-modal problems (Functions f_1–f_3) and multi-modal problem (Function f_6). It also dominated the initial stages of the competition phases of Function f_7 and Function f_8, but was superseded by Island 7 in the later stages. Interestingly, Island 7 that was constructed with a non-uniform problem decomposition strategy has shown be an efficient decomposition strategy as it was dominant in most of the multi-modal problems (Functions f_4, f_5, f_7 and f_8). This suggests that Island 2 and Island 7 have been contributing more towards the global fitness than the rest of the islands. In addition, Island 8 competed well on problems (Functions f_5 and f_6). This justifies the improvements in the solution quality obtained in Table 4 as we increase the number of islands in the existing MICCC which was limited to five islands [12].

Furthermore, the visualized competition analysis obtained from the bar charts of Figs. 2, 3 and 4 offer some interesting observations. Since the proposed MICCC algorithm incorporates solution *migration* phase during the course of the optimisation, it motivates the runner-up islands to catch up and compete over time. We observe intense competition between the all the nine different islands for quite a number of different problems. Intense competition is observed in some of the problems (Functions f_2,f_5, f_6 and f_8), whereby the islands that were losing initially were able to improve their performances over time. This notable trend is quite evident in most of the cases given in Figs. 2, 3 and 4. Considering the competition analyses for certain problems (Functions f_2,f_5 and f_6), we can observe that few other islands have also been contributing well during the evolutionary process.

The current MICCC algorithm splits the computational budget equally amongst all the islands. This gives motivation to divide the computational budget more wisely according to the contributions of each of the islands. The concept of contribution based cooperative coevolution (CBCC) [22] can be incorporated to enhance the performance of MICCC. In conducting such empirical analysis as above, it is possible to identify the stronger islands (implemented through uniform or non-uniform problem decomposition strategies) and accurately quantify the contribution of each of the islands towards the global fitness. Once the contribution information is available, the computational budget can be utilized more strategically. The stronger islands can be given more evolution time and the weakest ones can be eliminated during the course of optimisation.

4   Conclusion

In this paper, we scaled up MICCC by enforcing competition with a wider pool of islands. The experimental results show that enforcing competition with a wider pool of islands (more than five) can improve the performance during the course of the optimisation phase. The introduction of a hybrid pool of uniform and non-uniform problem decomposition strategies can substantially enhance the quality of the overall fitness solutions. More diversity is introduced as we increase the number of islands for competition which helps in escape the vulnerable fitness stagnation or local minimums of complex multimodal problems.

In future, a contribution based scheme can be developed that quantifies the contributions of each of the islands towards the global fitness. This can enhance the optimisation performance of MICCC whereby the stronger islands can be given more time (fitness evaluations) to compete and collaborate while the weaker ones can be eliminated. This extended multi-island algorithm can also be applied to large scale global optimisation as well as combinatorial optimisation problems.

References

1. Potter, M.A., De Jong, K.A.: A cooperative coevolutionary approach to function optimization. In: Davidor, Y., Männer, R., Schwefel, H.-P. (eds.) PPSN 1994. LNCS, vol. 866, pp. 249–257. Springer, Heidelberg (1994)
2. Yang, Z., Tang, K., Yao, X.: Large scale evolutionary optimization using cooperative coevolution. Inf. Sci. **178**(15), 2985–2999 (2008)
3. Bäck, T., Fogel, D.B., Michalewicz, Z. (eds.): Handbook of Evolutionary Computation. Institute of Physics Publishing/Oxford University Press, Bristol/New York (1997)
4. Omidvar, M., Li, X., Mei, Y., Yao, X.: Cooperative co-evolution with differential grouping for large scale optimization. IEEE Trans. Evol. Comput. **18**(3), 378–393 (2014)
5. Salomon, R.: Reevaluating genetic algorithm performance under coordinate rotation of benchmark functions - a survey of some theoretical and practical aspects of genetic algorithms. BioSystems **39**, 263–278 (1995)
6. Liu, Y., Yao, X., Zhao, Q., Higuchi, T.: Scaling up fast evolutionary programming with cooperative coevolution. In: Proceedings of the 2001 Congress on Evolutionary Computation, IEEE 2001, vol. 2, pp. 1101–1108 (2001)
7. Mahdavi, S., Shiri, M.E., Rahnamayan, S.: Cooperative co-evolution with a new decomposition method for large-scale optimization. In: Proceedings of the IEEE Congress on Evolutionary Computation, CEC 2014, pp. 1285–1292 (2014)
8. Chen, W., Weise, T., Yang, Z., Tang, K.: Large-scale global optimization using cooperative coevolution with variable interaction learning. In: Schaefer, R., Cotta, C., Kołodziej, J., Rudolph, G. (eds.) PPSN XI. LNCS, vol. 6239, pp. 300–309. Springer, Heidelberg (2010)
9. Omidvar, M.N., Li, X., Yao, X.: Cooperative co-evolution with delta grouping for large scale non-separable function optimization. In: Proceedings of IEEE Congress on Evolutionary Computation, pp. 1762–1769 (2010)
10. Omidvar, M.N., Li, X., Tang, K.: Designing benchmark problems for large-scale continuous optimization. Inf. Sci. **316**, 419–436 (2015)
11. Omidvar, M.N., Mei, Y., Li, X.: Effective decomposition of large-scale separable continuous functions for cooperative co-evolutionary algorithms. In: Proceedings of IEEE Congress on Evolutionary Computation, pp. 1305–1312 (2014)
12. Bali, K., Chandra, R.: Multi-island competitive cooperative coevolution for real parameter global optimization. In: International Conference on Neural Information Processing (ICONIP), Istanbul, Turkey, November 2015 (in press)

13. Chandra, R., Bali, K.: Competitive two island cooperative coevolution for real parameter global optimisation. In: IEEE Congress on Evolutionary Computation, Japan, Sendai, pp. 93–100 (2015)

14. Bali, K., Chandra, R., Omidvar, M.N.: Competitive island-based cooperative co-evolution for efficient optimization of large-scale fully-separable continuous functions. In: International Conference on Neural Information Processing (ICONIP), Istanbul, Turkey, November 2015 (in press)

15. Chandra, R.: Competition and collaboration in cooperative coevolution of Elman recurrent neural networks for time-series prediction. IEEE Trans. Neural Netw. Learn. Syst. (2015). doi:10.1109/TNNLS.2015.2404823. http://ieeexplore.ieee.org/stamp/stamp.jsp?tp=&arnumber=7055352&isnumber=6104215

16. Chandra, R.: Competitive two-island cooperative coevolution for training Elman recurrent networks for time series prediction. In: International Joint Conference on Neural Networks (IJCNN), Beijing, China, pp. 565–572, July 2014

17. Deb, K., Anand, A., Joshi, D.: A computationally efficient evolutionary algorithm for real-parameter optimization. Evol. Comput. 10(4), 371–395 (2002)

18. Van den Bergh, F., Engelbrecht, A.P.: A cooperative approach to particle swarm optimization. IEEE Trans. Evol. Comput. 8(3), 225–239 (2004)

19. Tang, K., Yao, X., Suganthan, P.N., MacNish, C., Chen, Y.P., Chen, C.M., Yang, Z.: Benchmark functions for the CEC 2008 special session and competition on large scale global optimization. Technical report, Nature Inspired Computation and Applications Laboratory, USTC, China (2007). http://nical.ustc.edu.cn/cec08ss.php

20. Li, X., Tang, K., Omidvar, M.N., Yang, Z., Qin, K.: Benchmark functions for the CEC 2013 special session and competition on large-scale global optimization. Technical report, RMIT University, Melbourne, Australia (2013). http://goanna.cs.rmit.edu.au/xiaodong/cec13-lsgo

21. Herrera, F., Lozano, M., Molina, D.: Test suite for the special issue of soft computing on scalability of evolutionary algorithms and other metaheuristics for large scale continuous optimization problems. Last accessed July 2010

22. Omidvar, M.N., Li, X., Yao, X.: Smart use of computational resources based on contribution for cooperative co-evolutionary algorithms. In: Proceedings of Genetic and Evolutionary Computation Conference, pp. 1115–1122. ACM (2011)

An Evolutionary Algorithm with Classifier Guided Constraint Evaluation Strategy for Computationally Expensive Optimization Problems

Kalyan Shankar Bhattacharjee[✉] and Tapabrata Ray

School of Engineering and Information Technology,
University of New South Wales, Canberra, ACT 2600, Australia
{k.bhattacharjee,t.ray}@adfa.edu.au

Abstract. Practical optimization problems often involve objective and constraint functions evaluated using computationally expensive numerical simulations e.g. computational fluid dynamics (CFD), finite element methods (FEM) etc. In order to deal with such problems, existing methods based on surrogates/approximations typically use cheaper and less accurate models of objectives and constraint functions during the search. Promising solutions identified using approximations or surrogates are only evaluated using computationally expensive analysis. In the event the constraints and objectives are evaluated using independent computationally expensive analysis (e.g. multi-disciplinary optimization), there exists an opportunity to only evaluate relevant constraints and/or objectives that are necessary to ascertain the utility of such solutions. In this paper, we introduce an efficient evolutionary algorithm for the solution of computationally expensive single objective constrained optimization problems. The algorithm is embedded with selective evaluation strategies guided by Support Vector Machine (SVM) models. Identification of promising individuals and relevant constraints corresponding to each individual is based on SVM classifiers, while partially evaluated solutions are ranked using SVM ranking models. The performance of the approach has been evaluated using a number of constrained optimization benchmarks and engineering design optimization problems with limited computational budget. The results have been compared with a number of established approaches based on full and partial evaluation strategies. Hopefully this study will prompt further efforts in the direction of selective evaluation, which so far had attracted little attention.

Keywords: Constraint handling · Selective evaluation · Approximation

1 Background

Constraint handling is an important area of research and several schemes have been proposed in the literature. Depending on the evaluation strategy, these schemes can be broadly categorized into two groups (a) full evaluation policy

© Springer International Publishing Switzerland 2015
B. Pfahringer and J. Renz (Eds.): AI 2015, LNAI 9457, pp. 49–62, 2015.
DOI: 10.1007/978-3-319-26350-2_5

(all constraints and objectives are evaluated), (b) partial evaluation policy (objective functions of feasible individuals are evaluated or selected constraints are only evaluated till a violation is encountered). The first group can be again sub categorized into several types depending on the constraint handling strategy: use of penalty functions which is an aggregate of constraint violations and the objective function [1], use of repair schemes [2] etc. All these schemes are based on *feasibility first* principle and they use objective function values to order feasible individuals and use sum of constraint violations to order the infeasible individuals in the population. Hence objective function values of infeasible individuals are essentially unused information for such approaches. However, methods such as stochastic ranking (SR) [3], infeasibility driven evolutionary algorithm (IDEA) [4], epsilon level comparison [5] etc. utilize objective function values of the infeasible solutions to order them. In the context of computationally expensive optimization problems, to reduce the computational cost, researchers have looked into the prospect of evaluating only a set of constraint(s) for an individual. In the *evaluate till you violate* strategy [6], constraints were evaluated in a sequence until a violation was encountered. This scheme needs the number of satisfied constraints and the violation measure of the first violated constraint to order the solutions. These schemes naturally fall into the second category of partial evaluation policy. There is another class of algorithms which uses surrogates to model the constraints and the objectives and uses actual evaluation for elite solutions to update the model [7,8]. The evaluation cost can be further reduced if one can (a) determine the constraint with the highest probability of being violated and (b) identify potentially promising offspring solutions. In order to identify the above, an efficient algorithm needs to continuously assess the trade off between *cost to evaluate* vs. *cost to learn*. In this paper, we use a support vector machine (SVM) classifier [9] to identity promising solutions and also for each solution, the constraint that is most likely to be violated. It is important to highlight that in the proposed approach, the classifier is used to predict a single class label for a solution and its associated confidence. The use of a SVM classifier to identify promising solutions appear in [10]. In this study, we used SVM ranking [11] models to predict the rank of a partially evaluated solution.

2   Proposed Approach

A generic single objective constrained optimization problem is defined as follows:

$$
\begin{aligned}
\underset{X}{\text{minimize}} \quad & f(\mathbf{x}) \\
\text{subject to} \quad & g_i(\mathbf{x}) \geq a_i, \ i = 1, 2, \ldots, q \\
& h_j(\mathbf{x}) = b_j, \ j = 1, 2, \ldots, r
\end{aligned}
\tag{1}
$$

where there are q inequality and r equality constraints, $\mathbf{x} = [x_1 \ x_2 \ \ldots x_n]$ is the vector of n design variables and a_i and b_j are constants.

Our approach is based on a generational model where an initial parent population is created using Latin Hypercube Sampling. All objectives and constraints are evaluated for all the solutions in the initial population and the information is stored in an *Archive*. Parents for recombination are identified using binary tournament (BT). The binary tournament is held between two sets of parents i.e. the first set of potential parents identified using roulette wheel selection and the second set of potential parents identified using a random selection. Offspring solutions are created using simulated binary crossover (SBX) and polynomial mutation (PM). These offspring solutions need to be screened prior to any further evaluation. Such a screening is performed using a two-class SVM classifier. The SVM classifier is trained using the data from the *Archive* with its inputs being the variables of the optimization problem $\mathbf{x}$ and the output being the rank of the solutions. Offspring solutions predicted with a class label of 1 are considered as potential solutions.

In the next step, for each potential offspring solution, we identify the constraint that is most likely to be violated. Here we calculate two metrics corresponding to each potential offspring solution i.e. *Feasibility_Index* and *Rank_index*. *Feasibility_index* of a solution is represented using a vector of same size as the number of constraints of the problem, wherein each element corresponds to the probability of satisfying a particular constraint (a value of 0 indicates that the solution would violate the constraint and 1 otherwise). The *Rank_Index* for a solution is again a vector of same size as *Feasibility_index*, with its elements providing the confidence information i.e. the confidence of the solution being ranked among the top 50 % for that particular constraint. The solutions are ranked based on each constraint with the information derived from *Archive*. For any constraint, a solution that satisfies it and is farthest from the boundary is placed at the top of the list, while the one which is infeasible and farthest from the boundary is placed at the bottom. In order to capture the local behavior, for each potential solution, the classifiers (one for each constraint) are trained using k closest (in variable space) neighbors from the *Archive*. The inputs to such classifiers are the variable values and the outputs are the corresponding ranks based on the particular constraint under consideration. For a potential offspring solution, the constraint associated with least *Feasibility_Index* and least *Rank_Index* will be evaluated first. However, in the situation where a solution has same *Feasibility_index* values for all its constraints, the order of evaluation is based on the following rule: (a) if all neighboring k solutions have all the constraints violated i.e. (*Feasibility_Index* $= 0$), the constraint having least *Rank_Index* is evaluated first and (b) if all neighboring k solutions have all the constraints satisfied i.e. (*Feasibility_Index* $= 1$), the objective function for this solution is only computed since it is most likely a feasible solution.

In the next step, we use SVM ranking model to predict rank of the potential offspring based on all other constraints, where it has not been evaluated. In this ranking scheme, a regression model is created using actual ranks of all the solutions from the *Archive* based on that particular constraint (where the potential offspring has been evaluated) as input and ranks based on other constraints or final rank as output. Hence, for any potential offspring solution, ranks based on all other constraints and its final rank in the population can be predicted

---

Algorithm 1. CGCSM

SET: FE_{max}{Total amount of evaluation cost unit allowed}, N{Size of population}, S_t{Confidence associated with SVM classifier (exponentially increases from 0 to 0.8 as cost of evaluation increases)}, M_g{Total number of constraints}, f{Objective}, g_j{j^{th} Constraint}, $Popbin${Repository of all solutions over the generation}, $Archive${Repository of fully evaluated solutions}

1: Initialize the population of N individuals
2: Evaluate $Pop_{1:N,g_{1:M_g},f}$ and order them according to their final ranks
3: $Popbin = Pop$
4: Update FE, Update $Archive$
5: Update S_t
6: **while** ($FE \leq FE_{max}$) **do**
7:     Generate offspring solutions using BT, SBX and PM from $Popbin_{1:N}$
8:     Construct a SVM classifier where top $100(1\text{-}S_t)$ percent of $Archive$ have class label 1 and rest have 0
9:     Determine unique $Childpop$ w.r.t the $Archive$ with a predicted class label 1, say C eligible offsprings
10:     **for** i = 1:C **do**
11:         Calculate the $Feasibility\_Index_{i,1:M_g}$ and $Rank\_Index_{i,1:M_g}$ based on k neighbors of $Childpop_i$ from the $Archive$
12:         **if** $Feasibility\_Index_{i,1:M_g} = 1$ **then**
13:             Only evaluate $Childpop_{i,f}$
14:         **else if** $Feasibility\_Index_{i,1:M_g} = 0$ **then**
15:             Find $g_{eval} \in g_j$ where $Rank\_Index_{i,g_{eval}}$ is least and evaluate $Childpop_{i,f}$ and $Childpop_{i,g_{eval}}$
16:         **else**
17:             Find $g_{eval} \in g_j$ where $Feasibility\_Index_{i,g_{eval}}$ and $Rank\_Index_{i,g_{eval}}$ are minimum and evaluate $Childpop_{i,f}$ and $Childpop_{i,g_{eval}}$
18:         **end if**
19:         Update FE
20:         Predict rank of $Childpop_i$ in other g_i's and its final rank in $Popbin$ using its rank in g_{eval} based on SVM ranking
21:         **if** (Final rank of $Childpop_i$ is within top $100(1\text{-}S_t)$ percent of $Popbin$) **then**
22:             Evaluate $Childpop_{i,g_{1:M_g}}$ and place it in $Popbin$ based on its actual final rank
23:             Update FE, Update $Archive$
24:         **else**
25:             Place the offspring in $Popbin$ according to its predicted final rank
26:         **end if**
27:         $Popbin = Pop + Childpop_i$
28:     **end for**
29: **end while**

*FE denotes the sum of objective and all individual constraint evaluation cost (1 unit cost for each objective and 1 unit for each constraint)

---

based on the rank of this particular solution in its evaluated constraint. In order to provide learning instances for the classifiers and to update the $Archive$, all constraints and objective function values of the top (within a threshold) set of solutions are evaluated. Please note that the final ranking of the whole population is carried out via *feasibility first* principle. Following is the pseudo-code for our proposed approach: Classifier Guided Constraint Selection Mechanism (CGCSM).

2.1   Test Function

The behavior of the proposed algorithm is illustrated using a single variable constrained test function shown in Eq. 2 with 5 nonlinear inequality constraints. The variable x is bounded between 0 and 4.5. The objective and the constraints are generalized as y, a function of x (Figs. 1, 2 and 3). The following parameters have been used: $k = 16.9$, $m = 0.7$, $c = 0.8$ and $d = 0.8$. In this example, we have used a population size of 8, neighborhood size of 4 with the maximum number of function evaluations set to 240 i.e. a total of 1440 evaluation cost units (1 function evaluation corresponds to evaluation of objective and all constraints for one solution). This test function has feasibility ratio (ρ) = 2.238 % computed based on 1,000,000 random points. The relative sizes of the feasible spaces corresponding to the constraints are 19.139 %, 9.672 %, 73.040 %, 25.174 % and 100 % respectively. Since the proposed approach will attempt to satisfy the most difficult constraint, we expect the second constraint to be satisfied first, followed

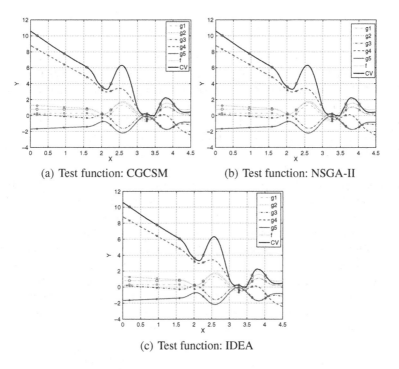

(a) Test function: CGCSM   (b) Test function: NSGA-II

(c) Test function: IDEA

Fig. 1. State of population at 1st gen

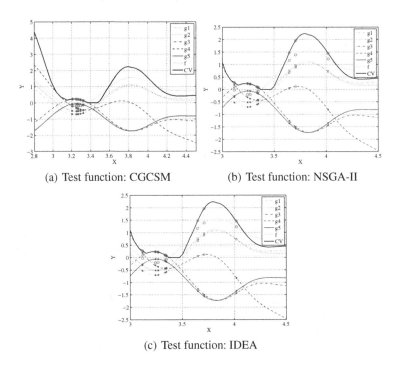

(a) Test function: CGCSM   (b) Test function: NSGA-II

(c) Test function: IDEA

Fig. 2. State of population at 3rd gen

by first, fourth and the third. The fifth constraint is feasible throughout the search space. The state of the population at the third generation is presented in Fig. 2(a). The corresponding states of IDEA and NSGA-II are presented in (Fig. 2(c) and (b)). One can note that all solutions of CGCSM have satisfied second constraint (equivalent to 144 function evaluations), while some solutions of IDEA and NSGA-II still could not satisfy the constraint.

The population of solutions after the 1st, and 7th generations for IDEA and NSGA-II are depicted in Figs. 1(c), 3(c), and 1(b), Fig. 3(b) respectively. Results obtained using CGCSM i.e. population of solutions after 48, and 336 evaluation cost units (equivalent to 1, 7 generations) are depicted in Figs. 1(a) and 3(a). It is important to analyse the effect of the confidence associated with the classifier, which has been varied between 0 and 0.8 in this study. A value of 0 indicates that we have no belief on the classifier and all solutions undergo evaluation. Evaluation of solutions would enrich the *Archive*, which in turn will allow the classifier to be trained better. To investigate this, we used a high value of confidence (0.8) for the classifier throughout the search process. The population of solutions after 336 evaluation cost units is presented in Fig. 4. Once can observe that there are no feasible solutions in the population. Since we placed a high confidence on

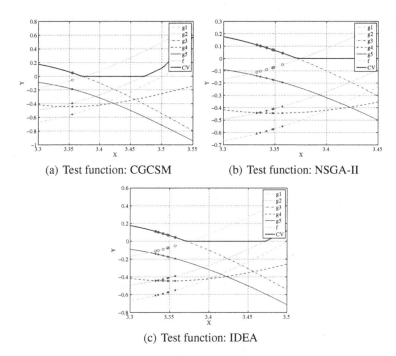

(a) Test function: CGCSM   (b) Test function: NSGA-II

(c) Test function: IDEA

Fig. 3. State of population at 7th gen

the classifier in early stages, the algorithm had limited opportunity to learn. On the contrary, if the confidence was allowed to vary between 0 and 0.8, feasible solutions emerge in the population as depicted in Fig. 3(a).

$$f(\mathbf{x}) = d + Ae^{-\zeta\omega|\mathbf{x}-3|^5} sin(\omega_d\mathbf{x} + \phi)$$

subject to the following constraints

$$0.5 - f \geq 0; \mathbf{x} - 2 - 4f \geq 0;$$

$$0.3\mathbf{x} - 1 + f \geq 0; 2.5\mathbf{x} - 8 - f \geq 0;$$

$$16f - 3\mathbf{x} + 14 \geq 0;$$

and bounds $0 \leq \mathbf{x} \leq 4.5$ 　　　　　　　　　　　　(2)

where

$$\omega = \sqrt{\frac{k}{m}}; A = \sqrt{\frac{(\zeta\omega)^2 + (\omega_d)^2}{(\omega_d)^2}};$$

$$\zeta = \frac{c}{2m\omega}; \omega_d = \omega\sqrt{1 - \zeta^2}; \phi = \tan^{-1}\frac{\omega_d}{\zeta\omega}$$

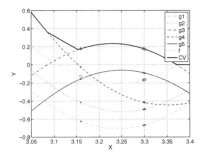

Fig. 4. State of population at generation 7 for test function using CGCSM with classifier threshold 0.2

3   Numerical Experiments

Having illustrated the search behavior of CGCSM, we objectively evaluate its performance and compare it with mIDEA and mNSGA-II using 15 well studied benchmark problems: G series (G1-G11) [12], Belleville Spring [13], Speed Reducer [14], Pressure Vessel [15] and Welded Beam [16]. These selected problems have variety of properties associated with their constraints: non-linearity, inequality, equality; different attributes associated with the objective functions: polynomial, linear, cubic, quadratic, non-linear etc.; different dimensions of variables and different feasibility ratios. The results obtained using the proposed algorithm CGCSM are compared with those delivered using infeasibility driven evolutionary algorithm (IDEA) [4] and non-dominated sorting genetic algorithm (NSGA-II) [17]. Since the proposed algorithm CGCSM utilizes a different scheme for offspring creation and retains unique solutions in the population, the baseline algorithms i.e. IDEA and NSGA-II have been modified to mIDEA and mNSGA-II for a fairer comparison. A one-to-one comparison of CGCSM with mIDEA and mNSGA-II would offer insights on the actual utility of the classifier. Results are presented in Table 1. The mean convergence plots are also presented for mIDEA and mNSGA-II and CGCSM.

4   Results and Discussion

The following parameters were used in this study: population size: 40; total cost of evaluation is 1000 times the total number of constraints and objective for the problem; crossover probability: 0.9; mutation probability: 0.1; distribution index for crossover: 20; distribution index of mutation: 30; confidence in the classifier varied exponentially from 0 to 0.8 and the number of neighbors (k) was set to 12. In our study, the standard SVM classifier of MATLAB toolbox was used with a Gaussian Radial Basis Function kernel with default settings and KKT violation level set as 0.05. The following subsection presents the results of all strategies i.e. CGCSM, IDEA, NSGA-II, mIDEA and mNSGA-II based on 30 independent runs for each problem. One can observe from Table 1 that the proposed approach is better in 2 out of 3 problems considering median statistics. The modified IDEA and NSGA-II reported better median objective value for the welded beam example, G11 and G4. However, baseline IDEA and NSGA-II has better performance in G7 and G2. One can also notice that at the end of 30 runs, none of the algorithms could identify feasible solutions for G5. The proposed approach is worse in G10, better in G6 and G11 and achieves competitive results for all other problems. It is also important to take note that our proposed approach performs much better in terms of convergence over others except for G10. This is probably due to the fact that all the constraints are active at the optimum for G10 which is beneficial to IDEA.

4.1 Performance on Problems

Figures 5(a), 6(a), and 7(a) show convergence plots of mean sum of constraint violations (CV) and average number of infeasible individuals over the evaluation

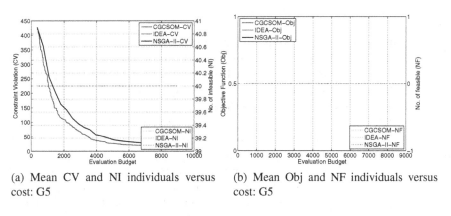

(a) Mean CV and NI individuals versus cost: G5

(b) Mean Obj and NF individuals versus cost: G5

Fig. 5. Convergence plots: G5

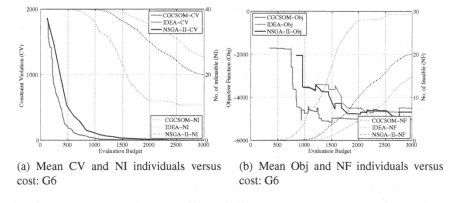

(a) Mean CV and NI individuals versus cost: G6

(b) Mean Obj and NF individuals versus cost: G6

Fig. 6. Convergence plots: G6

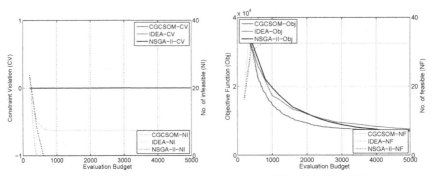

(a) Mean CV and NI individuals versus cost: Pressure Vessel

(b) Mean Obj and NF individuals versus cost: Pressure Vessel

Fig. 7. Convergence plots: Pressure Vessel

budget for problems: G5, G6, and Pressure Vessel. Similarly, mean objective value and average number of feasible individuals for these problems are shown in Figs. 5(b), 6(b), and 7(b).

The best, mean, worst, median and standard deviation measures of the best solution across 30 independent runs obtained using *CGCSM, mIDEA, IDEA, mNSGA-II, NSGA-II* for G5, G6 and Pressure Vessel problems are presented in Table 1. Here the last column success is defined as the number of runs which were successful to obtain a feasible solution at the end of evaluation budget. For all the other problems please refer to Supplementary where one can observe that proposed approach is better in 9 out of 14 problems considering the median statistics.

To compare any two stochastic algorithms a test based on statistical significance is performed. Wilcoxon Signed Rank Test [18] is used to judge the difference between paired scores when the assumptions required by the paired samples in t-test may not be valid, such as a normally distributed population. As a null hypothesis, it is assumed that there is no significant difference between the statistical measure of the two samples. Whereas the alternative hypothesis indicates a significant difference, with a significance level of 5 %. Based on the test results one of the three signs $(+,-,\approx)$ is assigned for the comparison of any two algorithms where the "+" sign indicates first algorithm is significantly better than the second algorithm, the "−" sign indicates first algorithm is significantly worse than the second algorithm and the "≈" sign indicates no significant difference between first algorithm and the second algorithm. Following Table 2 indicates the pairwise comparison of CGCSM with other state of the art algorithms:

The observations can be summarized as follows:
(a) At early stages of evolution, i.e. at lower evaluation budgets, our proposed approach provides better results than those obtained using IDEA and NSGA-II. This is evident from Figs. 5(a), 6(a), and 7(a). Similar observations were noticed

Table 1. Statistics for problems

Problem	Algorithms	Feasibility	Best	Mean	Median	Worst	Std	Success
G5	CGCSM	Feasible	NaN	NaN	NaN	NaN	NaN	0.000
		Infeasible	0.174	12.920	**6.266**	62.741	16.188	30.000
	mIDEA	Feasible	NaN	NaN	NaN	NaN	NaN	0.000
		Infeasible	0.373	21.507	9.612	182.587	33.759	30.000
	IDEA	Feasible	NaN	NaN	NaN	NaN	NaN	0.000
		Infeasible	0.015	13.098	7.833	68.571	14.345	30.000
	mNSGA-II	Feasible	NaN	NaN	NaN	NaN	NaN	0.000
		Infeasible	0.373	21.507	9.612	182.587	33.759	30.000
	NSGA-II	Feasible	NaN	NaN	NaN	NaN	NaN	0.0000
		Infeasible	0.015	13.098	7.833	68.571	14.345	30.000
G6	CGCSM	Feasible	−6785.808	−5041.048	**−5442.204**	−1228.347	1604.722	22.000
		Infeasible	0.0000	0.408	0.000	2.137	0.741	8.000
	mIDEA	Feasible	−6940.985	−4515.238	−4898.142	−1419.739	1858.821	16.000
		Infeasible	0.000	0.376	0.000	2.445	0.616	14.000
	IDEA	Feasible	−6862.173	−4584.322	−4551.042	−1301.353	1548.099	21.000
		Infeasible	0.000	0.338	0.000	2.074	0.617	9.000
	mNSGA-II	Feasible	−6941.706	−4690.694	−5114.001	−1543.642	1933.922	16.000
		Infeasible	0.000	0.376	0.000	2.445	0.616	14.000
	NSGA-II	Feasible	−6794.650	−4408.212	−4609.070	−1303.680	1593.556	21.000
		Infeasible	0.000	0.338	0.000	2.074	0.617	9.000
Vessel	CGCSM	Feasible	6106.624	7392.035	6930.623	15990.470	1855.454	30.000
		Infeasible	0.000	0.000	0.000	0.000	0.000	0.000
	mIDEA	Feasible	6364.481	7570.466	7368.356	10554.951	979.533	30.000
		Infeasible	0.000	0.000	0.000	0.000	0.000	0.000
	IDEA	Feasible	6472.820	7528.553	7389.138	11412.717	917.034	30.000
		Infeasible	0.000	0.000	0.000	0.000	0.000	0.000
	mNSGA-II	Feasible	6126.150	7015.135	**6854.071**	9961.452	712.944	30.000
		Infeasible	0.000	0.000	0.000	0.000	0.000	0.000
	NSGA-II	Feasible	6182.850	7152.707	7226.635	8551.650	541.266	30.000
		Infeasible	0.000	0.000	0.000	0.000	0.000	0.000

for other problems. (b) Since the best infeasible individual in IDEA and NSGA-II are the same, both have same convergence behavior in terms of mean sum of constraint violations. In the test problem, one can clearly observe the difference in search strategy between CGCSM, IDEA and NSGA-II. While all the strategies aim to reduce sum of constraint violations, CGCSM initially targets constraints that are difficult to satisfy. This can be evidenced from Figs. 1(a), 2(a), and 3(a). (c) In the illustrative example (test problem), one can clearly observe that the population of CGCSM reaches the feasible space much earlier than IDEA and NSGA-II. This is depicted in Fig. 3(c), (b) and (a). (d) Since only promising solutions are evaluated, use of classifiers would reduce the computational cost. However, the classifiers need to learn and their assessment needs to be reliable. The process of learning requires information from evaluated solutions. Although use of poorly trained classifiers would save computational cost, the search outcome may not be satisfactory. To achieve this balance, the confidence associated with the classifier is varied from 0 to 0.8 exponentially during the course of search. We have also presented the results when the trust

Table 2. Wilcoxon signed rank test for CGCSM

Comparison	Fitness	Better	Equal	Worse	Dec
CGCSM to IDEA	Median	10	0	4	+
	Mean	9	0	5	+
	Successful runs	2	11	1	≈
CGCSM to mIDEA	Median	12	0	2	+
	Mean	12	0	2	+
	Successful runs	2	11	1	≈
CGCSM to NSGA-II	Median	11	0	3	+
	Mean	10	0	4	+
	Successful runs	2	11	1	≈
CGCSM to mNSGA-II	Median	10	0	4	+
	Mean	10	0	4	+
	Successful runs	2	11	1	≈

associated with the classifier set to high (0.8) throughout the course of search. In such a scenario, very few solutions would be evaluated and in turn the classifier would not have the opportunity to learn from diverse solutions. This can be evidenced from Figs. 4 and 3(a). (e) Table 2 indicates the comparison of CGCSM with IDEA, NSGA-II and the proposed modified versions mIDEA, mNSGA-II. The table suggests significant improvement of CGCSM in terms of median and mean objective values considering all successful runs compared to other algorithms. However, in terms of number of successful runs where a feasible solution was obtained at the end of a limited computational budget, CGCSM performs at par with the other algorithms.

5   Summary and Conclusions

Real life optimization problems often involve objective and constraint functions that are evaluated using computationally expensive numerical simulations e.g. computational fluid dynamics (CFD), finite element methods (FEM) etc. In order to solve such classes of problems, surrogate assisted optimization (SAO) methods are typically used, wherein computationally cheap and less accurate surrogates/approximation models of objectives/constraints are used during the course of search. In this paper, we explore an alternative path i.e. one where promising solutions are identified using support vector machine (SVM) based models. The key difference being, SVM models are used to identify promising solutions without explicitly attempting to approximate objective and constraint functions. Furthermore, for every promising solution, the approach identifies the constraints that are most likely to be violated and evaluates them first. In the event the constraints and objectives are evaluated using independent computationally expensive analysis (e.g. multi-disciplinary optimization), such an approach would only evaluate relevant constraints and/or objectives that are necessary to ascertain the rank of the solutions. The differences in the search behavior of CGSCM, mNSGA-II and mIDEA are highlighted using a test function. The performance of the algorithm is further objectively assessed using a number of constrained optimization benchmarks and engineering design optimization problems with limited computational budget. The rate of convergence of CGCSM is better for most of the problems and the final set of results are clearly better on 9 out of 14 problems studied in this paper. We hope that this study would prompt design of efficient algorithms that selectively evaluate solutions and in particular selected set of constraints on the fly i.e. based on the trade-off between *need to learn/evaluate* and *cost to learn*.

References

1. Coit, D.W., Smith, A.E.: Penalty guided genetic search for reliability design optimization. Comput. Ind. Eng. **30**(4), 895–904 (1996)
2. FitzGerald, A., O'Donoghue, D.P.: Genetic repair for optimization under constraints inspired by *Arabidopsis Thaliana*. In: Rudolph, G., Jansen, T., Lucas, S., Poloni, C., Beume, N. (eds.) PPSN 2008. LNCS, vol. 5199, pp. 399–408. Springer, Heidelberg (2008)
3. Runarsson, T.P., Yao, X.: Stochastic ranking for constrained evolutionary optimization. IEEE Trans. Evol. Comput. **4**(3), 284–294 (2000)
4. Ray, T., Singh, H.K., Isaacs, A., Smith, W.: Infeasibility driven evolutionary algorithm for constrained optimization. In: Mezura-Montes, E. (ed.) Constraint-Handling in Evolutionary Optimization. SCI, vol. 198, pp. 145–165. Springer, Heidelberg (2009)
5. Takahama, T., Sakai, S.: Constrained optimization by the ϵ constrained differential evolution with gradient-based mutation and feasible elites. In: Proceedings of the IEEE Congress on Evolutionary Computation (CEC), pp. 1–8 (2006)
6. Asafuddoula, M., Ray, T., Sarker, R.: Evaluate till you violate: a differential evolution algorithm based on partial evaluation of the constraint set. In: Proceedings of the IEEE Symposium on Differential Evolution (SDE), pp. 31–37 (2013)
7. Regis, R.G.: Stochastic radial basis function algorithms for large-scale optimization involving expensive black-box objective and constraint functions. Comput. Oper. Res. **38**(5), 837–853 (2011)
8. Regis, R.G.: Evolutionary programming for high-dimensional constrained expensive black-box optimization using radial basis functions. IEEE Trans. Evol. Comput. **18**(3), 326–347 (2014)
9. Suykens, J.A.K., Van Gestel, T., De Brabanter, J., De Moor, B., Vandewalle, J.: Least Squares Support Vector Machines, vol. 4. World Scientific, Singapore (2002)
10. Loshchilov, I., Schoenauer, M., Sebag, M.: A mono surrogate for multiobjective optimization. In: Proceedings of the 12th Annual Conference on Genetic and Evolutionary Computation (GECCO), pp. 471–478. ACM (2010)
11. Chapelle, O., Keerthi, S.S.: Efficient algorithms for ranking with SVMs. Inf. Retrieval **13**(3), 201–215 (2010)
12. Michalewicz, Z., Schoenauer, M.: Evolutionary algorithms for constrained parameter optimization problems. Evol. Comput. **4**(1), 1–32 (1996)
13. Siddall, J.N.: Optimal Engineering Design: Principles and Applications. CRC Press, New York (1982)

14. Golinski, J.: Optimal synthesis problems solved by means of nonlinear programming and random methods. J. Mech. **5**(3), 287–309 (1970)
15. Coello Coello, C.A.: Use of a self-adaptive penalty approach for engineering optimization problems. Comput. Ind. **41**(2), 113–127 (2000)
16. Deb, K.: Multi-objective Optimization Using Evolutionary Algorithms. Wiley, Chichester (2001)
17. Deb, K., Pratap, A., Agarwal, S., Meyarivan, T.: A fast and elitist multiobjective genetic algorithm: NSGA-II. IEEE Trans. Evol. Comput. **6**(2), 182–197 (2002)
18. Corder, G.W., Foreman, D.I.: Comparing two related samples: the Wilcoxon signed ranks test. In: Nonparametric Statistics for Non-Statisticians: A Step-by-Step Approach, pp. 38–56. Wiley (2009)

Cost to Evaluate Versus Cost to Learn? Performance of Selective Evaluation Strategies in Multiobjective Optimization

Kalyan Shankar Bhattacharjee[(⊠)] and Tapabrata Ray

School of Engineering and Information Technology, University of New South Wales,
Canberra, ACT 2600, Australia
{k.bhattacharjee,t.ray}@adfa.edu.au

Abstract. Population based stochastic algorithms have long been used for the solution of multiobjective optimization problems. In the event the problem involves computationally expensive analysis, the existing practice is to use some form of surrogates or approximations. Surrogates are either used to screen promising solutions or approximate the objective functions corresponding to the solutions. In this paper, we investigate the effects of selective evaluation of promising solutions and try to derive answers to the following questions: (a) should we discard the solution right away relying on a classifier without any further evaluation? (b) should we evaluate its first objective function and then decide to select or discard it? (c) should we evaluate its second objective function instead and then decide its fate or (d) should we evaluate both its objective functions before selecting or discarding it? The last form is typically an optimization algorithm in its native form. While evaluation of solutions generate information that can be potentially learned by the optimization algorithm, it comes with a computational cost which might still be insignificant when compared with the cost of actual computationally expensive analysis. In this paper, a simple scheme, referred as Combined Classifier Based Approach (CCBA) is proposed. The performance of CCBA along with other strategies have been evaluated using five well studied unconstrained bi-objective optimization problems (DTLZ1-DTLZ5) with limited computational budget. The aspect of selective evaluation has rarely been investigated in literature and we hope that this study would prompt design of efficient algorithms that selectively evaluate solutions on the fly i.e. based on the trade-off between *need to learn/evaluate* and *cost to learn*.

Keywords: Multiobjective optimization · Approximation · Selective evaluation

1 Background

All population based stochastic optimization algorithms (e.g. evolutionary algorithms, particle swarm optimization, differential evolution etc.) typically work

© Springer International Publishing Switzerland 2015
B. Pfahringer and J. Renz (Eds.): AI 2015, LNAI 9457, pp. 63–75, 2015.
DOI: 10.1007/978-3-319-26350-2_6

with a set of solutions and evolve them over a number of generations. New solutions are constructed using stochastic recombination operators and the process continues until some prescribed termination condition is met. In the context of computationally expensive optimization problems, the termination condition is typically based on the allowable run time (several hours to several days depending upon the complexity of the problem and available computational resources). In mainstream evolutionary computation, evaluation of a solution typically refers to evaluating all the objective and constraint functions of a solution under consideration. This is quite a rigid assumption as in many practical problems, the objectives and constraints can be independently evaluated and each of such incurs a cost.

In the context of constrained single objective optimization, most existing population based stochastic algorithms (e.g. Nondominated sorting genetic algorithm (NSGA-II)) relying on *feasibility first* principles can easily save significant amount of computational time by only evaluating the objective function values for feasible individuals. Since such algorithms use objective function values to order feasible individuals and constraint violation values to order infeasible individuals, the computation of objective function values of infeasible solutions is redundant and is essentially unused information. However, methods such as stochastic ranking (SR) [1], infeasibility driven evolutionary algorithm (IDEA) [2], epsilon level comparison [3] etc. utilize information about the objective function values of the infeasible solutions to order them. Previous studies have highlighted benefits of using such information during the course of search. In an attempt to further contain the computational cost, one can look into possibilities of evaluating only a selected set of constraints for an individual. In the event *evaluate till you violate* strategy [4], constraints are evaluated in a sequence until a violation is encountered. Thus for an efficient algorithm design, it is important for us to continuously access the trade-off between *need to learn* and the *cost to learn*.

In the context of multiobjective optimization, the aspect of selective evaluation has rarely been studied. Selective evaluation is particularly important in the context of computationally expensive optimization problems, since every evaluation incurs significant cost. Canonical evolutionary algorithms rely on recombination schemes to construct solutions that are all considered promising and undergo full evaluation i.e. all objectives of all solutions in the offspring population are evaluated. In surrogate assisted optimization (SAO), approximations or surrogates are typically utilized in two different forms i.e. (a) used to identify promising solutions that are only evaluated i.e. a subset of offspring solutions undergo full evaluation or (b) performance of a subset of solutions are approximated and a mixed population (i.e. mix of individuals approximated and evaluated) is managed during the course of optimization. There is significant body of literature related to SAO and the readers are referred to [5] for further details.

It is clear from the above discussion that the overall performance of an approach would depend on the following (a) means to identify promising solutions (b) choice of model management i.e. evaluation of promising solutions versus managing a mixed pool of solutions and finally (c) the efficiency of the optimization algorithm. In this paper, we use a support vector machine classifier [6] to identity promising solutions. It is important to highlight that a classifier predicts a single class label for a solution, as opposed to surrogates that predict the objective function values. There are reports on the use of a SVM classifier to identify promising solutions [7], wherein a two-class classifier trained between variables and selected set of nondominated fronts was used to solve a number of multiobjective optimization problems. One class SVM classifiers are typically used when the data has a large imbalance i.e. number of non-promising solutions in the training data set is far more than number of promising solutions. While we agree that the data is likely to have significant imbalance [8], we focus our attention to several fundamental questions and understand the implications and their effects i.e. (a) can we discard a solution without even evaluating it based on the SVM classifier? (b) does the SVM classifier perform better if the first objective function is provided as an additional input although such a computation would incur cost? (c) should we evaluate its second objective function instead and then decide its fate based on the classifier? (d) should we evaluate both its objective functions and ignore use of classifier totally? We systematically conduct these experiments within the framework of Nondominated Sorting Genetic Algorithm (NSGA-II) [9].

2   Performance of Various Classifier Models

A generic bi-objective optimization problem is defined as follows:

$$\text{Minimize} \quad [f_1(\mathbf{x}), f_2(\mathbf{x})], \ \mathbf{x} \in \mathbb{R}^{nx} \tag{1}$$

where $f_1(\mathbf{x})$ and $f_2(\mathbf{x})$) are 2 objective functions, nx is the number of variables.

As in NSGA-II, a set of offspring solutions is generated using binary tournament among parents, followed by simulated binary crossover and polynomial mutation. In the context of computationally expensive optimization, we are interested to see if a subset of promising solutions can be identified using various selection strategies discussed earlier. The pseudo-code of the algorithm is presented below. In this paper we have used NSGA-II framework with a selective evaluation of potential child solutions. It is important to take note that the term, *number of function evaluations* used in this study refers to the sum of individual $f_1(\mathbf{x})$ and $f_2(\mathbf{x})$) evaluations, i.e. evaluating a solution would incur a cost of 2 units.

In the **Train** stage, all the nondominated solutions in the *Archive* are labelled with class 1 while the rest have been labelled with class 0. We study each of the possibilities in the **Train** and **Identify** stage i.e. (a) Ignore SVM in which case it is baseline NSGA-II (b) use SVM classifier using only X as inputs (c) evaluate f_1 for all child solutions and use X and f_1 as inputs to SVM (d) evaluate f_2 for all child solutions and use X and f_2 as inputs to SVM and finally (e) use any of the four strategies with equal probability for every child solution subsequently referred as *Random*. This process will try to identify potentially nondominated solutions using the above mentioned possibilities. We refer the readers to [6]

Algorithm 1. SEMO

Input: N = Population size, FE_{max} = Maximum number of function evaluations allowed, M_f= Number of Objectives (2 used in this study), $Archive$=Repository of all fully evaluated unique solutions.

1: $FE = 0, Gen = 0$
2: Initialize (pop); population of N individuals
3: Evaluate $(pop_{1:N, f_{1:M_f}})$
4: Update FE
5: Update $Archive$
6: **while** $(FE \leq FE_{max})\&(Gen \leq Gen_{max})$ **do**
7:    **Train** a two class SVM classifier
8:    $Gen = Gen + 1$
9:    $childpop$ = Generate N child solutions
10:   **Identify** C: set of promising child solutions
11:   Evaluate $childpop_{1:C, f_{1:M_f}}$
12:   Update FE
13:   Update $Archive$
14:   $S = pop + childpop$
15:   Order S using nondominated sorting and crowding
16:   $pop = S_{i=1:N, j=1:M_f)}$
17: **end while**

---

for the details on support vector machine classifiers. In our study, the standard SVM classifier of MATLAB toolbox was used with a Gaussian Radial Basis Function kernel with default settings and KKT violation level set as 0.05. One can observe from Fig. 1, that there are approximately 9 % class 1 solutions i.e. nondominated solutions at the initial stage of the search process for a bi-objective DTLZ1 problem. This justifies the use of a two-class classifier [8]. The percentage however reduces in later stages of the evolution process to less than 1 % which is ideally suited for one-class classifiers.

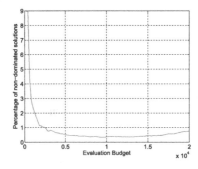

Fig. 1. Percentage of nondominated solutions in Archive versus cost: DTLZ1

3 Experimental Results

In this section, we would like to use the DTLZ1 [10] example to highlight some important observations. While this example is only used as an illustration and to lay the foundation of our proposed scheme, the observations are also true for other problems studied in the paper. The rationale behind the design of the proposed algorithm Combined Classifier Based Approach (CCBA) is presented in the next subsection. A complete benchmarking of all the above strategies and the newly proposed strategy is presented using two objective formulations of DTLZ1, DTLZ2, DTLZ3, DTLZ4 and DTLZ5 test functions. The number of variables nx is set to 6 for DTLZ1 and 11 for rest of the problems. The results are based on 30 independent runs and the hyper volume metric [11] is used as a measure of performance. The hypervolume is computed using true ideal point[0,0] as the reference and a lower HV is preferred. While, Nadir point is usually used as a reference for hypervolume computation (higher HV is better), use of true Nadir point to will lead to zero HV values for several initial generations which is not useful. The actual hypervolume measure is also given in Table 2 to provide more objective comparison. The performance of the approach is objectively assessed using inverted generational distance (IGD). The parameters used in this study include: Population size 100, Maximum allowable number of generations 500, Total allowable cost 20,000, probability of crossover of 1, crossover distribution index of 15, probability of mutation of $1/nx$ and a mutation distribution index of 20.

3.1 Illustrative Example: DTLZ1

The best, mean, worst, median and standard deviation of the hyper volume obtained using various selection strategies i.e. *NSGA-II, X, XF1, XF2 and Random* for the DTLZ1 problem are presented in Table 1.

Table 1. Hypervolume statistics for DTLZ1 problem for 20000 cost

Problems	Algorithms	Best	Mean	Worst	Median	Std
DTLZ1	NSGA-II	0.1277	0.3833	1.3051	0.2209	0.3359
	X	0.1289	1.6482	12.8327	0.6682	2.4747
	XF1	0.1330	1.4229	20.4741	0.5524	3.5965
	XF2	0.1366	1.8908	15.9307	0.8331	3.1704
	Random	0.1294	1.3630	4.1203	1.2384	1.0329

The progress plot of hypervolume with cost for the DTLZ1 problem is presented in Fig. 2(a), In order to more objectively observe the performance of the classifiers, the false positives and false negatives are analyzed. Take note that the definition of false positives and false negatives used in this study are different from those used in the domain of machine learning. The term false positive (FP)

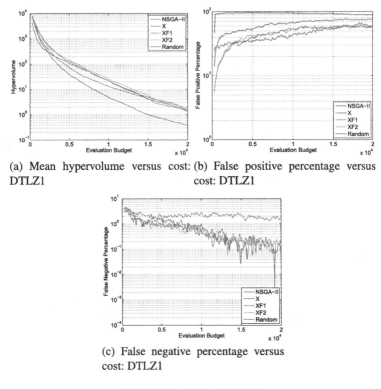

(a) Mean hypervolume versus cost: (b) False positive percentage versus
DTLZ1   cost: DTLZ1

(c) False negative percentage versus
cost: DTLZ1

Fig. 2. Base Plots-DTLZ1

Table 2. Actual hypervolume measure

DTLZ1	DTLZ2	DTLZ3	DTLZ4	DTLZ5
0.1250	0.7854	0.7854	0.7854	0.7854

would mean the classifier has identified a child solution to be nondominated
based on the archive, but in reality it is a dominated solution. The computation of FP considers one child solution at a time and if the child solution was
dominated with respect to the archive, it would be considered a false positive
i.e. the child was evaluated by the algorithm based on the classifier although it
was in reality a dominated solution. The term false negative (FN) would mean,
a nondominated child solution classified as a dominated solution and discarded
without further evaluation. High number of false positives will incur significant
computational cost, while a high number of false negatives would mean promising solutions discarded. As for NSGA-II, all child solutions are evaluated and
thus there are no false negatives. The false positives and false negatives of various
strategies for the DTLZ1 problem is presented in Fig. 2(b) and (c).

The preliminary observations can be summarized as follows: (a) Full evaluation of all child solutions i.e. NSGA-II offers the best performance for DTLZ1

based on final hypervolume. (b) Random model seems to offer an average performance. (c) In terms of false positives, XF1 and XF2 tend to have lower values. This supports the argument that the models are cautious i.e. they are selectively evaluating promising solutions. The false positives of model based on X alone is less strict in its evaluation policy when compared to XF1 and XF2 models. (d) It is interesting to note that the false negatives of all models are fairly low. This suggests, that all the models quite accurate in discarding i.e. very few around

Table 3. Hypervolume statistics for DTLZ problem for 20000 cost

Problems	Algorithms	Best	Mean	Worst	Median	Std
DTLZ1	NSGA-II	0.1277	0.3833	1.3051	0.2209	0.3359
	X	0.1289	1.6482	12.8327	0.6682	2.4746
	XF1	0.1330	1.4229	20.4741	0.5524	3.5965
	XF2	0.1366	1.8908	15.9307	0.8331	3.1704
	Random	0.1294	1.3630	4.1203	1.2384	1.0329
	CCBA	0.1315	1.1499	15.3663	0.5213	2.7715
DTLZ2	NSGA-II	0.7905	0.7910	0.7914	0.7910	0.0002
	X	0.7900	0.7903	0.7907	0.7903	0.0002
	XF1	0.7898	0.7904	0.7909	0.7905	0.0003
	XF2	0.7899	0.7906	0.7910	0.7906	0.0003
	Random	0.7906	0.7910	0.7915	0.7910	0.0002
	CCBA	0.7899	0.7903	0.7906	0.7902	0.0001
DTLZ3	NSGA-II	17.2377	158.3326	460.6635	111.9531	124.4603
	X	65.6367	847.7729	2280.8584	656.2069	641.5739
	XF1	469.8990	2166.9133	5148.9793	1782.7222	1122.1093
	XF2	377.6039	2109.2509	9849.1959	1809.4176	1870.4375
	Random	187.3159	691.4597	1984.1853	592.2983	415.63794
	CCBA	80.5458	537.4027	1730.7210	444.5855	356.0086
DTLZ4	NSGA-II	0.7907	0.7911	0.7918	0.7911	0.0002
	X	0.7900	0.7905	0.7911	0.7904	0.0003
	XF1	0.7899	0.7907	0.7914	0.7908	0.0003
	XF2	0.7901	0.7911	0.7987	0.7908	0.0017
	Random	0.7899	0.7910	0.7918	0.7910	0.0004
	CCBA	0.7901	0.7905	0.7909	0.7905	0.0002
DTLZ5	NSGA-II	0.7905	0.7910	0.7914	0.7910	0.0002
	X	0.7900	0.7903	0.7907	0.7903	0.0002
	XF1	0.7898	0.7904	0.7909	0.7905	0.0003
	XF2	0.7899	0.7906	0.7910	0.7906	0.0003
	Random	0.7906	0.7910	0.7915	0.7910	0.0002
	CCBA	0.7899	0.7903	0.7906	0.7902	0.0001

1 % of promising solutions are incorrectly eliminated. (e) There is no significant difference in the performance between XF1 and XF2 models. It is clear, that the number of FPs are extremely high and means to reduce that while maintaining similar levels of FN would be necessary to achieve any benefit. In an attempt to reduce FPs, Combined Classifier Based Approach (CCBA) is proposed which is same as Algorithm 1 with modifications in the **Train** and **Identify** stages. CCBA relies on the use of classifiers in two levels i.e. solutions identified promising based on X classifier is further analyzed using XF1 or XF2 classifier at random. Solutions identified as promising at the second stage via XF1 or XF2 classifier would only be evaluated. Take note that in the event, XF1 classifier is invoked, the f_1 is evaluated. This process is expected to be more *strict*, i.e., we would expect less false positives.

3.2   Performance on DTLZ Problem Suite

The best, mean, worst, median and standard deviation of the hyper volume measures for the problems using *NSGA-II, X, XF1, XF2, Random and CCBA* are presented in Table 3.

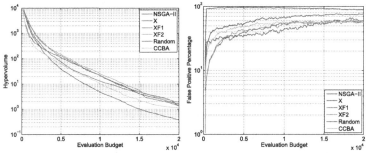

(a) Mean hypervolume versus cost: DTLZ1

(b) False positive percentage versus cost: DTLZ1

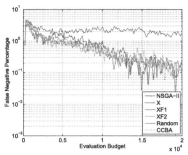

(c) False negative percentage versus cost: DTLZ1

Fig. 3. Plots-DTLZ1

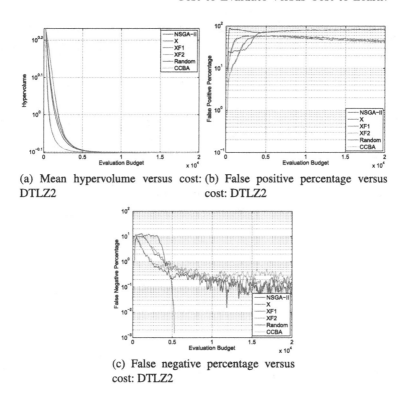

(a) Mean hypervolume versus cost: DTLZ2

(b) False positive percentage versus cost: DTLZ2

(c) False negative percentage versus cost: DTLZ2

Fig. 4. Plots-DTLZ2

The hypervolumes, false positives and false negatives of various strategies for DTLZ1, DTLZ2, DTLZ3, and DTLZ5 problems are presented in Figs. 3(a), 4(a), 5(a), 6(a), 3(b), 4(b), 5(b), 6(b), 3(c), 4(c), 5(c), 6(c).

The results listed in the above tables and plots can be summarized as follows: (a) Full evaluation of all child solutions i.e. NSGA-II offers the best performance for DTLZ1 and DTLZ3 problems. (b) Classification and selection based on CCBA and X only offers the best performance in DTLZ2 and DTLZ5. (c) Classification and selection based on CCBA and X only offers benefit in early phases. (d) In terms of false positives, the average values are 90 %, 70 %, 50 % for NSGA-II, X and CCBA models respectively and for rest of the models the values are at par which is around 40 % for DTLZ1 problem. Whereas, for DTLZ2 problem, the average values of FPs are 80 % for NSGA-II and Random models and 50 % for the rest. For DTLZ3 problem, NSGA-II has almost 100 % FPs, however X offers 50 %, CCBA offers 40 % and rest of the models offer around 30 %. Similarly for DTLZ5 problem, NSGA-II and Random offer around 80 % and rest of the models offer 50 %. However in terms of false negatives, NSGA-II offers best in all problems. Among the rest of the models, Random offers the highest in DTLZ1 and DTLZ3 and has less average value than others in DTLZ2, and DTLZ5 problems. X, XF1, XF2, CCBA models are approximately at par for

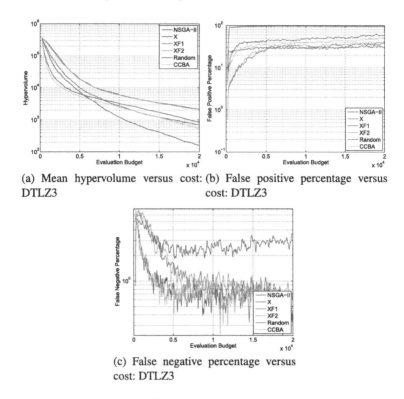

(a) Mean hypervolume versus cost: DTLZ3

(b) False positive percentage versus cost: DTLZ3

(c) False negative percentage versus cost: DTLZ3

Fig. 5. Plots-DTLZ3

all the problems in terms of FNs. These support the argument that the XF1 and XF2 models are cautious i.e. they are selectively evaluating promising solutions. The false positives of model based on X alone is less strict in its evaluation policy when compared to XF1 and XF2 models. However CCBA offers less or at par in terms of FPs and it has better or similar performance in terms of IGD and hypervolume in all the problems compared to X model. (e) Random model seems to offer an average performance across all problems. (f) There is no significant difference in the performance between XF1 and XF2 models in most problems except DTLZ4. One seems to more beneficial over another in different stages of the search process.

The last aspect seems counter-intuitive as we would expect providing additional information would allow the classifier to trained better. Since f_1 and f_2 are functions of X, it is difficult to generalize if evaluating them at a cost and subsequently using them to identify promising solutions would be better than using X alone without any additional cost. To examine this, we assumed zero cost associated in evaluating f_1 and f_2 of the child solutions during their classification phase using XF1 and XF1 models. We would certainly expect, XF1 and XF2 models to perform better or at par with X model. In all the cases, we noticed that XF1 and XF2 models performed better or at par with the X

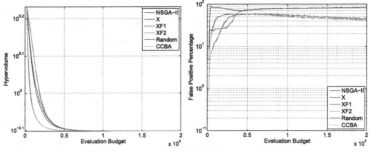

(a) Mean hypervolume versus cost: (b) False positive percentage versus
DTLZ5   cost: DTLZ5

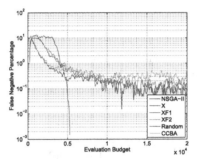

(c) False negative percentage versus
cost: DTLZ5

Fig. 6. Plots-DTLZ5

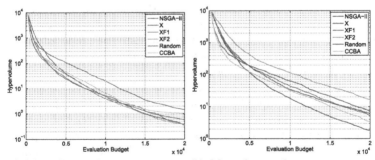

(a) Mean hypervolume versus cost: (b) Mean hypervolume versus cost:
DTLZ1: Zero cost at classification DTLZ1: Different cost
stage

Fig. 7. Plots-DTLZ1: Different Settings

model. In all cases, CCBA performed better than X model. A typical DTLZ1
plot is included to illustrate this. The above discussion is based on hypervolume
measure. The study has clearly indicated that the choice of X, XF1, XF2 or
CCBA classifier certainly improves the performance over a full evaluation policy

i.e. NSGA-II for all problems studied in the paper. We have also illustrated that CCBA can bring down FPs for all problems while maintaining similar levels of FNs and performance compared to X for all problems.

While in the above study, we considered equal cost of f_1 and f_2 evaluation i.e. each of 1 unit, the benefits of XF1 over XF2 or vice versa might be visible if the evaluation costs are different. We present the results of such an example, wherein f_1 evaluation costs 1 unit and f_2 evaluation costs 2 units. However the results clearly identify that CCBA model offers greater benefits as opposed to X, XF1 or XF2 models (Fig. 7).

4   Summary and Conclusions

In this paper, we investigated effects of classifiers and their utility for problems involving computationally expensive multiobjective optimization problems. There has been a renewed interest within the EMO community to investigate and develop classifier based approaches for the solution of such problems. The approach can be considered as a parallel to existing surrogate assisted optimization methods which independently evaluate the objectives. In classifier based approaches, a class label i,e, promising or non-promising is assigned to solutions. This paper attempts to investigate if partial evaluation is helpful i.e. can we evaluated one of the objectives and attempt to assess the quality of the solution. Results of classifiers being trained using X, XF1, XF2 and Random strategies have been used to for baseline investigation. The utility of X, XF1 and XF2 based classifiers have also been assessed with cost and zero cost models. The difference between XF1 and XF2 models have also been assessed to understand the utility of one over another. Furthermore, the results also suggest that some solutions in the population can be discarded based on its variable values only, while for others such a decision can only be made after evaluating its first or second or both its objectives or combining these two strategies. The results clearly indicate that development of strategies to exploit such schemes is extremely challenging.The observation of large number of false positives in all the above models prompted the design of Combined Classifier Based Approach (CCBA). The approach reduced the number of false positives and the false negatives were at par with other strategies resulting in competitive performance for all problems. While far from perfect, we hope that this study would prompt the design of efficient algorithms that selectively evaluate solutions on the fly i.e. based on the trade-off between *need to learn/evaluate* and *cost to learn*.

References

1. Runarsson, T.P., Yao, X.: Stochastic ranking for constrained evolutionary optimization. IEEE Trans. Evol. Comput. **4**(3), 284–294 (2000)
2. Ray, T., Singh, H.K., Isaacs, A., Smith, W.: Infeasibility driven evolutionary algorithm for constrained optimization. In: Mezura-Montes, E. (ed.) Constraint-Handling in Evolutionary Optimization. SCI, vol. 198, pp. 145–165. Springer, Heidelberg (2009)

 3. Takahama, T., Sakai, S.: Constrained optimization by the ϵ constrained differential evolution with gradient-based mutation and feasible elites. In: Proceedings of the IEEE Congress on Evolutionary Computation (CEC), pp. 1–8 (2006)
 4. Asafuddoula, M., Ray, T., Sarker, R.: Evaluate till you violate: a differential evolution algorithm based on partial evaluation of the constraint set. In: Proceedings of the IEEE Symposium on Differential Evolution (SDE), pp. 31–37 (2013)
 5. Jin, Y.: Surrogate-assisted evolutionary computation: recent advances and future challenges. Swarm Evol. Comput. 1(2), 61–70 (2011)
 6. Suykens, J.A.K., Van Gestel, T., De Brabanter, J., De Moor, B., Vandewalle, J.: Least Squares Support Vector Machines, vol. 4. World Scientific, Singapore (2002)
 7. Loshchilov, I., Schoenauer, M., Sebag, M.: A mono surrogate for multiobjective optimization. In: Proceedings of the 12th Annual Conference on Genetic and Evolutionary Computation (GECCO), pp. 471–478. ACM (2010)
 8. Chawla, N.V., Bowyer, K.W., Hall, L.O., Kegelmeyer, W.P.: SMOTE: synthetic minority over-sampling technique. J. Artif. Intell. Res. 16, 321–357 (2002)
 9. Deb, K., Pratap, A., Agarwal, S., Meyarivan, T.: A fast and elitist multiobjective genetic algorithm: NSGA-II. IEEE Trans. Evol. Comput. 6(2), 182–197 (2002)
10. Deb, K., Thiele, L., Laumanns, M., Zitzler, E.: Scalable Test Problems for Evolutionary Multiobjective Optimization. Springer, Heidelberg (2005)
11. Beume, N., Fonseca, C., Lopez-Ibanez, M., Paquete, L., Vahrenhold, J.: On the complexity of computing the hypervolume indicator. IEEE Trans. Evol. Comput. 13(5), 1075–1082 (2009)

A Propositional Plausible Logic

David Billington[✉]

School of Information and Communication Technology, Griffith University,
Nathan Campus, Brisbane, QLD 4111, Australia
d.billington@griffith.edu.au

Abstract. A new non-monotonic propositional logic called PPL — for Propositional Plausible Logic — is defined. An example is worked and four theorems about PPL are stated.

Keywords: Plausible logic · Defeasible logic · Non-monotonic reasoning

1 Introduction

We shall define a new non-monotonic propositional logic called PPL, for Propositional Plausible Logic. PPL belongs to the family of propositional non-monotonic logics called Defeasible Logics, first defined in [2]. Unlike PPL, previous Defeasible Logics [1] do not handle conjunctions properly.

For a given formula, a proof algorithm will satisfy exactly one of the following four conditions.

(i) not terminate,
(ii) terminate in a state indicating that the formula is proved,
(iii) terminate in a state indicating that the formula is not provable,
(iv) terminate in some other state.

A proof algorithm is said to be **decisive** if and only if for every formula, the proof algorithm satisfies either (ii) or (iii). The proof algorithms of all previous Defeasible Logics were not decisive. All the proof algorithms of PPL are decisive, and so PPL is decidable.

This article is organised into the following sections. The definition of PPL is in Sect. 2. In Sect. 3 we apply PPL to an example. Four important theorems about PPL are stated and discussed in Sect. 4. Section 5 is the conclusion.

2 PPL: A Propositional Plausible Logic

PPL uses a propositional language — with negation $\neg$, conjunction $\wedge$, and disjunction $\vee$ — to reason about facts and plausible information. The facts are represented by formulas that are converted into clauses called axioms, which are then converted into strict rules. The plausible information is represented by defeasible rules, warning rules, and a priority relation, $>$, on rules.

© Springer International Publishing Switzerland 2015
B. Pfahringer and J. Renz (Eds.): AI 2015, LNAI 9457, pp. 76–82, 2015.
DOI: 10.1007/978-3-319-26350-2_7

Definition 1. A **rule**, r, is any triple $(A(r), arrow(r), c(r))$ such that $A(r)$ is a finite set of formulas, $arrow(r) \in \{\rightarrow, \Rightarrow, \rightsquigarrow\}$, and $c(r)$ is a formula. If r is a rule then $A(r)$ is called the **set of antecedents** of r, and $c(r)$ is called the **consequent** of r.

Strict rules use the **strict arrow**, $\rightarrow$, and are written $A(r) \rightarrow c(r)$.

Defeasible rules use the **defeasible arrow**, $\Rightarrow$, and are written $A(r) \Rightarrow c(r)$.

Warning rules use the **warning arrow**, $\rightsquigarrow$, and are written $A(r) \rightsquigarrow c(r)$.

Intuitively, $A \rightarrow c$ means if every formula in A is true then c is true; whereas $A \Rightarrow c$ means if every formula in A is true then c is usually true. Roughly, $A \rightsquigarrow c$ means if every formula in A is true then c might be true. So $A \rightsquigarrow \neg c$ warns against concluding usually c, but does not support usually $\neg c$.

Definition 2. Let R be any set of rules. A binary relation, $>$, on R is **cyclic** iff there exists a finite sequence, $(r_1, r_2, ..., r_n)$ where $n \geq 1$, of elements of R such that $r_1 > r_2 > ... > r_n > r_1$.

A priority relation, $>$, on rules is used to indicate the more relevant of two rules. For instance the specific rule 'Quails usually do not fly' is more relevant than the general rule 'Birds usually fly' when reasoning about the flying ability of a quail. Hence 'Quails usually do not fly' > 'Birds usually fly'.

We shall now introduce some needed notation. The empty sequence is denoted by $()$. Let S be a sequence. If S is finite then $S + e$ denotes the sequence formed by just adding e onto the right end of S. Define $e \in S$ to mean e is an element of S, and $e \notin S$ to mean e is not an element of S. A formula is **contingent** iff it is not a tautology and it is not a contradiction. If a is an atom then $\sim a$ is $\neg a$, and $\sim \neg a$ is a. If L is a set of literals then $\sim L = \{\sim l : l \in L\}$.

Next we define, among other things, the conversion of a clause to strict rules.

Definition 3. Let R be a set of rules, F be a finite set of formulas, C be a set of contingent clauses, and c be a contingent clause.

(1) R_s is the set of strict rules in R.
(2) R_d is the set of defeasible rules in R.
(3) $c(R)$ is the set of consequents of the rules in R.
(4) $Ru(c) = \{ \{\} \rightarrow c \} \cup \{ \{ \wedge \sim (L-K) \} \rightarrow \vee K : c = \vee L \text{ and } \{\} \subset K \subset L \}$.
(5) $Ru(C) = \bigcup \{ Ru(c) : c \in C \}$.
(6) $Ru(C, F)$ is the set of rules in $Ru(C)$ whose set of antecedents is F.

If c has n literals then $Ru(c)$ has $2^n - 1$ strict rules. For example $Ru(\vee \{a, b, c\})$ $= \{ \ \{\} \rightarrow \vee \{a, b, c\}, \ \{\wedge \{\neg b, \neg c\}\} \rightarrow a, \ \{\wedge \{\neg a, \neg c\}\} \rightarrow b, \ \{\wedge \{\neg a, \neg b\}\} \rightarrow c,$ $\{\neg a\} \rightarrow \vee \{b, c\}, \ \{\neg b\} \rightarrow \vee \{a, c\}, \ \{\neg c\} \rightarrow \vee \{a, b\} \ \}$.

Definition 4. Let R be a finite set of rules. $(R, >)$ is a **plausible theory** iff there is a satisfiable set of contingent clauses, Ax, such that 1, 2, and 3 hold.

(1) $R_s = \{ A \rightarrow \wedge c(Ru(Ax, A)) : A \in \{ A(r) : r \in Ru(Ax) \} \}$.

(2) If $Ax \neq \{\}$ then $\mathbf{r}$ denotes the strict rule $\{\} \rightarrow \wedge Ax$.

(3) $>$ is a relation on R such that $> \subseteq R \times (R - \{\mathbf{r}\})$ and $>$ is not cyclic.

Space limitations mean that we have made R finite, although this is not necessary. The rules in R_s are the 'anding' together of the rules in $Ru(Ax)$ that have the same antecedent. The set Ax is the set of axioms of the plausible theory $(R, >)$ and is denoted by $Ax(R)$.

In the rest of this section we define how to prove formulas given a plausible theory. This is done by naming the proof algorithms (Definition 5), defining the evidence for and against a formula (Definition 6), defining a mechanism for preventing proofs from looping (Definition 7), and finally defining the proof function and the proof algorithms (Definition 9).

Definition 5. Define $Alg = \{\varphi, \pi, \psi, \beta, \beta', \psi', \pi'\}$ to be the **set of names of the proof algorithms**. Define $\varphi' = \varphi$. If $\alpha \in \{\pi, \psi, \beta\}$ then define $(\alpha')' = \alpha'' = \alpha$. If $\alpha \in Alg$, then the **co-algorithm** of α is α'.

The φ algorithm mimics classical provability, π and ψ propagating ambiguity, and β blocks ambiguity, see Sect. 3. The co-algorithms are used to evaluate evidence against a formula.

The evidence for and against a formula is now defined.

Definition 6. Let $(R, >)$ be a plausible theory, $R' \subseteq R$, $Ax = Ax(R)$, $\alpha \in Alg$, f is a formula, and $\{r, s\} \subseteq R$.

(1) $R_d^s = (R_s \cup R_d) - \{\mathbf{r}\}$.

(2) $R'[f] = \{r \in R' : Ax \cup \{c(r)\}$ is satisfiable and $Ax \cup \{c(r)\} \models f\}$.

(3) $R'[f; s] = \{t \in R'[f] : t > s\}$.

(4) If $\alpha \in \{\varphi, \pi'\}$ or $r = \mathbf{r}$ then $Foe(\alpha, f, r) = \{\}$.

$Foe(\psi', f, r) = \{s \in R[\neg f] : s > r\} = R[\neg f; r]$.

If $\alpha \notin \{\varphi, \pi', \psi'\}$ and $r \neq \mathbf{r}$ then $Foe(\alpha, f, r) = \{s \in R[\neg f] : s \not< r\}$.

Roughly $R'[f]$ is the set of rules in R' whose consequents and axioms imply f. If f is not a fact then $R_d^s[f]$ is all the evidence for f; and $Foe(\alpha, f, r)$ is the set of rules that α regards as the evidence against f that is not inferior to r.

A history of used rules prevents proofs from looping.

Definition 7. Suppose $(R, >)$ is a plausible theory and $\alpha \in Alg$. Define $\alpha R = \{\alpha r : r \in R\}$. Then H is an α-**history** iff H is a finite sequence of elements of $\alpha R \cup \alpha' R$ that has no repeated elements.

A formula is proved by evaluating its proof value, which is either $+1$ or -1; $+1$ indicates the formula is proved, and -1 indicates the formula cannot be proved. The arithmetic properties of the proof values are defined below. These are as expected, but note that $\max\{\} = -1$ and $\min\{\} = +1$.

Definition 8. Suppose $S \subseteq \{+1, -1\}$. Then $\min S = -1$ iff $-1 \in S$, $\min S = +1$ iff $-1 \notin S$, $\max S = +1$ iff $+1 \in S$, $\max S = -1$ iff $+1 \notin S$, $--1 = +1$, and $-+1 = -1$.

The proof function P evaluates the proof value of its three arguments: the proof algorithm α, the history of used rules H, and the formula f or set of formulas F to be evaluated. An explanation of each part of the following definition concludes this section.

Definition 9. Suppose $\mathcal{P} = (R, >)$ is a plausible theory, $Ax = Ax(R)$, $\alpha \in Alg$, H is an α-history, and f is a formula. The **proof function for** $\mathcal{P}$, P, and the proof algorithms are defined by P1 to P5.

(P1) If F is a finite set of formulas, then $P(\alpha, H, F) = \min\{P(\alpha, H, f) : f \in F\}$.

(P2) If $Ax \models f$ then $P(\alpha, H, f) = +1$. Also $P(\varphi, H, f) = +1$ iff $Ax \models f$.

(P3) If $Ax \not\models f$ and $\alpha \neq \varphi$ then
$$P(\alpha, H, f) = \max\{For(\alpha, H, f, r) : \alpha r \notin H \text{ and } r \in R_d^s[f]\}.$$

(P4) If $Ax \not\models f$, $\alpha \neq \varphi$, $\alpha r \notin H$, and $r \in R_d^s[f]$ then $For(\alpha, H, f, r) = \min[\{P(\alpha, H + \alpha r, A(r))\} \cup \{Dftd(\alpha, H, f, r, s) : s \in Foe(\alpha, f, r)\}]$.

(P5) If $Ax \not\models f$, $\alpha \notin \{\varphi, \pi'\}$, $\alpha r \notin H$, $r \in R_d^s[f]$, and $s \in Foe(\alpha, f, r)$ then
$$Dftd(\alpha, H, f, r, s) = \max[\{P(\alpha, H + \alpha t, A(t)) : \alpha t \notin H \text{ and } t \in R_d^s[f; s]\} \cup \{-P(\alpha', H + \alpha's, A(s)) : \alpha's \notin H\}].$$

If x is a formula or a finite set of formulas then define $(\alpha, H) \vdash x$ iff $P(\alpha, H, x) = +1$, and $\alpha \vdash x$ iff x is α-provable iff $P(\alpha, (), x) = +1$.

To prove a set F of formulas every element of F must be proved; hence P1. If $Ax \models f$ then f is a fact and so is declared proved. As φ only proves facts we have the second part of P2.

To prove f we need a rule for f, $r \in R_d^s[f]$, that has not been used before, $\alpha r \notin H$, such that $For(\alpha, H, f, r) = +1$; hence P3. $For(\alpha, H, f, r) = +1$ whenever the set of antecedents of r, $A(r)$, is proved; and when all the evidence against f, that is, rules in $Foe(\alpha, f, r)$, is defeated, that is, $Dftd(\alpha, H, f, r, s) = +1$. Whenever α uses a rule r, we must add αr to the history, H, of used rules giving $H + \alpha r$. Hence P4.

A rule s is defeated either by team defeat or by disabling s. The team of rules for f is $R_d^s[f]$. A rule s is defeated by team defeat if and only if there is a member t of the team for f such that $t > s$, that is, $t \in R_d^s[f; s]$, and every formula in the set of antecedents of t, $A(t)$, is proved. Of course we must be sure that α has not previously used t, $\alpha t \notin H$, and we must add αt to H. A rule s is disabled by showing that α' cannot prove $A(s)$. Again we must be sure that α' has not previously used s, $\alpha's \notin H$, and we must add $\alpha's$ to H. Hence P5.

Now that Plausible Logic is defined, it is time to apply it.

3   The Ambiguity Puzzle

Consider the following four statements.

(1) There is evidence that a is usually true.

(2) There is evidence that $\neg a$ is usually true.

(3) There is evidence that b is usually true.

(4) If a is true then $\neg b$ is usually true.

What can be concluded about b? The evidence for b is (3). The evidence against b comes from (1) and (4). If we knew that a was definitely true then the evidence for b and against b would be equal. But (1) and (2) means that the evidence against b has been weakened.

Since (1) and (2) give equal evidence for and against a, a is said to be **ambiguous**. If the evidence against b has been weakened sufficiently to allow b to be concluded, then b is not ambiguous. So the ambiguity of a has been blocked from propagating to b. An algorithm that can prove b (but not $\neg b$) is said to be **ambiguity blocking**.

If the evidence against b has not been weakened sufficiently to allow b to be concluded, then b is ambiguous. So the ambiguity of a has been propagated to b. An algorithm that cannot prove b (or $\neg b$) is said to be **ambiguity propagating**.

The plausible theory $(R, >)$ which models the Ambiguity Puzzle is defined as follows. The priority relation $>$ is empty, and $R = \{r_a, r_{na}, r_b, r_{anb}\}$, where r_a is $\{\} \Rightarrow a$, r_{na} is $\{\} \Rightarrow \neg a$, r_b is $\{\} \Rightarrow b$, and r_{anb} is $\{a\} \Rightarrow \neg b$.

Since $R_s = \{\}$, $Ax(R) = Ax = \{\}$. So $R[a] = \{r_a\}$, $R[b] = \{r_b\}$, $R[\neg a] = \{r_{na}\}$, and $R[\neg b] = \{r_{anb}\}$. If $l \in \{a, \neg a, b, \neg b\}$ and $s \in R$ then $R[l; s] = \{\}$.

We shall show that π and ψ are ambiguity propagating and that β is ambiguity blocking. In the following evaluations we shall use † and □.

†) $P(\alpha, H, \{f\}) = P(\alpha, H, f)$, by P1.

□) $P(\alpha, H, \{\}) = \min\{\} = +1$, by P1.

Evaluation E1. $\alpha \in \{\pi, \psi, \beta\}$
1α) $P(\alpha, (), b) = For(\alpha, (), b, r_b)$, by P3
2α) $= \min\{P(\alpha, (\alpha r_b), \{\}), Dftd(\alpha, (), b, r_b, r_{anb})\}$, by P4
3α) $= Dftd(\alpha, (), b, r_b, r_{anb})$, by □
4α) $= -P(\alpha', (\alpha' r_{anb}), a)$, by P5, †
5α) $= -For(\alpha', (\alpha' r_{anb}), a, r_a)$, by P3

Evaluation E2. $\alpha \in \{\pi, \psi\}$ and $\alpha \nvdash b$
5α) $P(\alpha, (), b) = -For(\alpha', (\alpha' r_{anb}), a, r_a)$, by E1
6α) $= -P(\alpha', (\alpha' r_{anb}, \alpha' r_a), \{\})$, by P4
7α) $= -1$, by □.

Evaluation E3. $\beta \vdash b$
5β) $P(\beta, (), b) = -For(\beta', (\beta' r_{anb}), a, r_a)$, by E1
6β) $= -\min\{P(\beta', (\beta' r_{anb}, \beta' r_a), \{\}), Dftd(\beta', (\beta' r_{anb}), a, r_a, r_{na})\}$, by P4
7β) $= -Dftd(\beta', (\beta' r_{anb}), a, r_a, r_{na})$, by □
8β) $= --P(\beta, (\beta' r_{anb}, \beta r_{na}), \{\})$, by P5
9β) $= +1$, by □.

By Evaluation E2 and Theorem 1 (Decisiveness), π and ψ cannot prove b and so they are ambiguity propagating. By Evaluation E3 and Theorem 3 (2-Consistency), β proves b and so is ambiguity blocking.

4 Theorems About Propositional Plausible Logic

A major property of PPL is that it is decisive and hence decidable.

Theorem 1 (Decisiveness). Suppose $\mathcal{P}$ is a plausible theory, $\alpha \in Alg$, H is an α-history, and x is either a formula or a finite set of formulas. Then either $P(\alpha, H, x) = +1$ or $P(\alpha, H, x) = -1$, but not both.

Right Weakening is the property that if a formula f is provable and f classically implies a formula g then g is provable. Thus Right Weakening is closure under classical inference. The following result shows that both Right Weakening and 'Modus Ponens (MP) for strict rules' hold for PPL.

Theorem 2 (Right Weakening). Suppose $(R, >)$ is a plausible theory, $Ax = Ax(R)$, $\alpha \in Alg$, H is an α-history, and both f and g are formulas.

(1) If $(\alpha, H) \vdash f$ and $Ax \cup \{f\} \models g$ then $(\alpha, H) \vdash g$.
(2) If $(\alpha, H) \vdash f$ and $f \models g$ then $(\alpha, H) \vdash g$. (Right Weakening)
(3) If $A \rightarrow g \in R_s$ and $(\alpha, H) \vdash A$ then $(\alpha, H) \vdash g$. (MP for strict rules)

The next theorem states that any two proved formulas are consistent with the axioms.

Theorem 3 (2-Consistency). Suppose $(R, >)$ is a plausible theory, $Ax = Ax(R)$, $\alpha \in \{\varphi, \pi, \psi, \beta, \beta'\}$, and both f and g are formulas.
If $\alpha \vdash f$ and $\alpha \vdash g$ then $Ax \cup \{f, g\}$ is satisfiable.

The final result shows that the proof algorithms form a hierarchy that is consistent with the intuition that ambiguity propagating proof algorithms are more cautious than ambiguity blocking algorithms.

Theorem 4 (Hierarchy). Suppose $\mathcal{P} = (R, >)$ is a plausible theory. Let $\mathcal{P}(\alpha)$ be the set of all formulas provable from $\mathcal{P}$ using the proof algorithm α.

(1) $\mathcal{P}(\varphi) \subseteq \mathcal{P}(\pi) \subseteq \mathcal{P}(\psi) \subseteq \mathcal{P}(\beta) = \mathcal{P}(\beta') \subseteq \mathcal{P}(\psi') \subseteq \mathcal{P}(\pi')$.
(2) If $>$ is empty then $\mathcal{P}(\varphi) \subseteq \mathcal{P}(\pi) = \mathcal{P}(\psi) \subseteq \mathcal{P}(\beta) = \mathcal{P}(\beta') \subseteq \mathcal{P}(\psi') = \mathcal{P}(\pi')$.

So β', the co-algorithm of β, proves exactly the same formulas as β. The set of formulas proved by π' is very similar to the union of all extensions of an extension based logic, like Default Logic or argumentation systems. The difference between ψ' and π' is that ψ' considers the priority relation $>$ whereas π' does not. Since both ψ' and π' can prove a formula and its negation, it is better to think of them as evidence finders rather than algorithms that prove formulas are true.

5 Conclusion

A new propositional non-monotonic logic, called PPL for Propositional Plausible Logic, was defined. An example was worked and four theorems about PPL were stated. PPL has been implemented by George Wilson under the direction of Dr. Andrew Rock, who has implemented other Defeasible Logics.

Future work on PPL could involve a complexity analysis and a study of PPL's implementation. It may be worthwhile relating PPL and argumentation systems. Adding variables to PPL would significantly increase its usefulness.

Acknowledgement. The author thanks Patrick Marchisella for many helpful discussions.

References

1. Billington, D.: A defeasible logic for clauses. In: Wang, D., Reynolds, M. (eds.) AI 2011. LNCS, vol. 7106, pp. 472–480. Springer, Heidelberg (2011)
2. Nute, D.: Defeasible reasoning. In: Proceedings of the 20th Hawaii International Conference on System Science, pp. 470–477 (1987)

Monte Carlo Analysis of a Puzzle Game

Cameron Browne[(✉)] and Frederic Maire

School of Electrical Engineering and Computer Science,
Science and Engineering Faculty, Queensland University of Technology,
Gardens Point, Brisbane 4000, Australia
{c.browne,f.maire}@qut.edu.au
http://www.qut.edu.au

Abstract. When a puzzle game is created, its design parameters must be chosen to allow solvable and interesting challenges to be created for the player. We investigate the use of random sampling as a computationally inexpensive means of automated game analysis, to evaluate the BoxOff family of puzzle games. This analysis reveals useful insights into the game, such as the surprising fact that almost 100 % of randomly generated challenges have a solution, but less than 10 % will be solved using strictly random play, validating the inventor's design choices. We show the 1D game to be trivial and the 3D game to be viable.

1 Introduction

Any newly designed game must undergo a process of playtesting and refinement to ensure that its equipment and rule set are optimally tuned to realise an interesting playing experience. This can be a painstaking and tedious process that may take years, but can be assisted by mathematical and/or computer modelling of the game in question [2]. However, analyses that rely on full game tree expansions or complete enumerations of the design space can be prohibitively expensive to compute for real-world cases of even modest complexity.

In this paper, we investigate ways in which random sampling can be used instead, to quickly give some insight into a game's inherent nature with less computational effort. We use as our test case a new puzzle game called BoxOff.

Monte Carlo approaches have had spectacular success in game AI over the last decade, especially *Monte Carlo tree search* (MCTS) methods, which now drive the world champion AI players of many games [5]. MCTS approaches have been especially successful in the related field of *general game playing*, i.e. the study of computer programs for playing a range of games well rather than specialising in any one particular game, as they allow the AI to make plausible moves for a given game without any domain-specific strategic or tactical knowledge [6].

However, we use Monte Carlo approaches for a different purpose in this study. We are less concerned with *how to play* the game than with *how well it plays*. This study can be phrased as an optimisation problem – given the basic design of a game, what are the parameters that provide the best experience for the player? – which places it within the remit of *procedural content generation* [13].

© Springer International Publishing Switzerland 2015
B. Pfahringer and J. Renz (Eds.): AI 2015, LNAI 9457, pp. 83–95, 2015.
DOI: 10.1007/978-3-319-26350-2_8

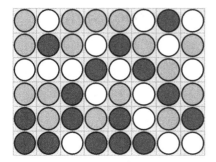
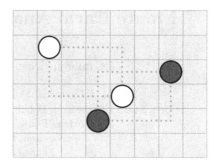

Fig. 1. A 6×8 challenge with three colours, and legal white and illegal black moves.

1.1 BoxOff

BoxOff[1] is a solitaire puzzle game invented in 2013 by American game designer Stephen Meyers [9]. The rules are as follows:

1. *Start:* The game starts with pieces in C colours randomly placed to cover all cells of a regular rectangular grid (the *board*). This defines the *challenge* to be solved. Figure 1 (left) shows a challenge on the standard 6×8 board with $C = 3$ colours.
2. *Play:* The player then makes a series of moves, each involving the removal of a pair of same-coloured pieces that occupy a *box* (rectangle) that includes no other pieces. For example, the two white pieces in Fig. 1 (right) can be removed because they occupy a box (dotted) that is otherwise empty, whereas the two black pieces are blocked from removal by the white piece.
3. *End:* The player wins by removing all pieces from the board, else loses if there are no legal moves at any point while pieces remain on the board.

In order to be solvable, each challenge must have an even number N of board cells, and for each colour c the number of pieces P_c must also be even. The standard game is played on a 6×8 board with 16 pieces in each of $C = 3$ colours. For aesthetic reasons, the board dimensions are typically chosen to be as square as possible for a given N, and the piece counts of each colour $P_1 \ldots P_C$ are typically chosen to be as similar as possible while summing to N.

1.2 Game Design Goals

The design parameters for BoxOff are therefore the board size N and number of colours C. We define the *design space* to be the set of valid combinations of N and C, the *challenge space* to be the set of possible challenges for a given design, and the *solution space* to be the set of possible solutions for a given challenge. We are interested in whether random sampling of the design space of BoxOff can shed some light on questions such as:

---

[1] The name "BoxOff" was coined by the first author.

*   – *What are good values for the board size N for this game?*
*   – *What are good values for the number of piece colours C?*
*   – *How likely is a randomly generated challenge to be solvable?*
*   – *How likely is a randomly generated challenge to be interesting?*

Solvability. The most important question regarding BoxOff – and indeed most puzzles – is that of *solvability*, namely how likely it is that a given challenge will actually have a solution. There are two relevant probabilities: $P(S_r)$ the probability that randomly sampled challenges will be solvable with random play, and $P(S_p)$ the probability that randomly sampled challenges will be solvable with perfect play.

It is easy to artificially construct unsolvable challenges. For example, Fig. 2 shows a 2×3 challenge with no opening moves (left), and a 3×4 challenge that allows several moves but never the removal of the lower left piece (right). Conversely, it is easy to artificially construct challenges that are guaranteed to be solvable, by starting with an empty board then adding same-coloured pairs such that no odd-sized regions of empty cells are created, until the board is full. Greg Schmidt uses this approach in his Axiom AI player for BoxOff [12].

However, a player using a physical version of the game does not want to worry about such niceties. A patient player can construct challenges guaranteed to be solvable if they wish, but most players will just want to place pieces randomly and as quickly as possible, hence the question of solvability becomes important.

Firstly, players must have confidence that the majority of challenges they construct will be solvable with perfect play, otherwise there is little point in playing the game. Knowing the likelihood that challenges are solvable also helps players gauge their progress based on win rate.

Secondly, it would be detrimental if most challenges could be solved by random play. Challenges that can be trivially solved without thought or forward planning will be of little interest to most players. Well designed puzzles tend to display structure or *dependency*, such that certain moves reveal key information that allows further moves to be made, and must be performed in a certain order [4]. Challenges that can be solved by making random moves without any planning indicate a lack of such dependency, and are described as being *susceptible to random play*.

Fig. 2. Unsolvable 2×3 and 4×3 challenges.

To answer the game design questions, we consider different complexity measures. *The majority of challenges should ideally be solvable by perfect play but not by random play.* We therefore want to maximise $P(S_p)$ while minimising $P(S_r)$. Section 2 discusses the complexity of BoxOff, Sect. 3 explores the solvability of randomly generated challenges, Sect. 4 looks briefly at the interestingness of randomly generated challenges, and Sect. 5 summarises our results.

2 Complexity

For this analysis, we implemented two types of AI solver for BoxOff:

1. S_r: Random solver that applies a random legal move each turn, until the game is won or lost.
2. S_p: Depth-first backtracking solver that returns the first valid solution found (if any). S_p recursively tries each available action of each state in order, using a *transposition table* to avoid repetition [11].

2.1 Challenge Space Complexity

The *challenge space complexity* of a given board size is the number of distinct challenges that it allows, not counting reflections, rotations and colour permutations.

Table 1 shows the number of distinct challenges found in complete enumerations of smaller board sizes up to 4×5 with $C = 3$ colours, the number of these challenges that are solvable with perfect play S_p, and the ratio of these two numbers in bold. The rightmost column of Table 1 shows the observed solvability ratios of 10,000 randomly generated challenges for the same board sizes, with 95 % confidence intervals. Randomly sampled challenges appear to offer a fair representation of the complete set of actual challenges. The number of challenges increases exponentially with board size, for example, there exist approximately 1.355×10^{21} challenges for the standard 6×8 game played with $C = 3$ colours, including reflections, rotations and colour transpositions. This makes exhaustive analysis of even the standard board size impractical.

Table 1. Solvability of complete and sampled challenge sets.

	N	Complete			Sampled
		Challenges	Solvable	Ratio	Ratio
3×4	12	1,523	682	**0.4478**	**0.4438** ± 0.0097
4×4	16	105,561	59,545	**0.5641**	**0.5592** ± 0.0097
4×5	20	13,098,310	9,036,038	**0.6998**	**0.6934** ± 0.0090

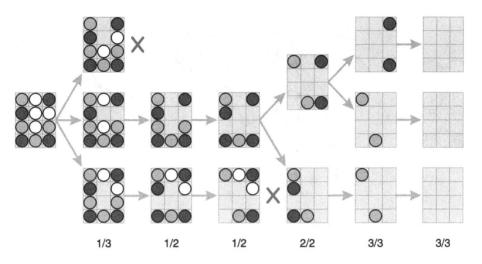

1/3   1/2   1/2   2/2   3/3   3/3

Fig. 3. Full game tree expansion, showing ratios of winning choices.

2.2 Game Tree Complexity

The *game tree complexity* of a challenge is defined as "the number of leaf nodes in the solution search tree of the initial position" [1, p. 160]. For example, Fig. 3 shows the full game tree expansion of a simple 4×3 challenge with $C = 3$ colours, showing the ratio of winning moves each turn (a statistic used later in Sect. 4). Obviously, no game of BoxOff on N cells can ever exceed $N/2$ moves, and a solution will always be of length $N/2$. For example, the 6th move wins in all cases on the $N = 12$ board shown in Fig. 3. Such complete game tree expansions are infeasible for larger boards, but game tree complexity can be estimated based on a challenge's *branching factor*, i.e. number of legal moves M_t for each turn t.

6×8 challenges tend to start with a branching factor of around $M_0 = 26$ legal opening moves, decreasing almost linearly to zero over the course of the game. The product of means $\prod_{t=1}^{N/2} M_t$ gives an estimated game tree complexity of approximately 2.86×10^{22} non-distinct board positions for the 6×8 board with $C = 3$ colours.

2.3 State Space Complexity

The *state space complexity* of a challenge is defined as "the number of legal game positions reachable from the initial position" [1, p. 158]. This is equivalent to the number of distinct board positions stored in the transposition table following a complete traversal of all possible lines of play.

A full BoxOff game tree expansion on a 6×8 board will typically involve less than 1×10^8 distinct board states. As a rough rule of thumb, BoxOff games tend to have state space complexity in the order of $2^{N/2}$. While this is a more manageable number than the game tree complexity, it is still prohibitively time

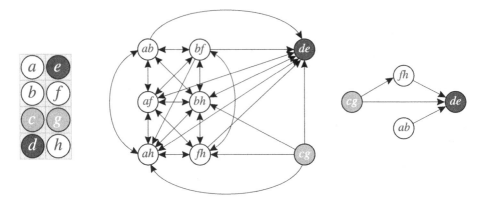

Fig. 4. A challenge, its digraph, and solution subgraph.

consuming to expand full game trees for even medium sized boards, hence our interest in analysing the game through random sampling alone.

2.4 Computational Complexity

We note that a given challenge can be represented as a directed graph G, in which each potential move corresponds to a vertex, and an arc connects vertex p_i to vertex p_j if move p_j cannot be played before move p_i. The game can then be reduced to the problem of finding the largest subgraph in G that does not contain a directed cycle. For example, Fig. 4 shows the reduction of the 4×2 challenge shown to its largest acyclic subgraph.

   A general version of this problem is equivalent to the *feedback vertex set* problem, which was among the first problems shown to be NP-complete by Karp [7]. However, we also note that graphs associated with BoxOff challenges may contain structural properties that allow the design of polynomial time algorithms for their solution. The complexity of the decision problem of the solvability of a given BoxOff puzzle is an open problem.

3 Experiments

As stated in the introduction, we now examine the solvability of 2D, 1D and 3D versions of the game. Recall from Sect. 1.2 (Solvability) that we especially want to maximise solvability while minimising susceptibility to random play.

3.1 2D Case

The standard 2D version of the game is the case we are most interested in. Figure 5 shows the observed solvability probabilities $P(S_r)$ and $P(S_p)$ for various 2D boards up to size $N = 64$ for $C = 2, 3$ and 4 colours. We only consider board sizes with at least $2C$ cells in each case, to allow at least one pair of each colour, and we only consider board sizes up to $N \leq 64$, in order to allow an efficient bitboard encoding with 64-bit long integers [3]. Board sizes tested: 2×2, 2×3, 3×4, 4×4, 4×5, 4×6, 5×6, 6×6, 6×7, 6×8, 7×8 and 8×8.

The dotted lines show the observed probabilities of success for the random solver $P(S_r)$ averaged over 10,000 randomly sampled challenges. The solid lines show the observed probabilities of success for the perfect solver $P(S_p)$ averaged over 1,000 randomly sampled challenges (decreasing to 100 for some of the larger board sizes, due to time constraints). The arrows indicate the disparity between $P(S_r)$ and $P(S_p)$ for the standard $N = 48$ (i.e. 6×8) case, which we want to maximise. Note that the random solver S_r solvability curves (dotted) tend to drop sharply, while the perfect solver S_p solvability curves trend upwards to plateau at almost 100 % for larger boards.

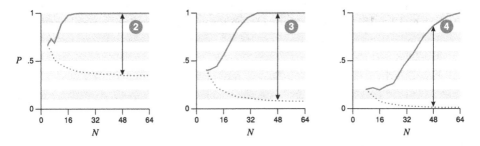

Fig. 5. Observed solvability probabilities $P(S_r)$ (dotted) and $P(S_p)$ (solid) for 2D cases, for $C = 2, 3$ and 4. Arrows show the disparity between $P(S_r)$ and $P(S_p)$ for the default $N = 6 \times 8 = 48$ case, which is the key measurement here.

Table 2. Solvability probabilities for the 6×8 (2D) case for $C = 2 \ldots 6$.

	C				
	2	3	4	5	6
$P(S_r)$	$.349 \pm .009$	$.083 \pm .005$	$.015 \pm .002$	$.002 \pm .001$	$.000 \pm .000$
$P(S_p)$	$.999 \pm .000$	$.999 \pm .000$	$.866 \pm .021$	$.293 \pm .028$	$.039 \pm .012$

Table 2 shows the exact solvability probabilities $P(S_r)$ and $P(S_p)$ for the standard 6×8 case, for $C = 2 \ldots 6$ colours. Using two colours, almost 100 % of randomly sampled challenges are solvable, although the high random solvability rate of around 35 % points to a lack of difficulty. Using three colours, almost 100 % of randomly sampled challenges are solvable, with a much lower susceptibility to random play of around 8 %. Using four colours, less than 87 % of randomly sampled challenges are solvable. Three colours show the greatest difference between $P(S_r)$ and $P(S_p)$ so are the optimal choice here.

In practice, only 1 in around 5,000 randomly sampled challenges prove to be unsolvable on the standard 6×8 board with three colours. Most players could spend their entire lives without constructing a single unsolvable challenge, while still having the luxury of being able to blame bad luck for any failure to solve a particular challenge.

For completeness, Fig. 6 shows the observed solvability probabilities $P(S_r)$ and $P(S_p)$ for various 2D boards up to size $N = 64$ for $C = 5$ and 6 colours. It can be seen that solvability by random player $P(S_r)$ drops quickly to almost 0% for both $C = 5$ and 6, which is good. However, solvability by perfect play $P(S_p)$ is in general much poorer than when using fewer colours. For example, less than 30% of randomly sampled challenges will be solvable on the standard 6×8 board using five colours, and less than 5% will be solvable using six. This means that most challenges that players set themselves will be unsolvable, which is very undesirable. These findings are consistent with an observation by the game's designer, Steve Meyers, that five and six colours may be suitable for very large boards, e.g. 12×15, but are a poor choice for small or medium sized boards.[2]

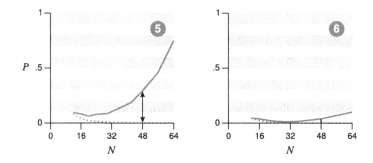

Fig. 6. Observed solvability probabilities $P(S_r)$ (dotted) and $P(S_p)$ (solid) for 2D cases, for $C = 5$ and 6.

Summary: The 6×8 board with three colours seems to be an astute design choice, which allows a good balance between high solvability and low susceptibility to random play.

3.2   1D Case

The simplest version of the game is the 1D case played on a $1 \times n$ board.[3] The pieces start in a line, and the player removes same-coloured pairs in clear line-of-sight of each other. Figure 7 shows the observed solvability probabilities $P(S_r)$ and $P(S_p)$ for various 1D boards up to size $N = 64$ for $C = 2, 3$ and 4 colours. Board sizes tested: 1×4, 1×8, 1×12, 1×16, 1×20, 1×28, 1×36, 1×42, 1×48,

---

[2] Personal correspondence.

[3] The 1D version of the game might be called "LineOff".

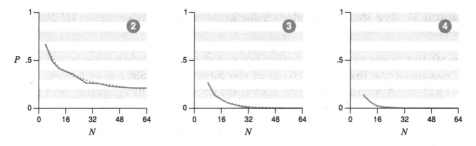

Fig. 7. Observed solvability probabilities $P(S_r)$ (dotted) and $P(S_p)$ (solid) for 1D cases, for $C = 2, 3$ and 4.

Table 3. Average solvability of the 1×48 (1D) case.

	C		
	2	3	4
1×48	0.2268 ± 0.0082	0.0042 ± 0.0013	0.0001 ± 0.0001

1×56 and 1×64. The first thing to note is that both curves are very similar for each value of C. There is no significant difference between the success rates of S_r and S_p at any point, and these curves would in fact be identical if they were measured on the same sample sets rather than being sampled independently. This is due to an unexpected anomaly that *for any solvable 1D position, no sequence of moves can ever lead to a loss*. This is proven below, and is an example of important knowledge about the game revealed through random sampling.

We characterise the winnable games in terms of a context-free grammar, and show that for such games, any sequence of legal moves wins the game. To compactly represent a challenge, we map a row of coloured pieces to a string of integers. For example, the string "1 2 2 1" codes a challenge where "1" corresponds to a *black* piece, and "2" corresponds to *white* a piece.

We will show that the context-free grammar $\mathcal{G}_1 = (\{S\}, \mathcal{P}, \{1, 2, \ldots, n\}, S)$ defined below generates exactly all winnable challenges. The grammar $\mathcal{G}_1$ has a unique non-terminal S. The set of terminals are the first n integers. The set $\mathcal{P}$ contains the production rules $\{S \to xSx, \ S \to SS, \ S \to \epsilon\}$ where x takes all values in $\{1, 2, \ldots, n\}$, and ϵ denotes the empty string.

We can consider that the rules of the form $S \to xx$ are also part of the grammar as they are obtained by applying the third rule after the first rule. We write $S \xrightarrow{\star} \alpha$, if the string α can be generated by repeated application of the production rules. By abuse of notation, we also write $\alpha \in \mathcal{G}_1$. For example, we have $S \xrightarrow{\star} 322113$, as "3 2 2 1 1 3" can be derived as:

$$S \to \mathbf{3S3} \to 3\mathbf{SS}3 \to 32\mathbf{S}2S3 \to 32\,S21\mathbf{S}13 \to 322113$$

where the **bold** substrings highlight the latest substitions of S. The grammar $\mathcal{G}_1$ is very similar to a *Balanced Parentheses* grammar.

Lemma 1. *Any string generated by the grammar $\mathcal{G}_1$ is a winnable 1D BoxOff challenge.*

Proof: In the derivation of a string generated by $\mathcal{G}_1$, the sub-sequence of substitions using the first rule of $\mathcal{G}_1$ is of the form $S \to x_1 S x_1$, $S \to x_2 S x_2$, ... , $S \to x_k S x_k$. A winning strategy is to contract $x_k x_k$, then $x_{k-1} x_{k-1}$, and continue the contractions until $x_1 x_1$. Reciprocally, we have:

Lemma 2. *If a string is a winnable 1D BoxOff challenge, then the string belongs to the language generated by $\mathcal{G}_1$.*

Proof: *Base case*: If the string is of length 2, then the string is of the form xx. Therefore it can be generated by the sequence, $S \to xSx$, and $S \to \epsilon$. *Induction case*: Without loss of generality, we assume that the first character of the string is contracted at step k of a winning sequence of moves. The string is of the form $x_k \alpha x_k \beta$, where β is possibly the empty string. The string α must be contracted before the pair $x_k x_k$. Therefore, by induction on the length of the string, we have $S \xrightarrow{*} \alpha$. Once the pair $x_k x_k$ is contracted, we are left with the winnable string β. Again, by induction we have $S \xrightarrow{*} \beta$. In summary, we can derive the initial string as $S \to SS \to x_k S x_k S \xrightarrow{*} x_k \alpha x_k S \xrightarrow{*} x_k \alpha x_k \beta$.

The 1D version of BoxOff is uninteresting for a human player because it does not require any forward thinking.

Theorem 1. *A 1D BoxOff challenge is winnable if and only if it belongs to the language generated by the grammar $\mathcal{G}_1$. Moreover, if the challenge is winnable, then any sequence of legal moves is a winning strategy.*

Proof: The first part of the theorem is a direct consequence of the two previous lemmas. We prove the second part of the theorem by induction on the length of the string. *Base case*: The result is trivial for a string of length 2. *Induction case*: Without loss of generality, assume that the winnable string is of the form $\alpha x x \beta$, where xx is a contractible pair that we arbitrarily choose as the first move. If there exists a derivation $S \xrightarrow{*} \alpha S \beta \to \alpha x x \beta$, then $\alpha \beta$ can be generated by $\mathcal{G}_1$. Indeed, we just have to replace $S \to xx$ with $S \to \epsilon$ as the last step of the derivation. Therefore, by induction, any sequence of legal moves on $\alpha \beta$ is winning. If there exists no derivation $S \xrightarrow{*} \alpha S \beta$, then the first x following α in $\alpha x x \beta$ must be generated as the right x of a production $S \to xSx$. Similarly, the x in front of β in $\alpha x x \beta$ must be generated as the left x of a production $S \to xSx$. Hence the string $\alpha x x \beta$ must be of the form $\alpha_1 x \alpha_2 x x \beta_1 x \beta_2$, with $\alpha_1 x \alpha_2 x = \alpha x$ and $x \beta_1 x \beta_2 = x \beta$. We therefore have $S \xrightarrow{*} \alpha_1 SS \beta_2 \xrightarrow{*} \alpha_1 \mathbf{x} \mathbf{S} \mathbf{x} S \beta_2 \xrightarrow{*} \alpha_1 x S x \mathbf{x} \mathbf{S} \mathbf{x} \beta_2 \xrightarrow{*} \alpha_1 x \alpha_2 x x \beta_1 x \beta_2$. This shows that $S \xrightarrow{*} \alpha_1 SS \beta_2$ and $S \xrightarrow{*} \alpha_2$ and $S \xrightarrow{*} \beta_1$ with $\alpha = \alpha_1 x \alpha_2$ and $\beta = \beta_1 x \beta_2$.

Recall that we want to prove that $\alpha \beta \in \mathcal{G}_1$. Starting from $S \xrightarrow{*} \alpha_1 SS \beta_2$ and using $S \to \epsilon$, we derive that $S \xrightarrow{*} \alpha_1 S \beta_2$. Applying the rules $S \to xSx$ and $S \to SS$, and using the fact that $S \xrightarrow{*} \alpha_2$ and $S \xrightarrow{*} \beta_1$, we derive that:

$$S \xrightarrow{*} \alpha_1 S \beta_2 \to \alpha_1 \mathbf{x} \mathbf{S} \mathbf{x} \beta_2 \xrightarrow{*} \alpha_1 x \mathbf{S} \mathbf{S} x \beta_2 \xrightarrow{*} \alpha_1 x \alpha_2 \beta_1 x \beta_2$$

We have just shown that $\alpha_1 x \alpha_2 \beta_1 x \beta_2 = \alpha\beta$, therefore $\alpha\beta \in \mathcal{G}_1$.

Solvability in the 1D case drops to around 25 % for larger boards with $C = 2$, and around 0 % for larger boards with $C = 3$ and $C = 4$. Table 3 shows the average solvability of the special 1×48 case, which has the same number of cells as the standard 6×8 board, with 95 % confidence intervals. Players can expect to win around 23 % of games on this board with 2 colours, but less than 1 % of games with 3 or 4 colours.

Summary: The 1D version is trivial – if a challenge is solvable, then any move is as good as any other – hence is of little interest to players.

3.3 3D Case

For completeness, we also consider the 3D version of the game in which piece pairs define 3D boxes that must otherwise be empty. However, such 3D boards would be difficult to make as physical sets so are mostly of academic interest only. Figure 8 shows the observed solvability probabilities $P(S_r)$ and $P(S_p)$ for various 3D boards up to size $N = 64$ for $C = 2, 3$ and 4 colours. Again, the arrows indicate the disparity between $P(S_r)$ and $P(S_p)$ for the $N = 48$ case.

The 3D solvability curves are strikingly similar to those of 2D case, although the solvability rates for both random and perfect play tend to be slightly higher in general. The exact values shown in Table 4 for the target case $N = 48$ ($4 \times 4 \times 4$) indicate that $C = 4$ colours is probably optimal for this board, giving an almost 100 % solvability rate with a low susceptibility to random solution of around 2 %.

Summary: The 3D case is a viable version of the game.

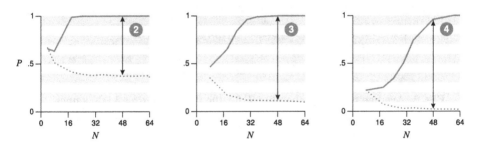

Fig. 8. Observed solvability probabilities $P(S_r)$ (dotted) and $P(S_p)$ (solid) for 3D cases, for $C = 2, 3$ and 4.

Table 4. Solvability probabilities of the $4 \times 4 \times 4$ (3D) case.

		C		
		2	3	4
$4 \times 4 \times 4$	$P(S_r)$	0.376 ± 0.009	0.097 ± 0.005	0.019 ± 0.003
	$P(S_p)$	0.999 ± 0.000	0.999 ± 0.000	0.999 ± 0.000

4   Tension

In this section, we use random sampling to evaluate the potential of BoxOff challenges to interest human players, based on estimated *tension*, i.e. the degree to which the players' decisions affect the outcome of the game. If the player can win by making random choices then the game is not tense, but if every decision is critical to success then the game is very tense. Kramer [8] and Rose [10] observe that well designed games tend to display points of high and low tension.

Tension T is measured as the average ratio of losing moves to total moves at each turn (reduced to 0 if all moves are losing). For example, the first move of the game shown in Fig. 3 is relatively tense, as 2 out of the 3 possible moves will lose. However, the remainder of this game lacks tension as every subsequent move leads to a win (or a loss, if the losing path is chosen).

Figure 9 shows relative tension per turn for the 6×8 case, averaged over 1,000 randomly sampled solvable games. Games typically start in a state of low tension that builds to a peak of almost 50 % in the mid game, followed by a quick dénouement in the end game. This tension curve is actually a good shape for this game. We want low tension (i.e. fewer losing moves) in the early game, as the repercussions of losing moves may not become obvious until say 20 turns later, which would be frustrating for the player and make such challenges intractable. We want higher tension in the middle-to-end game, where there are fewer move choices and the player can plan ahead with greater certainty, as found.

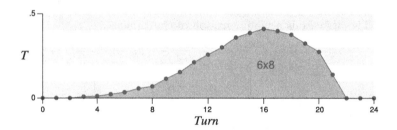

Fig. 9. Average tension probability T for the 6×8 case over 1,000 games.

5   Conclusion

Random sampling yielded useful insights into the BoxOff puzzle game, where analyses through complete game tree expansion would have been impractical. The inventor's default design parameter choices (three colours on a 6×8 board) appear to be optimal, as almost every challenge randomly constructed by the player will be solvable, while few will be susceptible to random play. The 1D version of the game is trivially solvable and hence of little interest to players, while the 3D version of the game appears to be viable. Our analysis also revealed that BoxOff challenges on the standard board tend to start with a low degree of tension, and build to a climax in the mid-to-late game, allowing them to be tractable but still demanding for players. Monte Carlo analysis proved useful in this case. The general nature of our method, which requires no domain-specific strategic or tactical knowledge, makes it potentially applicable to any domain with discrete actions and computable outcomes.

Acknowledgements. This work was supported by a QUT Vice-Chancellor's Research Fellowship, as part of the project *Games Without Frontiers*.

References

1. Allis, V.: Searching for solutions in games and artificial intelligence. Ph.D. dissertation, University of Limburg, Maastricht, Netherlands (1994)
2. Althöfer, I.: Computer-aided game inventing. Technical report, Friedrich-Schiller University, Faculty of Mathematics and Computer Science, Jena (2003)
3. Browne, C.: Bitboard methods for games. Int. Comput. Games Assoc. (ICGA) J. **37**(2), 67–84 (2014)
4. Browne, C.: The nature of puzzles. Game Puzzle Des. **1**(1), 23–34 (2015)
5. Browne, C., Powley, E., Whitehouse, D., Lucas, S., Cowling, P.I., Rohlfshagen, P., Tavener, S., Perez, D., Samothrakis, S., Colton, S.: A survey of monte carlo tree search methods. IEEE Trans. Comp. Intell. AI Games **4**(1), 1–43 (2012)
6. Finnsson, H., Björnsson, Y.: Simulation-based approach to general game playing. In: Proceedings of the Association for the Advancement of Artificial Intelligence, Chicago, Illinois, pp. 259–264 (2008)
7. Karp, R.M.: Reducibility among Combinatorial Problems. Springer, Berlin (1972)
8. Kramer, W.: What makes a game good? Games J. (2000)
9. Meyers, S.: BoxOff: a new solitaire board game. GAMES **37**(6), 12–13 (2013)
10. Rose, J.: Addressing conflict: tension and release in games. Gamasutra (2015). http://www.gamasutra.com
11. Russell, S.J., Norvig, P.: Artificial Intelligence - A Modern Approach (3rd internat edn.). Pearson Education, New York (2010)
12. Schmidt, G.: The axiom universal game system project. Mindsports (2012). http://www.mindsports.nl/index.php/axiom
13. Togelius, J., Yannakakis, G.N., Stanley, K.O., Browne, C.: Search-based procedural content generation: a taxonomy and survey. IEEE Trans. Comp. Intell. AI Games **3**, 172–186 (2011)

Tracking Drift Severity in Data Streams

Kylie Chen$^{(\boxtimes)}$, Yun Sing Koh, and Patricia Riddle

The University of Auckland, Auckland, New Zealand
kche309@aucklanduni.ac.nz, {ykoh,pat}@cs.auckland.ac.nz

Abstract. The evolution of data or concept drift is a common phenomena in data streams. Currently most drift detection methods are able to locate the point of drift, but are unable to provide important information on the characteristics of change such as the magnitude of change which we refer to as drift severity. Monitoring drift severity provides crucial information to users allowing them to formulate a more adaptive response. In this paper, we propose a drift detector, MagSeed, which is capable of tracking drift severity with a high rate of true positives and a low rate of false positives. We evaluate MagSeed on synthetic and real world data, and compare it to state of the art drift detectors ADWIN2 and DDM.

Keywords: Change detection · Severity detection · Drift detector

1 Introduction

One unique property of data streams and temporal data is the possibility of changes in the underlying model through time. This is referred to as concept drift. In data streams concept drift has been a well studied area. However most existing work in drift detection focuses on accurately and efficiently finding true drift points whilst minimising the delay time of detection, and are unable to detect the magnitude of concept drifts which we refer to as drift severity. There has been relatively little work that addresses the problem of drift severity, and how to accurately measure the magnitude of concept drift for streams with different characteristics. One of the challenges in measuring drift severity is that it is difficult to define a measure to directly compare classifiers before and after change.

The main motivation for detecting the severity of changes is that it gives an indication of the magnitude of concept change which allows responses to be better adapted to the current situation. For example, change mining techniques may be applied to data to identify the onset of viral outbreaks in a population. Such monitoring could be beneficial to public health by providing an indication of when episodes occur and how severe an outbreak is likely to be. Suppose we are interested in monitoring the outbreak of flu in a population of a million people. Drift detection techniques could potentially allow us to capture the onset of change (a drift point) in the population to identify when a flu outbreak occurs. Currently most techniques are unable to detect the magnitude of change within the population contracting the flu. Measuring the magnitude of change could

© Springer International Publishing Switzerland 2015
B. Pfahringer and J. Renz (Eds.): AI 2015, LNAI 9457, pp. 96–108, 2015.
DOI: 10.1007/978-3-319-26350-2_9

provide an estimate of the number of individuals affected by the flu (e.g. 30 % of the population), to give an early indication of how severe an outbreak may be. This allows preventative and responsive measures to be taken in a timely manner, such as stocking of medicine, alerting or vaccinating high risk groups in response to the severity of the outbreak. Furthermore, the ability to capture and monitor the magnitude of changes in a stream may be useful for magnitude prediction of future drifts and the visualisation of changes in streams.

Recently [10] introduced a metric for measuring drift severity and presented a technique for detectors that use statistical process control. It has been shown to be effective at tracking changes of large magnitudes, and at high speed but performs poorly for more gradual and less severe changes [10]. We will attempt to address these issues by developing a drift detector with capability for severity detection with greater sensitivity to smaller changes. Our main focus will be on streams with gradual change.

The main contribution of our work is the development of a drift detector, MagSeed, that is able to accurately capture drift severity. We compare our method to other drift detectors DDM [4] and ADWIN2 [1]. We evaluate our method using synthetic and real datasets and show that our algorithm detects different severity levels in both gradual and abrupt concept drifts with some robustness to noise.

The paper is organised as follows: Sect. 2 reviews the background, and Sect. 3 describes related work. Section 4 introduces our algorithm. Our experiments are presented in Sect. 5. Section 6 discusses future directions and concludes our paper.

2   Background

In this section, we give the formal definitions of the concept drift problem [5], and drift severity metric [10].

Concept Drift: Let a stream consist of labelled instances where each instance is a pair (X, y) of input feature values $X \in R^P$ and its class label $y \in R^1$, where R^P is the input feature space and R^1 is the class label space. The classification problem can be described as $p(y|X) = \frac{p(y)p(X|y)}{p(X)}$, where $p(y|X)$ is the concept function we are trying to learn. We follow the definition by [5] which defines concept drift as a change in the concept function $p(y|X)$.

Drift Severity: Let f_t and f_{t+1} represent two different consecutive concept functions. The drift severity $\in [0, 1]$ is defined as the percentage of the input space that has a different class label when the change from f_t to f_{t+1} is complete [10]. This is the amount of change between two concepts, where high severity represents a large change, and low severity represents a small change in concepts.

Suppose we are interested in monitoring flu trends in a population of 1000 individuals. Say there were 100 individuals with flu at time t, and at a later time $t + 1$ there were 600 affected individuals. The shift in the population of healthy and affected individuals could be quantified by dividing the difference in the number of affected individuals by the population size, severity $= (600 - 100)/1000 = 0.5$.

3   Related Work

Existing work in drift detection focuses on detectors. Gama et al.'s survey [5] groups detection methods based on: (1) sequential analysis [11], (2) statistical process control [4], (3) monitoring two distributions [1,7], and (4) contextual approaches [9]. Our work is most closely related to SEED [7], ADWIN2 [1], and DDM [4]. ADWIN2 [1] is an adaptive windowing technique that uses exponential histograms as a data structure to store classification errors, and a Hoeffding bound for drift detection. SEED [7] uses the same bound as ADWIN2, but stores data in blocks and utilises a block compression algorithm to minimise the number of boundary checks. DDM [4] detects drifts based on a significant increase in the mean classification errors by comparing the current cumulative mean and standard deviation with the minimum mean and standard deviation. It is effective for abrupt drifts [3] and has both a drift and warning threshold which allows it to monitor drift severity [10]. A comparison of different drift detectors can be found in [3,6,12]. In [3] Bifet et al. showed using prediction accuracy as the only measure of detector performance can be misleading and propose a new metric that can more accurately describe the trade-off between false positives and fast detection. Currently most drift detection methods are focused on supervised learning and only work with numerical data such as classification accuracy [8]. These techniques are able to monitor when a drop in accuracy occurs, but [8] point out that it is equally important to detect changes in categorical data and also in the unsupervised setting, and work in this direction.

4   MagSeed Drift Detector

Our detector, MagSeed extends the SEED algorithm [7] which uses classification error as input data. The novelty of our approach is that it allows the detection of drift severity. Our algorithm has two parts: drift detection, and magnitude tracking, each with three possible states: (1) no change - indicating there is no concept drift, (2) warning - in anticipation of a drift, and (3) drift - signalling a concept drift has occurred. First we describe the drift detection part, then we address the magnitude tracking component of the detector. As input data arrives, it is partitioned into blocks of size b where the block boundaries represent potential drift points. Drift detection is performed by testing for a significant difference in means between data on the left W_L and right W_R sides of the block boundaries based on the Hoeffding bound with Bonferroni correction. The condition for triggering a drift is

$$|\hat{\mu}_{W_L} - \hat{\mu}_{W_R}| > \epsilon_d, \tag{1}$$

$$\epsilon_d = \sqrt{\frac{2}{m} \cdot \sigma_W^2 \cdot ln\frac{2}{\delta'} + \frac{2}{3m}ln\frac{2}{\delta'}}, \quad \delta' = \frac{\delta}{n} \tag{2}$$

where $\hat{\mu}_{W_L}$ and $\hat{\mu}_{W_R}$ represent the mean error rate of data in W_L and W_R respectively, ϵ_d is the Hoeffding bound with Bonferroni correction using a confidence parameter $\delta_d \in (0,1)$, m is the harmonic mean of the lengths of W_L and W_R, and n is the length of W where $W = W_L + W_R$. When the drift threshold is surpassed, the detector enters a drift state.

When the detector is not in a warning or drift state, block compression is enabled to improve efficiency by removing potential drift points that have low probability of becoming actual drift points. We use the block compression algorithm detailed in [7] which uses the bound $\epsilon' = \hat{\epsilon} \cdot \alpha \cdot t$, where the parameters $\hat{\epsilon}$ is a base value, α is the linear growth term, and t is the relative arrival position.

To allow the anticipation of drift we introduced a more relaxed threshold for detecting warnings. The first warning test is similar to the drift detection trigger but uses the bound ϵ_w, where $\epsilon_w < \epsilon_d$. We use the condition below for warning detection

$$|\hat{\mu}_{W_L} - \hat{\mu}_{W_R}| > \epsilon_w, \tag{3}$$

where ϵ_w is the Hoeffding bound with Bonferroni correction using a confidence parameter $\delta_w \in (0,1)$, where $\delta_w > \delta_d$. By using only the first warning condition, it may cause warnings to be triggered too early. For example if the error rate increases and passes the ϵ_w threshold, followed by a decrease that falls below the ϵ_w threshold. This introduces a false warning. To address this, we used a second warning threshold. We monitor the current mean p and standard deviation s of the classification error over a sliding window of size 100, as well as the minimum p and s values denoted by p_{min} and s_{min}. Our second warning threshold is defined as

$$p + s > p_{min} + c_w \cdot s_{min}, \tag{4}$$

where $c_w \in (1,3)$ is parameter which defines the confidence level for warning detection. It is similar to the threshold used by DDM, but uses a sliding window rather than a cumulative one. This second threshold should allow our detector to pick up local changes and respond more quickly than the cumulative threshold used by DDM. Here the choice of the window size was arbitrarily chosen, but we note that a more adaptive approach would be more ideal. For example, monitoring the sensitivity of different window sizes to internal false warnings (such as when the error rate increases above the warning threshold then decreases) could help us adjust the window size adaptively. When either condition (3) or (4) is satisfied the detector enters a warning state. We also use them to control the rate of false warnings. When either (4) or (3) is not satisfied for all block boundaries then the warning is flagged as a false warning, and the detector enters the no change state. For computing drift severity we use the metric defined by [10] which we denote as *Rate*, that computes the error rate in the warning window.

$$Rate = \frac{\text{number of misclassifications in warning}}{\text{total instances seen in warning}} \tag{5}$$

Algorithm 1 shows the pseudocode for the MagSeed algorithm. Lines 1–3 initializes the change detector. The algorithm (lines 4–7) processes the input as

Algorithm 1. MagSeed algorithm

1: Initialize window W as blocks $\{B_0, ..., B_t\}$ each of size n
2: Initialize $state \leftarrow no\ change$
3: Initialize $p_{min} \leftarrow \infty$, $s_{min} \leftarrow \infty$

4: **for** $t > 0$ **do** where t is time
5:     SetInput(x_t, W) where x_t is classification error at t
6:         **return** $state$
7: **end for**

8: **function** SETINPUT(item k, List W)
9:     AddElement(k, W) adds element into tail block
10:     update p, s, p_{min}, s_{min}
11:     $warningFound \leftarrow false$
12:     **if** $p + s > p_{min} + c_w * s_{min}$ **then**
13:         $state \leftarrow warning$
14:     **else** $state \leftarrow no\ change$
15:     **end if**
16:     **for** every split of W into $W = W_L, W_R$ **do**
17:         **if** $|\mu_{W_L} - \mu_{W_R}| > \epsilon_d$ **then**
18:             $state \leftarrow drift$
19:             $p_{min} \leftarrow \infty$
20:             $s_{min} \leftarrow \infty$
21:             remove all blocks in W_L
22:         **else if** $|\mu_{W_L} - \mu_{W_R}| > \epsilon_w$ **then**
23:             $state \leftarrow warning$
24:             $warningFound \leftarrow true$
25:         **end if**
26:     **end for**
27:     **if** $warningFound = false$ **then** $state \leftarrow no\ change$
28:     **end if**
29:     **if** $state = no\ change$ **then**
30:         CompressionCheck(W) compresses blocks in W
31:     **end if**
32: **end function**

33: **function** ADDELEMENT(item k, List W)
34:     **if** Tail Block of W is full **then** create Block B with k
35:         $W \leftarrow W \cup \{B\}$
36:     **else** add k to tail block of W
37:     **end if**
38: **end function**

39: **function** COMPRESSIONCHECK(List W)
40:     compressCount++
41:     **if** compressCount = compressionInterval **then**
42:         **for** each two consecutive blocks B_t, B_{t+1} **do**
43:             **if** $|\mu B_t - \mu B_{t+1}| < \epsilon'$ **then** merge B_t, B_{t+1}
44:             **end if**
45:         **end for**
46:     **end if**
47: **end function**

each instance arrives by passing the classification output $x_t \in \{1, 0\}$ at time t to the *SetInput* function, and returns the state of the detector indicating whether a drift or warning has occurred at each time step t (line 6). The *SetInput* function (lines 8–32) contains the main algorithm used for change detection. Line 9 adds the classification output k to the window W of data blocks by the *AddElement* function. Line 10 updates the current p and s values by computing a rolling average of the classification error, and standard deviation of the error respectively, and the minimums p_{min}, s_{min} are updated when $p + s < p_{min} + s_{min}$. If the warning threshold (Eq. 4) is surpassed (line 12), the detector enters the warning state (line 13). Line 16 checks every block boundary in W against the drift threshold (line 17), and the warning threshold (Eq. 3) in line 22 which also acts as a false warning threshold. Lastly, if the detector is in the *no change* state after the data is processed, the blocks may be compressed by the *CompressionCheck* function (line 30). The *AddElement* function (lines 33–38) adds an element k that represents the classification error to the window W by appending k to the last block in W if the last block is not full (line 36), or by adding a new block with element k to the end of the window W (line 35). The *CompressionCheck* function (lines 39–47) merges consecutive homogeneous blocks in the window at set intervals using the ϵ' bound which is detailed above (line 43).

5   Experiments

Our evaluation has three parts. First we evaluate the effectiveness of drift detection. Second we evaluate the effectiveness of the warning technique, and its capability for tracking severity. Third we evaluate the performance and overall prediction accuracy of our algorithm on real data streams using a Hoeffding tree.

5.1   Experimental Setup

We evaluated the algorithms on synthetic streams over 100 trials, and also on real data streams. We test the robustness of the algorithms using synthetic streams with three noise levels: 0 %, 5 % and 10 %, but only present results for streams with 0 % and 10 % noise due to space constraints.

Synthetic and Real Streams: We use three different synthetic streams: Gradual Bernoulli [1], Abrupt SEA concepts [10], and Gradual CIRCLES [4] detailed below. We use three real streams from MOA: Forest Covertype [3], Pokerhand [2], and Airlines [8].

  i. **Gradual Bernoulli:** These streams are binary streams of length 1,000,000 where each data point represents the classification error (1 for error, and 0 for a correct prediction). Initially a stable period is generated according to a Bernoulli distribution with a mean error rate of $\mu = 0.2$. In the last 1,000 instances, gradual drift was simulated by linearly increasing the mean error rate μ by a slope value at each time step. Noise was controlled by flipping the classification error data point according to the defined noise probability.

ii. **Abrupt SEA concepts:** Each stream has a length of 90,000 instances with 2 balanced classes and a single point of abrupt drift. Instances have three attributes $f_1, f_2, f_3 \in [0, 10]$ of which the last two (f_2, f_3) are relevant. An instance has label class 1 if $f_2 + f_3 \leq \theta$, where θ is a threshold parameter. Abrupt drift was introduced by changing the class boundary threshold θ at the midpoint of the stream. Noise was controlled by flipping the class label of instances according to the defined noise probability.

iii. **Gradual CIRCLES:** Each stream has a length of 1,000,000 instances with 2 balanced classes and a single drift of low severity. Instances have two attributes $x, y \in [0, 1]$ which define the location of a point in a 2D space. The concept is defined by a circle in the 2D space, where instances inside the circle have a class label of 1, or label 0 otherwise. Drift is introduced by changing the radius of the circle concept, where a larger difference in radii corresponds to higher severity. Gradual drift was introduced by linearly increasing the probability of generating instances from the new concept by 0.1 % at each time step in the last 1000 instances. Noise was controlled by flipping the class label of instances according to the defined noise probability.

Parameter Selection: We evaluate the algorithms over a range of parameters, and present the best and worst performance for each detector. For synthetic streams we select the best and worst settings based on the number of true positive drifts and the ratio of true and false positives. For real data streams we use the best settings from the synthetic experiments.

Evaluation Metrics: We evaluate the algorithms using the rate of true positive drifts (RD), rate of false positive drifts (FP), detection delay, memory (in bytes), time (in milliseconds) used by the detector, computed severity (Rate), rate of true warnings (TW), and correlation between computed and actual severity. A high correlation would suggest that the computed severity reflects actual severity. A true warning (TW) is a warning that was triggered between the true point of drift and true detection by the detector.

5.2   Drift Detection

In these experiments, we test the drift detection capabilities of our algorithm compared to current drift detection techniques: ADWIN2 [1], and DDM [4]. We compare our drift detector to ADWIN2 as it is a state of the art detector that uses the same drift detection bound (the Hoeffding bound), and has been shown to be effective at detecting gradual drifts [12], which are the type of streams we will be focusing on. We also compare our detector with DDM, which is a technique that is effective at detecting abrupt drifts. DDM also has a warning threshold that would allow us to directly compare its performance in capturing the warning state against our technique. Thus it makes sense that we chose these two different detectors as our benchmark comparisons.

Table 1. Rate of true and false positives for drift detection

Noise	Detector	Slope	RD	FP	Delay (SD)	Memory (SD)	Time (SD)
					Best		
0 %	MagSEED	0.0001	**84**	0.001	759.00 ±(165.61)	2134.86 ±(388.56)	146.58 ±(5.77)
	$\delta_d = 0.05$	0.0002	**100**	0.001	543.32 ±(118.79)	2197.92 ±(316.59)	147.68 ±(5.64)
	$\delta_w = 0.1$	0.0003	**100**	0.001	433.24 ±(97.32)	1986.72 ±(328.89)	147.09 ±(5.09)
		0.0004	**100**	0.001	361.56 ±(77.94)	1848.96 ±(282.66)	148.32 ±(5.24)
	ADWIN2	0.0001	83	0.001	782.71 ±(162.45)	1770.31 ±(198.92)	353.65 ±(11.60)
	$\delta = 0.05$	0.0002	100	0.001	555.80 ±(113.10)	1565.92 ±(52.42)	353.08 ±(12.51)
		0.0003	100	0.001	439.96 ±(94.04)	1520.56 ±(79.39)	352.63 ±(11.78)
		0.0004	100	0.001	370.20 ±(74.79)	1478.56 ±(89.92)	352.00 ±(11.41)
	DDM	0.0001	1	<0.001	785.00 ±(0)	248.00 ±(0)	23.22 ±(15.10)
	$\alpha = 3$	0.0002	1	<0.001	560.00 ±(0)	248.00 ±(0)	23.77 ±(14.31)
	$\beta = 2$	0.0003	1	<0.001	456.00 ±(0)	248.00 ±(0)	22.53 ±(14.43)
		0.0004	1	<0.001	379.00 ±(0)	248.00 ±(0)	22.24 ±(14.78)
10 %	MagSEED	0.0001	**54**	0.001	852.63 ±(148.49)	2218.67 ±(431.11)	149.00 ±(5.78)
	$\delta_d = 0.05$	0.0002	**100**	0.001	658.84 ±(133.12)	2237.28 ±(330.23)	150.14 ±(6.29)
	$\delta_w = 0.1$	0.0003	**100**	0.001	517.72 ±(107.82)	2160.48 ±(284.63)	148.39 ±(6.15)
		0.0004	**100**	0.001	432.28 ±(87.17)	2044.80 ±(321.96)	150.85 ±(7.12)
	ADWIN2	0.0001	49	0.001	859.90 ±(148.31)	1918.86 ±(265.54)	353.80 ±(12.86)
	$\delta = 0.05$	0.0002	100	0.001	669.40 ±(136.56)	1641.52 ±(101.01)	354.37 ±(11.16)
		0.0003	100	0.001	527.32 ±(109.60)	1559.20 ±(50.65)	355.76 ±(10.96)
		0.0004	100	0.001	437.72 ±(91.47)	1532.32 ±(74.06)	354.11 ±(12.80)
	DDM	0.0001	0	<0.001	-	248.00 ±(0.00)	21.58 ±(15.66)
	$\alpha = 3$	0.0002	0	<0.001	-	248.00 ±(0.00)	22.17 ±(16.17)
	$\beta = 2$	0.0003	0	<0.001	-	248.00 ±(0.00)	23.21 ±(15.35)
		0.0004	0	<0.001	-	248.00 ±(0.00)	20.87 ±(15.25)
					Worst		
0 %	MagSEED	0.0001	84	0.002	754.43 ±(165.77)	2220.00 ±(654.54)	862.82 ±(52.31)
	$\delta_d = 0.05$	0.0002	100	0.002	539.48 ±(122.93)	2192.64 ±(315.08)	864.47 ±(50.79)
	$\delta_w = 0.2$	0.0003	100	0.002	431.96 ±(100.28)	1977.12 ±(318.59)	864.57 ±(52.90)
		0.0004	100	0.002	359.96 ±(78.05)	1857.60 ±(285.12)	862.89 ±(52.66)
	ADWIN2	0.0001	92	0.011	671.70 ±(196.98)	1627.13 ±(133.52)	302.51 ±(8.58)
	$\delta = 0.3$	0.0002	100	0.011	476.44 ±(140.49)	1512.16 ±(88.64)	303.96 ±(6.96)
		0.0003	100	0.011	372.12 ±(104.52)	1461.76 ±(98.11)	303.57 ±(6.16)
		0.0004	100	0.011	314.20 ±(82.87)	1414.72 ±(78.91)	301.33 ±(7.66)
	DDM	0.0001	1	<0.001	785.00 ±(0)	248.00 ±(0)	23.41 ±(16.04)
	$\alpha = 3$	0.0002	1	<0.001	560.00 ±(0)	248.00 ±(0)	23.79 ±(13.57)
	$\beta = 1$	0.0003	1	<0.001	456.00 ±(0)	248.00 ±(0)	23.70 ±(14.58)
		0.0004	1	<0.001	379.00 ±(0)	248.00 ±(0)	24.37 ±(14.47)
10 %	MagSEED	0.0001	53	0.002	842.62 ±(158.10)	2311.25 ±(604.04)	949.91 ±(76.14)
	$\delta_d = 0.05$	0.0002	99	0.002	652.49 ±(138.28)	2266.18 ±(406.15)	944.83 ±(78.22)
	$\delta_w = 0.2$	0.0003	100	0.002	515.48 ±(113.33)	2162.88 ±(300.25)	945.61 ±(75.83)
		0.0004	100	0.002	432.28 ±(92.35)	2021.76 ±(316.66)	940.68 ±(77.88)
	ADWIN2	0.0001	**74**	0.011	714.03 ±(231.13)	1719.03 ±(168.42)	300.72 ±(7.90)
	$\delta = 0.3$	0.0002	100	0.011	559.00 ±(181.27)	1564.24 ±(76.47)	300.85 ±(7.39)
		0.0003	100	0.011	425.56 ±(134.13)	1517.20 ±(84.00)	301.11 ±(7.44)
		0.0004	100	0.011	363.80 ±(115.99)	1456.72 ±(87.07)	302.68 ±(7.78)
	DDM	0.0001	0	<0.001	-	248.00 ±(0.00)	23.60 ±(16.96)
	$\alpha = 3$	0.0002	0	<0.001	-	248.00 ±(0.00)	20.57 ±(15.16)
	$\beta = 1$	0.0003	0	<0.001	-	248.00 ±(0.00)	19.64 ±(13.40)
		0.0004	0	<0.001	-	248.00 ±(0.00)	20.96 ±(16.63)

Table 1 shows the rate of true drift detection for gradual Bernoulli streams in the best (top) and worst (bottom) cases. MagSeed is comparable to ADWIN2 in terms of delay and true positive drift rate given the same confidence level δ. Both MagSeed and ADWIN2 outperform DDM on gradual Bernoulli streams in terms of true drift detection rate, but require more memory as they monitor the error distribution. In all cases, the rate of false positive drifts in MagSeed and ADWIN2 are below the theoretical upper bound for false positives δ. In terms of memory ADWIN2 outperforms MagSeed as the latter requires additional memory to monitor drift warnings.

5.3　Warning Detection and Severity Measure

In these experiments, we test the accuracy of warning detection and our proposed severity measure by using classification errors generated by the synthetic Bernoulli, SEA concept, and CIRCLES streams as input for the drift detectors. For the SEA concept, and CIRCLES streams the classification errors are generated by passing the data through a Hoeffding tree learner. We compare our

Table 2. Warning detection on Gradual Bernoulli

Noise	Detector	Best				Worst		
		Slope	RD	TW	Rate (SD)	RD	TW	Rate (SD)
0%	MagSEED	0.0001	84	84	0.36 ±(0.06)	84	84	0.34 ±(0.05)
	$\delta_d = 0.05$	0.0002	100	100	0.37 ±(0.06)	100	99	0.36 ±(0.06)
	$\delta_w = 0.1$ (Best)	0.0003	100	100	0.38 ±(0.06)	100	99	0.37 ±(0.06)
	$\delta_w = 0.2$ (Worst)	0.0004	100	100	0.39 ±(0.06)	100	99	0.38 ±(0.06)
Correlation (slope and computed severity)					**0.9973**			**0.9802**
	DDM	0.0001	1	1	0.26 ±(0.00)	1	0	0.23 ±(0.00)
	$\alpha = 3$	0.0002	1	1	0.30 ±(0.00)	1	0	0.24 ±(0.00)
	$\beta = 2$ (Best)	0.0003	1	1	0.34 ±(0.00)	1	0	0.24 ±(0.00)
	$\beta = 1$ (Worst)	0.0004	1	1	0.39 ±(0.00)	1	0	0.24 ±(0.00)
Correlation (slope and computed severity)					**0.9963**			**0.9876**
10%	MagSEED	0.0001	54	54	0.41 ±(0.05)	53	52	0.40 ±(0.05)
	$\delta_d = 0.05$	0.0002	100	100	0.42 ±(0.05)	99	98	0.41 ±(0.06)
	$\delta_w = 0.1$ (Best)	0.0003	100	100	0.46 ±(0.06)	100	99	0.44 ±(0.06)
	$\delta_w = 0.2$ (Worst)	0.0004	100	100	0.46 ±(0.07)	100	99	0.44 ±(0.06)
Correlation (slope and computed severity)					**0.9365**			**0.9844**
	DDM	0.0001	0	0	0	0	0	0
	$\alpha = 3$	0.0002	0	0	0	0	0	0
	$\beta = 2$ (Best)	0.0003	0	0	0	0	0	0
	$\beta = 1$ (Worst)	0.0004	0	0	0	0	0	0
Correlation (slope and computed severity)					**0.0000**			**0.0000**

Table 3. Warning detection on Abrupt SEA concepts

Noise	Detector	Actual	Best			Worst		
		Severity	RD	TW	Rate (SD)	RD	TW	Rate (SD)
0 %	MagSEED	0.03	100	100	0.08 ±(0.03)	100	99	0.09 ±(0.03)
		0.05	100	100	0.09 ±(0.04)	100	98	0.08 ±(0.03)
	$\delta_d = 0.05$ (Best)	0.09	100	100	0.13 ±(0.04)	100	100	0.12 ±(0.04)
	$\delta_w = 0.1$ (Best)	0.12	100	100	0.12 ±(0.04)	100	99	0.12 ±(0.04)
		0.13	100	99	0.16 ±(0.05)	100	98	0.16 ±(0.05)
	$\delta_d = 0.1$ (Worst)	0.15	100	88	0.20 ±(0.06)	100	76	0.20 ±(0.05)
	$\delta_w = 0.2$ (Worst)	0.16	100	92	0.21 ±(0.06)	100	85	0.20 ±(0.06)
		0.24	100	96	0.19 ±(0.06)	100	90	0.19 ±(0.06)
		0.27	100	91	0.21 ±(0.06)	100	84	0.21 ±(0.07)
Correlation (actual and computed severity)			**0.8670**			**0.8760**		
	DDM	0.03	90	90	0.04 ±(0.01)	90	55	0.04 ±(0.01)
		0.05	100	100	0.04 ±(0.01)	100	47	0.04 ±(0.00)
	$\alpha = 3$	0.09	100	100	0.10 ±(0.02)	100	51	0.09 ±(0.01)
		0.12	100	100	0.08 ±(0.01)	100	49	0.08 ±(0.01)
	$\beta = 2$ (Best)	0.13	100	100	0.14 ±(0.02)	100	43	0.12 ±(0.02)
	$\beta = 1$ (Worst)	0.15	100	100	0.20 ±(0.03)	100	54	0.18 ±(0.03)
		0.16	100	100	0.19 ±(0.02)	100	52	0.17 ±(0.03)
		0.24	100	100	0.16 ±(0.02)	100	40	0.14 ±(0.03)
		0.27	100	100	0.17 ±(0.03)	100	40	0.15 ±(0.03)
Correlation (actual and computed severity)			**0.7980**			**0.7710**		
10 %	MagSEED	0.03	29	28	0.14 ±(0.09)	70	69	0.16 ±(0.13)
		0.05	38	36	0.21 ±(0.09)	86	85	0.19 ±(0.08)
	$\delta_d = 0.05$ (Best)	0.09	100	100	0.24 ±(0.05)	100	99	0.24 ±(0.05)
	$\delta_w = 0.1$ (Best)	0.12	100	100	0.25 ±(0.06)	100	100	0.23 ±(0.05)
		0.13	100	99	0.26 ±(0.05)	100	98	0.25 ±(0.05)
	$\delta_d = 0.1$ (Worst)	0.15	100	100	0.29 ±(0.06)	100	99	0.29 ±(0.06)
	$\delta_w = 0.2$ (Worst)	0.16	100	100	0.30 ±(0.06)	100	99	0.29 ±(0.06)
		0.24	100	100	0.29 ±(0.06)	100	99	0.28 ±(0.06)
		0.27	100	100	0.30 ±(0.05)	100	99	0.30 ±(0.06)
Correlation (actual and computed severity)			**0.8216**			**0.8662**		
	DDM	0.03	0	0	0.00 ±(0.00)	0	0	0.00 ±(0.00)
		0.05	6	6	0.14 ±(0.01)	6	2	0.14 ±(0.01)
	$\alpha = 3$	0.09	99	99	0.17 ±(0.01)	99	17	0.16 ±(0.01)
		0.12	99	99	0.17 ±(0.01)	99	18	0.16 ±(0.01)
	$\beta = 2$ (Best)	0.13	100	100	0.20 ±(0.01)	100	15	0.19 ±(0.02)
	$\beta = 1$ (Worst)	0.15	98	98	0.24 ±(0.02)	98	23	0.23 ±(0.02)
		0.16	100	100	0.24 ±(0.02)	100	16	0.23 ±(0.02)
		0.24	98	97	0.22 ±(0.02)	98	10	0.19 ±(0.03)
		0.27	98	97	0.23 ±(0.02)	98	8	0.20 ±(0.04)
Correlation (actual and computed severity)			**0.7608**			**0.6732**		

method to the DDM method presented by Kosina et al. [10], as their technique has warning detection and can also capture drift severity. We did not compare our method with the PHT method [10] as it was shown to perform worse in terms of severity tracking. We did not include ADWIN2 in our comparison as the nature of the exponential histogram data. structure used in the detector makes it difficult to implement a warning threshold for computing severity.

Table 2 shows the rate of true warning detection on gradual Bernoulli streams with the correlation between average computed severity and slope of change highlighted in bold. MagSeed has a high rate of true warning detection (98–100 % given a true drift is detected) and is able to detect more true drifts than DDM. Both detectors have high correlations for noise free data which shows that they are capable of capturing the speed of concept drift. For noisy data (5 % or 10 % noise), MagSeed shows a clear advantage as it is able to detector more true drifts, true warnings and the measure correlates well with speed of change. This suggests that our detector is capable of tracking the speed of concept drift.

Table 3 shows results for abrupt SEA streams over a range of severity levels. The third column in the table (Actual Severity) is the theoretical severity of the streams computed as the area difference between two concepts. In these experiments, the severity measure of MagSeed shows high correlation with actual severity and consistently performed better than DDM in terms of true warning

Table 4. Warning detection on Gradual CIRCLES

Noise	Detector	Actual Severity	Best			Worst		
			RD	TW	Rate (SD)	RD	TW	Rate (SD)
0 %	MagSEED	0.07	100	100	0.08 ±(0.03)	100	98	0.07 ±(0.03)
	$\delta_d = 0.05, \delta_w = 0.1$ (Best)	0.16	100	100	0.09 ±(0.03)	100	99	0.07 ±(0.03)
	$\delta_d = 0.1, \delta_w = 0.3$ (Worst)	0.26	100	100	0.09 ±(0.03)	100	99	0.08 ±(0.03)
		0.38	100	100	0.09 ±(0.03)	100	99	0.08 ±(0.03)
Correlation (actual and computed severity)					**0.8482**			**0.9782**
	DDM	0.07	0	0	0.00 ±(0.00)	0	0	0.00 ±(0.00)
	$\alpha = 3, \beta = 2$ (Best)	0.16	8	8	0.25 ±(0.02)	8	8	0.15 ±(0.01)
	$\alpha = 3, \beta = 1$ (Worst)	0.26	97	97	0.26 ±(0.02)	97	62	0.15 ±(0.03)
		0.38	100	100	0.28 ±(0.02)	100	65	0.16 ±(0.03)
Correlation (actual and computed severity)					**0.7892**			**0.7514**
10 %	MagSEED	0.07	100	100	0.24 ±(0.05)	100	98	0.23 ±(0.05)
	$\delta_d = 0.05, \delta_w = 0.1$ (Best)	0.16	100	100	0.25 ±(0.05)	100	98	0.24 ±(0.05)
	$\delta_d = 0.1, \delta_w = 0.3$ (Worst)	0.26	100	100	0.26 ±(0.06)	100	97	0.25 ±(0.06)
		0.38	100	100	0.26 ±(0.06)	100	97	0.25 ±(0.06)
Correlation (actual and computed severity)					**0.9654**			**0.9374**
	DDM	0.07	0	0	0.00 ±(0.00)	0	0	0.00 ±(0.00)
	$\alpha = 3, \beta = 2$ (Best)	0.16	0	0	0.00 ±(0.00)	0	0	0.00 ±(0.00)
	$\alpha = 3, \beta = 1$ (Worst)	0.26	0	0	0.00 ±(0.00)	0	0	0.00 ±(0.00)
		0.38	0	0	0.00 ±(0.00)	0	0	0.00 ±(0.00)
Correlation (actual and computed severity)					**0.0000**			**0.0000**

rate and true drift rate. For the noise free streams as the severity increases there is also a decrease in delay which decreases the area for correct warning detection. Most of these incorrect warnings are detected too early and the increasing error rate prevents the warning from shifting forward in time. In contrast the noisy streams have greater fluctuations in the error rate and are more apt to recover from local maxima.

Table 4 shows results for gradual CIRCLES streams with low severity. The third column (Actual Severity) is the difference in area between the new and old circle concepts. The MagSeed detector shows high correlation between the computed and theoretical severity of the stream, and has higher true drift and true warning detection rates without compromising the false positive drift rate as $FP \leq 0.34\%$ (not shown in table).

5.4   Performance on Real World Data

We also examine the performance of our algorithm on real world data, and compare it to ADWIN2 and DDM using a Hoeffding tree learner which is retrained using examples from the warning period for detectors with warnings.

Table 5. Performance on real world data

Dataset	Detector	Accuracy	Drifts	Warnings	Rate (SD)
Forest covertype	MagSEED	0.84	2380	1395	0.49 ±(0.23)
	DDM	0.83	1942	1475	0.83 ±(0.27)
	ADWIN2	**0.85**	2492	0	-
Pokerhand	MagSEED	**0.75**	2032	1248	0.54 ±(0.20)
	DDM	0.73	1046	1007	0.63 ±(0.25)
	ADWIN2	**0.75**	2130	0	-
Airlines	MagSEED	**0.66**	173	131	0.40 ±(0.15)
	DDM	0.65	14	14	0.38 ±(0.07)
	ADWIN2	0.65	384	0	-

Table 5 shows the performance of MagSeed, DDM and ADWIN2 on three real world datasets: Forest Covertype [3], Pokerhand [2], and Airlines [8]. All the detectors are comparable in terms of overall prediction accuracy. However it is difficult to access the performance of the severity measure as we do not know the location or magnitude of the true drifts in these real world datasets. For the MagSeed detector there is a large number of drifts that did not trigger a warning, this may be caused by changes of large magnitudes which do not trigger the warning threshold prior to the drift threshold.

6   Conclusions

In this paper, we presented a drift detector that is capable of detecting drift severity. We experimentally showed that there is a strong correlation between the computed and real magnitudes in our synthetic datasets which suggests that the computed and real magnitudes are similar. One limitation of our detector is that it may be unable to capture the drift severity for changes of high magnitudes where there are drastic changes in concepts. This is often due to the small delay time between the true change and detected drift point which provides less opportunity for correct warning detection. In the future, we would like to extend MagSeed to adaptively determine the window size of our second warning threshold. We would also like to conduct experiments on a wider range of data streams and use drift severity for analysing characteristics of real data streams.

References

1. Bifet, A., Gavaldà, R.: Learning from time-changing data with adaptive windowing. In: Proceedings of the Seventh SIAM International Conference on Data Mining, 26–28 April 2007, Minneapolis, MN, USA, pp. 443–448 (2007)
2. Bifet, A., Holmes, G., Pfahringer, B.: Leveraging bagging for evolving data streams. In: Balcázar, J.L., Bonchi, F., Gionis, A., Sebag, M. (eds.) ECML PKDD 2010, Part I. LNCS, vol. 6321, pp. 135–150. Springer, Heidelberg (2010)
3. Bifet, A., Read, J., Pfahringer, B., Holmes, G., Žliobaitè, I.: CD-MOA: change detection framework for massive online analysis. In: Tucker, A., Höppner, F., Siebes, A., Swift, S. (eds.) Advances in Intelligent Data Analysis XII. LNCS, vol. 8207, pp. 92–103. Springer, Heidelberg (2013)
4. Gama, J., Medas, P., Castillo, G., Rodrigues, P.: Learning with drift detection. In: Bazzan, A.L.C., Labidi, S. (eds.) SBIA 2004. LNCS (LNAI), vol. 3171, pp. 286–295. Springer, Heidelberg (2004)
5. Gama, J., Žliobaitè, I., Bifet, A., Pechenizkiy, M., Bouchachia, A.: A survey on concept drift adaptation. ACM Comput. Surv. **45**(4), 44:1–44:37 (2014)
6. Gonçalves, P.M., de Carvalho, S.G., Santos, R., Barros, S., Vieira, D.C.: A comparative study on concept drift detectors. Expert Syst. Appl. **41**(18), 8144–8156 (2014)
7. Huang, D.T.J., Koh, Y.S., Dobbie, G., Pears, R.: Detecting volatility shift in data streams. In: 2014 IEEE International Conference on Data Mining, ICDM 2014, Shenzhen, China, 14–17 December 2014, pp. 863–868 (2014)
8. Ienco, D., Bifet, A., Pfahringer, B., Poncelet, P.: Change detection in categorical evolving data streams. In: Proceedings of the 29th Annual ACM Symposium on Applied Computing, pp. 792–797. ACM (2014)
9. Klinkenberg, R.: Learning drifting concepts: example selection vs. example weighting. Intell. Data Anal. **8**(3), 281–300 (2004)
10. Kosina, P., Gama, J., Sebastião, R.: Drift severity metric. In: 19th European Conference on Artificial Intelligence, Lisbon, Portugal, 16–20 August 2010, ECAI 2010, pp. 1119–1120 (2010)
11. Page, E.: Continuous inspection schemes. Biometrika **41**, 100–115 (1954)
12. Sebastião, R., Gama, J.: A study on change detection methods. In: 4th Portuguese Conference on Artificial Intelligence, Lisbon (2009)

Probabilistic Belief Contraction: Considerations on Epistemic Entrenchment, Probability Mixtures and KL Divergence

Kinzang Chhogyal[1,2]([✉]), Abhaya Nayak[2], and Abdul Sattar[1]

[1] Griffith University, Brisbane, Australia
a.sattar@griffith.edu.au
[2] Macquarie University, Sydney, Australia
{kin.chhogyal,abhaya.nayak}@mq.edu.au

Abstract. Probabilistic belief contraction is an operation that takes a probability distribution P representing a belief state along with an input sentence a representing some information to be removed from this belief state, and outputs a new probability distribution P_a^-. The contracted belief state P_a^- can be represented as a mixture of two states: the original belief state P, and the resultant state $P_{\neg a}^*$ of revising P by $\neg a$. Crucial to this mixture is the mixing factor ϵ which determines the proportion of P and $P_{\neg a}^*$ that are used in this process in a uniform manner. Ideas from information theory such as the *principle of minimum cross-entropy* have previously been used to motivate the choice of the probabilistic contraction operation. Central to this principle is the *Kullback-Leibler (KL) divergence*. In an earlier work we had shown that the KL divergence of P_a^- from P is fully determined by a function whose only argument is the mixing factor ϵ. In this paper we provide a way of interpreting ϵ in terms of a belief ranking mechanism such as epistemic entrenchment that is in consonance with this result. We also provide a much needed justification for why the mixing factor ϵ must be used in a uniform fashion by showing that the minimal divergence of P_a^- from P is achieved only when uniformity is respected.

1 Introduction

Cognitive agents use new information to form beliefs, and modify them. The field of *belief change* (Alchourrón et al. 1985) studies how a rational agent's set of beliefs, represented as sentences, may change when a piece of new information is acquired. It is convenient to view beliefs probabilistically when a finer grain of uncertainty is desired. The *belief state* of an agent is then represented by a probability distribution. The two main operations that are employed to represent change in a belief state are *contraction* and *revision*. Contraction removes sentences that represent beliefs where as revision accommodates information that is possibly inconsistent with existing beliefs. The result of both these operations are (usually) new belief states. One of the main guiding principles in belief change is that of *minimal information loss* which says that in the process of

© Springer International Publishing Switzerland 2015
B. Pfahringer and J. Renz (Eds.): AI 2015, LNAI 9457, pp. 109–122, 2015.
DOI: 10.1007/978-3-319-26350-2_10

belief change the loss of information should be minimised.[1] As belief states are probability distributions, researchers have resorted to the *principle of minimum cross-entropy* from information theory which is a technique that minimizes relative information loss and thus provides a way of selecting new belief states. The principle of minimum cross-entropy is based on the *Kullback-Leibler* divergence which measures the similarity between two probability distributions. Following (Gärdenfors 1988), we represent the contracted belief state P_a^- as the ϵ-mixture of two states: the original belief state P, and the resultant state $P_{\neg a}^*$ of revising P by $\neg a$. The factor ϵ determines the proportion of P and $P_{\neg a}^*$ that are to be used *in a uniform fashion* in this process.

In an earlier work, (Chhogyal et al. 2015), we showed a simple but somewhat surprising result that the Kullback-Leibler divergence of P_a^- from P is solely and completely determined by ϵ. This result is somewhat baffling since one would have thought that the divergence would at least depend on the two belief states P and $P_{\neg a}^*$. One of the things we do in this paper is take up this issue, and provide plausible ways for viewing the mixing factor ϵ that are consistent with this result. Of particular interest is one that connects ϵ with the entrenchment level of the belief a that is being discarded, and provides a way of obtaining the value of ϵ.

While the aforementioned contribution involves the *value* of the mixing factor ϵ, the other contribution deals with how ϵ is used. Although the factor ϵ determines what proportions of P and $P_{\neg a}^*$ are to be used in the process of computing P_a^-, in principle there are many ways of doing it. Following (Gärdenfors 1988), it has been customary to use ϵ as a uniform scaling factor for the probability mass assigned to the worlds that model the current beliefs, but no justification has been provided for this practice. We show that indeed, if the weight ϵ is given to the models of the current beliefs as a whole, but not uniformly (and hence violating the "ratio principle"), then the KL-divergence would be more than otherwise. Thus, in showing that the employment of ϵ as a uniform scaling factor is dictated by the principle of minimum cross-entropy, we provide a much needed justification for the way probabilistic belief contraction is conceived in the literature.

2   Background

Consider a finite set of propositional variables from which a language $\mathcal{L}$ is generated. The set of all possible worlds (interpretations) of $\mathcal{L}$ is Ω. Lower case Roman letters $a, b, \ldots$ are sentences in $\mathcal{L}$ and, ω with or without subscript represents worlds in Ω. An agent's belief state can be represented as a single sentence by the special symbol k or as a set of sentences K. Given a probability distribution P, the belief set K is the *top*[2] of P. Henceforth, P will be

---

[1] This principle is subject to debate and different interpretations; see for instance (Rott and Pagnucco 1999; Arló-Costa and Levi 2006). It has also been employed to provide accounts of iterated belief contraction, e.g. in (Nayak et al. 2007).

[2] Sentences that have a probability of 1.

referred to as the belief state. Given P, the probability of a sentence a is given by $P(a) = \sum_{\omega \in \Omega} P(\omega)$, where $\omega \models a$. If $P(a) = 1$, we say a is *belief*. If $P(a) = 0$, a is a *disbelief*, i.e. $\neg a$ is a belief. For all other cases, a is called a *non-belief*. The two belief change operations we are interested in are *probabilistic belief revision* and *probabilistic belief contraction*. Belief revision changes the status of a sentence from a belief to a disbelief (and *vice versa*) because the agent accepts information contrary to what it believes. Let P be the belief state and a be a belief. The agent upon receiving word that a is false *revises* P by $\neg a$ to transform P to a new belief state represented by $P^*_{\neg a}$ where a is a disbelief. Thus, $P(a) = 1$ and $P^*_{\neg a}(a) = 0$. Note that since $P(a) = 0$, Bayesian conditionalization cannot be used for revision because it runs into the zero prior problem. In fact, conditionalization is more appropriate for turning non-beliefs into beliefs.

Belief contraction changes the status of a sentence from a belief to a non-belief. The contraction of P by a is represented by the new belief state P^-_a. Thus, $P(a) = 1$ but $0 < P^-_a(a) < 1$. The semantics of revision and contraction of probabilistic belief states is given by the movement of probabilities between the worlds in Ω. When revising P by $\neg a$, we must ensure that all a-worlds have zero probability mass, i.e. $\sum_{\omega \in \Omega} P^*_{\neg a}(\omega) = 0$, where $\omega \models a$. This way $P^*_{\neg a}(a) = 0$ as required. On the other hand, in contracting P by a, some models of a must have non-zero probability mass but the sum of these masses should not equal 1, i.e. $0 < \sum_{\omega \in \Omega} P^-_a(\omega) < 1$, where $\omega \models a$. So, we have $0 < P^-_a(a) < 1$ as required for a non-belief. Revision and contraction are functions that map belief states and input sentences to belief states. Thus, given the set of all belief states $\mathbb{P}$, $* : \mathbb{P} \times \mathcal{L} \to \mathbb{P}$ and $- : \mathbb{P} \times \mathcal{L} \to \mathbb{P}$. These functions must of course be subject to some conditions. For instance, as we mentioned earlier, $P^*_{\neg a}(a) = 0$ for revision and $0 < P^-_a(a) < 1$ for contraction. The former constitutes the so called *revision postulates*, $P^*1 - P^*5$, where as the latter constitutes *contraction postulates*, $P^-1 - P^-5$ (Gärdenfors 1986; Gärdenfors 1988) provided below:

(P^-1) P^-_a is a probability function
(P^-2) $P^-_a(a) < 1$ iff not $\vdash a$.
(P^-3) If $\vdash a \leftrightarrow b$, then $P^-_a = P^-_b$.
(P^-4) If $P(a) < 1$, then $P^-_a = P$.
(P^-5) If $P(a) = 1$, then $(P^-_a)^+_a = P$.[3]

All the above postulates possibly except (P^-5) named *Recovery* are self-explanatory. *Recovery* captures the intuition that all the information lost in the process of giving up a belief a should be regained by reinstating a as a belief. It is a relatively controversial requirement, and discussion of it can be found, for instance, in (Hansson 1991).

Definition 1 *(Gärdenfors 1988). Given P with $P(a) = 1$, for all $x \in \mathcal{L}$:*

$$P^-_a(x) = \epsilon \cdot P(x) + (1 - \epsilon) \cdot P^*_{\neg a}(x)$$

for some ϵ, $0 < \epsilon \leq 1$.

---

[3] P^+_a is simply Bayesian conditioning.

Thus, P_a^- is a *mixture* of P and $P_{\neg a}^*$ and is also often written as $P\epsilon P_{\neg a}^*$. To keep in line with (P^-4), ϵ is taken to be 1 when $P(a) < 1$.[4] The following theorem guarantees that probabilistic contraction functions obtained via Definition 1 satisfy the contraction postulates.

Theorem 1 *(Gärdenfors 1988). If a revision function satisfies $P^*1 - P^*5$, then the contraction function generated by Definition 1 satisfies $P^-1 - P^-5$, where $P^*1 - P^*5$ are probabilistic revision postulates.*

3   What Is ϵ?

In (Gärdenfors 1988), it is claimed that P_a^- in Definition 1 is "a compromise between the states of belief represented by P and $P_{\neg a}^*$, where ϵ is a measure of degree of closeness to the beliefs in P." It is not clear from this statement what the nature of ϵ exactly is. Neither has there been any work that we know of which explicitly discusses ϵ. Hence the nature of ϵ merits more discussion. First, is ϵ sensitive to the input a in P_a^-? More precisely, if the current belief state is P and there are two beliefs b and b' such that $b \not\equiv b'$, should the value of ϵ be the same or different while constructing P_b^- and $P_{b'}^-$ respectively? Consider the example below for the purpose of illustration.

> Let's assume that John believes: *b: Betty's new car is green*, and *g: grass is green*. If someone raises doubt about the colour of Betty's new car, chances are that John, who always believed that Betty's favourite colour is red, will discard the belief b and the new probability he will assign to it will be relatively low. On the other hand, if John were to suspend the belief that grass is green, it is likely to be with much reluctance that he will do so, and assign relatively high probability to g.

This example indicates that ϵ should be sensitive to the input. Indeed, Gärdenfors himself has hinted support for it when he points out that $P_a^-(a) = \epsilon$, which is easily verified:

$$P_a^-(a) = \epsilon \cdot P(a) + (1 - \epsilon) \cdot P_{\neg a}^*(a)$$
$$= \epsilon \cdot 1 + (1 - \epsilon) \cdot 0$$
$$= \epsilon.$$

So ϵ is the probability that an agent would assign to a belief just discarded. Presumably, a belief that an agent very reluctantly discards will retain a higher degree of probability than one she happily discards. We next present two views on ϵ, both of which are sensitive to the input. One we call the *epistemic entrenchment view* and the other the *probabilistic view*.

---

[4] One might wonder if the value of ϵ is prefixed. We take the view that it is not, and is indeed sensitive to the information a that is being removed.

3.1   Two Views on ϵ

We may now ask whether there exists some sort of measure that could be used to represent John's reluctance (or willingness) to give up a belief as in the example above. An obvious candidate is the *epistemic entrenchment* (Gärdenfors 1988) which is the degree of resistance of a belief to change. The more highly entrenched a belief is the more reluctant an agent is to discard it. Epistemic entrenchment also serves as an extra-logical tool when having to choose between which beliefs to give up. Whilst more typically studied syntactically, for our purposes it helps to characterize it semantically via Groves' System of Spheres (Grove 1988; Nayak 1994).

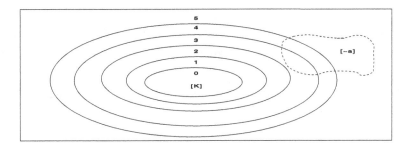

Fig. 1. Grove's system of spheres: Semantics for Epistemic Entrenchment. The inner most sphere $[K]$ represents the set of worlds considered most plausible by the agent and have an ordinal value of 0. The dotted line represents $[\neg a]$.

We give here a quick overview of entrenchment semantics using Fig. 1. An epistemic entrenchment relation, $\preceq$, can be viewed as inducing a system of spheres that is centered on $[K]$, where $[K]$ represents the worlds that have been assigned non-zero probability mass by the current belief state P. A system of spheres represents the relative plausibility of worlds in Ω. Each "solid" sphere may be viewed as consisting of a number of smaller spherical bands. Each band in this system consists of subsets of worlds that are considered equally plausible. The further out a band is from the centre, the less plausible the worlds contained in it are. Thus, the innermost sphere $[K]$ contains the most plausible worlds. Ordinal values, $0, 1, 2, \ldots$, are used to denote the relative plausibility of the spheres. Thus, $[K]$ being the set of most plausible worlds has an ordinal value of 0. Now in order to determine how epistemically entrenched a belief a is, we look at the innermost band that intersects $[\neg a]$. In the figure above, the smallest sphere intersecting $[\neg a]$ has an ordinal value of 3 which we assign to be the *entrenchment rank* of a and is denoted as $EE(a)$. Note that even though a is belief with $P(a) = 1$, it is still not maximally entrenched since $EE(a) \neq 5$.[5]

---

[5] Strictly speaking Gärdenfors epistemic entrenchment is completely relational, and using ordinals in this way is used for convenience only. Our approach may be taken to be closer to Spohn's degree of beliefs modeled via Ordinal Conditional Functions (Spohn 1988).

We would expect that epistemic entrenchment can help with the determination of ϵ when contracting a belief a from P. Since we argued above that both ϵ and epistemic entrenchment have to do with the reluctance to give up beliefs, it is reasonable to assume that ϵ should be determined by $EE(a)$. Given P and two beliefs a and b, what we are then aiming for is a relation between ϵ on the one hand, and $EE(a)$ and $EE(b)$ on the other that satisfies the following criteria. For all beliefs a and b,

1. If $EE(a) \leq EE(b)$, then $\epsilon_a \leq \epsilon_b$, where $P_a^- = P\epsilon_a P_{\neg a}^*$ and $P_b^- = P\epsilon_b P_{\neg b}^*$, and
2. $0 < \epsilon_a, \epsilon_b < 1$.

We know that $P_x^-(x) = \epsilon$, for any belief x in P. The first criteria says if belief a is less epistemically entrenched than belief b, then when we contract a from P the probability assigned to a should be less than the probability assigned to b when we contract b from P. The following definition satisfies the criteria above:

Definition 2. *Given an epistemic entrenchment relation, $\preceq$, and the corresponding epistemic ranking function, $EE(\cdot)$:*

$$\epsilon_a = 1 - \frac{1}{EE(a)}$$

where a is the belief to be contracted and $EE(a)$ is the epistemic rank of a.[6]

It is easily seen that the more entrenched a belief is, the closer will be the value of ϵ to 1 and the less entrenched a belief is, the closer ϵ will be to 0. It is important to note here that since a is a belief proper, it will never be the case that $EE(a) = 0$ and thus ϵ is always well defined. We state this as a simple proposition.

Proposition 1. *Given a consistent belief state P and a belief a, ϵ obtained via Definition 2 is always well defined.*

Proof. The only case where ϵ is undefined is the case when $EE(a) = 0$. We show that this case is not possible. Since P is consistent and $P(a) = 1$, each world in $[K]$, i.e. the set of worlds P assigns non-zero mass to, must also be models of a. This also means $[K] \cap [\neg a] = \emptyset$. Assume $EE(a) = 0$, this means the innermost band that intersects $[\neg a]$ is $[K]$ or in other words $[K] \cap [\neg a] \neq \emptyset$ but this contradicts the fact that $[K] \cap [\neg a] = \emptyset$. Hence, $EE(a) \neq 0$. □

On a final note, it is interesting to ask whether epistemic entrenchment is in someway connected to probabilities. If so, then ϵ may be possibly computed

---

[6] We assume that $a \not\equiv k$. The special case when the agent discards all that it believes will need special treatment, and will digress us to the discussion of special forms of belief contraction such as *pick contraction* and *bunch contraction* that are not directly relevant to the main contribution of this paper.

simply from the probabilities assigned by P. Thus, for two sentences, the one with a lower probability can be taken to be the one less epistemically entrenched. However, the problem with this reasoning is that all sentences of probability 1 should be maximally entrenched. According to (Gärdenfors 1988), one should find it harder to give up beliefs in natural laws than beliefs in single factual sentences. Thus, the former is more epistemically entrenched than the latter even though they may be both maximally probable. Indeed, as (Levi 1983) notes, "It is tempting to correlate these grades of corrigibility with grades of certainty or probability. According to the view I advocate, that would be a mistake. All items in the initial corpus L which is to be contracted are, from X's point of view, certainly and infallibly true. They all bear probability 1." What Levi describes as "grades of corrigibility" can be roughly seen as the degree of epistemic entrenchment. So, statically speaking, the connection between probability and epistemic entrenchment is rather tenuous since only sentences with maximal probability have a nontrivial entrenchment ranking. However, our approach indicates that the connection between probability and epistemic entrenchment is dynamic in nature – epistemic entrenchment is a prime driver of the probability assigned to non-beliefs.

We now depart from the epistemic entrenchment view of ϵ, and instead look at how ϵ might be determined from probabilities alone. This view is relatively straightforward. Since P_a^- is defined as $P \epsilon P_{\neg a}^*$, ϵ should at least partly be a function of P and $P_{\neg a}^*$. We have the following general definition:

Definition 3. *Given P and $P_{\neg a}^*$,*

$$\epsilon = f(P, P_{\neg a}^*, \Theta).$$

where Θ is contextually determined set of parameters that is possibly empty and f is a function with an appropriate signature.

We do not provide an explicit construction of the function f here since for our purpose the general definition will suffice. A possible way to construct f is provided in (Chhogyal et al. 2015) where arguments of varying degrees of strength are derived from P and $P_{\neg a}^*$, and then used to determine ϵ.

4   The Kullback-Leibler Divergence

In belief change, the Kullback-Leibler (KL) divergence (Kullback and Leibler 1951) has attracted some researchers to use it as tool for choosing between different probabilistic belief states when the current belief state is to undergo change. The KL divergence is central to the *principle of minimum cross-entropy* (Kullback and Leibler 1951), which in simple terms says that when faced with several probability distributions to move to from the current belief state, the probability distribution with the minimum KL divergence should be the one chosen. In this paper, we do not prescribe specific constructs that exploit the KL divergence in choosing a suitable belief state P_a^- given P and P_a^*. The interested reader is referred to works such as (Kern-Isberner 2008; Potyka et al. 2013; Ramachandran et al. 2012). Our main aim is to study the relation between ϵ and the KL divergence. This work builds on the work done in (Chhogyal et al. 2015) where a basic result which we present later was established. However, it was rather reticent on the implications of that basic result and this is what we hope to shed more light on. Recall from Definition 1 that $P_a^-(x) = \epsilon \cdot P(x) + (1 - \epsilon) \cdot P_{\neg a}^*(x)$. From here onwards, by P_a^-, we refer to the probability distribution obtained via Definition 1. We start with the definition of the KL divergence.

Definition 4. *Given two discrete probability distributions P and Q over the set of worlds Ω, the Kullback-Leibler (KL) divergence of Q from P is defined as:*[7]

$$D_{KL}(P \parallel Q) = \sum_{\omega \in \Omega} P(\omega) \ln \left(\frac{P(\omega)}{Q(\omega)} \right)$$

Intuitively, the KL divergence measures the difference between two probability distributions. If P is the real distribution then $D_{KL}(P \parallel Q)$ measures how good an approximation Q is of P, or alternatively how close Q is to P. It can also be viewed as measuring how much information is lost in moving from P to Q. The following are some important properties of the KL divergence:

1. $D_{KL}(P \parallel Q) \geq 0$,
2. $D_{KL}(P \parallel Q) = 0$ iff $P(\omega) = Q(\omega)$ for all $\omega \in \Omega$, and
3. $D_{KL}(P \parallel Q) \neq D_{KL}(Q \parallel P)$.

The first property says that the KL divergence is always non-negative. The second property says that the KL divergence between two distributions is 0 if and only if the distributions are equal. The third property says that in general the KL divergence is not symmetric. We will also adopt the convention that $0/0 = 0$. We have some simple observations:

Observation 1. *Given P, $P_{\neg a}^*$ and ϵ,*

1. *If $\epsilon = 1$, then $DL_{KL}(P \parallel P_a^-) = 0$,*
2. *If $0 < \epsilon < 1$, then $0 < DL_{KL}(P \parallel P_a^-) < \infty$ and,*
3. *If $\epsilon = 0$, then $DL_{KL}(P \parallel P_a^-)$ is undefined, i.e. it is ∞.*

Proof. The observation follows directly from Definition 1. First consider when $\epsilon = 1$ and we get $P_a^- = P$. From property 2 of KL divergence, we know $DL_{KL}(P \parallel P_a^-) = 0$. Next consider when $0 < \epsilon < 1$, by virtue of Definition 1, any world assigned non-zero probability mass by P is also assigned non-zero mass by P_a^-. It is easily seen then that: a) the ratio $\frac{P(\omega)}{P_a^-(\omega)}$ is either 0 or greater than 0, and b) there exists ω such that $\frac{P(\omega)}{P_a^-(\omega)}$ is positive and well-defined. It follows then that $0 < DL_{KL}(P \parallel P_a^-) < \infty$. For the final case when $\epsilon = 0$, $P_a^- = P_{\neg a}^*$ and there exists a world ω such that $P(\omega) > 0$ and $P_a^-(\omega) = 0$. This means $\frac{P(\omega)}{P_a^-(\omega)}$ is undefined. Thus, $DL_{KL}(P \parallel P_a^-) = \infty$. $\square$

---

[7] KL divergence is often defined only when $Q(w) = 0$ implies $P(w) = 0$, obviating the need for special conventions such as $0/0 = 0$.

When $\epsilon = 1$, contraction does not change the beliefs, so there should be no divergence. When $\epsilon = 0$, contraction results in a drastic change of beliefs so the divergence should be maximum.

Observation 2. *Given P, $P_{\neg a}^*$ and ϵ, $DL_{KL}(P_a^- \parallel P)$ is undefined, i.e. ∞.*

Proof. This proof is similar to case 3 above. There are worlds that P_a^- assign non-zero probability mass to but not by P.

Theorem 2 *(Chhogyal et al. 2015). Given P, $P_{\neg a}^*$, and ϵ, $DL_{KL}(P \parallel P_a^-) = ln(\frac{1}{\epsilon})$.*

Theorem 2 was established in (Chhogyal et al. 2015). As we can see, the claims made in Observation 1 agree with this theorem. Corollary 1 below is a simple consequence of Theorem 2 and the two will be central to our discussions later on.

Corollary 1. *Let belief state P, the mixing factor ϵ, belief a, and revision operators $*$ and $*'$ be given. Let $P_a^- = P\epsilon P_{\neg a}^*$ and $P_a^{-'} = P\epsilon P_{\neg a}^{*'}$. Then $DL_{KL}(P \parallel P_a^-) = DL_{KL}(P \parallel P_a^{-'})$.*

Proof. Since ϵ is fixed, it easily follows from Theorem 2, that $DL_{KL}(P \parallel P_a^-) = ln(\frac{1}{\epsilon})$ and $DL_{KL}(P \parallel P_a^{-'}) = ln(\frac{1}{\epsilon})$. □

Table 1. Two *partial* belief states P' and P'' that are obtained from P via scaling by $\epsilon = 0.1$. ϵ is applied uniformly to obtain P' and non-uniformly to obtain P''. Note that the probabilities assigned by P' and P'' sum to ϵ.

ω	P	P'	P"
ω_1	0.2	0.02	0.01
ω_2	0.3	0.03	0.06
ω_3	0.5	0.05	0.03
ω_4	0	0	0

We next look at the relation between the uniform scaling of probabilities and the KL divergence. For ease of explanation, let $P_a^- = Q + Q'$, where Q is $\epsilon \cdot P$ and Q' is $(1 - \epsilon) \cdot P_{\neg a}^*$. Thus, $DL_{KL}(P \parallel P_a^-)$ would be:

$$DL_{KL}(P \parallel P_a^-) = \sum_{\omega \in \Omega} P(\omega) \ln \left(\frac{P(\omega)}{Q(\omega) + Q'(\omega)} \right)$$

In both cases Q and Q' are obtained by scaling P and $P_{\neg a}^*$ uniformly and results in $\sum_{\omega \in \Omega} Q(\omega) = \epsilon$ and $\sum_{\omega \in \Omega} Q'(\omega) = 1 - \epsilon$. This raises an interesting question. Why should Q and Q' be obtained by uniform scaling? To make the idea of uniform scaling clearer, consider Table 1. We obtain the partial distribution P' by taking the probability of each world assigned by P and uniformly scaling it by ϵ. For P'', each world is scaled by a different factor but the sum of the probability mass in P'' still ϵ. What if we only adhere to the condition that $\sum_{\omega \in \Omega} Q(\omega) = \epsilon$ and $\sum_{\omega \in \Omega} Q'(\omega) = 1 - \epsilon$ but not bother that they are obtained by scaling P and $P^*_{\neg a}$ uniformly? Of course, we also require that $Q(\omega) \neq 0$ iff $P(\omega) \neq 0$, and similarly for Q' and $P^*_{\neg a}$. Would the divergence $DL_{KL}(P \parallel P^-_a)$ be less in this case when the scaling is non-uniform? A little reflection will show that in calculating the $DL_{KL}(P \parallel P^-_a)$, Q' really has no influence. This is because in the ratio $\frac{P(\omega)}{Q(\omega) + Q'(\omega)}$ when $P(\omega) \neq 0$ or equivalently $Q(\omega) \neq 0$, we have $Q'(\omega) = 0$ and vice versa. This means we only have to focus on the case where $\sum_{\omega \in \Omega} Q(\omega) = \epsilon$ and $Q \neq \epsilon \cdot P$. To help answer the question regarding $DL_{KL}(P \parallel P^-_a)$, we first introduce a result from information theory:

Theorem 3. *(Cover and Thomas 1991, Log Sum Inequality). For positive numbers, $a_1, \ldots, a_n$ and $b_1, \ldots, b_n$,*

$$\sum_{i=1}^{n} a_i \ln\left(\frac{a_i}{b_i}\right) \geq \left(\sum_{i=1}^{n} a_i\right) \ln\left(\frac{\sum_{i=1}^{n} a_i}{\sum_{i=1}^{n} b_i}\right)$$

with equality iff for all i, $\frac{a_i}{b_i} = constant$.

Theorem 4. *Given P, $P^*_{\neg a}$, and ϵ, let $[k]$ be the set of worlds such that $P(\omega) > 0$ and $\sum_{\omega \in [k]} P(\omega) = 1$. Let Q be such that a) $\sum_{\omega \in \Omega} Q(\omega) = \epsilon$, b) $Q(\omega) \neq \epsilon \cdot P(\omega)$ for some $\omega \in [k]$,[8] and c) $Q(\omega) \neq 0$ iff $P(\omega) \neq 0$. If $P^-_a(\omega) = \epsilon \cdot P(\omega) + (1 - \epsilon) \cdot P^*_{\neg a}(\omega)$ and $P^{-'}_a(\omega) = Q(\omega) + (1 - \epsilon) \cdot P^*_{\neg a}(\omega)$, then $DL_{KL}(P \parallel P^-_a) < DL_{KL}(P \parallel P^{-'}_a)$.*

Proof. We want to prove that $DL_{KL}(P \parallel P^-_a) < DL_{KL}(P \parallel P^{-'}_a)$. First, we compute $DL_{KL}(P \parallel P^-_a)$. Since P is scaled uniformly by ϵ, from Theorem 2 we know

$$DL_{KL}(P \parallel P^-_a) = \ln\left(\frac{1}{\epsilon}\right).$$

We next compute $DL_{KL}(P \parallel P^{-'}_a)$. By definition,

$$DL_{KL}(P \parallel P^{-'}_a) = \sum_{\omega \in \Omega} P(\omega) \ln\left(\frac{P(\omega)}{P^{-'}_a(\omega)}\right).$$

We know $P^{-'}_a(\omega) = Q(\omega) + (1 - \epsilon) \cdot P^*_{\neg a}(\omega)$, thus

$$DL_{KL}(P \parallel P^{-'}_a) = \sum_{\omega \in \Omega} P(\omega) \ln\left(\frac{P(\omega)}{Q(\omega) + (1 - \epsilon) \cdot P^*_{\neg a}(\omega)}\right).$$

Since $\Omega = [k] \cup \Omega \setminus [k]$, we can rewrite the equation above as follows:

---

[8] This is the same as saying *it is not the case that $Q(\omega) = \epsilon \cdot P(\omega)$ for all $\omega \in [k]$.*

$$DL_{KL}(P \parallel P_a^{-'}) = A + B$$

where $A = \sum_{\omega \in [k]} P(\omega) \ln \left(\frac{P(\omega)}{Q(\omega)+(1-\epsilon) \cdot P_{\neg a}^*(\omega)} \right)$ and $B = \sum_{\omega \in \Omega \setminus [k]} P(\omega) \ln \left(\frac{P(\omega)}{Q(\omega)+(1-\epsilon) \cdot P_{\neg a}^*(\omega)} \right)$.

Consider A first. Since $\omega \in [k]$, $P(\omega) > 0$, $Q(\omega) > 0$ and $P_{\neg a}^*(\omega) = 0$.

$$A = \sum_{\omega \in [k]} P(\omega) \ln \left(\frac{P(\omega)}{Q(\omega) + (1-\epsilon) \cdot 0} \right)$$

$$= \sum_{\omega \in [k]} P(\omega) \ln \left(\frac{P(\omega)}{Q(\omega)} \right).$$

Now consider B. Since $\omega \in \Omega \setminus [k]$, $P(\omega) = 0$ and $Q(\omega) = 0$.

$$B = \sum_{\omega \in \Omega \setminus [k]} 0 \cdot \ln \left(\frac{0}{0 + (1-\epsilon) \cdot P_{\neg a}^*(\omega)} \right)$$

$$= \sum_{\omega \in \Omega \setminus [k]} 0 \cdot \ln(0)$$

$$= 0.$$

Thus,

$$DL_{KL}(P \parallel P_a^{-'}) = \sum_{\omega \in [k]} P(\omega) \ln \left(\frac{P(\omega)}{Q(\omega)} \right).$$

This is the left side of the log-sum inequality. Thus,

$$DL_{KL}(P \parallel P_a^{-'}) \geq \left(\sum_{\omega \in [k]} P(\omega) \right) \ln \left(\frac{\sum_{\omega \in [k]} P(\omega)}{\sum_{\omega \in [k]} Q(\omega)} \right).$$

Now we know $\sum_{\omega \in [k]} P(\omega) = 1$ and $\sum_{\omega \in [k]} Q(\omega) = \epsilon$. Thus,

$$DL_{KL}(P \parallel P_a^{-'}) \geq \ln \left(\frac{1}{\epsilon} \right).$$

The equality in $\geq$ above only holds iff $\frac{P(\omega)}{Q(\omega)} = c$ or $Q(\omega) = \frac{1}{c} \cdot P(\omega)$, where c is a constant. Assume there is a constant c, we get

$$\sum_{\omega \in [k]} Q(\omega) = \sum_{\omega \in [k]} \frac{1}{c} \cdot P(\omega)$$

$$= \frac{1}{c} \sum_{\omega \in [k]} P(\omega)$$

$$= \frac{1}{c}$$

as we know $\sum_{\omega \in [k]} P(\omega) = 1$. We also know $\sum_{\omega \in [k]} Q(\omega) = \epsilon$ so,

$$\frac{1}{c} = \epsilon$$

This means that $Q(\omega) = \epsilon \cdot P(\omega)$ for all $\omega \in [k]$ which contradicts the assumption we made at the beginning that $Q(\omega) \neq \epsilon \cdot P(\omega)$ for some $\omega \in [k]$. It follows that there cannot be any constant c such that $\frac{P(\omega)}{Q(\omega)} = c$ and the equality in $\geq$ above no longer holds. Thus,

$$DL_{KL}(P \parallel P_a^{-'}) > \ln\left(\frac{1}{\epsilon}\right).$$

Since, $DL_{KL}(P \parallel P_a^{-}) = \ln\left(\frac{1}{\epsilon}\right)$, we have:

$$DL_{KL}(P \parallel P_a^{-}) < DL_{KL}(P \parallel P_a^{-'}).$$

$\square$

Theorem 4 gives us the answer that we were seeking, namely, when constructing P_a^{-} using Definition 1, scaling P *uniformly* is preferable since it produces a smaller divergence. This provides a much needed justification for the uniform scaling of P as conceived by Gärdenfors.

5   Conclusion

In this paper we provided a reasonable way of establishing the value of the mixing factor ϵ that is used for the computation of probabilistic belief contraction. We also showed that the uniform scaling of the belief state P used for this purpose can be formally justified by appealing to the principle of minimum cross-entropy.

As concluding remarks, let us consider the two competing views of ϵ provided in Sect. 3: the *epistemic entrenchment view* and the *probabilistic view*. Note that our results for the KL divergence in Theorem 2 does not specify which of these two views that ϵ is being interpreted under. That is, Theorem 2 may be interpreted under both these views. We consider these two cases in a little more detail.

Let us look at the epistemic entrenchment view first. Consider two revision operators $*$ and $*'$. Presumably they are associated with two different entrenchment ranking functions.[9] Let us further assume that both these ranking functions assign the same rank to the belief a whereby the mixing factor ϵ is identical in both the cases.[10] Thus, given an initial belief state P, an existing belief a, the two

---

[9] Strictly speaking relations, but we make appropriatete mental adjustments here.

[10] We note here in passing that even if a is assigned the same epistemic rank by the two ranking functions, and hence the revised belief sets $K_{\neg a}^{*}$ and $K_{\neg a}^{*'}$ are the same, the revised probabilistic states $P_{\neg a}^{*}$ and $P_{\neg a}^{*'}$ could be different. Support for this view can be obtained based on the accounts of probabilistic belief revision developed in (Chhogyal et al. 2014).

revision operators $*$, $*'$ and the mixing factor ϵ, we know from Corollary 1 that the following holds: $DL_{KL}(P \parallel P_a^-) = DL_{KL}(P \parallel P_a^{-'})$, where $P_a^- = P\epsilon P_{\neg a}^*$ and $P_a^{-'} = P\epsilon P_{\neg a}^{*'}$. Since the states P_a^- and $P_a^{-'}$ are both equally divergent from the initial state P, they are equally attractive belief states to move to after the contraction by a. Thus, according to the epistemic entrenchment view, ϵ is independent of different revisions of P by a as long as the presumed entrenchment ranking functions assign equal rank to a.

Let us now consider the probabilistic interpretation of ϵ. According to this view, ϵ is a function of P, $P_{\neg a}^*$ and possibly some set of other parameters Θ. Different revision operators $*1, *2, \ldots *n$ will produce different probability functions $P_{\neg a}^{*1}, P_{\neg a}^{*2}, \ldots, P_{\neg a}^{*n}$ with possibly different values for ϵ. One way of concretely establishing the value of ϵ in this way is given in (Chhogyal et al. 2015). Accordingly we will obtain different contracted states $P_a^{-1}, P_a^{-2}, \ldots, P_a^{-n}$. Since we know from Theorem 2 that $DL_{KL}(P \parallel P_a^-) = \left(\frac{1}{\epsilon}\right)$, unlike the case of the epistemic entrenchment view, some of the $P_a^{-1}, P_a^{-2}, \ldots, P_a^{-n}$ are almost guaranteed to be preferable to others. Thus, the probabilistic interpretation of ϵ leads to a more fine-grained account of probabilistic belief contraction than the epistemic entrenchment interpretation.

Thus we have provided a reason why the probabilistic interpretation is preferable to an epistemic entrenchment interpretation of the mixing factor. Nonetheless, it remains to be seen if the probabilistic interpretation of ϵ has the last word in this debate. We suspect a more informative measure for belief revision such as provided in the possibility theory (Dubois et al. 1994) will be a strong contender. But that is a topic for a future work.

References

Alchourrón, C.E., Gärdenfors, P., Makinson, D.: On the logic of theory change: partial meet contraction and revision functions. J. Symb. Log. **50**(2), 510–530 (1985)

Arló-Costa, H.L., Levi, I.: Contraction: on the decision-theoretical origins of minimal change and entrenchment. Synthese **152**(1), 129–154 (2006)

Chhogyal, K., Nayak, A.C., Sattar, A.: On the KL divergence of probability mixtures for belief contraction. In: Proceedings of the 38th German Conference on Artificial Intelligence (KI-2015) (to appear 2015)

Chhogyal, K., Nayak, A., Schwitter, R., Sattar, A.: Probabilistic belief revision via imaging. In: Pham, D.-N., Park, S.-B. (eds.) PRICAI 2014. LNCS, vol. 8862, pp. 694–707. Springer, Heidelberg (2014)

Chhogyal, K., Nayak, A. C., Zhuang, Z., Sattar, A.: Probabilistic belief contraction using argumentation. In: Proceedings of the Twenty-Fourth International Joint Conference on Artificial Intelligence, IJCAI 2015, Buenos Aires, Argentina, July 25–31, 2015, pp. 2854–2860 (2015)

Cover, T.M., Thomas, J.A.: Elements of Information Theory. John Wiley & Sons, New York (1991)

Dubois, D., Lang, J., Prade, H.: Possibilistic logic. In: Gabbay, D.M., Hogger, C.J., Robinson, J.A. (eds.) Handbook of Logic in Artificial Intelligence and Logic Programming, vol. 3, pp. 439–513. Clarendon Press, Oxford (1994)

Gärdenfors, P.: The dynamics of belief: contractions and revisions of probability functions. Topoi **5**(1), 29–37 (1986)

Gärdenfors, P.: Knowledge in Flux. Modelling the Dymanics of Epistemic States. MIT Press, Cambridge (1988)

Grove, A.: Two modellings for theory change. J. Philos. Logic **17**(2), 157–170 (1988)

Hansson, S.O.: Belief contraction without recovery. Stud. Logica **50**(2), 251–260 (1991)

Kern-Isberner, G.: Linking iterated belief change operations to nonmonotonic reasoning. In: Principles of Knowledge Representation and Reasoning: Proceedings of the Eleventh International Conference, KR 2008, pp. 166–176 (2008)

Kullback, S., Leibler, R.A.: On information and sufficiency. Ann. Math. Stat. **22**, 79–86 (1951)

Levi, I.: Truth, faillibility and the growth of knowledge in language, logic and method. Boston Stud. Philos. Sci. N.Y., NY **31**, 153–174 (1983)

Nayak, A.C.: Iterated belief change based on epistemic entrenchment. Erkenntnis **41**, 353–390 (1994)

Nayak, A.C., Goebel, R., Orgun, M.A.: Iterated belief contraction from first principles. In: IJCAI 2007, Proceedings of the 20th International Joint Conference on Artificial Intelligence, Hyderabad, India, January 6–12, 2007, pp. 2568–2573 (2007)

Potyka, N., Beierle, C., Kern-Isberner, G.: Changes of relational probabilistic belief states and their computation under optimum entropy semantics. In: Timm, I.J., Thimm, M. (eds.) KI 2013. LNCS, vol. 8077, pp. 176–187. Springer, Heidelberg (2013)

Ramachandran, R., Ramer, A., Nayak, A.C.: Probabilistic belief contraction. Mind. Mach. **22**(4), 325–351 (2012)

Rott, H., Pagnucco, M.: Severe withdrawal (and recovery). J. Philos. Logic **28**(5), 501–547 (1999)

Spohn, W.: Ordinal conditional functions: a dynamic theory of epistemic states. In: Harper, W., Skryms, B. (eds.) Causation in Decision, Belief Change, and Statistics, II', Kluwer, pp. 105–134 (1988)

DMAPP: A Distributed Multi-agent Path Planning Algorithm

Satyendra Singh Chouhan$^{(\boxtimes)}$ and Rajdeep Niyogi

Department of Computer Science and Engineering,
Indian Institute of Technology Roorkee, Roorkee 247667, India
{satycdec,rajdpfec}@iitr.ac.in

Abstract. Multi-agent path planning is a very challenging problem that has several applications. It has received a lot of attention in the last decade. Multi-agent optimal path planning is computationally intractable. Some algorithms have been suggested that may not return optimal plans but are useful in practice. These works mostly use centralized algorithms to compute plans. However in a multi-agent setting it would be more appropriate for the agents, with limited information, to compute the plans. In this paper, we suggest a distributed multi-agent path planning algorithm DMAPP, where all the phases are distributed. We have implemented DMAPP and have compared its performance with some existing algorithms. The results show the effectiveness of our approach.

Keywords: Multi-agent path planning · Distributed decision making · Plan restructuring

1 Introduction

Multi-agent path planning problem has been studied in several application scenarios like inventory management, warehouse management, games, and robotics [1–5]. The main objective of multi-agent path planning is to find a set of paths such that the paths avoid the obstacles in a domain and the paths do not conflict with each other. Multi-agent optimal path planning is PSPACE-hard [6].

Multi-agent path planning approaches can be broadly divided in to coupled and decoupled approaches. In the former, a centralized planner computes a path in the composite state space, which is a Cartesian product of the state space of the individual agents. A centralized planner can compute optimal plans [7,8]. However, this approach becomes computationally intensive, as the number of states and the number of agents increases. In fact, the computational complexity of the time taken to compute such a plan grows exponentially with the number of agents. It has been shown recently in [9] that suboptimal solutions for path finding with several agents can be found efficiently.

Decoupled approaches although not complete and not guaranteed to return optimal solutions, are, however, more practical in use [10,11]. It works in three

© Springer International Publishing Switzerland 2015
B. Pfahringer and J. Renz (Eds.): AI 2015, LNAI 9457, pp. 123–135, 2015.
DOI: 10.1007/978-3-319-26350-2_11

phases: (i) computing the plans of individual agents, (ii) deciding the priority of the agents for plan coordination (restructuring), and (iii) plan restructuring. Recent works [12–14] use only phases (i) and (iii), where the decision making for priority can be done in any of these two phases. Decoupled approaches scale well compared to coupled approaches. [15] suggest a decoupled approach, where a class of problems called SLIDEABLE has been identified for which planning is complete and it takes low polynomial time to compute plans.

The existing decoupled approaches [8,16–20] use centralized algorithms for the three phases. An exception being [19] where phase (ii) is computed in a distributed manner but the other two phases use centralized algorithms. In distributed multi-agent planning [21] the agents are involved in the planning process, there is no central agent to compute the plans. In [17] the agents are spatially distributed but the plan coordination is achieved by a centralized agent. Multi-agent systems are inherently distributed where each agent has limited or no knowledge about the environment and other agents. In this paper, we present a fully distributed multi-agent path planning algorithm, called DMAPP, where all the three phases are distributed.

The rest of this paper is organized as follows. In Sect. 2 we discuss some related work. DMAPP is given in Sect. 3. In Sect. 4 the simulation results are given. Conclusions are given in Sect. 5.

2   Related Work

A decoupled approach to solving a multi agent path planning problem typically consists of three phases: (i) computing the plans of individual agents, (ii) deciding the order (priority of the agents) in which the plans of individual agents would be restructured so as to obtain an overall collision free solution, (iii) restructuring the individual plans based on the order obtained in (ii).

In phase (i) individual plans are obtained with respect to static obstacles but not considering the plans of other agents. In phase (iii) the plan of the highest priority agent is first computed by looking at only the static obstacles. The plan of the next agent in the order of priority is computed by looking at the plan of the highest priority agent and the static obstacles. This continues till we get the agent with the least priority. Eventually the plan of the least priority agent is computed by considering the plans of all the higher priority agents and the static obstacles. We review some works that adopt different strategies for the phases (ii) and (iii).

In [10], a multi agent path planning algorithm is suggested. The phases (ii) and (iii) are combined together by first assigning a priority to the agents, which is obtained in a centralized manner. The combined phases (ii) and (iii) is also computed by a centralized algorithm.

In [5], a path finding problem for real-time strategy games (RTS) is considered. The environment is an 8-connected deterministic grid world. The paths of the individual agents, without considering collision with other agents, are obtained by using the A* algorithm. The priority of the agents in such game settings is predetermined. A centralized algorithm is suggested (phase (iii)) to remove the collisions, if any, in the individual plans, using the priority. In order to resolve collisions, a biased cost function is defined on collision points for all the colliding agents, except the agent with the highest priority.

In [15, 22], a multi-agent path planning problem is considered in a 4-connected grid world. In phase (i), plans of individual agents are obtained together with alternate plans. In phase (ii) the priority of the agents is computed using some heuristics. A centralized algorithm is suggested to find an overall collision free solution. A class of problems in grid world has been identified called SLIDEABLE for which the overall algorithm MAPP is complete. Moreover, MAPP has low polynomial complexity for computing plans.

In [19], a path planning problem for multiple robots is considered. In phase (i) the plans of the individual robots are obtained. In phase (ii) a priority value is obtained by each robot based on the static obstacles and the kinematic constraints of the robot. Then each robot broadcasts its priority value and the number of obstacles that it senses to all the other robots. These values are compared to obtain the order of priority. This phase is thus distributed. In phase (iii) a centralized algorithm is used to restructure the plans.

In the light of the above works, we now present some features of the proposed algorithm in this paper, DMAPP. Phase (i) of DMAPP is distributed since each agent computes its plans by considering the static obstacles, whereas the plans of the agents are computed by a centralized agent in [5, 15, 19, 22]. Phase (ii) in [5, 15, 22] is computed in a centralized manner whereas it is computed in a distributed manner in [19]. However, in [19] it is assumed that each agent knows the total number of agents in the system, which makes it possible to broadcast the messages. In DMAPP, this assumption is relaxed, where an agent does not know the total number of agents, which is a standard assumption, is most distributed systems [23]. Phase (iii) in all the above works is computed by a centralized algorithm. In DMAPP phase (iii) is computed in a distributed manner by the agents by sending messages. Thus, DMAPP is a fully distributed multi-agent path planning algorithm.

3   A Distributed Multi-agent Path Planning Algorithm

In this section, we present the DAMPP algorithm. It works in three phases: path planning by an agent, Distributed decision-making, and plan restructuring. The algorithm uses the following data structures as listed in Table 1.

3.1   Path Planning

In this phase, each agent computes a path corresponding to its initial-goal states by using the search technique used in the FF planning system [24]. In each iteration [24] performs a complete breadth-first-search to find a state with strictly better evaluation. It [24] evaluates all the successor states of a state s. If a state s' with better heuristic value than that of s is found, add the path up to s' and

Table 1. Data Structures and variables used in DAMPP

Notation	Meaning
i, n	Agent i, where $1 \leq i \leq n$, n: number of agents
P_i	Plan of agent i
$|P_i|$	Length of the plan of agent i
T_i	$\langle id_i, |P_i| \rangle$, id is an identifier
MA-Plan$[1 \ldots n][1 \ldots m]$	A cell (i, k) of MA-Plan contains the kth action of agent i
I_i, G_i	Initial and goal states of agent i

make s' the current state. We have used Euclidean distance as the heuristic. The path is computed by considering only the static obstacles in the environment; the presence of other agents is, however, not considered. At the end of this phase, each agent comes up with an individual plan.

Pseudo code of phase (i) for agent i is as follows:

> **Input**: Initial and goal state of each agent
> **Output**: plan (P_i) of agent i
> $P_i := ComputePlan(I_i, G_i, null)$
> $T_i := \langle id_i, |P_i| \rangle$
> insert T_i in set S_i

Algorithm 1. Phase (i): individual path planning

3.2 Priority Decision-Making

After first phase, each agent has its own plan. The individual plans may be conflicting with each other. Therefore the agents need to restructure their plans by considering the plans of the other agents. Thus the order in which the restructuring will be done has to be decided. In the literature the following strategies have been adopted: distance between initial and goal states [19], a random sequence of agents [10], individual plan length of the agents as obtained in phase (i) [16,18].

In this paper, we decide priority based on individual plan length of the agents as obtained in phase (i). Highest priority is given to the agent that has the longest plan length. The advantages of using this measure are:

- The overall plan restructuring time would be reduced compared to that obtained by assigning a random priority as in [16].
- The number of problem instances solvable is more than that obtained by assigning a random priority as in [16].

Priority of the agents can easily be achieved by a central agent that can compare the individual plan length of the agents or distance between initial and goal states of the agents. In a distributed approach, the agents can exchange messages (that contain information on plan length, kinematic constraints) with all the other agents to come up with a priority as in [19]. This method [19] assumes that an agent knows the total number of agents in the system, which makes it possible to broadcast the messages. In a multi-agent setting it would be more appropriate to assume that an agent has limited information about the environment and other agents. Typically an agent does not know the total number of agents in the system. We suggest a distributed decision making procedure.

3.2.1   Distributed Priority Decision-Making

The proposed algorithm is based on the classical synchronous leader election LCR algorithm [23].

Problem (Priority of Plan Restructuring): The underlying network topology for message passing is a unidirectional ring consisting of n nodes, numbered 1 to n in the clockwise direction. Counting is modulo n. A ring topology is used to illustrate the main idea of the algorithm. However, an arbitrary graph structure can also be used [23]. The messages can only be sent in a clockwise direction. The agents are associated with the nodes. The agents do not know their indices, nor those of their neighbors. Each agent can distinguish its clockwise neighbor from its counterclockwise neighbor. The number n of nodes in the ring is unknown to each agent. Each agent has a unique identifier chosen from the set of positive integers. All the agents communicate and compute in synchronous rounds. At each round all the agents send a message at the same time to its clockwise neighbor; after receiving a message every agent performs some computation. Then the next round begins. The communication system is lossless.

Each agent j has a set that is initialized to a tuple $\langle id_j, plan - length_j \rangle$. The goal of each agent is to know the plan length of every other agent. This allows an agent to determine the priority of the agents for plan restructuring. The algorithm terminates when each agent knows the plan length of every other agent.

Priority of Agents (Pa) Algorithm (Informal): Each agent sends a message that consists of a pair (identifier, plan length) around the ring. When an agent receives a message, it compares the incoming identifier to its own. If the incoming identifier is not equal to its own, it first updates its set with the pair and then passes the message; if it is equal to its own, it comes to know the plan length of every other agent and thus the priority of the agents.

For each agent j, the states in $states_j$ consist of the following components:

u_j, a tuple $\langle id, Plan\_length \rangle$ initially $\langle id_j, |P_j| \rangle$
S_j is a set, initially, $S_j = \{\langle id_j, |P_j| \rangle\}$
$send_j$, a tuple, initially $\langle id_j, |P_j| \rangle$,
$status_j$, with values in $\{unknown, known\}$, initially $unknown$.

When the status becomes '*known*' for all the agents, the algorithm terminates.

send Function for Agent j
    send the current value of $send_j$ to agent $j + 1$
Receive function for agent j
    $send_j := null$
    if *the incoming message is v, a tuple, (id, plan length)* **then**
    case:
    $v.id \neq u.id$: $S_j := S_j \cup \{v\}$; $send_j := v$
    $v.id = u.id$: $status := known$
    endcase

Algorithm 2. Phase (ii): Distributed priority decision making

Lemma 1. *Every agent j outputs 'known' by the end of round 'n'.*

Proof. Note that u_j is the initial value of j. Also note that the value of u_j is not changed by the code. Therefore, by the code, it is suffices to show Assertion 1.

Assertion 1. After n rounds, $status_j =$ known for all agents.
To prove this assertion, we need a preliminary invariant that says something about the situation after smaller number of rounds. Therefore, we add Assertion 2.

Assertion 2. For $0 \leq r \leq n - 1$, after r rounds, $send_{j+r} = u_j$ for all agents.
    This says that the initial value of u for the agent j appears in the send component of the agent $(j + r)$ after r rounds. It is straightforward to prove by induction on r. For $r = 0$, $send_j = u_j$. The inductive step is based on the fact that, every agent other than j will accept the tuple and places it into its send component. Hence, Assertion 2 is proved.
    Having proved Assertion 2, we use its special case for $r = n - 1$, and one more argument about what happens in a single round to show that Assertion 1 holds. The key fact here is that process j accepts u_j as a signal to set its status '*known*'.   ☐

Lemma 2. *After n rounds, Set_j is same for all the agents.*

Proof. Consider an agent j. By Lemma 1 for agent j, $status = known$ after n rounds. By the code of the Receive function for an agent, it means that the condition $v.id \neq u.id$ holds for $n - 1$ rounds. This implies that the set S_j has been updated $n - 1$ times. But at each round it is updated with a unique value since the ids are distinct for each agent. Since the initial value of S_j is the tuple of j ($\langle id_j, plan_j \rangle$), so after n rounds [$(n - 1) + 1$] the set contains n distinct values. Since the choice of j is arbitrary, so S_j is same for all agents.   ☐

Lemmas 1 and 2 together imply the following:

Theorem 1. The Pa algorithm solves the problem of priority of plan restructuring.

Time complexity of the distributed decision making algorithm is $O(n)$ since after n rounds the algorithm terminates. Message complexity of the algorithm is $O(n^2)$ since in each round n messages are sent and the algorithm takes n rounds. The message complexity can be reduced to $O(nlogn)$ by suitably modifying the HS algorithm [23].

Example. Consider a network of 4 agents in a unidirectional ring, numbered 1 to 4 in the clockwise direction. Plan length of the agents (1,2,3,4) with identifiers (11,22,33,44) are 20, 10, 15 and 25 respectively. The distributed priority decision-making process is shown in Fig. 1.

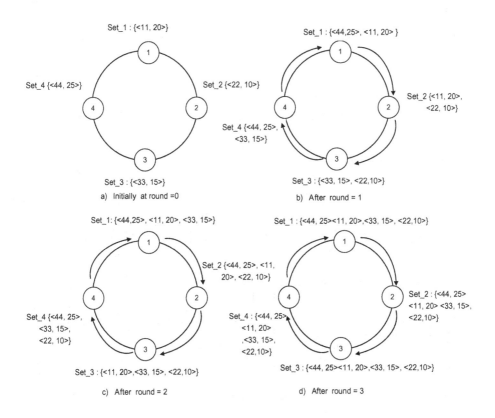

Fig. 1. Distributed priority decision making

Figure 2(a) shows the initial tuple in the sets of agents. After each round, a new tuple is added in to the set (Fig. 2(b)–(d)). Figure 2 shows that after round 3, all the agents have same tuples in the set. The algorithm runs for 4 rounds. In the last (fourth) round every agent receives its own tuple. There is no change in the sets of the agents in the fourth round.

The set $S_j = \{\langle 44, 25 \rangle, \langle 11, 20 \rangle, \langle 33, 15 \rangle, \langle 22, 10 \rangle\}$ is same for all the agents. Set of each agent is sorted according to plan length of the agents. Hence, all agents know the order in which the plan restructuring will be done. For this example plan restructuring sequence will be $4 \rightarrow 1 \rightarrow 3 \rightarrow 2$.

In addition, each agent knows for the restructuring phase, who would be the communicating agents. For example, agent 1 knows that it will receive a message from agent 4, then it would restructure its plan, and then it would send a message to agent 3.

3.3   Plan Restructuring

The highest priority agent will not restructure its plan and least priority agent will restructure the plan according to the plans of all the other agents. After phase (ii):

- Every agent knows the value of n.
- Every agent knows the sequence (of agents) in which restructuring will be done.

Plan restructuring for the example illustrated in Fig. 2 will be done as follows. Plan restructuring sequence is agent 4 $\rightarrow$ agent 1 $\rightarrow$ agent 3 $\rightarrow$ agent 2.

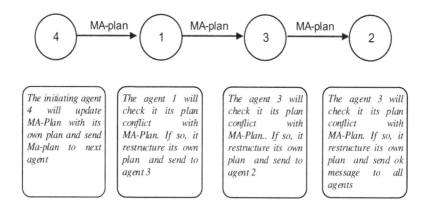

Fig. 2. Plan restructuring for the example illustrated in Fig. 1

Pseudo code for plan restructuring is as follows:

send function for agent j

    send MA-plan to agent k

Receive function for agent k, where k is not the last agent to restructure

    if *the plan of k is conflicting with MA-Plan* **then**

        $P_k :=$ ComputePlan(I_k, G_k , MA-Plan)

        if $P_k \neq null$ **then**

            update MA-Plan With P_k; send MA-Plan

        else

            send 'fail' to all the other agents.

Receive function for agent k, where k is the last agent to restructure

    if *the plan of k is conflicting with MA-Plan* **then**

        $P_k :=$ ComputePlan(I_k, G_k , MA-Plan)

        if $P_k \neq null$ **then**

            update MA-Plan With P_k; send 'ok' to all the other agents

        else

            send 'fail' to all the other agents.

Algorithm 3. Phase (iii): plan restructuring

Lemma 3. *Plan restructuring terminates in at most O(nT) time, O(T) is time complexity of FF.*

Proof. Initially highest priority agent will send the plan to the next agent. It is clear from the algorithm that at most *n-1* communication will be done between the agents. In worst case, apart from highest priority agent, every agent will restructure its plan. Hence, it will take $O(nT)$ time.   □

Lemma 4. *MA-plan obtained by the restructuring phase is correct.*

Proof. For $n = 1$ (basis of induction), MA-plan consists of only one plan from highest priority agents. Since there is only one plan, therefore it is coordinated and valid plan. Let the Lemma be true for the *ith* agent in the priority. To prove that if the *(i + 1)th* agent in the priority finds a plan then it is correct. For this we need the following cases.

Case 1: The plan of the *(i + 1)th* agent in the priority is not conflicting with the remaining plans. In this case no restructuring is to be done. So the overall plan is correct.

Case 2: The plan of the *(i + 1)th* agent in the priority is conflicting with the remaining plans. The FF algorithm computes a conflict free plan by looking at all the other plans and static obstacles. Thus the overall plan is correct. Thus if the restructuring phase obtains a plan it is correct.   □

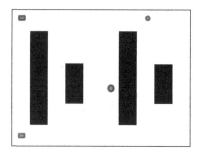

Fig. 3. A simple warehouse domain (represented as a grid structure) with two agents. Initial state of an agent is shown using a rectangle and the final state using a circle.

4   Implementation

We have implemented the DAMPP algorithm in JAVA using NetBeans 8.0.2 IDE. Experiments were performed on a workstation with 2.53GHz processor with 4 GB of RAM. Simulation of the overall implementation is done using Java threads (*Using Thread Class*). Each agent (thread) runs FF algorithm [19] to compute its plan (Phase 1). To ensure synchronization between agents in phase2, we used inbuilt semaphore class of JAVA (*java.util.concurrent.Semaphore*). Synchronization between all three phases is done using valid checkpoints. Experiments were run for several times for each instance and average running time is reported.

Abbreviations used in Tables.
$\#n$: Number of Agents
T1: Average individual Planning Time (in millisecond)
T2: Average time for Distributed Priority decision making (in millisecond)
T3: Average time for plan restructuring (in millisecond)

4.1   Warehouse Domain

We have evaluated our algorithm in the warehouse domain, by considering a grid of size 50×50. It can be easily extended to larger grid structure. Some cells contain obstacles. An agent can occupy a cell if there is no obstacle or another agent. Objective is to find a coordinated path for all the agents. Figure 3 shows a simple warehouse domain having two agents with their initial and goal locations respectively (Fig. 3).

We have compared the performance of DMAPP with the randomized prioritized planning approach [10] and the priority scheme in [19]. We have implemented these approaches in java to compare the peformance. In [10] priority of the agent are pre-decided. In [19] the priority of agents is decided based on a distance heuristic whereas in our work priority of agents is based on plan length.

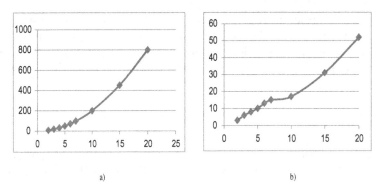

a)   b)

Fig. 4. (a) Number of messages communicated vs Number of agents (b) Number of agents vs average time for Distributed Priority decision-making (T2) in DMAPP

Table 2. Comparison results with Random prioritized planning and Path planning using distance heuristic [19]

Sr. no	n	DMAPP		Random prioritized planning		Path planning using distance heuristic	
		Solved?	(T1 + T2 + T3)	Solved?	Time (ms)	Solved?	Time (ms)
1	2	Yes	15	Yes	13	Yes	16
2	3	Yes	16	yes	15	Yes	18
3	5	Yes	53	Yes	48	Yes	42
4	5	Yes	43	yes	39	Yes	43
5	6	Yes	35	No	-	Yes	56
6	7	Yes	77	No	-	Yes	79
7	10	Yes	166	No	-	No	-
8	15	Yes	221	No	-	No	-

Table 3. Experimental results of DMAPP algorithm

Sr. no	#n = 5				#n = 10				#n = 15				#n = 20			
	Solved?	T1	T2	T3	Solved?	T1	T2	T3	Solved?	T1	T2	T3	Solved?	T1	T2	T3
1	Yes	44	9	2	yes	72	13	3	Yes	176	27	167	Yes	202	44	36
2	Yes	36	6	1	Yes	134	15	9	Yes	220	30	160	Yes	179	67	49
3	Yes	44	5	2	Yes	126	14	9	Yes	136	35	166	Yes	154	50	32
4	Yes	65	8	2	Yes	132	18	16	Yes	209	21	54	Yes	148	50	118
5	Yes	63	10	3	Yes	132	12	63	Yes	180	27	54	Yes	201	41	107
6	Yes	39	16	2	Yes	117	16	149	Yes	173	34	42	Yes	169	46	68
7	Yes	53	9	3	No	130	20	-	Yes	164	26	31	Yes	188	43	19
8	Yes	64	8	4	Yes	130	15	109	Yes	170	47	12	Yes	170	50	30
9	Yes	53	9	3	Yes	133	17	109	Yes	150	33	15	Yes	180	60	20
10	Yes	64	8	4	Yes	124	13	142	Yes	179	22	32	No	201	47	-

The experimental results given in Table 2 show the difference between using these two measures for priority decision making. It can be seen from the table, that DMAPP solves large problem instances that are not solvable by the other approaches. In Table 3, we have reported the average running time of DMAPP for different sized problem instances.

5   Conclusions

In this paper we considered the multi-agent path planning problem. Most of the existing works have used a centralized approach for solving the problem. We feel that in multi agent settings it would be more appropriate for the agents to compute the plans, rather than relying on a centralized unit. Moreover it is typical in multi-agent systems that the agents have limited information. Motivated by these aspects, we have developed a distributed multi-agent path planning algorithm DMAPP. The salient feature of DMAPP is that all the three phases are computed in a distributed manner. For determining the priority for plan restructuring we have developed a distributed algorithm by modifying a classical leader election algorithm [23]. The implementation of a distributed algorithm is seemingly more challenging. We have implemented DMAPP in Java and the results are quite promising.

As part of future work, we would like to determine the class of problems solvable by DMAPP and compare it with the class SLIDEABLE [15, 22] and examine the two classes from both theoretical and practical perspectives. Another line of work would be to see the scope of DMAPP for some application scenarios.

Acknowledgements. The authors thank the anonymous reviewers of AI-2015 for their valuable comments and suggestions for improving the paper.

References

1. Parker, L.E.: Distributed intelligence: overview of the field and its application in multi-robot systems. J. Phys. Agents **2**(1), 5–14 (2008)
2. Bernardini, S., Fox, M., Long, D.: Planning the behaviour of low-cost quadcopters for surveillance missions. In: Proceedings of ICAPS (2014)
3. Wurman, P.R., D'Andrea, R., Mountz, M.: Coordinating hundreds of cooperative autonomous vehicles in warehouses. AI Mag. **29**(1), 9 (2008)
4. Cirillo, M., Pecora, F., Andreasson, H., Uras, T., Koenig, S.: Integrated motion planning and coordination for industrial vehicles. In: Proceedings of the 24th International Conference on Automated Planning and Scheduling, vol. 2126 (2014)
5. Geramifard, A., Chubak, P., Bulitko, V.: Biased cost pathfinding. In: AIIDE, pp. 112–114 (2006)
6. Hopcroft, J.E., Schwartz, J.T., Sharir, M.: On the complexity of motion planning for multiple independent objects; PSPACE-hardness of the warehouseman's problem. Int. J. Robot. Res. **3**(4), 76–88 (1984)
7. Standley, T.S.: Finding optimal solutions to cooperative pathfinding problems. In: AAAI, vol. 1, pp. 28–29 (2010)

8. Sharon, G., Stern, R., Felner, A., Sturtevant, N.R.: Conflict-based search for optimal multi-agent pathfinding. Artif. Intell. **219**, 40–66 (2015)
9. Cohen, L., Uras, T., Koening, S.: Feasibility study: using highways for bounded-suboptimal multi-agent path finding. In: International Symposium on Combinatorial Search (SOCS) (2015)
10. Erdmann, M., Lozano-Perez, T.: On multiple moving objects. Algorithmica **2**(1–4), 477–521 (1987)
11. Silver, D.: Cooperative pathfinding. In: AIIDE, pp. 117–122 (2005)
12. Surynek, P.: A novel approach to path planning for multiple robots in bi-connected graphs. In: IEEE International Conference on Robotics and Automation, 2009, pp. 3613–3619 (2009)
13. Botea, A., Surynek, P.: Multi-agent path finding on strongly biconnected digraphs. In: Twenty-Ninth AAAI Conference on Artificial Intelligence (2015)
14. de Wilde, B., Ter Mors, A.W., Witteveen, C.: Push and rotate: a complete multi-agent pathfinding algorithm. J. Artif. Intell. Res. **51**, 443–492 (2014)
15. Wang, K.-H.C., Botea, A.: Tractable multi-agent path planning on grid maps. In: IJCAI, pp. 1870–1875 (2009)
16. Van Den Berg, J.P., Overmars, M.H.: Prioritized motion planning for multiple robots. In: International Conference on Intelligent Robots and Systems (IROS 2005), pp. 430–435 (2005)
17. Wilt, C., Botea, A.: Spatially distributed multiagent path planning. In: Twenty-Fourth International Conference on Automated Planning and Scheduling (2014)
18. Liu, S., Sun, D., Zhu, C.: A dynamic priority based path planning for cooperation of multiple mobile robots in formation forming. Robot. Comput. Integr. Manuf. **30**(6), 589–596 (2014)
19. Yu, W., Peng, J., Zhang, X.: A prioritized path planning algorithm for MMRS. In: 33rd Chinese Control Conference (CCC 2014), pp. 966–971 (2014)
20. Wagner, G., Choset, H.: Subdimensional expansion for multirobot path planning. Artif. Intell. **219**, 1–24 (2015)
21. De Weerdt, M.M., Clement, B.: Introduction to planning in multiagent systems. Multiagent Grid Syst. **5**(4) (2009) (preprint)
22. Wang, K.-H.C., Botea, A.: Mapp: a scalable multi-agent path planning algorithm with tractability and completeness guarantees. J. Artif. Intell. Res. **42**, 55–90 (2011)
23. Lynch, N.A.: Distributed Algorithms. Morgan Kaufmann, San Francisco (1996)
24. Hoffmann, J., Nebel, B.: The FF planning system: fast plan generation through heuristic search. J. Artif. Intell. Res. **14**, 253–302 (2001)

Graph-Based Collaborative Filtering Using Rating Nodes: A Solution to the High Ratings/Low Ratings Problem

Alphan Culha[✉] and Andrew Skabar

Department of Computer Science and Information Technology, La Trobe University,
Melbourne, Australia
{a.culha,a.skabar}@latrobe.edu.au

Abstract. Graph-based random walk models have recently become a popular approach to collaborative filtering recommendation systems. Under the conventional graph-based approach, a user node and item node are connected if the user has rated the item, and the value of the rating is represented as the weight of the connection. Commencing from some target user, a random walk is performed on the graph, and the results used to perform useful tasks such as ranking items in order of their importance to the user. Because random walk favors large-weighted connections, walk is more likely to proceed through two users that share a high rating for some item, than through users who share a low rating. This is a problem because there are similarity relations implicit in the data that are not being captured under this representation. We refer to this as the 'High Ratings/Low Ratings' problem. This paper proposes a novel graph representation scheme in which item ratings are represented using multiple nodes, allowing flow of information through both low-rating and high-rating connections. Empirical results on the MovieLens dataset show that recommendation rankings made using the proposed scheme are much better correlated with results in the test ratings, and that under a top-k evaluation, there is an improvement of up to 15 % in precision and recall. An attractive feature of the approach is that it also associates a confidence value with a recommendation.

1 Introduction

Recommender Systems are defined as software tools and techniques that provide suggestions for items to be of use to a target user [14]. Although recommender systems have been an important research area since the mid-1990s, interest in this area has increased dramatically as a result of the vast amounts of online data, information, and options that are now available to users [1, 6]. Recommender systems are now a vital component of many online services and retail e-commerce sites including Amazon (general products), Netflix (movies), YouTube (videos) and Pandora (music).

Collaborative filtering has been the most popular and successful approach in building recommender systems [12, 17]. In contrast to content-based filtering, in which the list of recommendations is based on the similarity of an item to items the user has previously purchased, collaborative filtering bases the recommendations on the item similarities as well as the purchases of users who have similar tastes and preferences; for example, if a target user had bought books on recommendation systems in the past, then recommend

© Springer International Publishing Switzerland 2015
B. Pfahringer and J. Renz (Eds.): AI 2015, LNAI 9457, pp. 136–148, 2015.
DOI: 10.1007/978-3-319-26350-2_12

books that are preferred by other users who have also bought books on recommendation systems. One of the main problems suffered by collaborative filtering approaches is the sparsity problem; i.e., because the number of items is so large, the average user will only have rated an extremely small proportion of these items, meaning that even the most popular items will have very few ratings.

Graph-based random walk models have recently become a popular approach to collaborative filtering [2, 4, 12, 16, 17]. These models represent users and items as the nodes of a graph. A user node and item node are connected if the user has rated the item, purchased the item, or displayed interest in the item in some other way. In the case of ratings data (which we focus on in this paper), the value of the rating is usually represented as the weight of the edge connecting the user and item. Commencing from the node for some target user, a random walk is performed on the graph, and the results are then used to perform useful tasks such as ranking items in order of their importance to the user. Representing the relationships between users and items as a graph helps overcome the data sparsity problems inherent in more traditional approaches to collaborative filtering. Graph-based models also offer the advantage of being easily able to incorporate social network data, as recently demonstrated in [2, 16].

A problem with the graph-based representation described above, which we will refer to as the 'weighted-edge representation', is that it is unable to fully capture the similarity relations that are implicit in the data. The premise of collaborative filtering is to make item recommendations based on user similarities; however, because random walk favors large-weighted connections, walk is more likely to proceed through two users that share a high rating for some item than through users who share a low rating. That is, two users who are similar by virtue of rating the same items highly will be more influential in the construction of the recommendation list than two users who have rated the same items poorly. We refer to this as the 'High Ratings/Low Ratings' problem. The problem arises as a result of representing ratings using a single weight.

In this paper we introduce a novel scheme for representing recommendation data as a graph. Unlike the weighted-edge approach, in which a user and an item are each represented by a single node, under the proposed scheme item ratings are represented by multiple nodes. This enables the explicit representation of low ratings, thus facilitating the flow of information through both low-rating and high-rating connections. Empirical results on the MovieLens dataset show that recommendations made using the proposed scheme are much better correlated with results in the test ratings set, and that under a top-k evaluation, there is an improvement of up to 15 % in precision and recall. An attractive feature of the approach is that it also associates a confidence value with a recommendation.

The remainder of the paper is structured as follows. Section 2 describes the weighted-edge graph representation, and how random walk commencing from some target user can be used to produce a ranking of nodes according to their importance to that user. The section also provides a simple example demonstrating the inability of the weighted-edge representation to fully capture and utilize similarities based on low rankings. Section 3 presents the proposed approach, which we refer to as the 'rating-nodes representation'. Section 4 presents results comparing the performance of the two approaches on the well-known MovieLens dataset. Section 5 provides further discussion of the rating-nodes approach, and concludes the paper.

2 Graph-Based Recommender Models

We define a weighted-edge recommendation graph as an undirected graph $G = \{V, E\}$, in which the set of nodes V is the union of a set of user nodes U and a set of item nodes I. A user node is connected to an item node if the user has rated the item; i.e., for some user $u \in U$ and item $i \in I$, edge $(u, i) \in E$ only if $r_{ui} \neq 0$. Associated with the edge will be a weight w_{ui}, based on some normalization of r_{ui}. We assume for simplicity that the recommendation graph is bipartite, meaning that edges may exist only between user nodes and item nodes, but not between nodes of the same type; however, this assumption does not limit the generality of the approach, which can easily be extended to deal with connections between nodes of the same type, as well as the introduction of other node types (e.g., nodes representing social network data, as in [16]).

The Markov Chain describing the sequence of nodes visited by a random walker is called a 'random walk'. The concept of random walk is important, because the amount of time a random walker spends visiting some node provides a measure of the relative importance of that node. The probability of moving from node i to node j is calculated by dividing the edge weight w_{ij} by the sum of the weights of all of i's outgoing edges:

$$p_{ij} = \frac{w_{ij}}{\sum_{k \in N(i)} w_{ik}} \tag{1}$$

where $N(i)$ denotes the nodes that are the immediate neighbors of i. The larger the value of an outgoing weight, the larger is the probability of traversing that edge as opposed to the other outgoing edges.

The relative importance of nodes can be determined by finding the stationary distribution. For a graph with adjacency matrix $W = \{w_{ij}\}$, the stationary distribution can be found by solving the eigenvector equation $x = Wx$, in which case the dominant eigenvector x will represent the stationary distribution. Page-Rank [3] is a variation of this in which at each step the walker is teleported with probability $(1 - d)$ to a random node rather than following an edge (typically $d \cong 0.85$). Personalized PageRank [8] is a further variation in which the user is teleported not to a random node, but to the node corresponding to that user. The eigenvector equation in this case is

$$x = dWx + (1 - d)\theta \tag{2}$$

where θ is a vector in which $\theta = 1$ for the node corresponding to the target user, and 0 for all others. The components of x are the PageRank values. Each node will have a PageRank value.

In order to find the dominant eigenvector x, the above eigenvector equation needs to be solved. A general and robust approach is power iteration, which begins with a random vector x_k, and iterates the step $x_{k+1} = dWx_k + (1 - d)\theta$ until convergence, when x will be the dominant eigenvector [13]. Note that the dominant eigenvector will not be unique, since any linear scaling of this eigenvector will also satisfy the eigenvector equation. Therefore it is the relative, not absolute, scores which are important. It is common to normalize the eigenvector such that its components sum to one.

The PageRank values obtained by performing personalized PageRank for some target user provide a measure of the relative importance of those nodes to the user. A recommendation list for that user can be constructed by simply sorting the user's unrated items in decreasing order of PageRank.

2.1 The High Ratings/Low Ratings Problem

When a random walk is applied to a weighted-edge graph as described above, the PageRank value of an item node provides a relative measure of the strength of the recommendation of that item to the target user. The PageRank for an item ultimately depends on how similar users rated the item, and we now demonstrate through a simple example that the weighted-edge representation is deficient in fully capturing the similarity relations that may be present in ratings data.

Consider the graph shown in Fig. 1, which represents recommendation data for four users and three items. The edge weightings represent the ratings, and can take an integer value from 1 to 5, with a 1 representing a low rating and a 5 representing a high rating. The absence of an edge indicates that the user has not rated the item. Table 1 is the adjacency matrix for the graph, where a value of 0 indicates no edge (i.e., no rating).

Table 1. Adjacency matrix for Fig. 1 graph

		USERS				ITEMS		
		U1	U2	U3	U4	I1	I2	I3
USERS	U1	0	0	0	0	5	5	0
	U2	0	0	0	0	5	0	0
	U3	0	0	0	0	1	0	5
	U4	0	0	0	0	1	0	0
ITEMS	I1	5	5	1	1	0	0	0
	I2	5	0	0	0	0	0	0
	I3	5	0	0	0	0	0	0

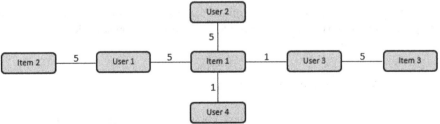

Fig. 1. Sample recommendation graph for 4 users and 3 items.

User 1 and User 2 gave a rating of 5 to Item 1, while User 3 and User 4 gave a rating of 1. Users 1 and 2 liked the item as much as each other, and Users 3 and 4 (dis)liked it as much as each other. In addition to this, User 1 gave a rating of 5 to Item 2, while User 3 gave a rating of 5 to Item 3. We want to find out how User 2 would have rated Item 2 and how User 4 would have rated Item 3. That is, we need Item 2's PageRank for target User 2, and Item 3's PageRank for User 4.

Table 2 shows the PageRank values for the three items for separate random walks commencing from each user. Note that the PageRank values in each row do not sum to unity. This is because some of the PageRank will be assigned to user nodes. The percentages indicate the proportion of *item* PageRank assigned to each item node. The PageRank of Item 2 for target User 2 is 0.0795 (17 %), while the PageRank of Item 3 for User 4 is significantly lower than this at 0.0425 (9 %). This is despite the fact that, on the basis of the ratings data available, User 1 and User 2 are as similar to each other (i.e., one rating in common) as User 3 and User 4 are to each other (also one rating in common). Also problematic is the fact that for User 4, Item 2 actually receives a higher PageRank than Item 3, which is also at odds with what the similarities in the ratings data suggest.

Table 2. PageRank values for Fig. 1 graph

	I1	I2	I3
U1	0.264 (58%)	0.162 (35%)	0.033 (7%)
U2	0.337 (74%)	**0.079 (17%)**	0.042 (9%)
U3	0.141 (31%)	0.033 (7%)	0.284 (62%)
U4	0.337 (74%)	0.079 (17%)	**0.042 (9%)**

We refer to this problem as the 'High ratings/Low ratings' problem. The problem occurs because the weighted-edge representation causes random walk to favor high-weighted edges. Item 1 has four outgoing links: edges of weight 1 to each of Users 3 and 4; and edges of weight 5 to Users 1 and 2. Any walk passing through Item 1, will be more likely to proceed to Item 2 (via User 1) than to Item 3 (via User 3), thus explaining the observations from Table 2. Under the weighted-edge representation, two users who are similar by virtue of rating the same items highly will be more influential in the construction of the recommendation list than two users who have rated the same items poorly. In the next section we provide an alternative representation scheme that avoids these problems.

3   Rating Node Representation

In order to overcome the High/Low Ratings problem, we propose in this section a novel graph representation scheme which we refer to as the 'rating-nodes representation'.

Whereas in the weighted-edge representation each item is represented by a single node, under the rating-nodes representation each item is represented by multiple nodes, which we refer to as 'rating nodes'. That is, $G = \{V, E\}$ is an undirected graph in which $V := U \cup I_R$, where U is the set of users, and $I_R := I_1 \cup I_2 \cup \ldots \cup I_N$ is the set of item rating nodes. Corresponding to each item there are N rating nodes, not just one. This allows the representation of an item rating to be distributed over a collection of rating nodes, and it is this distributed representation that allows a fuller representation of the similarity relationships implicit in the ratings dataset.

There is some scope in how an item rating may be represented using such a collection of nodes. In the following, each of the rating nodes for an item corresponds to one of the possible ratings that can be given to an item. For example, in the MovieLens dataset, ratings can take an integer value from 1 to 5; thus ratings will be represented using five rating nodes, each corresponding to each of the possible rating values. If a user $u \in U$ gave movie i a rating of n (where n is a number between 1 and 5), then there will be an edge (of weight 1) between user node u and item rating node i_n, but no connection (i.e., weight 0) between u and the other rating nodes for item i. Suppose that a user has given an item a rating of 5, and that another user has given the same item a rating of 1. Figure 2a shows the weighted-edge representation, and Fig. 2b shows the rating nodes representation.

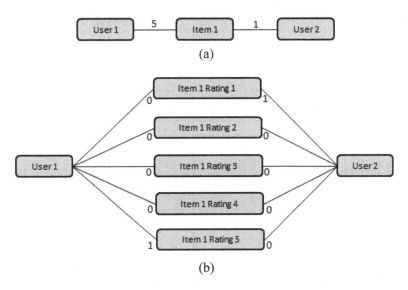

Fig. 2. Recommendation graph for 2 users and 1 item (a) weighted-edge representation; (b) ranking-node representation.

The rating–nodes representation does not change the nature of the random walk, and personalized PageRank can be applied exactly as described in Sect. 2. At convergence, the PageRank for a particular item will be distributed over the rating nodes for that item. Prior to describing how these PageRank values can be collapsed to provide a single recommendation value, we return to the example from Sect. 2.1 to demonstrate how the rating-nodes representation overcomes the High Ratings/Low Ratings problem.

3.1 High Ratings/Low Ratings Revisited

Under the ranking-nodes representation, the rating information in Table 1 would be represented as per the adjacency matrix shown in Table 3. Running personalized Page-Rank from each of the four users, results in the PageRank values shown in Table 4. There are two things to note. Firstly, the PageRank of Item 2 for target User 2 (0.101) is now the same as the PageRank value of Item 3 for target User 4. Secondly, the PageRank value of Item 3 for User 4 (0.101) is now equal to the PageRank value of Item 2 for User 2. The rating-nodes representation has solved the High Ranking/Low Ranking problem: similarity is equally propagated through both low and high ratings.

Table 3. Adjacency matrix for Fig. 1 graph (ranking-nodes representation)

			USER				ITEM														
							I1					I2					I3				
			U1	U2	U3	U4	R1	R2	R3	R4	R5	R1	R2	R3	R4	R5	R1	R2	R3	R4	R5
USER		U1									1					1					
		U2									1										
		U3					1														1
		U4					1														
ITEM	I1	R1			1	1															
		R2																			
		R3																			
		R4																			
		R5	1	1																	
	I2	R1																			
		R2																			
		R3																			
		R4																			
		R5	1																		
	I3	R1																			
		R2																			
		R3																			
		R4																			
		R5			1																

Table 4. PageRank values for Fig. 1 graph (ranking-nodes representation)

		ITEMS														
		Item 1					Item 2					Item 3				
		R1	R2	R3	R4	R5	R1	R2	R3	R4	R5	R1	R2	R3	R4	R5
USERS	U1	0	0	0	0	0.280	0	0	0	0	0.179	0	0	0	0	0
	U2	0	0	0	0	0.358	0	0	0	0	0.101	0	0	0	0	0
	U3	0.280	0	0	0	0	0	0	0	0	0	0	0	0	0	0.179
	U4	0.358	0	0	0	0	0	0	0	0	0	0	0	0	0	0.101

3.2 Recommendation Value

The recommendation value rec_{ui} of an item i for some target user u is simply the PageRank-weighted sum of the rating values represented by the rating nodes; i.e.

$$rec_{ui} = \sum_{j=1}^{n} x_i^j r_j \Big/ \sum_{j=1}^{n} x_i^j \qquad (3)$$

where x_i^j is the PageRank score for the jth rating node for item i and r_j is the rating value represented by the rating node (1 through to 5 for the MovieLens dataset). While rec_{ui} will always take a real value on the interval [1 5], this value should not be interpreted as being on the same scale as the ratings in the dataset, since by the nature of the averaging process the rec_{ui} values will rarely take extreme values of 1 or 5. Note also that the same value for rec_{ui} may result from more than one combination of values for the x_i^j. For example, multiplying each of the x_i^j by a factor of 2 will not affect the value of rec_{ui}. The total PageRank associated with an item i (i.e., the denominator in the above equation) can be interpreted as the strength of the recommendation. That is, some rec_{ui} value may be predicted strongly or weakly.

3.3 Additional Comments

One potential issue that can be identified from Table 4 concerns the disconnection between nodes. For example, random walk starting at User 3 now gives a PageRank of 0 for Item 2. This might be considered reasonable on the grounds that there is no longer a path from User 3 to Item 2 (because Users 3 and 1 disagree on their rating of Item 1, which has a pivotal position in this graph). However the fact that Users 1 and 3 even reviewed the same movie (despite disagreeing about their ratings of it) suggests that there should perhaps be some connection between the two users. This issue of disconnection is not a problem under the weighted-edge representation since any item that was reviewed by the same two reviewers will result in weights of at least 1 connecting the two users with that item node. This problem can be solved in the rating-nodes approach by inserting edges, weighted with some small positive value α (e.g., 0.05), between user nodes and item rating nodes corresponding to recommendations not made by the user. However we will see from the results in Sect. 4 that in the case of large graphs, disconnection is not an issue, as there are many other pathways through which random walk can proceed. While the rating-nodes representation that we have described in this section has assumed that nodes are of two types—user nodes and item nodes—the representation can easily be extended to incorporate social attributes. In this paper we are primarily interested in determining whether the rating-nodes representation leads to improved performance, and we leave the investigation of incorporating social attributes to future work.

4 Experiments

This section presents empirical results comparing the performance of the rating-nodes and weighted-edge representations on the well-known MovieLens dataset (www.movielens.umn.edu) [5]. The version of the dataset used in this research contains 943 users,

1682 movies and 100,000 ratings in the scale of 1 to 5. The data is organized into five 80 %-20 % training/test splits, for use in five-fold cross-validation. The dataset has been collected on the MovieLens web site and made available by GroupLens Research and has been used in [6, 10, 11, 15, 16].

4.1 Correlation-Based Evaluation

The recommendation value rec_{ui} defined in Sect. 3.2 provides a measure which can be used to produce a ranked item recommendation list for the target user. This suggests an evaluation based on correlation; i.e., how well does the ranking of items in the recommendation list correlate with the ranking of items by their star-rating value in the test set?

This correlation can be measured using Spearman's rank correlation coefficient, defined as

$$\rho = \frac{\sum_i \left(x_i - \bar{x}\right)\left(y_i - \bar{y}\right)}{\sum_i \left(x_i - \bar{x}\right)^2 \sum_i \left(y_i - \bar{y}\right)^2} \tag{4}$$

where x_i and y_i are ranks of the scores represented by variables X_i and Y_i. In our case, one of these variables corresponds to the recommendation values; the other to the star-ratings in the test set. Since the star-ratings in the test set can only take an integer value in the set $\{1, 2, ..., 5\}$, the test ratings will involve many ties. Ties are assigned a rank equal to the average of their position in the descending order of values. For example, assuming that six movies are sorted according to their star-ratings of 5, 4, 4, 3, 2, 1 (second and third movies tied with a rating of 4), the rankings given to the movies would be 1, 2.5, 2.5, 4, 5, 6 (The second and third movie both receive a rating of 2.5 = (2 + 3)/2).

Table 5 shows the Spearman rank correlations between test rankings and recommendation list for the weighted-edge and rating-nodes representations. For the latter we used both $\alpha = 0$ and $\alpha = 0.1$. The correlations were obtained by averaging the correlations across all users and all 5 train/test splits. The recommendation values resulting from the rating-nodes approach are clearly much more highly correlated with the test set ratings than are the recommendation values from the weighted-edge approach.

Table 5. Spearman rank correlations between test ratings and recommendation values

Graph representation	Correlation
Weighted-edge representation	0.221
Rating-nodes representation ($\alpha = 0$)	0.389
Rating-nodes representation ($\alpha = 0.1$)	0.381

To explore this further, Fig. 3 shows scatterplots of recommendation rankings (vertical axis) versus test set rankings (horizontal axis) for a typical user (User 1, Test Set 1). The test set for this user consisted of 137 movie reviews. Higher-rated movies appear to the left of the plot; lower-rated movies appear to the right. Thus the first vertical bar corresponds to movies with a 5-star rating; the second to movies with a 4-star rating, etc. The correlation coefficient for the weighted-edge representation is 0.331, and that for the rating-nodes representation is 0.578. This difference in correlation can be clearly discerned through visual inspection, and is particularly apparent for both higher-rated and lower-rated movies, and to a lesser degree for middle-rated movies.

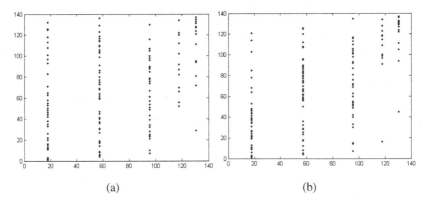

(a)   (b)

Fig. 3. Recommendation rankings (vert.) versus test rankings (horz.) for User 1, Test_Set 1. (a) weighted-node representation ($\rho = 0.331$); (b) rating-nodes representation ($\rho = 0.578$).

4.2   Top-K Evaluation

A common approach to evaluating recommendations on the MovieLens dataset is Top-k evaluation (used, for example in [7, 9, 16]). Under this style of evaluation, a recommendation set is formed by selecting the k most highly recommended items, and a test set is formed by selecting the most highly rated items from some test data. Measures such as Precision and Recall are then calculated based on these lists. It is not surprising that graph-based approaches such as the weighted-edge approach typically perform well under this type of evaluation, since these type of recommendation systems are designed to predict higher-rated items. Given the correlation results above, it is interesting to compare the performance of the rating-nodes approach with the weighted-edge approach under this style of evaluation.

To evaluate the predictions for some user u, we select the k most highly recommended movies from the sorted recommendation list, and place these in the Recommendation Set (*recset*). Following Shang *et al.* [16], we use only the test set movies with a 5-star rating. We place all of the movies to which user u has given a 5-star rating in the Test Set (*testset*). Any movie which appears in both the Test Set and the Recommendation Set is a *hit*. Recall, Precision and F1-measure are then defined as follows:

$$Recall = \frac{|recset \cap testset|}{|testset|}$$

$$Precision = \frac{|recset \cap testset|}{|recset|}$$

$$F1 - measure = \frac{2 \cdot Precision \cdot Recall}{Precision + Recall}$$

Figure 4 shows the Precision, Recall, and F1 curves corresponding to the weighted-edge and rating-nodes representations for $k = 5, 10, \ldots, 50$. (For the rating-nodes representation, curves for $\alpha = 0$ and $\alpha = 0.1$ were identical). The values are averaged across all users and all 5 training/test splits. As can be seen from the curves, the Precision and Recall for low values of k are approximately 15 % higher for the rating-nodes representation than they are for the weighted-edge representation. As the value of k increases, the difference in performance becomes less significant.

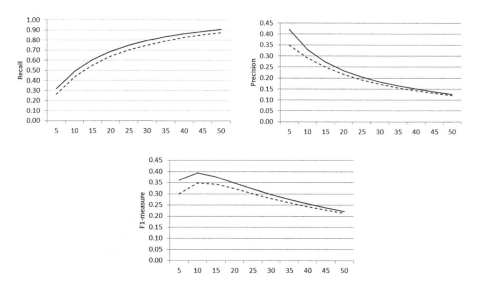

Fig. 4. Recall, Precision and F1-measure for Top-k analysis.

5   Conclusion

Graph-based approaches provide a powerful means of combining collaborative and content-based filtering. Users are similar if they like the same items, and items are similar if they are liked by the same users. This provides a powerful mechanism by which similarity relations can be propagated through such networks. In Sect. 2.1 we demonstrated that while the weighted-edge representation is able to capture similarities between users who share high-ratings for products, it is not able to adequately capture relationships between users who share a low-rating for some item, and we referred to this as the high-ratings/low-ratings problem. The rating-nodes representation proposed in Sect. 3 was designed to overcome this problem.

Results on the MovieLens dataset have shown that recommendation rankings made using the rating-nodes representation are much better correlated with results in the test ratings than are the rankings produced using the weighted-edge approach. However, what is surprising about this result is that the correlation improved so dramatically over the full range of ratings; i.e., for high ratings as well as low ratings. While we expected a significant increase in correlation for low ratings, there is less room for improvement in the case of high rankings since the weighted-edge approach is designed to predict high ratings. The improved performance in predicting high ratings was corroborated by the results of the top-k analysis, in which the rating-nodes approach displayed a clear improvement in precision and recall over the weighted-edge approach.

The graphs we used in this paper were bipartite; i.e., edges exist only between user nodes and item nodes, but not between nodes of the same type. However, both the weighted-edge and rating-nodes representations are general, and can easily represent user-user links, as well as item-item links. There has been considerable recent interest in incorporating social network data into graph-based recommender systems. For example Bogers [2] and Shang *et al.* [16] have recently incorporated social network data into graph-based recommender systems using the weighted-edge representation. With the increasing amount and variety of social networking data that is becoming available, this area of research will no doubt continue to grow, and we believe that the distributed, multiple-node representation scheme that we have proposed in this paper will provide a powerful method of incorporating such data, thus allowing a richer representation of similarities than is possible through weighted-edge representations.

References

1. Adomavicius, G., Tuzhilin, A.: Toward the next generation of recommender systems: a survey of the state-of-the-art and possible extensions. IEEE Trans. Knowl. Data Eng. **17**(6), 734–749 (2005)
2. Bogers, T.: Movie recommendation using random walks over the contextual graph. In: Proceedings of the 2nd International Workshop on Context-Aware Recommender Systems (2010)
3. Brin, S., Page, L.: The anatomy of a large-scale hypertextual Web search engine. Comput. Netw. ISDN Syst. **30**(1), 107–117 (1998)
4. Fouss, F., Pirotte, A., Renders, J.M.: Random-walk computation of similarities between nodes of a graph with application to collaborative recommendation. IEEE Trans. Knowl. Data Eng. **19**(3), 355–369 (2007)
5. Gabrielsson, S., Gabrielsson, S.: The Use of self-organizing maps in recommender systems. Master's thesis, Department of Information Technology at the Division of Computer Systems, Uppsala University (2006)
6. Herlocker, J.L., Konstan, J.A.: Content-independent task-focused recommendation. IEEE Internet Comput. **5**(6), 40–47 (2001)
7. Herlocker, J.L., Konstan, J.A., Terveen, L.G.: Evaluating collaborative filtering recommender systems. ACM Trans. Inf. Syst. (TOIS) **22**(1), 5–53 (2004)

8. Jeh, G., Widom, J.: Scaling personalized web search. In: Proceedings of the 12th Wold Wide Web Conference (WWW), pp. 271–279 (2003)

9. Karypis, G.: Evaluation of item-based top-n recommendation algorithms. In: Proceedings of the Tenth International Conference on Information and Knowledge Management, pp. 247–254. ACM (2001)

10. Massa, P., Avesani, P.: Trust-aware recommender systems. In: Proceedings of the 2007 ACM Conference on Recommender Systems, pp. 17–24. ACM (2007)

11. Miller, B.N., Albert, I., Lam, S.K., Konstan, J.A., Riedl, J.: Movielens unplugged: Experiences with a recommender system on four mobile devices. In: Proceedings of the 2003 Conference on Intelligent User Interfaces (2003)

12. Mohsen, J., Ester, M.: TrustWalker: a random walk model for combining trust-based and item-based recommendation. In: Proceedings of the 15th ACM SIGKDD International Conference on Knowledge Discovery and Data Mining, pp. 397–406. ACM (2009)

13. Newman, M.E.J.: Networks: An Introduction. Oxford University Press, Oxford (2010)

14. Ricci, F., Rokach, L., Shapira, B., Kantor, P.: Recommender Systems Handbook. Springer, New York (2011)

15. Sarwar, B., Karypis, K., Konstan, J., Riedl, J.: Item-based collaborative filtering recommendation algorithms. In: Proceedings of the 10th International Conference on World Wide Web, pp. 285–295. ACM (2001)

16. Shang, S., Kulkarni, S.R., Cuff, P.W., Hui, P.: A random walk based model incorporating social information for recommendation. International IEEE Workshop Machine Learning for Signal Processing, pp. 1–6 (2012)

17. Yildirim, H., Krishnamoorthy, M.S.: A random walk method for alleviating the sparsity problem in collaborative filtering. In: Proceedings of the 2008 ACM Conference on Recommender Systems, pp. 131–138. ACM (2008)

A Comparative Study on Vector Similarity Methods for Offer Generation in Multi-attribute Negotiation

Aodah Diamah[1(✉)], Michael Wagner[1,2], and Menkes van den Briel[2,3]

[1] Faculty of ESTeM, University of Canberra, Canberra, Australia
{aodah.diamah,michael.wagner}@canberra.edu.au
[2] College of Engineering and Computer Science,
Australian National University, Canberra, Australia
menkes@nicta.com.au
[3] National ICT Australia, Eveleigh, Australia

Abstract. Offer generation is an important mechanism in automated negotiation, in which a negotiating agent needs to select bids close to the opponent preference to increase their chance of being accepted. The existing offer generation approaches are either random, require partial knowledge of opponent preference or are domain-dependent. In this paper, we investigate and compare two vector similarity functions for generating offer vectors close to opponent preference. Vector similarities are not domain-specific, do not require different similarity functions for each negotiation domain and can be computed in incomplete-information negotiation. We evaluate negotiation outcomes by the joint gain obtained by the agents and by their closeness to Pareto-optimal solutions.

Keywords: Multi-attribute negotiation · Offer generation · Vector similarity · Cosine distance · Euclidean distance · Pareto-optimal solutions

1 Introduction

Automated multi-attribute negotiations have been studied extensively in recent years because of their potential applications for e-commerce. One of the research foci in this field is the negotiation strategy to achieve high benefit for the negotiating parties. The negotiation strategy can be decomposed into three key components: bidding (which includes conceding and offer generation), opponent modeling, and acceptance conditions [1, 2]. Much of the research has been applied to opponent modeling and conceding, and less attention has been placed on offer generation. Many negotiating agents choose their offers randomly when they are indifferent to a set of iso-utility offers [3–5]. Faratin et al. in [6] introduced a fuzzy similarity function to choose an offer similar to the opponent's last offer from a set of iso-utility offers, since that increases the chance of the offer being accepted by the opponent. That approach, however, requires a similarity function that is defined for every issue/attribute being negotiated, which makes the approach domain dependent. Moreover, to perform well, that similarity function requires information on the weights placed by the opponent on each negotiation issue. To address this, we investigate the use of Euclidean and cosine vector

© Springer International Publishing Switzerland 2015
B. Pfahringer and J. Renz (Eds.): AI 2015, LNAI 9457, pp. 149–156, 2015.
DOI: 10.1007/978-3-319-26350-2_13

similarities to choose a bid similar to the opponent's last offer from a set of iso-utility offers. Cosine similarity has been used for text and image information retrieval [7], face verification [8] and computer virus detection [9]. However, it has not been used before for the offer generation mechanism in multi attribute negotiation. Vector similarity eliminates the need to define a similarity function for each attribute, and no information on the opponent's weights is needed[1]. Because the attributes of a negotiation rarely fall in the same scale (e.g. price may be $1000–5000, while delivery time may be 3–10 days), we are also interested to see if attribute scaling has any effect on the similarity computation and thus on the negotiation outcome.

The structure of the paper is as follows: First we address domain-dependent similarity modeling by using vector similarities. Second, we compare how two vector similarities (Euclidean and cosine) perform on (1) unscaled and (2) normalized attributes. This comparison will allow agents to decide whether they should normalize attributes before performing similarity computation.

2 Negotiation Settings

We define the negotiation model between two agents, buyer A and seller B, as follows.

Utility function: In evaluating incoming offers, agents need to calculate how closely the offer matches their own preference. Utility functions have been used to score offers based on negotiator preferences [10]. In multi-attribute negotiation, each attribute value is assigned a score by an evaluation function that falls into a normalized interval of [0,1]. Each attribute has a weight attached to it, representing the importance the agent places on that attribute. In the literature, the common assumption is that the utilities of the attributes are positive and mutually independent [10, 11]. For n attributes, the sum of the weights $\sum_{i=1}^{n} w^i$ must be one.

The total utility score of an offer vector $x = (x_1 \cdots x_n)$ is:

$$V_m(x) = \sum_{i=1}^{n} w_m^i u_m^i(x_i) \tag{1}$$

where u_m^i is the utility of attribute i for agent m, and V_m is the total utility for agent m.

Concession strategy: We assume that agents propose offers with high utility at the beginning of the negotiation and lower their utility (i.e. concede) as the negotiation proceeds. We use an alternating-offer protocol, in which agents in each round make offers in turn until either an agreement is reached or the maximum number of rounds. We use the monotonic, time-dependent concession strategy from [12] to determine the target utility in each round:

$$V_m^k = 1 + \left(V_m^{min} - 1\right) \left(\frac{k}{k_{max}}\right)^{\alpha} \tag{2}$$

---

[1] In this paper, agents do not reveal their utility function to their opponent.

where k is the current round, k_{max} is the maximum round, V_m^{min} is the minimum acceptable utility for agent m and α is the rate at which agent m is conceding. A larger value of α represents a slower conceding strategy.

3 Vector Similarities for Offer Generation

Suppose agent A receives agent B's offer x_B^k in round k and that offer does not meet agent A's acceptable minimum utility. In generating a counter offer, agent A has to satisfy two objectives: (1) an acceptable utility for itself and (2) an acceptable utility for its opponent. Objective 1 is relatively easy to achieve. The agent calculates its target utility for a given round using Eq. (2) and any bid with that utility can be offered. The set of those iso-utility offers was defined by [6] as

$$iso_A(\theta) = \{x | V_A(x) = \theta\} \tag{3}$$

where θ is the acceptable utility for agent A and $V_A(x)$ is agent A's total utility function. As any two offers $x, y \in iso_A(\theta)$ may have different utilities to agent B, i.e. $V_A(x) = V_A(y)$, but $V_B(x) \neq V_B(y)$, agent A needs to find an offer from $iso_A(\theta)$ with maximum utility for agent B to increase the chance that objective (2) is met and an agreement can be reached. Because the agent does not know the opponent's utility function, it can approximate the opponent's partial preference from the opponent's last offer and use that to find the offer $x \in iso_A(\theta)$ that is maximally similar to the opponent's last offer. This offer generating mechanism is expressed as:

$$x_A^{k+1} = \text{argmax}_{x \in iso_A(\theta)} \left\{ sim\left(x, x_B^{best}\right) \right\} \tag{4}$$

where x_A^{k+1} is the offer agent A is about to propose at round $k + 1$, x_B^k is the opponent's last offer, and $S(x, x_B^k)$ is the similarity of x and x_B^k.

Similarity between two offers is calculated using Euclidean and cosine distances $S_E(x, y) = 1/x - y$ and $S_C(x, y) = x \cdot y/(|x||y|)$, respectively.

4 Experiment and Results

4.1 Negotiation Domain

We evaluate cosine and Euclidean similarity with two negotiation scenarios taken from [13] and [14] for ease of comparison with the results in those papers. The domain for each attribute and the corresponding evaluation function for negotiation settings 1 and 2 are shown in Table 1. Each agent knows its own utility function, but not its opponent's utility function. However the domain knowledge is common to both agents.

There are many possible different weight combinations that an agent may attach to the negotiation attributes and it is not feasible to try them all. However, to allow us to see how cosine and Euclidean similarity perform under different possible preferences, we define several weight sets for agents A and B, as seen from Tables 2 and 3.

Table 1. Domain and utility functions for negotiation setting 1 and negotiation setting 2

Domain	Negotiation setting 1			Negotiation setting 2			
	Attribute 1	Attribute 2	Attribute 3	Attribute 1	Attribute 2	Attribute 3	Attribute 4
	[5000, 10,000]	[30,90]	[0,3]	[0,5]	[0,4]	[30,70]	[5,15]
u_A^i	$\frac{(x_1-5000)}{5000}$	$\frac{(90-x_2)}{60}$	$\frac{x_3}{3}$	$\frac{x_1}{5}$	$\frac{x_2}{4}$	$\frac{70-x_3}{40}$	$\frac{15-x_4}{10}$
u_B^i	$\frac{(10000-x_1)}{5000}$	$\frac{(x_2-30)}{60}$	$\frac{(3-x_3)}{3}$	$\frac{(5-x_1)}{5}$	$\frac{(4-x_1)}{4}$	$\frac{x_3-30}{40}$	$\frac{x_4-5}{10}$

Table 2. Weights for negotiation setting 1

Set	w_A	w_B
1	0.2, 0.4, 0.4	0.1,5 0.6, 0.25
2	0.15, 0.6, 0.25	0.2, 0.4, 0.4
3	0.3, 0.55, 0.15	0.5, 0.2, 0.3
4	0.5, 0.2, 0.3	0.3, 0.55, 0.15
5	0.5, 0.3, 0.2	0.15, 0.45, 0.4
6	0.15, 0.45, 0.4	0.5, 0.3, 0.2
7	0.8, 0.1, 0.1	0.33, 0.33, 0.33
8	0.33, 0.33, 0.33	0.8, 0.1, 0.1
9	0.3, 0.1, 0.6	0.4, 0.5, 0.1
10	0.4, 0.5, 0.1	0.3, 0.1, 0.6

Table 3. Weights for negotiation setting 2

Set	w_A	w_B
1	0.35, 0.15, 0.45, 0.05	0.1, 0.15, 0.4, 0.35
2	0.1, 0.15, 0.4, 0.35	0.35, 0.15, 0.45, 0.05
3	0.7,0.1, 0.1, 0.1	0.25,0.25, 0.25, 0.25
4	0.25, 0.25, 0.25, 0.25	0.7, 0.1, 0.1, 0.1
5	0.4, 0.05, 0.1, 0.45	0.1, 0.45, 0.4, 0.05
6	0.1, 0.45, 0.4, 0.05	0.4, 0.05, 0.1, 0.45
7	0.15, 0.15, 0.3, 0.4	0.4, 0.4, 0.1, 0.1
8	0.4, 0.4, 0.1, 0.1	0.15, 0.15, 0.3, 0.4
9	0.05, 0.6, 0.3, 0.05	0.1, 0.3, 0.4, 0.2
10	0.1, 0.3, 0.4, 0.2	0.05, 0.6, 0.3, 0.05

The Pareto front will be further away from the origin if agents have very different weight rankings, i.e. low opposition, while identical weight rankings mean high opposition, thus increasing the possibility of a high-gain agreement for both agents.

The attributes in the two negotiation settings are of different scales. It is likely that attributes with a larger scale dominate the similarity calculation (e.g. attribute 1 and attribute 3 are dominating similarity calculation between offers in negotiation setting 1 and 2, respectively). We hypothesize that the effect of dominating attributes will be removed if attributes are normalized to the range [0,1], and that the agents can therefore improve their offers and thus the negotiation outcome.

For all negotiations, the reservation utility for both agents is set to be 0.5 so that a zone of agreement exists between them. The agents are assumed to be "tough" negotiators with a conceding rate of $\alpha = 1.3$. We are aware that in real life agents may negotiate differently, e.g., one agent may be hard to budge and the other generous, and thus have different conceding rates α. However, in this negotiation we deliberately posit agents of equal toughness to eliminate an imbalanced negotiation dyad affecting the negotiation outcome. The maximum number of rounds is varied from 10 to 15. For each maximum, we run the negotiations ten times (totaling 50 negotiations in each set) and calculate the average joint utility for all deals reached.

4.2   Negotiation Outcome Evaluation

For every deal x reached, the evaluation is carried out with the following metrics:

Joint utility reflects the total value of a negotiation agreement for the negotiating parties. It is the sum of the gains received by each agent and is calculated as

$$V_J(x) = V_A(x) + V_B(x) \tag{5}$$

Distance to Maximum Joint Utility on the Pareto Front. The utilities $V_A(x)$ and $V_B(x)$ each possible solution x have coordinates $(V_A(x), V_B(x))$ in the two-dimensional utility space. A solution is Pareto-optimal when utility improvement in one direction is impossible without lowering utility in the other direction. We find the maximum joint utility V_J^{max} of all Pareto-front solutions, which is the metric used to calculate the distance from a deal. The closer a deal to the maximum joint utility the higher is the gain for both agents. That distance for a vector x is

$$D_J^{max}(x) = V_J^{max} - V_J(x) \tag{6}$$

4.3   Results and Discussion

We compare the average joint utility for all sets in negotiation settings 1 and 2 in Tables 4 and 5, respectively. The average joint utility was computed from 50 negotiation deals in each set. Because each set has a different Pareto front, the maximum joint utility the agents can obtain for each set is different. The results indicate that for the majority of sets, a better joint utility can be achieved when the domains are normalized. For both Euclidean and cosine similarities, agents obtain higher joint utility and the deals are closer to maximum joint utility.

This confirms our hypothesis that scaling the attributes removes the effect of large-scale attributes dominating the similarity computation. When the attributes are normalized, Euclidean similarity gives higher joint utility in 12 out of 20 negotiation

Table 4. Negotiation setting 1: Outcomes

	Joint utility				Distance to max JU			
	Attributes are not normalised		Attributes are normalised		Attributes are not normalised		Attributes are normalised	
	EUC	COS	EUC	COS	EUC	COS	EUC	COS
Set 1	1.0430	1.0979	1.1506	1.1868	0.1570	0.1021	0.0494	0.0132
Set 2	1.0596	1.0958	1.1478	1.1525	0.1404	0.1042	0.0522	0.0475
Set 3	1.2101	1.2583	1.2770	1.2070	0.1399	0.0917	0.0730	0.1430
Set 4	1.1845	1.2709	1.2755	1.2854	0.1655	0.0791	0.0745	0.0646
Set 5	1.1448	1.2378	1.2672	1.2558	0.2052	0.1122	0.0828	0.0942
Set 6	1.1328	1.2484	1.2640	1.2208	0.2172	0.1016	0.0860	0.1292
Set 7	1.2292	1.2152	1.4081	1.3942	0.2375	0.2515	0.0586	0.0725
Set 8	1.2235	1.2424	1.4013	1.2326	0.2432	0.2243	0.0654	0.2341
Set 9	1.3594	1.3609	1.4404	1.4304	0.1406	0.1391	0.0596	0.0696
Set 10	1.3353	1.3645	1.4501	1.4433	0.1647	0.1355	0.0499	0.0567

Table 5. Negotiation setting 2: Outcomes

	Joint utility				Distance to max JU			
	Attributes are not normalised		Attributes are normalised		Attributes are not normalised		Attributes are normalised	
	EUC	COS	EUC	COS	EUC	COS	EUC	COS
Set 1	1.2484	1.1999	1.2766	1.2601	0.0516	0.1001	0.0234	0.0399
Set 2	1.2778	1.1927	1.2765	1.2513	0.0222	0.1073	0.0235	0.0488
Set 3	1.2621	1.3951	1.3660	1.2513	0.1879	0.0549	0.0840	0.1988
Set 4	1.2498	1.3605	1.3353	1.4103	0.1903	0.0795	0.1047	0.0297
Set 5	1.5592	1.5817	1.6667	1.6667	0.1408	0.1183	0.0333	0.0333
Set 6	1.5852	1.4133	1.6692	1.6733	0.1148	0.2868	0.0308	0.0267
Set 7	1.3135	1.3904	1.3971	1.3988	0.1865	0.1096	0.1029	0.1013
Set 8	1.2913	1.4338	1.3875	1.4225	0.2088	0.0663	0.1125	0.0775
Set 9	1.1725	1.2525	1.2795	1.2154	0.0516	0.1001	0.0234	0.0399
Set 10	1.1725	1.1888	1.2803	1.2198	0.1275	0.0475	0.0205	0.0846

sets compared to cosine similarity. For unnormalised attributes, the cosine similarity performs better in 16 out of 20 negotiation sets. This suggests that the agents may want to choose cosine similarity for their offer generation mechanism when negotiating in unscaled domains. As an example, agents may find themselves in a situation where they know the minimum and the maximum value of the price they are going to offer, but they do not know these values for their opponent. Hence, scaling the attribute values is not possible in this situation and the agents may choose cosine similarity with higher chance for high joint utility.

Further investigation is needed to find out which factors cause the different similarity measures to perform better in the different domains. With those factors identified, agents could then choose the appropriate similarity confidently to achieve higher joint utility.

5 Conclusion

We have presented an experiment to compare two vector similarity measures for generating offers in multi-attribute negotiation. Vector similarities eliminate the need to redefine similarity functions when the negotiation domain changes. In addition, vector similarities in this experiment do not require information on the opponent's weight values or ranking. In this study we also compare the results of using the two vector similarities on the original and normalized domains.

The results indicate that cosine similarity performs better for unscaled domains while Euclidean distance gives better joint utility in normalized domains. In real-life applications there are situations when negotiating parties are supplied with information on allowed value ranges to bid. In such situation agents can normalize the domain. However when the negotiators have different ranges of attribute values and that information is not shared, it is not possible for agents to normalize the domain. The results of our study can give options on what similarity measure to use when negotiating in either scenario. However, further experiments are required to identify what factors contribute to these results so as to allow agents negotiate for higher joint utility confidently with either similarity method.

References

1. Baarslag, T., Hindriks, K., Hendrikx, M., Dirkzwager, A., Jonker, C.: Decoupling negotiating agents to explore the space of negotiation strategies. Novel Insights in Agent-Based Complex Automated Negotiation, pp. 61–83. Springer, Berlin (2014)
2. Dirkzwager, A.S.Y.: Towards understanding negotiation strategies: analyzing the dynamics of strategy components (2013)
3. Chen, S., Weiss, G.: A novel strategy for efficient negotiation in complex environments. In: Timm, I.J., Guttmann, C. (eds.) MATES 2012. LNCS, vol. 7598, pp. 68–82. Springer, Heidelberg (2012)
4. Williams, C.R., Robu, V., Gerding, E.H., Jennings, N.R.: Iamhaggler 2011: a gaussian process regression based negotiation agent. Complex Automated Negotiations: Theories, Models, and Software Competitions, pp. 209–212. Springer, Berlin (2013)
5. Baarslag, T., Hindriks, K., Jonker, C., Kraus, S., Lin, R.: The first automated negotiating agents competition (ANAC 2010). In: Ito, T., Zhang, M., Robu, V., Fatima, S., Matsuo, T. (eds.) New Trends in Agent-Based Complex Automated Negotiations. SCI, vol. 383, pp. 113–135. Springer, Heidelberg (2012)
6. Faratin, P., Sierra, C., Jennings, N.R.: Using similarity criteria to make issue trade-offs in automated negotiations. Artif. Intell. **142**, 205–237 (2002)
7. Korenius, T., Laurikkala, J., Juhola, M.: On principal component analysis, cosine and Euclidean measures in information retrieval. Inf. Sci. **177**, 4893–4905 (2007)
8. Nguyen, H.V., Bai, L.: Cosine similarity metric learning for face verification. In: Kimmel, R., Klette, R., Sugimoto, A. (eds.) ACCV 2010, Part II. LNCS, vol. 6493, pp. 709–720. Springer, Heidelberg (2011)
9. Karnik, A., Goswami, S., Guha, R.: Detecting obfuscated viruses using cosine similarity analysis. In: Modelling and Simulation. In: First Asia International Conference on AMS 2007, pp. 165–170. IEEE (2007)

10. Ragone, A., Di Noia, T., Di Sciascio, E., Donini, F.M.: Propositional-logic approach to one-shot multi issue bilateral negotiation. ACM SIGecom Exch. **5**, 11–21 (2006)
11. Coehoorn, R.M., Jennings, N.R.: Learning on opponent's preferences to make effective multi-issue negotiation trade-offs. Presented at the (2004)
12. Jazayeriy, H., Azmi-Murad, M., Sulaiman, N., Izura Udizir, N.: The learning of an opponent's approximate preferences in bilateral automated negotiation. J. Theor. Appl. Electron. Commer. Res. **6**, 65–84 (2011)
13. Ros, R., Sierra, C.: A negotiation meta strategy combining trade-off and concession moves. Auton. Agent. Multi-Agent Syst. **12**, 163–181 (2006)
14. Cheng, C.-B., Chan, C.-C.H., Lin, K.-C.: Intelligent agents for e-marketplace: negotiation with issue trade-offs by fuzzy inference systems. Decis. Support Syst. **42**, 626–638 (2006)

Analytical Results on the BFS vs. DFS Algorithm Selection Problem. Part I: Tree Search

Tom Everitt[(⊠)] and Marcus Hutter

Australian National University, Canberra, Australia
tom.everitt@anu.edu.au

Abstract. Breadth-first search (BFS) and depth-first search (DFS) are the two most fundamental search algorithms. We derive approximations of their expected runtimes in complete trees, as a function of tree depth and probabilistic goal distribution. We also demonstrate that the analytical approximations are close to the empirical averages for most parameter settings, and that the results can be used to predict the best algorithm given the relevant problem features.

1 Introduction

A wide range of problems in artificial intelligence can be naturally formulated as *search problems* (Russell and Norvig 2010; Edelkamp and Schrödl 2012). Examples include planning, scheduling, and combinatorial optimisation (TSP, graph colouring, etc.), as well as various toy problems such as Sudoku and the Towers of Hanoi. Search problems can be solved by exploring the space of possible solutions in a more or less systematic or clever order. Not all problems are created equal, however, and substantial gains can be made by choosing the right method for the right problem. Predicting the best algorithm is sometimes known as the *algorithm selection problem* (Rice 1975).

A number of studies have approached the algorithm selection problem with machine learning techniques (Kotthoff 2014; Hutter et al. 2014). While demonstrably a feasible path, machine learning tend to be used as a *black box*, offering little insight into *why* a certain method works better on a given problem. On the other hand, most existing analytical results focus on worst-case big-O analysis, which is often less useful than average-case analysis when selecting algorithm. An important worst-case result is Knuth's (1975) simple but useful technique for estimating the depth-first search tree size. Kilby et al. (2006) used it for algorithm selection in the SAT problem. See also the extensions by Purdom (1978), Chen (1992), and Lelis et al. (2013). Analytical IDA* runtime predictions based on problem features were obtained by Korf et al. (2001) and Zahavi et al. (2010). In this study we focus on *theoretical analysis* of *average runtime* of breadth-first search (BFS) and depth-first search (DFS). While the IDA* results can be interpreted to give rough estimates for average BFS search time, no similar results are available for DFS.

© Springer International Publishing Switzerland 2015
B. Pfahringer and J. Renz (Eds.): AI 2015, LNAI 9457, pp. 157–165, 2015.
DOI: 10.1007/978-3-319-26350-2_14

To facilitate the analysis, we use a probabilistic model of goal distribution and graph structure. Currently no method to automatically estimate the model parameters is available. However, the analysis still offers important theoretical insights into BFS and DFS search. The parameters of the model can also be interpreted as a *Bayesian prior belief* about goal distribution. A precise understanding of BFS and DFS performance is likely to have both practical and theoretical value: Practical, as BFS and DFS are both widely employed; theoretical, as BFS and DFS are two most fundamental ways to search, so their properties may be useful in analytical approaches to more advanced search algorithms as well.

Our main contributions are estimates of average BFS and DFS runtime as a function of tree depth and goal distribution (goal quality ignored). This paper focuses on the performance of *tree search* versions of BFS and DFS that do *not* remember visited nodes. Graph search algorithms are generally superior when there are many ways to get to the same node. In such cases, tree search algorithms may end up exploring the same nodes multiple times. On the other hand, keeping track of visited nodes comes with a high prize in memory consumption, so graph search algorithms are not always a viable choice. BFS tree search may be implemented in a memory-efficient way as iterative-deepening DFS (ID-DFS). Our results are derived for standard BFS, but are only marginally affected by substituting BFS with ID-DFS. The technical report (Everitt and Hutter 2015a) gives further details and contains all omitted formal proofs. Part II of this paper (Everitt and Hutter 2015b) provides a similar analysis of the *graph search* case where visited nodes are marked. Our main analytical results are developed in Sects. 3 and 4, and verified experimentally in Sect. 5. Part II of this paper offers a longer discussion of conclusions and outlooks.

2   Preliminaries

BFS and DFS are two standard methods for uninformed graph search. Both assume oracle access to a *neighbourhood function* and a *goal check function* defined on a *state space*. BFS explores increasingly wider neighbourhoods around the start node. DFS follows one path as long as possible, and *backtracks* when stuck. The *tree search* variants of BFS and DFS do not remember which nodes they have visited. This has no impact when searching in trees, where each node can only be reached through one path. One way to understand tree search in general graphs is to say that they still *effectively* explore a tree; branches in this tree correspond to paths in the original graph, and copies of the same node v will appear in several places of the tree whenever v can be reached through several paths. DFS tree search may search forever if there are cycles in the graph. We always assume that path lengths are bounded by a constant D. Figure 1 illustrates the BFS and DFS search strategies, and how they (initially) focus on different parts of the search space. Russell and Norvig (2010) has further details.

The *runtime* or *search time* of a search method (BFS or DFS) is the number of nodes explored until a first goal is found (5 and 6 respectively in Fig. 1). This simplifying assumption relies on node expansion being the dominant operation,

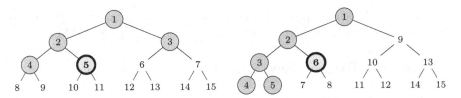

Fig. 1. The difference between BFS (left) and DFS (right) in a complete binary tree where a goal is placed in the second position on level 2 (the third row). The numbers indicate traversal order. Circled nodes are explored before the goal is found. Note how BFS and DFS explore different parts of the tree. In bigger trees, this may lead to substantial differences in search performance.

consuming similar time throughout the tree. If no goal exists, the search method will explore all nodes before halting. In this case, we define the runtime as the number of nodes in the search problem plus 1 (i.e., 2^{D+1} in the case of a binary tree of depth D). Let Γ be the event that a goal exists, Γ_k the event that a goal exists on level k, and $\bar{\Gamma}$ and $\bar{\Gamma}_k$ their complements. Let $F_k = \Gamma_k \cap (\cap_{i=0}^{k-1} \bar{\Gamma}_i)$ be the event that level k has the *first* goal.

A random variable X is *geometrically distributed* Geo(p) if $P(X = k) = (1-p)^{k-1}p$ for $k \in \{1, 2, \dots\}$. The interpretation of X is the number of trials until the first success when each trial succeeds with probability p. Its cumulative distribution function (CDF) is $P(X \le k) = 1 - (1-p)^k$, and its *average* or *expected value* $\mathbb{E}[X] = 1/p$. A random variable Y is *truncated geometrically distributed* $X \sim \text{TruncGeo}(p, m)$ if $Y = (X \mid X \le m)$ for $X \sim \text{Geo}(p)$, which gives

$$P(Y = k) = \begin{cases} \frac{(1-p)^k p}{1-(1-p)^m} & \text{for } k \in \{1, \dots, m\} \\ 0 & \text{otherwise.} \end{cases}$$

$$\text{tc}(p, m) := \mathbb{E}[Y] = \mathbb{E}[X \mid X \le m] = \frac{1 - (1-p)^m(pm+1)}{p(1-(1-p)^m)}.$$

When $p \gg \frac{1}{m}$, Y is approximately Geo(p), and $\text{tc}(p, m) \approx \frac{1}{p}$. When $p \ll \frac{1}{m}$, Y becomes approximately uniform on $\{1, \dots, m\}$ and $\text{tc}(p, m) \approx \frac{m}{2}$.

A random variable Z is *exponentially distributed* Exp(λ) if $P(Z \le z) = 1 - e^{-\lambda z}$ for $z \ge 0$. The expected value of Z is $\frac{1}{\lambda}$, and the probability density function of Z is $\lambda e^{-\lambda z}$. An exponential distribution with parameter $\lambda = -\ln(1-p)$ might be viewed as the continuous counterpart of a Geo(p) distribution. We will use this approximation in Sect. 4.

Lemma 1 (Exponential approximation). *Let $Z \sim \text{Exp}(-\ln(1-p))$ and $X \sim \text{Geo}(p)$. Then the CDFs for X and Z agree for integers k, $P(Z \le k) = P(X \le k)$. The expectations of Z and X are also similar in the sense that $0 \le \mathbb{E}[X] - \mathbb{E}[Z] \le 1$.*

We will occasionally make use of the convention $0 \cdot undefined = 0$, and often expand expectations by conditioning on disjoint events:

Lemma 2. *Let X be a random variable and let the sample space $\Omega = \dot{\bigcup}_{i \in I} C_i$ be partitioned by mutually disjoint events C_i. Then $\mathbb{E}[X] = \sum_{i \in I} P(C_i)E[X \mid C_i]$.*

3   Complete Binary Tree with a Single Goal Level

Consider a binary tree of depth D, where solutions are distributed on a single *goal level* $g \in \{0, \ldots, D\}$. At the goal level, any node is a goal with iid probability $p_g \in [0, 1]$. We will refer to this kind of problems as *(single goal level) complete binary trees with depth D, goal level g and goal probability p_g* (Sect. 4 generalises the setup to multiple goal levels).

As a concrete example, consider the search problem of solving a Rubik's cube. There is an upper bound $D = 20$ to how many moves it can take to reach the goal, and we may suspect that most goals are located around level 17 (± 2 levels) (Rokicki and Kociemba 2013). If we consider search algorithms that do not remember where they have been, the search space becomes a complete tree with fixed branching factor 3^6. What would be the expected BFS and DFS search time for this problem? Which algorithm would be faster?

The probability that a goal exists is $P(\Gamma) = P(\Gamma_g) = 1 - (1 - p_g)^{2^g}$. If a goal exists, let Y be the position of the first goal at level g. Conditioned on a goal existing, Y is a truncated geometric variable $Y \sim \text{TruncGeo}(p_g, 2^g)$. When $p_g \gg 2^{-g}$ the goal position Y is approximately $\text{Geo}(p_g)$, which makes most expressions slightly more elegant. This is often a realistic assumption, since if $p \not\gg 2^{-g}$, then often no goal would exist.

Proposition 1 (BFS runtime Single Goal Level). *Let the problem be a complete binary tree with depth D, goal level g and goal probability p_g. When a goal exists and has position Y on the goal level, the BFS search time is $t_{\text{SGL}}^{\text{BFS}}(g, p_g, Y) = 2^g - 1 + Y$, with expectation, $t_{\text{SGL}}^{\text{BFS}}(g, p_g \mid \Gamma_g) = 2^g - 1 + \text{tc}(p_g, 2^g) \approx 2^g - 1 + \frac{1}{p_g}$. In general, when a goal does not necessarily exist, the expected BFS search time is $t_{\text{SGL}}^{\text{BFS}}(g, p_g) = P(\Gamma) \cdot (2^g - 1 + \text{tc}(p_g, 2^g)) + P(\bar{\Gamma}) \cdot 2^{D+1} \approx 2^g - 1 + \frac{1}{p_g}$. The approximations are close when $p_g \gg 2^{-g}$.*

Proposition 2. *Consider a complete binary tree with depth D, goal level g and goal probability p_g. When a goal exists and has position Y on the goal level, the DFS search time is approximately $\tilde{t}_{\text{SGL}}^{\text{DFS}}(D, g, p_g, Y) := (Y - 1)2^{D-g+1} + 2$, with expectation $\tilde{t}_{\text{SGL}}^{\text{DFS}}(D, g, p_g \mid \Gamma_g) := (1/p_g - 1)\, 2^{D-g+1} + 2$. When $p_g \gg 2^{-g}$, the expected DFS search time when a goal does not necessarily exist is approximately*

$$\tilde{t}_{\text{SGL}}^{\text{DFS}}(D, g, p_g) := P(\Gamma)((\text{tc}(p_g, 2^g) - 1)2^{D-g+1} + 2) + P(\bar{\Gamma})2^{D+1} \approx \left(\frac{1}{p_g} - 1\right)2^{D-g+1}.$$

The proofs only use basic counting arguments and probability theory. A less precise version of Proposition 1 can be obtained from (Korf et al. 2001, Theorem 1). Full proofs and further details are provided in (Everitt and Hutter 2015a). Figure 2 shows the runtime estimates as a function of goal level. The runtime estimates can be used to predict whether BFS or DFS will be faster, given the parameters D, g, and p_g, as stated in the next Proposition.

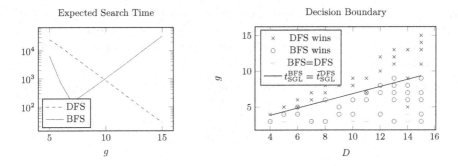

Fig. 2. Two plots of how expected BFS and DFS search time varies in a complete binary tree with a single goal level g and goal probability $p_g = 0.07$. The left depicts search time as a function of goal level in a tree of depth 15. BFS has the advantage when the goal is in the higher regions of the graph, although at first the probability that no goal exists heavily influences both BFS and DFS search time. DFS search time improves as the goal moves downwards since the goal probability is held constant. The right graph shows the decision boundary of Proposition 3, together with 100 empirical outcomes of BFS and DFS search time according to the varied parameters $g \in [3, D] \cap \mathbb{N}$ and $D \in [4, 15] \cap \mathbb{N}$. The decision boundary gets 79 % of the winners correct.

Proposition 3. *Let* $\gamma_{p_g} = \log_2 \left(tc(p_g, 2^g) - 1\right)/2 \approx \log_2 \left(\frac{1-p_g}{p_g}\right)/2$. *Given the approximation of DFS runtime of Proposition 2, BFS wins in expectation in a complete binary tree with depth D, goal level g and goal probability p_g when* $g < \frac{D}{2} + \gamma_{p_g}$ *and DFS wins in expectation when* $g > \frac{D}{2} + \gamma_{p_g} + \frac{1}{2}$.

The term γ_{p_g} is in the range $[-1, 1]$ when $p_g \in [0.2, 0.75]$, $g \geq 2$, in which case Proposition 3 roughly says that BFS wins (in expectation) when the goal level g comes before the middle of the tree. BFS benefits from a smaller p_g, with the boundary level being shifted $\gamma_{p_g} \approx k/2$ levels from the middle when $p_g \approx 2^{-k} \gg 2^{-g}$. Figure 2 illustrates the prediction as a function of goal depth and tree depth for a fixed probability $p_g = 0.07$. The technical report (Everitt and Hutter 2015a) gives the full proof, which follows from the runtime estimates Propositions 1 and 2.

It is straightforward to generalise the calculations to arbitrary branching factor b by substituting the 2 in the base of $t_{\mathrm{SGL}}^{\mathrm{BFS}}$ and $\tilde{t}_{\mathrm{SGL}}^{\mathrm{DFS}}$ for b. In Proposition 3, the change only affects the base of the logarithm in γ_{p_g}. See (Everitt and Hutter 2015a) for further details.

4   Complete Binary Tree with Multiple Goal Levels

We now generalise the model developed in the previous section to problems that can have goals on any number of levels. For each level $k \in \{0, \ldots, D\}$, let p_k be the associated *goal probability*. Not every p_k should be equal to 0. Nodes on level k have iid probability p_k of being a goal. We will refer to this kind of problems as *(multi goal level) complete binary trees with depth D and goal probabilities p.*

DFS Analysis To find an approximation of goal DFS performance in trees with multiple goal levels, we approximate the geometric distribution used in Proposition 2 with an exponential distribution (its continuous approximation by Lemma 1).

Proposition 4 (Expected Multi Goal Level DFS Performance). *Consider a complete binary tree of depth D with goal probabilities $\mathbf{p} = [p_0, \ldots, p_D] \in [0, 1)^{D+1}$. If for at least one j, $p_j \gg 2^{-j}$, and for all k, $p_k \ll 1$, then the expected number of nodes DFS will search is approximately $\tilde{t}_{\mathrm{MGL}}^{\mathrm{DFS}}(D, \mathbf{p}) := 1/\sum_{k=0}^{D} \ln(1 - p_k)^{-1} 2^{-(D-k+1)}$.*

The proof (available in Everitt and Hutter 2015a) constructs for each level k an exponential random variable X_k that approximates the search time before a goal is found on level k (disregarding goals on other levels). The minimum of all X_k then becomes an approximation of the search time to find a goal on any level. The approximations use exponential variables for easy minimisation.

In the special case of a single goal level, the approximation of Proposition 4 is similar to the one given by Proposition 2. When $\mathbf{p}$ only has a single element $p_j \neq 0$, the expression $\tilde{t}_{\mathrm{MGL}}^{\mathrm{DFS}}$ simplifies to $\tilde{t}_{\mathrm{MGL}}^{\mathrm{DFS}}(D, \mathbf{p}) = -2^{D-j+1}/\ln(1 - p_j)$. For p_j not close to 1, the factor $-1/\ln(1 - p_j)$ is approximately the same as the corresponding factor $1/p_j - 1$ in Proposition 2 (the *Laurent expansion* is $-1/\ln(1 - p_j) = 1/p_j - 1/2 + O(p_j)$).

BFS Analysis The corresponding expected search time $t_{\mathrm{MGL}}^{\mathrm{BFS}}(D, \mathbf{p})$ for BFS requires less insight and can be calculated exactly by conditioning on which level the first goal is. The resulting formula is less elegant, however. The same technique cannot be used for DFS, since DFS does not exhaust levels one by one.

The probability that level k has the first goal is $P(F_k) = P(\Gamma_k) \prod_{j=0}^{k-1} P(\bar{\Gamma}_j)$, where $P(\Gamma_i) = (1 - (1 - p_i)^{2^i})$. The expected BFS search time gets a more uniform expression by the introduction of an extra *hypothetical level $D + 1$* where all nodes are goals. That is, level $D + 1$ has goal probability $p_{D+1} = 1$ and $P(F_{D+1}) = P(\bar{\Gamma}) = 1 - \sum_{k=0}^{D} P(F_k)$.

Proposition 5 (Expected Multi Goal Level BFS Performance). *The expected number of nodes $t_{\mathrm{MGL}}^{\mathrm{BFS}}(p)$ that BFS needs to search to find a goal in a complete binary tree of depth D with goal probabilities $\mathbf{p} = [p_0, \ldots, p_D]$, $\mathbf{p} \neq \mathbf{0}$, is $t_{\mathrm{MGL}}^{\mathrm{BFS}}(\mathbf{p}) = \sum_{k=0}^{D+1} P(F_k) t_{\mathrm{SGL}}^{\mathrm{BFS}}(k, p_k \mid \Gamma_k) \approx \sum_{k=0}^{D+1} P(F_k)\left(2^k + \frac{1}{p_k}\right)$.*

See (Everitt and Hutter 2015a) for a proof. For $p_k = 0$, the expression $t_{\mathrm{CB}}^{\mathrm{BFS}}(k, p_k)$ and $1/p_k$ will be undefined, but this only occurs when $P(F_k)$ is also 0. The approximation tends to be within a factor 2 of the correct expression, even when $p_k < 2^{-k}$ for some or all $p_k \in \mathbf{p}$. The reason is that the corresponding $P(F_k)$'s are small when the geometric approximation is inaccurate. Both Propositions 4 and 5 naturally generalise to arbitrary branching factor b. Although their combination does not yield a similarly elegant expression as Proposition 3, they can still be naively combined to predict the BFS vs. DFS winner (Fig. 3).

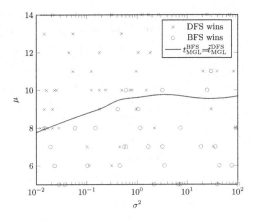

Fig. 3. The decision boundary for the Gaussian tree given by Propositions 4 and 5, together with empirical outcomes of BFS vs. DFS winner. The scattered points are based on 100 independently generated problems with depth $D = 14$ and uniformly sampled parameters $\mu \in [5, 14] \cap \mathbb{N}$ and $\log(\sigma^2) \in [-2, 2]$. The most deciding feature is the goal peak μ, but DFS also benefits from a smaller σ^2. The decision boundary gets 74 % of the winners correct.

5   Experimental Verification

To verify the analytical results, we have implemented the models in Python 3 using the `graph-tool` package (Peixoto 2015)[1]. The data reported in Tables 1

Table 1. BFS and DFS performance in the single goal level model with depth $D = 14$, where g is the goal level and p_g the goal probability. Each box contains empirical average/*analytical expectation*/error percentage.

$g \backslash p_g$	0.001	0.01	0.1	$g \backslash p_g$	0.001	0.01	0.1
5		46.33	40.01	5		14 680	8206
		46.64	*39.86*			*15 000*	*8053*
		0.7 %	0.4 %			2.2 %	1.9 %
8	369.5	332.8	264.6	8	14 530	9833	1105
	378.0	*333.9*	*265.0*		*15 620*	*9967*	*1154*
	2.3 %	0.3 %	0.2 %		7.5 %	1.4 %	4.5 %
11	2748	2143	2057	11	11 200	1535	152.3
	2744	*2147*	*2057*		*11 140*	*1586*	*146.0*
	0.1 %	0.2 %	0. %		0.5 %	3.4 %	4.1 %
14	17 360	16 480	16 390	14	1971	208.8	30.57
	17 380	*16 480*	*16 390*		*2000*	*200.0*	*20.00*
	0.1 %	0. %	0. %		1.4 %	4.2 %	35 %

(a) BFS single goal level   (b) DFS single goal level

---

[1] Source code for the experiments is available at http://tomeveritt.se.

Table 2. BFS and DFS performance in Gaussian binary trees with depth $D = 14$. Each box contains empirical average/*analytical expectation*/error percentage.

$\mu\backslash\sigma$	0.1	1	10	100	$\mu\backslash\sigma$	0.1	1	10	100
5	37.24	43.75	90.87	225.1	5	5374	8572	3405	385.8
	37.04	*41.55*	*83.72*	*210.8*		*5949*	*10 070*	*3477*	*379.1*
	0.5%	5.0%	7.9%	6.4%		11%	18%	2.1%	1.7%
8	261.2	171.9	119.6	212.0	8	677.3	1234	454.6	252.6
	261.3	*173.4*	*119.8*	*211.0*		*743.6*	*1259*	*473.6*	*260.0*
	0.%	0.9%	0.2%	0.5%		9.8%	2.1%	4.2%	2.9%
11	2049	953.0	303.9	249.5	11	97.38	168.1	117.4	210.0
	2050	*953.0*	*305.0*	*247.5*		*92.95*	*157.4*	*106.7*	*211.7*
	0.%	0.%	0.3%	0.8%		4.5%	6.4%	9.1%	0.8%
14	16 210	5159	968.5	332.9	14	24.00	43.38	81.75	213.6
	16 150	*5136*	*960.6*	*329.7*		*11.62*	*32.89*	*74.46*	*205.0*
	0.4%	0.4%	0.8%	0.9%		52%	24%	8.9%	4.0%

(a) BFS multi goal level	(b) DFS multi goal level

and 2 is based on an average over 1000 independently generated search problems with depth $D = 14$. The first number in each box is the empirical average, the second number is the analytical estimate, and the third number is the percentage error of the analytical estimate.

For certain parameter settings, there is only a small chance ($< 10^{-3}$) that there are no goals. In such circumstances, all 1000 generated search graphs typically inhabit a goal, and so the empirical search times will be comparatively small. However, since a tree of depth 14 has about $2^{15} \approx 3 \cdot 10^5$ nodes (and a search algorithm must search through all of them in case there is no goal), the rarely occurring event of no goal can still influence the *expected* search time substantially. To avoid this sampling problem, we have ubiquitously discarded all instances where no goal is present, and compared the resulting averages to the analytical expectations *conditioned on at least one goal being present*.

To develop a concrete instance of the multi goal level model we consider the special case of *Gaussian goal probability vectors*, with two parameters μ and σ^2. For a given depth D, the goal probabilities are given by $p_i = \min \left\{ \frac{1}{20\sqrt{\sigma^2}} e^{(i-\mu)^2/\sigma^2}, \frac{1}{2} \right\}$. The parameter $\mu \in [0, D] \cap \mathbb{N}$ is the *goal peak*, and the parameter $\sigma^2 \in \mathbb{R}^+$ is the *goal spread*. The factor $1/20$ is arbitrary, and chosen to give an interesting dynamics between searching depth-first and breadth-first. No p_i should be greater than $1/2$, in order to (roughly) satisfy the assumption of Proposition 5. We call this model the *Gaussian binary tree*.

The accuracy of the predictions of Propositions 1 and 2 are shown in Table 1, and the accuracy of Propositions 4 and 5 in Table 2. The relative error is always small for BFS ($< 10\%$). For DFS the error is generally within 20%, except when the search time is small (< 35 probes), in which case the absolute error is always small. The decision boundary of Proposition 3 is shown in Fig. 2, and the decision boundary of Propositions 4 vs. 5 is shown in Fig. 3. These boundary plots show that the analysis generally predict the correct BFS vs. DFS winner (79 % and 74 % correct in the investigated models).

6 Conclusions and Outlook

Part II of this paper (Everitt and Hutter 2015b) generalises the setup in this paper, analytically investigating search performance in general graphs. Part II also provides a more general discussion and outlook on future directions.

References

Chen, P.C.: Heuristic sampling: a method for predicting the performance of tree searching programs. SIAM J. Comput. **21**(2), 295–315 (1992)

Edelkamp, S., Schrödl, S.: Heuristic Search. Morgan Kaufmann Publishers Inc., San Francisco (2012)

Everitt, T., Hutter, M.: A topological approach to meta-heuristics: analytical results on the BFS vs. DFS algorithm selection problem. Technical report, Australian National University (2015a). arXiv:1509.02709 [cs.AI]

Everitt, T., Hutter, M.: Analytical results on the BFS vs. DFS algorithm selection problem. Part II: graph search. In: 28th Australian Joint Conference on Artificial Intelligence (2015b)

Hutter, F., Xu, L., Hoos, H.H., Leyton-Brown, K.: Algorithm runtime prediction: methods & evaluation. Artif. Intell. **206**(1), 79–111 (2014)

Kilby, P., Slaney, J., Thiébaux, S., Walsh, T.: Estimating search tree size. In: Proceedings of the 21st National Conference of Artificial Intelligence. AAAI, Menlo Park (2006)

Knuth, D.E.: Estimating the efficiency of backtrack programs. Math. Comput. **29**(129), 122–136 (1975)

Korf, R.E., Reid, M., Edelkamp, S.: Time complexity of iterative-deepening-A*. Artif. Intell. **129**(1–2), 199–218 (2001)

Kotthoff, L.: Algorithm selection for combinatorial search problems: a survey. AI Mag., 1–17 (2014)

Lelis, L.H.S., Otten, L., and Dechter, R.: Predicting the size of Depth-first Branch and Bound search trees. In: IJCAI International Joint Conference on Artificial Intelligence, pp. 594–600 (2013)

Peixoto, T.P.: The graph-tool python library. figshare (2015)

Purdom, P.W.: Tree size by partial backtracking. SIAM J. Comput. **7**(4), 481–491 (1978)

Rice, J.R.: The algorithm selection problem. Adv. Comput. **15**, 65–117 (1975)

Rokicki, T., Kociemba, H.: The diameter of the rubiks cube group is twenty. SIAM J. Discrete Math. **27**(2), 1082–1105 (2013)

Russell, S.J., Norvig, P.: Artificial Intelligence: A Modern Approach, 3rd edn. Prentice Hall, Englewood Cliffs (2010)

Zahavi, U., Felner, A., Burch, N., Holte, R.C.: Predicting the performance of IDA* using conditional distributions. J. Artif. Int. Res. **37**, 41–83 (2010)

Analytical Results on the BFS vs. DFS Algorithm Selection Problem: Part II: Graph Search

Tom Everitt[(✉)] and Marcus Hutter

Australian National University, Canberra, Australia
tom.everitt@anu.edu.au

Abstract. The algorithm selection problem asks to select the best algorithm for a given problem. In the companion paper (Everitt and Hutter 2015b), expected runtime was approximated as a function of search depth and probabilistic goal distribution for tree search versions of breadth-first search (BFS) and depth-first search (DFS). Here we provide an analogous analysis of BFS and DFS graph search, deriving expected runtime as a function of graph structure and goal distribution. The applicability of the method is demonstrated through analysis of two different grammar problems. The approximations come surprisingly close to empirical reality.

1 Introduction

Search is a fundamental problem of artificial intelligence (Russell and Norvig 2010), and a sizeable list of search algorithms with different pros and cons can be found in the literature (Edelkamp and Schrödl 2012). Examples of search tasks include combinatorial optimisation problems and planning, and core search algorithms include BFS, DFS, A*, simulated annealing, and genetic algorithms. Techniques for selecting the best algorithm for a given problem is of obvious importance (Rice 1975; Kotthoff 2014; Hutter et al. 2014).

Tightly related to the algorithm selection problem is the problem of predicting algorithm runtime; in particular expected runtime. In the companion paper (Everitt and Hutter 2015b) we gave a brief survey of related work and derived (approximations of) expected runtime for breadth-first search (BFS) and depth-first search (DFS) in trees. The results also applied to search in general graphs for the variants of BFS and DFS that do *not* remember visited nodes; so called *tree search algorithms*. Their name does not stop them from being used also in other types of graphs, but they then run the risk of spending most of the time searching the same nodes many times. This paper analyses expected runtime of *graph search* variants of BFS and DFS that do remember which nodes they have visited. Although graph search variants usually are more efficient in the sense that they search fewer nodes, the extra memory overhead means that they are not always applicable.

© Springer International Publishing Switzerland 2015
B. Pfahringer and J. Renz (Eds.): AI 2015, LNAI 9457, pp. 166–178, 2015.
DOI: 10.1007/978-3-319-26350-2_15

Our main contributions are estimates of expected BFS and DFS graph search runtime as a function of graph structure and distributions of goals (Sect. 3). Note that we focus solely on the time it takes to find a goal, and ignore aspects such as solution quality. We demonstrate the relevance of the results by applying them to two different grammar problems (Sect. 4). Setup and background are described in Sect. 2, and experimental verification in Sect. 5. Finally, conclusions and outlooks come in Sect. 6. The technical report (Everitt and Hutter 2015a) offers a greatly extended discussion about the setup.

2   Preliminaries

The graph search variants of BFS and DFS are two standard methods for uninformed graph search. Both BFS and DFS assume oracle access to a *neighbourhood function* and a *goal check function* defined on a *state space*. BFS explores increasingly wider neighbourhoods around the start node. DFS follows one path as long as possible, and *backtracks* when stuck. Figure 1 illustrates the different search strategies, and how they (initially) focus on different parts of the search space. Please refer to Russell and Norvig (2010) for details.

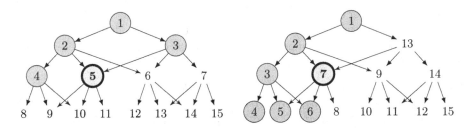

Fig. 1. The difference between BFS (left) and DFS (right) in a directed graph where a goal is placed in the second position on level 2 (the third row). The numbers indicate traversal order. Circled nodes are explored before the goal is found. Note how BFS and DFS explore different parts of the tree. In bigger search spaces, this may lead to substantial differences in search performance.

The *runtime* or *search time* of a search method (BFS or DFS) is the number of nodes explored until a first goal is found (5 and 7 respectively in Fig. 1). This simplifying assumption relies on node expansion being the dominant operation, requiring similar time throughout the tree. If no goal exists, the search method will explore all nodes before halting. In this case, we define the runtime as the number of nodes in the search problem plus 1. Let Γ be the event that a goal exists, Γ_k the event that a goal exists on level k, and $\bar{\Gamma}$ and $\bar{\Gamma}_k$ their complements. Let $F_k = \Gamma_k \cap (\bigcap_{i=0}^{k-1} \bar{\Gamma}_i)$ be the event that level k has the *first* goal.

A random variable X is *geometrically distributed* $\mathrm{Geo}(p)$ if $P(X = k) = (1-p)^{k-1}p$ for $k \in \{1, 2, \dots\}$. The interpretation of X is the number of trials until the first success when each trial succeeds with probability p. Its cumulative distribution function (CDF) is $P(X \leq k) = 1 - (1-p)^k$, and its *average* or *expected value* $\mathbb{E}[X] = 1/p$. A random variable Y is *truncated geometrically distributed* $X \sim \text{TruncGeo}(p, m)$ if $Y = (X \mid X \leq m)$ for $X \sim \text{Geo}(p)$, which gives

$$P(Y = k) = \begin{cases} \frac{(1-p)^k p}{1-(1-p)^m} & \text{for } k \in \{1, \ldots, m\} \\ 0 & \text{otherwise.} \end{cases}$$

$$\text{tc}(p, m) := \mathbb{E}[Y] = \mathbb{E}[X \mid X \leq m] = \frac{1 - (1-p)^m(pm+1)}{p(1-(1-p)^m)}.$$

When $p \gg \frac{1}{m}$, Y is approximately $\text{Geo}(p)$, and $\text{tc}(p, m) \approx \frac{1}{p}$. When $p \ll \frac{1}{m}$, Y becomes approximately uniform on $\{1, \ldots, m\}$ and $\text{tc}(p, m) \approx \frac{m}{2}$.

We will occasionally make use of the convention $0 \cdot \textit{undefined} = 0$, and often expand expectations by conditioning on disjoint events:

Lemma 1. *Let X be a random variable and let the sample space $\Omega = \dot{\bigcup}_{i \in I} C_i$ be partitioned by mutually disjoint events C_i. Then $\mathbb{E}[X] = \sum_{i \in I} P(C_i)E[X \mid C_i]$.*

3   Colliding Branches

The companion paper (Everitt and Hutter 2015b) explores a model of *tree search*, where path redundancies are not recognized by the search algorithms. In this section we develop a similar model for *graph search* performance. The abstract results of this section are applied to two grammar problems in the next section.

Definition 1. *For a given search problem: Let the* level *of a node v, level(v), be the length of a shortest path from the start node to v. Let $D = \max_v \text{level}(v)$ be the (generalised)* depth *of the search graph. Let δ_n be the first node on level n reached by DFS, $0 \leq n \leq D$. Any node reachable from v is a* descendant *of v.*

The descendant counter *L plays a central role in the analysis. For a given search problem, let*

$$L(n, d) = |\{v : \text{level}(v) = d, v \in \text{descendants}(\delta_n)\}|$$

count the number of nodes on level d that are reachable from δ_n.

As in the companion paper, we assume that goals are distributed by level in an iid manner according to a goal probability vector **p**. We will also assume that the probability of DFS finding a goal before finding δ_D is negligible. We will refer to this kind of problems as *search problems with depth D, goal probabilities* **p** *and descendant counter L*. The rest of this section justifies the following proposition.

Proposition 1. *The DFS and BFS runtime of a search problem can be roughly estimated from the descendant counter L, the depth D and the goal probabilities* $\mathbf{p} = [p_0, \ldots, p_D]$ *when the probability of DFS finding a goal before δ_D is negligible.*

The assumption of DFS not finding a goal before δ_D is not always realistic, but is for example satisfied in the grammar problems considered in Sect. 4 below.

3.1   DFS Analysis

The nodes $\delta_0, \ldots, \delta_D$ play a central role in the analysis of DFS runtime, since all the descendants of δ_{n+1} will be explored before the descendants of δ_n (excluding the δ_{n+1} descendants). We say that *DFS explores from* δ_n after DFS has explored all descendants of δ_{n+1} and until all descendants of δ_n have been explored. The general idea of the DFS analysis will be to count the number of nodes under each δ_n, and to compute the probability that any of these nodes is a goal.

   Some notation for this:

- Let the δ_n-*subgraph* $S_n = \{v : v \in \text{descendants}(\delta_n)\}$ be the set of nodes reachable from δ_n, with cardinality $|S_n| = \sum_{i=0}^{D} L(n, i)$, $0 \le n \le D$. Let $S_{D+1} = \emptyset$ and let S_{-1} be a set of cardinality $|S_{-1}| = |S_0| + 1 = \sum_{i=0}^{D} L(0, i) + 1$.
- Let the δ_n-*explorables* $T_n = S_n \setminus S_{n+1}$ be the nodes explored from δ_n.
- Let the *number of level-d* δ_n-*explorables* $A_{n,d} = L(n, d) - L(n + 1, d)$ be the number of level d descendants of δ_n that are not descendants of δ_{n+1} for $0 \le n, d \le D$. The relation between T_n and $A_{n,d}$ is the following: $|T_n| = \sum_{i=n}^{D} A_{n,i}$.

Let $q_k = 1 - p_k$ for $0 \le k \le D$.

Lemma 2. *Consider a search problem with depth D, goal probabilities $\mathbf{p}$, and descendant counter L. The probability that the δ_n-explorables T_n contains a goal is $\tau_n := 1 - \prod_{k=0}^{D} q_k^{A_{n,k}}$, and the probability that T_n contains the first goal is $\phi_n := \tau_n \prod_{i=n+1}^{D}(1 - \tau_i)$.*

Proof. τ_n is 1 minus the probability of *not* hitting a goal at any level d, $n \le d \le D$, since at each level d, $A_{n,d}$ probes are made when exploring from δ_n.

Proposition 2 (Colliding Branches Expected DFS Search Time). *The expected DFS search time $t_{\mathrm{CB}}^{\mathrm{DFS}}(D, \mathbf{p}, L)$ in a search problem with depth D, goal probabilities $\mathbf{p}$, and descendant counter L is bounded by*

$$t_{\mathrm{CBL}}^{\mathrm{DFS}}(D, \mathbf{p}, L) := \sum_{n=-1}^{D} |S_{n+1}| \phi_n \le t_{\mathrm{CB}}^{\mathrm{DFS}}(D, \mathbf{p}, L) \le \sum_{n=-1}^{D} |S_n| \phi_n := t_{\mathrm{CBU}}^{\mathrm{DFS}}(D, \mathbf{p}, l)$$

where $\phi_{-1} = \bar{\Gamma} = 1 - \sum_{n=0}^{D} \phi_n$ is the probability that no goal exists.

   The arithmetic mean $\tilde{t}_{\mathrm{CB}}^{\mathrm{DFS}}(D, \mathbf{p}, L) := (t_{\mathrm{CBL}}^{\mathrm{DFS}}(D, \mathbf{p}, L) + t_{\mathrm{CBU}}^{\mathrm{DFS}}(D, \mathbf{p}, L))/2$ of the bounds can be used for a single runtime estimate.

Proof. Let X be the DFS search time in a search problem with the features described above. The expectation of X may be decomposed as

$$\mathbb{E}[X] = P(\bar{\Gamma})\mathbb{E}[X \mid \bar{\Gamma}] + \sum_{n=0}^{D} P(\text{first goal in } T_n) \cdot \mathbb{E}[X \mid \text{first goal in } T_n]. \quad (1)$$

The conditional search time (X | first goal in T_n) is bounded by $|S_{n+1}| \leq (X$ | first goal in $T_n) \leq |S_n|$ for $0 \leq n \leq D$, since to find a goal DFS will search the entire δ_{n+1}-subgraph S_{n+1} before finding it when searching the δ_n-explorables T_n, but will not need to search more than the δ_n-subgraph $S_n = S_{n+1} \cup T_n$ (disregarding the few probes made 'on the way down to' δ_n (i.e. to T_n); these probes were assumed negligible). The same bounds also hold with S_0 and S_{-1} when no goal exists (recall that $|S_{-1}| := |S_0| + 1$). Therefore the conditional expectation satisfies

$$|S_{n+1}| \leq \mathbb{E}[X \mid \text{first goal in } T_n] \leq |S_n| \qquad (2)$$

for $-1 \leq n \leq D$. By Lemma 2, the probability that the first goal is among the δ_n-explorables T_n is ϕ_n, and the probability $P(\bar{\Gamma})$ that no goal exists is ϕ_{-1} by definition.

Substituting ϕ_n and (2) into (1) gives the desired bounds for expected DFS search time $\tilde{t}_{CB}^{DFS}(D, \mathbf{p}, L) = \mathbb{E}[X]$.

The informativeness of the bounds of Proposition 2 depends on the dispersion of nodes between the different T_n's. If most nodes belong to one or a few sets T_n, the bounds may be almost completely uninformative. This happens in the special case of complete trees with branching factor b, where a fraction $(b-1)/b$ of the nodes will be in T_0. The companion paper (Everitt and Hutter 2015b) derives techniques for these cases. The grammar problems investigated in Sect. 4 below show that the bounds may be relevant in more connected graphs, however.

3.2   BFS Analysis

The analysis of BFS only requires the descendant counter $L(0, \cdot)$ with the first argument set to 0, and follows the same structure as the BFS analysis in (Everitt and Hutter 2015b). In contrast to the DFS bounds above, this analysis gives a precise expression for the expected runtime. The idea is to count the number of nodes in the upper k levels of the tree (derived from $L(0,0), \ldots, L(0,k)$), and to compute the probability that they contain a goal. Let the *upper subgraph* $U_k = \sum_{i=0}^{k-1} L(0, i)$ be the number of nodes above level k. When there is only a single goal level, the following expression for BFS runtime may be readily derived.

Lemma 3 (BFS Runtime Single Goal Level). *For a search problem with depth D and descendant counter L, assume that the problem has a single goal level g with goal probability p_g, and that $p_j = 0$ for $j \neq g$. When a goal exists and has position Y on the goal level, the BFS search time is:*

$$t_{CB}^{BFS}(g, p_g, L, Y) = U_g + Y, with\ expected\ value$$
$$t_{CB}^{BFS}(g, p_g, L \mid \Gamma_g) = U_g + \mathrm{tc}(p_g, L(0, g))$$

Proof. When a goal exists, BFS will explore all of the top of the tree until depth $g - 1$ (that is, U_g nodes) and Y nodes on level g before finding the first goal. The expected value of Y is $\mathrm{tc}(p_g, L(0, g))$.

The probability that level k has a goal is $P(\Gamma_k) = 1 - q_k^{L(0,k)}$, and the probability that level k has the first goal is $P(F_k) = P(\Gamma_k) \prod_{i=0}^{k-1} P(\bar{\Gamma}_i)$. To BFS, only the first goal level matter. This allows BFS runtime to be expanded over the F_k events as in Lemma 1. For greater uniformity, a *hypothetical level* $D + 1$ only containing goals is introduced to handle the event of no goal in the first D levels.

Proposition 3 (Branch Colliding Expected BFS Performance). *The expected number of nodes that BFS needs to search to find a goal in a search problem with depth D, goal probabilities $\mathbf{p} = [p_0, \dots, p_D]$, $\mathbf{p} \neq \mathbf{0}$, and descendant counter L is*

$$t_{\mathrm{CB}}^{\mathrm{BFS}}(\mathbf{p}, L) = \sum_{k=0}^{D+1} P(F_k) t_{\mathrm{CB}}^{\mathrm{BFS}}(k, p_k, L \mid \Gamma_k)$$

where the goal probabilities have been extended with an extra element $p_{D+1} = 1$, and $F_{D+1} = \bar{\Gamma}$ is the event that no goal exists.

For $p_k = 0$, $t_{\mathrm{CB}}^{\mathrm{BFS}}$ will be undefined, but this only occurs when $P(F_k)$ is also 0. Propositions 2 and 3 give (rough) estimates of average BFS and DFS graph search time given the goal distribution $\mathbf{p}$ and the structure parameter L. The results can be combined to make a decision whether to use BFS or DFS (Fig. 3).

4 Grammar Problems

We now show how to apply the general theory of Sect. 3 to two concrete grammar problems. A *grammar problem* is a search problem where nodes are strings over some finite alphabet B, and the neighbourhood relation is given by a set of production rules. *Production rules* are mappings $x \to y$, $x, y \in B^*$, defining how strings may be transformed. For example, the production rule $S \to Sa$ permits the string aSa to be transformed into $aSaa$. A grammar problem is defined by a set of production rules, together with a *starting string* and a *set of goal strings*. A *solution* is a sequence of production rule applications that transforms the starting string into a goal string. Many search problems can be formulated as grammar problems, with string representations of states modified by production rules. Their generality makes it *computably undecidable* whether a given grammar problem has a solution or not. We here consider a simplified version where the search depth is artificially limited, and goals are distributed according to a goal probability vector p.

4.1 Binary Grammar

Let ϵ be the empty string. The *binary grammar* consists of two production rules, $\epsilon \to a$ and $\epsilon \to b$ over the alphabet $B = \{a, b\}$. The starting string is the empty string ϵ. A maximum depth D of the search graph is imposed, and strings on level k are goals with iid probability p_k, $0 \leq k \leq D$. Since the left hand substring

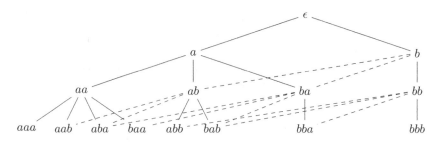

Fig. 2. Graph of binary grammar problem with max depth $D = 3$. Contiguous lines indicate first discovery by DFS, and dashed lines indicate rediscoveries.

of both production rules is the empty string, both can always be applied at any place to a given string. The resulting graph is shown in Fig. 2.

The first node on level n that DFS reaches in the binary grammar problem is $\delta_n = a^n$ for $0 \le n \le D$, assuming that the production rule $\epsilon \to a$ is always used first by DFS. The following lemma derives an expression for the descendant counter L^{BG} required by Proposition 2. Incidentally, the number of level-d δ_n explorables $A_{n,d}$ (Sect. 3.1) gets an elegant form in the binary grammar problem.

Lemma 4. *For $n < d$, let $L^{\mathrm{BG}}(n, d) = |\{v : \mathrm{level}(v) = d, v \in \mathrm{descendants}(a^n)\}|$ be the number of nodes reachable from a^n, and let $A_{n,d} = L^{\mathrm{BG}}(n, d) - L^{\mathrm{BG}}(n + 1, d)$ be the number of descendants of a^n that are not descendants of a^{n+1}. Then $L^{\mathrm{BG}}(n, d) = \sum_{i=0}^{d-n} \binom{d}{i}$, and $A_{n,d} = \binom{d}{d-n}$.*

Proof. The reachable nodes on level d that we wish to count are $d - n$ levels below a^n. To reach this level we must add $i \le d - n$ number of b's and $d - n - i$ number of a's to a^n. The number of length d strings containing exactly i number of b's is $\binom{d}{i}$ (we are choosing positions for the b's non-uniquely with repetition among $d - i + 1$ possible positions). Summing over i, we obtain $L^{\mathrm{BG}}(n, d) = \sum_{i=0}^{d-n} \binom{d}{i}$, and $A_{n,d} = L^{\mathrm{BG}}(n, d) - L^{\mathrm{BG}}(n + 1, d) = \binom{d}{d-n}$. $\square$

Corollary 1 (Expected Binary Grammar BFS Search Time). *The expected BFS search time $\tilde{t}_{\mathrm{BG}}^{\mathrm{DFS}}(\mathbf{p})$ in a Binary Grammar Problem of depth D with goal probabilities $\mathbf{p} = [p_0, \ldots, p_D]$ is*

$$t_{\mathrm{BG}}^{\mathrm{BFS}}(\mathbf{p}) = t_{\mathrm{CB}}^{\mathrm{BFS}}(\mathbf{p}, L^{\mathrm{BG}}).$$

Corollary 2 (Expected Binary Grammar DFS Search Time). *The expected DFS search time $\tilde{t}_{\mathrm{BG}}^{\mathrm{DFS}}(D, \mathbf{p})$ in a binary grammar problem of depth D with goal probabilities $\mathbf{p} = [p_0, \ldots, p_D]$ is bounded between $t_{\mathrm{BGL}}^{\mathrm{DFS}}(D, \mathbf{p}) := t_{\mathrm{CBL}}^{\mathrm{DFS}}(D, \mathbf{p}, L^{\mathrm{BG}})$ and $t_{\mathrm{BGU}}^{\mathrm{DFS}}(D, \mathbf{p}) := t_{\mathrm{CBU}}^{\mathrm{DFS}}(D, \mathbf{p}, L^{\mathrm{BG}})$, and is approximately*

$$\tilde{t}_{\mathrm{BG}}^{\mathrm{DFS}}(D, \mathbf{p}) := \tilde{t}_{\mathrm{CB}}^{\mathrm{DFS}}(D, \mathbf{p}, L^{\mathrm{BG}}).$$

Proof (Proof of Corollaries 1 and 2). Direct application of Lemma 4, and Propositions 2 and 3 respectively. $\square$

The bounds are plotted for a single goal level in Figs. 3 and 4.

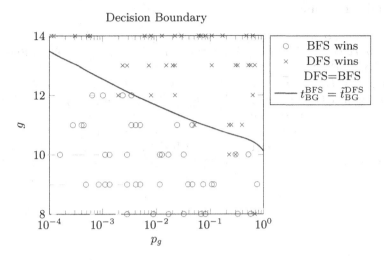

Fig. 3. The decision boundary predicted by Corollaries 1 and 2, together with empirical outcomes of BFS and DFS search time. The scattered points are based on 100 independently generated binary grammar problems of depth $D = 14$ with uniformly sampled (single) goal level $g \in [8, 14] \cap \mathbb{N}$ and $\log(p_g) \in [-4, 0]$. DFS benefits from a deeper goal level and higher goal probability compared to BFS. The decision boundary gets 87 % of the instances correct.

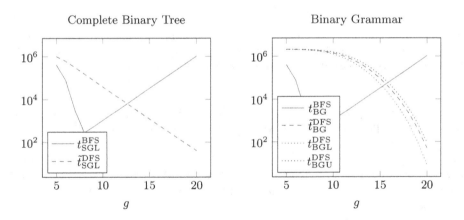

Fig. 4. The expected search time of BFS and DFS as a function of a single goal level g with goal probability $p_g = 0.05$ in a tree of depth $D = 20$. BFS has the advantage when the goal is in the higher regions of the graph, although at first the probability that no goal exists heavily influences both BFS and DFS search time. The greater connectivity of the graph in the binary grammar problem permits DFS to spend more time in the lower regions before backtracking, compared to the complete binary tree analysed in the companion paper (Everitt and Hutter 2015b). This penalises DFS runtime when the goal is not in the very lowest regions of the tree. BFS behaviour is identical in both models.

4.2   Full Grammar

The *full grammar problem* has alphabet $B = \{S, a, b\}$ and start string S. The *production rules* are $S \to \epsilon$ (with ϵ the empty string) plus the *adding rules* $S \to Sa$, $S \to aS$, $S \to Sb$, and $S \to bS$, and the *moving rules* $Sa \to aS$, $aS \to Sa$, $Sb \to bS$, and $bS \to Sb$. Only S-less strings can be goal nodes. As usual, a maximum depth D and a goal probability vector $\mathbf{p} = [p_0, \ldots, p_D]$ are given.

For simplified analysis, we will abuse notation the following way. We will consider S-less nodes to be one level higher than they actually are. For example, a would normally be on level 2 (e.g. reached by the path $S \to Sa$, $S \to \epsilon$), but we will consider it to be on level 1. A slight modification of BFS and DFS makes them always check the S-less child first (which is always child-less in turn), which means the change will only slightly affect search time. We will still consider $\delta_n = Sa^n$ whenever $S \to Sa$ is among the production rules, however.

The search graph of the full grammar problem is shown in Fig. 5.

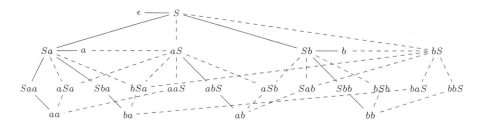

Fig. 5. Search graph for the Grammar problem until level 2. Connections induced by moving rules are not displayed. Contiguous lines indicate the first discovery of a child by DFS and dashed lines indicate rediscoveries.

The problem can be analysed by a reduction to a binary grammar problem with the same parameters D and $\mathbf{p}$. Assign to each string v of the binary grammar problem the set of strings that only differ from v by (at most) an extra S. We call such sets *node clusters*. For example, $\{a, Sa, aS\}$ constitutes the node cluster corresponding to a. Due to the abusing of levels for the S-less strings, all members of a cluster appear on the same level (the level is equal to the number of a's and b's). The level is also the same as the corresponding string in the binary grammar problem.

Lemma 5 (Binary Grammar Reduction). *For every* $n, d, n \le d$, *the descendant counter* L^{FG} *of the full grammar problem is* $L^{FG}(n, d) = (d+2)L^{BG}(n, d)$.

Proof. $L^{BG}(n, d)$ counts the level d descendants of a^n in the binary grammar problem (BGP), and $L^{FG}(n, d)$ counts the level d descendants of Sa^n in the full grammar problem (FGP). The node u is a child of v in BGP iff the members of the u node cluster are descendants of Su. Therefore the node clusters on level d

descending from Sa^n in FGP correspond to the BGP nodes descending from a^n. At level d, each node cluster contains $d + 2$ nodes.

Corollary 3 (Expected Full Grammar BFS Search Time). *The expected BFS search time* $\tilde{t}_{FG}^{DFS}(\mathbf{p})$ *in a full grammar problem of depth D with goal probabilities* $\mathbf{p} = [p_0, \ldots, p_D]$ *is*

$$t_{FG}^{BFS}(\mathbf{p}) := t_{CB}^{BFS}(\mathbf{p}, L^{FG}).$$

Corollary 4 (Expected Full Grammar DFS Search Time). *The expected DFS search time* $\tilde{t}_{FG}^{DFS}(D, p)$ *in a full grammar problem of depth D with goal probabilities* $\mathbf{p} = [p_0, \ldots, p_D]$ *is bounded between* $t_{FGL}^{DFS}(D, \mathbf{p}) := t_{CBL}^{DFS}(D, \mathbf{p}, L^{FG})$ *and* $t_{FGU}^{DFS}(D, \mathbf{p}) := t_{CBU}^{DFS}(D, \mathbf{p}, L^{FG})$, *and is approximately*

$$\tilde{t}_{FG}^{DFS}(D, \mathbf{p}) := \tilde{t}_{CB}^{DFS}(D, \mathbf{p}, L^{FG}).$$

Proof (Proof of Corollaries 3 and 4). Direct application of Lemma 5, and Propositions 2 and 3 respectively.

5   Experimental Verification

To verify the analytical results, we have implemented the binary grammar in Python 3 using the `graph-tool` package (Peixoto 2015)[1]. The data reported in Table 1 is based on an average over 1000 independently generated search problems with depth $D = 14$. The first number in each box is the empirical average, the second number is the analytical estimate, and the third number is the percentage error of the analytical estimate.

For certain parameter settings, there is only a small chance $(< 10^{-3})$ that there are no goals. In such circumstances, all 1000 generated search graphs typically inhabit a goal, and so the empirical search times will be comparatively small. However, since a binary grammar of depth 14 has about $2^{15} \approx 3 \cdot 10^5$ nodes (and a search algorithm must search through all of them in case there is no goal), the rarely occurring event of no goal may still influence the expected search time substantially. To avoid this sampling problem, we have ubiquitously discarded all instances where no goal is present, and compared the resulting averages to the analytical expectations *conditioned on at least one goal being present*.

The binary grammar model of Sect. 4.1 serves to verify the general estimates of Propositions 2 and 3. The results are shown in Table 1. The estimates for BFS are accurate $(< 3\%$ error). With few exceptions, the lower and the upper bounds t_{BGL}^{DFS} and t_{BGU}^{DFS} of Corollary 2 for DFS differ by at most 50 % on the respective sides from the true (empirical) average. The arithmetic mean $\tilde{t}_{BG}^{DFS}$ often gives surprisingly accurate predictions $(< 4\%)$ except when t_{BGL}^{DFS} and t_{BGU}^{DFS} leave wide margins as to the expected search time (when $g = 14$, the margin is up to 84 % downwards and 125 % upwards). Even then, the $\tilde{t}_{BG}^{DFS}$ error remains within 30 %.

---

[1] Source code for the experiments is available at http://tomeveritt.se.

Table 1. Comparison of analytical estimates with empirical averages for BFS and DFS in binary grammars of depth $D = 14$. Goals are distributed on a single goal level g with goal probability p_g. The BFS estimates t_{BG}^{BFS} are highly accurate, and the averaged DFS estimates $\bar{t}_{BG}^{DFS}$ are mostly accurate. Each box contains empirical average/*analytical expectation*/error percentage.

$g\backslash p_g$	0.001	0.01	0.1
5		46.74	40.53
		46.64	*39.86*
		0.2%	1.7%
8	375.7	332.5	265.7
	378.0	*333.9*	*265.0*
	0.6%	0.4%	0.3%
11	2751	2145	2058
	2744	*2147*	*2057*
	0.3%	0.1%	0.%
14	17370	16480	16390
	17 380	*16 480*	*16 390*
	0.1%	0.%	0.%

(a) BFS t_{BG}^{BFS}

$g\backslash p_g$	0.001	0.01	0.1
5		30910	27840
		31 370	*30 190*
		1.5%	8.4%
8	28000	25160	15490
	27 410	*24 420*	*15 200*
	2.1%	2.9%	1.9%
11	17280	5932	1815
	16 790	*5806*	*1788*
	2.9%	2.1%	1.5%
14	1304	122.1	25.60
	1522	*164.6*	*20.06*
	17%	35%	22%

(b) Average DFS $\bar{t}_{BG}^{DFS}$

$g\backslash p_g$	0.001	0.01	0.1
5		30910	27840
		30 710	*29 080*
		0.7%	4.5%
8	28000	25160	15490
	25 740	*22 150*	*12 070*
	8.1%	12%	22%
11	17280	5932	1815
	14 160	*3822*	*918.6*
	18%	36%	49%
14	1304	122.1	25.60
	808.8	*54.12*	*3.990*
	38%	56%	84%

(c) Lower DFS t_{BGL}^{DFS}

$g\backslash p_g$	0.001	0.01	0.1
5		30910	27840
		32 020	*31 290*
		3.6%	12%
8	28000	25160	15490
	29 080	*26 690*	*18 340*
	3.8%	6.1%	18%
11	17280	5932	1815
	19 410	*7790*	*2657*
	12%	31%	46%
14	1304	122.1	25.60
	2236	*275.1*	*36.12*
	72%	125%	41%

(d) Upper DFS t_{BGU}^{DFS}

6   Discussion

Search and optimisation problems appear in different flavors throughout the field of artificial intelligence; in planning, problem solving, games, and learning. Therefore even minor improvements to search performance can potentially lead to gains in many aspects of intelligent systems. It is even possible to equate intelligence with (Bayesian expectimax) optimisation performance (Legg and Hutter 2007).

Summary. In this paper and Part I (Everitt and Hutter 2015b) we have derived analytical results for expected runtime performance. Part I focused on BFS and DFS *tree search* where explored nodes were not remembered. A vector $\mathbf{p} = (p_1, \ldots, p_D)$ described *a priori* goal probabilities for the different levels of the tree. This concrete but general model of goal distribution allowed us to calculate approximate closed-form expression of both BFS and DFS average runtime. Earlier studies have only addressed *worst-case* runtimes: Knuth (1975) and followers for DFS; Korf et al. (2001) and followers for IDA*, effectively a generalised version of BFS.

This paper generalised the model of Part I to non-tree graphs. In addition to the goal probability vector $\mathbf{p}$, the graph search analysis required additional structural information in the form of a descendant counter L. The graph search estimates for DFS also took the form of less precise bounds. Even so, the arithmetic mean of the lower and the upper bound often came close the empirical average. The analysis of this paper does not supersede the results of Part I, as the bounds become uninformative when the graph is a tree. Overall, the analytical approximations derived in both papers were generally consistent with experimental outcomes.

Conclusions and Outlook. The value of the results are at least twofold. They offer a concrete means of deciding between BFS and DFS given some rough idea of the location of the goal (and the graph structure). To make the results more generally usable, automatic inference of model parameters would be necessary; primarily of goal distribution p and graph structure L. (The depth D will often be set by the searcher itself, and perhaps be iteratively increased.) There is good hope that the descendant counter L can be estimated online from the local sample obtained during search, similar to Knuth (1975). The goal distribution is likely to prove more challenging, but resembles the automatic creation of heuristic functions, so techniques such as *relaxed problems* could well prove useful (Pearl 1984). Estimates of goal distribution could possibly also be inferred from a heuristic function.

The results also offer theoretical insight into BFS and DFS performance. As BFS and DFS are in a sense the most fundamental search operations, we have high hopes that our results and techniques will prove useful as building blocks for analysis of more advanced search algorithms. For example, A* and IDA* may be viewed as a generalisations of BFS, and Beam Search and Greedy Best-First as generalisations of DFS.

Acknowledgements. Thanks to David Johnston for proof reading final drafts of both papers.

References

Edelkamp, S., Schrödl, S.: Heuristic Search. Morgan Kaufmann Publishers Inc, San Francisco (2012)

Everitt, T., Hutter, M.: A topological approach to Meta-heuristics: analytical results on the BFS vs. DFS algorithm selection problem. Technical report, Australian National University. arXiv:1509.02709[cs.AI] (2015a)

Everitt, T., Hutter, M.: Analytical results on the BFS vs. DFS algorithm selection problem. In: 28th Australian Joint Conference on Artificial Intelligence, Part I: Tree Search (2015b)

Hutter, F., Xu, L., Hoos, H.H., Leyton-Brown, K.: Algorithm runtime prediction: methods and evaluation. Artif. Intell. **206**(1), 79–111 (2014)

Knuth, D.E.: Estimating the efficiency of backtrack programs. Math. Comput. **29**(129), 122–122 (1975)

Korf, R.E., Reid, M., Edelkamp, S.: Time complexity of iterative-deepening-A*. Artif. Intell. **129**(1–2), 199–218 (2001)

Kotthoff, L.: Algorithm selection for combinatorial search problems: a survey. AI Magazine, pp. 1–17 (2014)

Legg, S., Hutter, M.: Universal intelligence. Minds Mach. **17**(4), 391–444 (2007)

Pearl, J.: Heuristics: Intelligent Search Strategies for Computer Problem Solving. Addison-Wesley, Boston (1984)

Peixoto, T.P.: The graph-tool python library. figshare (2015)

Rice, J.R.: The algorithm selection problem. Adv. Comput. **15**, 65–117 (1975)

Russell, S.J., Norvig, P.: Artificial Intelligence: A Modern Approach, 3rd edn. Prentice Hall, Upper Saddle River (2010)

Region-Growing Planar Segmentation
for Robot Action Planning

Reza Farid[(✉)]

Institute for Integrated and Intelligent Systems, Griffith University,
170 Kessels Rd, Nathan, QLD 4111, Australia
r.farid@griffith.edu.au

Abstract. Object detection, classification and manipulation are some of
the capabilities required by autonomous robots. The main steps in object
classification are: segmentation, feature extraction, object representation
and learning. To address the problem of learning object classification
using multi-view range data, we used a relational approach. The first
step of our object classification method is to decompose a scene into
shape primitives such as planes, followed by extracting a set of higher-
level, relational features from the segmented regions. In this paper, we
compare our plane segmentation algorithm with state-of-the-art plane
segmentation algorithms which are publicly available. We show that our
segmentation outperforms visually and also produces better results for
the robot action planning.

Keywords: Object classification · Robot action planning · Planar
segmentation · Point cloud · Range data

1 Introduction

A considerable amount of research has been devoted to generic object recognition
(Opelt, 2006; Vasudevan et al., 2007; Shin, 2008; Endres, 2009), which is required
by robots in many tasks. For instance, in service robotics applications, such as a
catering or a domestic robot (Shin, 2008), the robot must recognise specific kinds
of tableware, while the robot's ability to distinguish a set of products is necessary
in industrial applications (Endres, 2009). We are mostly interested in urban
search and rescue; where a team of robots are sent to a post-disaster environment.
The robot's mission is to traverse the arena, to search for victims while making
a map of the area. Rescue robots may be tele-operated or autonomous. When
running autonomously, classification of objects is useful for reporting to human
rescuers what is in the environment as well as determining the robot's behaviour.
For example, recognising a *staircase* can provide necessary information to a
wheeled robot (Fig. 1a) to avoid that object, whereas a tracked robot (Fig. 1b) is
capable of climbing stairs but it must reconfigure its flippers (Fig. 2) to be able
to climb (Kalantari et al., 2009) successfully as shown in Fig. 3. Another example
is to use the relation between surfaces to grasp objects (Prankl et al., 2013).

© Springer International Publishing Switzerland 2015
B. Pfahringer and J. Renz (Eds.): AI 2015, LNAI 9457, pp. 179–191, 2015.
DOI: 10.1007/978-3-319-26350-2_16

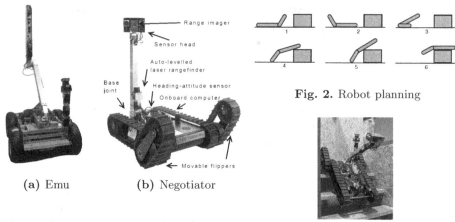

(a) Emu   (b) Negotiator

Fig. 2. Robot planning

Fig. 1. Some autonomous robots (Team CASuality in RoboCup 2011) (McGill et al., 2012)

Fig. 3. Robot climbing stairs

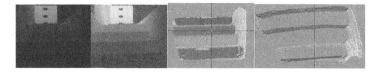

Fig. 4. Range image and corresponding point cloud from front and top view (Farid and Sammut, 2014a)

Range cameras have become popular in robotics because they are small, light, consume relatively little power and have the ability to produce range measurements of up to several metres, making them suitable for indoor use. Range images are acquired by these 3D range/depth cameras, such as the Microsoft Kinect and ASUS Xtion PRO LIVE. These images are like grey scale images in which the value of each pixel represents the distance of the sensor to the surface of an object in a scene from a specific viewpoint (Gächter et al., 2006), and can be used to infer the shape of the object (Hegazy and Denzler, 2009). The Kinect and Xtion sensors also provide a colour video image. However, in this research, only the depth image is used for object recognition as it is capable of operating in the dark, which is often required in search and rescue operations. Furthermore, colour calibration under different lighting conditions is troublesome (Opelt, 2006).

A range image can be transformed into a point cloud by converting each pixel of the image into 3D coordinates. Figure 4 shows a range image of a *staircase* with four steps. The image was taken by a robot positioned in front of the staircase. In the leftmost image (the grey scale), darker colours represents closer surfaces. The next image (a colour-mapped version) is presented for clarity. The next two images are front and top views for the same point cloud, in which the point cloud is segmented into planes that are identified by unique colours. Since a range image is taken from one viewpoint, it only provides a partial view of a scene.

An object class describes a set of instances that share common properties, such as shape or function. A common starting point for finding the properties to be used in similarity matching is to segment the image into different regions and to characterise the relationships between those regions. Segmenting a point cloud can be viewed as the process of assigning each point to a region, with an accompanying label.

In previous work (Farid and Sammut, 2012b; 2014a), we extracted planes from a 3D point cloud based on a region growing plane segmentation algorithm (Farid and Sammut, 2012a; 2014c) and used them as primitives for object categorisation. In this paper, we show that our plane segmentation algorithm outperforms state-of-the-art plane segmentation methods which are publicly available. For this purpose, in the following sections, we compare the methods based on their visual results and the suitability of the plane features for robot action planning.

2   Background Work

Planes are useful features in built environments, including urban search and rescue for identifying floors, walls, stairs, ramps and other terrain that the robot is likely to encounter. Modelling a scene from planar patches is used in computer vision, robotics and augmented reality (Prankl et al., 2013). For example, it has been used for scene understanding (Bartoli, 2007; Xu and Petrou, 2011), localisation (Mohr et al., 1992) and 3D virtual reconstruction of the environment (McGill et al., 2012).

Our earlier approach (Farid and Sammut, 2012b; 2014a) was most closely related to Shanahan (2002) and Shanahan and Randell (2004) who used a logic program as a relational representation for 3D objects in 2D line drawings, while abduction is used in object recognition. We extended this representation, replacing the 2D lines with 3D planes. Furthermore, we used ALEPH (Srinivasan, 2002) to learn the logic programs from instances obtained by a robot equipped with a depth camera.

The fact that all points belonging to the same plane are supposed to have approximately the same normal vector, formed the core of the our segmentation algorithm. We introduced a region-growing plane segmentation algorithm based on neighbourhood normal vector similarity to segment an object into a set of planar surfaces. The method starts with a point and traverses the other points through the neighbourhood structure. To decide if the point can be added to the planar surface, it must satisfy the planar surface criteria, which determine when to add a point to a region. Our algorithm is based on using neighbouring points to grow the region. This is where the distance threshold, δ, can be used to decide whether a point is too far away to be accepted as a neighbour for a point. If a point is not too far, it can be included in the *not visited neighbours* list, *candidates*, as shown in the Algorithm 1. We have used the below values for input variables:

```
min_neighbour_num = 4          base_update_step = 8
num_initial_points = 16        θ = 15° − 20°
```

Features of these planar regions and their relationships were generated to form a planar description for an object class. The segmentation result and features were used for learning. We also showed that the learning system was able to use other primitives such as cylinders and spheres for the same purpose (Farid and Sammut, 2014b). A relational representation is useful in this application because it is our interest to recognise objects that are characterised by relationships between its parts, as in the steps constituting a staircase, and the number of parts may not be fixed, as the number of steps in a staircase can vary.

Ideally, an off-the-shelf segmentation method would have been used to decompose a scene into shape primitives. Several methods claim to provide good plane segmentation. However, they are not publicly available or they are not usable as claimed (Farid, 2014). Two algorithms are provided by PCL (Rusu and Cousins, 2011) using the RANdom SAmple Consensus (RANSAC) (Fischler and Bolles, 1981) algorithm. We use these state-of-the-art plane segmentation methods especially because they are publicly available:

– One of the PCL algorithms, setting the model type as SACMODEL_PLANE, uses 3D points belonging to the point cloud[1] without considering normals or any additional constraints. We call it as SP.
– The other algorithm, using SACMODEL_NORMAL_PLANE for model type, has an additional constraint similar to the method used in our research. We call it as SNP. It assumes the normal of each point must be parallel to the output plane normal within a maximum angular difference[2]. The use of SNP has been shown as a part of PCL's tutorial for cylinder model segmentation[3].

These two methods require a few parameters such as "Distance Threshold" and "Angle Threshold" to decide whether a point must be added to a plane. We will discuss these parameters later.

PCL has an algorithm as region growing segmentation. However, this algorithm merges the points to form a segment considering a smoothness constraint. The output clusters can be considered as smooth surfaces, not primitives such as planes, spheres and cylinders. This algorithm can be used to cluster the point cloud before passing each cluster to other segmentation algorithms such as SP and SNP. That is why we will not consider this algorithm for comparison in this paper. Our algorithm will be compared with SP and SNP visually and based on the quality of the segmented planes.

3   Experimental Evaluation

3.1   Dataset

Figure 1 shows several robots used for urban search and rescue. These ground robots were designed to participate in the RoboCup Rescue Robot competition,

---

[1] http://pointclouds.org/documentation/tutorials/planar_segmentation.php.

[2] http://docs.pointclouds.org/1.7.0/group__sample__consensus.html.

[3] http://pointclouds.org/documentation/tutorials/cylinder_segmentation.php.

Algorithm 1. Region growing plane segmentation algorithm using normal vectors

---

Input: *PointCloud, normal vector for all points in PointCloud*
Input: $min\_neighbour\_num > 0$, $base\_update\_step > 0$
Input: $num\_initial\_points > 0$, min_region_size
Input: θ // angle threshold
Input: δ // distance threshold
Input: $angle\_mf < 1$ // angle modifying factor

1: $R \leftarrow \{\}$ // output: Regions
2: **for all** p in the PointCloud **do**
3:     **if** *p is visited* $\vee$ *p is rejected* **then**
4:         continue
5:     **else if** $number\_of\_usable\_neighbour(p) < min\_neighbour\_num$ **then**
6:         continue
7:     **end if**
8:     $C_R \leftarrow p$
9:     $Base\_normal \leftarrow get\_normal\_vector(p)$
10:     $candidates \leftarrow get\_not\_visited\_neighbours(p, \delta)$
11:     **for all** q in candidates **do**
12:         **if** $Size(C_R) < num\_initial\_points \vee mod(Size(C_R), base\_update\_step) = 0$ **then**
13:             $Base\_normal \leftarrow get\_average\_normal\_vectors(C_R)$
14:         **end if**
15:         $current\_angle \leftarrow get\_angle(Base\_normal, get\_normal\_vector(q))$
16:         $accepted \leftarrow$ **false**
17:         **if** $Size(C_R) < num\_initial\_points$ **then**
18:             **if** $current\_angle < \theta$ **then**
19:                 $accepted \leftarrow$ **true**
20:             **end if**
21:         **else if** $current\_angle < \theta * angle\_mf$ **then**
22:             $accepted \leftarrow$ **true**
23:         **end if**
24:         **if** $accepted$ **then**
25:             $C_R \leftarrow C_R \cup q$
26:             $set\_visited(q)$
27:             $candidates \leftarrow candidates \cup get\_not\_visited\_neighbours(q, \delta)$
28:         **end if**
29:     **end for**
30:     **if** $Size(C_R) > min\_region\_size$ **then**
31:         $set\_final\_normal\_vector(C_R)$
32:         $R \leftarrow R \cup C_R$
33:     **end if**
34: **end for**
35: **return** R

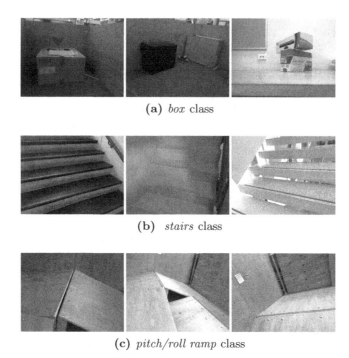

(a) *box* class

(b) *stairs* class

(c) *pitch/roll ramp* class

Fig. 5. Examples of instances used in this research

held annually. The competition arena uses elements developed by the US National Institute of Standards and Technology (NIST, 2010) to certify robots for emergency operations. These elements simulate typical hazards that might be encountered in buildings damaged by a disaster such as an earthquake. We captured data during RoboCup Rescue competitions, as well as from rescue laboratories and other indoor locations. In this paper, we use a subset (45 images) of such data which we used for learning classes such as *box* (12 images), *stairs* (15 images) and *pitch/roll ramp* in a maze (18 images). Since it is difficult to comprehend the range image, the corresponding colour (RGB) image of the scene will be shown in the rest of the paper. For each class, different multi-view data are chosen. For example, Fig. 5 shows one view of some of the examples in this research which previously were used for training *box*, *stairs* and *pitch/roll ramp* classes respectively.

3.2   Parameters

Distance Threshold. SP and SNP are using the distance threshold parameter to limit the maximum acceptable distance of a point to the plane model. If the point is further, it will not be considered as an inlier for the plane. In the PCL tutorial, this value is set to 0.01 for SP. However, for SNP, this value is set to 0.03. Due to this difference, we use more than one value as the distance threshold in our experiments. For example, for experiments regarding SP, we configured four experiments by using 0.005, 0.01, 0.03 and 0.05 as distance thresholds.

Angle Threshold. SNP also considers the surface normal of each point and employs a weight value to determine the surface normal influence. We used the suggested value by the PCL tutorial as 0.1. We also have used similar threshold (15) in our region-growing algorithm to accept or reject adding a point to the current plane. In other words, if the angle of the current plane and the candidate point is more than 15, the point can not be added to the region.

Minimum Region Size. All methods use a value as the minimum size for the plane. If the number of points belonging to the region is less than this value, the region will be rejected. We employed 90 for this purpose.

3.3   Data Preparation

Before applying PCL plane segmentation algorithms, we must prepare our data. Since our range images have 640 × 480 pixels, we sub-sample them to 160 × 120, while we converting them to point clouds. All data and the result of experiments are available via http://rfarid.altervista.org/plane_seg_compare/index.html.

3.4   Evaluating SP

The first experiment set is based on using the first PCL plane segmentation algorithm, called as SP. We applied SP on our data four times by using 0.005, 0.01, 0.03 and 0.05 as distance thresholds.

Considering the Number of Planes. Table 1 shows the total and average number of planes produced per each class. This table indicates that the number of planes is closer and more reliable using distance thresholds 0.03 and 0.05.

Table 1. Total and average number of planes using SP

	Sum				Average			
	Distance threshold				Distance threshold			
Class	**0.005**	**0.01**	**0.03**	**0.05**	**0.005**	**0.01**	**0.03**	**0.05**
box	182	108	67	60	15.17	9	5.58	5
pitch/roll ramp	493	281	144	116	27.39	15.61	8	6.44
stairs	329	226	124	84	21.93	15.07	8.27	5.6
Total	1004	615	335	260				

Table 2. Distribution of segmentation quality using SP

Distance threshold	Segmentation quality level			
	H	**MH**	**ML**	**L**
0.005	2.22 %	11.11 %	8.89 %	77.78 %
0.01	11.11 %	6.67 %	53.33 %	28.89 %
0.03	0.00 %	33.33 %	62.22 %	4.44 %
0.05	0.00 %	24.44 %	71.11 %	4.44 %

Table 3. Distribution of segmentation quality per object class using SP for distance threshold as 0.03

Class	Seg. quality level			
	H	**MH**	**ML**	**L**
box	0 %	83 %	17 %	0 %
pitch/roll ramp	0 %	28 %	61 %	11 %
stairs	0 %	0 %	100 %	0 %

Fig. 6. Example of SP segmentation result for *stairs* using dis. thr. as 0.03

Visual Quality of the Output. We defined four levels of segmentation quality as H, MH, ML and L indicating high, mid to high, mid to low and low respectively. We went through all visual results and scored them based on a human-manual segmentation expectation. Table 2 shows the percentage of the images per each distance threshold and segmentation quality level. It illustrates that although we get high segmentation quality around 2 % and 11 % of times for using distance threshold as 0.005 and 0.01, these thresholds cause less quality level of segmentation. In contrast, using threshold as 0.03 and 0.05 produces results with the mid to low and mid to high level of segmentation quality.

Table 3 shows the same distribution per class while we used 0.03 as distance threshold. It indicates that the 83 % of images containing *box* class are segmented with the mid to high level of segmentation quality, while 61 % of *pitch/roll ramp* images have mid to low level of segmentation quality and all the *stairs* class images are segmented with a mid to low level of segmentation quality. Figure 6 shows three examples of *stairs* using SP segmentation (with distance threshold as 0.03) corresponding to the scenes shown in Fig. 5b. Each plane is coloured differently. All the segmentation results are available in the experiment website.

3.5   Evaluating SNP

The second experiment set is based on using the second PCL plane segmentation algorithm, called as SNP, which employs normal vectors in its process. Since using 0.005 as distance threshold caused a major low quality segmentation for SP, we avoided using 0.005 and applied SNP on our data three times by using 0.01, 0.03 and 0.05 as distance threshold values.

Table 4. Total and average number of planes using SNP

Class	Sum			Average		
	Distance threshold					
	0.01	**0.03**	**0.05**	**0.01**	**0.03**	**0.05**
box	235	98	89	19.58	8.17	7.42
pitch/roll ramp	456	269	177	25.33	14.94	9.83
stairs	174	205	191	12.43	13.67	12.73
Total	865	572	457	19.22	12.71	10.16

Table 5. Distribution of segmentation quality using SNP

Distance threshold	Segmentation quality level			
	H	**MH**	**ML**	**L**
0.01	0.0 %	0.0 %	2.2 %	97.8 %
0.03	0.0 %	51.1 %	48.9 %	0.00 %
0.05	4.4 %	80.0 %	15.6 %	0.00 %

Table 6. Distribution of segmentation quality per class using SNP

Dist. Thr.=0.05	Seg. quality level			
Class	**H**	**MH**	**ML**	**L**
box	8.3 %	75.0 %	16.7 %	0.00 %
pitch/roll ramp	0.0 %	72.2 %	27.8 %	0.00 %
stairs	6.7 %	93.3 %	0.0 %	0.00 %

Considering the Number of Planes. Table 4 shows the total number and average number of planes produced using SNP per each class. This table indicates that the number of planes are closer and more reliable using distance thresholds as 0.03 and 0.05.

Visual Quality of the Output. For analysing the visual quality of the output, we used the same approach employed for SP. Table 5 shows the percentage of the images per each distance threshold and segmentation quality level. It illustrates that the distance threshold set at 0.05 produces more mid to high quality segmentation.

Table 6 shows the same distribution per each class while we used 0.05 as distance threshold. It indicates that for each class images, the majority of segmentation quality are mid to high. The visual comparison between SP and SNP results shows that SNP outperforms SP. Figure 7 shows the corresponding version of Fig. 6 using SNP segmentation (with distance threshold as 0.05), where each plane is coloured differently. All the segmentation results are available in the experiment website.

3.6   Comparing SNP and Our Method

We applied our region-growing plane segmentation algorithm on the same data. As shown before, SNP outperformed SP, so we just compared the result of segmentations between ours and SNP (using distance threshold 0.05) as follows:

Fig. 7. Example of SNP segmentation result for *stairs* using dis. thr. as 0.05

Considering the Number of Planes. The average number of planes is 9.44 for ours while this average is 10.16 for SNP.

Visual Quality of the Output. To compare the results visually, we split a score 100 between the result of our segmentation method and SNP. We scored SNP as 2138 totally, which means 47.51 on average. Additionally, we also asked some people to do the same. We provided a web-page[4] showing the RGB version of the scene and the results of segmentation for method 1 and 2. The participants did not know which method was which. They were asked to split the score 100 between the two methods based on their expectations of the correct manual segmentation. SNP was scored 46.86 on average, while our algorithm was scored 53.14 on average. This comparison shows that the our segmentation algorithm outperforms the SNP visually.

Comparison Based on the Quality of the Features. Visual comparison might not be good enough to compare two segmentation methods. Since the result of segmentation can be passed to a robot as features for action planning, it is important to evaluate the correctness of these features, which is not possible to do just by visual comparison. In this case, a plane can be represented by a point belonging to the plane, its normal vector and its boundaries. The boundary can be represented by a convex hull (Farid and Sammut, 2012b). That is where SNP fails. SNP uses RANSAC and produces planes that cover many sparse points. As a result, two set of points, which are very far from each other, are put together in the same plane, while there is no such planar surface in the reality. These virtual planes can interfere with robot action planning, since there is no planar surface where the robot expects one based on the features provided. Figure 8 shows an example of such situation. Using the colour legend provided, the figure shows that our segmentation produces 10 planes, while SNP produces 12 planes in which planes coloured as regions 1, 3, 8, 9 and 10 are sparse and the corresponding features will be problematic. Figure 9 shows another example based on the leftmost *stairs* instance in Fig. 5b. SNP produces regions 8, 9, 12 and 13 by putting edges of steps together as planes, which cause trouble when the robot uses these planes for actions such as climbing.

---

[4] http://rfarid.altervista.org/plane_seg_compare/comp.html.

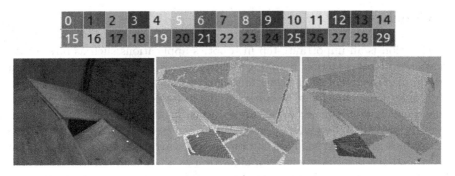

Fig. 8. Colour legend for segmentation, RGB version of the scene and results of segmentations using our algorithm (left) and SNP (right) (Color figure online).

Table 7. Distribution of sparse planes using SNP (distance threshold = 0.05)

Number of sparse planes	Frequency
0	5
1	13
2	9
3	9
4	7
5	2

Fig. 9. Segmented planes using our algorithm (left) and SNP (right)

Considering this sparse issue, we counted the number of sparse planes for the segmentation results by SNP. The details are available in the experiment website. Table 7 shows the numerical results of this evaluation and how the number of sparse regions per images are distributed. Based on these numbers, we can say 96 planes of total 457 planes for 45 images had this issue due to SNP segmentation. That is, there is an average of 2.13 planes per image affected by this issue.

Comparison on Distance Threshold. As shown before, SP and SNP are sensitive to the value chosen for the distance threshold. Some images might work well with one value and another value might produces better results on another subset of images. SP and SNP do not suggest any systematic way to define the distance threshold. Our algorithm calculates the distance threshold based on each image automatically by using the minimum distance between each point and its adjacent neighbours and finding the average of them as the base. So, this distance threshold is also reliable in existence of noisy data. The detail and the relevant experiments are provided in Farid and Sammut (2014c).

4   Conclusion

Segmentation is an important step in robotics applications such as object classification and robot action planning. In this paper, we compared our region-growing plane segmentation algorithm with two state-of-the-art plane segmentation algorithms which are publicly available by PCL. We showed that the visual quality of our segmentation outperforms the others. We also showed that those RANSAC based segmentation algorithm can create planes with very sparse points which provide wrong information for robot action planning. We are planning to add our plane segmentation algorithm as a new method to PCL and/or ROS. The URL http://rfarid.altervista.org/plane_seg_compare/index.html provides the data and the detailed results.

Acknowledgement. We thank the people who kindly participated on visual comparison between our method and SNP.

References

Bartoli, A.: A random sampling strategy for piecewise planar scene segmentation. Comput. Vis. Image Underst. **105**(1), 42–59 (2007)

Endres, F.L.: Scene Analysis from Range Data. Master thesis, Albert-Ludwigs-University Freiburg, Faculty of Applied Sciences (2009)

Farid, R.: Generic 3D Object Recognition Using Multi-view Range Data. Ph.D. thesis (2014). URL http://handle.unsw.edu.au/1959.4/53848

Farid, R., Sammut, C.: A relational approach to plane-based object categorisation. In: RSS 2012 Workshop on RGB-D: Advanced Reasoning with Depth Cameras (2012a). http://mobilerobotics.cs.washington.edu/rgbd-workshop-2012/papers/farid-rgbd12-object-categorization.pdf

Farid, R., Sammut, C.: Plane-based object categorisation using relational learning. In: Online Proceedings of ILP 2012 (2012b). URL http://ida.felk.cvut.cz/ilp2012/wp-content/uploads/ilp2012_submission_6.pdf

Farid, R., Sammut, C.: Plane-based object categorisation using relational learning. Mach. Learn. **94**(1), 1–21 (2014a). doi:10.1007/s10994-013-5352-9

Farid, R., Sammut, C.: Region-based object categorisation using relational learning. In: Pham, D.-N., Park, S.-B. (eds.) PRICAI 2014. LNCS, vol. 8862, pp. 357–369. Springer, Heidelberg (2014). doi:10.1007/978-3-319-13560-1_29

Farid, R., Sammut, C.: Plane-based object categorisation using relational learning: implementation details and extension of experiments. Technical Report UNSW-CSE-TR-201416, School of Computer Science and Engineering, The University of New South Wales (2014c). URL ftp://ftp.cse.unsw.edu.au/pub/doc/papers/UNSW/201416.pdf

Fischler, M.A., Bolles, R.C.: Random sample consensus: a paradigm for model fitting with applications to image analysis and automated cartography. Comm. ACM **24**(6), 381–395 (1981). doi:10.1145/358669.358692

Gächter, S., Nguyen, V., Siegwart, R.: Results on range image segmentation for service robots. In: Proceedings of IEEE International Conference on Computer Vision Systems, pp. 53–53 (2006). doi:10.1109/ICVS.2006.54

Hegazy, D., Denzler, J.: Generic 3D object recognition from time-of-flight images using boosted combined shape features. In: Ranchordas, A., Araújo, H., (eds.) Proceedings of the 4th International Conference on Computer Vision, Theory and Applications, vol. 2, pp. 321–326. INSTICC Press (2009)

Kalantari, A., Mihankhah, E., Moosavian, S.A.A.: Safe autonomous stair climbing for a tracked mobile robot using a kinematics based controller. In: IEEE/ASME International Conference on Advanced Intelligent Mechatronics (AIM2009), pp. 1891–1896 (2009). doi:10.1109/AIM.2009.5229765

McGill, M., Salleh, R., Wiley, T., Ratter, A., Farid, R., Sammut, C., Milstein, A.: Virtual reconstruction using an autonomous robot. In: Proceedings of the International Conference on Indoor Positioning and Indoor Navigation (IPIN2012), pp. 1–8 (2012). doi:10.1109/IPIN.2012.6418851

Mohr, R., Morin, L., Grosso, E.: Relative positioning with uncalibrated cameras. In: Mundy, J.L., Zisserman, A. (eds.) Geometric Invariance in Computer Vision, pp. 440–460. MIT Press, Cambridge (1992). http://dl.acm.org/citation.cfm?id= 153634.153656, ISBN 0-262-13285-0

NIST: The national institute of standards and technology; test methods. Retrieved 14–02-2014 (2010). URL http://www.nist.gov/el/isd/test-methods.cfm

Opelt, A.: Generic Object Recognition. Ph.D. thesis, Graz University of Technology (2006)

Prankl, J., Zillich, M., Vincze, M.: Interactive object modelling based on piecewise planar surface patches. Comput. Vis. Image Underst. 117(6), 718–731 (2013). doi:10. 1016/j.cviu.2013.01.010. ISSN 1077–3142

Rusu, R.B., Cousinsm, S.: 3D is here: point cloud library (PCL). In: Proceedings of ICRA 2011, pp. 1–4 (2011). doi:10.1109/ICRA.2011.5980567

Shanahan, M.: A logical account of perception incorporating feedback and expectation. In: Proceedings of 8th International Conference on Principles of Knowledge Representation and Reasoning, pp. 3–13. Morgan Kaufmann, Toulouse, France (2002)

Shanahan, M., Randell, D.: A logic-based formulation of active visual perception. In: Dubois, D., Welty, C.A., Williams, M.-A., (eds.) Proceedings of KR 2004, pp. 64–72. AAAI Press (2004)

Shin, J.: Parts-Based Object Classification for Range Images. Ph.D. thesis, Swiss Federal Institute of Technology Zurich (2008)

Srinivasan, A.: The Aleph Manual (Version 4 and above). Technical report, University of Oxford (2002)

Vasudevan, S., Gächter, S., Nguyen, V., Siegwart, R.: Cognitive maps for mobile robots-an object based approach. Robot. Auton. Syst. (From Sensors to Human Spatial Concepts) 55(5), 359–371 (2007). doi:10.1016/j.robot.2006.12.008

Xu, M., Petrou, M.: 3D scene interpretation by combining probability theory and logic: the tower of knowledge. Comput. Vis. Image Underst. 115(11), 1581–1596 (2011). doi:10.1016/j.cviu.2011.08.001

A Differentially Private Random Decision Forest Using Reliable Signal-to-Noise Ratios

Sam Fletcher[✉] and Md Zahidul Islam

School of Computing and Mathematics, Charles Sturt University, Bathurst, Australia
{safletcher,zislam}@csu.edu.au

Abstract. When dealing with personal data, it is important for data miners to have algorithms available for discovering trends and patterns in the data without exposing people's private information. Differential privacy offers an enforceable definition of privacy that can provide each individual in a dataset a guarantee that their personal information is no more at risk than it would be if their data was not in the dataset at all. By using mechanisms that achieve differential privacy, we propose a decision forest algorithm that uses the theory of Signal-to-Noise Ratios to automatically tune the algorithm's parameters, and to make sure that any differentially private noise added to the results does not outweigh the true results. Our experiments demonstrate that our differentially private algorithm can achieve high prediction accuracy.

Keywords: Differential privacy · Noise · Decision tree · Data mining

1 Introduction

Collecting data about people–whether it be for economic, medical, political, militaristic or academic purposes–is becoming increasingly commonplace. With it, comes the question of privacy: what safeguards are data collectors required to enforce to protect the privacy of those whose data they are collecting? The world at large has agreed that privacy is a human right [18]; fortunately, this right can still be upheld without forgoing the benefits of data collection wholesale. Instead, privacy preservation techniques can be employed to remove or distort identifying markers about individuals in the data without ruining the underlying trends and patterns in the data.

Balancing the loss of information with the gain in privacy has been an active area of research for nearly two decades [11]. In the early years, a privacy-preservation technique known as generalization was at the forefront of research, with k-anonymity [17] gaining a lot of support and leading to the development of machine learning techniques that minimized the loss of information in the data while still preserving privacy [11]. One thing these approaches lacked however was a strong definition of "privacy". This was rectified by differential privacy, proposed in 2006 [4–6,15,16]. It made the following promise to each individual in the data: "Any information that could be discovered about you with your data

© Springer International Publishing Switzerland 2015
B. Pfahringer and J. Renz (Eds.): AI 2015, LNAI 9457, pp. 192–203, 2015.
DOI: 10.1007/978-3-319-26350-2_17

in the dataset could also, with high probability, be discovered *without* your data in the dataset". In other words, the output of any query Q performed on dataset D will be indistinguishable from the output of the same query Q performed on dataset D', where D' differs from D by one record (the record of an individual).

Using differential privacy, we propose a data mining algorithm that builds an ensemble of randomized decision trees (i.e. a forest), queries the data in a differentially private manner, and outputs a classifier capable of high accuracy even with very high privacy requirements. We approach the problem by phrasing it in terms of the Signal-to-Noise Ratio of our queries, and using signal averaging to reduce the noise in the queries.

1.1   Problem Statement

Dataset D is a two-dimensional matrix of rows and columns, where each row (i.e. record) $r \in D$ describes a single individual, and each column is an attribute a in the set of attributes A. Each r possesses one discrete value $v \in a; \forall a \in A$. We symbolize that record r has value v for attribute a by writing $r_a = v$. Each r also has a class value c, from the class attribute C. The aim of a decision forest is to correctly predict r_C (the class value c of record r) for records $r \in B : B \cap D = \emptyset$, where B and D are drawn from the same population.

A user is given limited access to D, in which they are allowed to query D in an ϵ-differentially private way. For any given query Q, the value of ϵ can be equal to or less than the total privacy budget β provided to the user by the data owner. We will be dividing β into smaller parts for each query Q. Our aim is to build τ decision trees by only submitting ϵ-differentially private queries Q to D, and without exceeding our total budget β. The decision trees need to have acceptably high prediction accuracy in order to be valuable.

Decision Trees [13] work by iteratively selecting attributes in a dataset that can most accurately classify a class attribute.[1] When an attribute is selected, the records in the dataset are split up according to what value they have for the chosen attribute. For each of these partitions, the process is then repeated until a user-defined termination condition is met.

1.2   Our Contributions

Our novel contributions can be summarized as the following:

- We re-phrase the problem of making a differentially private data mining algorithm in a novel way, by using Signal-to-Noise Ratio theory to assess the noise added to differentially private queries (Sect. 3).
- We present a differentially-private randomized decision forest algorithm (referred to as DP-RF) in which the structure of the decision trees is decided before querying the dataset D at all (Sect. 4).

---

[1] The class attribute is the attribute that the user wishes to accurately predict the value of for future records, where the value is not known.

- Our algorithm automatically tunes all parameters, with the only inputs being: access to the secure dataset D, the domains of the attributes A, and the privacy budget β (Sects. 4.3, 4.4 and 4.5).
- We take full advantage of the benefits of randomly built decision trees, while identifying the assumptions usually made about decision trees that no longer hold, and providing solutions (Sect. 4.6).

We also provide code for our algorithm online.[2]

2   Previous Work

2.1   Differential Privacy

We provide a brief summary of the main components of differential privacy that we use in this paper. We refer the reader to [6] for a more thorough introduction. Differential privacy can be formally defined as follows:

Definition 1 (Differential Privacy [4]). *A query $Q : Q(D) \rightarrow Y$ satisfies ϵ-differential privacy if for all datasets D and D' differing by at most one record,*

$$Pr(Q(D) = y \in Y) \leq e^{\epsilon} \times Pr(Q(D') = y \in Y). \tag{1}$$

Common values for ϵ range from 0.005 to 0.1. This definition allows a data collector to make a strong promise to each individual in D: that for any query Q, the output observed is $1/e^{\epsilon}$ as likely to occur even if they had not been in D. It does not promise that a malicious user cannot find out any information about them, but it does promise that any information they can find, they could have found without the individual even being in D.

In order for Definition 1 to be possible for query Q to achieve, there must be a randomized component in Q, preventing any output y from being 100% likely. One mechanism commonly used to inject randomness into queries is the Laplace Mechanism. Before we define this mechanism, we first need to define the "sensitivity" of Q:

Definition 2 (Sensitivity [5]). *A query Q has sensitivity $\Delta(Q)$, where:*

$$\Delta(Q) = \max_{K,K'} |Q(K) - Q(K')| \tag{2}$$

and K and K' are any datasets that differ by at most one record.

Using Definition 2, we now define:

Definition 3 (The Laplace Mechanism [5,6]). *A query Q satisfies ϵ-differential privacy if it outputs $y + Lap(\frac{\Delta(Q)}{\epsilon})$, where $y \in Y : Q(D) \rightarrow Y$ and $Lap(x)$ is an i.i.d. random variable drawn from the Laplace distribution with mean 0 and scale x (i.e. variance $2x^2$).*

---

[2] Our code can be found at http://csusap.csu.edu.au/zislam/, or you can email us.

We will later take advantage of two more theorems that have been proven about differential privacy:

Definition 4 (The Composition Theorem [16]). *The application of queries Q_i, each satisfying ϵ_i-differential privacy, satisfies $\sum_i \epsilon_i$-differential privacy.*

Definition 5 (The Parallel Composition Theorem [15]). *Let D_i be a disjoint subset of dataset D, and let $Q_i(D_i)$ satisfy ϵ-differential privacy; then $\sum_i Q_i(D_i)$ also satisfies ϵ-differential privacy.*

2.2   Previous Differentially Private Decision Trees

The work most closely related to ours is the differentially private random decision forest proposed by [14] in 2012. The main differences between our proposed algorithm and theirs stem from our re-framing of the scenario in terms of Signal-to-Noise Ratios. Our algorithm automatically tunes the required parameters, while the work in [14] requires the user to manually set parameters, or otherwise uses heuristics. Their heuristics are based on the combinatorial reasoning used in an earlier randomized decision tree algorithm [7], which demonstrated that randomly built trees can actually produce high quality classifiers, even in scenarios without privacy restrictions. Other work since then has supported their findings [12]. Outside of randomly built decision trees, another differentially private data mining algorithm was proposed in 2010 [10], in which one decision tree was built using the Gini Index in the same way that CART does [3]. This work has since been improved upon [8].

3   Signal-to-Noise Ratio

Signal-to-Noise Ratio is a comparison between some sort of measurement (signal) and the background noise accompanying that measurement [19]. The ratio of the "proper" or "real" signal to the noise is the Signal-to-Noise Ratio, or SNR. The mathematics underpinning the SNR concept is applicable in any scenario where the following assumptions are met: the signal and the noise are uncorrelated; the signal would be the same if it was measured again (ignoring the noise); and the noise is random, with a mean of 0 and a constant variance for repeated measurements of the signal. Differential privacy meets these assumptions: noise added with the Laplace Mechanism is only dependent on ϵ and Δ, neither of which are affected by the signal; the number of records meeting a query's criteria will not change if the same dataset is queried again; and the Laplace distribution has a mean of 0 and a known, constant variance of $2x^2$.

Expressing the SNR mathematically is quite simple:

Definition 6 (Signal-to-Noise Ratio). *The Signal-to-Noise Ratio of a measurement can be expressed as $\frac{signal}{noise}$, where the signal and noise are expressed in the same units. An alternate way of writing this is:*

$$SNR = \frac{\mu}{\sigma} \tag{3}$$

where μ is the signal mean or expected value and σ is the standard deviation of the noise.

We take advantage of Definition 6 in Sect. 4.

3.1 Signal Averaging

Once a problem has been phrased in terms of the Signal-to-Noise ratio, there is additional property we are able to take advantage of: signal averaging. Signal averaging provides an explicit definition of the intuition that if noise can increase or decrease a signal with equal probability, then summing multiple signals together will result in a total that is less noisy:

$$\text{SNR} = \frac{\mu}{\sigma} = \frac{\sum_x^X \mu_x}{\sqrt{|X|\sigma^2}} \tag{4}$$

where X is the set of signals, and $|X|$ is the size of that set.

4 Our Differentially Private Random Decision Forest

It has been demonstrated in the past that randomized decision trees can have surprisingly high performance, and often have much lower computational complexity compared to their less random counterparts [7,12]. They gain their computational efficiency from the fact that they do not need to use the training data D in order to build a tree. This is a valuable advantage when trying to achieve differential privacy, as the less times we need to query the training data the better. The overall design philosophy of a differentially private machine learning algorithm is to be as efficient at spending the privacy budget β as possible. Below, we present our novel algorithm for building a differentially private random forest (DP-RF), and achieve better accuracy than any differentially private decision tree/forest algorithm proposed before it, even with very low privacy budgets.

4.1 Overview of Our Algorithm, DP-RF

Our algorithm, described in detail in the sections below, can be summarized as the following steps:

1. Based off the size of the privacy budget β, the size of the dataset D and the domain sizes of the attributes A, automatically tune the following parameters:
   - τ, the number of trees in the decision forest, which in turn dictates the ϵ spent per tree (Sect. 4.4).
   - θ, the minimum support threshold for nodes in the trees (Sect. 4.3).
2. Build a randomized decision forest using τ and θ, and no user-inputted parameters (Sect. 4.3).

3. Query the dataset using the Laplace Mechanism to learn the class counts in each leaf (Sect. 4.2).
4. Prune away nodes with SNR < 1, using signal averaging for non-leaf nodes (Sect. 4.5).
5. Find the node with the highest confidence in each path from the root node to a leaf node, in each tree (Sect. 4.6).
6. Predict the class value of future records by voting on the most confident predictions made by each tree (Sect. 4.7).

4.2 Querying the Leafs of a Tree

By following any chain of directed edges from a decision tree's root to a leaf, we have what is called a "decision rule". Each decision rule is a collection of attribute values, defined by the attributes from A that each node splits on, with each directed edge having one value a_v from the attribute a in the node it came from. Every record r in the training data D possesses a value $a_v; \forall a \in A$ that make it match one and only one decision rule in the decision tree. We therefore say that each record "fits into" or "belongs to" the leaf at the end of the chain of nodes its values match.

The Parallel Composition Theorem (Definition 5) means that for a decision tree–where every record appears in one and only one leaf–we can perform an ϵ-differentially private query Q on every leaf and use a total of ϵ out of the privacy budget β [14].

This is precisely what we will be doing: performing a single query Q on each leaf of a decision tree. That single query will be, "How many records have each class value?". In other words, we will get a histogram of the class counts in each leaf. The sensitivity Δ of this query Q is $\Delta(Q) = 1$, because the bins in a histogram are disjoint–the removal or addition of a single record can affect at most one bin of the histogram. Using the Laplace Mechanism (Definition 3) we can achieve ϵ-differential privacy by adding $\text{Lap}(1/\epsilon)$ to each class count, in each leaf of a decision tree.

4.3 Building a Random Tree

The building of a random decision tree is straight-forward, and does not require the training data D at all. An attribute a is chosen as the root node of the tree, with a directed edge (i.e. a branch of the tree) being created for each value $a_v \in a$. This process is recursively performed for the nodes at the end of each directed edge. The chosen attribute a at each node is selected from the set of attributes not chosen previously in the recursion chain, so that for any chain of nodes (i.e. path) from the root node to the bottom of the tree, no attribute appears twice.

The recursion ends for a given chain in the tree when either of the following termination criteria are met: (1.) there are no attributes left that have not been used previously in the recursion chain; or (2.) The estimated support of the current node is so low that the estimated SNR is below 1 (i.e. the noise outweighs the signal).

Criteria 2 uses the Signal-to-Noise Ratio, described in Sect. 3, and refers to the "estimated support" of the node. We define "estimated support" as the number of records that are estimated to match the attribute values $a_v; \forall a$ on the directed edges that led to the current node. The number of records is predicted by using the assumption that the subset of records D_i in a given node will be divided equally amongst all the directed edges leaving that node. In other words, starting from the root node and working our way down the tree, we make a rough estimate of the support of each node by dividing it by the domain size of each attribute used earlier in the chain.

This has the effect of not restricting the entire tree to the same maximum depth, unlike the differentially-private random decision tree algorithm proposed by [14]. Assuming that each value of an attribute has an equal portion of the records in D is clearly not an accurate assumption, but it does not need to be for our purposes. While rough, we mostly need the order of magnitude of a node's support, and we apply pruning later (see Sect. 4.5) to clean up the rough estimates. By estimating the support of nodes without using the training data, we avoid having to spend any of the privacy budget β on queries.

In the case of differential privacy, there is an additional reason to define a minimum support threshold: the larger the support, the less likely it is that noise will disrupt the signal. If we apply a query Q to a leaf with very small support, the noise could easily completely overwhelm the signal.

In order for the SNR to be above 1, the estimated support of a node must be larger than the minimum support threshold. We define the minimum support threshold as:

$$\theta = |C| \times \sqrt{2} \times \frac{1}{\epsilon}. \tag{5}$$

This is derived from the Laplace Distribution's variance, $\sigma^2 = 2x^2$ (see Definition 3), and the Signal-to-Noise Ratio, μ/σ. Because the scale of the noise we are adding is $1/\epsilon$ we can substitute that in for x, and by taking the square root of the variance we are left with the standard deviation, as seen in the formula for the SNR (Definition 6). Combining these observations leaves us with:

$$\frac{\mu}{\sigma} = \frac{\mu}{\sqrt{2 \times (1/\epsilon)^2}} = \frac{\epsilon\mu}{\sqrt{2}}. \tag{6}$$

A simple rearrangement lets us see that the signal μ (i.e. the support of a node) must be larger than $\mu > \sqrt{2}/\epsilon$ in order for $\mu/\sigma > 1$. Note that this assumes that $Lap(1/\epsilon)$ is only being added to the support of a node once. However, it as actually being added to each node a number of times equal to the size of the class attribute: $|C|$. We therefore need to increase the noise component of the SNR accordingly:

$$\mu > \frac{|C|\sqrt{2}}{\epsilon}, \tag{7}$$

and thus we arrive at (5): the estimated support μ must be larger than θ.

After recursively building upon each node until the estimated support of each node is less than the minimum support threshold θ, we can then learn how many records in the training data D fit into each leaf. Learning about the records in each leaf requires querying the dataset, and spending ϵ amount of the privacy budget β. This process is described in full in Sect. 4.2.

Thus we have built a random decision tree. We then repeat this tree-building process a number of times equal to τ to end up with a forest of random decision trees. The parameter τ is automatically defined by our algorithm, and is described below in Sect. 4.4. τ defines the fraction of the privacy budget that each decision tree can spend. When differentially-privately querying the leafs of one decision tree, ϵ is defined as:

$$\epsilon = \frac{\beta}{\tau}. \tag{8}$$

4.4   Defining the Number of Trees

To promote as diverse a collection of random trees as possible, we can make the root node of each tree unique by making sure a different attribute is chosen each time. Diversity is well known to be an advantageous property of decision forests [2]–hence why randomly built trees are valuable even when there are no privacy restrictions involved [7,12]. Extending this idea, we can define the number of trees as being equal to the number of attributes–thus the dataset will be partitioned based off the values of every attribute, giving us a chance to learn how predictive each attribute can be on their own. Any more trees and we would start having redundant root nodes, and thus be getting less value for the privacy budget spent on it than we did for the earlier trees.

Limiting the number of trees is only necessary because we have a privacy budget we must adhere to. As explained in Sect. 4.2, we can query the dataset about the (disjoint) leafs in a tree using the same portion of the privacy budget. However we cannot use the same portion of the privacy budget for multiple trees–each tree is a re-partitioning of the same data, and so they are not disjoint. This means that the more trees we have, the more we have to divide the privacy budget among them due to Definition 4 (see (8)).

It is therefore possible that if there is a large number of attributes, the budget may be spread too thin to output useful query results. In our case, "too thin" can be interpreted as "the minimum support threshold is too high to build a tree". We address this by reducing the number of trees to a point where the minimum support threshold becomes acceptable. We do so by imagining a scenario in which the dataset D is split into a number of partitions D_i equal to the average domain size of the attributes A–we now have the "stump" of a decision tree. We then repeat this, dividing each node on the next level of the tree by the average domain size of the attributes. We now have a tree with a depth of 3: the dataset is divided up twice by attributes (with average domain sizes), with the last level of the tree being leafs. We consider this to be the minimally acceptable tree size.

For this minimally acceptable tree size to be possible (at least in the average case), the minimum support threshold must be as follows:

$$\theta < \frac{|D|}{\delta^2}, \tag{9}$$

where δ is the average domain size of the attributes in A, and θ is defined by (5). Recall that there is no maximum depth defined in our algorithm: some parts of the decision tree might have a depth shorter than 3, while other parts of the same tree might have a much larger depth. Thus we define τ as the largest number, up to the number of attributes $|A|$, that satisfies (9) and (5).

4.5   Pruning to a Reliable Depth

Once the forest of random decision trees has been built, the next step in our algorithm is to prune each decision tree. This is somewhat similar to the pruning done in other decision tree algorithms [13], including the pruning done in Private-RDT [14]. The only pruning done in Private-RDT is to remove leafs with no records in them. We propose a more sophisticated approach, but first a remark on their approach. It is important to note that even if there are 0 records in a leaf, Laplace noise still needs to be added to each class count. While of course a count cannot go below 0, a count that was originally 0 can be raised above 0 by the noise. This drastically reduces the amount of pruning that is done in Private-RDT, and keeps leafs that most likely have a Signal-to-Noise Ratio of 0 (i.e. they are 100 % noise). Our proposed approach solves this problem.

Earlier, in Sect. 4.3, we estimated the support of each node in order to estimate its Signal-to-Noise Ratio and decide whether to branch from the node further. Now that we have queried the server and have the (noisy) class counts of each leaf, we no longer have to use such rough estimates for each node's support–we can re-check which nodes have acceptable SNRs, and which do not. We can also take advantage of signal averaging, described in Sect. 3.1. The idea is simple: by summing together the class counts of all the leafs that have the same parent node, we know the class counts of the parent node. This can be repeated up the tree until we know the class counts of every node in the tree. Not only that, but the class counts of the parent node are actually *less noisy* than those of the child nodes. We can see this in (4), as well as the intuition behind it in Sect. 3.1. The noise is reduced by a factor equal to the square root of the number of leaf nodes being summed together.

Using this knowledge, we now not only have more accurate SNRs for the leafs than we did in Sect. 4.3, but we also have increasingly more accurate SNRs for the nodes above the leafs, the further we traverse up the tree. Our pruning is simple: we remove any nodes with SNR < 1. Because of the reduction in noise in nodes using the sum of class counts in its children, it becomes easier and easier for nodes higher up the tree to have SNR > 1. When re-written for this scenario, (4) becomes:

$$\text{SNR} = \frac{\epsilon \sum_L^{\text{Leafs}} \left(\sum_c^C L_c \right)}{|C|\sqrt{2 \times |\text{Leafs}|}}, \tag{10}$$

where *Leafs* is the set of all leaf nodes that can be reached by traversing down from the current node (not all leafs in the entire tree), and L_c is the class count for $c \in C$ in leaf L. Observe that (10) reduces down to (6) when considering one leaf and one class value – they both originate from the SNR theory.

4.6   Finding the Most Confident Prediction in Each Rule

Once unreliable leafs have been pruned away, we are left with only nodes where the true counts of the class values outweigh the noise. Our aim now is to use these nodes to predict the class values of future records. For a given future record r and a given decision tree, we can find the leaf that r belongs to, and know the decision rule that it obeys.

Usually in non-randomly built decision trees, a node is only split into more nodes if an attribute can be found that increases the confidence of the decision rule's prediction of r's class value [3]. The predicted class value of r is therefore often defined as the most common class value in the leaf of the decision rule (i.e. the majority class value). The "confidence" of the prediction is then the fraction of the records in the leaf that have the majority class value. However in the case of a randomly built decision tree, there is nothing preventing a node above the leaf from having a higher confidence than the leaf.

We therefore find the most confident prediction in each decision rule, and then use that prediction for any future records that obey that rule.

4.7   Voting

When predicting the class value of a future record r, the process of finding the most confident prediction in the decision rule that r obeys is repeated for each decision tree in the forest. r obeys one decision rule per tree, and our algorithm selects one node out of each decision rule with the highest confidence, as described in Sect. 4.6. It is unlikely that all the predictions are predicting the same class value, so a voting mechanism is required to decide which class value is the decision forest's final prediction.

Our algorithm votes using the following process: (1.) The final prediction is the class value with the highest confidence; if two different class values are predicted with equal confidence, then (2.) sum the confidences of each class value from the nodes selected in Sect. 4.6 (including the non-majority class values) and (3.) the final prediction is the class value with the highest summed confidence.

5   Experiments

We test our proposed algorithm (DP-RF) using 9 datasets from the UCI Machine Learning Repository [1]. We use 10-fold stratified cross-validation repeated 30 times to calculate the average prediction accuracy [9] of our algorithm, using various privacy budgets. We test the following privacy budgets: $\beta = 0.01, 0.05,$ 0.1, 0.25, 0.5, 1.0, 2.0. We compare our technique to Private-RDT, proposed by

[14] (Sect. 2.2). For continuous attributes, we discretize them by splitting their domain into 5 bins of equal range. We use the default parameters recommended in [14] for Private-RDT. Note that high values of β (especially values ≥ 1) are unlikely in real-world situations, and are presented to demonstrate the trends.

Our results are presented in Fig. 1. For all datasets, our algorithm has higher prediction accuracy than Private-RDT on average. Out of the 63 measurements (7 values for β per dataset) our algorithm out-performs Private-RDT in 55 cases, sometimes only under-performing by $< 1\,\%$ (i.e. 0.01 on the y axis). At its best our algorithm can beat Private-RDT by 45 %; at its worst it loses by 7 %.

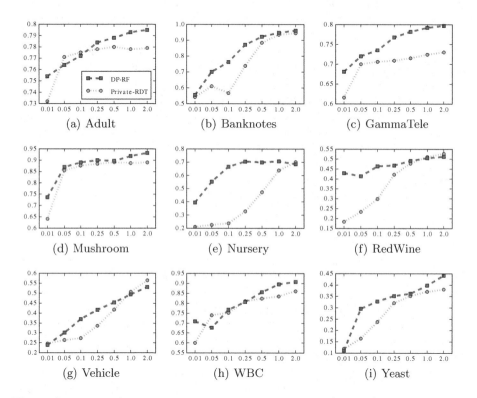

Fig. 1. Comparing the prediction accuracy of our technique (DP-RF) to Private-RDT [14] at different ϵ values, using 9 different datasets from [1]. The y axis represents the prediction accuracy of the classifier and the x axis represents β.

6   Conclusion

By re-framing the problem of building a differentially private decision forest in terms of the Signal-to-Noise Ratios, we are able to propose intuitive methods for automating the tuning of necessary parameters. We are also able to guarantee that any predictions made about future records are made using class counts that not only have high confidence, but also outweigh any noise that might have been added to them. Our results prove the success of this approach, and pave the way for extending the application of Signal-to-Noise Ratio theory to other implementations of differential privacy.

References

1. Bache, K., Lichman, M.: UCI Machine Learning Repository (2013). http://archive.ics.uci.edu/ml/
2. Breiman, L.: Random forests. Mach. Learn. **45**, 5–32 (2001)
3. Breiman, L., Friedman, J., Stone, C., Olshen, R.: Classification and Regression Trees. Chapman & Hall/CRC, Boca Raton (1984)
4. Dwork, C.: Differential privacy. In: Bugliesi, M., Preneel, B., Sassone, V., Wegener, I. (eds.) ICALP 2006. LNCS, vol. 4052, pp. 1–12. Springer, Heidelberg (2006)
5. Dwork, C., McSherry, F., Nissim, K., Smith, A.: Calibrating noise to sensitivity in private data analysis. In: Halevi, S., Rabin, T. (eds.) TCC 2006. LNCS, vol. 3876, pp. 265–284. Springer, Heidelberg (2006)
6. Dwork, C., Roth, A.: The Algorithmic Foundations of Differential Privacy. Now Publishers, Hanover (2013)
7. Fan, W., Wang, H., Yu, P., Ma, S.: Is random model better? On its accuracy and efficiency. In: Third IEEE International Conference on Data Mining (2003)
8. Fletcher, S., Islam, M.Z.: A differentially private decision forest. In: Proceedings of the 13th Australasian Data Mining Conference, Sydney, Australia (2015)
9. Fletcher, S., Islam, M.Z.: Quality evaluation of an anonymized dataset. In: 22nd International Conference on Pattern Recognition. IEEE, Stockholm (2014)
10. Friedman, A., Schuster, A.: Data mining with differential privacy. In: 16th SIGKDD Conference on Knowledge Discovery and Data Mining, pp. 493–502. ACM, Washington, DC, USA (2010)
11. Fung, B., Wang, K., Chen, R., Yu, P.: Privacy-preserving data publishing: a survey of recent developments. ACM Comput. Surv. (CSUR) **42**(4), 14 (2010)
12. Geurts, P., Ernst, D., Wehenkel, L.: Extremely randomized trees. Mach. Learn. **63**(1), 3–42 (2006)
13. Han, J., Kamber, M., Pei, J.: Data Mining: Concepts and Techniques. Morgan Kaufmann Publishers, San Francisco (2006)
14. Jagannathan, G., Pillaipakkamnatt, K., Wright, R.: A practical differentially private random decision tree classifier. Trans. Data Priv. **5**(1), 273–295 (2012)
15. McSherry, F.: Privacy integrated queries: an extensible platform for privacy-preserving data analysis. In: Proceedings of the 35th SIGMOD International Conference on Management of Data, pp. 19–30. ACM, Providence, USA (2009)
16. McSherry, F., Talwar, K.: Mechanism design via differential privacy. In: 48th Annual IEEE Symposium on Foundations of Computer Science, pp. 94–103 (2007)
17. Sweeney, L.: Achieving k-anonymity privacy protection using generalization and suppression. Int. J. Uncertainty Fuzziness Knowl. Based Syst. **10**(5), 571–588 (2002)
18. UN General Assembly: Universal Declaration of Human Rights (1948)
19. Van Drongelen, W.: Signal processing for Neuroscientists: An Introduction to the Analysis of Physiological Signals. Academic Press, Burlington (2006)

Learning from Demonstration Using GMM, CHMM and DHMM: A Comparison

Fenglu Ge[(✉)], Wayne Moore, and Michael Antolovich

Charles Sturt University, Bathurst 2795, Australia
fge@csu.edu.au

Abstract. Greater production and improved safety in the mining industry can be enhanced by the use of automated vehicles. This paper presents results in applying Learning from Demonstration (LfD) to a laboratory semi-automated mine inspection robot following a path through a simulated mine. Three methods, Gaussian Mixture Model (GMM), Continuous Hidden Markov Model (CHMM), and Discrete Hidden Markov Model (DHMM) were used to implement the LfD and a comparison of the implementation results is presented. The results from the different models were then used to implement a novel, optimised path decomposition technique that may be suitable for possible robot use within an underground mine.

Keywords: Mining · Robot · Autonomous · Path · LfD · Markov Model · Gaussian

1 Introduction

Mine tunnels are dangerous areas for mining employees. During production, the working environment is dusty and muddy. When workers walk across mine tunnels, while Load Haul Dump (LHD) and other vehicles travel through tunnels, it increases the probability of human-truck collisions (Burgess-Limerick and Steiner 2006; Tyson 2006). However, there are still many tasks that require human intervention, including inspections, solving unexpected problems, and so on. Previously, humans have been hired to perform tasks in mining tunnels, which required special training and specialised equipments (Gunningham 2007; Dra 2012; Ven 2011). For instance, to inspect muck piles, an inspector needs to periodically go to many (>400) underground positions, carrying a monitoring device in complex mining tunnels. Heavy equipment plus distractions from vehicles, influence the working efficiency. Therefore, the nature and the speed required for these tasks make the use of humans problematic.

Based on the above, it is necessary to reduce the need for a human presence in the mine environment.

1.1 Safety and Efficiency

Robots have been introduced into tunnel tasks (González et al. 2009; Yu et al. 2007; Zhuang et al. 2008; Tang et al. 2011), which can reduce accidents, human error and increase efficiency.

© Springer International Publishing Switzerland 2015
B. Pfahringer and J. Renz (Eds.): AI 2015, LNAI 9457, pp. 204–217, 2015.
DOI: 10.1007/978-3-319-26350-2_18

One potential solution is a pre-programmed robot, which has the drawback of heavy coding work to adapt to different environments.

Another potential solution is Simultaneous Localization and Mapping (SLAM). However, due to the numerous uncertainties and computational complications, there isn't an established solution to mapping and localization when dealing with large areas (Aulinas et al. 2008). Furthermore, it can only be applied to trajectory related tasks, reducing the boundary of robot applications in mining.

Hence, it's necessary to find a suitable solution, which can replace humans without compromising safety and efficiency.

2 Current Robot Learning

Robot Learning is a potential solution to replace humans. Current robot learning has been classified into two groups, Reinforcement Learning (RL) and Learning from Demonstration (LfD).

2.1 Reinforcement Learning

Reinforcement Learning is a trial-and-error technique, finding a policy via maximizing a reward function or functions. The learning process can be modelled as a Markov Decision Process (MDP), and Expectation Maximization (EM) is often used to find final solutions by updating parameters in the previously built MDP. RL has the advantage of making a robot learn a task autonomously. However, it is difficult to give a suitable definition of reward in RL (Kaelbling et al. 1996), which is very important to guide it to achieve the most appropriate behaviour for the task to be learned. Moreover, building a policy (Kaelbling et al. 1996; Gosavi 2009; Kober et al. 2013) for RL also requires state information of the ambient environment so as to receive rewards, which costs time and can be computationally expensive in the real world.

2.2 Learning from Demonstration

LfD (Argall et al. 2009; Bruno and Oussama 2008) is an unsupervised technique, where a demonstrator can teach and guide a robot's behaviours during a training period. Compared to RL, LfD is a faster learning method since it doesn't need an unpredictable reward, saving time and computation. However, it's not adaptable to changes in the environment, so it would need retraining. Therefore, to make a robot adapt to changes in the environment is still a big challenge.

The basic concept of LfD is to set up a mapping between the environment and a robot's actions. Trajectory reconstruction and modelling is a way to relate the environment and corresponding actions (Calinon and Billard 2007b, 2009). These methods can build mappings between the environment and related actions using time labels explicitly, leading to unexpected errors because of time mismatching in actual operations. As an alternative, hidden time-space encoding is a popular way to map a robot's actions and external information (Ogata et al. 2005;

Kulic et al. 2008; Calinon et al. 2010). However, heavy computation is still a big issue, leading to impractically long learning times for reproduction of tasks. An improved version has been developed in this paper, where a modified DHMM method was used to help a robot to learn a task quickly. The experiment details will be explained later.

In this paper, an inspection task is used as a template for comparing different learning methods. Performance of experiments based on the following methods will be evaluated for the purpose of choosing a suitable strategy for mining tasks.

Gaussian Mixture Model (GMM) Based Method

GMM. Expectation-Maximization (EM) (Dempster et al. 1977; Bilmes 1997) is initialized to obtain the initial parameters of a Gaussian Mixture Model (GMM). Input datasets are modelled by a mixture of K components, where K is the number of clusters (actions).

$$p(X_i) = \sum_{k=1}^{K} p(k)p(X_i|k) \tag{1}$$

where $p(k)$ is the prior Probability Density Function (PDF), X_i is a d-dimensional vector and $p(X_i|k)$ is the posterior PDF. For every cluster, a Gaussian model is used as the PDF of the data distribution in a cluster, which can be denoted by:

$$
\begin{aligned}
p(X_i|k) &= N\left(\mu_k, \Sigma_k\right) \\
&= \frac{1}{\sqrt{(2\pi)^d |\Sigma_k|}} e^{-\frac{1}{2}(X_i - \mu_k)' \Sigma_k^{-1}(X_i - \mu_k)}
\end{aligned}
\tag{2}
$$

where μ_k is a d-dimensional vector of X_{set}. Σ_k is a $k \times k$ covariance matrix of X_{set}.

$$X_{set} = (X_i)_{i=1}^{N} \tag{3}$$

In this paper, X_{set} was the data collected by a robot in one task. Parameters of GMM were calculated by EM. After the initialization step, E-step and M-step were used to retrieve μ_k and Σ_k.

E-step:

$$p^t(k|X_i) = \frac{p^t(X_i|k)p^t(k)}{p^t(X_i)} \tag{4}$$

M-step:

$$\mu_k^{t+1} = \frac{\sum_{i=1}^{M} p^t(k|X_i)X_i}{\sum_{i=1}^{M} p^t(k|X_i)} \tag{5}$$

$$\Sigma_k^{t+1} = \frac{\sum_{i=1}^{M} p^t(k|X_i)(X_i - \mu_k^{t+1})(X_i - \mu_k^{t+1})^T}{\sum_{i=1}^{M} p^t(k|X_i)} \tag{6}$$

$$p^{t+1}(k) = \frac{1}{N} \sum_{i=1}^{M} p^t(k|X_i) \tag{7}$$

where t indexes the iteration of the EM algorithm.

Gaussian Mixture Regression(GMR). GMR is an algorithm for data reconstruction (Calinon and Billard 2007a). It is used to estimate a variable Y given X on the basis of a series of previous observations $\{X, Y\}$. Here we define:

$$X_{in} = X \tag{8}$$

$$X_{out} = Y \tag{9}$$

For every Gaussian component k, we define

$$\mu_k = \{\mu_{in,k}, \mu_{out,k}\} \tag{10}$$

$$\Sigma_k = \begin{pmatrix} \Sigma_{inin,k} & \Sigma_{inout,k} \\ \Sigma_{outin,k} & \Sigma_{outout,k} \end{pmatrix} \tag{11}$$

and the following equation can be inferred:

$$p(X_{out,k}|X_{in,k}) = N(\mu_{GMR}, \Sigma_{GMR}) \tag{12}$$

$$\mu_{GMR} = \mu_{out,k} + \Sigma_{outin,k}(\Sigma_{inin})^{-1}(X_{in} - \mu_{in,k}) \tag{13}$$

$$\Sigma_{GMR} = \Sigma_{outout,k} - \Sigma_{outin,k}(\Sigma_{inin,k})^{-1}\Sigma_{inout,k} \tag{14}$$

Therefore, it can be deduced that

$$p(X_{out}|X_{in}) = \sum_{k=1}^{K} \gamma_k N(\mu_{GMR}, \Sigma_{GMR}) \tag{15}$$

$$\gamma_k = \frac{p(k)p(X_{in}|k)}{\sum_{i=1}^{K} p(i)p(X_{in}|i)} \tag{16}$$

Thus, the value of X_{out} can be estimated by X_{in} via the following equations:

$$X_{out} = \sum_{k=1}^{K} \gamma \mu_{GMR} \tag{17}$$

$$\Sigma_{outout} = \sum_{k=1}^{K} \gamma^2 \Sigma_{GMR} \tag{18}$$

Continuous Hidden Markov Model (CHMM) Based Method. CHMM is another potential solution for LfD. CHMM sets up a mathematical model by encoding time and space together. Given inputs, all parameters in that model are used to retrieve outputs during the regression stage (Calinon et al. 2010). However, the high computational cost is a great challenge for real time applications (Rabiner 1989; Hwamg et al. 1994). In this task, there are 2 states, and 1 cluster for each state.

Discrete Hidden Markov Model (DHMM) Based Method. The authors revised the traditional DHMM batch based method (Yang et al. 1994; Vakanski et al. 2012) to a sequential processing method , making single training enough for most tasks. The theory part can be summarized as follows:

Learning Stage. The learning stage can be summarized in Fig. 1.

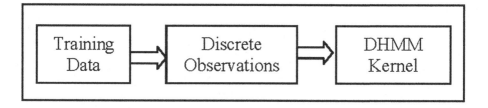

Fig. 1. Scheme in learning stage.

In this section, the training data consists of T 5-dimension vectors $\mathbf{V} = [ld\ rd\ fd\ lsp\ rsp]$ (Table 1), where every vector consists of two parts, including a distance vector $x_t = ld, rd, fd$ and a corresponding speed vector $y_t = lsp, rsp$. The training data x_t, y_t was converted to discrete observations via K-means, and a codebook was also generated including distinct observations $o_k{}_{k=1}^{16}$. The observations were used to generate a DHMM learning kernel of 2 states, including the following elements:

1. N: The number of distinct states
2. M: The number of distinct observations
3. π_i: Initial probability of state i
4. A: The transition probability matrix of states, where $A = a_{ij}$ the transition probability from state i to state j
5. B: The emission probability matrix, $B = b_i(O_k)$ where the probability of observing symbol $b_i(O_k)$ given state i

Here, the states can be understood as moving forward, turning, and stop.

Regression Stage. The Traditional DHMM regression algorithm, Viterbi algorithm, finds a sequence of actions with the most likelihood. Then a robot will perform this sequence without interactions with environment. If the length of one part of an experimental environment is changed, the robot will still perform the predefined sequence of actions, leading to potential conflicts.

The only difference with traditional Viterbi happens in the Regression Stage. In this DHMM, an action is generated based on the environmental data, then the robot will perform the action generated. Given updated environmental data, the robot will perform corresponding action as well. Therefore, this method has the privilege of interacting with surrounding environment against Viterbi algorithm.

The target for regression is to get the estimated speed $\dot{y}_t$ given x_t in real-time. Because the estimated speed is contributed from different states, Eq. 19 can be derived:

$$\dot{y}_t = \sum_{i=1}^{N}[F(x_t|i)(y|x_t,i)] \tag{19}$$

where $F(x_t|i)$ is the DHMM forward variable.

In the GMM (Calinon et al. 2007), the x_t is a measured value at time t, and $F(x_t|i)$ is the influence of state i given x_t, without considering temporal influence. However, the temporal influence and the spatial influence are both considered in DHMM. Therefore, the influence of state i has a relationship to the spatial information x_t and temporal information $x_1, x_2, ...x_{t-1}$, which is the forward variable of DHMM.

The actual procedure of implementation can be summarized as follows:

1. convert x_t to the corresponding observation O_t at state I using the codebook generated by K-means,
2. get speed Y given O_t and the state based on the codebook,
3. calculate Forward variable F, and
4. calculate the estimated speed using Eq. 19.

The forward variable can be calculated using Eq. 20, where the denominator is used for the purpose of normalization.

$$F_i(x_t) = \frac{(\sum_{j=1}^{N} F_j(x_{t-1})a_{ji})b_i(O_t)}{\sum_{i=1}^{N}(\sum_{j=1}^{N} F_j(x_{t-1})a_{ji})b_i(O_t)} \tag{20}$$

3   Environment and Metrics

3.1   Experimental Setup

Figure 2 shows the floorplan of the training environment. The concrete simulated mine tunnel shown in Figs. 3 and 4 was designed to reflect a more realistic environment for the robot. This was achieved by moving some of the concrete blocks in such a way as to make a very uneven wall contour. This allowed the

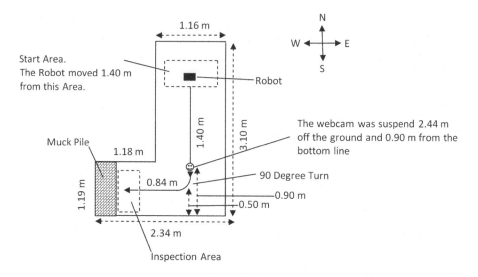

Fig. 2. Floorplan of the training environment.

Fig. 3. Original robot training environment with smooth walls.

full potential of LfD to be used in learning to traverse a very uneven, irregular environment. A P3-DX robot (mob)[1] was used for experiments, and a webcam was used to capture the robot motion.

3.2 Training

A teleoperated robot was allowed to run through the training environment as shown in Fig. 3. The robot acquired data from its on-board sonar sensors while traversing segment A, segment B and segment C. These segments were labelled A, B and C in Fig. 3. It can be seen that segment A was a straight line from the starting position to a point at B where the robot needed to turn to the right.

---

[1] Adept robot. http://www.mobilerobots.com/. Accessed: 2014-10-18.

Fig. 4. Randomly placed obstacles (round tubes) to simulate a noisy environment.

Table 1. Variables collected

ld	Distance to the nearest left object
rd	Distance to the nearest right object
fd	Distance to the nearest front object
lsp	Linear speed of a robot
asp	Angle speed of a robot

Table 2. Kernel generation.

First demonstration	=>	First learning Kernel
First two demonstrations	=>	Second learning Kernel
...	...	...
First ten demonstrations	=>	Tenth learning Kernel

Segment B was the rotation through 90 degrees to face the simulated muck pile which was the desired end position for the robot. After turning, the robot then traversed segment C to end up facing the muck pile.

A vector $V = \begin{bmatrix} ld\ rd\ fd\ lsp\ rsp \end{bmatrix}$ was generated at every time interval as shown in Table 1. Due to noise and general interference from the environment, a vector X was formed from the average of three consecutive V vectors. A single demonstration/training session generated a number (T) of these X vectors. A training set consisted of N demonstrations with a value of lsp of 0.1 m/s and an rsp of 10 degrees/s.

After 10 separate data acquisitions for each of the three different trajectories (A, B and C), learning kernels (Herbrich 2001) were generated based on strategies as listed in Table 2. These were generated for each of the LfD methods (GMM, CHMM, and DHMM) used in this paper The robot was then allowed to perform each task independently. At every time interval it recorded a distance array

$Y = \begin{bmatrix} fd & ld & rd \end{bmatrix}$ and processed it. After processing, outputs were produced which were the actions the robot needed to take to carry out its task.

3.3   Environmental Modifications

The training environment was as described in Sect. 3.1, a simple tunnel constructed with smooth blocks. The environment was modified to include noise by placing cylindrical tubes at uneven spacings. This was to represent a more realistic space for the robot and allowed the authors to see the results of changing the environment. Training was done without the presence of the cylindrical tubes. The modified space can be seen in Fig. 4. All kernels in Table 2 were applied to both the simple environment as well as the modified, noisy environment. The results of these runs are given in Figs. 5, 6 and 7. The aim of this exercise was to determine which of the three kernels (GMM, CHMM, DHMM) were the best for each of the three segments. The best kernels would be chosen and run for each of the segments independently. This would allow the robot to complete the overall task of moving from the start position to the "muck pile" using a decomposition of the task into three segments using the best LfD kernel for each section.

Besides the above mentioned, the following performances were also recorded after every robot autonomous run:

1. Distance Error (DE), which is the absolute distance between the average training stop point and the free run stop point in training (segment A, B, C) in cm
2. Time, which is time span to run a trajectory.
3. Degree, which is the actural turn in degrees for segment B.

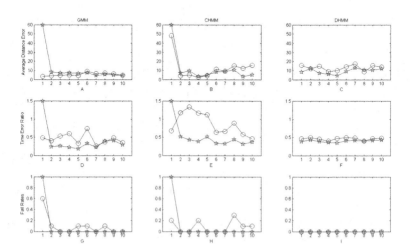

Fig. 5. Performance Comparison for Segment A under both environments. The circles represent performance in a low noise environment. The stars represent performance in a noisy environment. X-axes represent training numbers.

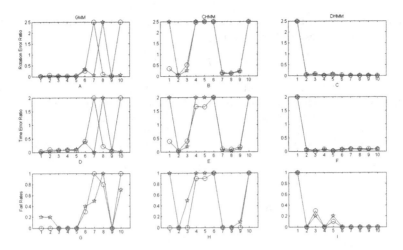

Fig. 6. Segment B: Performance Comparison for Segment B under both environments. The circles represent performance in a low noise environment. The stars represent performance in a noisy environment. X-axes represent training numbers.

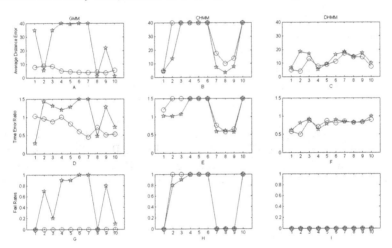

Fig. 7. Segment C: Performance Comparison for Segment C under both environments. The circles represent performance in a low noise environment. The stars represent performance in a noisy environment. X-axes represent training numbers.

3.4 Metrics

Each learning kernel was run 10 times to improve reliability. The average performances are shown in Figs. 5, 6 and 7.

For Segment A and C, the following performances were used and are displayed in Figs. 5 and 7.

1. Average Distance Error (ADE) which is an average of 10 DEs.

2. Time Error Ratio (TER) which shows time error between an average running time and the corresponding training time.

$$TER_i = (T_f - T_t)/T_t \tag{21}$$

where T_f is a running time and T_t is a related training time.

3. Fail Rates which show the number of failures for a learning kernel (e.g., wall strike).

For Segment B, the following performances were displayed (see Fig. 6).

1. Rotation Error Ratio (RER) which shows degree error between an average stop spot and the ending point in training.

$$RER_i = (R_f - R_t)/R_t \tag{22}$$

where R_f is a stop spot during free running, while R_t is a related stop spot during training.

2. TER which shows time error between an average running time and the corresponding training time.

3. Fail Rates (FR) which show the number of failures for a learning kernel (e.g., kept spinning).

In every plot, circles represent performance under the low noise environment, while stars represent performance in the noisy environment. The horizontal coordinates are the number of training times. As shown in Figs. 5, 6 and 7, a variable has different variations on a section based on different methods and situations. For the purpose of comparing values of these variables, the values were graphed on the same scale. All variables with FR equal to 1 were assigned the maximum value in a corresponding scale. ADE was assigned to 60 and TER was assigned to 1.5 for any training with FR equal to 1 in Fig. 5. RER was assigned to 2.5 and TER was assigned to 2 for any training with FR equal to 1 in Fig. 6. ADE was assigned to 40 and TER was assigned to 1.5 for any training with FR equal to 1 in Fig. 7.

4   Experiment Results and Discussion

4.1   Segment A

Figure 5 shows performances of the three algorithms under both environments.

1. For the GMM based method, errors were down to a lower level after 2 trainings. Kernels achieved better performance in the noisy environment. Overtraining occurred in the runs shown in Fig. 5A, D and G.

2. For the CHMM based method, errors were down to a lower level after 2 trainings. Kernels also achieved better performance in the noisy environment. Overtraining occurred in the runs shown in Fig. 5B, E and H. The CHMM based method has the highest time variation in the low noise environment.

3. For the DHMM based method, performances were stable, and errors varied in a small range after 1 training.

4.2 Segment B

Figure 6 shows performances of the three algorithms in both environments.

1. For the GMM based method, performances were similar in both environments.
2. For the CHMM based method, performances were unstable across all training sets in both environments.
3. For the DHMM based method, performances became stable after 2 trainings. Noise didn't show any significant impact on a kernel.

4.3 Segment C

Figure 7 shows performances of the three algorithms in both environments, which was the shortest trajectory.

1. For the GMM based method, noise had negative impacts on the performance of a kernel. Performances became unstable in the noisy environment.
2. For the CHMM and DHMM based methods, noise didn't play an important role in the performance for both methods.
3. The DHMM based method was the most stable across the three methods with the lowest variation after 1 training.

5   Conclusion

In this paper, three methods were compared based on three trajectories in the low noise and noisy environments. The GMM based method had the best performance in a low noise environment. In practice, there's always unexpected noise around a robot, implying the GMM based method was not practical for real environments. The CHMM based method was suitable for turning trajectories, while The DHMM based method was more robust for straight trajectories.

References

Ventilation standards in underground mines. (mines safety bulletin no. 95). Technical report, Department of Mines and Petroleum, WA, Australia, 2011

Draft guide for tunnelling work: Technical report. Safe Work Australia, Canderra, Australia (2012)

Argall, B.D., Chernova, S., Veloso, M., Browning, B.: A survey of robot learning from demonstration. Robot. Autonom. Syst. **57**(5), 469–483 (2009)

Aulinas, J., Petillot, Y., Salvi, J., Lladó, X.: The slam problem: a survey. In: Proceedings of the 2008 Conference on Artificial Intelligence Research and Development: Proceedings of the 11th International Conference of the Catalan Association for Artificial Intelligence, pp. 363–371, Amsterdam, The Netherlands, 2008. IOS Press. ISBN 978-1-58603-925-7

Bilmes, J.: A gentle tutorial of the em algorithm and its application to parameter estimation for gaussian mixture and hidden markov models. Technical report, University of Berkeley (1997)

Siciliano, B., Khatib, O. (eds.): Springer Handbook of Robotics. Springer, Heidelberg (2008)

Burgess-Limerick, R., Steiner, L.: Injuries associated with continuous miners, shuttle cars, load-haul-dump and personnel transport in new south wales underground coal mines. Technical report, Office of Mine Safety and Health Research (2006)

Calinon, S., Billard, A.: What is the teacher's role in robot programming by demonstration? - toward benchmarks for improved learning. Interaction Studies. Special Issue on Psychological Benchmarks in Human-Robot Interaction 8(3) (2007a)

Calinon, S, Billard, A.: Incremental learning of gestures by imitation in a humanoid robot. In: HRI, pp. 255–262 (2007b)

Calinon, S., Billard, A.: Statistical learning by imitation of competing constraints in joint space and task space. Adv. Robot. 23(15), 2059–2076 (2009)

Calinon, S., Guenter, F., Billard, A.: On learning, representing, and generalizing a task in a humanoid robot. IEEE Trans. Syst. Man Cybern. Part B 37(2), 286–298 (2007)

Calinon, S., D'halluin, F., Sauser, E.L., Caldwell, D.G., Billard, A.: Learning and reproduction of gestures by imitation. IEEE Robot. Autom. Mag. 17(2), 44–54 (2010)

Dempster, A.P., Laird, N.M., Rubin, D.B.: Maximum likelihood from incomplete data via the em algorithm. J. Roy. Stat. Soc. Ser. B 39(1), 1–38 (1977)

González, J.C., Martínez, S., Jardón, A., Balaguer, C.: Robot-aided tunnel inspection and maintenance system. In: 26th International Symposium on Automation and Robotics in Construction (ISARC), Austin, TX, USA (2009)

Gosavi, A.: Reinforcement learning: a tutorial survey and recent advances. INFORMS J. Comput. 21(2), 178–192 (2009)

Gunningham, N.: Mine Safety: Law Regulation Policy. Federation Press, Annandale (2007)

Herbrich, R.: Learning Kernel Classifiers: Theory and Algorithms. Adaptive Computation and Machine Learning. The MIT Press, Cambridge (2001)

Hwamg, M., Rosenfeld, R., Theyer, E., Mosur, R., Chase, L., Weide, R., Huang, X., Alleva, F.: Improving speech recognition performance via phone-dependent VQ codebooks and adaptive language models in SPHINX-II. In: IEEE International Conference on Acoustics, Speech, and Signal Processing, ICASSP-94 1994 , vol. 1, pp. 549–552, April 1994

Kaelbling, L.P., Littman, M.L., Moore, A.W.: Reinforcement learning: a survey. J. Artif. Intell. Res. (JAIR) 4, 237–285 (1996)

Kober, J., Bagnell, J.A., Peters, J.: Reinforcement learning in robotics: a survey. Int. J. Robot. Res. 32(11), 1238–1274 (2013)

Kulic, D., Takano, W., Nakamura, Y.: Incremental learning, clustering and hierarchy formation of whole body motion patterns using adaptive hidden Markov chains. I. J. Robot. Res. 27(7), 761–784 (2008)

Ogata, T., Sugano, S., Tani, J.: Open-end human-robot interaction from the dynamical systems perspective: mutual adaptation and incremental learning. Adv. Robot. 19(6), 651–670 (2005)

Rabiner, L.R.: A tutorial on hidden markov models and selected applications in speech recognition. Proc. IEEE 77(2), 257–286 (1989)

Tang, S., Chen, S., Liu, Q., Wang, B., Guo, X.: A small tracked robot for cable tunnel inspection. In: Lee, G. (ed.) Advances in Automation and Robotics, Vol.1. LNEE, vol. 122. Springer, Heidelberg (2011). (ISSN 1876–1119)

Jonathan, T.: Designing to maximize operator visibility in lhd equipment. Technical report, Ontario Natural Resources Safety Association, North Bay, Ontario, Canada (2006)

Vakanski, A., Mantegh, I., Irish, A., Janabi-Sharifi, F.: Trajectory learning for robot programming by demonstration using hidden markov model and dynamic time warping. IEEE Trans. Syst. Man Cybern. Part B: Cybern. **42**(4), 1039–1052 (2012)

Yang, J., Yangsheng, X., Chen, C.S.: Hidden Markov model approach to skill learning and its application to telerobotics. IEEE Trans. Robot. Autom. **10**(5), 621–631 (1994)

Seung-Nam, Y., Jang, J.-H., Han, C.-S.: Auto inspection system using a mobile robot for detecting concrete cracks in a tunnel. Autom. Constr. **16**(3), 255–261 (2007)

Zhuang, F., Zupan, C., Zheng, C., Zhao, Y.: A cable-tunnel inspecting robot for dangerous environment. Int. J. Adv. Robot. Syst. **5**(3) (2008)

Implementing Modal Tableaux
Using Sentential Decision Diagrams

Rajeev Goré[1]([⊠]), Jason Jingshi Li[1,2], and Thomas Pagram[1]

[1] Research School of Computer Science, RSISE Building,
The Australian National University, Canberra, ACT 0200, Australia
{Rajeev.Gore,us163504}@anu.edu.au
[2] ShoreTel, Canberra Technology Park, 49 Phillip Avenue,
Watson, ACT 2602, Australia

Abstract. A Sentential Decision Diagram (SDD) is a novel representation of a boolean function which contains a Binary Decision Diagram (BDD) as a subclass. Previous research suggests that BDDs are effective in implementing tableaux-based automated theorem provers. We investigate whether SDDs can offer improved efficiency when used in the same capacity. Preliminarily, we found that SDDs compile faster than BDDs only on large CNF formulae. In general, we found the BDD-based modal theorem prover still outperforms our SDD-based modal theorem prover. However, the SDD-based approach excels over the BDD-based approach in a select subset of benchmarks that have large sizes and modalities that are less nested or fewer in number.

1 Introduction

Propositional modal logics are of fundamental importance to various fields of Artificial Intelligence (AI). Modal logics are extensions of propositional logic by the modal quantifier $\Box$ [2], which in AI is most commonly interpreted as epistemic knowledge [8]. Under this interpretation the statement $\Box\varphi$ is read as "φ is known", and a second modal quantifier $\Diamond\varphi$, defined by convenience as $\neg\Box\neg\varphi$, is read as "not φ is not known". Other common interpretations are doxastic (φ is believed), deontic (there is an obligation for φ) or temporal (always φ).

Many important problems that arise from the use of modal logics as the basis for knowledge representation or agent specification can be reduced to the problem of determining satisfiability in modal logic. Automated Theorem Provers (ATPs) are software used to automate the process of determining satisfiability in a particular logic [14]. Arguably, the most widely used technique for modal ATPs is that of semantic tableaux [9].

The current obstacle for Automated Theorem Proving, and thereby also theorem proving in modal logic, is the speed at which a set of formulae can be proven satisfiable. Theorem proving in propositional logic is NP-complete by the Cook-Levin Theorem [4]. Since most modal logics of interest will be extensions of propositional logic, modern computers struggle to handle most logics. Making theorem provers more efficient is therefore an active field of research [13].

© Springer International Publishing Switzerland 2015
B. Pfahringer and J. Renz (Eds.): AI 2015, LNAI 9457, pp. 218–228, 2015.
DOI: 10.1007/978-3-319-26350-2_19

Specifically, we investigate the potential for improvements in efficiency of the method of semantic tableau by using a novel data structure called a Sentential Decision Diagram (SDD). SDDs are used to represent sets of logical formulae upon which the rules of tableau can be applied. Previous research into the use of Binary Decision Diagrams (BDDs) as the basis for modal tableau has established positive results [11]. Since SDDs are proven to be at least as compact as BDDs, while also maintaining polytime operations [5], it is plausible that SDDs will also be competitive with state-of-the-art modal theorem provers.

2   Preliminaries

2.1   Motivation for SDDs

An SDD is a novel representation of a boolean function $f : \mathbb{B}^n \longrightarrow \mathbb{B}$ proposed in 2011 by Darwiche [5]. It is closely related to two popular representations: BDDs and decomposable negation normal form (d-DNNF). Together, these three forms share the relationship:
$$BDD \subset SDD \subset \text{d-DNNF} \tag{1}$$
When comparing representations of boolean functions, there are two important properties to consider. The first, succinctness, is the size of the boolean function when compiled into the form. The other, tractability, is the set of operations on the compiled representation that can be performed in polynomial time. In general, there appears to be an inverse relationship between succinctness and tractability: i.e. more succinct representations will tend to be less tractable [5].

BDDs and d-DNNF follow this trade-off between succinctness and tractability. Although BDDs are less succinct, they maintain the ability to perform boolean operations such as negation, conjunction and disjunction in polynomial-time. On the other hand, a boolean function can be compressed into a much smaller size using d-DNNF, but without polynomial-time operations [5]. The decision of whether to use BDDs or d-DNNF within a software system hence depends entirely on the context and needs of the system.

SDDs are aimed at being a middle-of-the-road approach between these two forms. They have the ability to perform boolean operations in polynomial-time, while also being more succinct than a BDD. This has raised the possibility for SDDs to supersede the use of BDDs in practical applications [5].

Like BDDs, the final size and speed of compilation of SDDs is influenced by heuristics and minimisation algorithms used during compilation. Using a dynamic minimisation algorithm and tested over a series of combinatorial circuit benchmarks, it was demonstrated that SDDs can compile faster and to a more compressed representation than BDDs, often by an order of magnitude [3].

The speed of compilation provides the motivation for SDDs to be used as the fundamental data structure of a modal theorem prover. There are many other modal theorem provers to compare with - the two most state-of-the-art provers are likely InKreSAT [13] and FaCT++ [15]. InKreSAT works by incrementally reducing the modal satisfiability problem to a boolean satisfiability problem, while FaCT++ is a heavily optimised tableaux-based approach. Prior research conducted by Gore et al. [11] has a shown a tableaux-based approach using BDDs in place of the usual Davis-Putnam-Logemann-Loveland (DPLL) algorithm can be promising. At this point in time, we are aware of no other research published on the efficacy of SDDs for use in any theorem prover, modal or otherwise.

2.2   Constructing SDDs

Figure 1a shows a graphical depiction of an SDD. A circle node ○ represents a disjunction of its children, while a pair of boxes $\boxed{q\,|\,p}$ represents a conjunction of q and p. We can therefore read this diagram as:

$$f = (\neg A \wedge ((B \wedge C) \vee (\neg B \wedge \bot))) \vee (A \wedge B). \tag{2}$$

Which, via De Morgan's laws, can be reduced to:

$$f = (A \wedge B) \vee (B \wedge C). \tag{3}$$

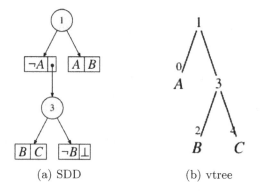

(a) SDD   (b) vtree

Fig. 1. An SDD and vtree for the boolean expression $(A \wedge B) \vee (B \wedge C)$

Each SDD is based uniquely off a seperate structure called a vtree, or vertical tree [5]. A vtree is a full binary tree using the variables of a given boolean function. Figure 1b shows the vtree corresponding to the SDD in Fig. 1a. Each non-leaf node of a vtree forms a variable split of the variables found in the left subtree and those in the right subtree. For example, node 1 corresponds to a variable split $\mathbf{X} = \{A\}$, $\mathbf{Y} = \{B, C\}$.

For each variable split, a boolean function f using these variables can be written as what is called a compressed $(\mathbf{X}, \mathbf{Y})$-partition:

$$f = [p_1(\mathbf{X}) \wedge s_1(\mathbf{Y})] \vee [p_2(\mathbf{X}) \wedge s_2(\mathbf{Y})] \vee \dots \vee [p_n(\mathbf{X}) \wedge s_n(\mathbf{Y})]. \tag{4}$$

Here each p_i expression is known as a prime, while each s_i expression is known as a sub. A given function $f(\mathbf{X}, \mathbf{Y})$ has a unique compressed $(\mathbf{X}, \mathbf{Y})$ partition [5]. These partitions have the following constraints:

1. Each pair of primes is mutually exclusive. $(\forall i, j : i \neq j)\ p_i \wedge p_j = \text{false}$.
2. All primes are consistent. $(\forall i)\ p_i \neq \text{false}$.
3. The disjunction of all primes is valid. $\bigvee_i p_i = \text{true}$.
4. All subs are distinct. $(\forall i, j : i \neq j)\ s_i \neq s_j$.

The procedure to construct an SDD given a vtree and a boolean function is given as follows [5]: first, the boolean function is expressed as an $(\mathbf{X}, \mathbf{Y})$-partition for the variable split imposed by the root node of the vtree. For the expression $(A \wedge B) \vee (B \wedge C)$ and the vtree in Fig. 1b, we would have:

$$[A, B], [\neg A, B \wedge C] \tag{5}$$

As usual, a literal is either an atomic formula or its negation. Each non-literal prime of the partition is then recursively expressed as a $(\mathbf{X}, \mathbf{Y})$-partition for the variable split of the child node in the left subtree, and each sub is expressed as a partition for the child node in the right subtree. This recursive procedure continues until each prime is a literal. For our expression, we need only to express $B \wedge C$ as a $(\mathbf{X}, \mathbf{Y})$-partition for the variables split $\mathbf{X} = \{B\}$, $\mathbf{Y} = \{C\}$ imposed by node 3. This gives us:

$$[A, B], [\neg A, ([B, C], [\neg B, \bot])] \tag{6}$$

This is equivalent to expression 2 and the SDD in Fig. 1a.

2.3 Modal Tableaux

The method of semantic tableaux is arguably the most widely used technique for deciding satisfiability in non-classical logics [9]. The tableaux procedure, as presented by Goré [10], creates a tree structure called a tableau. Each node of the tableau contains a set of modal formula, while each branch represents the application of a particular rule on a parent node to create a child node. The finite set of these rules is called the tableaux calculus. The tableaux calculus for modal logic K is as follows:

Static rules:

$$(id)\frac{\varphi; \neg\varphi; X}{\times} \qquad (\wedge)\frac{\varphi \wedge \psi; X}{\varphi; \psi; X} \qquad (\vee)\frac{\varphi \vee \psi; X}{\varphi; X \mid \psi; X} \tag{7}$$

Transitional rule:

$$(\Diamond K)\frac{\Diamond\varphi; \Box\psi; X}{\varphi; \psi; X} \tag{8}$$

To test whether a given set Y of formulae is K-satisfiable, we build an upside down tree of nodes, each obtained from the root node Y by applying these rules. A branch of the tree closes if the rule (id) is applied, else it is open. If all branches close, the tableaux is said to be closed. A set Y is shown unsatisfiable if it has a closed tableau, and shown satisfiable otherwise.

The static rules can be safely applied at any time. Usually, we apply them in the left-to-right order shown in (7) as this closes a branch as soon as possible, or extends it linearly with only one child, or extends it with two children. The transitional rule forces us to choose a particular $\Diamond\varphi$ from the node and different choices give different instances of this rule. To ensure completeness, we have to try all possible instances until one of them closes, or all of them are open.

This gives us a sound and complete strategy for rule applications: apply the static rules as much as possible, then apply the transitional rule in all possible ways.

The first part is called the saturation phase, while the second is called a transitional jump. These phases alternate, with all possible static rules being applied after a single transitional rule application. The paper showing the efficacy of BDD-based modal tableaux prover [11] used an arrangement where BDDs were used to automatically implement the saturation phase, while transitional rules were applied to the BDDs by a master program.

3   Method

3.1   Preliminary Experiment: Propositional Compilation

Before developing a modal theorem prover using SDDs, we conducted a preliminary experiment to test the speed at which a statement in propositional logic can be compiled into an SDD. Compiling classical propositional logic (CPL) is equivalent to the initial saturation phase of the tableaux method - where one applies only the rules of propositional logic to a modal formula to find any satisfying evaluations. It should be noted that successive saturations will be generally less intensive than the initial saturation, as they will be integrating a smaller number of propositions into an existing SDD. However, the speed of the initial saturation phase will be indicative of the speed of successive saturation phases, as they use the same SDD operations. It is therefore integral for the success of an SDD-based modal theorem prover that this initial saturation phase be efficient.

The compilation from propositional logic to an SDD was performed in a bottom-up fashion, illustrated by the following example:

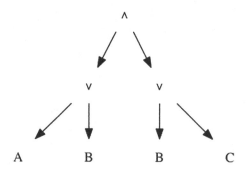

Fig. 2. Parse tree for boolean expression $(A \lor B) \land (B \lor C)$.

Each formula of propositional logic has a corresponding parse tree. Here Fig. 2 is the parse tree corresponding to the formula $(A \vee B) \wedge (B \vee C)$. A parse tree in this context will have all leaf nodes denoting a boolean variable, and each non-leaf node denoting an operation to be performed upon its children.

To bottom-up construct an SDD for a given parse tree, an SDD is created for each leaf node of the tree, then for each non-leaf node the corresponding operation is performed upon its children, working from the bottom of the tree up to the root node.

In Fig. 3, an SDD is created for each leaf A, B and C. The SDDs representing A and B are disjoined to create an SDD for $A \vee B$, and equivalently for $B \vee C$. Finally, the SDDs representing $A \vee B$ and $B \vee C$ are conjoined to create an SDD representing the entire formula $(A \vee B) \wedge (B \vee C)$.

The bottom-up compiler was written in C++ using the SDD Package [16], a C library for constructing and manipulating SDDs.

3.2 Theorem Proving

A modal theorem prover called SDDtab was developed in C++, using the SDD Package [16]. In order to compare the efficacy of BDDs and SDDs for modal theorem proving, SDDtab is based on the same algorithm as BDDtab [11], a modal theorem prover using BDDs. We prioritised ensuring a fair comparison between SDDtab and BDDtab. Optimisations present in BDDtab, such as SAT caching, were implemented in SDDtab when suitable, else they were disabled within BDDtab while benchmarking.

We will only summarise the method used for both SDDtab and BDDtab. Further details are presented by Gore et al. [11].

SDDs are used to implement the saturation phase of the tableau procedure, as described in Sect. 2.3. Modal components $\Diamond \varphi$ and $\Box \varphi$ in the input formula are expressed as atomic propositional variables $v_{\Diamond \varphi}$ and $v_{\Box \varphi}$ throughout the saturation phase, allowing the SDDs to complete the static tableaux rules.

After an initial saturation phase, the SDD structure representing the formula returns a list of possible evaluations of the variables that would make the formula true. In a traditional tableaux proof, this represents a list of open branches in the tableau. The prover then performs transitional tableaux rules on one of these branches in an iterative, depth first fashion, interrupting each transitional jump with a saturation phase.

If the prover encounters a contradiction, it closes the branch. In a typical tableaux procedure, the prover would now choose another open branch to explore. In the case of SDDtab, the SDD representing the formula is appended with the negation of a set of variables that were responsible for the branch closing. This effectively assists in closing other branches that would otherwise need to be explored.

For example, a branch with variables $\Box(\neg a) \wedge a \wedge b$ would find the contradiction $a \wedge \neg a$ after extracting $\neg a$ from the boxed variable $\Box(\neg a)$. We would then append $\neg((\Box \neg a) \wedge a)$ to the SDD. This would, for instance, close a branch with variables $(\Box \neg a) \wedge a \wedge c$.

Using our strategy, if the prover finds the tableau closed then the prover has found the initial set of formulae K-unsatisfiable, else it has found the set of formulae K-satisfiable.

4   Results

4.1   Preliminary Results

We evaluated the SDD-based bottom-up compiler (sddtab) against two state-of-the-art SAT solvers, MiniSat [7] and Z3 [6], as well a BDD-based bottom-up compiler (bddtab) and a CNF-to-SDD compiler (sdd-cnf) provided with the SDD library [16].

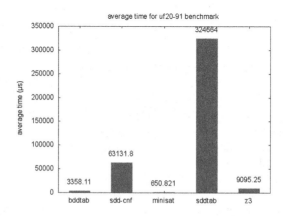

Fig. 3. Average compilation time from CPL for uf20-91 benchmarks.

We ran each method over a series of propositional benchmarks, then averaged to find the mean time per benchmark. We have included two graphs corresponding to two sets from the SATLIB Benchmark Suite [12]: uf20-91 and uf50-218. Each provides 1000 propositional benchmarks of 20 and 50 variables respectively.

SDD bottom-up compilation is at least two orders of magnitude slower than bottom-up BDD compilation at all variable sizes. At 50 variables the compiler fails to terminate in under 3 min. From this we are forced to conclude that this current implementation of SDD bottom-up compilation is likely to be unsuitable for use in modal theorem proving (Fig. 4).

However, the CNF-to-SDD compiler performs better than BDD bottom-up compilation as the variable size increases. Thus SDDs may still have potential as the basis for efficient modal tableux if the speed of the CNF-to-SDD compiler is reproducible for propositional logic outside of the CNF form.

Although SAT solvers MiniSat and Z3 outperformed all other approaches, BDD-based modal theorem provers [11] have been shown to compete with SAT-based modal theorem provers such as InKreSAT [13].

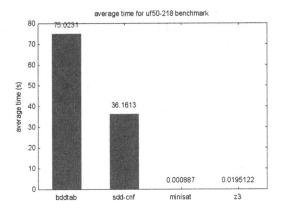

Fig. 4. Average time of compilation from propositional logic for the uf50-218 benchmark set. The bottom-up sdd compiler failed to terminate, and hence is not included.

4.2 Theorem Proving Results

SDDtab[1] was evaluated against BDDtab [11] using benchmarks from the Logic Work Bench [1] (LWB) for modal logics K and S4. Each formula set contains 21 numbered benchmarks, sorted by size and difficulty to solve. Both programs were given 600 seconds to solve each problem - the number of solved instances for each set is listed in Table 1. All benchmarking was performed on an Intel 3.5 GHz CPU with 32 GB of memory. The benchmark instances, source, and scripts are all publicly available in a single repository[2].

Table 2 shows us the average size of the constructed logical form for each benchmark set. BDDs and SDDs appear similarly matched in this regard. On both the K and S4 benchmarks, half of all benchmarks are compiled to a smaller BDD than SDD, while the other half are compiled to a smaller SDD than BDD. This suggests the speed of operations rather than the succinctness of compilation is determining the speed of each modal logic prover.

In Table 1, we can see that the SDD-based prover completes similar to BDD-based prove for the majority of benchmarks. It solved fewer problems to BDD-based solver for some problems, but outperforms the BDD-based prover for the benchmarks $s4\_ph\_n$, $s4\_ipc\_n$, k_ph_n, and k_ph_p. Furthermore, we observed that for many benchmarks, the SDD-based prover may timeout for instances that are considered relatively easier, but successful in solving other harder instances that belong to the same class of problems.

We believe the success on these benchmarks suggests the SDD-based approach is more suited to solving instances that are large in size, yet contain modalities that are less nested or fewer in number. As an illustrating example we examine the benchmark k_ph_p, in which SDDtab outperforms BDDtab:
$$k\_ph\_p(n) :\equiv \Diamond left(n) \rightarrow \Diamond right(n)$$

---

[1] https://github.com/tpagram/sddtab.
[2] https://github.com/jasonjli/ai2015-benchmarks.

Table 1. Results on the LWB benchmark for K and S4 modal logics. Entries represent the number of problem instances solved by the program, where the program was given a time limit of 600 s per instance.

S4 Benchmark	BDDtab	SDDtab	K Benchmark	BDDtab	SDDtab
s4_45_n	**21**	**21**	k_branch_n	**21**	15
s4_45_p	**21**	**21**	k_branch_p	**21**	**21**
s4_branch_n	**21**	16	k_d4_n	**21**	19
s4_branch_p	**21**	**21**	k_d4_p	**21**	**21**
s4_grz_n	**21**	**21**	k_dum_n	**21**	**21**
s4_grz_p	**21**	20	k_dum_p	**21**	**21**
s4_ipc_n	13	**15**	k_grz_n	**21**	**21**
s4_ipc_p	**21**	12	k_grz_p	**21**	**21**
s4_md_n	**16**	13	k_lin_n	**21**	7
s4_md_p	**21**	**21**	k_lin_p	**21**	18
s4_path_n	**3**	**3**	k_path_n	**21**	15
s4_path_p	**5**	3	k_path_p	**21**	16
s4_ph_n	5	**6**	k_ph_n	11	**12**
s4_ph_p	**10**	6	k_ph_p	10	**12**
s4_s5_n	**21**	**21**	k_poly_n	**21**	**21**
s4_s5_p	**21**	**21**	k_poly_p	**21**	**21**
s4_t4p_n	**21**	**21**	k_t4p_n	**21**	**21**
s4_t4p_p	**21**	**21**	k_t4p_p	**21**	**21**

$$left(n) :\equiv \bigwedge_{i=1,\ldots,n+1}\left(\bigvee_{j=1,\ldots,n}(l(i,j))\right)$$
$$right(n) :\equiv \bigvee_{j=1,\ldots,n;i_1=1,\ldots,n+1;i_2=i_1+1,\ldots,n+1}(l(i_1,j) \wedge l(i_2,j))$$
$$l(i,j) :\equiv \begin{cases} \Box p_{100i+j} & i < j \\ p_{100i+j} & \text{otherwise} \end{cases}$$

Other than two diamond quantifiers surrounding the left and right components of the formula, k_ph_p contains only modalities that apply to literals. This effectively reduces the amount of branching within the solver and consequently the number of operations being performed upon the underlying decision diagram. Given this, we would expect the conciseness of the initial compiled diagram to have a greater influence on the result for k_ph_p than for benchmarks with modalities that are more nested. Table 2 shows us that SDDtab compiles to a more concise diagram for k_ph_p, which may explain its success.

We hypothesise that the comparative immaturity of vtree heuristics for SDDs compared with the more established variable order heuristics for BDDs results in slowness when applying operations. This gives SDDs a disadvantage on benchmarks that require excessive branching.

Table 2. The average number of nodes of the initial BDD/SDD constructed over the first 20 benchmarks of each set from the LWB benchmark. Dashed entries denote sets containing a benchmark that failed to construct in under a minute.

S4 Benchmark	BDDtab	SDDtab	K Benchmark	BDDtab	SDDtab
s4_45_n	**56.50**	83.75	k_branch_n	**29.00**	38.80
s4_45_p	**46.00**	63.20	k_branch_p	**30.00**	39.55
s4_branch_n	3.00	**2.00**	k_d4_n	**65.00**	92.85
s4_branch_p	4.00	**3.00**	k_d4_p	**54.50**	74.20
s4_grz_n	**29.00**	37.40	k_dum_n	**11.60**	12.90
s4_grz_p	95.65	**80.15**	k_dum_p	**11.60**	12.85
s4_ipc_n	9.50	**9.00**	k_grz_n	45.00	**37.95**
s4_ipc_p	10.50	**10.05**	k_grz_p	65.85	**59.40**
s4_md_n	**11.40**	13.80	k_lin_n	**150.20**	267.00
s4_md_p	**30.40**	46.60	k_lin_p	183.00	**91.75**
s4_path_n	**171.50**	277.25	k_path_n	**174.00**	258.50
s4_path_p	**156.50**	254.85	k_path_p	**158.60**	248.45
s4_ph_n	-	-	k_ph_n	3.00	**1.00**
s4_ph_p	-	-	k_ph_p	2.85	**0.95**
s4_s5_n	3.00	**1.00**	k_poly_n	5.00	**3.40**
s4_s5_p	3.00	**1.00**	k_poly_p	5.00	**3.25**
s4_t4p_n	9.00	**8.85**	k_t4p_n	8.00	**7.75**
s4_t4p_p	**9.00**	9.05	k_t4p_p	8.00	**7.90**

Overall, these results suggests that SDD-based approaches are promising for application in modal tableaux. However, more progress on vtree heuristics are essential.

5 Conclusions and Further Work

Sentential Decision Diagrams are shown to be a promising data structure for use in modal theorem proving. Comparison with Binary Decision Diagrams shows SDDs currently have a slower compilation speed for propositional logic outside of CNF, and a comparable for modal theorem proving capacity when used in a similar algorithmic capacity. In the empirical evaluation carried out in this paper, the SDD-based approach solved as many instances as the BDD-based approach for over half of the benchmark, and even outperformed the BDD-based approach for some subset of instances with large sizes and modalities that are less nested or fewer in number. This also indicates that SDDs may be promising for theorem proving in other extensions of propositional logic, such as intuitionistic logic. Since the alleged benefits of SDDs are reproducible only on large CNF formulae, further work could include developing an SDD-based modal theorem prover that uses CNF at each stage of the tableaux method. However, it is important that further work is done on developing heuristics for formulae in a general form before SDDs can find success in theorem proving.

Acknowledgements. We thank the anonymous reviewers for their reviews and their suggestions to improve the paper.

References

1. Balsiger, P., Heuerding, A., Schwendimann, S.: A benchmark method for the propositional modal logics K, KT, S4. J. Automat. Reason. **24**(3), 297–317 (2000)
2. Blackburn, P., De Rijke, M., Venema, Y.: Modal Logic. Cambridge University Press, Cambridge (2002)
3. Choi, A., Darwiche, A.: Dynamic minimization of sentential decision diagrams. In: AAAI (2013)
4. Cook, S.A.: The complexity of theorem-proving procedures. In: ACM (1971)
5. Darwiche, A.: SDD: a new canonical representation of propositional knowledge bases. In: IJCAI (2011)
6. de Moura, L., Bjørner, N.S.: Z3: an efficient SMT solver. In: Ramakrishnan, C.R., Rehof, J. (eds.) TACAS 2008. LNCS, vol. 4963, pp. 337–340. Springer, Heidelberg (2008)
7. Een, N., Sörensson, N.: MiniSat: a SAT solver with conflict-clause minimization. In: SAT (2005)
8. Fagin, R., Moses, Y., Halpern, J.Y., Vardi, M.Y.: Reasoning About Knowledge. MIT Press, Cambridge (2003)
9. Girle, R.: Modal Logics and Philosophy (2000)
10. Goré, R.: Tableau methods for modal and temporal logics. In: D'Agostino, M., Gabbay, D.M., Hähnle, R., Posegga, J. (eds.) Handbook of Tableau Methods, pp. 297–396. Springer, Amsterdam (1999)
11. Goré, R., Olesen, K., Thomson, J.: Implementing tableau calculi using BDDs: BDDTab system description. In: Demri, S., Kapur, D., Weidenbach, C. (eds.) IJCAR 2014. LNCS, vol. 8562, pp. 337–343. Springer, Heidelberg (2014)
12. Hoos, H., Stiitzle, T.: Satlib: an online resource for research on SAT (2000)
13. Kaminski, M., Tebbi, T.: InKreSAT: modal reasoning via incremental reduction to SAT. In: Bonacina, M.P. (ed.) CADE 2013. LNCS, vol. 7898, pp. 436–442. Springer, Heidelberg (2013)
14. Loveland, D.W.: Automated Theorem Proving: A Logical Basis. Elsevier, Toronto (2014)
15. Tsarkov, D., Horrocks, I.: FaCT++ description logic reasoner: system description. In: Furbach, U., Shankar, N. (eds.) IJCAR 2006. LNCS (LNAI), vol. 4130, pp. 292–297. Springer, Heidelberg (2006)
16. UCLA: The SDD Package 1.1.1 (2014). http://hreasoning.cs.ucla.edu/sdd/

Decision Making Strategy Based on Time Series Data of Voting Behavior

Shogo Higuchi[✉], Ryohei Orihara, Yuichi Sei,
Yasuyuki Tahara, and Akihiko Ohsuga

Graduate School of Information Systems, The University of Electro-Communications,
Tokyo, Japan
higuchi.shogo@ohsuga.is.uec.ac.jp, ryohei.orihara@toshiba.co.jp,
{sei,tahara}@is.uec.ac.jp, ohsuga@uec.ac.jp

Abstract. In gambling such as horse racing, we are sometimes able to peep peculiar voting behavior by a punter with the advantageous information closely related to the results. The punter is often referred as an *insider*. In this study, our goal is to propose a reasonable investment strategy by peeping *insiders'* decision-making based on the time series odds data in horse racing events held by JRA. We have found the conditions that the rate of return is more than 642 % for races whose winner's prize money is 20 million yens or more. That suggests the possibility of *Knowledge Peeping*.

Keywords: Horse racing prediction · Time series data · Optimization · Decision tree · Knowledge peeping

1 Introduction

In gambling there is a family of betting where participants cannot be involved in the outcome. In such betting, the support rate is determined by voting of punters who have various decision-making strategies. The payout is determined based on the support rate. Horse racing, roulette and lottery are examples of such betting.

In horse racing, interesting voting behaviors have been reported. In autumn 2014, at Kikka Sho and Tenno Sho, two of the most prestigious and the highest-ranked (Grade-1) events held by Japan Racing Association (JRA), horses with little track record have won, after mass votes from a punter. These phenomena have been observed in the past. It is possible to find similar cases in articles on the *nikkansports* newspaper site [1] by Google search. The oldest one of what we are aware is the 2003 Takarazuka Kinen (Grade-1). A punter who invested 13 million yens on the racehorse "Hishimirakuru", which had shown poor performance in past races, gained approximately 200 million yens of payout by the victory of the racehorse [2]. It is famous among horse race enthusiasts in Japan, and the punter is called as "Mirakuru Ojisan" (Mister of miracle in English). In the betting for the United States presidential election in 2008 held by a bookmaker, a punter who bet €100,000 on Barack Obama was reported to receive a large amount of payout [3].

© Springer International Publishing Switzerland 2015
B. Pfahringer and J. Renz (Eds.): AI 2015, LNAI 9457, pp. 229–241, 2015.
DOI: 10.1007/978-3-319-26350-2_20

Although there are researches analyzing peculiar voting behavior, whose typical embodiment is a mass vote as described above, to predict the outcome of the horse racing around the world, academic research in Japan, where bookmakers on horse racing are banned, has not yet been done. With ideas gained from generalization of these events, we explore the possibility to incorporate them into a reasonable investment strategy in gambling.

An unnatural odds formed by mass votes is regarded as an example of traces left by a punter with a peculiar voting behavior which may be peeped from other punters. In this study, we defined an *insider* as a voter who has the advantageous information closely related to the results and that is not known to the most of punters. While a term *insider* means an illegal voter in the stock investment, we want to note that according to our definition as *insider* could mean a voter with legally obtained special knowledge.

The structure of this paper is as follows. In Sect. 2, we explain the goal of the study. We introduce studies on investment strategy at horse racing around the world, and the studies of the horse racing market in Japan in Sect. 3. In Sect. 4, we explain the existing method to extract *insiders'* voting behavior which is likely to prompt a change in the odds, and a method to decide parameters for applying it to the time series data of events held by JRA. In Sect. 5, we experimentally evaluate the proposed parameter sets. In Sect. 6, we discuss the result of the experiment, and future challenges. We summarize this paper in Sect. 7.

2   Our Goal in This Study

In this study, our goal is to propose a reasonable investment strategy for getting a stable profit by peeping *insiders'* peculiar voting behavior based on the publicly available information in horse racing events held by JRA. The information we use is the odds, which is shared in the odds board. With permission by JRA, we analyze the time series of the odds in horse racing events and ultimately propose a strategy for voting in race-track betting. In the future, we aim to get a reasonable decision-making strategy that can be applied to general voting behavior by means of the concept *Knowledge Peeping*.

We peep knowledge to maximize the profit and make a reasonable investment for punters by comprehensively applying techniques such as machine learning to large amounts of voting data. The application domain is not only limited to a market such as gambling, but also more actions where we can perceive decisions of others, for example, stock market and political election. It has the potential to provide a variety of suggestions.

3   Related Work

We show the research on voting behavior in horse racing below. There are a number of studies targeted on gambling in the field of social intelligence science [4].

3.1 Studies on Horse Racing Investment Strategy Around the World

Comparing the expected profit rate of each horse, which is defined by multiplication of the objective probability of winning and the payout ratio against the wager, it is observed that the rates of favorite horses are large and ones of *dark horses* are small in horse racing markets around the world. The excessive preference for *dark horses* is called "Favorite-Longshot Bias", which many studies have been made to explain. In addition, there is a study that obtained empirical result whose expected profit rate is positive by the investment strategies based on the Favorite-Longshot Bias [5].

3.2 Studies on Horse Racing Market in Japan

There are some works dealing with horse racing market in Japan. From the voting data of races held by JRA, the Favorite-Longshot Bias is observed and analysis on the bias is performed from the viewpoint of behavioral economics [6]. There is also a study reported that *Accuracy Ratio* of the final odds is more than that of the logit model learned from the track record of racehorses and jockeys to perform a prediction. A neural network model to predict the clock of a race is also mentioned in the study, and it is outperformed by the final odds [7]. The final odds is an embodiment of the collective knowledge of punters. From these studies, the collective intelligence by punters in Japan are getting a positive evaluation. Moreover, a model for a situation where multiple investment strategies exist has been proposed and the estimation of the ratio of each strategy has been performed [8].

Although the investigation of the horse racing market as a whole has been carried out as described above, there are no study to propose an investment strategy that focuses on the individual decision-making. Therefore this study can be considered to be a new type of research in social intelligence field. We defined *Knowledge Peeping* in this paper as analyzing *insiders'* trace of voting from the publicly available information, and acquiring the useful information to decision-making.

4 Knowledge Peeping in the Horse Racing

4.1 *Knowledge Peeping* from *Insiders'* Votes

We try to maximize the profit in horse racing betting by peeping knowledge from *insiders'* peculiar voting behavior. In this study, we apply the method proposed by Law and Peel [9] in order to extract racehorses whose odds may be altered by *insider's* vote.

In the horse racing in Japan, only "pari-mutuel betting" is allowed. The payout in pari-mutuel betting is determined depending on the reciprocal of the final odds. On the other hand, "Bookmaker betting" accounts for the majority of votes in the race of Europe, the United States and Australia. The payout in bookmaker betting is determined according to the odds posed by a bookmaker at the time of voting.

Law and Peel are dealing with the odds in the United Kingdom horse racing which is based on the bookmaker betting. Therefore it may be difficult to apply this method directly to the odds on pari-mutuel betting.

4.2 Conditions of Racehorse Extraction Proposed by Law and Peel

Law and Peel defined two parameters *mop* and *dz* to extract racehorses which are likely to be voted by an *insider*. They are able to get the positive expected rate of return and get profit when the parameters are in certain ranges. First, *mop* is a parameter representing the difference between opening odds and starting odds. Here, opening odds are the first odds posted by a bookmaker, and starting odds are the final odds before the race starts. When the opening odds of horse number i is $\omega_i(o)$ and the starting odds of horse number i is $\omega_i(s)$, the formula for calculating the *mop* is shown in Eq. (1).

$$mop = \log\left(\frac{\omega_i(s) + 1}{\omega_i(s)}\right) - \log\left(\frac{\omega_i(o) + 1}{\omega_i(o)}\right) \tag{1}$$

Second, *dz* is a parameter based on the model proposed by Shin [10]. Shin has shown z, the ratio of *insider* voting in the total voting behavior, is described as Eq. (2), where p_i shows objective probability of the i-th horse is to win in the number of horses participated in the race n and β shows maximum sales allowed for the bookmaker.

$$z = \left(\frac{1}{\beta}\sum_{i=1}^{n}\left[\frac{1}{\omega_i + 1}\right]^2 - \sum_{i=1}^{n}p_i^2\right)\bigg/\left(1 - \sum_{i=1}^{n}p_i^2\right) \tag{2}$$

dz is defined by the difference of between z in the opening odds and z in starting odds. Because p_i and β can be treated as constant unique to each race, *dz* can be calculated as shown in Eq. (3). Here, $\beta > 0$ and $C < 1$.

$$dz = \frac{1}{\beta(1 - C)}\sum_{i=1}^{n}\left(\left[\frac{1}{\omega_i(s) + 1}\right]^2 - \left[\frac{1}{\omega_i(o) + 1}\right]^2\right) \tag{3}$$

Law and Peel observed the expected rate of return is over 100 % where $dz > 0$ and *mop* > 0.05. The condition implies the odds of the entire horses becomes more uniform at starting odds compared to the opening odds, and individual odds has dropped significantly. Regarding this factor, Law and Peel explain that the bookmakers tend to shift the odds of *dark horses* towards the popularity side to avoid the damage upon receiving a large amount of voting to *dark horses* by *insiders*.

4.3 The Method of Determining the Opening Odds

We evaluate the expected rate of return of the method with data drawn from racing events held by JRA while changing the parameters to verify whether it is possible to peep knowledge in the pari-mutuel betting. The opening odds in pari-mutuel betting cannot be determined uniquely because the odds in pari-mutuel betting is dependent on the inverse of the rates of votes. Therefore the odds when it reaches certain rate of the total number of votes is used as the opening odds. We evaluate the rate of return while changing the rates of votes that determines the opening odds.

5   Experiments

5.1   Application to the Racing Events Held by JRA

First, we evaluate the expected rate of return of the method introduced in the previous section for some of the *mop* and *dz* values, changing the rates of votes that determines the opening odds. Our dataset is time-series odds in the Win[1] market of 36357 races, which are every racing events held by JRA from 2009 to 2014. Note that the time series odds are recorded every 5–10 min until the votes are closed.

Table 1. Results in each parameter sets (all races)

	mop	*dz*	Number of Horses	Rate of return
1	*mop* $\geq$ 0.05	*dz* $\geq$ 0	~14000	46–78 %
2		*dz* < 0	~9600	41–88 %
3	0 $\leq$ *mop* < 0.05	*dz* $\geq$ 0	~520000	72–82 %
4		*dz* < 0	~86000	67–96 %
5	−0.05 $\leq$ *mop* < 0	*dz* $\geq$ 0	~350000	60–71 %
6		*dz* < 0	~64000	59–78 %
7	*mop* < −0.05	*dz* $\geq$ 0	~8600	33–98 %
8		*dz* < 0	~12000	58–95 %

Table 2. Results in each parameter sets (prize money: 20 million yens or more)

	mop	*dz*	Number of Horses	Rate of return
1	*mop* $\geq$ 0.05	*dz* $\geq$ 0	~490	65–132 %
2		*dz* < 0	~420	66–139 %
3	0 $\leq$ *mop* < 0.05	*dz* $\geq$ 0	~24000	75–89 %
4		*dz* < 0	~8500	58–108 %
5	−0.05 $\leq$ *mop* < 0	*dz* $\geq$ 0	~22000	60–83 %
6		*dz* < 0	~4600	38–99 %
7	*mop* < −0.05	*dz* $\geq$ 0	~270	28–220 %
8		*dz* < 0	~740	31–105 %

---

[1] A type of betting offered by JRA where a punter selects one horse to win.

To calculate the return rate, we extract horses who meet the condition specified by the parameters, then assume to have bought their Win tickets. The return rate is calculated by their outcomes. The rate of return over all races are shown in Table 1. The ranges of the parameters are ones used in Law and Peel's paper. In parameter sets 3 through 6, where the absolute value of the *mop* is small, the number of extracted horses is larger compared to other cases because the extraction conditions are met even with a slight change in the odds. Although parameter set 1, which is the Law and Peel's proposed condition, got poor expected rate of return, parameter sets 7 and 8 where the *mop* is over 0.05, got good expected rate such as 95–98 %. Note that it is not easy to get the rate more than 80 % of return because deduction rate of the Win market in JRA is about 20 %. However, the rate of return in Table 1 does not exceed 100 % and the strategy cannot be seen as a profitable betting policy.

Because the method is trying to peep *insider*'s voting in comparison with the *outsider*'s voting, the result could be explained by a lack of quality information from *outsiders*. The vast majority of our data is drawn from low-ranked races, whose prize money is low and hardly gets huge attentions. In such events, the number of *outsider* participants is assumed to be small and it could be insufficient to form a reliable collective voting behavior. Therefore we investigate the parameter sets for the subset of the dataset, where the events are limited to *big races* that more *outsider* punters are expected to participate.

Among the 36357 races, 2200 races are events whose winner's prize money is 20 million yens or more. For the 2200 races we calculate the expected rate of return. The results are shown in Table 2. There are five parameter sets whose rates are more than 100 %. The better result comparing to Table 1 could be explained by the presence of more *outsiders*.

Then, focusing on the *mop*, we draw graphs of the expected rate of return and the number of extracted horses against the rates of votes used to determine the opening odds, in the three conditions on the *mop*, that is, $mop \geq 0.05$, $0 \leq mop < 0.05$ and $mop < -0.05$. The graphs are shown in Fig. 1 through Fig. 3.

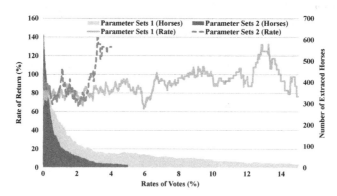

Fig. 1. Rates of return and number of extracted horses with $mop \geq 0.05$

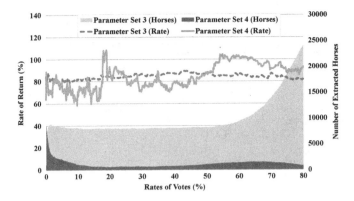

Fig. 2. Rates of return and number of extracted horses with $0 \leq mop < 0.05$

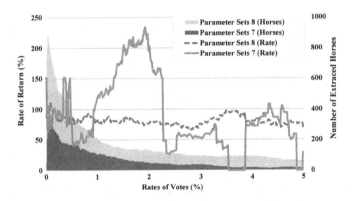

Fig. 3. Rates of return and number of extracted horses with $mop < -0.05$

First, Fig. 1 is the transition of the expected rate of return and the number of extracted horses with $mop \geq 0.05$ (parameter sets 1 and 2). The line graph (left axis) is the expected rate of return, the bar graph (right axis) is the number of racehorses that have been extracted by the specified conditions. The calculation of the expected rate of return is made if more than 20 horses are extracted. Although the both parameter sets have yielded approximately 130 % as maximum of the expected rate of return when the number of extracted horses is barely more than 20, the range in which the strategy is profitable remains negligible. We cannot say that they are rational investment strategies.

Then, Fig. 2 is the transition with $0 \leq mop < 0.05$ (parameter sets 3 and 4). Throughout the entire voting period, the method has extracted racehorses that meet the conditions. We have decided to consider cases where the rates of votes is less than 80 % only, because the time when the rates of votes becomes 80 % is immediately before the vote closure and it would be unrealistic to make a decision in such last minute. Although parameter set 4 has yielded 108 % as maximum of expected rate of return when the number of extracted horses is about 570, the range in which the strategy is profitable remains negligible. We cannot say that it is a reasonable investment strategy.

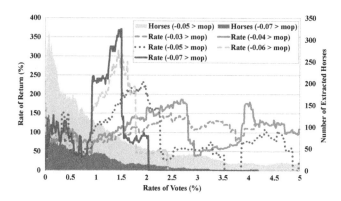

Fig. 4. Rates of return and number of extracted horses added with the Vicinity Conditions of *mop* in Fig. 3.

Finally, Fig. 3 is the transition with *mop* < -0.05 (parameter sets 7 and 8). It is also plotted for cases where more than 20 horses are extracted. In parameter set 7, the expected rate of return is 234 % as maximum when the number of extracted horses is 88, and the strategy remains profitable in the range of 1.0–2.2 % of x-axis, which is significantly wider than cases for parameter sets 1 through 4. It can be said that with the parameter set there is a condition for a stable profits. Furthermore, we investigate the effect of *mop* on the expected rate of return more closely. Figure 4 is the transition where the *mop*'s thresholds are changed from -0.03 to -0.07. We can see a trend that the higher the expected rate of return is obtained when the larger the change in the odds is, namely the larger the absolute value of the *mop* is. The maximum expected rate of return 368 % has been achieved with *mop* < -0.07 when the rates of votes is 1.5 %, where the number of extracted horses is about 20. From this result, it has suggested that it is possible to peep knowledge in pari-mutuel betting.

In pari-mutuel betting, upon receiving a large amount of voting to dark horse by *insiders*, its odds first soar and then shift towards the less popular side. The overall odds are also influenced by the vote. This can explain why it has been possible to peep *insiders*' knowledge under the conditions of $dz \geq 0$ and *mop* < -0.05.

Note that the approach by Law and Peel is based on the assumption that bookmakers set reasonable odds. Therefore, it is impossible to carry out the *Knowledge Peeping* when the assumption does not hold. On the other hand, because the assumption always holds in the pari-mutuel betting, it is expected to perform *Knowledge Peeping* with more occasions.

5.2   Expected Rate of Return Without *Knowledge Peeping*

In order to evaluate the expected rate of return in previous section, we compare the investment decision suggested in Sect. 5.1 with another one that does not rely on the concept of *Knowledge Peeping*. In this section, we build a classification model using decision tree, and calculate the expected rate of return. The decision tree has been chosen because we were interested in which explanatory variables would be included in the trained model.

We draw data from 2200 races whose winner's prize money is 20 million yens or more, where 34888 racehorses have participated. We define the objective variables as the results of the racehorses whose values are 1 if the horse wins and 0 otherwise. We use 38 explanatory variables related to the performance and attributes of racehorses. However, explanatory variables involved in the constructed decision tree was only 6 variables. We show the explanatory variables in Table 3, and the decision tree in Fig. 5. Please refer to the JRA website [12] for definitions of terms in Table 3.

Table 3. Explanatory variables used in the decision tree classification

Continuous values		Discrete values	
Prize money	Horse's weight	Horse Name	Weight limits
Earning money	Change of weight	Trainer	Race conditions
Career of racing	Weight	Owner	Affiliation limits
Interval of races	Weight ratio	Jockey	Sex limits
Race number	Odds in win	Affiliation	Specified limits
Distance	Odds in place (max)	Sex	Track condition
Bracket number	Odds in place (min)	Coat Color	Weather
Horse number	Popularity ranking	Class of race	Surface
Horse's age	Mining score	Demotedrace	
Jockey's age	Mining ranking	Blinkers	

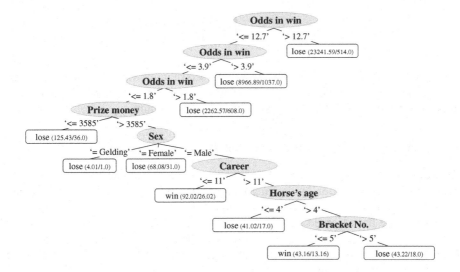

Fig. 5. The constructed decision tree

We employ J4.8 in the Weka [11] as the classification algorithm. The parameters C and M are set to 0.25 and 40 respectively. We have evaluated the accuracy through 10–fold cross validation, and have obtained 48.7 % precision and 3.2 % recall. Here, the poor recall can be overlooked because its influences to the expected rates are limited, for example, the profit will not be negative when the winning horse is missed.

Then, we calculate the expected rate of return with the investment decision based on Fig. 5. The result is 108.0 %. The fact that *Knowledge Peeping* is superior to the decision from *outsiders* suggests that *Knowledge Peeping* is effective at decision making at horse race investment.

5.3 Optimization of Parameters Related to *Knowledge Peeping*

In order to maximize the expected rate of return, we optimize the parameters related to extraction of the racehorses based on the discussion in Sect. 5.1. We employ the steepest descent method to search the conditions where the expected rate of return gets the maximum value. We apply the steepest descent method to $f(x)$ which represents the expected rate of return as a function of the rates of votes x_1, $x_2 = mop$, and $x_3 = dz$. We update the values as shown in Eq. (4) in k-th iteration. Here, α determines the amount of the values updated each time.

$$x^{(k+1)} = x^{(k)} - \alpha \; \text{grad} \, f\left(x^k\right)$$

$$= x^{(k)} - \alpha \begin{bmatrix} \dfrac{\partial f\left(x^k\right)}{\partial x_1^{(k)}} \\[2mm] \dfrac{\partial f\left(x^k\right)}{\partial x_2^{(k)}} \\[2mm] \dfrac{\partial f\left(x^k\right)}{\partial x_3^{(k)}} \end{bmatrix} \tag{4}$$

Although we have experimented with several sets of initial values and learning rates, we could not observe the convergence to the optimal solution. The algorithm has prematurely stopped because it has been trapped in a local minimum.

5.4 Visualization of the Expected Rate of Return

Before proceeding to methods to globally optimize the expected rate of return, we study the property of the function by drawing its graph as a 3D surface chart over the plane defined by two most important parameters *mop* and rates of votes.

Because we have learned that the cases with $dz > 0$ are more promising than the cases with $dz \le 0$ from Tables 1 and 2, we draw the 3D surface chart of the function fixing dz positively. We show the 3D graph for races whose winner's prize money is 20 million yens or more in Fig. 6.

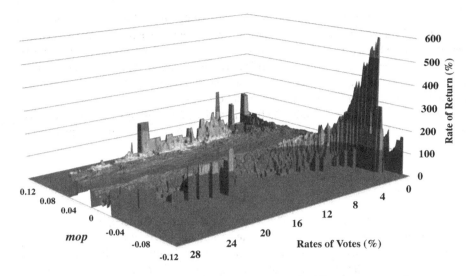

Fig. 6. Expected rates of return (prize money: 20 million yens or more)

From Fig. 6, it has been clearly indicated that the expected rates are 0 % in the area where *mop*'s absolute value is large and the rates of votes is large. This is because the large rates of votes means the small difference between the starting odds and the opening odds. In those cases little voting behaviors meet the conditions and therefore little race-horses are extracted.

There are sporadic peaks around the area where the rates of votes is approximately 18 %. These could be results of mass votes made at the time frames correspond to the rates of votes.

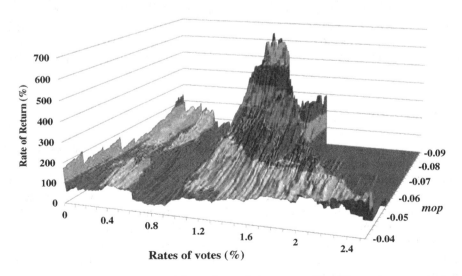

Fig. 7. Enlarged view around the peak area (prize money: 20 million yens or more)

Figure 7 is obtained by enlarging the peak area ($-0.09 \leq mop \leq -0.04$, $0 \leq$ *rates of votes* ≤ 2.4) of Fig. 6. There are spots where the expected rates exceed the best value in the Sect. 5.1. As the maximum, we have obtained the 641.8 % of the expected rate of return where *rates of votes* $= 1.01\%$ and $mop = -0.09$. Although the surface looks generally monotonous, a lot of local maximums exist. That is why we could not optimize the function by the simple algorithm shown in Sect. 5.3.

Although we have investigated the function fixing dz positively here, it is necessary to find the solution including the optimal condition for dz. We plan to do so by applying methods like simulated annealing in the future.

6   Discussion

As shown in the experiments, peculiar voting behaviors we can peep are ones made at relatively early stages where the rates of votes take values around 1 to 3 %. When a mass vote is made by an *insider* at a relatively early stage, its effect on the overall odds is significant because the total number of votes at that time is relatively small. The mass vote changes the odds of favorite horses in the opening odds as if they are underrated. It motivates *outsiders* to vote to favorite ones. As a result, favorite horses are overrated at starting odds to give the *insider* greater profit. This is the rationale of *insiders* to perform mass votes at early stage.

However, there is a reason for *insiders* to make mass votes just before races start. First, peculiar votes made at the final moments are unlikely to be spotted because they tend to be buried in a large amount of other votes at that time. For an *insider*, this means less possibility to be accused as an illegal voter. Second, *insiders* may want to prevent *free riding* behaviors from *outsiders* by making their voting behavior obscure. Here, free riding means that the profit *insiders* are supposed to get is shared by *outsiders* by imitating *insiders'* peculiar voting behaviors.

From the second point of view, it is desirable for us to be able to peep knowledge from voting behaviors just before races start. Right now it is difficult for our proposed parameter sets. The method assumes that there are time frames when enough *insider* votes exist against *outsiders* votes. The assumption is not satisfied if an *insider* decides to go with the second policy. We leave this issue for future work.

7   Conclusions

In this paper, we have proposed *Knowledge Peeping* and investigated extraction of *insiders'* voting behaviors which are likely to prompt a change in the time series odds data drawn from horse racing events held by JRA. As a result, we have found the condition where the rate of return is more than 642 % for races whose winner's prize money is 20 million yens or more. That suggests the potential of *Knowledge Peeping*. Furthermore, we have evaluated the results by comparing them with one obtained from a classification model built by a decision tree learner to show the former is better. This suggests that *Knowledge Peeping* is effective at decision making at horse race investment.

Although we have tried to obtain an investment strategy to constantly get profit by optimizing parameters related to extraction of votes, it has been difficult to find the optimal solution in this paper because the target function is complex.

In future, we will use methods like simulated annealing to globally optimize parameters. Further, we will evaluate the method by comparing it with other baselines, such as improved classification models and applications of past racetrack forecasting studies to data drawn from JRA events.

Ultimately, we will apply the method to other domains where voting data are publicly available, such as investment decision on stock market only from turnover volumes, and demonstrate the general effectiveness of this proposed parameter sets.

Acknowledgements. This work was supported by JSPS KAKENHI Grant Numbers 24300005, 26330081, 26870201. The horse racing data was supplied by Japan Racing Association (JRA). We would like to thank everyone that has helped us.

References

1. Nikkansports.com: http://www.nikkansports.com/ (in Japanese)
2. Wikipedia: Hishimirakuru (Mister of miracle), https://ja.wikipedia.org/wiki/%E3%83%92% E3%82%B7%E3%83%9F%E3%83%A9%E3%82%AF%E3%83%AB (in Japanese)
3. NBC NEWS.com: Irish bookie say race over, pays off on Obama, http://www. nbcnews.com/id/27236818/ns/world_news-europe/t/irish-bookie-say-race-over-pays-obama/ #.VZzm5vntlBc
4. Williams, V., Siegel, S.: The Oxford Handbook of The Economics of Gambling, 752 p. Oxford University Press, Oxford (2014). ISBN: 978-0-19-979791-2
5. Ashiya, M.: The Survey on Favorite-Longshot Bias, Kokumin Keizai Zasshi, Kobe University, vol. 202(2), pp. 13–28 (2010)
6. Obata, S., Dazai, H.: Horse racing and prospect theory: an empirical analysis of overweighting of low probabilities. J. Behav. Econ. Finan. **7**, 1–18 (2014)
7. Mori, S., Hisakado, M.: Accuracy and fine structure of predictions in a racetrack betting. JWEIN: Special Interest Group on Emergent Intelligence on Network (2010)
8. Mori, S., Hisakado, M.: Component Ratios of Independent and Herding Betters in a Racetrack Betting Market, preprint arXiv:1006.4884 (2010)
9. Law, D., Peel, D.A.: Insider trading, herding behaviour and market plungers in the british horse-race betting market. Economica **69**, 327–338 (2002)
10. Shin, H.S.: Measuring the incidence of insider trading in a market for state-contingent claims. Econ. J. **103**(420), 1141–1153 (1993)
11. Hall, M., Frank, E., Holmes, G., Pfahringer, B., Reutemann, P., Witten, I.H.: The WEKA data mining software: an update. SIGKDD Explor. **11**(1), 10–18 (2009)
12. Japan Racing Association (JRA): Terminology on Horse Racing. http://jra.jp/kouza/yougo/

Finding Within-Organisation Spatial Information on the Web

Jun Hou[✉], Ruth Schulz, Gordon Wyeth, and Richi Nayak

Queensland University of Technology (QUT), Brisbane, Australia
{j1.hou,ruth.schulz,gordon.wyeth,r.nayak}@qut.edu.au

Abstract. Information available on company websites can help people navigate to the offices of groups and individuals within the company. Automatically retrieving this within-organisation spatial information is a challenging AI problem This paper introduces a novel unsupervised pattern-based method to extract within-organisation spatial information by taking advantage of HTML structure patterns, together with a novel Conditional Random Fields (CRF) based method to identify different categories of within-organisation spatial information. The results show that the proposed method can achieve a high performance in terms of F-Score, indicating that this purely syntactic method based on web search and an analysis of HTML structure is well-suited for retrieving within-organisation spatial information.

Keywords: HTML pattern · Search engine · Spatial information extraction

1 Introduction

When navigating to new locations, such as an office within a large organisation, people typically use many different cues to help them and often visit web search engines for extra information prior to performing navigation in the real world. Company websites are typically the first place people will look for information about where an organisation is physically located, as well as where groups of people and individuals within the organisation are located, and any extra information about how to get there. Automatically retrieving information from such an unstructured source is a challenging AI problem, which could be useful in autonomous navigation systems for robots or mobile apps that provide navigation directions for pedestrians.

Typically, address and location information is formatted or tagged in a particular way to make it easier for people looking at the website to find, but this is not always the case. Other sources of navigation information include GPS, which can benefit large-scale navigation through road networks, but is less useful at a building scale, and spatial databases like GeoName (www.geonames.org), which include high level spatial information, including country, city, and suburb, but not within-organisation location information such as campus, building, floor, and room information.

While traditional robot navigation involves constructing detailed maps that can be used for navigation to known locations, robotic systems have recently been developed

© Springer International Publishing Switzerland 2015
B. Pfahringer and J. Renz (Eds.): AI 2015, LNAI 9457, pp. 242–248, 2015.
DOI: 10.1007/978-3-319-26350-2_21

that use symbolic spatial information such as gestures [1], route directions [2, 3], and floor plans [4] to aid navigation to new locations. As yet, automated navigation systems do not take advantage of the spatial information available on the web, including within-organisation location information, maps such as floor plans and campus maps, and natural language directions, partially due to a lack of automated methods to extract this information.

In this paper we introduce a novel unsupervised pattern-based method to extract within-organisation spatial information by taking advantage of HTML structure patterns, together with a novel Conditional Random Fields (CRF) based method to identify different categories of within-organisation spatial information. The methodology (see Fig. 1) includes HTML structure pattern learning (described in Sect. 2), spatial information extraction (described in Sect. 3), and spatial information tagging (described in Sect. 4). We present the results of a study extracting and tagging office locations of academics from two different universities in Sect. 5, and finish the paper with our conclusions and plans for future work in Sect. 6.

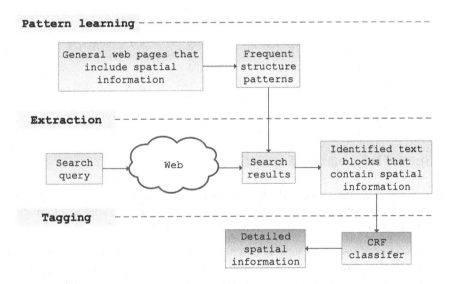

Fig. 1. Overview of the locating within-organisation spatial information methodology.

2   HTML Structure Pattern Learning

Previous location information extraction methods either use text-features including Name Entity Recognition (NER) and address keywords [5], or the visual or layout difference between two text blocks [6–9] to find relevant text blocks. However, the address keywords are domain-dependent and NER can produce incorrect results, while visual or layout features fail when there is not much visual or layout difference between two text blocks, for example, similar background colour and font size, or the web page mainly contains plain text. The proposed pattern-based extraction method described in this section is unsupervised by using frequent HTML structures rather than the content of a web page or any pre-specified user pattern.

The common HTML structure patterns in which spatial information is organised on general web pages were found using the statistics of the HTML patterns in the training data set (the top-100 search results by issuing a query "allinurl:/com.au/contact" to Google, see Table 1). Most spatial information appears in the form of multiple lines. Among multiple lines, the most frequent pattern is the "<p>-
" HTML structure. The second most frequent pattern is the table structure, that is, a HTML tag <table> and sub-tags <tr> (table rows) and <td> (table data). Lastly, "Others" involves the situations where the spatial information is displayed with visual difference, for example, different font sizes or background colours from context, rather than frequent structure patterns. Similar patterns are observed in the single line scenario.

Consequently, the two frequent HTML patterns (<p>-
 and table) are used as frequent structure patterns for the methods described in the following sections.

Table 1. Summary of the HTML patterns in the training data set.

Multiple lines	<p>- 	58.5 %	73.9 %	90.8 %
	Table	15.4 %		
	Others	16.9 %		
Single line	9.2 %			

3 Spatial Information Extraction

There are two steps involved in the within-organisation spatial information extraction from the web: (1) retrieving relevant web pages and (2) extracting spatial information. Two set of queries are constructed to retrieve relevant web pages for a person:

1. General query – a person's name or a person's name + university name. This type of query is to retrieve general web pages related to a person's name.
2. Spatial query – a person's name + spatial cue. The spatial cue is selected from the set "address, location, contact, office, about". A spatial query aims to find web pages containing the within-organisation spatial information.

For each name, the two general queries and five spatial queries are issued to the Google Search API. Top-50 results are collected for each query. As there can be overlap between search results of different queries on the same name, it is necessary to remove duplicates and re-rank the search results. Two factors are considered: (1) relevance in individual search results and (2) popularity among different search results. For calculating relevance, we assign a score for each search result based on its rank position with the web page at the rank position 1 assigned a score of 5.0. We reduce the score by an interval 0.1 for each ranked search result, from the second ranked result, which is assigned a score of 4.9, to the 50th rank search result, which is assigned a score of 0.1.

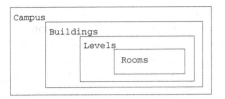

Fig. 2. Label scheme hierarchy.

Popularity among different search results in related queries is measured by merging individual search results and adding together the scores.

The two frequent HTML structure patterns identified in the previous section are used to extract spatial information from the collected web pages. For the "<p>-
" pattern, there are four variations based on Jsoup selector: (1) div[class~=.?content.?] p:has(br); (2) div[class~=.?contact.?] p:has(br); (3) p:has(br); and (4) address:has(br).

Patterns 1 and 2 are designed to find "<p>-
" structures nested inside the HTML tag <div> that have an attribute "class" with its value containing word "content" or "contact". The attribute "class" with value "content" or "contact" often indicates that the <div> tag defines the main content or the contact section of a web page. Thus, patterns 1 and 2 ensure only selecting "<p>-
" structures located in the main content or the contact section of a web page. Pattern 3 can select "<p>-
" structures that are not nested inside another HTML tag. It can capture "<p>-
" structures without embedded design. Pattern 4 can extract "<p>-
" structures inside the special tag <address>, which is used to store location or address information.

For the "table" pattern, two instances are designed and used: (5) dl; and (6) table:has(tr). Pattern 5 is designed to extract any description list (DL) structure inside a web page. Pattern 6 can select any table structure within a web page. Patterns 5 and 6 will cover both flat and embedded DL and table structure.

4   Spatial Information Tagging

Classification using Conditional Random Fields (CRF) [10] can use a domain-dependent ontology [11, 12] or use many NER and address keywords for training a classifier [5]. The proposed CRF-based method presented in this section for assigning categories to the extracted spatial information uses the CRF model [10, 13] but utilises only a few low-level lexical and syntactical features.

The extracted spatial information is presented as input to the CRF in the form of a set of sequential words $X = \{X_i\}$ ($i = 1,2,...,|X|$). The CRF is trained to find the correct category Y_i for each X_i. In order to train CRF, a label scheme was designed to identify categories in extracted spatial information with four types of category label (see Fig. 2): campus (CA), building (BU), level (LE), and room (RO). For example, for the extracted spatial information "Gardens Point, O Block Podium Level 4 453" the following fine-grained categories should be identified: Gardens Point – > *Campus*; O Block Podium – > *Building*; Level 4 – > *Level*; 453 – > *Room*.

As labels may contain multiple words, the IOB2 notation with two position labels is used, that is, begin label (B) and inside label (I). For example, a phrase of two words of category campus (CA) is represented by the beginning of the chunk (B-CA) and the inside of the chunk (I-CA) to indicate two positions. Words that do not belong to any of the four category labels are tagged by O labels. With the designed labels, extracted spatial information is manually tagged for training and testing CRF, in order to generate the datasets.

To identify different categories of spatial information, rich features, including transition features and syntactic features, are utilised in the CRF model. A *transition* feature indicates label transition between adjacent states in the CRF. We only use first-order transition features in this work. A syntactic feature can be either lexical or Part-of-Speech (PoS). A *lexical* feature is a binary feature that indicates if a specific word co-occurs with a label. We generalise this feature to n-grams by applying a sliding window. Each word of the input sequence $w_{1:n}$ is sequentially viewed as the centre of a window with size n (we have used $n = 2$ in the models used in this paper). The *PoS* feature indicates whether a label occurs depending on the part-of-speech of the current word. The PoS feature is extended from the current word to its neighbourhood with a size of n (we use in this paper, n = 2).

5   Experiments

In the study presented in this paper, academic's names from two universities (QUT and UQ) were used to retrieve relevant webpages, extract the related within-organisation spatial information using the proposed spatial information extraction method, and the outputs were tagged using the proposed CRF-based tagging method.

5.1   Data Set

The names of 80 academics were used to build a data set, 60 from QUT Gardens Point Campus and 20 from UQ St Lucia Campus. The academics were located on various levels of four buildings on QUT Gardens Point Campus (O, P, S, and Y) and two buildings on UQ St Lucia Campus (47 and 78). To evaluate the pattern-based extraction method, true positive search results were identified. A search result was judged as true positive only if it contained the correct organisation-level room information with clear context information. As only a few websites provide the required spatial information (typically personal homepages or official university staff pages), there was little within-organisation spatial information available on the web: 0.621 % of 9667 pages retrieved for the 60 QUT academics and 0.526 % of the 3419 pages retrieved for the 20 UQ academics were true positives. For two of the UQ academics none of the webpages retrieved contained any relevant spatial information. We use the true positives as our ground truth to evaluate the proposed method in the following section.

5.2 Results

For the pattern-based extraction method, very high precision was achieved, with all extracted spatial information containing relevant within-organisation spatial information (see Table 2). The high precision demonstrates that the HTML structure patterns are very effective for extracting spatial information. High recall was also obtained for names from QUT. The reason for lower recall for names from UQ is that within-organisation spatial information for some UQ academics appears as unstructured sentences of paragraphs and does not follow one of the common patterns.

Table 2. Precision, recall and F-Score of the proposed pattern-based extraction method and CRF-based tagging method.

	Extraction			Tagging		
	Precision	Recall	F-Score	Precision	Recall	F-Score
QUT	1	0.984	0.992	0.889	0.889	0.889
UQ	1	0.833	0.909	0.454	0.519	0.462
All	1	0.956	0.978	0.951	0.914	0.926

For the CRF-based tagging method, higher precision and recall is achieved for names from QUT compared to names from UQ (see Table 2). This occurs due to the availability of more training data for QUT entities than the UQ data set. Furthermore, combining QUT and UQ data (All) achieves the best performance with very high precision, recall and F-Score. These results show the effectiveness of the proposed CRF-based tagging method.

6 Conclusions and Future Work

In this paper, we proposed a pattern-based extraction method to identify within-organisation information from the search results returned by a search engine in response to user queries. In order to identify different types of spatial information, a CRF-based tagging method is proposed by assigning different labels to the extracted spatial information. Experiments using data from two different universities have shown that high accuracy performance (i.e. F-Score) can be achieved for the proposed pattern-based extraction method and CRF-based tagging method. The results have shown that a purely syntactic method based on web search and an analysis of HTML structure is well-suited for retrieving within-organisation spatial information.

Our previous robot work used floor plans and door label information to guide a robot to a goal location specified by room number [4]. In future work, the goal could be a person's name, with the system determining the campus, building, level, and room using the methods described here. Future work also includes investigating different ways to extract within-organisation information on the web that will be applicable to a large variety of organisations.

Acknowledgements. This research was supported under Australian Research Council's Discovery Projects funding scheme (project number DP140103216).

References

1. Bauer, A., et al.: The autonomous city explorer: towards natural human-robot interaction in urban environments. Int. J. Soc. Robot. **1**(2), 127–140 (2009)
2. Hemachandra, S., et al.: Learning models for following natural language directions in unknown environments. In: 2015 IEEE International Conference on Robotics and Automation. IEEE. Seattle, WA (2015)
3. Kollar, T., et al.: Toward understanding natural language directions. In: 5th ACM/IEEE International Conference on Human-Robot Interaction (HRI), pp. 259–266. IEEE (2010)
4. Schulz, R., et al.: Robot navigation using human cues: a robot navigation system for symbolic goal-directed exploration. In: 2015 IEEE International Conference on Robotics and Automation. IEEE. Seattle, WA (2015)
5. Chia Hui, C., Shu-Ying, L.: MapMarker: extraction of postal addresses and associated information for general web pages. In: 2010 IEEE/WIC/ACM International Conference on Web Intelligence and Intelligent Agent Technology (WI-IAT), pp. 105–111 (2010)
6. Lin, C., et al.: Postal address detection from web documents. In: 2005 Proceedings of the International Workshop on Challenges in Web Information Retrieval and Integration, (WIRI 2005) pp. 40–45 (2005)
7. Cai, D., Yu, S., Wen, J., Ma, W.-Y.: Extracting content structure for web pages based on visual representation. In: Zhou, X., Zhang, Y., Orlowska, M.E. (eds.) APWeb 2003. LNCS, vol. 2642, pp. 406–417. Springer, Heidelberg (2003)
8. Asadi, S., Yang, G.-W., Zhou, X., Shi, Y., Zhai, B., Jiang, W.W.-R.: Pattern-based extraction of addresses from web page content. In: Zhang, Y., Yu, G., Bertino, E., Xu, G. (eds.) APWeb 2008. LNCS, vol. 4976, pp. 407–418. Springer, Heidelberg (2008)
9. Liu, Y., Liu, W., Jiang, C.: User interest detection on web pages for building personalized information agent. In: Li, Q., Wang, G., Feng, L. (eds.) WAIM 2004. LNCS, vol. 3129, pp. 280–290. Springer, Heidelberg (2004)
10. Lafferty, J., McCallum, A., Pereira, F.C.: Conditional random fields: probabilistic models for segmenting and labeling sequence data. In: Proceedings of the 18th International Conference on Machine Learning 2001 (ICML 2001), pp. 282–289 (2001)
11. Cai, W., Wang, S., Jiang, Q.: Address extraction: extraction of location-based information from the web. In: Zhang, Y., Tanaka, K., Yu, J.X., Wang, S., Li, M. (eds.) APWeb 2005. LNCS, vol. 3399, pp. 925–937. Springer, Heidelberg (2005)
12. Borges, K.A.V., et al.: Discovering geographic locations in web pages using urban addresses. In: Proceedings of the 4th ACM Workshop on Geographical Information Retrieval, pp. 31–36. ACM, Lisbon, Portugal (2007)
13. Okazaki, N., CRFsuite: a fast implementation of Conditional Random Fields (CRFs). http://www.chokkan.org/software/crfsuite/

Knowledge Sharing in Coalitions

Guifei Jiang[1,2(✉)], Dongmo Zhang[1], and Laurent Perrussel[2]

[1] AIRG, Western Sydney University, Penrith, Australia
jgfei1987@gmail.com
[2] IRIT, University of Toulouse 1, Toulouse, France

Abstract. The aim of this paper is to investigate the interplay between knowledge shared by a group of agents and its coalition ability. We characterize this relation in the standard context of imperfect information concurrent game. We assume that whenever a set of agents form a coalition to achieve a goal, they share their knowledge before acting. Based on this assumption, we propose a new semantics for alternating-time temporal logic with imperfect information and perfect recall. It turns out this semantics is sufficient to preserve all the desirable properties of coalition ability in traditional coalitional logics. Meanwhile, we also show that the fixed-point characterisations of coalition operators which normally fail in the context of imperfect information can be recovered through the interplay of epistemic and coalition modalities. This work provides a partial answer to the question: which kind of group knowledge is required for a group to achieve their goals in the context of imperfect information.

1 Introduction

Reasoning about coalitional abilities and strategic interactions is fundamental in analysis of multiagent systems (MAS). Among many others [8,14,15,25,27], Coalition Logic (CL) [22] and Alternating-time Temporal Logic (ATL) [2] are typical logical frameworks that allow to specify and reason about effects of coalitions [16]. In a nutshell, these logics express coalition ability using a modality in the form, say $\langle\langle G \rangle\rangle\varphi$, to mean coalition G (a set of agents) can achieve a property φ, regardless what the other agents do. ATL/CL assume that each agent in a multi-agent system has complete information about the system at all states (perfect information). Obviously this is not always true in the real world. Different agents might own different knowledge about their system. To model the systems in which agents have imperfect information, a few attempts have been made in the last few years by extending ATL with epistemic operators [11,17,19,26]. With the extensions, agents' abilities are associated with their knowledge. For instance, assuming a few agents are trying to open a safe, only the ones who know the code have the ability to open the safe.

One difficulty of ATL with imperfect information is how to model knowledge sharing among a coalition. In other words, if a group of agents form a coalition, whether their knowledge will be shared and be contributed to the group

ⓒ Springer International Publishing Switzerland 2015

B. Pfahringer and J. Renz (Eds.): AI 2015, LNAI 9457, pp. 249–262, 2015.
DOI: 10.1007/978-3-319-26350-2_22

abilities [13]? For simplicity, all the existing epistemic ATL-style logics do not assume that members of a coalition share knowledge unless the information is general knowledge [7,24] or common knowledge [11] to a group or a system. However, most of the time when a set of agents form a coalition, their cooperation is not merely limited to acting together, but, more importantly, sharing their knowledge when acting. Safe opening is an example.

This paper aims to take the challenge of dealing with knowledge sharing among coalitions. By a coalition we mean a set of agents that can not only act together to achieve a goal, but also share their knowledge when acting. We say that a coalition can ensure φ if the agents in the coalition distributedly know that they can enforce φ. Based on this idea, we provide a new semantics for the coalition operator in ATL with imperfect information. It turns out that this semantics is sufficient to preserve desirable properties of coalition ability [12,22]. More importantly, we provide characterization results of coalition operators through the interplay of distributed knowledge and coalition ability. Our contribution is twofold: firstly, this work can be seen as an attempt towards the difficulty: which kind of group knowledge is required for a group to achieve some goal in the context of imperfect information; secondly, these results show that the fixed-point characterizations of coalition operators which normally fail in the context of imperfect information [3,4] can be recovered by the interplay of epistemic and coalitional operators.

The rest of this paper is structured as follows. Section 2 introduces a motivating example for our new semantics. Section 3 provides the new semantics and investigates its properties. Section 4 explores the characterizations for the interplay of epistemic and coalitional operators. Section 5 discusses related work. Finally we conclude the paper with future work.

2   A Motivating Example

Let's consider the following example which highlights our motivation to study coalition abilities under the assumption of knowledge sharing within coalitions.

Example 1. Figure 1 depicts a variant of the shell game [7] with three players: the shuffler s, the guessers g_1 and g_2. Initially the shuffler places a ball in one of the two shells (the left (L) or the right (R)). The guesser g_1 can observe which action the shuffler does, while the other guesser g_2 can't. A guesser or a coalition of two wins if he picks up the shell containing the ball. We assume that the guesser g_1 takes no action (n) and the guesser g_2 chooses the shell (the left (l) or the right (r)).

Clearly, g_1 knows the location of the ball but cannot choose. Instead g_2 does not know where the ball is, though he has right to choose the shell. It's easy to see that neither g_1 nor g_2 can win this game individually. But if g_1 and g_2 form a coalition, it should follow that by sharing their knowledge they can cooperate to win. However, according to the existing semantics for ATL with imperfect information including the latest one, called truly perfect recall (also referred as no-forgetting semantics) [7], the coalition of g_1 and g_2 does not have such ability to win since they claim that coalition abilities require general knowledge or even common knowledge.

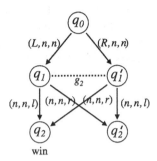

Fig. 1. The model M_1

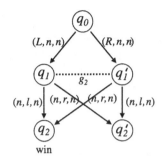

Fig. 2. The model M_2

(The tuple $(\alpha_1, \alpha_2, \alpha_3)$ represents an action profile, i.e., action α_1 of player s, action α_2 of player g_1, and action α_3 of player g_2. The dotted line represents g_2's indistinguishability relation: reflexive loops are omitted. State q_2 is labelled with the proposition win.)

Moreover, these semantic variants fail to preserve the coalition monotonicity which is a desirable property for coalition ability in coalitional logics [12,22], that is, if a coalition can achieve some goal, then its superset can achieve this goal as well. For instance, the model M_2 in Fig. 2 depicts a variant game of Example 1 by just switching the available actions of two guessers. The guesser g_1 chooses the shell and the guesser g_2 takes no action. Then it's clear that the guesser g_1 can win no matter what the others do, As he sees the location of the ball and can pick up the right shell. It should follow that as a group, the guessers g_1 and g_2 can win this game. However, according to most existing semantics, though the guesser g_1 has the ability to win, this ability no longer holds once he forms a coalition with guesser g_2. These counterintuitive phenomena motivate our new semantics for ATL with imperfect information and perfect recall.

3 The Framework

In this section, we provide a new semantics for ATL with imperfect information and perfect recall based on the assumption of knowledge sharing in coalitions, and then investigate logical properties of ATL under this semantics.

3.1 Syntax of ATL

Let Φ be a countable set of atomic propositions and N be a finite nonempty set of agents. The language of ATL, denoted by $\mathcal{L}$, is defined by the following grammar:

$$\varphi := p \mid \neg\varphi \mid \varphi \wedge \varphi \mid \langle\langle G \rangle\rangle \bigcirc \varphi \mid \langle\langle G \rangle\rangle \square \varphi \mid \langle\langle G \rangle\rangle \varphi \mathcal{U} \varphi$$

where $p \in \Phi$ and $\emptyset \neq G \subseteq N$.

A coalition operator $\langle\langle G \rangle\rangle \varphi$ intuitively expresses that the group G can cooperate to ensure that φ. The temporal operator $\square$ means "from now on (always)" and other temporal connectives in ATL are $\mathcal{U}$ ("until") and $\bigcirc$ ("in the next state"). The dual operator $\Diamond$ of $\square$ ("either now or at some point in the future") is defined as $\Diamond \varphi =_{def} \top \mathcal{U} \varphi$. Moreover, the standard epistemic operators can be defined as follows: $K_i \varphi =_{def} \langle\langle i \rangle\rangle \varphi \mathcal{U} \varphi$ and $D_G \varphi =_{def} \langle\langle G \rangle\rangle \varphi \mathcal{U} \varphi$. As we will show in the semantics, these abbreviations capture their standard intuitions, i.e., "$K_i \varphi$" says the agent i knows φ, and $D_G \varphi$ means it is distributed knowledge among the group G that φ.

3.2 Semantics of ATL

The semantics is built upon the *imperfect information concurrent game structure (iCGS)* [17,24].

Definition 1. *An iCGS is a tuple* $M = (N, \Phi, W, \mathcal{A}, \pi, d, \delta, \{R_i\}_{i \in N})$ *where*

- $N = \{1, 2, \cdots, k\}$ *is a nonempty finite set of* players;
- Φ *is a set of* atomic propositions;
- W *is a nonempty finite set of* states;
- $\pi : \Phi \mapsto \wp(W)$ *is a* valuation function;
- $\mathcal{A}$ *is a nonempty finite set of* actions;
- $d : N \times W \mapsto \wp(\mathcal{A})$ *is a mapping specifying nonempty sets of actions available to agents at each state. We will write* $d_i(w)$ *rather* $d(i, w)$. *The set of joint actions at* w *for* N *is denoted as* $D(w) = d_1(w) \times \cdots \times d_k(w)$;
- $\delta : W \times D(W) \mapsto W$ *is the* transition function *from every pair* $(w \in W, \alpha \in D(w))$ *to an outcome state* $\delta(w, \alpha) \in W$.
- $R_i \subseteq W \times W$ *is an equivalence relation for agent* i *indicating the states that are indistinguishable from her viewpoint. For consistency, we assume that each agent knows which actions are available for her, i.e.,* $d_i(w) = d_i(w')$ *whenever* $w R_i w'$.

A *path* λ is an infinite sequence of states and actions $w_0 \overset{\alpha_1}{\to} w_1 \overset{\alpha_2}{\to} w_2 \cdots$, where for each $j \geq 1$, $\alpha_j \in D(w_{j-1})$ and $\delta(w_{j-1}, \alpha_j) = w_j$. Any finite segment $w_k \overset{\alpha_{k+1}}{\to} w_{k+1} \overset{\alpha_{k+2}}{\to} \cdots \overset{\alpha_l}{\to} w_l$ of a path is called a *history*. The set of all histories for M is denoted by H. We use $\lambda[j]$ to denote the j-th state on path λ, $\lambda[j, k]$ $(0 \leq j \leq k)$ to denote the segment of λ from the j-th state to the k-th state, and $\lambda[j, \infty]$ to denote the subpath of λ starting from j. The length of history h, denoted by $|h|$, is defined as the number of actions.

The following definition specifies what a player with perfect reasoning capabilities can in principle know at a special stage of an imperfect information game.

Definition 2. *Two histories* $h = w_0 \overset{\alpha_1}{\to} w_1 \overset{\alpha_2}{\to} \cdots \overset{\alpha_m}{\to} w_m$ *and* $h' = w'_0 \overset{\alpha'_1}{\to} w'_1 \overset{\alpha'_2}{\to} \cdots \overset{\alpha'_n}{\to} w'_n$ *are equivalent for agent* $i \in N$, *denoted by* $h \approx_i h'$, *iff*

1. $m = n$,

2. $w_j R_i w'_j$ for any $0 \leq j \leq m$, and
3. $\alpha_k(i) = \alpha'_k(i)$ for any $1 \leq k \leq m$.

where $\alpha_k(i)$ is the i-th component of α_k.

Intuitively, two histories are indistinguishable for an agent if (1) they have the same length, (2) their corresponding states are indistinguishable for this agent, and (3) the agent takes the same action at the each corresponding stage. Our notion of perfect recall is more like GDL perfect recall [11,25] as well as perfect recall in extensive games [21] by requiring that an agent remember the past states as well as its own actions. This is stronger than the one in most epistemic ATL-style logics which often use the state-based equivalence without taking the actions into consideration, that is, a (truly) perfect recall agent just remembers the past states. Our version has the advantage to deal with situations where different actions may have the same effects. For instance, consider two histories $q_0 \xrightarrow{a} q_1$ and $q_0 \xrightarrow{b} q_1$ with a single agent. According to the state-based equivalence, the agent cannot distinguish the two histories, but actually they are different from his view since he takes different actions at state q_0[1]. Note that the perfect recall agent does not observe or remember other agents' actions.

In particular, we say two paths λ and λ' are equivalent up to stage $j \geq 0$ for agent $i \in N$, denoted by $\lambda \approx_i^j \lambda'$, iff $\lambda[0,j] \approx_i \lambda'[0,j]$. As mentioned before, we assume that whenever a set of agents form a coalition to achieve their goals, the agents share their own knowledge before acting. To make this idea precise, we extend the indistinguishability relation $\approx_i^j$ to groups as the intersection of all its members' individual equivalence relation, i.e., $\approx_G^j = \bigcap_{i \in G} \approx_i^j$. Let $\approx_G^j (\lambda)$ denote the set of all paths that are indistinguishable from λ up to stage j for coalition G, i.e., $\approx_G^j (\lambda) = \{\lambda' \mid \lambda \approx_G^j \lambda'\}$.

A *strategy* is a plan telling one agent what to do at each stage of a given game. We say a strategy of agent $i \in N$ is *uniform* if the strategy specifies the same action for agent i at all indistinguishable histories.

Definition 3. *A uniform perfect recall strategy for agent i is a function $f_i :$ $H \rightarrow A$ such that for any history $h, h' \in H$,*

1. $f_i(h) \in d_i(last(h))$, and
2. if $h \approx_i h'$, then $f_i(h) = f_i(h')$,

where $last(h)$ denotes the last state of h.

Intuitively, a uniform perfect recall strategy for an agent tells one of her legal actions to take at each history and specifies the same action for her indistinguishable histories. In the rest of paper, we simply call a uniform perfect recall strategy a strategy. A *joint strategy* for group $\emptyset \neq G \subseteq N$, denoted by F_G, is a vector of its members' individual strategies, i.e., $\langle f_i \rangle_{i \in G}$. Function $\mathcal{P}(h, f_i)$ returns the set of all paths that can occur when agent i's strategy f_i executes

---

[1] It is worth to mention that [23] proposed a way to embed actions to a state so that the state-based equivalence can achieve the same meaning.

after an initial history h. Formally, $\lambda \in \mathcal{P}(h, f_i)$ iff $\lambda[0, |h|] = h$ and for any $j \geq |h|$, $f_i(\lambda[0, j]) = \theta_i(\lambda, j)$ where $\theta_i(\lambda, j)$ is the action of agent i taken at stage j on path λ. Obviously, the set of all paths complying with joint strategy F_G after h is defined as $\mathcal{P}(h, F_G) = \bigcap_{i \in G} \mathcal{P}(h, f_i)$.

We are now in the position to introduce the new semantics for ATL. Formulae are interpreted over triples consisting of a model, a path and an index which indicates the current stage on the path.

Definition 4. *Let M be an iCGS. Given a path λ of M and a stage $j \in \mathbb{N}$ on λ, the satisfiability of a formula φ wrt. M, λ and j, denoted by $M, \lambda, j \models \varphi$, is defined as follows:*

$$
\begin{array}{lll}
M, \lambda, j \models p & \textit{iff} & p \in \pi(\lambda[j]) \\
M, \lambda, j \models \neg\varphi & \textit{iff} & M, \lambda, j \not\models \varphi \\
M, \lambda, j \models \varphi_1 \wedge \varphi_2 & \textit{iff} & M, \lambda, j \models \varphi_1 \text{ and } M, \lambda, j \models \varphi_2 \\
M, \lambda, j \models \langle\langle G \rangle\rangle \bigcirc \varphi & \textit{iff} & \exists F_G \; \forall \lambda' \in \approx_G^j (\lambda) \; \forall \lambda'' \in \mathcal{P}(\lambda'[0, j], F_G) \\
& & M, \lambda'', j+1 \models \varphi \\
M, \lambda, j \models \langle\langle G \rangle\rangle \Box \varphi & \textit{iff} & \exists F_G \; \forall \lambda' \in \approx_G^j (\lambda) \; \forall \lambda'' \in \mathcal{P}(\lambda'[0, j], F_G) \\
& & \forall k \geq j \; M, \lambda'', k \models \varphi \\
M, \lambda, j \models \langle\langle G \rangle\rangle \varphi_1 \mathcal{U} \varphi_2 & \textit{iff} & \exists F_G \; \forall \lambda' \in \approx_G^j (\lambda) \forall \lambda'' \in \mathcal{P}(\lambda'[0, j], F_G) \\
& & \exists k \geq j, \; M, \lambda'', k \models \varphi_2, \text{ and} \\
& & \forall j \leq t < k, \; M, \lambda'', t \models \varphi_1
\end{array}
$$

The interpretation for the coalition operator $\langle\langle G \rangle\rangle \varphi$ captures its precise meaning that the coalition G by sharing knowledge can cooperate to enforce that φ. Alternatively, the agents in G distributedly know that they can enforce that φ. A formula φ is *valid* in an iCGS M, written as $M \models \varphi$, if $M, \lambda, j \models \varphi$ for all paths $\lambda \in M$ and every stage j on λ. A formula φ is *valid*, denoted by $\models \varphi$, if it is valid in every iCGS M.

We first show that, as we claimed before, the abbreviations capture the intended meanings of the epistemic operators.

Proposition 1. *Given an iCGS M, a path λ of M and a stage $j \in \mathbb{N}$ on λ,*

- $M, \lambda, j \models K_i \varphi$ *iff* *for all $\lambda' \approx_i^j \lambda$, $M, \lambda', j \models \varphi$.*
- $M, \lambda, j \models D_G \varphi$ *iff* *for all $\lambda' \in \approx_G^j (\lambda)$, $M, \lambda', j \models \varphi$.*

Proof. We just prove the first clause, and the second one is proved in a similar way. It suffices to show that $M, \lambda, j \models \langle\langle i \rangle\rangle \varphi \mathcal{U} \varphi$ iff for all $\lambda' \approx_i^j \lambda$, $M, \lambda', j \models \varphi$. The direction from the right to the left is straightforward according to the truth condition for $\mathcal{U}$. We next show the other direction. Suppose $M, \lambda, j \models \langle\langle i \rangle\rangle \varphi \mathcal{U} \varphi$ and for all $\lambda' \approx_i^j \lambda$, then there is f_i such that for any $\lambda'' \in \mathcal{P}(\lambda'[0, j], f_i)$, $M, \lambda'', j \models \varphi$. Thus, $M, \lambda', j \models \varphi$ by $\lambda'[0, j] = \lambda''[0, j]$. ($\Box$)

We demonstrate with the variant shell game that the new semantics justifies our intuitions that the coalition of two guessers by sharing their knowledge can win the game.

Example 1 (*continued*). *Consider the model* M_1 *in Fig. 1. It's easy to check that at the stage 1 on the left path* $\lambda_1 := q_0 q_1 q_2 \cdots$, *neither guesser* g_1 *nor guesser* g_2 *has the ability to win at the next stage, i.e.,* $M_1, \lambda_1, 1 \not\models \langle\langle g_1 \rangle\rangle \bigcirc win$ *and* $M_1, \lambda_1, 1 \not\models \langle\langle g_2 \rangle\rangle \bigcirc win$. *Instead when* g_1 *and* g_2 *form a coalition, after sharing knowledge, the guesser* g_2 *is able to distinguish the history* $q_0 q_1$ *from the history* $q_0 q_1'$, *then they can cooperate to win, i.e.,* $M_1, \lambda_1, 1 \models \langle\langle \{g_1, g_2\} \rangle\rangle \bigcirc win$.

For the coalition monotonicity property, consider the model M_2 *in Fig. 2. It's easy to check that at the stage 1 on the left path* $\lambda_1 := q_0 q_1 q_2 \cdots$, *guesser* g_1 *has the ability to win at the next stage by choosing the left shell, i.e.,* $M_2, \lambda_1, 1 \models \langle\langle g_1 \rangle\rangle \bigcirc win$. *Moreover, when* g_1 *and* g_2 *form a coalition, then they can cooperate to win, i.e.,* $M_2, \lambda_1, 1 \models \langle\langle \{g_1, g_2\} \rangle\rangle \bigcirc win$.

At last, we would like to say that the reason why alternative semantics [7, 11, 24] fail to keep the coalition monotonicity property is that their interpretations of coalition operators $\langle\langle G \rangle\rangle \varphi$ use either the union of each member's equivalence relation or its transitive reflexive closure. This means that the coalition ability implicitly requires general knowledge or common knowledge of the group, while neither of them is coalitionally monotonic. Instead distributed knowledge is sufficient for coalition ability under our assumption of knowledge sharing in coalitions. Clearly, distributed knowledge preserves the coalition monotonicity property.

3.3 Properties of the New Semantics

We first show that the new semantics satisfied the desirable properties of coalition ability in traditional coalitional logics [12, 22].

Proposition 2. *For any* G, G_1, $G_2 \subseteq N$ *and any* $\varphi, \psi \in \mathcal{L}$,

1. $\models \neg \langle\langle G \rangle\rangle \bigcirc \bot$
2. $\models \langle\langle G \rangle\rangle \bigcirc \top$
3. $\models \langle\langle G \rangle\rangle \bigcirc (\varphi \wedge \psi) \rightarrow \langle\langle G \rangle\rangle \bigcirc \varphi$
4. $\models \langle\langle G_1 \rangle\rangle \bigcirc \varphi \rightarrow \langle\langle G_2 \rangle\rangle \bigcirc \varphi$ *where* $G_1 \subseteq G_2$
5. $\models \langle\langle G_1 \rangle\rangle \bigcirc \varphi \wedge \langle\langle G_2 \rangle\rangle \bigcirc \psi \rightarrow \langle\langle G_1 \cup G_2 \rangle\rangle \bigcirc (\varphi \wedge \psi)$ *where* $G_1 \cap G_2 = \emptyset$
6. $\models \langle\langle G \rangle\rangle \bigcirc \varphi \rightarrow \neg \langle\langle N \backslash G \rangle\rangle \bigcirc \neg \varphi$

Similarly for the $\square$ *and* $\mathcal{U}$ *operators.*

Clause 1 says that no coalition G can enforce the falsity while 2 states every coalition G can enforce the truth. 3 and 4 specify the outcome-monotonicity and the coalition-monotonicity, respectively. 5 is the superadditivity property specifying disjoint coalitions can combine their strategies to achieve more. 6 is called G-regularity specifying that it is impossible for a coalition and its complementary set to enforce inconsistency.

The next proposition provides interesting validities about epistemic and coalitional operators.

Proposition 3. *For any* $G \subseteq N$ *and any* $\varphi, \psi \in \mathcal{L}$,

1. $\models \langle\langle G\rangle\rangle \bigcirc \varphi \leftrightarrow \langle\langle G\rangle\rangle \bigcirc D_G\varphi$
2. $\models \langle\langle G\rangle\rangle \bigcirc \varphi \leftrightarrow D_G \langle\langle G\rangle\rangle \bigcirc \varphi$
3. $\models \langle\langle G\rangle\rangle\Box\varphi \leftrightarrow \langle\langle G\rangle\rangle\Box D_G\varphi$
4. $\models \langle\langle G\rangle\rangle\Box\varphi \leftrightarrow D_G \langle\langle G\rangle\rangle\Box\varphi$
5. $\models \langle\langle G\rangle\rangle D_G\varphi \mathcal{U} D_G\psi \rightarrow \langle\langle G\rangle\rangle\varphi\mathcal{U}\psi$
6. $\models \langle\langle G\rangle\rangle\varphi\mathcal{U}\psi \leftrightarrow D_G \langle\langle G\rangle\rangle\varphi\mathcal{U}\psi$

Proof. We only give proof for the first two clauses and the proof for $\Box$, $\mathcal{U}$ is similar.

1. For every iCGS M, every path λ of M and every stage $j \in \mathbb{N}$ on λ, assume $M, \lambda, j \models \langle\langle G\rangle\rangle \bigcirc \varphi$, then there is $F_G = \langle f_i\rangle_{i\in G}$ such that for all $\lambda' \in\approx_G^j (\lambda)$, for all $\lambda'' \in \mathcal{P}(F_G, \lambda'[0,j])$, $M, \lambda'', j+1 \models \varphi$. We next show that F_G is the joint strategy to verify $\langle\langle G\rangle\rangle \bigcirc D_G\varphi$. Suppose for a contradiction that there is $\lambda_1 \in\approx_G^j (\lambda)$, there is $\lambda_2 \in \mathcal{P}(F_G, \lambda_1[0,j])$, there is $\lambda_3 \in\approx_G^{j+1} (\lambda_2)$ such that $M, \lambda_3, j+1 \not\models \varphi$. Then $\lambda_3 \in\approx_G^j (\lambda)$ and $\theta_i(\lambda_3, j) = \theta_i(\lambda_2, j) = f_i(\lambda_2[0,j])$ for every $i \in G$, so there is some $\lambda^* \in \bigcup_{\lambda'\in\approx_G^j(\lambda)} \mathcal{P}(F_G, \lambda'[0,j])$ such that $\lambda^*[0,j+1] = \lambda_3[0,j+1]$. And by assumption we have $M, \lambda^*, j+1 \models \varphi$. It follows that $M, \lambda_3, j+1 \models \varphi$: contradiction. Thus, $M, \lambda, j \models \langle\langle G\rangle\rangle \bigcirc D_G\varphi$. The other direction is straightforward.

2. For every iCGS M, every path λ of M and every stage $j \in \mathbb{N}$ on λ, assume $M, \lambda, j \models \langle\langle G\rangle\rangle \bigcirc \varphi$, then there is $F_G = \langle f_i\rangle_{i\in G}$ such that for all $\lambda' \in\approx_G^j (\lambda)$, for all $\lambda'' \in \mathcal{P}(F_G, \lambda'[0,j])$, $M, \lambda'', j+1 \models \varphi$. We next prove that $M, \lambda^*, j \models \langle\langle G\rangle\rangle \bigcirc \varphi$ for any $\lambda^* \in\approx_G^j (\lambda)$. We consider the strategy F_G and it is easy to check that for all $\lambda_1 \in\approx_G^j (\lambda^*)$, for all $\lambda_2 \in \mathcal{P}(F_G, \lambda_1[0,j])$, $M, \lambda_2, j+1 \models \varphi$ as $\approx_G^j (\lambda^*) =\approx_G^j (\lambda)$. Thus, $M, \lambda, j \models D_G \langle\langle G\rangle\rangle \bigcirc \varphi$. The other direction is straightforward.   ($\Box$)

Note that it is not generally the case that $\models \langle\langle G\rangle\rangle\varphi\mathcal{U}\psi \rightarrow \langle\langle G\rangle\rangle D_G\varphi\mathcal{U} D_G\psi$. Here is a counter-example. Consider the model M_3 in Fig. 3 with two agents 1 and 2 and states $\{q_0, q_1, q_1', q_2, q_2'\}$, where $q_1 R_1 q_1'$, but not for 2, and all the other states can be distinguished by both agents. There are two propositions p, q, and $\pi(p) = \{q_1\}$, $\pi(q) = \{q_1', q_2\}$. The transitions are depicted in Fig. 3. Consider

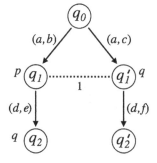

Fig. 3. The counter-model M_3

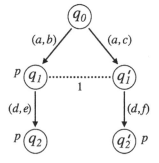

Fig. 4. The counter-model M_4

the left path $\lambda_1 := q_0 q_1 q_2 \cdots$. It is easy to check that $M_3, \lambda_1, 1 \models \langle\langle 1 \rangle\rangle p \mathcal{U} q$, but $M_3, \lambda_1, 1 \not\models \langle\langle 1 \rangle\rangle K_1 p \mathcal{U} K_1 q$.

It follows from Proposition 3 that the distributed knowledge operator and the coalition operator are interchangeable w.r.t temporal operators $\bigcirc$ and $\square$.

Corollary 1. *For any $G \subseteq N$ and any $\varphi \in \mathcal{L}$,*

$$- \models \langle\langle G \rangle\rangle \bigcirc D_G \varphi \leftrightarrow D_G \langle\langle G \rangle\rangle \bigcirc \varphi$$
$$- \models \langle\langle G \rangle\rangle \square D_G \varphi \leftrightarrow D_G \langle\langle G \rangle\rangle \square \varphi$$

4   The Fixed-Point Characterization

In this section, we will investigate the fixed-point characterizations of coalition operators through the interplay between knowledge shared by a group of agents and its coalition ability in ATL with imperfect information and perfect recall. We first show that similar to [3,4], the standard fixed-point characterizations of coalition operators for ATL [12] fail under our new semantics.

Proposition 4. *For any $G \subseteq N$ and any $\varphi, \psi \in \mathcal{L}$,*

$$- \not\models \varphi \wedge \langle\langle G \rangle\rangle \bigcirc \langle\langle G \rangle\rangle \square \varphi \rightarrow \langle\langle G \rangle\rangle \square \varphi$$
$$- \not\models \varphi \vee \langle\langle G \rangle\rangle \bigcirc \langle\langle G \rangle\rangle \Diamond \varphi \rightarrow \langle\langle G \rangle\rangle \Diamond \varphi$$
$$- \not\models \psi \vee (\varphi \wedge \langle\langle G \rangle\rangle \bigcirc \langle\langle G \rangle\rangle \varphi \mathcal{U} \psi) \rightarrow \langle\langle G \rangle\rangle \varphi \mathcal{U} \psi$$

Here is a counter-example for the first one. Consider the model M_4 in Fig. 4 which is obtained from M_3 by just changing the valuations. There is one proposition p, and $\pi(p) = \{q_1, q_2, q_2'\}$. Consider $\varphi := p$ and the left path $\lambda_1 := q_0 q_1 q_2 \cdots$. Then it is easy to check that $M_4, \lambda_1, 1 \models p$ and $M_4, \lambda_1, 1 \models \langle\langle 1 \rangle\rangle \bigcirc \langle\langle 1 \rangle\rangle \square p$, but $M_4, \lambda_1, 1 \not\models \langle\langle 1 \rangle\rangle \square p$. Thus, $M_4, \lambda_1, 1 \not\models p \wedge \langle\langle 1 \rangle\rangle \bigcirc \langle\langle 1 \rangle\rangle \square p \rightarrow \langle\langle 1 \rangle\rangle \square p$.

On the other hand, different from them, we have the following proposition showing that all the converse directions hold under the new semantics.

Proposition 5. *For any $G \subseteq N$ and any $\varphi, \psi \in \mathcal{L}$,*

$$- \models \langle\langle G \rangle\rangle \square \varphi \rightarrow \varphi \wedge \langle\langle G \rangle\rangle \bigcirc \langle\langle G \rangle\rangle \square \varphi$$
$$- \models \langle\langle G \rangle\rangle \Diamond \varphi \rightarrow \varphi \vee \langle\langle G \rangle\rangle \bigcirc \langle\langle G \rangle\rangle \Diamond \varphi$$
$$- \models \langle\langle G \rangle\rangle \varphi \mathcal{U} \psi \rightarrow \psi \vee (\varphi \wedge \langle\langle G \rangle\rangle \bigcirc \langle\langle G \rangle\rangle \varphi \mathcal{U} \psi)$$

Proof. We only give the proof for the $\mathcal{U}$ operator, and the proof for $\square$ and $\Diamond$ is similar. For every iCGS M, every path λ of M and every stage $j \in \mathbb{N}$ on λ, assume $M, \lambda, j \models \langle\langle G \rangle\rangle \varphi \mathcal{U} \psi$, then there is $F_G = \langle f_i \rangle_i \in G$ such that for all $\lambda' \in \approx_G^j (\lambda)$, for all $\lambda'' \in \mathcal{P}(F_G, \lambda'[0, j])$, there is $k \geq j$ such that $M, \lambda'', k \models \psi$ and for all $j \leq t < k$, $M, \lambda'', t \models \varphi$. We next prove by two cases: either $k = j$ or $k > j$.

If $k = j$, then by assumption for all $\lambda' \in \approx_G^j (\lambda)$, for all $\lambda'' \in \mathcal{P}(F_G, \lambda'[0, j])$, $M, \lambda'', j \models \psi$. In particular, $\lambda \in \approx_G^j (\lambda)$, then for all $\lambda^* \in \mathcal{P}(F_G, \lambda[0, j])$, $M, \lambda^*, j \models \psi$, so by $\lambda[0, j] = \lambda^*[0, j]$ we have $M, \lambda, j \models \psi$.

If $k > j$, then by assumption for all $\lambda' \in\approx_G^j (\lambda)$, for all $\lambda'' \in \mathcal{P}(F_G, \lambda'[0,j])$, there is $k > j$ such that $M, \lambda'', k \models \psi$ and for all $j \leq t < k$, $M, \lambda'', t \models \varphi$. Then $M, \lambda'', j \models \varphi$. In particular, $\lambda \in\approx_G^j (\lambda)$, so for all $\lambda^* \in \mathcal{P}(F_G, \lambda[0,j])$, $M, \lambda^*, j \models \varphi$, so by $\lambda[0,j] = \lambda^*[0,j]$ we have $M, \lambda, j \models \varphi$. We next prove that $M, \lambda, j \models \langle\langle G \rangle\rangle \bigcirc \langle\langle G \rangle\rangle \varphi \mathcal{U} \psi$. It suffices to show that F_G is just the joint strategy for the both coalition operators. That is, for all $\lambda_1 \in\approx_G^j (\lambda)$, for all $\lambda_2 \in \mathcal{P}(F_G, \lambda_1[0,j])$, for all $\lambda_3 \in\approx_G^{j+1} (\lambda_2)$, for all $\lambda_4 \in \mathcal{P}(F_G, \lambda_3[0,j+1])$, we want to prove that there is some $r \geq j+1$ such that $M, \lambda_4, r \models \psi$ and for all $j+1 \leq s < r$, $M, \lambda_4, s \models \varphi$. As $\lambda_4 \in \mathcal{P}(F_G, \lambda_3[0,j+1])$, then $\lambda_4[0,j+1] = \lambda_3[0,j+1]$, then $\lambda_4 \in\approx_G^{j+1} (\lambda_2)$, then $\lambda_4 \in\approx_G^j (\lambda_2)$ and $\lambda_4 \in \bigcup_{\lambda_1 \in\approx_G^j(\lambda)} \mathcal{P}(F_G, \lambda_1[0,j])$. So by the assumption we have that there is $k > j$ such that $M, \lambda_4, k \models \psi$ and for all $j \leq t < k$, $M, \lambda_4, t \models \varphi$, so $M, \lambda, j \models \langle\langle G \rangle\rangle \bigcirc \langle\langle G \rangle\rangle \varphi \mathcal{U} \psi$. Thus, in this case we obtain $M, \lambda, j \models \varphi \wedge \langle\langle G \rangle\rangle \bigcirc \langle\langle G \rangle\rangle \varphi \mathcal{U} \psi$.

Therefore, in both cases, $M, \lambda, j \models \psi \vee (\varphi \wedge \langle\langle G \rangle\rangle \bigcirc \langle\langle G \rangle\rangle \varphi \mathcal{U} \psi)$.   ($\square$)

We now present the main result about the fixed-point characterizations of coalition ability for ATL with imperfect information and perfect recall.

Theorem 1. *For any $G \subseteq N$ and for any $\varphi \in \mathcal{L}$,*

$$\models \langle\langle G \rangle\rangle \Box \varphi \leftrightarrow D_G \varphi \wedge \langle\langle G \rangle\rangle \bigcirc \langle\langle G \rangle\rangle \Box \varphi$$

$$\models \langle\langle G \rangle\rangle \Diamond \varphi \leftrightarrow D_G \varphi \vee \langle\langle G \rangle\rangle \bigcirc \langle\langle G \rangle\rangle \Diamond \varphi$$

$$\models \langle\langle G \rangle\rangle \varphi \mathcal{U} \psi \leftrightarrow D_G \psi \vee (D_G \varphi \wedge \langle\langle G \rangle\rangle \bigcirc \langle\langle G \rangle\rangle \varphi \mathcal{U} \psi)$$

Proof. We only give the proof for the $\mathcal{U}$ operator, the proof for $\Box$ and $\Diamond$ is similar. We first show the right to left direction. For every iCGS M, every path λ of M and every stage $j \in \mathbb{N}$ on λ, assume $M, \lambda, j \models D_G \psi \vee (D_G \varphi \wedge \langle\langle G \rangle\rangle \bigcirc \langle\langle G \rangle\rangle \varphi \mathcal{U} \psi)$, want $M, \lambda, j \models \langle\langle G \rangle\rangle \varphi \mathcal{U} \psi$. We next prove this by two cases: either $M, \lambda, j \models D_G \psi$ or $M, \lambda, j \models D_G \varphi \wedge \langle\langle G \rangle\rangle \bigcirc \langle\langle G \rangle\rangle \varphi \mathcal{U} \psi$.

If $M, \lambda, j \models D_G \psi$, then for all $\lambda' \in\approx_G^j (\lambda)$, $M, \lambda', j \models \psi$, then we have that for any F_G, for all $\lambda' \in\approx_G^j (\lambda)$, for all $\lambda'' \in \mathcal{P}(F_G, \lambda'[0,j])$, $M, \lambda'', j \models \psi$ by $\lambda''[0,j] = \lambda'[0,j]$. Thus, $M, \lambda, j \models \langle\langle G \rangle\rangle \varphi \mathcal{U} \psi$.

If $M, \lambda, j \models D_G \varphi \wedge \langle\langle G \rangle\rangle \bigcirc \langle\langle G \rangle\rangle \varphi \mathcal{U} \psi$, then $M, \lambda, j \models D_G \varphi$ and $M, \lambda, j \models \langle\langle G \rangle\rangle \bigcirc \langle\langle G \rangle\rangle \varphi \mathcal{U} \psi$. By the latter, we get that there is $F_G^1 = \langle f_i^1 \rangle_{i \in G}$ such that for all $\lambda_1 \in\approx_G^j (\lambda)$, for all $\lambda_2 \in \mathcal{P}(F_G^1, \lambda_1[0,j])$, $M, \lambda_2, j+1 \models \langle\langle G \rangle\rangle \varphi \mathcal{U} \psi$. It follows that there is $F_G^{2 \cdot x} = \langle f_i^{2 \cdot x} \rangle_{i \in G}$ where $x = \lambda_2[0,j+1]$ such that for all $\lambda_3 \in\approx_G^{j+1} (\lambda_2)$, for all $\lambda_4 \in \mathcal{P}(F_G^{2 \cdot x}, \lambda_3[0,j+1])$, there is $k \geq j+1$ such that $M, \lambda_4, k \models \psi$ and for all $j+1 \leq t < k$, $M, \lambda_4, t \models \varphi$. We next construct a new joint strategy $F_G = \langle f_i \rangle_{i \in G}$ based on F_G^1 and $F_G^{2 \cdot x}$. In order to define F_G, we first need the following notation. Let $X = \{\lambda'[0,j] \xrightarrow{\alpha} w \mid \lambda' \in\approx_G^j (\lambda), \forall i \in G, \alpha(i) = f_i^1(\lambda'[0,j])$ and $w = \delta(\lambda'[j], \alpha)\}$. Intuitively, X is the set of all possible outcomes generated by the agents in G taking the next actions specified by F_G^1 from a history that is indistinguishable from history $\lambda[0,j]$. We can now define the strategy $F_G = \langle f_i \rangle_{i \in G}$ as follows: For all $h \in H(M)$ and for all $i \in G$,

$$f_i(h) = \begin{cases} f_i^{2\cdot l}(h) & \text{if } \exists l \in X \text{ such that } l \text{ is a segment of } h \\ f_i^1(h) & \text{otherwise} \end{cases}$$

Note that this strategy is well defined, because if a history h has a segment in X, there is only one such segment due to the fact that all histories in X has the same length according to the definition for equivalence relation.

We next show that F_G is just the joint strategy we need to verify $\langle\langle G \rangle\rangle \varphi \mathcal{U} \psi$. That is, for all $\lambda' \in \approx_G^j (\lambda)$, for all $\lambda'' \in \mathcal{P}(F_G, \lambda'[0, j])$, we want to prove that there is $r \geq j$, $M, \lambda'', r \models \psi$ and for all $j \leq s < r$, $M, \lambda'', s \models \varphi$. As for any $l \in X$, $|l| > |\lambda'[0, j]|$, then there is no $l \in X$ such that l is a segment of $\lambda'[0, j]$, then by the definition of F_G, $F_G(\lambda'[0, j]) = F_G^1(\lambda'[0, j])$, so $\lambda''[0, j+1] \in X$ and $M, \lambda'', j+1 \models \langle\langle G \rangle\rangle \varphi \mathcal{U} \psi$. From the later, we get that for all $\lambda^\bullet \in \approx_G^{j+1} (\lambda'')$, for all $\lambda^* \in \mathcal{P}(F_G^{2\cdot y}, \lambda^\bullet[0, j+1])$ where $y = \lambda''[0, j+1]$, there is $k \geq j+1$ such that $M, \lambda^*, k \models \psi$ and for all $j+1 \leq t < k$, $M, \lambda^*, t \models \varphi$. Since $\lambda''[0, j+1] \in X$, then $\lambda^\bullet[0, j+1] \in X$. And by the definition of F_G and the assumption $\lambda'' \in \mathcal{P}(F_G, \lambda'[0, j])$, we get $\lambda'' \in \mathcal{P}(F_G^{2\cdot y}, \lambda^\bullet[0, j+1])$, so there is $r \geq j+1$ such that $M, \lambda'', r \models \psi$ and for all $j+1 \leq s < r$, $M, \lambda'', s \models \varphi$. And by the assumption $M, \lambda, j \models D_G \varphi$, we get $M, \lambda'', j \models \varphi$. So there is $r \geq j$, $M, \lambda'', r \models \psi$ and for all $j \leq s < r$, $M, \lambda'', s \models \varphi$, so $M, \lambda, j \models \langle\langle G \rangle\rangle \varphi \mathcal{U} \psi$.

Thus, in both cases $M, \lambda, j \models \langle\langle G \rangle\rangle \varphi \mathcal{U} \psi$.

The other direction is proved by a similar method in Proposition 5. $(\square)$

The first clause says that a coalition by sharing their knowledge can cooperate to maintain φ iff the coalition distributedly knows φ at the current stage and there is a joint strategy for this coalition to possess this ability at the next stage. The intuitions behind the other two clauses are similar. In particular, we have the following result for a single agent.

Corollary 2. *For any $i \in N$ and any $\varphi, \psi \in \mathcal{L}$,*

$$\models \langle\langle i \rangle\rangle \square \varphi \leftrightarrow K_i \varphi \wedge \langle\langle i \rangle\rangle \bigcirc \langle\langle i \rangle\rangle \square \varphi$$

$$\models \langle\langle i \rangle\rangle \Diamond \varphi \leftrightarrow K_i \varphi \vee \langle\langle i \rangle\rangle \bigcirc \langle\langle i \rangle\rangle \Diamond \varphi$$

$$\models \langle\langle i \rangle\rangle \varphi \mathcal{U} \psi \leftrightarrow K_i \psi \vee (K_i \varphi \wedge \langle\langle i \rangle\rangle \bigcirc \langle\langle i \rangle\rangle \varphi \mathcal{U} \psi)$$

Finally, we would like to remark that on the one hand, such fixed-point characterizations shed light on the interplay of group knowledge and coalition ability under imperfect information and perfect recall; on the other hand, as pointed by [4], the interest of such fixed-point characterizations lies in the fact that similar characterizations feature prominently in decision procedures for the satisfiability and model checking problems of temporal logics [5]. We envisage they may play a similar role for ATL with imperfect information and perfect recall.

5 Related Work

In recent years, there are many logical formalisms for reasoning about coalition abilities and strategic interactions in MAS. [10,13] provide a latest survey of this topic. In this following, we will review several works which are most related to ATL with imperfect information and perfect recall.

In the context of imperfect information, several semantic variants have been proposed for ATL based on different interpretations of agents' ability [1,17,19,24]. In particular, [6,18] provide formal comparisons of validity sets for semantic variants of ATL. Similar to Bulling et al.'s no forgetting semantics [20], our semantics is also history-based w.r.t a path and an index on the path, but there are fundamental differences. First of all, we consider a finer notion of perfect recall by taking both past states and actions into considerations to deal with situations where different actions may have the same effects. Secondly, our notion of group uniform strategies is defined in terms of distributed knowledge instead of general knowledge as we assume that when a set of agents form a coalition, they are able to share their knowledge before cooperating to ensure a goal.

Several epistemic-ATL style logics have been proposed to investigate the interaction of group knowledge and coalition ability [6,11,24]. In particular, the most relevant work is [11] where van Ditmarsch et al. propose three types of coalition operators to specify different cases of how all agents in the coalition cooperate to enforce a goal. Among them, the communication strategy operator $\langle\langle G \rangle\rangle_c$ captures the intuition behind our coalition operator. Specifically, we have the following correspondence.

Proposition 6. *Given an iCGS M, a path λ of M and a stage $j \in \mathbb{N}$ on λ, let φ be any formula of the form $\bigcirc\psi$, $\square\psi$ or $\psi_1\mathcal{U}\psi_2$, $M, \lambda, j \models \langle\langle G \rangle\rangle\varphi$ iff M, $\lambda[0, j] \models_{euATL} \langle\langle G \rangle\rangle_c\varphi$*

However, their work is different from ours in the following aspects: firstly, they propose two epistemic versions of ATL, namely uATL and euATL, to address the issue of uniformity of strategies in the combination of strategic and epistemic systems, while we introduce a new semantics without adding new operators to the language to explore the interplay of epistemic and coalitional operators; secondly, their results mainly focus on the relations and logical properties of three coalition ability operators, while we investigate fixed-pointed characterizations for the interplay of distributed knowledge and coalition operators which is not involved in [11]; thirdly, their meaning by coalition is more subtle than ours. Except the communication strategy operator, the comparison with the other two strategy operators is less straightforward since they are based on assumptions of coalitions without sharing knowledge. We hope to understand them better in the future.

Finally, it is also worth mentioning that [15] adopts a similar meaning of coalition so as to capture the notion of "knowing how to play". Besides the different motivations, that work is based on STIT framework and just considers one-step uniform strategies without investigating the interplay of epistemic and coalitional operators.

6   Conclusion

In this paper, we have proposed a new semantics for ATL with imperfect information and perfect recall to explore the interplay of the knowledge shared by a group of agents and its coalitional ability. Compared to existing alternative semantics, we have showed that our semantics can not only preserve the desirable properties of coalition ability in traditional coalitional logics, but also provide a finer notion of perfect recall requiring an agent remembers the past states as well as the past actions. More importantly, we have recovered the fixed-point characterizations of coalition operators which normally fail under imperfect information through the interplay of epistemic and coalitional operators.

In the future we intend to study the computational complexity of ATL with this new semantics, such as the model-checking problem. Based on previous results [9,24], we expect similar undecidability result hold under our setting. If it is the case, it would also be interesting to explore the decidable segment of this variant.

In addition, we have investigated how knowledge sharing within a group of agents contributes to its coalitional ability. This work can be seen as an attempt towards the question: which kind of group knowledge is required for a group to achieve some goal in the context of imperfect information. We believe that it is an interesting question for further investigation by considering other cases such as group without knowledge sharing or with partial knowledge sharing.

Acknowledgments. We are grateful to Heng Zhang for his valuable help, and special thanks are due to three anonymous referees for their insightful comments. This research was partially supported by A key project of National Science of China titled with A study on dynamic logics for games (15AZX020).

References

1. Ågotnes, T., Goranko, V., Jamroga, W.: Alternating-time temporal logics with irrevocable strategies. In: TARK 2007, pp. 15–24. ACM (2007)
2. Alur, R., Henzinger, T.A., Kupferman, O.: Alternating-time temporal logic. J. ACM **49**(5), 672–713 (2002)
3. Belardinelli, F.: Reasoning about knowledge and strategies: epistemic strategy logic. In: SR 2014, pp. 27–33 (2014)
4. Belardinelli, F.: A logic of knowledge and strategies with imperfect information. In: LAMAS 2015 (2015)
5. Bolander, T., Braüner, T.: Tableau-based decision procedures for hybrid logic. J. Logic Comput. **16**(6), 737–763 (2006)
6. Bulling, N., Jamroga, W.: Comparing variants of strategic ability: how uncertainty and memory influence general properties of games. Auton. Agent. Multi-Agent Syst. **28**(3), 474–518 (2014)
7. Bulling, N., Jamroga, W., Popovici, M.: Agents with truly perfect recall in alternating-time temporal logic. In: Proceedings of AAMAS 2014, pp. 1561–1562 (2014)
8. Chatterjee, K., Henzinger, T.A., Piterman, N.: Strategy logic. Inf. Comput. **208**(6), 677–693 (2010)

9. Diaconu, R., Dima, C.: Model-checking alternating-time temporal logic with strategies based on common knowledge is undecidable. Appl. Artif. Intell. **26**(4), 331–348 (2012)
10. van Ditmarsch, H., Halpern, J.Y., van der Hoek, W.: Handbook of Epistemic Logic. College Publications, London (2015)
11. van Ditmarsch, H., Knight, S.: Partial information and uniform strategies. In: Bulling, N., van der Torre, L., Villata, S., Jamroga, W., Vasconcelos, W. (eds.) CLIMA 2014. LNCS, vol. 8624, pp. 183–198. Springer, Heidelberg (2014)
12. Goranko, V., Van Drimmelen, G.: Complete axiomatization and decidability of alternating-time temporal logic. Theoret. Comput. Sci. **353**(1), 93–117 (2006)
13. Herzig, A.: Logics of knowledge and action: critical analysis and challenges. In: Autonomous Agents and Multi-Agent Systems, pp. 1–35 (2014)
14. Herzig, A., Lorini, E.: A dynamic logic of agency I: STIT, capabilities and powers. J. Logic Lang. Inform. **19**(1), 89–121 (2010)
15. Herzig, A., Troquard, N.: Knowing how to play: uniform choices in logics of agency. In: AAMAS 2006, pp. 209–216. ACM (2006)
16. van der Hoek, W., Pauly, M.: Modal logic for games and information. In: Handbook of Modal Logic, vol. 3, pp. 1077–1148 (2006)
17. van der Hoek, W., Wooldridge, M.: Cooperation, knowledge, and time: alternating-time temporal epistemic logic and its applications. Studia Logica **75**(1), 125–157 (2003)
18. Jamroga, W., Bulling, N.: Comparing variants of strategic ability. In: IJCAI 2011, pp. 252–257 (2011)
19. Jamroga, W., van der Hoek, W.: Agents that know how to play. Fundamenta Informaticae **63**(2), 185–219 (2004)
20. Jamroga, W.: Some remarks on alternating temporal epistemic logic. In: FAMAS 2003, pp. 133–140. Citeseer (2003)
21. Kuhn, H.W.: Extensive games and the problem of information. Contrib. Theor. Games **2**(28), 193–216 (1953)
22. Pauly, M.: A modal logic for coalitional power in games. J. Logic Comput. **12**(1), 149–166 (2002)
23. Ruan, J., Thielscher, M.: Strategic and epistemic reasoning for the game description language GDL-II. In: ECAI, pp. 696–701 (2012)
24. Schobbens, P.Y.: Alternating-time logic with imperfect recall. Electron. Notes Theoret. Comput. Sci. **85**(2), 82–93 (2004)
25. Thielscher, M.: A general game description language for incomplete information games. In: AAAI 2010, pp. 994–999 (2010)
26. Van Otterloo, S., Jonker, G.: On epistemic temporal strategic logic. Electron. Notes Theoret. Comput. Sci. **126**, 77–92 (2005)
27. Zhang, D., Thielscher, M.: A logic for reasoning about game strategies. In: AAAI 2015, pp. 1671–1677 (2015)

Possibilistic Inferences
in Answer Set Programming

Yifan Jin[(✉)], Kewen Wang, and Zhe Wang

School of Information and Communication Technology,
Griffith University, Nathan, Australia
yifan.jin@griffithuni.edu.au

Abstract. Answer set programming (ASP) has been extended to possibilistic ASP (PASP), in which the notion of possibilistic stable models is defined for possibilistic logic programs. However, possibilistic inferences that correspond to the three inferences in ordinary possibilistic logic have not been explored in PASP yet. In this paper, based on the skeptical reasoning determined by possibilistic stable models, we define three inference relations for PASP, provide their equivalent characterisations in terms of possibility distribution, and develop algorithms for these possibilistic inferences. Our algorithms are achieved by generalising some important concepts (Clarke's completion, loop formulas, and guarded resolution) and properties in ASP to PASP. Besides their theoretical importance, these results can be used to develop efficient implementations for possibilistic reasoning in ASP.

Keywords: Possibilistic inference · Loop formula · Guarded resolution

1 Introduction

Answer set programming (ASP) (Baral 2003, Gelfond et al. 1988) is currently one of the most widely used nonmonotonic reasoning systems due to its simple syntax, precise semantics and importantly, the availability of efficient ASP solvers, such as clasp (Gebser et al. 2009), dlv (Leone et al. 2006), and smodels (Syrjanen et al. 2001). On the other hand, a variety of practical applications require the ability of ASP to represent and reason about uncertain and (partially) inconsistent knowledge. As Zadeh (1999) argued, "when our main concern is with the meaning of information-rather than with its measure, the proper framework for information analysis is possibilistic rather than probabilistic in nature". We are not going to discuss details of this statement but point out that 'possibilistic' information is more on the side of representing priority of formulas and rules. Moreover, it is easier for the user to manage 'possibilistic' information than 'probabilistic' information since probability axioms are not enforced in possibilistic reasoning (Dubois and Prade 2001). Zadeh's observation motivated the introduction of possibilistic logic (Dubois et al. 1994), which has been applied in several application domains (Alsinet et al. 2008, Benferhat et al. 2013, Dubois

© Springer International Publishing Switzerland 2015
B. Pfahringer and J. Renz (Eds.): AI 2015, LNAI 9457, pp. 263–275, 2015.
DOI: 10.1007/978-3-319-26350-2_23

et al. 1999; 2001). According to Dubois and Prade, possibilistic approach has its advantages over probabilistic approach in dealing with logical entailment.

Thus, it is natural and useful to combine possibilistic reasoning and ASP, and as a result, a possibilistic extension of ASP (PASP) is proposed in Nicolas et al. (2005), Bauters et al. (2015), and Nieves et al. (2007). The semantics of a possibilistic program in Nicolas et al. (2006) is defined by its *possibilistic stable models*, which are obtained from a generalisation of Gelfond-Lifschitz reduction. This semantics is a natural generalisation of stable models for normal logic programs in the sense that if each normal logic program P is regarded as a possibilistic program P' whose rules have the same necessity degree 1, the possibilistic stable models of P' are exactly answer sets of P. The skeptical reasoning determined by this semantics is actually not inconsistency-tolerant like other possibilistic logics. For instance, given a simple possibilistic logic program $\{(a \leftarrow, 0.8), (b \leftarrow not\ b, 0.1)\}$. The semantics proposed in Nicolas et al. (2006) will not have any possibilistic stable model even the rule that causes inconsistency has a low necessity degree. We will discuss this further in Sect. 3.

In fact, several issues in PASP are still unexplored. First, possibilistic inferences, similar to those in ordinary possibilistic logic, have not been investigated in the literature; Second, the counterparts of some important concepts in ASP, such as Clarke's completion (Clark 1978), loop formulas (Lin et al. 2004) and resolution (Marek et al. 2011), have not been adapted to PASP. Moreover, how can such generalised concepts and algorithms be used in developing algorithms for possibilistic inferences in ASP?

In this paper, we aim to provide a solution to these open problems. We first introduce three possibilistic inferences in PASP. Moreover, we will show that, these inferences can be fully characterized by possibility distribution defined in Nicolas et al. (2006). This is not the case in many previous proposed semantics like Nicolas et al. (2006) and Bauters et al. (2012; 2014). We also develop algorithms for these possibilistic inferences. Our algorithms are achieved by generalising some important concepts (Clarke's completion, loop formulas, and guarded resolution) and their properties in ASP to PASP.

Specifically, major contributions of this work are summarised as follows:

1. We define three forms of possibilistic reasoning in PASP and provide characterisations of these possibilistic inferences in terms of possibility distributions.
2. We provide a translation from possibilistic logic programs to possibilistic propositional theories. This is achieved by generalising Clarke's completion and loop formulas from standard ASP to PASP. As a result, possibilistic completion and loop formulas allow us to compute the least specific distribution of a possibilistic program by just computing that of the corresponding possibilistic theory. Moreover, this result provides an algorithm for all of our three possibilistic inferences in PASP through computing inconsistency degrees.
3. We propose possibilistic guarded resolution for PASP, which generalises the guarded resolution for ordinary ASP in 2011. We show that the new resolution algorithm is sound and complete with respect to possibilistic stable models. This result provides an algorithm for two of our possibilistic inferences in PASP.

2   Preliminaries

2.1   Possibilistic Logic

In possibilistic propositional logic, a (possibilistic) formula is a pair of the form (ϕ, α) where ϕ is a propositional formula and α is an element of a totally ordered set, called the necessity degree of ϕ. In this paper, we assume that the necessity degrees are taken from a fixed finite subset of the interval $[0, 1]$. Informally, the formula (ϕ, α) expresses that ϕ is certain at least to the level α. A possibilistic knowledge base is a finite set of possibilistic formulas $\Sigma = \{(\phi_i, \alpha_i), i = 1, \ldots, n\}$. Given a possibilistic knowledge base Σ we use Σ^* to denote its classical part that is obtained by ignoring necessity degrees of all formulas in Σ.

The semantics of possibilistic logic is based on the notion of possibility distributions, which are mappings from the set Ω of all propositional interpretations to $[0, 1]$. Given a possibility distribution π and an interpretation $I \in \Omega$, $\pi(I)$ represents the degree of compatibility of the interpretation I with the available beliefs about the real world. By convention, $\pi(I) = 0$ means that I is impossible, and $\pi(I) = 1$ means that nothing prevents I from being the real world. When $\pi(I) > \pi(I')$, I is a preferred candidate to I' for being the real state of the world. Given two possibility distributions π and π', π is said to be less specific than π' if for all interpretation I, $\pi(I) \geq \pi'(I)$.

Given a possibility distribution π, two measures can be defined for possibilistic formulas. That is, the possibility degree $\Pi_\pi(\phi) = max\{\pi(I) : I \models \phi\}$ and the necessity degree $N_\pi(\phi) = 1 - max\{\pi(I) : I \not\models \phi\}$. We say a possibility distribution π satisfies a possibilistic formula (ϕ, α), denoted $\pi \models \phi$ if $N_\pi(\phi) \geq \alpha$.

For each possibilistic knowledge base Σ, the least specific possibility distribution, denoted π_Σ, is defined as, for each $I \in \Omega$, $\pi_\Sigma(I) = 1$ if $I \models \phi_i$ for all $(\phi_i, \alpha_i) \in \Sigma$ and $\pi_\Sigma(I) = 1 - max\{\alpha \mid I \not\models \phi, (\phi, \alpha) \in \Sigma\}$ otherwise.

In possibilistic logic, when a knowledge base is inconsistent, we are able to extract certain consistent sub-knowledge bases based on the priority level determined by necessity degrees of formulas and to use inconsistency degree to characterise the inconsistency of a possibilistic knowledge base. In possibilistic logic, each priority level is referred to as a *cut*. Formally, given a necessity degree α, an α-cut (resp. strict α-cut) of Σ, denoted $\Sigma_{\geq \alpha}$ (resp. $\Sigma_{> \alpha}$), is the set of formulas in Σ having a necessity degree greater than (resp. strictly greater than) α. The inconsistency degree of Σ is defined as $Inc(\Sigma) = max\{\alpha \mid \Sigma_{\geq \alpha}$ is inconsistent$\}$. This inconsistency degree defines a plausibility level under which information is no more pertinent. The inconsistency degree of Σ is actually determined by its least specific possibility distribution. Specifically, $Inc(\Sigma) = 1 - max\{\pi_\Sigma(I) \mid I \in \Omega\}$.

There are three major inferences in possibilistic logic. Here $(\Sigma_{\geq \alpha})^*$ is denoted $\Sigma^*_{\geq \alpha}$.

1. A formula ϕ is said to be a plausible consequence of Σ, denoted $\Sigma \models_p \phi$, if $\Sigma^*_{> Inc(\Sigma)} \models \phi$.
2. A possibilistic formula (ϕ, α) is a consequence of Σ, denoted by $\Sigma \models (\phi, \alpha)$, if $\alpha > Inc(\Sigma)$ and $\Sigma^*_{\geq \alpha} \models \phi$.

3. A formula ϕ is said to be a possibilistic consequence of Σ to degree α, denoted by $\Sigma \models_{\pi} (\phi, \alpha)$, if the following conditions are satisfied:(1) $\Sigma_{\geq \alpha}$ is consistent, (2) $\Sigma^{*}_{\geq \alpha} \models \phi$, and (3) $\forall \beta > \alpha$, $\Sigma^{*}_{\geq \beta} \not\models \phi$.

The second and third inferences attach to the consequences degrees α that are at least equal to the inconsistency degree of the possibilistic knowledge base Σ, yet they deal with different consequences. The second inference checks if a possibilistic formula (ϕ, α) can be inferred from Σ, whereas the third checks whether a formula ϕ can be inferred from Σ and to what degree it can be inferred. Clearly, $\Sigma \models_{\pi} (\phi, \alpha)$ implies $\Sigma \models (\phi, \alpha)$, which in turn implies $\Sigma \models_{p} \phi$. In (Dubois and Prade 2004), it is proven that $\Sigma \models_{p} \phi$ if and only if $N_{\pi_{\Sigma}}(\phi) > Inc(\Sigma)$, $\Sigma \models (\phi, \alpha)$ if and only if $N_{\pi_{\Sigma}}(\phi) \geq \alpha$ and $\alpha > Inc(\Sigma)$, and $\Sigma \models_{\pi} (\phi, \alpha)$ if and only if $N_{\pi_{\Sigma}}(\phi) = \alpha$ and $\alpha > Inc(\Sigma)$. We can see that meaningful consequences can be inferred from an inconsistent possibilistic knowledge base and thus possibilistic logic is inconsistency tolerant.

2.2   Possibilistic ASP (PASP)

In this subsection, we briefly introduce Nicolas et al.'s approach to extending ASP by allowing possibilistic reasoning proposed in (Nicolas et al. 2006). In their PASP framework, each rule is associated with a level of priority like formulas in possibilistic logic. Formally, an ASP rule r is of the form $a \leftarrow a_1, \ldots, a_m, not\ b_1, \ldots, not\ b_n$ where not is the default negation, a, a_i's and b_j's are propositional atoms. A rule in PASP is a pair (r, α) where r is an ASP rule and α is the necessity degree of r. Informally, the possibilistic rule (r, α) expresses that the rule r is certain at least to the necessity degree $n(r) = \alpha$. A possibilistic logic program is a finite set of possibilistic rules. Rules with zero degree are not explicitly represented in the knowledge base. Given a possibilistic program P, we use P^{*} to denote the answer set program of a PASP obtained by ignoring all necessity degrees in rules.

Nicolas et al. (2006) defined an immediate consequence operator for possibilistic programs without default negation as follows.

Let r be a possibilistic rule of the form $a \leftarrow a_1, \ldots, a_n, \alpha$ and A be a set of possibilistic atoms. We say that r is β-applicable in A if $\{(a_1, \alpha_1), \ldots, (a_n, \alpha_n)\} \subseteq A$ and $\beta = min\{\alpha, \alpha_1, \ldots, \alpha_n\}$. Given an atom a, define
$AP(P, A, a) = \{r \in P \mid head(r) = a, r$ is β-applicable in $A, \beta > 0\}$.

The immediate consequence operator $\mathcal{T}_P$ for possibilistic program P is then defined by, for each set A of possibilistic atoms,

$$\mathcal{T}_P(A) = \left\{ (a, \alpha) \left| \begin{array}{l} a \in head(P^{*}), AP(P, A, a) \neq \emptyset, \\ \alpha = max\{\beta \mid r \text{ is } \beta\text{-applicable in } A, r \in AP(P, A, a)\} \end{array} \right. \right\} \quad (1)$$

This operator is monotonic and thus has the least fixpoint $L(\mathcal{T}_P)$.

Given a possibilistic program P and a set of possibilistic atoms A, A is a *possibilistic stable model* of P if $A = L(\mathcal{T}_{P^A})$, where $P^A = \{r^A \mid r \in P, body^{-}(r) \cap A^{*} = \emptyset\}$ and r^A is the possibilistic rule $head(r) \leftarrow body^{+}(r), n(r)$.

A possibilistic program P is *consistent* if P has at least one possibilistic stable model; otherwise P is inconsistent.

Example 1. *Consider possibilistic program* $P = \{(a \leftarrow, 0.9), (b \leftarrow not\ c, 0.7),$ $(c \leftarrow a, not\ b, 0.8)\}$. *If* $A = \{(a, 0.9), (b, 0.7)\}$, *then* $A^* = \{a, b\}$ *and* $P^A = \{(a \leftarrow, 0.9), (b \leftarrow, 0.7)\}$. *It is easy to see that* $L(T_{P^A}) = A$. *Thus,* $\{(a, 0.9), (b, 0.7)\}$ *is a possibilistic stable model of* P. *Similarly,* $\{(a, 0.9), (c, 0.8)\}$ *is another possibilistic stable model of* P.

Given a possibilistic program P, let Ω be the collection of all Herbrand interpretations of P^* (i.e., all subsets of the atoms in P^*). For a Herbrand interpretation I and a rule r, $I \models r$ denotes that I is a model of r. r is *applicable w.r.t.* I if $body^+(r) \subseteq I$ and $body^-(r) \cap I = \emptyset$. The set of all applicable rules in P^* *w.r.t.* I is denoted $app(P^*, I)$. P^* is *grounded* if it can be ordered as a sequence $r_1, \ldots, r_n$ such that for each i $(1 \le i \le n)$, $r_i \in app(P^*, head(\{r_1, \ldots, r_{i-1}\}))$. $(P^*)^I$ is the GL-reduction of P^* *w.r.t.* I defined by $(P^*)^I = \{head(r) \leftarrow body^+(r) \mid r \in P^*,$ $body^-(r) \cap I = \emptyset\}$. A possibilistic distribution π on Ω satisfies P, denoted $\pi \models P$, if for each $I \in \Omega$, the following conditions are satisfied:

$$\pi(I) = 0, \text{ if } I \not\subseteq head(app((P^*)^I, I)) \text{ or } app((P^*)^I, I) \text{ is not grounded} \quad (2)$$

$$\pi(I) = 1, \text{ if I is a model of } (P^*)^I \quad (3)$$

$$\pi(I) \le 1 - max\{\alpha \mid I \not\models r \text{ and } (r, \alpha) \in P^I\}, \text{ otherwise} \quad (4)$$

Given two possibility distributions π and π' on Ω, we say that π is less specific than π' if $\pi(I) \ge \pi'(I)$ for each $I \in \Omega$. The *least specific* possibility distribution π_P for P satisfies for each $I \in \Omega$, the above conditions (2), (3), and otherwise $\pi_P(I) = 1 - max\{\alpha \mid I \not\models r \text{ and } (r, \alpha) \in P^I\}$. The inconsistency degree, α-cut, possibility degree and necessity degree of an atom a under the possibility distribution can all be similarly defined as in ordinary possibilistic logic. In particular, a possibility distribution π satisfies a possibilistic atom (a, α), denoted $\pi \models (a, \alpha)$, if $N_\pi(a) \ge \alpha$.

3 Possibilistic Inferences in ASP

In this section, we first introduce three forms of possibilistic reasoning for logic programs based on possibilistic stable models. Our possibilistic inferences are inconsistency-tolerant. We then show that they can be characterised in terms of possibility distributions defined in Nicolas et al. (2006).

Unlike in ordinary possibilistic logic, given a possibilistic logic program P and a necessity degree α such that $\alpha \ge Inc(P)$, $P_{>\alpha}$ may not be consistent as shown in Example 9 in (Nicolas et al. 2006). This difference is mainly due to the fact that PASP is nonmonotonic in general. So, some changes have to be made when we adapt the three major forms of possibility inferences from propositional possibilistic logic to PASP.

Recall that an atom a is a skeptical consequence of a normal program Q under stable model semantics, denoted $Q \models_s a$, if a is in every stable model of P. For simplicity, we use $P^*_{\ge\alpha}$ to denote $(P^*)_{\ge\alpha}$.

Definition 1. *Let P be a possibilistic program.*

1. *An atom a is a* plausible consequence *of P, denoted $P \models_p a$, if there exists $\alpha \in [0,1]$ such that $P^*_{\geq \alpha}$ is consistent and $P^*_{\geq \alpha} \models_s a$.*
2. *An atom a is a* possibilistic consequence *of P to degree α, written $P \models_\pi (a, \alpha)$, if $P^*_{\geq \alpha}$ is consistent, $P^*_{\geq \alpha} \models_s a$, and for all $\beta > \alpha$, $P^*_{\geq \beta}$ is inconsistent or $P^*_{\geq \beta} \not\models_s a$.*
3. *A possibilistic atom (a, α) is a* consequence *of P, written $P \models (a, \alpha)$ if $P^*_{\geq \alpha}$ is consistent and $P^*_{\geq \alpha} \models_s a$.*

Example 2. *Consider possibilistic program $P = \{(\leftarrow b, 0.9), (b \leftarrow not\ c, 0.8), (c \leftarrow not\ a, 0.7), (a \leftarrow, 0.1)\}$. Since $P_{\geq 0.7}$ is consistent and $P^*_{\geq 0.7} \models_s c$, we have that $P \models (c, 0.7)$.*

These inference services allow us to deal with uncertainty. For example, possibilistic inference with degree allows us to infer to what degree an individual atom can be nontrivially inferred from a possibilistic program. These inference services are also heavily related to possibilistic distribution, as we noted before, a possibility distribution π satisfies (a, α), denoted $\pi \models (a, \alpha)$ if and only if $N_\pi(a) \geq \alpha$. In the following, we will discuss the relation between possibility distribution and possibilistic inference services.

The above three forms of possibilistic reasoning for ASP can be equivalently characterized in terms of possibility distributions. Unlike other semantics proposed before, in Nicolas et al. (2006) and Nieves et al. (2007) the possibility distribution can only be used to characterize possibilistic answer sets in definite programs and in Bauters et al. (2012; 2014) the possibility distribution is defined over a sub set of a program, not over the interpretations of a program.

Proposition 1. *Let P be a possibilistic program and a be an atom. Then $P \models (a, \alpha)$ if and only if $P^*_{\geq \alpha}$ is consistent, and $\pi \models P$ implies $\pi \models (a, \alpha)$ for each possibility distribution π.*

By Proposition 1, it is straightforward to show the following result for plausible consequences.

Proposition 2. *Let P be a possibilistic program and a be an atom. Then $P \models_p a$ if and only if there exists $\alpha \in [0,1]$ s.t. $P^*_{\geq \alpha}$ is consistent and for each possibility distribution π, if $\pi \models P$, then $\pi \models (a, \alpha)$.*

While the above two properties are similar to those in possibilistic logic in form, the characterisation of possibilistic inference in terms of possibility distributions is quite different from that in possibilistic logic. In possibilistic logic, if a formula ϕ is a possibilistic inference in possibilistic knowledge base Σ with the necessity degree α, then $\alpha = N_{\pi_\Sigma}(\phi)$. This is not the case in possibilistic programs as the following example shows.

Example 3 (*example 2 continued*). *Since for all $\beta > 0.7$ we have $P^*_{\geq \beta} \not\models c$, thus $P \models_\pi (c, 0.7)$. But, $N_{\pi_P}(c) = 0.8$ and $P^*_{\geq 0.8}$ is not consistent.*

However, we have the following characterisation of possibilistic inference, which is significantly different from its counterpart in possibilistic logic.

Proposition 3. *Let P be a possibilistic program and a be an atom. Then $P \models_\pi$ (a, β) if and only if $P^*_{\geq \beta}$ is consistent, $\beta \leq N_{\pi_P}(a)$, and there is no $\gamma > \beta$ s.t. $\gamma \leq N_{\pi_P}(a)$ and $P^*_{\geq \gamma}$ is consistent.*

In possibilistic logic, reasoning services can be easily reduced to computing the inconsistency degree of a possibilistic knowledge base. However, this is not straightforward in PASP due to its nonmonotonicity. Moreover, inconsistency degree in possibilistic programs is now only defined under possibilistic distribution, there is no syntactic definition of inconsistency degree in possibilistic programs. We will address these issues in subsequent sections.

4   Completion and Loop Formulas for PASP

There are several reasons for studying the concept of loop formulas for possibilistic programs. First, it is interesting to extend the definitions of completion and loop formulas to possibilistic programs. Moreover, loop formulas provide a elegant way to compute inconsistency degrees and the least specific possibility distribution, which are important for developing inference algorithms for possibilistic programs.

Given a possibilistic program rule $r = (a \leftarrow a_1, \ldots, a_m, not\ b_1, \ldots, not\ b_n, \alpha)$. Denote $G(r) = a_1 \wedge \cdots \wedge a_m \wedge \neg b_1 \wedge \cdots \wedge \neg b_n$. The completion of possibilistic program P is defined by $Comp(P) = Comp^+(P) \cup Comp^-(P)$:

(i) $G(r) \rightarrow a, n(r)$ is in $Comp^+(P)$, for each rule $r = (a \leftarrow body(r), n(r))$ in P.
(ii) For each atom a in P, let $r_1 = (a \leftarrow body(r_1), n(r_1))$, $\ldots$, $r_n = (a \leftarrow body(r_n), n(r_n))$ be all the rules with head a in P, then $(a \rightarrow G(r_1) \vee \cdots \vee G(r_n), 1)$ is in $Comp^-(P)$.

We note that due to the involvement of necessity degrees, the definition of completion for possibilistic programs looks different from the standard one but it is obviously a generalisation of the latter. Given a possibilistic program P, the positive dependency graph G_P of P is the directed graph whose set of nodes is the set of all atoms in P, and whose set of edges contains an arc from p to q if there is a rule of the form $r = (p \leftarrow body(r), n(r))$ in P such that $q \in body(r)$, for each pair of nodes p and q. Recall that a directed graph is said to be strongly connected if for any two nodes in the graph there is a directed path from one node to the other node. Given a directed graph, a strongly connected component is a set of nodes such that for any two nodes u and v in that set, there is a path from u to v and that set is not a subset of any other such sets.

Given a finite possibilistic program P, a non-empty subset L of all atoms in P is called a loop of P if for any u and v in L, there is a path of length > 0 from u to v in G_P such that all the vertices in the path are in L. This means if L is non-empty, then L is a loop if and only if the subgraph of G_P induced by L is strongly connected.

Given a possibilistic program P, and a loop L in it, $R^-(L, P) = \{G(r) \to p \mid r \in P, head(r) = p, p \in L$, and there does not exist q s.t. $q \in body(r) \cap L\}$. Without loss of generality, suppose that $R^-(L, P)$ consists of the following formulas:

$$G_{r_{11}} \to p_1, \ldots, G_{r_{1k_1}} \to p_1,$$

$$\ldots$$

$$G_{r_{n1}} \to p_n, \ldots, G_{r_{nk_n}} \to p_n.$$

Then the possibilistic loop formula $LF(L, P)$ associated with L is

$$(\neg[G_{r_{11}} \lor \cdots \lor G_{r_{1k_1}} \lor \cdots \lor G_{r_{n1}} \lor \cdots \lor G_{r_{nk_n}}] \to \bigwedge_{p \in L} \neg p, 1).$$

Let $F(P) = Comp(P) \cup LF$ where LF is the set of loop formulas associated with the loops of P. Then the least specific possibilistic distributions of P and $F(P)$ coincide.

The major result of this section is stated as follows.

Theorem 1. *If P is a possibilistic program and A is a set of atoms, then $\pi_P(A) = \pi_{F(P)}(A)$.*

By Theorem 1, we have the following useful result.

Corollary 1. *If P is a possibilistic program, then we have $Inc(P) = Inc(F(P))$ and $Inc(P) = max\{\alpha \mid F(P) \vdash (\bot, \alpha)\}$.*

The above corollary shows that the inconsistency degree of possibilistic program P can be obtained from those of the corresponding possibilistic theory $F(P)$. This allows us to deal with inconsistency degree of a possibilistic program in a syntactic way. As a result, the loop formulas can be used to reduce possibilistic reasoning in ASP to the task of computing inconsistency degree of the corresponding possibilistic theory.

Proposition 4. *Given a possibilistic program P, then $P \models (a, \alpha)$ if and only if $Inc(F(P_{\geq \alpha})) = 0$ and $Inc(F(P_{\geq \alpha}) \cup \{(\neg a, 1)\}) > 0$.*

Plausible consequences of a possibilistic program can be obtained by computing inconsistency degrees of possibilistic theories.

Proposition 5. *If P is a possibilistic program and A is a set of atoms, then $P \models_p a$ if and only if there exists α such that $Inc(F(P_{\geq \alpha})) = 0$ and $Inc(F(P_{\geq \alpha}) \cup \{(\neg a, 1)\}) > 0$.*

Moreover, possibilistic consequences of a possibilistic program can also be obtained by computing inconsistency degrees of possibilistic theories.

Proposition 6. *If P is a possibilistic program and A is a set of atoms, then $P \models_\pi (a, \alpha)$ if and only if the following three conditions are satisfied:*

1. *$Inc(F(P_{\geq \alpha})) = 0$,*
2. *$Inc(F(P_{\geq \alpha}) \cup \{(\neg a, 1)\}) > 0$, and*
3. *If $\alpha \neq N_{\pi_{F(P)}}(a)$, then $\alpha < N_{\pi_{F(P)}}(a)$ and $Inc(F(P_{\geq \beta})) > 0$ for each β with $\alpha < \beta \leq N_{\pi_{F(P)}}(a)$.*

By Corollary 1, a straightforward approach to computing the inconsistency degree of a possibilistic program is to compute its loop formulas. Yet similar to the case of ASP, such a naive approach may not be practical, as there can be an exponential number of loops in a possibilistic program. For this reason, it would be more efficient if only necessary loop formulas are included into the completion. This motivates the following method of computing its inconsistency degree when a possibilistic program is inconsistent.

Algorithm 1. Computing Inconsistency Degree
Input: Possibilitic logic program P
Output: Inconsistency degree of P

1. Set $T_P := Comp(P)$.
2. If $(T_P)^*$ has no models, then terminate and return $Inc(T_P)$.
3. Find a model M of $(T_P)^*$.
4. If M is a stable model of P^*, then terminate and return 0.
5. If M is not a stable model of P^*, then find a loop L of P such that its loop formula $LF(L, P)$ is not satisfied by M.
6. Set $T_P := T_P \cup \{LF(L, P)\}$ and go back to step 2.

It is obvious that the above algorithm will terminate in finite number of steps. We can show that the algorithm is sound.

Theorem 2. *Let P be an inconsistent possibilistic program P, then Algorithm 1 outputs $Inc(P)$.*

The notions of completion and loop formulas for possibilistic programs can be used to compute possibility distribution and inconsistency degree of a possibilistic program. In addition, it can be used to reduce the task of computing the inconsistency degree, like the three inferences for PASP and consistency restore investigated in (Nicolas et al. 2006). In the next section, we will introduce a resolution procedure for possibilistic programs, which provides a sound and complete procedure with respect to possibilistic stable models.

5   Guarded Resolution for PASP

In this section, we first present a resolution procedure for possibilistic reasoning in logic programs, which is a generalisation of the resolution rule defined in (Marek et al. 2011), and then show that our procedure is sound and complete with respect to possibilistic stable models (Nicolas et al. 2006). Moreover, as a useful application of the guarded resolution, we show that it can provide alternative way of reducing an inference task in a possibilistic program to that of computing inconsistency degrees of the possibilistic program and its cuts.

Given a possibilistic rule $r = (a \leftarrow a_1, \ldots, a_m, not\ b_1, \ldots, not\ b_n, \alpha)$, a guarded clause $pg(r)$ for r is defined as

$$(a \leftarrow a_1, \ldots, a_m : \{b_1, \ldots, b_n\}, \alpha). \tag{5}$$

If a possibilistic rule r has no positive atoms in its body (that is, $m = 0$), then $pg(r)$ is also denoted $(a : \{b_1, \ldots, b_n\}, \alpha)$, called a possibilistic guarded atom and $\{b_1, \ldots, b_n\}$ is its guard. For a possibilistic program P, define $pg(P) = \{pg(r) : r \in P\}$.

The possibilistic guarded resolution rule is defined as the following inference rule:

$$\frac{(a \leftarrow A : B, \alpha). \quad (a_j : C, \beta).}{(a \leftarrow A \setminus \{a_j\} : B \cup C, min\{\alpha, \beta\})} \tag{6}$$

where $A = \{a_1, \ldots, a_j, \ldots, a_m\}$, $B = \{b_1, \ldots, b_n\}$, $C = \{c_1, \ldots, c_h\}$

The guarded resolution rule naturally leads to the notion of a guarded resolution proof $\mathcal{P}$ of a possibilistic guarded atom $(a : S, \alpha)$ from the program $pg(P)$. A guarded resolution proof of $(a : S, \alpha)$ is a labelled tree such that every node that is not a leaf has two parents, the two parents are the upper part of Eq. (6). Each leaf is either $(a \leftarrow a_1, \ldots, a_m : \{b_1, \ldots, b_n\}, \alpha)$. such that $(a \leftarrow a_1, \ldots, a_m : not\ b_1, \ldots, not\ b_n, \alpha)$ is in P or $(a : \{b_1, \ldots, b_m\}, \alpha)$ such that $(a \leftarrow not\ b_1, \ldots, not\ b_n, \alpha)$ is in P. Note that in a guarded resolution proof, guards only grow as proceeding down the tree. Thus, the root of the proof contains the guards of every label in the tree. In the following we use $(a : S, \alpha)$ to denote both a possibilistic guarded atom as well as the possibilistic guarded resolution proofs associated with it.

We say a set A of possibilistic atoms admits a possibilistic guarded atom $(a : S, \alpha)$ if $A^* \cap S = \emptyset$, and that A admits a guarded resolution proof $\mathcal{P}$ if it admits the label of the root of $\mathcal{P}$.

The following result shows that possibilistic guarded resolution is sound and complete with respect to possibilistic stable models.

Proposition 7. *Let P be a possibilistic program and A be a set of possibilistic atoms. Then A is a possibilistic stable model of P iff the following two conditions are satisfied:*

(1) for every $(a, \alpha) \in A$, for some set S of atoms, there is a guarded resolution proof of $(a : S, \alpha)$ from $pg(P)$ that is admitted by A and $\alpha \geq \beta$ for all other guarded resolution proof of $(a : S', \beta)$ from $pg(P)$ admitted by A.

(2) for every $a \notin A^*$ and each S, there is no guarded resolution proof of $(a : S, \alpha)$ from $pg(P)$ that is admitted by A.

Based on the soundness and completeness of possibilistic guarded resolution, we are able to transform a possibilistic program P into possibilistic propositional theory $E(P)$ such that both possibilistic and plausible consequences of P can be obtained from the inconsistency degrees of $E(P)$ and its cuts.

Given an atom a, assume that $(a : S_1, \alpha_1)$, $\ldots$, and $(a : S_n, \alpha_n)$ consist of all possibilistic guarded resolution proofs from $pg(P)$ for every atom a appearing in P. Then $eq_P(a)$ is defined as conjunction of the following possibilistic propositional formulas:

$(a \Leftrightarrow \neg S_1 \vee \ldots \neg S_n, max(\alpha_1, \ldots, \alpha_n))$

If every S_i is empty, $eq_P(a)$ is defined as the possibilistic atom $(a, max(\alpha_1, \ldots, \alpha_n))$ and if there is no possibilistic guarded resolution proof for a, $eq_P(a)$ is $(\neg a, 1)$.

Let $E(P)$ denote the possibilistic propositional theory consisting of all $eq_P(a)$ for a in P. Then $E(P)$ has the following property that is useful for reducing the tasks of computing plausible consequences and consequences to the task of computing inconsistency degrees of $E(P)$ and its cuts.

Lemma 1. *Let P be a possibilistic program and A be a set of atoms. Then $\pi_P(A) = 1$ if and only if $\pi_{E(P)}(A) = 1$.*

The task of computing consequences of a possibilistic program can be reduced to that of computing inconsistency degrees of the corresponding possibilistic theory and its cuts.

Proposition 8. *Given a possibilistic program P, $P \models (a, \alpha)$ if and only if $Inc(E(P_{\geq \alpha})) = 0$ and $Inc(E(P_{\geq \alpha}) \cup \{(\neg a, 1)\}) > 0$.*

Similarly, the task of deciding plausible consequence can also be reduced to that of computing inconsistency degrees as the following property shows.

Proposition 9. *Given a possibilistic program P, $P \models_p a$ if and only if there exists an α such that $Inc(E(P_{\geq \alpha})) = 0$ and $Inc(E(P_{\geq \alpha}) \cup \{(\neg a, 1)\}) > 0$.*

However, it is unclear to us whether the task of computing possibilistic consequences of P can be reduced to the problem of computing inconsistency degrees of $E(P)$ and its cuts.

6   Discussions and Conclusion

Based on the framework of possibilistic answer set programming (PASP) in Nicolas et al. (2006), we have introduced three forms of possibilistic reasoning in ASP, provided their characterizations in terms of possibility distributions. Our possibilistic inferences are inconsistency-tolerant. In order to develop inference algorithms for these possibilistic inferences, the notions of possibilistic loop formulas and possibilistic guarded resolution are proposed, which extends their counterparts in standard ASP. We have demonstrated how these techniques can be used to perform possibilistic reasoning in a possibilistic program by computing inconsistency degrees of the corresponding possibilistic theory and its cuts. Besides their theoretical importance, our results provide promising inference algorithms for PASP.

Besides Nicolas et al.'s definition, Bauters et al. (2010; 2015) have proposed an alternative definition of possibilistic stable models. Their new semantics is based on a different view over negation as failure atoms. However, as shown in Bauters et al. (2014), their semantics may give ambiguous result for atoms in some possibilistic logic programs. Another approach similar to ours was proposed in Bauters et al. (2012; 2014). They treat the necessity degree of a rule as the uncertainty as to whether the rule is valid. However, their semantics cannot be equivalently characterized in terms of possibility distribution defined in Nicolas et al. (2006).

Several issues are interesting for future research. First, an efficient implementation of the three forms of reasoning in PASP is still missing; Second, it is unclear to us if both possibilistic loop formulas and guarded resolution provide the same translation to a possibilistic program; Third, we will investigate the possibility of adapting our results to the framework of PASP proposed in Bauters et al. (2015); Finally, we will extend our result to possibilistic disjunctive programs (Nieves et al. 2007).

References

Alsinet, T., Chesñevar, C.I., Godo, L., Simari, G.R.: A logic programming framework for possibilistic argumentation: Formalization and logical properties. Fuzzy Sets Syst. **159**(10), 1208–1228 (2008)

Baral, C.: Knowledge Representation, Reasoning and Declarative Problem Solving. Cambridge University Press, Cambridge (2003)

Bauters, K., Schockaert, S., Cock, M.D., Vermeir, D.: Possibilistic answer set programming revisited. In: Proceedings of the 26th Conference on Uncertainty in Artificial Intelligence (UAI 2010), pp. 48–55 (2010)

Bauters, K., Schockaert, S., Cock, M.D., Vermeir, D.: Possible and necessary answer sets of possibilistic answer set programs. In: Proceedings of the 24th International Conference on Tools with Artificial Intelligence (ICTAI 2012), pp. 836–843 (2012)

Bauters, K., Schockaert, S., Cock, M.D., Vermeir, D.: Semantics for possibilistic answer set programs: uncertain rules versus rules with uncertain conclusions. Int. J. Approximate Reasoning **55**(2), 739–761 (2014)

Bauters, K., Schockaert, S., Cock, M.D., Vermeir, D.: Characterizing and extending answer set semantics using possibility theory. Theory Pract. Logic Program. **15**(1), 79–116 (2015)

Benferhat, S., Bouraoui, Z.: Possibilistic DL-lite. In: Liu, W., Subrahmanian, V.S., Wijsen, J. (eds.) SUM 2013. LNCS, vol. 8078, pp. 346–359. Springer, Heidelberg (2013)

Clark, K.: Negation as failure. In: Ginsberg, M. (ed.) Readings in Nonmonotonic Reasoning, pp. 311–325. Morgan Kaufmann, San Francisco (1987)

Dubois, D., Berre, D.L., Prade, H., Sabbadin, R.: Using possibilistic logic for modeling qualitative decision: ATMS-based algorithms. Fundamenta of Informatica **37**(1–2), 1–30 (1999)

Dubois, D., Lang, J., Prade, H.: Possibilistic logic. In: Handbook of Logic in Artificial Intelligence and Logic Programming, vol. 3, pp. 439–513. Oxford University Press, New York (1994)

Dubois, D., Prade, H.: Possibility theory, probability theory and multiple-valued logics: a clarification. Ann. Math. Artif. Intell. **32**(1–4), 35–66 (2001)

Dubois, D., Prade, H.: Possibilistic logic: a retrospective and prospective view. Fuzzy Sets Syst. **144**(1), 3–23 (2004)

Gebser, M., Kaufmann, B., Schaub, T.: The conflict-driven answer set solver clasp: Progress report. In: Erdem, E., Lin, F., Schaub, T. (eds.) LPNMR 2009. LNCS, vol. 5753, pp. 509–514. Springer, Heidelberg (2009)

Gelfond, M., Lifschitz, V.: The stable model semantics for logic programming. In: ICLP/SLP 1988, pp. 1070–1080 (1988)

Leone, N., Pfeifer, G., Faber, W., Eiter, T., Gottlob, G., Perri, S., Scarcello, F.: The DLV system for knowledge representation and reasoning. ACM Trans. Comput. Logic **7**(3), 499–562 (2006)

Lin, F., Zhao, Y.: ASSAT: Computing answer sets of a logic program by SAT solvers. J. Artif. Intell. **157**(1), 115–137 (2004)

Marek, V.W., Remmel, J.B.: Guarded resolution for answer set programming. Theory Pract. Logic Program. **11**(1), 111–123 (2011)

Nicolas, P., Garcia, L., Stéphan, I., Lefèvre, C.: Possibilistic uncertainty handling for answer set programming. Ann. Math. Artif. Intell. **47**(1–2), 139–181 (2006)

Nicolas, P., Garcia, L., Stéphan, I.: A possibilistic inconsistency handling in answer set programming. In: Godo, L. (ed.) ECSQARU 2005. LNCS (LNAI), vol. 3571, pp. 402–414. Springer, Heidelberg (2005)

Nieves, J.C., Osorio, M., Cortés, U.: Semantics for possibilistic disjunctive programs. In: Baral, C., Brewka, G., Schlipf, J. (eds.) LPNMR 2007. LNCS (LNAI), vol. 4483, pp. 315–320. Springer, Heidelberg (2007)

Syrjänen, T., Niemelä, I.: The smodels system. In: Eiter, T., Faber, W., Truszczyński, M. (eds.) LPNMR 2001. LNCS (LNAI), vol. 2173, pp. 434–438. Springer, Heidelberg (2001)

Zadeh, L.A.: Fuzzy sets as a basis for a theory of possibility. Fuzzy Sets Syst. **100**, 9–34 (1999)

A Two Tiered Finite Mixture Modelling Framework to Cluster Customers on EFTPOS Network

Yuan Jin and Grace Rumantir[✉]

Faculty of Information Technology Caulfield, Monash University,
Melbourne, Australia
{Yuan.Jin,Grace.Rumantir}@monash.edu

Abstract. This paper proposes a framework to build a clustering model of customers of the retailers on the EFTPOS network of a major bank in Australia. The framework consists of two clustering tiers using Finite Mixture Modelling (FMM) that segments customers based on their probabilities of generating transactions of different categories. The first tier generates the transaction categories and the second tier segments the customers, each with a vector of the fractions of their transaction categories as parameters. For each tier, we determine the optimal number of clusters based on the Minimum Message Length (MML) criterion. With the premise that the most valuable customer segment is one that is most likely to generate the most valuable transaction category, we rank the customer segments based on their respective joint probabilities with the most valuable transaction category. By doing so, we are able to reveal the relative value of each customer segment.

Keywords: Customer segmentation · EFTPOS transactions · Finite mixture modelling · Minimum message length

1 Introduction

Electronic Funds Transfer at Point Of Sale (EFTPOS) is a payment system that transfers money from a customer account to the retailer account at the conclusion of a transaction. In 2014, over 2.4 billion transactions took place on the EFTPOS system in Australia with approximately 5.6 % growth in volume compared to the previous year [1]. This substantial number of EFTPOS transactions can be mined to gain insights into the characteristics of the retailers and the customers on the system. These insights can then be translated into business strategies by the bank that manages the EFTPOS system. Despite this potential, data mining projects on EFTPOS transaction data is still in its infancy. The lack of research in this area may be due to the difficulty in data collection in the first place. There is only one paper found in the literatures on the use of data mining techniques on EFTPOS data obtained from a small bank in Iran [2]. For our research, we have secured a set of EFTPOS transaction data from one of the major banks in Australia. Previous work that has used the same data set over a different time period to perform segmentation of EFTPOS retailers using the RFM (Recency, Frequency and Monetary values) analysis has been proposed in [3–5].

© Springer International Publishing Switzerland 2015
B. Pfahringer and J. Renz (Eds.): AI 2015, LNAI 9457, pp. 276–284, 2015.
DOI: 10.1007/978-3-319-26350-2_24

In this paper, we take a different angle in the research using the EFTPOS data in that we are particularly interested in finding out the types of customers that have generated the transactions, which may indicate their potential values to the bank. Insights into the types of businesses the customers are spending money on, the methods of payment most commonly used, etc., are of particular interest to the bank. To this end, the way customers are represented is of prime importance. We propose to segment the customers using information captured in a few transaction categories.

2   Customer Segmentation Using Finite Mixture Modelling

Finite Mixture Modelling (FMM) is advocated as a formal statistical modelling approach [6] that incorporates mixtures of parametric distributions to discover the true underlying clusters shaped by significant changes in data density. Each cluster in a mixture model consists of a number of probability distributions, each of which is assumed for a particular attribute of the given dataset. The goal of FMM is to find a parameterisation for the mixture of clusters that is most likely to generate the given data. The usual approach to achieve this is the maximum (log) likelihood estimation via the EM algorithm [7]. An important notion in EM is the "fraction" $r_{c_k|\mathbf{x}_n}$ of the multidimensional datum $\mathbf{x}_n$ assigned to Cluster c_k, $1 \leq k \leq K$ as shown in Eq. (1). Essentially, $r_{c_k|\mathbf{x}_n}$ is equivalent to the posterior probability of picking Cluster c_k to generate data given $\mathbf{x}_n$ observed. This allows $\mathbf{x}_n$ to be partially assigned to all the clusters in a probabilistic manner.

$$
\begin{aligned}
r_{c_k|\mathbf{x}_n} \equiv P\left(c_k|\mathbf{x}_n, a_{c_k}, \theta_{c_k}\right) &= \frac{P\left(c_k\right)p\left(\mathbf{x}_n|\theta_{c_k}\right)}{\sum_{j=1}^{K} P\left(c_j\right)p\left(\mathbf{x}_n|\theta_{c_j}\right)} \\
&= \frac{a_{c_k}p\left(\mathbf{x}_n|\theta_{c_k}\right)}{\sum_{j=1}^{K} a_{c_j}p\left(\mathbf{x}_n|\theta_{c_j}\right)}
\end{aligned}
\tag{1}
$$

3   Proposed Customer Segmentation Framework

Our framework segments customers based on their respective probabilities of generating different categories of transactions. Two issues arise naturally; the derivation of different transaction categories and the definition of a customer's probability of generating each category of transactions. To address the first issue, our framework starts with performing FMM clustering on the EFTPOS transactions with the resultant clusters being interpreted as the transaction categories. As for the second issue, in the context of FMM, the probability of a customer generating a particular transaction category is calculated as the total fraction of the transactions made by the customer assigned to the category. The definition of this fraction is given in Eq. (1).

With each customer being represented as a single vector whose components are the total fractions of the individual's transactions assigned to the corresponding transaction categories, our framework performs customer segmentation based on FMM, followed by the labelling of the segments. Figure 1 shows the scientific workflow of our framework. Four stages comprise this framework. The first stage is the clustering of the processed EFTPOS data, which yields a number of transaction clusters. The immediate stage arises as the construction of the customer representation data by following the aforementioned definition of a customer's probability of generating a transaction category. The third stage is the segmentation of the customer data, which yields unlabelled customer segments. These segments are then interpreted and labelled in the segment-labelling stage, which finalises the entire workflow. The MML criterion, in this case, evaluates every mixture model examined in the first and the third stages to determine the best one in each stage in terms of the trade-off between the model complexity and the goodness of the data fit.

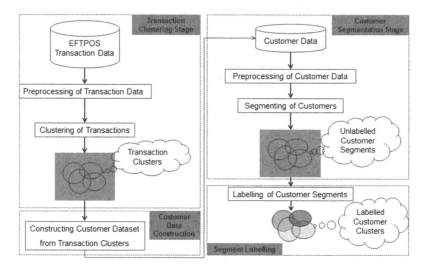

Fig. 1. Scientific workflow of the proposed customer segmentation framework

4   EFTPOS Data Exploration

The EFTPOS transaction data is provided by one of the "Big Four" banks in Australia. The raw data comprises records of transactions made by customers across the nation who have made purchases by means of EFTPOS payment. Banks do not store raw EFTPOS data for a long time on their data warehouse, but for the purpose of this project, the data has been collected on a daily basis with volume varying from 1.5 GB to 2.5 GB. This project uses 46 days of EFTPOS data collected from 1st September 2013 to 17th October 2013.

The EFTPOS data contains 55 variables, some of which are confidential and sensitive in nature. These attributes, e.g. retailer names and ids, credit/debit card numbers, pin numbers, customer's billing address, etc., have been removed, masked or changed.

These operations have been done in a consistent manner to ensure the bank can convert the data back to the original form once the market segmentation modeling has been completed. In this research, we only use three attributes of the EFTPOS data to carry out transaction clustering. They are the attributes "MV", "SIC" and "Acc_Type", which respectively records the monetary value, the type of business and the type of account associated with each transaction. For Attribute "MV", we firstly take the natural logarithm of its data to eliminate the skewedness in the monetary value distribution, which ends up into a bell shape, and then round off the log monetary values to two decimal places. For Attribute "SIC", since its values (i.e. SIC codes) can be organised into a conceptual hierarchy of business sectors by recursive prefix subsumption, we use only the first two digits of these codes in transaction clustering. The round-off operation for Attribute "MV" and the subsumption operation for Attribute "SIC" are both data reduction techniques we employ to accelerate the transaction clustering.

5   Experimental Design

5.1   Design of Transaction Clustering and Customer Segmentation

Both transaction clustering and customer segmentation can be viewed as a process of inferring a mixture of clusters that is able to achieve the optimal trade-off between the model complexity and the goodness of the data fit. The inference process is essentially a heuristic search that explores the space of all possible mixture models to find the optimal one within a specific timeframe. In the model space, any transition from one state to another is triggered by one of the atomic model-changing operations conducted on the current model.

6   Experimental Design

6.1   Design of Transaction Clustering and Customer Segmentation

Both transaction clustering and customer segmentation can be viewed as a process of inferring a mixture of clusters that is able to achieve the optimal trade-off between the model complexity and the goodness of the data fit. The inference process is essentially a heuristic search that explores the space of all possible mixture models to find the optimal one within a specific timeframe. In the model space, any transition from one state to another is triggered by one of the atomic model-changing operations conducted on the current model.

In this experiment, the Wallace-Freeman MML87 criterion [13] is applied to the search heuristics to either accept or reject a state transition. According to [15], we encode the following aspects of a mixture of transaction clusters or a mixture of customer segments into the assertion part of the MML message. The first aspect is the current number of clusters K in the mixture. Its corresponding assertion length A_K is $-\log 2^K$. Secondly, the cluster/segment proportions $\{a_{c_k}|c_k = c_1, \dots, c_K\}$ need to be encoded and the assertion length A_a is calculated as:

$$A_a = -\log \frac{h\left(a_{c_1}, \ldots, a_{c_K}\right)}{\sqrt{\left|F\left(a_{c_1}, \ldots, a_{c_K}\right)\right|}} = \frac{C \prod_{k=1}^{K} a_{c_k}^{\beta_{c_k}-1}}{N^{K-1}/\prod_{k=1}^{K} a_{c_k}} \tag{2}$$

where $|\cdot|$ is the determinant function, $F\left(a_{c_1}, \ldots, a_{c_K}\right)$ is the Fisher information matrix for the multinomial distribution with respect to the parameter vector a, C is the normalisation constant of the Dirichlet distribution as the prior distribution over a, and β_{c_k} is the hyper-parameter governing the proportion of Cluster c_k in some pseudo data population whose size is $\sum_{k=1}^{K} \beta_{c_k}$. In the experiment, we assume a non-informative prior on the cluster proportions by setting each β_{c_k} equal 1, thereby transforming the Dirichlet distribution into a uniform distribution whose density $h(a_{c_1}, \ldots, a_{c_K})$ is just the constant C. Thirdly, for each cluster c_k, we encode its parameters θ_{c_k} for the probability distributions over different attributes. For a nominal attribute, we assume a multinomial distribution over its data. In this case, the assertion length for encoding its parameters is exactly the same form as that for encoding the cluster proportions specified in Eq. (2). For a numeric attribute, in this research, it is either the only numeric attribute (i.e. Attribute "MV") of the EFTPOS data or a transaction category. In the first case, we assume non-informative priors independently over the mean μ_{tc_k} and the log standard deviation $\log \sigma_{tc_k}$ of Attribute "MV" of each transaction cluster tc_k. They are $1/R_{\mu_{tc_k}}$ and $1/(\log b - \log a)$, where $R_{\mu_{tc_k}}$ and $(\log b - \log a)$ are respectively the sizes of the ranges over μ_{tc_k} and $\log \sigma_{tc_k}$. Then, the prior density over μ_{tc_k} and σ_{tc_k} is just:

$$h\left(\mu_{tc_k}, \sigma_{tc_k}\right) = h\left(\mu_{tc_k}\right) h\left(\sigma_{tc_k}\right) = \frac{1}{R_{\mu_{tc_k}}}\left(\frac{1}{R_{\sigma_{tc_k}} \sigma_{tc_k}}\right)$$

In this experiment, we set $R_{\mu_{tc_k}}$ and $R_{\sigma_{tc_k}}$ equal 1 and $(\log 2 - \log 0.1)$ respectively. Thus, the assertion length for encoding μ_{tc_k} and σ_{tc_k} is:

$$\begin{aligned} A_{\theta_{tc_k}} &= -\log \frac{h\left(\mu_{tc_k}, \sigma_{tc_k}\right)}{\sqrt{\left|F\left(\mu_{tc_k}, \sigma_{tc_k}\right)\right|}} \\ &= \log R_{\mu_{tc_k}} R_{\sigma_{tc_k}} + \frac{1}{2}\log a_{tc_k} - \log \sigma_{tc_k} + \log N + \frac{1}{2}\log 2 \end{aligned}$$

In the second case, we assume the following priors respectively over the mean vector μ_{cs_k} and the precision matrix $\Sigma_{cs_k}^{-1}$ of each customer segment cs_k:

$$\mu_{cs_k} \sim \text{Gaussian}(\mu_0, \Sigma_{cs_k}/m_1), \quad \Sigma_{cs_k}^{-1} \sim \text{Wishart}(\Phi, m_0)$$

where $\mathbf{Wishart}(\cdot)$ is the Wishart distribution. The hyper-parameters μ_0, Φ, m_0 and m_1 are fixed in the experiment with $\mu_0 = \vec{0}$, $\Phi = \mathbb{I}$ (the identity matrix), $m_0 = d$ and $m_1 = 1$.

$$\left| F\left(\mu_{cs_k} \right) \right| = \left(Na_{cs_k} \right)^d \left| \Sigma_{cs_k}^{-1} \right| \qquad \left| F\left(\Sigma_{cs_k}^{-1} \right) \right|$$

$$= \left(Na_{cs_k} \right)^{d(d+1)/2} \left| \Sigma_{cs_k}^{-1} \right|^{-(d+1)}$$

Therefore, the assertion length for encoding μ_{cs_k} and $\Sigma_{cs_k}^{-1}$ is:

$$A_{\theta_{cs_k}} = -\log \frac{h\left(\mu_{cs_k} | \Sigma_{cs_k}^{-1} \right) h\left(\Sigma_{cs_k}^{-1} \right)}{\sqrt{\left| F\left(\mu_{cs_k} \right) \right| \left| \Sigma_{cs_k}^{-1} \right|}}$$

The total message length for encoding a mixture of clusters or segments is:

$$ML(\Theta, X) = A_K + A_a + \sum_{k=1}^{K} A_{\theta_{tc_k or cs_k}} + \sum_{n=1}^{N} \left(A_{x_n} + L_{x_n} \right) + c_d - \log K! \tag{3}$$

By setting the partial derivative of Eq. (3) to zero with respect to each parameter of each cluster, we are able to derive the MML parameter estimates that minimise the message length for encoding a mixture model.

We apply the simulated annealing (SA) algorithm [18] to the search where the candidate model is always accepted to be the next state to visit if its MML score is less than that of the current model, and is accepted with certain probability if its MML score is greater or equal to it. The initial temperature of the SA system is set to be 10 with the cooling rate being 0.999.

Apart from the model inference, the customer data construction, which bridges the transaction clustering and the customer segmentation, is majorly important to the framework. Specifically, after the optimal mixture of transaction clusters is obtained, we treat each cluster as a unique category of transactions that acts as an attribute of the customers in customer segmentation. Thus, each customer is represented by a single vector of the quantities of transactions he/she has made from the respective transaction categories.

6.2 Experimental Result Analysis

The result analysis answers the question of which customer segments are valuable to the bank. Here, we start with identifying the tendency for each segment of customers to generate each category of transactions. Here, the tendency for the ith segment of customers to generate the kth category of transactions is measured by the joint probability of Segment cs_i and Cluster tc_k, which is $P\left(TC = tc_k, CS = cs_i \right)$. This probability is the product of the conditional probability $P\left(TC = tc_k | CS = cs_i \right)$ and the prior probability $P\left(CS = cs_i \right)$.

7   Results and Discussions

The transaction clustering stage of our framework converges to 22 transaction clusters and the customer segmentation stage converges to 9 customer segments. The proportion of each of these segments is respectively: 0.383, 0.212, 0.126, 0.098, 0.080, 0.043, 0.026, 0.014, 0.011. The sum of these proportions is smaller than 1 ($\approx$ 0.993). This is due to the presence of some outliers in the customer data that fail to be fit by the mixture. In this research, we have decided to ignore these outliers, but they warrant further investigation in our future work.

Table 1 shows the rank of the 9 segments in descending order with respect to their joint probabilities by percentage with the most valuable transaction category tc_{22}. It further shows that Segment cs_1 manages to achieve a joint probability (2.31 %) higher than those of the other segments. This means it is the most valuable customer segment in our analysis.

Table 1. Rank of the 9 customer segments in descending order of joint probability by percentage with Cluster tc_{22}

%	tc_{21}	tc_7	tc_2	tc_{16}	tc_{11}	tc_{12}	tc_8	tc_{15}	tc_1	tc_9	tc_{17}
cs_1	0.00	0.00	0.00	0.01	0.00	0.00	0.51	1.93	0.00	0.36	1.00
cs_5	0.00	0.00	0.48	0.03	0.00	0.00	0.88	1.95	0.00	0.42	0.75
cs_7	0.00	0.00	0.48	0.04	0.00	0.00	0.14	0.65	0.00	0.08	0.20
cs_6	0.00	0.00	0.00	0.02	00.0	0.00	0.37	1.29	1.71	1.18	0.79
cs_3	0.00	0.00	0.00	0.01	0.00	0.00	0.30	1.35	1.25	0.84	0.75
cs_9	0.00	0.00	0.00	0.03	0.00	0.00	0.25	0.49	0.38	0.30	0.21
cs_8	0.00	0.00	0.00	0.02	0.00	0.00	0.01	0.05	0.23	0.10	0.06
cs_4	0.00	0.00	0.00	0.01	0.00	0.00	0.00	0.00	2.56	1.04	0.49
cs_2	0.00	0.00	0.00	0.00	0.00	0.00	0.00	0.00	1.15	0.48	0.25

%	tc_3	tc_{18}	tc_{13}	tc_6	tc_{14}	tc_4	tc_{20}	tc_{19}	tc_{10}	tc_5	tc_{22}
cs_1	7.09	0.00	0.00	0.00	2.00	2.21	0.00	0.00	1.58	0.00	2.31
cs_5	5.41	0.00	0.58	0.00	1.31	1.51	0.00	0.00	1.13	0.00	1.75
cs_7	1.67	0.00	0.08	0.00	0.49	0.56	0.27	0.04	0.44	0.00	0.64
cs_6	2.65	4.34	0.29	0.43	1.06	0.55	0.24	0.00	1.09	0.51	0.54
cs_3	3.50	3.42	0.23	0.24	1.11	0.63	0.00	0.00	1.05	0.34	0.42
cs_9	0.75	0.77	0.10	0.11	0.27	0.24	0.07	0.02	0.26	0.11	0.33
cs_8	0.14	0.56	0.02	0.04	0.11	0.03	0.01	0.04	0.14	0.05	0.02
cs_4	0.08	6.12	0.22	0.57	0.77	0.00	0.00	0.00	0.99	0.65	0.00
cs_2	0.03	3.60	0.24	0.17	0.41	0.00	0.09	0.00	0.55	0.24	0.00

Table 2 shows the rank of the 9 customer segments in descending order of their total joint probabilities with the transaction categories with the highest level of value. It further shows that Segment cs_1 manages to achieve the highest total joint probability.

This displays the overall inclination of this segment of customers towards generating more valuable categories of transactions than the other segments. Moreover, we find that some of the customer segments, including Segments cs_7, cs_8 and cs_9, are more keen on making on average the most expensive category of transactions, as their ranks are higher in Table 1 than in Table 2, while others, including Segments cs_2, cs_3, cs_4, and cs_6, show greater general interests in their purchases.

Table 2. Rank of the 9 segments in descending order of total joint probability by percentage over the highest level of transaction category value MV_5

%	MV_1	MV_3	MV_4	MV_5
cs_1	0.00	0.00	10.89	8.10
cs_5	0.00	0.00	9.41	6.29
cs_6	0.00	0.00	12.33	4.71
cs_3	0.00	0.00	11.41	4.01
cs_4	0.00	0.00	0.00	3.21
cs_7	0.00	0.00	2.75	0.00
cs_2	0.00	0.00	0.00	1.71
cs_9	0.00	0.00	0.00	1.50
cs_8	0.00	0.00	0.00	0.00

8   Conclusion and Future Work

This project proposes customer segmentation using FMM based on MML criterion using a relatively long term EFTPOS data. The first contribution of this work is the development of a novel framework that segments EFTPOS customers by leveraging fine-grained transactional information. This framework consists of two unsupervised clustering tiers, the transaction clustering and the customer segmentation. The results of the first tier, as transaction categories, are used to construct the representation vector for each customer, whose components accommodate total fractions of transactions made by the customer for the corresponding transaction categories. In the experiment, both tiers are implemented as inference of the optimal mixture model evaluated by the MML criterion.

The second contribution is our derivation of the value of each customer segment to the bank. Specifically, we rank the customer segments with respect to their joint probabilities with the most valuable transaction category, which is the one with the highest log monetary value mean.

The future directions of this research involve (1) handling outliers in the customer data and (2) comparing our framework with some baseline methods, such as the RFM analysis, to see whether it is better in terms of the clustering performance.

References

1. EFTPOSAustralia. http://www.eftposaustralia.com.au/wp-content/uploads/2015/01/eftpos-2014-annual-report.pdf
2. Bizhani, M., Tarokh, M.J.: Behavioural rules of bank's point-of-sale for segments description and scoring prediction. Int. J. Ind. Eng. Comput. **2**(2), 337–350 (2011)
3. Singh, A., Rumantir, G., South, A.: Market segmentation of EFTPOS retailers. In: Nayak, R., Li, X., Liu, L., Ong, K.-L., Zhao, Y., Kennedy, P. (eds.) Proceedings of the Twelfth Australasian Data Mining Conference, Brisbane, Conferences in Research and Practice in Information Technology, vol. 158 (2014)
4. Singh, A., Rumantir, G., South, A., Bethwaite, B.: Clustering experiments on big transactional data for market segmentation. In: Proceedings of the Third ASE International Conference on Big Data Science and Computing, Beijing. ACM (2014). http://dx.doi.org/10.1145/2640087.2644161. 978-1-4503-2891-3/14/08
5. Singh, A., Rumantir, G., South, A.: Two-tiered clustering classification experiments for market segmentation of EFTPOS retailers. Australas. J. Inf. Syst. Spec. Issue Bus. Analytics (accepted, 2015)
6. McLachlan, G., Peel, D.: Finite Mixture Models. Wiley, New York (2004)
7. Neal, R.M., Hinton, G.E.: A view of the EM algorithm that justifies incremental, sparse, and other variants. In: Jordan, M.I. (ed.) Learning in Graphical Models, vol. 89, pp. 355–368. Springer, Netherlands (1998)
8. Akaike, H.: A new look at the statistical model identification. IEEE Trans. Autom. Control, **19**(6), 716–723 (1974)
9. Schwarz, G.: Estimating the dimension of a model. Ann. Stat. **6**(2), 461–464 (1978)
10. Barron, A., Rissanen, J., Yu, B.: The minimum description length principle in coding and modelling. IEEE Trans. Inf. Theory **44**(6), 2743–2760 (1998)
11. Wallace, C.S., Boulton, D.M.: An information measure for classification. Comput. J. **11**(2), 185–194 (1968)
12. Wallace, C.S.: Statistical and inductive inference by minimum message length. Springer, New York (2005)
13. Wallace, C.S., Freeman, P.: Estimation and inference by compact coding. J. Royal Stat. Soc. Ser. B (Methodological) **49**, 240–265 (1987)
14. Conway, J., Sloane, N.: On the Voronoi regions of certain lattices. SIAM J. Algebraic Discrete Methods **5**(3), 294–305 (1984)
15. Dowe, D.L.: MML, hybrid Bayesian network graphical models, statistical consistency, invariance and uniqueness, pp. 901–982. Citeseer (2011)
16. Dempster, A.P., Laird, N.M., Rubin, D.B.: Maximum likelihood from incomplete data via the EM algorithm. J. Roy. Stat. Soc. Ser. B (Methodological) **39**, 1–38 (1977)
17. Thiesson, B., Meek, C., Heckerman, D.: Accelerating EM for large databases. Mach. Learn. **45**(3), 279–299 (2001)
18. Bertsimas, D., Tsitsiklis, J.: Simulated annealing. Stat. Sci. **8**(1), 10–15 (1993)

A Dual Network for Transfer Learning with Spike Train Data

Keith Johnson[✉] and Wei Liu

School of Computer Science and Software Engineering, The University of Western
Australia, Crawley, Australia
keith@csse.uwa.edu.au, wei.liu@uwa.edu.au
http://www.csse.uwa.edu.au

Abstract. A massive amount of data is being produced in a continual
stream, in real time, which renders traditional batch processing based
data mining and neural network techniques as incapable. In this paper we
focus on transfer learning from Spike Train Data, for which traditional
techniques often require tasks to be distinctively identified during the
training phase. We propose a novel dual network model that demonstrates
transfer learning from spike train data without explicit task specification.
An implementation of the proposed approach was tested experimentally
to evaluate its ability to use previously learned knowledge to improve the
learning of new tasks.

1   Introduction

Humans can intelligently apply previously learned knowledge to the learning of
new tasks. However, exploiting such knowledge in machine learning is not trivial
and remains an open problem. Motivated by this pursuit, the field of "transfer
learning" considers the scenario of learning multiple tasks and focuses on the
problem of extracting knowledge from one or more source tasks (i.e. previously
learned tasks) to improve the learning of a target task (i.e. a new task) [5].

There are a wide variety of approaches to knowledge transfer in neural networks and these are often categorised as either representational or functional
transfer [6,11]. In representational transfer, the target network is initialised based
on learned representations (usually weights) from the source tasks. This biases
the search trajectory in the weight space by influencing its starting point. Functional transfer on the other hand does not modify the initial conditions of the
target network, it instead makes use of implicit pressures via a number of mechanisms. These include the modification of training parameters such as the learning rate or the error gradient, provision of additional training examples or the
simultaneous training of several tasks whilst sharing a common representation.

One type of representational transfer can occur in growing or constructive
networks by utilising the output of previously learned feature detectors as input
to new feature detectors (without allowing previously learned feature detectors
to be catastrophically interfered). It is this type of transfer that we will focus on
in this paper. Some examples of networks that utilise this type of construction

© Springer International Publishing Switzerland 2015
B. Pfahringer and J. Renz (Eds.): AI 2015, LNAI 9457, pp. 285–297, 2015.
DOI: 10.1007/978-3-319-26350-2_25

include Cascade Correlation (CC) [1] and incremental Feature Dependency Discovery (iFDD) [2], though they are not designed specifically for transfer learning. Knowledge Based Cascade Correlation (KBCC) [10], on the other hand, is an extension of CC that is designed specifically for transfer learning.

In general, machine learning approaches designed for transfer learning require a pre-defined distinction of the tasks, either by having the inputs separated by task or by informing the system of which task the current input relates to (because for example, some methods use a different network to learn each task). Furthermore, existing approaches applied to transfer learning have not been designed to learn directly from or represent continuous time information. The type of continuous time data we are interested in is the *Spike Train Data*, which is often generated by recording neuronal activity in the brain. All variables are in an "off" state, except when they are instantaneously activated (i.e. they spike). However we may also translate other data types into this format, including many real world data such as Internet traffic, financial data and web server logs. Few systems, such as Spiking Neural Networks (SNNs) [4], attempt to learn directly from this type of continuous time data.

To address these two issues we propose a novel dual network connectionist approach. The approach separates state information and learned knowledge into two distinct connectionist networks, named the sate model and environment model respectively. The dedicated state model, a novel way of representing state information, allows a more expressive representation of state instances including the representation of instances that occur over any continuous time duration. The environment model stores learned knowledge via its structure and parameters, as per the traditional single model, however its structure is dynamic and it is capable of representing continuous time features and probabilistic associations. It is a growing model, in which new feature detectors are added iteratively via incremental learning. When learning new feature detectors (e.g. for a new task), inputs may be taken from the output of any existing feature (e.g. that was learned in a previous task). When used appropriately, this can result in transfer learning, by reducing the time taken to learn the new task. This method of transfer learning does not require the distinction of tasks by the experimenter or the system to be informed of which task the current input relates to.

An implementation of the proposed approach was tested experimentally to evaluate its basic ability to learn suitable representations directly from continuous time spike train and whether it exhibited transfer learning. The results show that the system as a whole is capable of forming accurate model representations in a small test environment. We also demonstrate that the proposed system is capable of unsupervised transfer learning without the need for task distinction.

The paper is structured as follows. Section 2 reviews the relevant literature on the various techniques dealing with transfer learning. Section 3 details the proposed connectionist state and environment models. Section 4 uses an example operation to illustrate how pattern search and construction work in the environment model and recognition for the construction of the state model. Section 5 provides experimental results on transfer learning from spike train data. The paper concludes in Sect. 6 with an outlook to future work.

2   Existing Approaches to Transfer Learning

As mentioned in Sect. 1, transfer learning can be broadly categorised into two camps, representational and functional transfer. The type of learning our dual network model exhibits is closer to representational than functional. Hereby we give a more detailed review on representational transfer learning.

The simplest method of representational transfer is to copy the network parameters learned from a source task to the initial state of network learning the target task. While in some cases this may reduce the training time of the target task it may also increase it [7]. The benefit of the transfer depends on how related the tasks are and without an attempt to measure this, this methods usefulness is limited.

One approach, often known as "task decomposition" aims to improve learning on complex problems by breaking a large task into a set of smaller tasks. Tasks are typically decomposed by limiting each sub-task to learning only a subset of the outputs. The smaller tasks are first learned by individual networks and then combined into one large network. The combined network is then refined by training on the whole task. In this form of transfer, the parameters learned in the smaller tasks are directly transferred to the combined network, as initial weight values. This approach has been used successfully by Waibel et al. [12] and Pratt et al. [9] in speech recognition tasks.

Cascade Correlation [1], when trained in a sequential regime is another form of representational transfer. After training the network on the first task, the input to hidden layer weights are frozen, leaving the nodes as permanent feature detectors in the network. For each subsequent task, a set of hidden nodes are added to the network and trained on the new data set. The feature detectors learned in all previous tasks are effectively transferred to the network used to learn the current task. Weights to the output layer are modified in each new task and as a result, catastrophic interference can occur. However, because of the persistence of the feature detectors added for each task, the network can rapidly recover when retrained on previous tasks. Knowledge Based Cascade Correlation (KBCC) [10] extends Cascade Correlation to improve transfer learning by employing separate sub-networks for each task. Output from sub-networks learned in previous tasks provide input to new hidden units trained during subsequent tasks.

In the representational transfer methods described so far, the parameters from previous learning tasks are transferred to subsequent tasks with no intermediate modification. In some cases, this may either hinder the subsequent learning tasks (negative transfer), when compared to randomly initialised parameters. Rather than copy parameters literally, the Discriminality Based Transfer (DBT) [7,8] algorithm first analyses the discriminality of each hyperplane (each node in the hidden layer) in the source network. The analysis measures the information theoretic value, based on Shannon's information theory, in order to determine the contribution of the node in separating the target tasks training data. If the hyperplane is found to be useful in the new task, then the node's weights are scaled up in order to help preserve them during the next learning phase. Otherwise, they are reduced or initialised to random weights. This algorithm is applied to initialisation of weights for the target task and once complete, training takes place using standard back-propagation. Pratt's results demonstrate the improvement of this method over literal transfer or random initial weights in several real world learning problems.

A common requirement of the current transfer learning networks is to pre-define a set of distinctive tasks or sub-tasks, which sometimes can be counter-intuitive if not impossible. In the rest of the paper, we demonstrate a dual network model for transfer learning without explicit task identification.

3   State-Based Dual Network Connectionist Model

Most, if not all, connectionist approaches combine the learned knowledge about an environment and the current interpreted state within a single model. As shown on the left-hand-side of Fig. 1b, in traditional ANNs the knowledge about an environment is stored via a combination of nodes, full connection between layers, connection weights and activation rules. Information about the current state is stored in the same model, as a set of activation values, one for each node. In the approach proposed in this paper, environment knowledge and state information are represented separately by two different connectionist models — the environment and the state model. The system, illustrated in Fig. 1a, shows that the two connectionist models, environment and state, are connected via data associations. Pattern search algorithms construct the environment model. The recognition process constructs the state model.

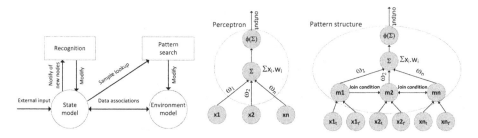

Fig. 1. (a) Diagram of the proposed system. (b) (Left) A perceptron in traditional ANN; (Right) A pattern perceptron in the proposed environment model.

3.1   Environment Model

Similar to ANNs, the environment model is a connectionist model that stores knowledge in a set of nodes, connections, connection parameters and activation rules. However, the simple perceptron is now augmented by a graph structure that represents patterns between inputs. The perceptron with *pattern structures* as depicted on the right-hand-side of Fig. 1b forms the basic component of our environment model.

One of the primary motivations behind the use of pattern structures, rather than the traditional layers of fully connected nodes to facilitate unsupervised transfer learning without the need to notify the system of changes in task. The transfer learning can occur when learning a new feature detector (i.e. pattern structure). The environment model is not layered, instead each feature can potentially take input from a combination of the outputs of any other nodes. Thus features learned in previous tasks may be used as building blocks for constructing new features when learning a new task. A key aspect of the pattern structure is that its parameters are frozen after construction, ensuring that knowledge learned in previous tasks stays available (i.e. unchanged) when learning future tasks. If the increased representation, made available from previous learning, can appropriately be re-used and leveraged to reduce the learning time of the new task, then transfer learning will occur. An illustrative hypothetical example of this is given in Sect. 5.1 as well as results for transfer learning in a synthetic environment.

General to all pattern structures (ps_{env})

$$ps_{env} = \langle v_{env}^{root}, V_{env}^{mem}, E_{env} \rangle$$

is that they have one unique root node (v_{env}^{root}), a set of member nodes (V_{env}^{mem}) and a set of environment model connections (E_{env}). The root node v_{env}^{root} can be thought of as a perceptron in the traditional sense of ANN, referring to the aggregation and activation function. Each member node v_{env}^{mem} connects to the v_{env}^{root} with two edges, one in each direction. v_{env}^{mem} also connect to one *original* node through two edges, one in each direction. All edges store dynamically updated conditional probabilities, $p_{s \to d} = P(d|s) = \frac{n_d}{n_s}$, where n_s is the number of times v_{env}^s has been activated and n_d is the number of times v_{env}^s has been activated and connected to an activated v_{env}^d. In addition, the member to root edges also store weights ω and root nodes store an activation threshold ε, in the same way as traditional ANNs.

The original node is not a special type of node, it is a relative term that refers to a node that provides input to a pattern structure. A member node represents a specific subset of the original node's instances, those that are only activated when both the original node (the input) and the root node (which represents the pattern as a whole) are activated.

The V_{env}^{mem} may connect to each other via any number of member to member connections (e_{env}^{m2m}), how they are connected is determined by the pattern search algorithms. Each e_{env}^{m2m} defines a *join conditions* that determines which state instances of the original nodes are allowed to join together to form a pattern instance. In this paper, we focus primarily on one type of pattern structure — the timespan pattern structure (ps_{env}^{ts}) generated by the timespan search algorithm reported in Sect. 4.1, in which the e_{env}^{m2m} join condition is a continuous time delay interval. Other pattern generation algorithms will be reported in future work.

3.2 State Model

The state model is a connectionist representation of state instances over continuous time, containing not only instances of raw input but also instances of nodes activated by recognition (see Sect. 4.2). The main benefits of the proposed separate state model state model is its increased expressiveness (including continuous time durations, multiple overlapping instances of the same node and binding connections). The state model is a connectionist model

$$G_{state} = \langle V_{state}, E_{state} \rangle$$

where each state node ($v_{state} \in V_{state}$) is an instance of a v_{env} and are connected together by directional state connections ($e_{state} \in E_{state}$).

All state nodes

$$v_{state} = \langle ref, ps_{state}, [t_{start}, t_{end}] \rangle$$

maintain a reference (ref) to their corresponding v_{env} and belong to one pattern structure instance (ps_{state}) — a unique identifier for each instance of an ps_{env}. The state node also stores an observation time range, with a start (t_{start}) and end time (t_{end}).

State model connections

$$e_{state}^{s \rightarrow d} = \langle ref, v_{state}^{s}, v_{state}^{d} \rangle$$

connect a source node (v_{state}^{s}) to a destination node (v_{state}^{d}). There are two types of state model connections — *state time connections* (e_{state}^{time}) and *state instance connections* (e_{state}^{ins}). e_{state}^{time} represent the observed time difference (Δt) between two v_{state}. e_{state}^{ins} are instances of environment model connections (e_{env}) and maintain a reference (ref) to their corresponding e_{env}. These connections facilitate the binding of state nodes based on recognised instances of relationships modelled by environment model connections.

4   Operation of the Dual Network Model

In order to illustrate how the system represents continuous time information and learned features within the two models, while introducing the *timespan search algorithm* (for constructing the environment model) and the *recognition process* (for constructing the state model), consider the following example. Suppose we have a simple environment with 3 discrete state (binary) sensors. The input from these can be translated into spike train data (as shown in Fig. 2) by recording each instantaneous state change event as a spike. For binary state data, we can assign two input nodes per sensor, one to record the change from off to on (on sensor) and the other from on to off (off sensor).

Figure 3 illustrates and describes, step by step, an example of how the proposed system operates for the first 2 rounds based on the example environment described above. For this example and the experiments presented in this paper, a sequential schedule is used to run the various operations (in theory they can be run in parallel). The observations from the environment and operations run by the proposed system are divided into rounds. In each round, the sequential schedule used for all experiments, is (1) Receive a set of input spikes from the environment (2) Perform recognition (3) Run one instance of the timespan search algorithm (4) Cull state model. The timespan search algorithm and recognition process are given below in Sects. 4.1 and 4.2, respectively.

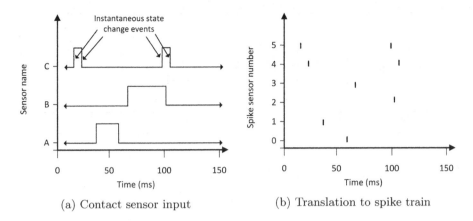

(a) Contact sensor input   (b) Translation to spike train

Fig. 2. (a) Example input from 3 continuous time, binary state sensors A, B and C.
(b) Translation of contact sensor input to spike train data. For each of the sensors
A, B and C, two spike sensors are used. One that generates a spike for instantaneous
changes from off to on (1, 3, 5 for A, B, C respectively) and the other from on to off
(0, 2, 4 for A, B, C respectively).

4.1   Environment Model Construction through Timespan Search

The timespan pattern search algorithm performs unsupervised learning via cluster analysis. Together with the environment model, it facilitates transfer learning.
The algorithm searches for patterns of consistent timing between two environment model nodes.

The timespan search algorithm can be summarised as follows. The algorithm
collects state node samples from the state model and performs a cluster analysis,
using the Self-Organising Map (SOM) algorithm [3], over sets of time difference
values between the sample nodes. The resulting clusters are converted into time
intervals to be used as the join condition (i.e. if both nodes are activated within
the relative time interval, the join condition is satisfied) in candidate timespan
structures. A multi-objective score is assigned to each and the highest scoring
candidate is constructed in the environment model. An example generation of
timespan candidates using a SOM is illustrated in Fig. 4.

Referring to the running example shown in Fig. 2, if we assume that the
same spike train is observed repeatedly, some timespan pattern structures that
may result from a timespan search include S5 to S4 (i.e the two original nodes
of the structure) with time range condition [7.0, 7.2], S1 to S0 with time range
[19.7, 20.1] or S4 to S0 with time range [31.1, 33.5].

Many of the existing clustering or density estimation techniques could be
employed for this cluster analysis. For the purposes of this paper, it is not considered particularly important exactly which technique is used to generate the
clusters. The k-means algorithm, Jenks natural breaks and the Self-Organising
Map (SOM) [3] are some common clustering algorithms used, where k-means
and Jenks natural breaks are particularly popular choices for single dimension

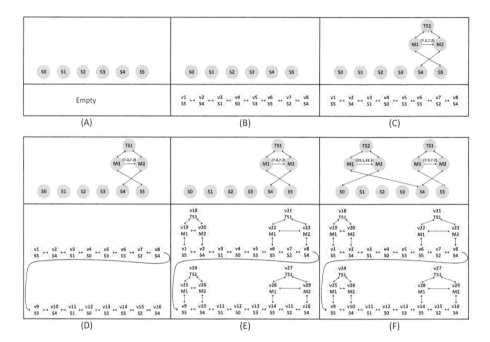

Fig. 3. An example operation of the system for the first 2 rounds using the environment shown in Fig. 2. The environment model is shown in the top half of each box with grey filled nodes, the state model in the bottom half with white filled nodes (the label below each state node v_i denotes the environment node that its an instance of). (A) Initialise the models. The environment model has 6 sensor nodes, the state model is empty. (B) Observe one round of data, creating 8 state nodes and 7 state connections. (C) Run one instance of the timespan search algorithm, resulting in the construction of TS1. (D) Observe second round of data, creating 8 state nodes and 8 state connections. (E) Perform recognition, resulting in the recognition of 4 instances of TS1. (F) Run one instance of the timespan search algorithm, resulting in the creation of TS2

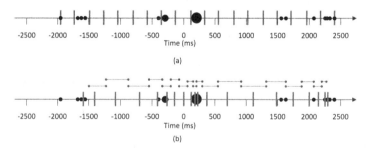

Fig. 4. Two time lines showing the results of SOM clustering to produce time intervals. (a) A set of sample time difference values between two environment nodes, where the larger points depict more data values at the same point on the time line. The initialised SOM centres in red, evenly distributed between the minimum and maximum points. (b) SOM centres after learning and candidate timespan intervals shown in blue (Color figure online).

data. However, we have chosen to use SOM for clustering in the timespan search algorithm because of its neighbourhood function (which draws non-winning cluster centres towards winning clusters), resulting in more clusters occupying the more dense areas of the sample space.

4.2   State Model Evolution through Recognition

The state model (G_{state}) is a growing dynamic model that is initialised as empty (i.e. $V_{state} = \{\}$ and $E_{state} = \{\}$). When a node in the environment model (v_{env}) is activated, a state instance of the node (v_{state}) is created in the state model and connected, via state model connections (e_{state}), to other state nodes (v_{state}). A state sensor node (v_{state}^{sen}) is created in the state model when a spike input is received (externally from the environment) by its corresponding v_{env}. v_{state}^{sen} are connected together in a chain via state time connections (e_{state}^{time}), in order of observation time. If multiple are observed simultaneously, the connection order is chosen randomly. The relative time difference between the connected v_{state}^{sen} is stored in each e_{state}^{time}. An example of this can be seen in Fig. 3B. Note that inactivity is not modelled by the state model thus all nodes in the state model represent an active instance of a v_{env}.

Recognition of a pattern structure ps_{env} (and the resulting creation of a pattern structure instance ps_{state} in the state model, if found to be active) occurs when one or more original node instances satisfy the join conditions defined by the ps_{env}. When this occurs, the activation rule for the root node (v_{env}^{root}) of the ps_{env} is evaluated (see the pattern perceptron in Fig. 1b). If the sum of weighted inputs (an input value of 1 if an instance of the original node is joined, otherwise 0) is greater or equal to activation threshold $\varepsilon_{activation}$, then the v_{env}^{root} is activated, otherwise it is not. Upon activation, an instance of v_{root} is created in the state model along with a member node instance for each original node instance that was joined. The new instances are connected via state instance connections (e_{state}^{ins}) that reflect the connectivity in the ps_{env} (which act as binding connections). Figure 3D and E demonstrate a clear transition of the state model before and after the recognition process.

5   Experimentation and Results

As with any new architecture, it is typical to do the proof-of-concept with toy examples, which not only provides clear abstraction but also allows for incremental and controlled testings. Having said that, the system is capable of handling more complex data sets, for example smart home sensor data or raw pixel input from classic arcade games, which we will report in a later work. In this paper, we will focus on demonstrating transfer learning in a small test environment.

5.1   Transfer Learning in a Moving Shape Environment

Transfer learning is demonstrated by learning a sequence of tasks. If the tasks share something in common that is transferrable, then subsequent tasks should be able to be learned faster, by taking advantage of knowledge learned in the previous tasks. The experiments carried out in this section use the moving shapes environment, where each shape is considered a different task. The system is given no information about where one task ends and the next begins.

Each task is presented to the proposed system until a representation score of 1, the maximum, is reached. To gain the maximum score, the state representations of a pattern structure must exactly match the spike inputs generated by an environment cause (e.g. a moving shape). The formula cannot be given in this paper, due to space limitations, however it is available on demand. At this point the next task is presented and so forth until all tasks have been learned. The number of learning rounds required for each task is recorded. The effectiveness of transfer learning can then be assessed by examining the average time taken to learn each new task sequentially. The expectation is that, if transfer learning is occurring then each new task should require less, on average, than the previous.

In the first transfer learning experiment, 6 shapes are generated randomly, by randomly filling 5 squares within a 4×4 pixel grid. The learning of each shape is considered one task. The shapes all move in the same direction, from left to right, across a field of view that is divided into a 4×4 sensor grid (total of 32 sensors because each grid square has an on and off sensor). The results, averaged over 30 experiment instances, are shown in Fig. 5a and indicate that transfer learning occurs. The average time for learning the first shape, starting from an environment model with no prior knowledge (i.e. a "blank" environment model with sensor nodes only) is approximately 32 learning rounds. The number of learning rounds required for each subsequent shape decreases, requiring an average of approximately 12 rounds for the final task.

Due to the fact that the shapes are randomly generated according to a grid formation and move in the same direction and at the same speed, it is likely that many of the smaller features that form the building blocks for the shape representations, might be common across several shapes, facilitating the re-use of representations that result in reduced learning time. In order to test how reliant the transfer learning is on these conditions, further experiments are carried out

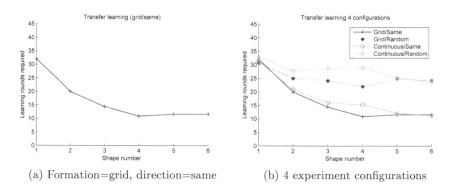

(a) Formation=grid, direction=same   (b) 4 experiment configurations

Fig. 5. Transfer learning with 4 different experiment configurations.

with shapes moving in random directions (either right, left, up or down) and where the position of the 5 randomly generated shape parts are not constrained to a grid formation (i.e. continuous formation), by assigning a random continuous x and y coordinate for each part. Three further experiments are carried out with the following configurations; (1) Shapes generated with grid formation, moving in a random direction (2) Shapes generated with continuous formation, moving in the same direction (3) Shapes generated with continuous formation, moving in a random direction.

The results, averaged over 30 experiment instances, are shown in Fig. 5b and appear to show that transfer learning occurs in all 4 experiment configurations. However, the effect is more pronounced in the two configurations for which the shapes are moving in the same direction (Grid/Same and Continuous/Same).

5.2   Example to Demonstrate How Transfer Learning Occurs

Suppose we have an environment configuration with two shapes; a vertical line with filled 2×2 grid square coordinates of (0,0),(0,1) and a backslash with (0,0),(1,1), both moving across a 2×2 visual field. If the system first learns the vertical line, followed by the backslash, we would expect that some of the pattern structures learned for the first shape would be re-usable for the second. For example, if we isolate each row of the visual field, the observation of each of the shapes, from the perspective of one row only, will look the same. Figure 6a illustrates an example of a possible environment model after learning the vertical line. The timespan pattern structure, named TS5, represents the movement observed in the first row of the field of view only and TS6 in the second row

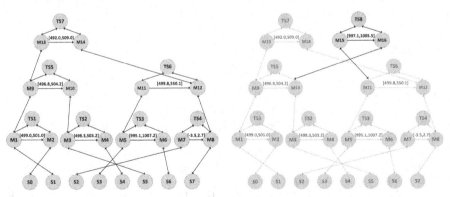

(a) Example environment model after learning the vertical line

(b) Example environment model after learning the backslash shape

Fig. 6. Transfer learning example. Shows the environment model after learning the movement of two shapes. Each timespan structure (TS1, ..., TS8) is hypothetically discovered and constructed by the timespan search algorithm. (a) The environment model after learning the vertical line, represented by TS7. (b) The environment model after learning the backslash shape, represented by TS8.

(note that this is just one possible way that the timespan search algorithm might form representations of the shape). TS7, a temporal combination of TS5 and TS6, provides a representation of the vertical line's movement as a whole. Figure 6b shows that by re-using the existing pattern structures, the learning of the backslash shape can be completed with the construction of only one timespan pattern structure, TS8, which is a different temporal combination of TS5 and TS6.

6   Conclusion

In this paper we introduced a novel state-based connectionist model, which is capable of transfer learning, while learning directly from spike train data and forming continuous time representations as well as probabilistic associations. The key idea for learning from and representing continuous time information is the use of a dedicated state model to represent more expressive state information. Transfer learning is facilitated by the re-use of feature detectors learned in previous tasks, when faced with new tasks.

Experiments were carried out to demonstrate the model's basic learning ability from spike train data as well as its capability in transfer learning. The model is also capable of knowledge persistence and learning hierarchical representations, which due to space limit, wasn't reported here. The detailed algorithms for pattern search, construction, recognition and calculation of representation score are available on request. Future work will include pruning of the environment model (so that continual learning can be more sustainable) and new pattern search types such as learning by structural analogy. Furthermore the various operations (e.g. recognition and pattern search) will be implemented as concurrent operations so that the system can run online in real-time.

References

1. Fahlman, S.E., Lebiere, C.: The cascade-correlation learning architecture. In: Touretzky, D.S. (ed.) Advances in Neural Information Processing Systems, vol. 2, pp. 524–532, Denver 1989. Morgan Kaufmann, San Mateo (1990)
2. Geramifard, A., Doshi, F., Redding, J., Roy, N., How, J.: Online discovery of feature dependencies. In: Proceedings of the 28th International Conference on Machine Learning (ICML 2011), pp. 881–888 (2011)
3. Kohonen, T.: Self-organized formation of topologically correct feature maps. Biol. Cybern. **43**(1), 59–69 (1982)
4. Maass, W.: Networks of spiking neurons: the third generation of neural network models. Neural Netw. **10**(9), 1659–1671 (1997)
5. Mitchell, T.: The need for biases in learning generalizations. In: Readings in Machine Learning, pp. 184–191 (1980)
6. Pratt, L., Jennings, B.: A survey of transfer between connectionist networks. Connection Sci. **8**(2), 163–184 (1996)
7. Pratt, L.Y.: Discriminability-based transfer between neural networks. In: Advances in Neural Information Processing Systems 5, [NIPS Conference], pp. 204–211. Morgan Kaufmann Publishers Inc., San Francisco (1993)

8. Pratt, L.Y.: Transferring previously learned back-propagation neural networks to new learning tasks. Ph.D. thesis, Rutgers University, Department of Computer Science (1993)
9. Pratt, L.Y., Mostow, J., Kamm, C.A.: Direct transfer of learned information among neural networks. In: Proceedings of the Ninth National Conference on Artificial Intelligence, AAAI 1991, vol. 2, pp. 584–589. AAAI Press (1991)
10. Shultz, T.R., Rivest, F.: Knowledge-based cascade-correlation: using knowledge to speed learning. Connection Sci. **13**(1), 43–72 (2001)
11. Silver, D.L.: Selective transfer of neural network task knowledge. Ph.D. thesis, The University of Western Ontario, AAINQ58190 (2000)
12. Waibel, A., Sawai, H., Shikano, K.: Modularity and scaling in large phonemic neural networks. IEEE Trans. Acoust. Speech Signal Process. **37**(12), 1888–1898 (1989)

Stable Feature Selection with Support Vector Machines

Iman Kamkar[✉], Sunil Kumar Gupta, Dinh Phung, and Svetha Venkatesh

Centre for Pattern Recognition and Data Analytics, Deakin University,
Geelong, Australia
ikamkar@deakin.edu.au

Abstract. The support vector machine (SVM) is a popular method for classification, well known for finding the maximum-margin hyperplane. Combining SVM with l_1-norm penalty further enables it to simultaneously perform feature selection and margin maximization within a single framework. However, l_1-norm SVM shows instability in selecting features in presence of correlated features. We propose a new method to increase the stability of l_1-norm SVM by encouraging similarities between feature weights based on feature correlations, which is captured via a feature covariance matrix. Our proposed method can capture both positive and negative correlations between features. We formulate the model as a convex optimization problem and propose a solution based on alternating minimization. Using both synthetic and real-world datasets, we show that our model achieves better stability and classification accuracy compared to several state-of-the-art regularized classification methods.

1 Introduction

High dimensional datasets have become increasingly popular in many real-world applications. However, it is generally believed that in these datasets often only a small number of features are informative and the remaining features are either noisy or contain irrelevant information. Hence, selecting truly informative features is essential for many real applications [8] and improves the prediction accuracy of the model.

One of the important attributes of feature selection methods is their "stability" in selecting informative features. The feature stability is defined as the variation in obtained feature sets due to small changes in dataset [18] and is crucial in applications where selected features are used for knowledge discovery and decision makings [12]. For example, in clinical domain, explaining the risk factors in prognosis is as important as the prognosis itself. Consequently, stable features in spite of data resampling, are critical for clinical adoption [7].

A widely used strategy for feature selection that imposes sparsity on regression or classification coefficients is l_1-norm regularization. Perhaps the most well-known example is Lasso that minimizes the sum of squared errors while penalizing the l_1-norm of the regression coefficients [13]. The idea of using l_1-norm penalty to automatically select features has also been extended to classification problems.

© Springer International Publishing Switzerland 2015
B. Pfahringer and J. Renz (Eds.): AI 2015, LNAI 9457, pp. 298–308, 2015.
DOI: 10.1007/978-3-319-26350-2_26

Zhu et al. in [21] proposed l_1-norm support vector machine that can perform feature selection and binary classification, simultaneously. Although using l_1-norm regularization has shown success in many applications and has been generalized for different settings [21], it shows instability in presence of correlated features. The reason for such instability is that it tends to assign a nonzero weight to only a single feature among a group of correlated features [19,20].

Different methods have been proposed to address the instability of l_1-norm methods. Many of these methods try to find the groups of correlated features because these groups are consistent to the variation of training data. In presence of feature grouping information, groups Lasso [19] can be used and if features have an intrinsic hierarchical structure, tree-Lasso can be considered as a solution for stabilizing Lasso [10]. When there are ordering between features which imposes correlation among them, fused-Lasso can be used as a remedy to increase the stability of Lasso by selecting neighboring features [14]. Use of these methods requires that we know the structure of the data. However, such a structure is not available in many applications, which renders these methods inapplicable. There are limited works that try to solve the instability of l_1-norm methods in general context by incorporating feature similarities. Elastic net [22] is one of these methods that assigns comparable weights to similar features by using a combination of l_1 and l_2 penalty. However, it results in a longer lists of features compared to Lasso. Another method is Oscar that performs feature grouping and feature selection, simultaneously [1]. This method uses a combination of l_1 and pairwise l_∞ norm penalties to impose sparsity and equal feature weights for highly correlated features. The features with equal weights automatically form a group. Although Oscar tends to increase the stability of Lasso by grouping correlated features, assigning equal weights to features that are partially correlated may degrade the performance of the method [3].

All the methods discussed above, are proposed to increase the feature stability of Lasso, where its loss function is residual sum of squares or the logit function. As mentioned before, l_1-norm penalty terms are also combined with support vector machines to encourage sparsity. However, limited research has been done to address the instability of l_1-norm in these methods. To the best of our knowledge the only work done to address the instability in l_1-norm support vector machines is combining SVM with elastic net penalty [16,17]. However, this method does not properly exploit the feature correlations.

To address the instability in l_1-norm SVM, we propose a regularization formulation that encourages the similarities between features based on their relatedness. In our formulation, the relatedness between features is captured through a feature covariance matrix. Our method can perform feature selection and capture both positive and negative correlations between features through a convex objective function. In summary, our contributions are as follows:

– Proposal of a new model aimed to improve the stability of l_1-norm support vector machines by capturing the similarities between features based on their relatedness via a feature covariance matrix.

- Proposal of a convex optimization formulation for the model and a solution based on an alternating optimization.
- Demonstration of improved feature stability in terms of two stability measures, Jaccard similarity measure and Spearman's rank correlation coefficient in comparison with several baseline methods namely, Lasso, l_1-SVM and Elastic net SVM.
- Demonstration of improved classification accuracy of the model in comparison with the above baseline methods.

2   Framework

We propose a new model to address the instability of l_1-SVM in selecting informative features. We consider a binary classification problem with training data $\{\mathbf{x}_i, y_i\}_{i=1}^n$, where $\mathbf{x}_i = (x_{i1}, \ldots, x_{ip})^T$ is the feature vector and $y_i \in \{-1, 1\}$ is the class label. In general, we make two assumptions: (1) We are dealing with high dimensional but sparse setting. By sparsity we mean that the majority of the features are not predictive of the outcome. (2) Among the features, there are sets of features with high levels of correlations. In this context, l_1-SVM shows instability in selecting informative features because it randomly assigns a nonzero weight to a single feature among a group of correlated features and so with small changes in dataset, another feature maybe selected from the correlated group. To overcome this problem, the similarities between the features can be encouraged based on their relatedness. To this end, we use a feature covariance matrix to capture relationships between features. Our proposed model, is the solution to the following optimization problem:

$$\arg\min_{\beta_0, \beta, \Omega} \frac{1}{n}(1 - y_i(\beta_0 + \mathbf{x}_i^T \beta))_+ + \lambda\|\beta\|_1 + \frac{\eta}{2}\beta^T \Omega^{-1}\beta \qquad (1)$$

$$\text{s.t.} \qquad \Omega \succeq 0, \operatorname{tr}(\Omega) = 1,$$

where β is the vector of feature weights and β_0 is the intercept. Also, Ω is the covariance matrix that models the relationships between features, λ and η are the tuning parameters and $(1 - T)_+ = \max(T, 0)$ is the hinge loss. The term $\beta^T \Omega^{-1}\beta$ ensures that feature weights follow the feature correlations, i.e. if two features are highly correlated their feature weights would become very high. We refer to the above model as **Covariance SVM (C-SVM)**.

2.1   Algorithm for Covariance-SVM

Although the objective function in (1) is convex with respect to all variables, it is not straight forward due to the non-smooth convexity. To solve this problem, we introduce an iterative algorithm that alternatively updates β and Ω as follows:

Optimizing w.r.t. β when Ω is Fixed: In this situation, the objective function can be stated as:

$$\arg\min_{\beta_0,\beta} \frac{1}{n}(1 - y_i(\beta_0 + \mathbf{x}_i^T\beta))_+ + \lambda\|\beta\|_1 + \frac{\eta}{2}\beta^T\Omega^{-1}\beta. \tag{2}$$

This problem can be solved using the alternate direction method of multipliers (ADMM), which has recently become a method of choice for solving many large-scale problems [2]. Because of the nondifferentiability of the hinge loss and l_1 norm term in (2), we introduce some auxiliary variables to handle these two nondifferentiable terms. Suppose $X = (x_{ij})_{i=1,j=1}^{n,p}$ and Y be a diagonal matrix, where its diagonal elements are the vector $y = (y_1\ldots,y_n)^T$. So the problem in (2), can be reformulated as:

$$\arg\min_{\beta_0,\beta} \frac{1}{n}\sum_{i=1}^{n}(a_i)_+ + \lambda\|z\|_1 + \frac{\eta}{2}\beta^T\Omega^{-1}\beta \tag{3}$$

$$\text{s.t.} \quad \mathbf{a} = 1 - Y(X\beta + \beta_0\mathbf{1}), \ z = \beta,$$

where $\mathbf{a} = (a_1,\ldots,a_n)$ and $\mathbf{1}$ is a column vector of 1's with length n. The augmented Lagrangian function of (3) is

$$L(\beta_0,\beta,a,z,u,v) = \frac{1}{n}\sum_{i=1}^{n}(a_i)_+ + \lambda\|z\|_1 + \frac{\eta}{2}\beta^T\Omega^{-1}\beta \tag{4}$$

$$+ \langle u, 1 - Y(X\beta + \beta_0\mathbf{1}) - a\rangle + \langle v, \beta - z\rangle,$$

where $u \in \mathbb{R}^n$ and $v \in \mathbb{R}^p$ are dual variables corresponding to the first and the second constraints in Eq. (3), respectively. $\langle .,.\rangle$ is the inner product in the Euclidean space and μ_1 and μ_2 control the convergence behavior and are usually set to 1. By solving the above equation w.r.t u,v, (β_0,β), a and z we have:

$$\begin{cases} (\beta_0^{k+1},\beta^{k+1}) = \arg\min_{\beta_0,\beta} L(\beta_0,\beta,a^k,z^k,u^k,v^k), \\ a^{k+1} = \arg\min_a L(\beta_0^{k+1},\beta^{k+1},a,z^k,u^k,v^k), \\ z^{k+1} = \arg\min_c L(\beta_0^{k+1},\beta^{k+1},a^{k+1},z,u^k,v^k), \\ u^{k+1} = u^k + \mu_1(1 - Y(X\beta^{k+1} + \beta_0^{k+1}\mathbf{1}) - a^{k+1}), \\ v^{k+1} = v^k + \mu_2(\beta^{k+1} - z^{k+1}). \end{cases} \tag{5}$$

The first term in (5) is a quadratic and differentiable objective function, so its solution can be found by solving a set of linear equations:

$$\begin{pmatrix} \lambda_2\Omega^{-1} + \mu_2\mathbf{I} + \mu_1 X^T X & \mu_1 X^T\mathbf{I} \\ \mu_1\mathbf{1}^T X & \mu_1 n \end{pmatrix}\begin{pmatrix} \beta^{k+1} \\ \beta_0^{k+1} \end{pmatrix} \tag{6}$$

$$= \begin{pmatrix} X^T Y u^k - \mu_1 X^T Y(a^k - 1) - v^k + \mu_2 z^k \\ \mathbf{1}^T Y u^k - \mu_1\mathbf{1}^T Y(a^k - 1) \end{pmatrix}.$$

The second term in (5) can be solved by using Proposition 1.

Proposition 1. *Let $h_\lambda(w) = \arg\min_x \lambda x_+ + \frac{1}{2}\|x - w\|_2^2$. Then $h_\lambda(w) = w - \lambda$ for $w > \lambda$, $h_\lambda(w) = 0$ for $0 \le w \le \lambda$ and $h_\lambda(w) = w$ for $w < 0$.*

So the second term in (5), can be written as

$$\frac{\|u\|_2^2}{2\mu_1} + \frac{\mu_1}{2}\|\mathbf{1} - Y(X\beta^{k+1} + \beta_0^{k+1}\mathbf{1}) - a\|_2^2 + \langle u^k, \mathbf{1} - Y(X\beta^{k+1} + \beta_0^{k+1}\mathbf{1}) - a\rangle$$
$$= \frac{\mu_1}{2}\|a - (\mathbf{1} + \frac{u}{\mu_1} - Y(X\beta^{k+1} + \beta_0^{k+1}\mathbf{1}))\|_2^2.$$

From above equation and Proposition 1, we can update a^{k+1} as follows:

$$a^{k+1} = H_{\frac{1}{n\mu_1}}(\mathbf{1} + \frac{u^k}{\mu_1} - Y(X\beta^{k+1} + \beta_0^{k+1}\mathbf{1})), \tag{7}$$

where $H_\lambda(w) = (h_\lambda(w_1), h_\lambda(w_2), \ldots, h_\lambda(w_n))^T$.

The third equation in (5) can be solved using soft thresholding. So we have

$$z^{k+1} = S_{\frac{\lambda}{\mu_2}}\left(\frac{v^k}{\mu_2} + \beta^{k+1}\right), \tag{8}$$

where S_λ is the soft threshold operator defined on vector space and $S_\lambda(w) = (s_\lambda(w_1), \ldots, s_\lambda(w_p))$, where $s_\lambda(w_i) = sgn(w_i)\max\{0, |w_i| - \lambda\}$.

By combining (5)–(8), we obtain the ADMM algorithm for solving the objective function (1) with respect to β when Ω is fixed.

Optimizing w.r.t Ω when β is fixed: In this situation, the optimization problem for finding Ω becomes

$$\min_\Omega \beta^T \Omega^{-1}\beta \text{ such that } \Omega \succeq 0, \text{tr}(\Omega) = 1$$

Let $B = \beta\beta^T$, as $\beta^T\Omega^{-1}\beta = \text{tr}(\beta^T\Omega^{-1}\beta) = \text{tr}(\Omega^{-1}\beta\beta^T)$ and $\text{tr}(\Omega) = 1$, so

$$\text{tr}(\Omega^{-1}B) = \text{tr}(\Omega^{-1}B)\text{tr}(\Omega) = \text{tr}((\Omega^{-\frac{1}{2}}B^{\frac{1}{2}})(B^{\frac{1}{2}}\Omega^{-\frac{1}{2}}))\text{tr}(\Omega^{\frac{1}{2}}\Omega^{\frac{1}{2}})$$
$$\ge (\text{tr}(\Omega^{-\frac{1}{2}}B^{\frac{1}{2}}\Omega^{\frac{1}{2}}))^2 = (\text{tr}(B^{\frac{1}{2}}))^2.$$

The inequality holds because of Cauchy-Schwarz inequality for the Frobenius norm. From this inequality, we can say that $\text{tr}(\Omega^{-1}B)$ achieves its minimum value $(\text{tr}(B^{\frac{1}{2}}))^2$ if and only if $\Omega^{-\frac{1}{2}}B^{\frac{1}{2}} = \zeta\Omega^{\frac{1}{2}}$ for some constant ζ and $\text{tr}(\Omega) = 1$. So Ω can be obtained from $\Omega = \frac{(\beta\beta^T)^{\frac{1}{2}}}{\text{tr}((\beta\beta^T)^{\frac{1}{2}})}$.

3   Experiments

In this section, we perform experiments using both synthetic and real datasets and compare the classification accuracy and feature stability of the C-SVM with several baselines that deemed to be closest to our work, namely Lasso [13], l_1-norm SVM [21] and Elastic net SVM (ENSVM) [16].

3.1 Tuning Parameter Selection

In case of synthetic data set, we use a validation set to select the tuning parameters λ and η. We train each model on the training set and use the validation set to select the best tuning parameter for the final model. The performance of each model is evaluated using the test set. In case of real data sets, we use the 5-fold cross validation to select the best tuning parameters.

3.2 Performance Metrics

Feature Stability Measures. To compare the feature stability of C-SVM with other methods we use two similarity measures, Jaccard similarity measure (JSM), which considers the indices of the selected features in its evaluation process and Spearman's rank correlation coefficient (SRCC), which considers the rank of the selected features for evaluating stability. Jaccard measures the similarities between any two sets of selected features S_q and $S_{q'}$ as $JSM(S_q, S_{q'}) = \frac{|S_q \cap S_{q'}|}{|S_q \cup S_{q'}|}$, where JSM$\in [0, 1]$ and 0 means there are no similarities between the two sets and 1 means the two sets are identical. SRCC measures similarity between two rankings r and r' as $SRCC(r, r') = 1 - 6 \sum_j \frac{(r_j - r'_j)}{p(p^2 - 1)}$, where SRCC$\in [-1, 1]$ and 1 shows the two rankings are identical, 0 shows there is no correlation between two rankings and -1 shows that rankings are in inverse order. In our experiments we generate M sub-samples of the training set and apply each algorithm to each sub-sample to obtain its selected feature set. We use JSM and SRCC to evaluate the similarity between each pair of selected features and finally, average similarities over all pairs to obtain the stability of each algorithm.

Classification Accuracy. To compare the classification accuracy of C-SVM with other baselines, F-measure and AUC score are used [9].

3.3 Simulation Results

We consider a binary classification problem in a p dimensional space where only the first 50 features are relevant for classification and the remaining features are noise. To this end, we generate n instances where half of them belong to $+1$ class and the other half belong to -1 class. Instances in positive class are i.i.d drawn from a normal distribution with mean $\mu_+ = (\underbrace{1, \ldots, 1}_{50}, \underbrace{0, \ldots, 0}_{p-50})^T$ and covariance

$$\Sigma = \begin{pmatrix} \Sigma^*_{50 \times 50} & \mathbf{0}_{50 \times (p-50)} \\ \mathbf{0}_{(p-50) \times 50} & \mathbf{I}_{(p-50) \times (p-50)} \end{pmatrix},$$

where in Σ^* the diagonal elements are 1 and others are all equal to ρ. The mean for negative class is $\mu_- = (\underbrace{-1, \ldots, -1}_{50}, \underbrace{0, \ldots, 0}_{p-50})$. In this situation, the Bayes optimal classification rule depends on $x_1, \ldots, x_{50}$, which are highly correlated if ρ is large.

Table 1. Stability performance of C-SVM compared to other baselines for Synthetic dataset. Means and standard error over 50 iterations are reported.

Synthetic data		Lasso	l_1-SVM	ENSVM	C-SVM
$\rho = 0$	JSM	0.652 (0.015)	0.649 (0.019)	0.655 (0.026)	**0.662** (0.031)
	SRCC	0.452 (0.031)	0.448 (0.037)	0.463 (0.027)	**0.487** (0.025)
$\rho = 0.8$	JSM	0.447 (0.027)	0.510 (0.021)	0.571 (0.034)	**0.603** (0.027)
	SRCC	0.305 (0.032)	0.319 (0.018)	0.368 (0.032)	**0.407** (0.034)

We explore two values of ρ, 0 and 0.8, where $\rho = 0$ simulates the situation that informative features are uncorrelated to each other and $\rho = 0.8$, simulates the situation that those features are highly correlated. The stability performance of each method, measured in terms of JSM and SRCC, is shown in Table 1. The high value of SRCC implies that ranks of features do not vary a lot for different training sets and high value of JSM means that the selected features do not change significantly when there is a slight change in the training set. As the table implies, when there is no correlation among variables ($\rho = 0$), the stability of C-SVM is comparable to other baselines. However, when the correlation among features is high ($\rho = 0.8$) Lasso and l_1-SVM show low stability performance in terms of both JSM and SRCC. However, ENSVM that incorporates l_2-norm penalty shows better stability performance compared to Lasso and l_1-SVM. For C-SVM, we can see that as this model encourages similarities between features based on their relatedness, it shows the best stability compared to the baselines in terms of both JSM and SRCC.

Figure 1 shows the classification performance of C-SVM in terms of two classification measures, F-measure, and AUC and compares them with other baselines. As shown, the classification performance of C-SVM outperforms other baselines in terms of the both classification measures.

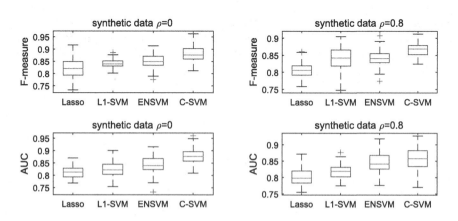

Fig. 1. Classification performance of C-SVM and baseline methods in terms of F-measure and AUC for Synthetic dataset.

3.4   Application on Real Datasets

In this section, we evaluate the performance of C-SVM on real datasets and compare it with other baselines. The datasets used are as follows:

Breast Cancer Dataset: This dataset is compiled by [15] and consists of gene expression data for 8141 genes in 295 breast cancer tumors (87 metastatic and 217 non-metastatic). As the dataset is very imbalanced, we balance it by using 3 replicates of each positive (metastasis) sample while keeping all replicates in the same fold during cross validation.

Cancer Dataset: This dataset is obtained from a large regional hospital in Australia. There are eleven different cancer types in this data recorded from patients visiting the hospital during 2010–2012. Patient data is acquired from Electronic Medical Records (EMR). The dataset consists of 4293 patients with 3867 variables including International Classification of Disease 10 (ICD-10), procedure and diagnosis related Group (DRG) codes of each patient as well as demographic data (age, gender and postcode). Using this dataset, our goal is to predict 1 year mortality of patients while ensuring the stable feature sets.

AMI Dataset: This dataset is also obtained from the same hospital in Australia. It involves patients admitted with AMI conditions and discharged later between 2007–2011. The task is to predict if a patient will be re-admitted to the hospital within 30 days after discharge. The dataset consists of 2941 patients with 2504 variables include International Classification of Disease 10 (ICD-10), procedure and diagnosis-related Group (DRG) codes of each admission; details of procedures; and departments involved in the patient's care.

Experimental Results

Stability Performance. The comparison between stability performance of C-SVM and other baselines in terms of JSM and SRCC for real datasets are presented in Table 2. For Breast cancer dataset, C-SVM shows the best stability performance in terms of JSM (0.620). However, in terms of SRCC, ENSVM represents the best stability (0.512), which is closely followed by C-SVM (0.509). For Cancer dataset, C-SVM shows the best stability performances with $JSM = 0.631$ and $SRCC = 0.518$. In terms of both JSM and SRCC, C-SVM is followed by ENSVM with $JSM = 0.568$ and $SRCC = 0.427$. For AMI dataset, again C-SVM shows the best stability in terms of both JSM (0.572) and SRCC (0.509), which is followed by ENSVM with $JSM = 0.516$ and $SRCC = 0.457$. As seen, stability performances of Lasso and l_1-SVM are close to each other and these methods show the least stability, the reason for which is that these methods use only l_1 regularization term which is unstable in selecting correlated features.

Classification Performance. Figure 2 shows the classification performance of C-SVM and other baselines in terms of F-measure and AUC for real-world datasets. For Breast cancer dataset, we can see that the classification performance of C-SVM outperforms other methods in terms of both F-measure and

Table 2. Stability performance of C-SVM compared to other baselines for real data sets. Means and standard error over 50 iterations are reported.

Real data		Lasso	l_1-SVM	ENSVM	C-SVM
Breast cancer	JSM	0.352 (0.028)	0.348 (0.025)	0.551 (0.021)	**0.620** (0.031)
	SRCC	0.237 (0.028)	0.235 (0.019)	**0.512** (0.026)	0.509 (0.023)
Cancer	JSM	0.420 (0.019)	0.423 (0.025)	0.568 (0.018)	**0.631** (0.026)
	SRCC	0.273 (0.021)	0.276 (0.030)	0.427 (0.022)	**0.518** (0.015)
AMI	JSM	0.372 (0.026)	0.368 (0.033)	0.516 (0.036)	**0.572** (0.036)
	SRCC	0.268 (0.031)	0.270 (0.024)	0.457 (0.028)	**0.509** (0.021)

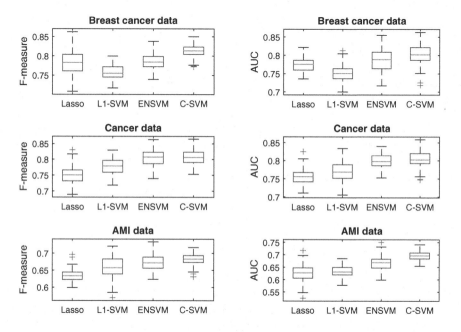

Fig. 2. Classification performance of C-SVM and other baselines in terms of accuracy, F-measure and AUC for Cancer dataset.

AUC. For Cancer data, we can see that C-SVM along with ENSVM show the best classification performance in terms of both F-measure and AUC. Turning to the AMI data, again C-SVM shows the best accuracy among other baselines in terms of the two classification measures.

Estimated Covariance Matrix. As feature names for AMI and Cancer datasets are available, we show the estimated covariance matrix for these datasets in Fig. 3 and we further discuss about some of the correlated features estimated in their Ω matrix. For better representation, we show the correlation matrix computed from Ω matrix by standardizing its values as $\Omega_{st}(i,j) = \frac{\Omega(i,j)}{\sqrt{\Omega(i,j)\Omega(i,j)}}$. In Ω matrix of

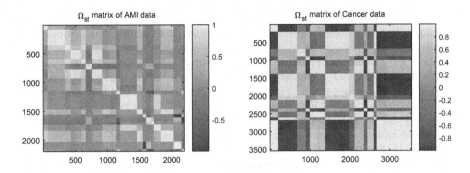

Fig. 3. The pictorial representation of estimated covariance matrix for real data sets. For better representation, we show the correlation matrix computed from Ω matrix by standardizing its values.

AMI dataset, the first group are the features related to cardiac troponin and the last group are features related to discharge sodium values. Both of these features are reported as important risk factors for Mayocardial infarction [6,11]. In Ω matrix obtained for Cancer dataset, the first group are the features related to diabetes mellitus and the last group are the features related to anemia, where both of these features are important risk factors for cancer survival prediction [4,5].

4   Conclusion

In this paper, we propose a method that can increase feature stability of l_1-norm SVM in presence of highly correlated features. The method can capture both the positive and negative relations between features using a covariance matrix, therefore the highly correlated features could be selected or rejected together by the model. We propose a convex formulation for the model that can be solved using an alternating optimization algorithm. We show the proposed method is more stable and more accurate than many existing methods.

References

1. Bondell, H.D., Reich, B.J.: Simultaneous regression shrinkage, variable selection, and supervised clustering of predictors with oscar. Biometrics **64**(1), 115–123 (2008)
2. Boyd, S., Parikh, N., Chu, E., Peleato, B., Eckstein, J.: Distributed optimization and statistical learning via the alternating direction method of multipliers. Found. Trends Mach. Learn. **3**(1), 1–122 (2011)
3. Bühlmann, P., Rütimann, P., van de Geer, S., Zhang, C.H.: Correlated variables in regression: clustering and sparse estimation. J. Stat. Planning Infer. **143**(11), 1835–1858 (2013)
4. Caro, J.J., Salas, M., Ward, A., Goss, G.: Anemia as an independent prognostic factor for survival in patients with cancer. Cancer **91**(12), 2214–2221 (2001)

5. Coughlin, S.S., Calle, E.E., Teras, L.R., Petrelli, J., Thun, M.J.: Diabetes mellitus as a predictor of cancer mortality in a large cohort of us adults. Am. J. Epidemiol. **159**(12), 1160–1167 (2004)

6. Eapen, Z.J., Liang, L., Fonarow, G.C., Heidenreich, P.A., Curtis, L.H., Peterson, E.D., Hernandez, A.F.: Validated, electronic health record deployable prediction models for assessing patient risk of 30-day rehospitalization and mortality in older heart failure patients. JACC Heart Fail. **1**(3), 245–251 (2013)

7. Ein-Dor, L., Kela, I., Getz, G., Givol, D., Domany, E.: Outcome signature genes in breast cancer: is there a unique set? Bioinformatics **21**(2), 171–178 (2005)

8. Fan, J., Li, R.: Statistical challenges with high dimensionality: feature selection in knowledge discovery (2006). arXiv preprint math/0602133

9. Han, J., Kamber, M., Pei, J.: Data Mining: Concepts and Techniques. Elsevier, Massachussets (2011)

10. Kamkar, I., Gupta, S.K., Phung, D., Venkatesh, S.: Stable feature selection for clinical prediction: exploiting ICD tree structure using tree-lasso. J. Biomed. Inf. **53**, 277–290 (2015)

11. Mair, J., Artner-Dworzak, E., Lechleitner, P., Smidt, J., Wagner, I., Dienstl, F., Puschendorf, B.: Cardiac troponin T in diagnosis of acute myocardial infarction. Clin. Chem. **37**(6), 845–852 (1991)

12. Saeys, Y., Abeel, T., Van de Peer, Y.: Robust feature selection using ensemble feature selection techniques. In: Daelemans, W., Goethals, B., Morik, K. (eds.) ECML PKDD 2008, Part II. LNCS (LNAI), vol. 5212, pp. 313–325. Springer, Heidelberg (2008)

13. Tibshirani, R.: Regression shrinkage and selection via the lasso. J. Roy. Stat. Soc. Ser. B (Methodological) **58**(1), 267–288 (1996)

14. Tibshirani, R., Saunders, M., Rosset, S., Zhu, J., Knight, K.: Sparsity and smoothness via the fused lasso. J. Roy. Stat. Soc. Ser. B (Statist. Method.) **67**(1), 91–108 (2005)

15. Van De Vijver, M.J., He, Y.D., van't Veer, L.J., Hart, A.A., Voskuil, D.W., Schreiber, G.J., Peterse, J.L., Roberts, C., Marton, M.J., Marton, M.J., et al.: A gene-expression signature as a predictor of survival in breast cancer. N. Engl. J. Med. **347**(25), 1999–2009 (2002)

16. Wang, L., Zhu, J., Zou, H.: The doubly regularized support vector machine. Stat. Sinica **16**(2), 589 (2006)

17. Ye, G.B., Chen, Y., Xie, X.: Efficient variable selection in support vector machines via the alternating direction method of multipliers. In: International Conference on Artificial Intelligence and Statistics, pp. 832–840 (2011)

18. Yu, L., Ding, C., Loscalzo, S.: Stable feature selection via dense feature groups. In: Proceedings of the 14th ACM SIGKDD International Conference on Knowledge Discovery and Data Mining, pp. 803–811. ACM (2008)

19. Yuan, M., Lin, Y.: Model selection and estimation in regression with grouped variables. J. Roy. Stat. Soc. Ser. B (Stat. Method.) **68**(1), 49–67 (2006)

20. Zhao, P., Yu, B.: On model selection consistency of lasso. J. Mach. Learn. Res. **7**, 2541–2563 (2006)

21. Zhu, J., Rosset, S., Hastie, T., Tibshirani, R.: 1-norm support vector machines. Adv. Neural Inf. Process. Syst. **16**(1), 49–56 (2004)

22. Zou, H., Hastie, T.: Regularization and variable selection via the elastic net. J. Roy. Stat. Soc. Ser. B (Stat. Method.) **67**(2), 301–320 (2005)

Gene Transfer: A Novel Genetic Operator for Discovering *Diverse-Frequent* Patterns

Shanjida Khatun[1], Hasib Ul Alam[1], Mahmood A. Rasid[3], and Swakkhar Shatabda[2]([✉])

[1] Department of CSE, Ahsanullah University of Science and Technology, Dhaka, Bangladesh
[2] Department of CSE, United International University, Dhaka, Bangladesh
[3] CIS, University of Massachussettes Dartmouth, North Dartmouth, MA, USA
swakkhar@cse.uiu.ac.bd

Abstract. Genetic algorithm (GA) based on evolution principles has found its strong base in pattern set mining. GA has proved to generate more accurate results when compared to other formal methods available in the past years. In this paper, we present a new genetic operator called *gene transfer* within the GA framework. Genes are part of the genetic code that represents a partial solution to the problem. The proposed GA operator creates a pool of *gene*s or partial solutions which are of good health and tries to replace genes of similar size in the population to improve the fitness of the individuals. This operation applied along with other traditional genetic operators like crossover and mutation results in speed up in convergence and finding individuals with better fitness function. On a set of standard benchmark dataset, we experimentally show that our new genetic operator improves the performance of a genetic algorithm consistently.

Keywords: Pattern set mining · Genetic algorithm · Gene transfer

1 Introduction

In last few years, many algorithms [1,5,8] have failed to give a small set of diverse-frequent patterns in an efficient way for several reasons. First of all, large data leads to enormous hypothesis space, making exhaustive search infeasible. Secondly, they perform some kind of greedy or local search and differ widely in the heuristics and search orders used. Thirdly, due to attributes of high cardinality and correlated attributes many variants of essentially the same pattern exist.

To address this issue, in our recent work [7], we investigated a technique to find a small set of diverse-frequent patterns using genetic algorithm (GA) which has several novel components such as relative encoding technique, twin removal technique and random restart technique. This algorithm is capable of handling the search space in more effective way with compared to large neighborhood search, hill climbing and random walk.

© Springer International Publishing Switzerland 2015
B. Pfahringer and J. Renz (Eds.): AI 2015, LNAI 9457, pp. 309–316, 2015.
DOI: 10.1007/978-3-319-26350-2_27

In this paper, we present two variants of genetic algorithms that individually and in a combined way use four different enhancement techniques: (i) a relative encoding technique; (ii) a twin removal technique; (iii) a random-walk based stagnation recovery technique; and (iv) gene transfer technique. GA with *geneTransfer* that use a combination of all the four enhancements significantly outperforms all current approaches of genetic algorithm. We applied the *gene transfer* operator only when the search faces stagnation or fails to improve for a number of iterations.

The paper is furnished as follows: Sect. 2 explains all the necessary definitions to understand the paper; Sect. 3 reviews the related work; Sect. 4 explains the algorithms we used; Sect. 5 discusses and analyzes the experimental results; and then, Sect. 6 presents our conclusions with a discussion and possible outlines for future work.

2   The Problem Model

In our recent work [7], we described all concepts to understand the *diverse-frequent* pattern set mining problems.

Dispersion score is the score of the frequent pattern sets based on the items categories within it. We used XOR based dispersion score to measure the diversity of a pattern set. The objective function that we wish to maximize is defined as follows:

$$obj Dispersion = \sum_{i=1}^{k=|\Pi|} \sum_{j=1}^{i-1} xor Dispersion(T^i, T^j). \tag{1}$$

Given this model, the problem can be defined as follows: Given a set of items $\mathcal{I}$ and a database, $\mathcal{D}$ of transactions $\mathcal{T}$, find a pattern set Π of size k such that the dispersion score defined in Eq. 1 is maximized.

3   Related Work

Many algorithms have been proposed as a general framework [3,5] in last few years to find patterns which are correlated [9], discriminative [11], contrast [4] and diverse [10]. Most of these methods are not only exhaustive in nature but also take huge amount of time and performed poor for large datasets. Guns et al. [5] investigated a technique to solve both of the concept learning task and the diverse pattern set mining problem by applying Large Neighborhood Search using constraint programing platform. The algorithms that they used only worked for small dataset. In a recent work, Hossain et al. [6] and Shanjida et al. [7] explored the use of genetic algorithms and other stochastic local search algorithms to solve the concept learning and the diverse pattern set mining tasks.

4 Our Approach

In this section, first we describe the GA framework and then we describe our gene transfer operator within the GA framework.

4.1 The Genetic Algorithm (GA) Framework

Algorithm 1. GAFramework()

---

1 $percentChange$: input parameter
2 $p = populationSize$
3 $\mathcal{P} =$ generate p valid pattern sets
4 $\mathcal{P}_b = \{\}$
5 **while** $timeout$ **do**
6      $\mathcal{P}_m =$ simpleMutation($\mathcal{P}$)
7      $\mathcal{P}_c =$ uniformCrossOver($\mathcal{P}$)
8      $\mathcal{P}_* =$ select($\mathcal{P} \cup \mathcal{P}_m \cup \mathcal{P}_c$)
9   **if** $stagnation$ **then**
10          $\prod =$ findBest($\mathcal{P}_*$)
11          $\mathcal{P}_b = \mathcal{P}_b \cup \{\prod\}$
12          $\mathcal{P}_* =$ randomRestart($percentChange, \mathcal{P}_*$)
13      $\mathcal{P} = \mathcal{P}_*$
14 $\prod^* =$ findBest($\mathcal{P}_b$)
15 return $\prod^*$

Algorithm 2. randomRestart($perChange, \mathcal{P}$)

---

1 $toChange = (perChange * \text{sizeOf}(\mathcal{P}))/100$
2 keep $toChange$ best $\prod$ in $\mathcal{P}$ and remove rest
3 $i = 1$
4 **while** $i \leq toChange$ **do**
5      $\prod =$ randomly create a valid patten set with k-items
6   **while** $\prod \in \mathcal{P}$ **do**
7          $\prod =$ randomly create a valid patten set with k-items
8      $\mathcal{P} = \mathcal{P} \cup \{\prod\}$
9      $i + +$
10 return $\mathcal{P}$

In Algorithm 1, we created a population in $\mathcal{P}_m$ using mutation and $\mathcal{P}_c$ using cross over [7]. Then we selected best population from $\mathcal{P}$, $\mathcal{P}_m$ and $\mathcal{P}_c$ into $\mathcal{P}_*$ where size of $\mathcal{P}_*$ will be same as population size. We repeated the procedure through several generation. Stagnation occurs if the quality of the best solution achieved does not change for a number of iterations. We call a procedure called $randomRestart()$. This method alters a percentage of the population by replacing them with randomly generated individuals. In each generation, the best individual or the pattern set with maximum fitness is saved in $\mathcal{P}_b$. Then we copied the value of $\mathcal{P}_*$ into $\mathcal{P}$. We got a new population in the next generation. We repeated the procedure until timeout occurs.

In Algorithm 2, we used random restart to avoid stagnation based on two variables. One, when it will be restarted and how much change will be done in the list of population. We created a new population where $\mathcal{P}$ represents the pattern set in which we have to change and $perChange$ represents how much patterns that we have to change. For example, if $perChange = 90$ that means 90 % value will be replaced by new value. As it saved only top 10 % score and other 90 % will be used to create new population. In our algorithm, we experimented with different values of $perChange$.

Algorithm 3. geneTransfer($perChange, \mathcal{P}$)

1 $toChange = (perChange * \text{sizeOf}(\mathcal{P}))/100$
2 keep $toChange$ best $\prod$ in $\mathcal{P}$ and remove the rest
3 randomly create n gene for each $\prod$ remains in $\mathcal{P}$
4 $i = 1$
5 **while** $i \leq toChange$ **do**
6 $\quad \prod = $ randomly create a valid patten set with k-items
7 $\quad$ **while** $\prod \in \mathcal{P}$ **do**
8 $\quad\quad \prod = $ randomly create a valid patten set with k-items
9 $\quad\quad \prod = $ add each gene individually into $\prod$ and keep the best one
10 $\quad \mathcal{P} = \mathcal{P} \cup \{\prod\}$
11 $\quad i + +$
12 return $\mathcal{P}$

4.2 Gene Transfer Operation

We randomly created n genes for each pattern set, $\Pi \in \mathcal{P}$ where n is the number of genes. After creating a valid pattern set with k-items randomly as previously done in Algorithm 2, we added each gene individually into pattern sets. Then we

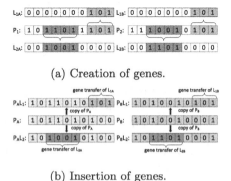

(a) Creation of genes.

(b) Insertion of genes.

Fig. 1. Gene transfer technique.

calculated objective score as describe earlier for each pattern sets and kept the best one. In our implementation, we randomly generated *geneSize* for each gene. We described gene transfer technique for pattern set size 2 in Fig. 1. The gene-Transfer algorithm is given in Algorithm 3 and used in Algorithm 1 at Line 12 instead of Algorithm 2.

In Fig. 1(a), there are two patterns (P_1, P_2) in pattern set(P). We created two genes (L_1, L_2) from each patterns. We copied three consecutive items from the item position 8 to 10 of P_1 into same position of L_{1A} for gene size 3. Other positions of L_{1A} holds 0. In similar way, we created L_{1B} from P_2. Again, we created L_{2A} from P_1 and L_{2B} from P_2 in similar way for gene size 4. In Fig. 1(b), we randomly generated a pattern set which has two patterns P_A and P_B. We created two copy of each pattern because in earlier, the total number of genes $(L_{1A}, L_{2A}, L_{1B}, L_{2B})$ was four. Then we transfered genes $(L_{1A}, L_{2A}, L_{1B}, L_{2B})$ into patterns (P_A, P_B) to create new patterns $(P_A L_1, P_A L_2, P_B L_1, P_B L_2)$. After calculating the objective score of $P_A L_1, P_B L_1, P_A L_2$ and $P_B L_2$, we took the best one among them.

5   Experimental Results

We have implemented our algorithms in JAVA language and have ran our experiments on an Intel core i3 2.27 GHz machine with 4 GB ram running 64bit Windows 7 Home Premium.

5.1   Dataset

In this paper, the datasets named tic-tac-toe, hypothyroid, mushroom and kr-vs-kp that we used are taken from UCI Machine Learning repository [2] and originally used in [5]. The datasets are available to download freely from the website: https://dtai.cs.kuleuven.be/CP4IM/datasets/. The datasets are given in this site with their properties.

5.2   Results

In our experiment, we implemented two version of the genetic algorithms (one with the geneTransfer and one without it) and calculated the objective score by running each algorithm for 1 min. using four datasets. We examined our result with various number of genes (5, 10 and 15), population size (10, 20, 30, 40, 50, 75 and 100), generation before random restart (50, 100 and 200) and percent of change for random restart (50 and 90) for $k = 3, 6, 9, 15$. We ran the experiments five times and took the best score and the average score for each test case where total number of test cases for each dataset was 42 differing the population size. In Table 1, we showed the average objective score for population size 30 and number of genes 15. The best objective scores in each row are shown in bold faced fonts.

Table 1. Objective score achieved by genetic algorithms for various datasets.

Data set	Pattern set size k	GA						GA with geneTransfer					
		Number of generation before random restart						Number of generation before random restart					
		50		100		200		50		100		200	
		Percent of Change		Percent of Change		Percent of Change		Percent of Change		Percent of Change		Percent of Change	
		50%	90%	50%	90%	50%	90%	50%	90%	50%	90%	50%	90%
Tic-tac-toe	3	1916	1916	1916	1916	1916	1916	1916	1916	1916	1916	1916	1916
	6	7892.2	7818	7695.6	7884.4	**7697.2**	7783.8	**7914**	**7951.8**	**7951.4**	**7925.6**	7693	**7890.2**
	9	**17699.2**	17520.4	16488.4	17255.6	**16903.2**	16754.8	16977.6	**18027.6**	17262	17458	16792.8	17460.8
	15	**43588.8**	45539.6	44120.4	43386.4	45460	45086	43095.2	**47950**	44801.6	45490	46370.8	45689.6
Hypothyroid	3	3903.2	4622.8	3375.6	3992	**4016.8**	3842.8	4062.4	**6459.2**	5611.6	**5727.2**	3287.6	5890
	6	6227	**13907.6**	8490	12273	8013.2	**9863.6**	9761.4	13118	10042.4	**12391.8**	6465	9256.4
	9	17567.6	18957.2	**15908.8**	15895.6	16602.8	14584.8	13581.6	**27992.4**	15480	21560	18661.6	25095.2
	15	**33617.2**	29517.2	34991.6	35694.8	16993.6	27356.8	15246	**38000.4**	26089.2	**38943.6**	31734.4	22190
Mushroom	3	13068.8	15979.2	16248	15993.6	12356.8	16248	**16248**	16171.2	16248	16164	**16248**	16248
	6	45795.6	**59755.6**	49703.2	**52431.6**	40180.8	37974.8	**54460.8**	58871.6	**63718.4**	49841.2	**45161.6**	55332
	9	90650.4	**93904**	91235.2	**99552**	87131.2	88196.8	82876.8	83490.4	90468	90091.2	77681.6	**90200.8**
	15	146877.6	162858.8	189993.6	162803.2	134732	**182298.4**	**189194.4**	164743.2	**190962.4**	172888	170994.4	157809.6
Kr-vs-kp	3	4852.4	6148.4	4888.8	**6078**	3679.2	6003.2	5268.4	**6366.8**	5750	6063.2	**3689.6**	6104.8
	6	**16294**	17922.8	**13275**	**17470**	13501.2	18107.4	14544.8	**18659.6**	12243.6	16657.2	**14865**	18276.6
	9	**29638.4**	23279.6	15323.6	**27568.8**	19716.8	27200.4	20184.8	**29472.8**	**23797.6**	26168.8	19932	**29232.8**
	15	29534.4	46320.4	31848.8	34736.8	37123.6	**42602.4**	44463.6	**54268.4**	39506	**55578.4**	37756.8	41171.6

5.3 Analysis

From Table 1, we find that GA with geneTransfer gives best in 60/96 and GA gives best in 28/96. We investigated that GA with gene transfer works well for other population size (10, 20, 40, 50, 75 and 100) and number of genes (5 and 10) but works comparatively very well when number of genes equals to 15 and population size equals to 30 or 50. We saw that when the number of itemset becomes greater, GA with geneTransfer performs very well within a short period of time for all the datasets. But too many or too less population size or number of genes may gives unsatisfactory result.

In Fig. 2, we only depict the performance of GA with geneTransfer for the dataset *tic-tac-toe* where pattern set size is 6, number of genes is 15, generation before random restart is 100 and percent of change for random restart is 90. The performance of GA with geneTransfer remains same for all the other values and datasets. Each figure shows the result for the average objective score and the best objective score. Figure 2(a) and (b) are shows the performance of the algorithms based on their objective score with respect to population size. We find that when population size is in 20–800, GA with geneTransfer gives best result where GA works well only when population size is in 40–500. After that when population size is increasing, the objective score is decreasing. So, when the size of population is too big, it performs not well in allocated time because the calculations become too expensive. In Fig. 2(c), we find that GA with geneTransfer performs well within a very short period of time with compared to GA. In Fig. 2(d), GA with geneTransfer gives high performance than GA.

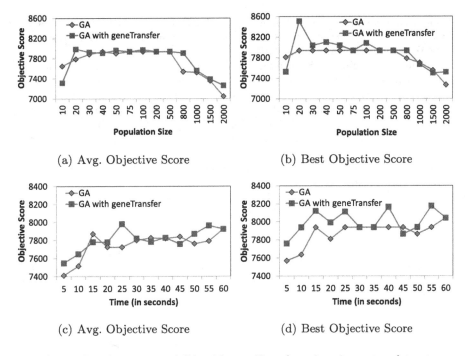

Fig. 2. Search progress of GA with geneTransfer using tic-tac-toe dataset.

6   Conclusion

In this paper, we proposed a new genetic operator called gene transfer within
the genetic algorithm framework to solve the task of mining diverse-frequent
pattern sets. The gene transfer technique plays a major role in reducing the execution time by selecting right patterns for reproduction. The right representation
scheme adopted tends to speed up the system and gave satisfactory result within
a short period of time. In the future, we would like to improve the performance
of the search techniques of GA for large population size within the framework
by using gene transfer, a genetic operator, and applying it in a similar way of
crossover and mutation. Also maintaining a pool of genes through all generations
of the genetic algorithm might result in better convergence.

References

1. Bringmann, B., Nijssen, S., Tatti, N., Vreeken, J., Zimmerman, A.: Mining sets of
   patterns. In: Tutorial at ECMLPKDD (2010)
2. Frank, A., Asuncion, A., et al.: UCI machine learning repository (2010)
3. Guns, T., Nijssen, S., De Raedt, L.: Itemset mining: a constraint programming
   perspective. Artif. Intell. **175**(12), 1951–1983 (2011)
4. Guns, T., Nijssen, S., Raedt, L.D.: K-pattern set mining under constraints. IEEE
   Trans. Knowl. Data Eng. **25**(2), 402–418 (2013)

5. Guns, T., Nijssen, S., Zimmermann, A., De Raedt, L.: Declarative heuristic search for pattern set mining. In: 2011 IEEE 11th International Conference on Data Mining Workshops (ICDMW), pp. 1104–1111. IEEE (2011)

6. Hossain, M., Tasnim, T., Shatabda, S., Farid, D.M.: Stochastic local search for pattern set mining. In: Proceedings of the 8th International Conference on Software, Knowledge, Information Management and Applications (SKIMA) (2014)

7. Khatun, S., Alam, H.U., Shatabda, S.: An efficient genetic algorithm for discovering diverse-frequent patterns. arXiv preprint arXiv:1507.05275 (2015)

8. van Leeuwen, M., Knobbe, A.: Diverse subgroup set discovery. Data Min. Knowl. Disc. **25**(2), 208–242 (2012)

9. Rossi, F., Van Beek, P., Walsh, T.: Handbook of Constraint Programming. Elsevier, Amsterdam (2006)

10. Rückert, U., Kramer, S.: Optimizing feature sets for structured data. In: Kok, J.N., Koronacki, J., Lopez de Mantaras, R., Matwin, S., Mladenič, D., Skowron, A. (eds.) ECML 2007. LNCS (LNAI), vol. 4701, pp. 716–723. Springer, Heidelberg (2007)

11. Shaw, P.: Using constraint programming and local search methods to solve vehicle routing problems. In: Maher, M.J., Puget, J.-F. (eds.) CP 1998. LNCS, vol. 1520, pp. 417–431. Springer, Heidelberg (1998)

Task Allocation Using Particle Swarm Optimisation and Anomaly Detection to Generate a Dynamic Fitness Function

Adam Klyne[✉] and Kathryn Merrick

University of New South Wales - Canberra, Canberra, ACT 2612, Australia
adam.klyne@student.adfa.edu.au, k.merrick@adfa.edu.au

Abstract. In task allocation a group of agents perform search and discovery of tasks, then allocate themselves to complete those tasks. Tasks are assumed to have a strong signature by which they can be identified. This paper considers task allocation in environments where the definition of a task is weak and can change over time. Specifically, we define tasks as environmental anomalies and present a new optimisation-based task allocation algorithm using anomaly detection to generate a dynamic fitness function. We present experiments in a simulated environment to show that agents using this algorithm can generate a dynamic fitness function using anomaly detection. They can then converge on optima in this function using particle swarm optimisation. The demonstration is conducted in a workplace hazard identification simulation.

1 Introduction

During the 2012–2013 period falls, trips, and slips of a person directly accounted for 22.2 % of serious injury claims in Australia and indirectly influenced many more [1]. The cause of such incidents are often the result of human interaction with hazardous situations. A hazard, in this context, may be defined as a particular substance or situation that has the potential to cause harmful effects [10]. Hazards may include liquid spills, fixed or temporary path obstructions or even the natural degradation of walkways and\or thoroughfares from tree roots distorting pavement. In a given environment these dangers occur in a dynamic way with respect to spatial and temporal parameters and are innumerate in their potential combinations. Hence, it is desirable to be able to autonomously identify and inspect unpredictable hazards and allocate resources to mitigate those hazards.

Many industries have been identified as assigning a high level of importance toward good hazard identification in the workplace [3]. However, many risk management processes make the assumption that employees have the requisite knowledge and skills to successfully identify workplace hazards. Bahn [2] found that when employees and managers of an underground mine in Western Australia were surveyed, experience was not a predetermining factor in staff ability to identify hazards across four categories (obvious hazards, trivial hazards, emerging hazards and hidden hazards). This indicates the deficiencies of humans at identifying hazards within workplaces in spite of training and experience.

© Springer International Publishing Switzerland 2015
B. Pfahringer and J. Renz (Eds.): AI 2015, LNAI 9457, pp. 317–329, 2015.
DOI: 10.1007/978-3-319-26350-2_28

We propose using an autonomous, multi-agent system to identify and allocate resources for mitigating physical hazards in the workplace. Our approach will specifically focus on integrating an online K-means clustering algorithm with particle swarm optimisation (PSO). The K-means component discovers tasks as anomalies, while the PSO component allocates resources to tasks. We analyse the performance of the new algorithm in terms of ability to detect anomalies and allocate resources to them.

This paper will be broken up into the following sections: Sect. 2 provides the background and literature in the areas of anomaly detection and PSO which underpin this research. Section 3 outlines the proposed new architecture, Sect. 4 presents our quantitative analysis of our simulation results, and Sect. 5 will conclude the paper and present a brief discussion on the future research issues and directions regarding the development and testing of our algorithm.

2 Background and Literature Review

2.1 Anomaly Detection

Anomaly detection is the process of finding patterns that deviate from the known or expected behaviour of a monitored system [15]. Although anomaly detection has a tendency to issue false-positives, it is versatile in that it has the potential to adapt models of normal and abnormal scenarios. This is a desirable trait as it removes the need for preprogrammed rules [17]. There are typically two learning paradigms associated with anomaly detection [12]; supervised and unsupervised learning. As we will be working with tasks whose nature is unknown, we will use unsupervised learning. In particular, we will use a clustering approach. We chose a neural network based clustering technique over other methods, such as statistical approaches [11], due to the advantage of not having to make a priori assumptions on the data being analysed. This is important as our system will be processing data which may contain anomalies(tasks) whose signature is weak, that is not defined prior to initialisation.

Clustering lends itself to making a determination of *how* similar one data-point is to another, a feature which is important in anomaly detection. As a starting point for our work, we will use a neural network derivation of the K-means clustering technique [13]. This is achieved by mapping cluster centres from traditional K-means, to locations in weight space. The path between the input and output neurons is weighted and undergoes reinforcement, given by (1), which is similar to shifting the cluster centroid locations in the solution space. This process is explained as follows:

$$w_t^{ibest} = w_{t-1}^{ibest} + \eta(g_t^i - w_t^{ik}), \tag{1}$$

where η is the learning rate, g^i is the ith attribute of data point G and w^{ik} weight connecting g^i to the kth output neuron O^k, and w^{ibest} is the ith attribute of $O^{best} = (w^{1best}, w^{2best}, \cdots, w^{kbest})$ as given by (3). The input data is fed through an input layer of neurons, g^i, each one of which are connected to each

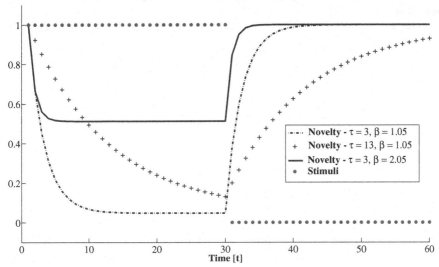

Fig. 1. Impact of parameter values on the shape of the habituation function

of the K output neurons, O^k (see Fig. 2 for further details). The output neurons compete to fire in a *winner takes all* situation, the winning neuron being the one being most similar to the input, using (2) and (3) as outlined below:

$$dist(G_t, O^k) = \sqrt{\sum_i \left(g_t^i - w^{ik} \right)^2}, \tag{2}$$

$$O^{best} = \operatorname*{argmin}_{k=1..K} dist(G_t, O^k), \tag{3}$$

where G_t is the input data vector$(g_t^1, g_t^2, \cdots, g_t^i)$, O^k is the kth output node $(w^{1k}, w^{2k}, \cdots, w^{ik})$. $dist(G_t, O^k)$ is the Euclidean distance between the input and O^k. This function will return a lower value for pairs of objects that are more similar to one another.

paginationUsing just an input and output layer means that, not only will the error associated with normal data tend toward zero (a desirable trait) so to will the error associated with anomalous data over a great enough time span. To combat this we add a habituation layer to the neural network. Habituation is a neurological phenomena that is one of the simplest forms of brain plasticity. Synaptic plasticity is the ability of synapses[1] to strengthen or weaken over time in response to increases or decreases in their activity. When an animal is repeatedly exposed to a stimuli it will eventually stop responding to that signal. That is, it becomes normal to the animal, which is then said to have become habituated to

---

[1] Synapses in our neural network are represented by the weight paths that link out input and output nodes.

the signal. This behaviour has been observed in a range of animals from humans down to sea slugs [14].

Marsland et al. [14] suggest using (4) to dis-habituate the system as follows:

$$\frac{dN_{t-1}^k}{dt} = \frac{\beta[N_0^k - N_{t-1}^k] - \zeta_t}{\tau}, \tag{4}$$

$$N_t^k = N_{t-1}^k + \frac{dN_{t-1}^k}{dt},$$

$$\zeta_t = \begin{cases} 1, & \text{if } O^k \text{ fired} \\ 0, & \text{otherwise} \end{cases},$$

where N_t^k is the kth synaptic efficacy (novelty synapses linking the output node layer to the habituation layer) at time t, τ and β are constants governing the rate of habituation and recovery respectively.

Using (4), we can see from Fig. 1 that the error associated with an output node will decrease in a non-linear fashion whilst being subjected to stimuli. When the stimuli ceases we see that the error increases, in a manner inverse to its decrease, over time. Also from this figure we can see that by varying the parameters β and τ the response of the novelty filter may be tailored to suit different criteria.

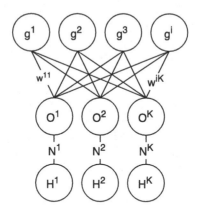

Fig. 2. Online K-means with Habituation Layer

Figure 2 shows the online K-means anomaly detection model with habituation layer. This approach is important as it allows us to detect *similar* anomalies repeatedly with a lower risk of anomalies being mislabelled as *normal* data.

As anomaly detection can only highlight regions of interest within our search space we will now discuss the PSO technique that will be used to move the agents to the detected hazard sites.

2.2 Particle Swarm Optimisation

PSO is a stochastic meta-heuristic algorithm that is inspired by emerging social behaviours found in nature, such as flocking birds or schools of fish [5]. Each member (particle) of the PSO population, known as a cloud or swarm, is a candidate solution to the problem at hand. The algorithm begins by randomly initialising the particle swarm within an n-dimensional search space. As the particles move throughout the space their individual positions ($\overrightarrow{x}_a(t)$) are updated using (5); this is calculated by adding a velocity ($\overrightarrow{v}_a(t)$) vector to its own current position. The velocity of a particle is calculated by using its own previous best position and the location within the search space of the globally best found solution. Traditionally (5) is defined as:

$$\begin{aligned}
\overrightarrow{x}_a(t) &= \overrightarrow{x}_i(t-1) + \overrightarrow{v}_a(t), \\
\overrightarrow{v}_a(t) &= \omega \overrightarrow{v}_a(t-1) + c_1 r_1 (p_a^{best}(t) - \overrightarrow{x}_a(t-1)) \\
&\quad + c_2 r_2 (s^{best}(t) - \overrightarrow{x}_a(t-1)),
\end{aligned} \tag{5}$$

where a constitutes the ath agent in the set of agents A, $\overrightarrow{v}_a(t)$ is the ath particle velocity at time t, $\overrightarrow{x}_a(t)$ is the ath particle position (solution) at time t, $p_a^{best}(t)$ is the best solution (fitness) achieved so far by the ath particle, $s^{best}(t)$ is the global best fitness achieved so far, r_1 and r_2 are random numbers, $U(0,1)$, between 0 and 1 with uniform distribution, ω is the inertial weight of the ath particle, and c_1 and c_2 are acceleration coefficients. The convergence behaviour of PSO algorithms is dependent on the values of the inertial weight and the acceleration coefficients, which are also known as the cognitive and social weights respectively.

The standard PSO algorithm was designed for optimisation of a static fitness function. Later variants propose different approaches to overcome its shortfalls in dynamic environments [4,7,9]. Whilst each of the adapted algorithms is successful in improving upon the efficiency of PSO in dynamic environments, they do not explicitly incorporate a mechanism for generating a dynamic fitness function.

The contribution of this paper is an algorithm for integrating anomaly detection with PSO to achieve a task allocation algorithm when the signature of a task is not well understood in advance. We assume that more than one task (characterised as an anomaly) may arise over time, but that the environment changes slowly enough for us to apply a standard PSO approach for converging agents around anomalies.

3 Integrating Anomaly Detection for Fitness Function Generation in PSO

Our new algorithm is shown in Algorithm 1. From lines 1–3, we can see that the system begins each simulation by initialising both the PSO agents and anomaly detector. The agents are initialised as per the canonical PSO discussed in Sect. 2.2 and the anomaly detector generates its neural network structure by randomly assigning values to the weight paths w^{ik}. It should be noted that our

K-means neural network does not have specific training\testing phases. Once initialised the synaptic weights are continually updated, as required, throughout the system's life-span. Once initialised the system moves into its main intelligence loop, occurring from line 4.

Algorithm 1. Task allocation using particle swarm optimisation and anomaly detection to generate a dynamic fitness function – a centralised control strategy.

1: **for** each agent **do**
2:     Initialise with random $\overrightarrow{x}_i$ and $\overrightarrow{v}_i$.
3: Initialise anomaly detector
4: **repeat**
5:     Obtain sensor data
6:     Perform anomaly detection on sensor data using (1) – (4)
7:     Generate fitness function using (6)
8:     **repeat**
9:         Perform PSO (5)
10:         Move agents to new positions
11:     **until** further anomaly detection required
12: **until** termination condition

Sensor data is sent to the anomaly detector in line 5. As outlined in Sect. 2.1 the data is analysed for anomalies (line 6) and a fitness function generated (line 7) in the form of an intensity field using (6), defined as:

$$I_{new} = \sum_{j=1}^{J} I_{current} + \frac{\varphi(j)}{1 + \gamma(j)(L_x(j)^2 + L_y(j)^2)}. \tag{6}$$

This fitness function was chosen for its ability to form gradient slopes decreasing from the environmental Cartesian coordinates(L_x and L_y) indicated by Algorithm 1. This is achieved by not only increasing the magnitude of the coordinate associated with the anomalous data-point, the rate of which is governed by φ, but also by slightly increasing neighbourhood coordinate points. The affected neighbourhood area is tuned via the γ parameter. This feature is beneficial for our agent swarm as it allows for the swarm to discover a gradient and follow it to it's summit. Additionally it should be noted that the fitness function parameter $I_{current}$ is initialised as zero.

Once the fitness function is generated our agent swarm will be driven about the environment using PSO (lines 9 and 10). This is in a nested loop as we do not anticipate that the anomaly detector will need to run at the same rate as the PSO component. This is investigated and discussed in Sect. 4.1.

4   Experimental Evaluation in a Workplace Hazard Detection Simulation

Our domain of interest in this paper is workplace hazard detection and mitigation. That is, we wish to use a swarm of agents to detect and either clear up, or warn passers by, of a hazard. We assume that we do not have a strong task signature for hazards, and instead identify hazards as anomalies. We further assume that our system has access only to image frames (such as those collected from a camera) from which to detect anomalies. This section presents three experiments. The first experiment in Sect. 4.1 uses randomly generated colour data to investigate the processing time characteristics of our anomaly detector using different neural network sizes. The second experiment in Sect. 4.2 uses one of the feasible network sizes identified in Experiment 1 to test the anomaly detection component on three simulated room images. The third experiment in Sect. 4.3 tests the end-to-end performance of the system using both the anomaly detection and PSO components. The three experiments were implemented using MATLAB 2012b running on a machine with an Intel 3.20 GHz CPU and 16.0 GB RAM within a Windows 7 operating environment.

4.1 Experiment 1: Anomaly Detection Processing Time

The first experiment examines only the anomaly detection component of the algorithm. The aim of this experiment was to investigate what values are appropriate for the anomaly detection parameters, in particular, the number of output neurons in the K-means network. This value of K was chosen using an *elbow method* [8, p. 519]. We generated 1×10^5 points of random colour data, this value was chosen as it is within the same order of magnitude as the input data, in the form of a VGA camera image (640×480 pixels) that we used in Experiments 2 and 3. We fed this data into the anomaly detector and computed the average network learning error against computation time for various numbers of output nodes (10, 100, 1000, and 1×10^4). It was found that there came a point where the reduction of error was outweighed by the increased computation time. We ran this experiment five times with the averaged results available in Fig. 3. In this experiment the learning rate of $\eta = 0.25$ was obtained from [13, p. 203]. The novelty function parameters (see 4) were fixed as shown in Table 2. The effect of altering β and τ on a small data set may be seen in Fig. 1. Parameter values for (6) were also fixed for Experiment 1.

4.2 Experiment 2: Minimum Anomaly Size

Our test environment for this experiment is a simulated 2-Dimensional quantised space, that we assume to represent a 640×480 pixel (VGA sized) image frame. We assume the space represents a room with a textured floor colour. Three floor colours were simulated: mottled grey, mottled orange and a brown\maroon checkered pattern(each measuring 5×5 pixels). In order to gain some appreciation of how the algorithm may perform in a physical environment we consider each pixel to be a $0.00156\,\text{m} \times 0.00156\,\text{m}$ square which equates to an equivalent environment size of $10\,\text{m} \times 7.5\,\text{m}$. Example situations where such large, open spaces, may be found are shopping centres, offices or warehouses. A random circularly shaped colour patch of a different colour to the floor was inserted into

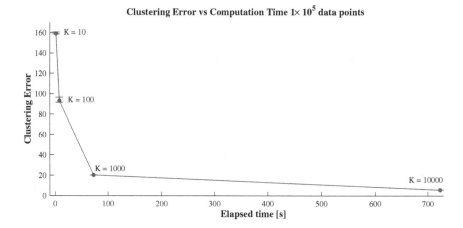

Fig. 3. Average clustering error vs computation time. The computation time is the average time over five simulations and the error bars indicate the error variance (Color figure online)

the environment representing the hazard to be detected. The aim of this experiment was to investigate the minimum size of a detectable anomaly in a single pass over a single image frame required to detect it. The system was first fed a *clean* image frame, that is no anomalies were present, and then a frame with a randomly generated anomaly. When one of the habituation layer neurons novelty score falls within the predefined novelty range(0.3–0.8) the associated data is flagged as anomalous. The aforementioned range was fixed for Experiment's 2 and 3.

Using these parameters we were able to detect objects in the simulated environment down to an average size of 0.16 % of the environment area in a single pass of the anomaly detector over an image. This correlates to a pixel density of 500 pixels in a 26×26 pixel area or approximately $0.035\,\mathrm{m} \times 0.035\,\mathrm{m}$ square, i.e. a standard cup of coffee worth of liquid. We found that the pixel density of the object had an effect on the systems ability to detect it. This is primarily an artefact of τ. For the value used in this experiment we found that 104 sequential node activations was required to drop the novelty value below 0.8, from 1, but 18182 activations was required to raise the novelty value by the same amount. From Table 1 it can be seen that environmental surface colour did not meaningfully effect the results.

4.3 Experiment 3: Anomaly Detection and PSO

Our test environment for this experiment is using the same series of simulated quantised spaces as in Experiment 2. Additionally, we set the agent maximum velocity to be 64 pixels iteration or $1\,\mathrm{m/s}$ equivalent physical speed. We ran the simulation 15 times, five times against each of the three background types as

Table 1. Minimum anomaly size

Environment	Average area		Std dev
	(% of environment)	(Pixels)	
Checker	0.16	504	12.87
Grey	0.16	502	12.62
Orange	0.16	506	13.08

outlined in Experiment 2. This time, each simulation runs for 1800 iterations, representing 30 min of wall clock time – assuming each iteration, or time-step, is 1 s. Randomised spacial and temporal anomalies were generated and cleared in 300 time unit cycles; i.e. an anomaly will spawn at $t = 300$, is cleared(whether detected or not) at $t = 600$, the second anomaly spawns at $t = 900$ and cleared at $t = 1200$, and the final anomaly spawns at $t = 1500$. As we are using the canonical PSO to drive our agents we only allow one anomaly to be present within the environment at any one time as it is suited to only locating single optima. Anomalies spawn as a coloured (various shades of blue) circularly shaped 'blob' and will represent a hazard that may be a spill requiring 'cleaning up' by one of our agents or a trip hazard that requires guarding against; that is our system will be required to identify anomalies, characterise them as tasks and then allocated resources in order to fulfil task requirements. Within the simulation this is represented by the agents changing colour when they perceive they are close to an anomaly.

In this experiment, to measure the end-to-end performance of our algorithm we employed measures of swarm diversity and error associated with the PSO perceived best fitness function result and the average swarm best.

In order to measure diversity we use a method known as the moment of inertia [16], given as:

$$J = \sum_{i=1}^{N} \sum_{a=1}^{|A|} (x_{ia} - \bar{x}_i)^2,$$
$$\bar{x}_i = \frac{1}{|A|} \sum_{a=1}^{|A|} x_{ia}, \tag{7}$$

where x_{ia} is the ith coordinate of the ath agent and $\bar{x}_i$ is the ith coordinate of the average agent position. N and $|A|$ denote the number of 'spatial' dimensions and size of the agent population respectively. It should be noted that for scalability we normalise the result of (7). This allows us to evaluate the ability of algorithm in maintaining diversity to deal with environmental dynamics such as the introduction of new tasks into the problem space as well as to indicate how well the agent swarm reacts to identified anomalies. We consider convergence occurs when the normalised moment of inertia is zero, this behaviour is evident in Fig. 4.

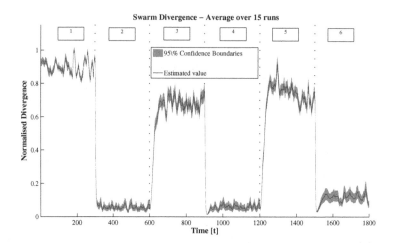

Fig. 4. Average simulation swarm divergence for $K = 1000$ over 15 simulations with a 95 % confidence interval. The numbered regions indicate when the anomaly detector is run (at the start of each region) and distinctly show that swarm divergence is a function of anomalies being detected within the environment (Color figure online)

For the simulations conducted as a part of Experiment 3 we used the parameter values in Table 2, the same as those used in Experiment 1 and 2, with the addition of the PSO parameter values obtained from [6].

Table 2. Simulation parameters

PSO			NN		Novelty				Fitness function	
c_1	c_2	ω	K	η	β_{decay}	τ_{decay}	β_{growth}	τ_{growth}	γ	φ
1.49	1.49	0.729	1000	0.25	0.03	500	0.09	750	0.05	1

From Experiment 1, we found that when using $K = 1000$ output nodes the computation time of the anomaly detector over 15 runs, five each of our three environment backgrounds, with an average time of 4 min 02.422 s. For this reason we chose to run the anomaly detector once at the start of the simulation with no anomaly's present to give the system a sense of normalcy. The anomaly detector was then run every 5 min of wall clock time, i.e. every 300 time units. It is felt that an update to the fitness function once every 5 min will to allow enough time for the anomaly detection to complete an entire iteration and for the agents to respond to any change in the fitness function. On the surface the processing time may seem excessive. If we consider a physical situation in a shopping centre, for example, it is not unreasonable for it to take longer than five for a member of staff to become aware of a spill, barricade it, and then fetch the required equipment to clean it up.

The system's ability to successfully allocate resources to detected anomalies was measured using the error, calculated as the difference between the global best fitness function value and the average swarm best. The value of the global best directly correlates to the size of the detected anomaly, with larger values indicating spatially larger anomalies. Figure 4 shows that once an anomaly is detected by the swarm, indicated by the divergence tending toward zero, the swarm converges on the location of the detected anomaly. Typically, the swarm converges on the anomaly in an average of 16 time units. Similarly, once the system detects an absence of anomalies we can see that the swarm immediately begins to diverge and move about the environment until a new anomaly is detected. It should be noted that the swarm best and each agent best fitness value is reset each time the anomaly detector is run. This behaviour indicates that the system is well placed to cope with slowly occurring anomalies. The data in Table 3, showing true\false anomaly detections, corroborates with the average swarm divergent behaviour as seen in Fig. 4. We can also see that whilst the system detects 95.56 % of the spawned anomalies it also 'found' non-existent anomalies in 11.11 % cases indicating the system may require additional settling time, i.e. multiple detector runs between anomalous events occurring in the environment.

Table 3. Simulation results over 15 iterations, $K = 1000$ - true vs false anomaly detections

Anomaly detector iteration	1 (False)	2 (True)	3 (False)	4 (True)	5 (False)	6 (True)
Number detected	0	14	3	15	2	14
Standard deviation	0.00	0.58	1	0.00	1.154	0.58
95 % Confidence interval	0.00	0.46	0.80	0.00	0.92	0.46

A second series of simulations was run to discover the effect of reducing the number of anomaly detector output nodes to $K = 100$. The computation time reduced to an average time of 31.29 s confirming the expectation set from Experiment 1. Overall, the system performance improved slightly in that all anomalies were detected and false positives were lower as seen in Table 4. However, further testing will be required in order to establish the statistical significance of the result. As would be expected from using a traditional PSO there was no significant change in ability of the swarm to converge on the globally indicated optima (see Table 4), except for the 'grey' case. As the global optima is a product of the 'shape' of the fitness function if multiple peaks are present the swarm will settle on the first peak that is found. In all cases, where the agent swarm focussed on a local optima it was in the vicinity of the detected anomaly.

Table 4. Simulation results over 15 iterations

Environment	Nodes	True anomalies detected (%)	False positive (%)	Global optima (%)
Checker	100	100.00	0.00	86.67
	1000	100.00	6.67	86.67
Grey	100	100.00	0.00	80.00
	1000	86.67	26.67	93.33
Orange	100	100.00	0.00	93.33
	1000	100.00	0.00	86.67

5   Conclusion and Future

This paper has described a method of autonomously detecting and mitigating against hazards in the workplace. We outlined the framework within which an optimising task allocation algorithm is produced which is able to elicit it's own fitness function from environmental stimuli which has not been predefined.

It was shown, by using an *elbow method*, that 1000 output nodes was appropriate for the system under test. However, it was later discovered that this particular system is somewhat tolerant to error and hence a smaller number of output nodes could be used. The result was faster computation time and comparable performance with respect to the system detecting anomalies and allocating resources to the indicated location.

Initial experimental results conclude that the system is able to detect anomalies down to a size of 0.16 % of the environment but is sensitive to the pixel density and location of the object.

Finally, we have shown that a fitness function may be elicited from an environment containing weak task signatures in the majority of cases and further, that this fitness function is able to provide the required impetus for allocating resources to the discovered task.

Future work will involve testing the anomaly detection sensitivity, that is how different an anomaly can we detect, and running simulations with real world camera data. Further investigation is required to ascertain the performance and stability implications of Algorithm 1 with respect to adjusting the various parameter values outlined in Table 2. Another area of interest will be investigating the performance of less centralised system architectures. Such architectures may involve equipping each of the agents with their own 'sensors' or perhaps have each agent operate autonomously.

We have not considered the implication of humans in the environment being potentially flagged as moving anomalies, as such we have only simulated static anomalous objects and will increase the complexity of further simulations in the future. One potential is that the interaction between agents and simulated humans in the environment will be such that if humans are detected they are treated, by the agents, as impenetrable obstacles to be avoided.

References

1. Australia, S.W.: Australian Workers Compensation Statistics, 2012–13 (2013). http://www.safeworkaustralia.gov.au/sites/SWA/about/Publications/Documents/897/australian-workers-compensation-statistics-2012-13.pdf
2. Bahn, S.: Workplace hazard identification and management: the case of an underground mining operation. Saf. Sci. **57**, 129–137 (2013)
3. Biggs, H.C., Sheahan, V.L., Dingsdag, D.P.: Improving industry safety culture: the tasks in which safety critical positions holders must be competent. In: 2006 CIB99 International Conference on Global Unity for Safety & Health in Construction, pp. 181–187 (2006)
4. Blackwell, T.: Particle swarm optimization in dynamic environments. Intelligence (SCI) **51**, 29–49 (2007)
5. Eberhart, R., Kennedy, J.: A new optimizer using particle swarm theory. In: MHS 1995, Proceedings of the Sixth International Symposium on Micro Machine and Human Science, vol. 1, pp. 39–43 (1995)
6. Eberhart, R., Shi, Y.: Comparing inertia weights and constriction factors in particle swarm optimization. In: Proceedings of the 2000 Congress on Evolutionary Computation, vol. 1, pp. 84–88. IEEE (2000)
7. Fernandez-Marquez, J., Arcos, J.: Adapting particle swarm optimization in dynamic and noisy environments. In: 2010 IEEE Congress on Evolutionary Computation (CEC), pp. 1–8 (2010)
8. Hastie, T., Tibshirani, R., Friedman, J.: The Elements of Statistical Learning. Springer Series in Statistics, vol. 18. Springer, New York (2001)
9. Karimi, J., Nobahari, H., Pourtakdoust, S.H.: A new hybrid approach for dynamic continuous optimization problems. Appl. Soft Comput. J. **12**(3), 1158–1167 (2012)
10. Manuele, F.: Acceptable risk - time for SH&E professionals to adopt the concept. Prof. Saf. **55**(May), 30–38 (2010)
11. Markou, M., Singh, S.: Novelty detection: a review - part 1: statistical approaches. Sig. Process. **83**(12), 2481–2497 (2003)
12. Markou, M., Singh, S.: Novelty detection: a review - part 2: neural network based approaches. Sig. Process. **83**(12), 2499–2521 (2003)
13. Marsland, S.: Machine Learning: An Algorithmic Perspective, 1st edn. CRC Press, New York (2011)
14. Marsland, S., Nehmzow, U., Shapiro, J.: A Real-Time Novelty Detector for a Mobile Robot, p. 8 (2000). arXiv preprint cs/0006006
15. Miljkovic, D.: Review of novelty detection methods. In: MIPRO, 2010 Proceedings of the 33rd International Convention, pp. 593–598 (2010)
16. Morrison, R.W., De Jong, K.A.: Measurement of population diversity. In: Collet, P., Fonlupt, C., Hao, J.-K., Lutton, E., Schoenauer, M. (eds.) EA 2001. LNCS, vol. 2310, pp. 31–41. Springer, Heidelberg (2002)
17. Shafi, K., Merrick, K.: A curious agent for network anomaly detection. In: The 10th International Conference on Autonomous Agents and Multiagent Systems, vol. 33, pp. 1075–1076 (2011)

Evolving High Fidelity Low Complexity Sheepdog Herding Simulations Using a Machine Learner Fitness Function Surrogate for Human Judgement

Erandi Lakshika[1]([✉]), Michael Barlow[1], and Adam Easton[2]

[1] School of Engineering and IT, The University of New South Wales,
Canberra, Australia
erasuru@gmail.com
[2] Simcentric Technologies, Oxford, UK

Abstract. Multi-agent simulations facilitate modelling the complex behaviours through simple rules codified within the agents. However, the manual exploration of effective rules is generally prohibitive or at least sub-optimal in these types of simulations. Evolutionary techniques can effectively explore the rule and parameter space for high fidelity simulations; however, the pairing of these two techniques is challenging in classes of problems where the evaluation of simulation fidelity is reliant on human judgement. In this work we present a machine learning approach to evolve high fidelity low complexity sheepdog herding simulations. A multi-objective evolutionary algorithm is applied to evolve simulations with both high fidelity and low complexity as the two objectives to be optimised. Fidelity is measured via a machine learning system trained using a small set of training samples of human judgements; whereas the complexity is measured using the number of rules and parameters used to codify the agents in the system. The experimental results demonstrate the effectiveness of the approach in evolving high fidelity and low complexity multi-agent simulations when the fidelity cannot be measured objectively.

Keywords: Multi-agent simulations · Machine learning · Complexity · Fidelity · Evolutionary computing · Multi-objective optimisation

1 Introduction

Multi-agent simulations are a type of simulations which facilitate modelling the complex behaviours of real world entities through agents codified with simple rules. The decentralised nature, autonomous operation and ability to capture complex macro level emergent phenomena through micro level interactions make them desirable in simulating complex systems that are otherwise difficult to understand or simulate. High fidelity in simulation (in this case, highly realistic behaviours) is desirable; however high fidelity often results in high computational

© Springer International Publishing Switzerland 2015
B. Pfahringer and J. Renz (Eds.): AI 2015, LNAI 9457, pp. 330–342, 2015.
DOI: 10.1007/978-3-319-26350-2_29

cost and complexity. Therefore computational complexity and simulation fidelity are often two conflicting objectives that need to be optimised based on the overall goals in simulation usage and available resources.

Multi-objective evolutionary algorithms provide a sophisticated mechanism to optimise two or more conflicting objectives simultaneously. Employing evolutionary algorithms to evolve high fidelity multi-agent simulations is challenging when the fidelity of the simulations depends on human judgement of the perceived realism of the synthesised behaviours rather than empirical data. In this work we overcome this challenge by training a machine learning system based on human evaluations, which is then employed as a fitness function to the evolutionary algorithm.

We present a case study of evolving high fidelity low complexity sheepdog herding dynamics to show the effectiveness of the approach by employing the Nondominated Sorting Genetic Algorithm II (NSGA-II) [1]. Sheepdog herding dynamics exhibit an interesting form of interaction between sheep and the dog. Sheepdog herding is extensively used in agriculture, and simulation of such behaviours has been studied in the literature due to its many application areas [2, 3]. In this study we employ the conceptual rule set presented in [3] for codifying sheep's behaviour as the objective of this study is evolve an existing rule set towards high fidelity and low complexity sheepdog herding simulations rather than devising a completely new set of rules. The said rule set has been chosen due to its simplicity and similarity to the simple rule based multi-agent approach we consider in this work. Moreover, the concepts associated with the rule set are based on the literature and observations of real sheepdog herding trials.

In this work, our primary contributions are: (1) introducing a novel approach to evolve sheepdog herding simulations when the fidelity is measured based on the 'look and feel' of the simulations (human judgement); (2) introducing multi-objective optimisation as a framework to evolve high fidelity low complexity sheepdog herding simulations; (3) developing an understanding of which rules are important in generating high fidelity sheepdog herding dynamics; and (4) developing an understanding of the properties of the optimal solution sets in order to understand a relationship between simulation fidelity and complexity of sheepdog herding dynamics.

The rest of the paper is organised as follows. The next section introduces related work in the problem area. The subsequent section presents the sheepdog herding behaviour model. This is followed by the experimental approach. Thereafter, the experimental results are presented together with a detailed discussion. Finally, the conclusions and future work are presented.

2   Related Work

The seminal boid simulation [4] demonstrated the effectiveness of the simple rule based approach in synthesising realistic collective behaviours. Following this a number of works in the literature applied the same approach to simulate realistic behaviour dynamics [3, 5–8]. Sheepdog herding behaviours demonstrate an interesting form of interaction between a flock of sheep and a herding dog and simulating such behaviours has a number of applications, especially in agriculture [2]. As such numerous works in the literature have attempted to synthesise sheepdog herding behaviours through various approaches [2,3,9–11]. The work presented in [3] describes a set of simple spatial and temporal rules to synthesise high fidelity sheepdog herding behaviours. This work presents six (6) spatial rules and one (1) temporal rule for synthesising the behaviour of a flock of sheep and four (4) simple spatial rules and one (1) temporal rule for the dog. The work also highlights the difficulty of deriving an appropriate objective function to co-evolve the behaviours towards high fidelity.

The work presented in [7] demonstrates the effectiveness of training a machine learning system to re-create human aesthetic judgement of the behavioural fidelity of the simulations. The work applies this approach to evolve high fidelity standing conversation group dynamics with the intent of exploring the relationship of simulation model complexity and fidelity. However, the approach presented in [7] requires multiple evolutionary runs (to explore simulations with different number of rules) to evolve high fidelity low complexity simulations due to the use of the standard single objective evolution.

In this work we present complexity and fidelity as two conflicting objectives that need to be optimised when developing multi-agent simulations. The preferable approach to multi-objective optimisation problems are to derive a set of optimal solutions that can be examined for different trade-offs with additional information. This type of optimal solution set is known as the Pareto-optimal solutions. A set of solutions is said to be Pareto-optimal if they are non-dominated with respect to each other [12]. In other words, no solution in the Pareto-optimal set can be treated as better than any other solution with respect to all the objectives. In this work we apply NSGA-II [1] in order to evolve high fidelity and low complexity simulations: i.e.: to generate a set of Pareto-optimal solutions that can be examined for complexity-fidelity trade-off. NSGA-II is a well-tested single parameter algorithm which has proven to be capable of finding a good spread of solutions closer to the true Pareto front. Further, it has been successfully employed in deriving the Pareto-optimal solutions in complex landscapes in different problem domains such as combat scenario planning [13], air traffic control [14], and simulation optimisation [15].

3   Sheepdog Herding Behaviour Model

The sheepdog herding behaviour model presented in this work is adapted from [3]. The objective of the simulation is to realistically simulate the behaviour of a flock of three sheep when they are herded by a herding dog (sheepdog). Initially, the flock is placed at the far end of an oval shaped ground- directly opposite to the starting location of the dog. The task of the dog is to bring the flock back to a target located at the dog's own starting position as quickly and as directly as possible (a kind of retrieval task). This task is illustrated in Fig. 1.

In this work, we only explore the spatial rules for sheep as the objective of the work is to explore the effectiveness of our proposed approach in evolving high

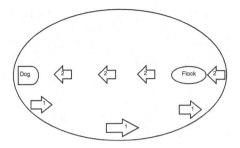

Fig. 1. An illustration of the herding scenario. The dog is required to bring the flock of sheep placed at the far end of the ground to its starting location as directly as possible. The flock is required to appropriately respond to the dog while maintaining their flocking behaviours.

fidelity and low complexity simulations. Therefore the rules for the dog were kept fixed. In doing so, we selected the two best performing rule sets (together with their weights and parameters) found for the dog by co-evolving the rule sets for the dog and sheep in the pilot studies. Both the best two rule sets for the dog consist of the two rules: *repel from the goal* and *attract to the flock*, but with different weights. The rule *attract to the flock* forces the dog to move towards the flock whereas the rule *repel from the goal* enables it to turn the flock towards the goal when combined with the rule *attract to the flock*. In the first rule set, the weight for the rule *repel from the goal* is 0.42 and the weight for the rule *attract to the flock* is 0.58. In the second rule set they are 0.28 and 0.72 respectively.[1]

The rule space for the sheep we explore in this work consists of 6 rules. The six spatial rules are, *separation, alignment, cohesion, repel from the dog, repel from walls* and *wandering*. The rule *repel from the dog* has two implementations due to different parameter choices and the two implementations are referred to as *repel from the dog-A* and *repel from the dog-B*. The rule *separation* forces the flock members to move away from each other if they are within a threshold distance. The rule *alignment* enables them to move in the same direction. The rule *cohesion* forces them to maintain the flock cohesion by moving towards each other. The rule *repel from the dog-A* generates a repulsion force from the dog with a speed inversely proportional to the separation distance and the rule *repel from the dog-B* generates a repulsion force from the dog only if the distance between the dog and the flock members is within a certain threshold. The rule *repel from walls* forces the flock members to move away from the walls if they are within a threshold distance to the wall. The rule *wandering* adds some randomness to the movements of the flock members.

---

[1] Note that these weights were decided by co-evolving the dog and sheep in our pilot experiments. Although we have provided the weights with a high precision, that precision has no significant impact on the behaviours. A higher weight for the rule *attract to the flock* makes the dog more aggressive.

Table 1. A list of parameters for each rule for sheep agents

Rule	Parameter	Conceptual function
Separation	Separation distance	Keep sheep separate
Alignment		Move together as herd
Cohesion		Stay together as herd
Repel from the dog-A	Speedup parameter	Move away when dog approaches
	Repulsion threshold	
Repel from the dog-B	Speedup parameter	Move away when dog approaches
	Activation distance	
Repel from walls	Activation threshold	Do not collide with walls
Wandering	Wandering angle	Add variability & individuality to motion
	Activation threshold	
	Speedup parameter	

The rules have parameters that need to be set appropriately for high-fidelity behaviours. A list of parameters for each rule is shown in Table 1. As the behaviours codified in the rules are sensitive to the parameters it is important to tune them correctly to obtain the desired behaviours. More rules in the system means that more parameters must be tuned. Further, as the number of rules increases the amount of computational resources required to run the model also increases. Therefore, the objective of this work is to automatically explore the rules space while automatically tuning the parameters to generate high fidelity simulations with low complexity (i:e: in this case low number of rules).

4   Experimental Approach

We propose a multi-objective evolutionary approach to explore the rule space (of sheep) towards generating high fidelity, low complexity sheepdog herding simulations. The two objectives that need to be optimised are complexity and fidelity. In this case we consider that when the number of rules in the system increases, the complexity of the simulation also increases. This appears to be natural in this case, because, as the number of rules increases the non-linear interactions happen between the rules and their parameters also increase. These interactions give rise to emergent behaviours that demonstrate the real life complexity- as such high fidelity behaviours. Therefore, in order to measure the complexity of the simulation we use two simple measures (1) rule count (2) parameter count.

Measuring fidelity of the simulations is not straight forward, especially in situations where the fidelity depends on the look and feel to the human eye. In this work, we address the problem of measuring the fidelity in cases where no empirical data is available or no objective forms for fitness functions are available. Interactive Evolutionary Algorithms have been proposed in the literature [16–18] to address the cases where the fitness depends upon human judgement. One of the common problems in an interactive evolutionary system is user fatigue [19]. Further, there is a practical need to keep the population size low. Therefore a more sophisticated mechanism is required to handle such situations.

In this work we apply the bootstrapping approach presented in [7], i.e.: training a machine learning system to recreate human aesthetic judgement. The approach we have taken can be described as follows: (1) Generate a sample of simulations using different permutations of rules and parameters to sufficiently capture the dynamics of the sheepdog herding simulations (2) Present them to human subjects and obtain their score for the fidelity of each of the simulations (3) Collect all potential features that the human subjects used to determine their scores for the simulations (4) Extract such features from the simulations and train a machine learning system to generate scores for unseen simulations (5) Employ NSGA-II [1] to explore the rule and the parameter space of the sheepdog herding model to generate high fidelity low complexity simulations with minimising the rule count as one objective and maximising fidelity (measured through the trained machine learning system) as the second objective.

5   Experimental Results and Discussion

5.1   Obtaining Human Evaluations

First, we generated 17 simulations using different permutations of the rules and parameter combinations. These permutations were selected out of all possible permutations based on pilot studies with the goal of sufficiently capturing the different aspects of dynamics that could occur in the simulations. Each simulation was generated as a video of about 30 s in length representing the scenario illustrated in Fig. 1- i.e.: a scenario where a dog is herding 3 sheep in a flock to a goal.

20 subjects were recruited on a voluntarily basis. Demographic information concerning the subjects was collected at the beginning of the survey and also information about whether they have watched any real sheepdog herding before was collected for further analysis.

After collecting the demographic information, the subjects were shown a video of sheepdog herding filmed at the 2012 Australian National Sheepdog Trials in order to make them familiar with the problem domain. This video was about 5 min in length. The subjects were then asked to list down the features they employ to determine the naturalness of sheepdog herding dynamics in a simulation. Afterward, the subjects were shown the 17 videos (each is less than 30 s length) one at a time in a randomised order and asked to rate the naturalness (based on aesthetic judgment) of the scenario in a range of 0–9, 0 being not realistic at all and 9 being as realistic as the real video shown. For each scenario the justification for their rating was also obtained. Finally, having watched the videos, the subjects were again asked to list down the features that they employed in determining their ratings.

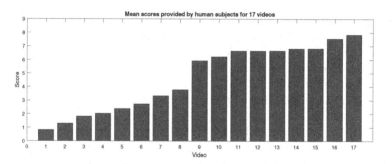

Fig. 2. Mean of the scores provided by human subjects for each video. Each video is a simulation of the herding task illustrated in Fig. 1. The videos are sorted from the lowest mean score to the highest mean score.

5.2 Training a Machine Learning System

Data collected through the process described in the previous section was utilised to explore the possibility of training a machine learning system to capture human aesthetic judgment. Based on the features listed by the subjects and by analysing the footage filmed at the 2012 Australian National Sheepdog Trials, a number of simulation features (per agent) were considered to capture the aesthetics as follows: *Distance to the centre of the flock in the current simulation tick, Average distance to the other sheep in the flock in the current simulation tick, Distance to the closest sheep in the flock in the current simulation tick, Distance to the furthest sheep in the flock in the current simulation tick, Distance the sheep moved, Simulation ticks the sheep kept moving, Simulation ticks the sheep did not move, Number of collisions with the flock members, Number of collisions with the dog, Simulation ticks the sheep went outside, Distance maintained to the flock by the dog in the current simulation tick, Simulation ticks the dog kept moving, Simulation ticks the dog did not move* and *Distance maintained to the goal by the dog in the current simulation tick.*

Since most of the features are recorded per simulation tick (and also per agent), the feature recording resulted in a large amount of data per feature. Therefore, in order to summarise the data obtained for a particular feature, the mean and the standard deviation of the recorded values were obtained. In addition, the change (delta) of the feature values between two consecutive time ticks were calculated to capture the temporal variability and mean and the standard deviation of those delta features were obtained. For example, for any single feature listed above, this process creates 4 features per simulation (a training example), i:e: $mean(feature)$, $standardDev(feature)$, $mean(deltaFeature)$ and $standardDev(deltaFeature)$.

Although a large number of features were considered, obviously, not all the features equally contributed to capturing the human aesthetic judgement. Some of the features are more important than the others and some features replicate the same concept. For example, the features, *Distance to the centre of the flock,*

Average distance to the other sheep, *Distance to the closest sheep and Distance to the furthest sheep* capture the same concept, i:e: the maintenance of the flock cohesion. Therefore, some of the repeating concepts were removed from the feature set and several experiments were conducted to analyse the key features that output the best performance based on the features used by the humans and the machine learner performance (This was not an exhaustive search on all the possible combination to get the best performance, as finding such a feature set is beyond the scope of this study. The idea was to find a well performing feature set that can capture the key properties to bootstrap from human judgment). The key features examined were related to the concepts such as flock cohesion, flock resting, appropriate response to the dog and reach of the goal, because these were the key features/concepts highlighted in human responses. The addition of the features such as total distance travelled did not improve the machine learner performance, perhaps because human evaluations are based on aesthetic judgement rather than efficiency of the herding task performed by the dog. The features, *mean (distanceToFurthestSheep)*, *mean (simulationTicksSheepKeptMoving / simulationTicksSheepDidnotMove)*, *mean (deltaDistaneToGoal)*, *total (collisionsWithFlockMembers)*, *total (numberOfCollisionsWithTheDog)*, *total (simulationTicksSheepWentOutside*, *mean (distanceToFlockFromTheDog)*, and *mean (distanceMaintainedToTheGoalByTheDog)* performed well on the data set, i:e: are able to capture the properties used by humans to judge the aesthetic quality of the scenarios.

The machine learning algorithm (SMOreg) [20] was trained based on (1) the mean of the participant scores given for the 17 simulations (Fig. 2) (17 training examples); (2) by inputting all the participant scores for a training example (this creates 17 * 20 = 340 training examples). In the latter experiment, when evaluating the performance of the machine learner, the performance was compared against the mean score given by the human subjects for a scenario (a training example), because evaluating the performance against all participant scores does not give a meaningful result. The performances of the two approaches are shown in Table 2.

The Mean Squared Error (MSE), Root Mean Squared Error (RMSE) and Correlation co-efficient (Corr) were used to compare the performance of the two approaches. The second approach (using all participant scores) outperformed the first approach (using only the mean scores of the participants) in all the criteria. This better performance describes that the usage of the mean of the participant scores for a given scenario (a training example) conceals some information for the machine learning algorithm. This is because the mean is sensitive to the outliers. Hence, approach 2 is applied in the subsequent experiments.

A pilot study was conducted to understand the capability of the machine learning system to score unseen solutions. It was noted that for some of the edge cases which were not included in the training set, the machine learner was giving unrealistic scores. To train the machine learner to understand such cases, another 4 training examples (altogether 21 training examples) were added to the training set. These 4 training examples included such scenarios in which the dog goes out

Table 2. Performance of the Machine Learning System. (1) based on the mean of the participant scores given for the 17 simulations; (2) by inputting all the participant scores for a simulation ($17 * 20 = 340$ training examples)

Approach	MSE	RMSE	Corr
1	0.62	0.83	0.95
2	0.41	0.50	0.99

of the arena to herd the sheep or the sheep go out of the area when herded by the dog. While the herding dynamics were realistic, the scenarios needed to be penalised for going out of the arena. Given that these are edge cases, a separate user study was not conducted for these 4 training examples as it is infeasible to conduct such a separate study. Therefore, a random set of scores was assigned with an upper bound of 7 to determine the scores. In that way, the scenarios were assigned with the scores in the range of not realistic to somewhat realistic without a bias of the investigator. After adding the new training examples, the performance of the machine learning system was slightly increased in terms of MSE and RMSE and decreased in terms of correlation co-efficient (MSE = 0.4, RMSE = 0.49 Corr = 0.98) when compared with the performance showed for $17 * 20$ training examples. In general, the changes are negligible, however the pilot study showed that the scores generated by the machine learning system after the addition of new training examples are more generalised as it recognises previously unknown regions of the fitness landscape.

5.3   Evolving Sheepdog Herding Dynamics

We employed NSGA-II [1] to evolve high fidelity low complexity sheepdog herding simulations. The complexity of the simulation was determined by the number of rules the agents are codified with and the number of parameters that must be tuned. The fidelity was determined by the machine learning system described above. Since the algorithm was implemented to minimise the objectives the fidelity value was multiplied by -1. The parameters used for NSGA-II (determined based on the pilot experiments) are as follows: Population size = 100, Number of Generations = 100, Cross over rate = 0.9 & Mutation rate = 0.1. The chromosome consists of both booleans and real values. Boolean genes determine whether a particular rule is on or off whereas real values represented the weights and parameters of the rule. The crossover and mutation mechanisms for the chromosome are described in [7]. As new Pareto-optimal solutions were not encountered after evolving for about 100 generations, the evolution was stopped after 100 generations.

When evolving the sheepdog herding behaviours using NSGA-II [1] we only evolved the rule space of the sheep. When determining the fidelity, for a considered rule combination of sheep (i.e.: one individual in a population), we evaluated two scenarios using two sets of rules for the dog (the best performing rule sets for dog found through pilot studies) described in Sect. 3. The fitness function which is the machine learning system trained based on human aesthetic judgement evaluated fidelity of the behavioural dynamics of the two scenarios as the herding dynamics occur and the average was obtained. The complexity was simply determined based on (1) the number of rules in the considered rule combination for sheep (rule count) (2) number of parameters that must be tuned (parameter count). For example if the sheep is encoded with the two rules, *cohesion* and *wandering*, the complexity of the simulation is 2 according to the complexity measure rule count, whereas if the sheep is encoded with the rules *separation*, *cohesion* and *alignment* the complexity of the simulation is 3 for the same complexity measure. According to the complexity measure parameter count, if the sheep is encoded with the two rules *cohesion* and *wandering* the complexity is 4 (three parameters and 1 weight to combine the rules). Similarly, if the sheep is encoded with the rules *separation*, *cohesion* and *alignment* the complexity of the simulation is 3 under the parameter count complexity measure (1 parameter and 2 weights to combine the rules - if N rules are used then a minimum of $(N-1)$ weights are required). We obtained the Pareto optimal frontiers of 5 evolutionary runs for each complexity measure and the average across the runs were obtained to identify the high fidelity low complexity sheepdog herding simulations in the Pareto frontier.

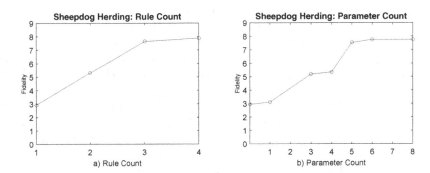

Fig. 3. Complexity versus fidelity of sheepdog herding dynamics: (a) Rule Count Versus Fidelity (b) Parameter Count Versus Fidelity

The rule space of the sheepdog herding model comprises of 6 rules. The exploration of this rule space utilising the rule count complexity measure is shown in Fig. 3 - (a). The maximum number of rules in the Pareto frontier is 4 indicating that the solutions with 5-rule and 6-rule combinations are dominated by the other solutions in the solution space. The lowest complexity Pareto-optimal solution in the Pareto-frontier has an average fidelity of 2.9. The Pareto-optimal single rule in the majority of runs was the rule *repel from the dog A/B*. This appears to show that a single rule is insufficient to generate acceptable realistic behaviours (as per human judgement at least a fidelity level of 5 in the range of 0–10 was required to be rated as somewhat realistic behaviours).

The next solution in the Pareto-frontier (Complexity = 2 fidelity = 5.3) was the combination *cohesion* and *repel from the dog-A*. This is about a 2.4 units increase in fidelity. This increase is due to that fact that the property cohesion is one of the key factors used by the dog to control the flock, and when cohesion is combined with repulsion from the predator the dynamics become somewhat natural.

The combination *cohesion, separation* and *repel from the dog (A/B)* is the next Pareto-optimal solution in the Pareto-frontier with a complexity of 3 and a fidelity of 7.7 showing a further increase in fidelity by 2.4. This combination has the key properties to become a high fidelity simulation - separation to avoid collisions, cohesion to exhibit flocking and response for the predator. Thereafter, adding more complexity showed only marginal improvements; for example the 4-rule combination in the Pareto-frontier (complexity of 4) shows an increase in fidelity by only 0.2 units (fidelity of 7.9).

The result obtained for the parameter count complexity measure (shown in Fig. 3 - (b)) is quite similar to the results obtained for the rule count complexity measure. The rule *cohesion* has no parameters (a parameter count of 0) and achieve a fidelity of 2.9. Because, there is no other rule with 0 complexity that can achieve a fidelity of 2.9, it has become a Pareto-optimal solution. The rules *cohesion* and *alignment* both do not have parameters. However, the weight by which they are combined becomes a parameter to the model. This combination which has a parameter count of 1 becomes the next Pareto-optimal solution with a fidelity of 3.1. The combination, *cohesion* and *repel from predator-A* has a parameter count of 3 and it achieves a fidelity of 5.2 which is in line with the result of the rule count complexity measure.

A new combination, *cohesion, alignment* and *repel from predator (A or B)* have been encountered in the Pareto-optimal frontier with a complexity of 4 and fidelity of 5.3 which is dominated by other 3-rule combinations when optimised with the rule count complexity measure. The combination *cohesion, separation* and *repel from the dog (A/B)* with a complexity of 5 and fidelity of 7.5 becomes the next Pareto-optimal solution with an increase of 2.2 in fidelity with 1 unit increase in complexity. Again, this is the combination which has the key properties to make the herding process look natural. Thereafter adding more complexity did not provide significant advantages in terms of fidelity; for example the combination *cohesion, separation, alignment* and *repel from the dog (A or B)* with a complexity of 6 provides only 7.7 units of fidelity and the same combination with the addition of the rule repel from walls which has a complexity of 8, provides the same amount of fidelity. The results indicate that once the key properties are present adding extra complexity does not provide significant advantages in terms of fidelity.

6   Conclusions and Future Work

We presented a machine learning approach to evolve high fidelity low complexity sheepdog herding simulations in cases where the fidelity is measured based on human judgement rather than empirical data. We trained a machine learning system by bootstrapping human judgement of the fidelity of the simulations and used this machine learning system as a fitness function to NSGA-II [1] to measure the fidelity (fitness) of the solutions encountered during the evolution. Complexity was measured using (1) the number of rules and (2) the number of parameters in the simulation.

The results indicate the effectiveness of the proposed approach in evolving high fidelity and low complexity sheepdog herding simulations, in particular when the fidelity depends on the aesthetics of the simulations. The solutions in the Pareto frontier can be analysed to identify which rules are more important to generate high fidelity sheepdog herding dynamics. Further, it is useful to generate simulations based on the required level of fidelity and available computational resources. More importantly, the results indicated that after a certain level, adding more complexity to the system only resulted in marginal improvements in fidelity (diminishing returns). This could open up interesting insights on a relationship between simulation fidelity and complexity.

In future work, we intend to validate the rule set and the machine learning system with empirical data. It is also interesting to explore the change of dynamics by adding more sheep in the flock. We also intend to extend this framework into several other problem domains to analyse the common properties in the Pareto frontiers in order to understand a relationship of simulation fidelity and complexity in general rather limiting to a particular problem area. Further, we intend to explore more complexity measures to characterise simulation model complexity, as the number of rules and parameters only provide an abstract view of the complexity.

References

1. Deb, K., Pratap, A., Agarwal, S., Meyarivan, T.A.M.T.: A fast and elitist multi-objective genetic algorithm: NSGA-II. IEEE Trans. Evol. Comput. **6**(2), 182–197 (2002)
2. Lien, J.M., Bayazit, B., Sowell, R.T., Rodriguez, S.: Shepherding behaviors. In: 2004 IEEE International Conference on Robotics and Automation, pp. 4159–4164. IEEE Press, New York (2004)
3. Lakshika, E., Barlow, M., Easton, A.: Co-evolving semi-competitive interactions of sheepdog herding behaviors utilizing a simple rule-based multi agent framework. In: 2013 IEEE Symposium on Artificial Life (ALIFE), pp. 82–89. IEEE Press, New York (2013)
4. Reynolds, C.W.: Flocks, herds and schools: a distributed behavioral model. ACM Siggraph Comput. Graph. **21**, 25–34 (1987)
5. Barlow, M., Easton, A.: Crocadile-an open, extensible agent-based distillation engine. Inf. Secur. **8**(1), 17–51 (2002)

6. Pan, X., Han, C.S., Dauber, K., Law, K.H.: A multi-agent based framework for the simulation of human and social behaviors during emergency evacuations. AI Soc. **22**(2), 113–132 (2007)
7. Lakshika, E., Barlow, M., Easton, A.: Fidelity and complexity of standing group conversation simulations: a framework for the evolution of multi agent systems through bootstrapping human aesthetic judgments. In: 2012 IEEE Congress on Evolutionary Computation, pp. 1044–1051. IEEE (2012)
8. Guy, S.J., Chhugani, J., Kim, C., Satish, N., Lin, M., Manocha, D., Dubey, P.: Clearpath: highly parallel collision avoidance for multi-agent simulation. In: 2009 ACM SIGGRAPH/Eurographics Symposium on Computer Animation, pp. 177–187. ACM (2009)
9. Vaughan, R., Sumpter, N., Henderson, J., Frost, A., Cameron, S.: Experiments in automatic flock control. Robot. Auton. Syst. **31**(1), 109–117 (2000)
10. Schultz, A., Grefenstette, J., Adams, W.: Robo-shepherd: learning complex robotic behaviors. In: Robotics and Manufacturing: Recent Trends in Research and Applications, pp. 763–768. ASME Press (1996)
11. Strömbom, D., Mann, R.P., Wilson, A.M., Hailes, S., Morton, A.J., Sumpter, D.J., King, A.J.: Solving the shepherding problem: heuristics for herding autonomous, interacting agents. J. R. Soc. Interface **11**(100), 20140719 (2014)
12. Konak, A., Coit, D.W., Smith, A.E.: Multi-objective optimization using genetic algorithms: a tutorial. Reliab. Eng. Syst. Safe. **91**(9), 992–1007 (2006)
13. Yang, A., Abbass, H.A., Sarker, R.A.: Land combat scenario planning: a multiobjective approach. In: Wang, T.-D., Li, X., Chen, S.-H., Wang, X., Abbass, H.A., Iba, H., Chen, G.-L., Yao, X. (eds.) SEAL 2006. LNCS, vol. 4247, pp. 837–844. Springer, Heidelberg (2006)
14. Alam, S., Shafi, K., Abbass, H.A., Barlow, M.: An ensemble approach for conflict detection in free flight by data mining. Transp. Res. part C: Emerging Technol. **17**(3), 298–317 (2009)
15. Rodrguez, D., Ruiz, M., Riquelme, J.C., Harrison, R. : Multiobjective simulation optimisation in software project management. In: 13th Annual Conference on Genetic and Evolutionary Computation, pp. 1883–1890. ACM (2011)
16. Aupetit, S., Bordeau, V., Monmarché, N., Slimane, M., Venturini, G.: Interactive evolution of ant paintings. In: 2003 IEEE Congress on Evolutionary Computation, pp. 1376–1383. IEEE Press, New York (2003)
17. den Heijer, E., Eiben, A.E.: Evolving pop art using scalable vector graphics. In: Machado, P., Romero, J., Carballal, A. (eds.) EvoMUSART 2012. LNCS, vol. 7247, pp. 48–59. Springer, Heidelberg (2012)
18. Chen, Y.-W., Kobayashi, K., Kawabayashi, H., Huang, X.: Application of interactive genetic algorithms to boid model based artificial fish schools. In: Lovrek, I., Howlett, R.J., Jain, L.C. (eds.) KES 2008, Part II. LNCS (LNAI), vol. 5178, pp. 141–148. Springer, Heidelberg (2008)
19. Takagi, H.: Interactive evolutionary computation: fusion of the capabilities of EC optimization and human evaluation. Proc. IEEE **89**(9), 1275–1296 (2001). IEEE, New York
20. Witten, I.H., Frank, E.: Data Mining: Practical Machine Learning Tools and Techniques. Morgan Kaufmann, San Francisco (2005)

An Episodic Memory Retrieval Algorithm for the Soar Cognitive Architecture

Francis Li[(⊠)], Jesse Frost, and Braden J. Phillips

School of Electrical and Electronic Engineering,
University of Adelaide, Adelaide, SA, Australia
{francis.li,jesse.frost,braden.phillips}@adelaide.edu.au

Abstract. Episodic memory in cognitive architectures allows intelligent agents to query their past experiences to influence decision making. Episodic memory in the Soar cognitive architecture must efficiently encode, store, search and reconstruct episodes as snapshots of short term working memory. The performance of the current search algorithm can be improved by doing a structural match phase first on all stored data and enforcing arc consistency in the first stage of the structural match. We demonstrate experimentally the performance of the improved search algorithm in two Soar environments.

1 Introduction

1.1 The Soar Cognitive Architecture

The Soar cognitive architecture is a model of cognition based on a symbolic production rule system. Soar models problem solving by moving an agent through a problem state-space in a sequence of *decision cycles* — wherein each cycle it uses a decision making algorithm to decide how to modify the problem state [5].

Soar internally represents its current knowledge of the problem state in temporary memory called *working memory* (WM), which consists of a set of 3-tuples called *working memory elements* (WMEs). The elements of each 3-tuple are called the identifier, attribute and value. The agent moves through the problem state-space by adding, removing and changing WMEs. Soar's WM is usually visualised as a directed graph (see Fig. 1).

1.2 Soar Episodic Memory

In addition to WM, a Soar agent can make use of *Episodic Memory* (EpMem) to assist its decision making process. Soar defines EpMem as an automatic architectural mechanism [4] that:

- Stores the state of WM at a specified phase in each decision cycle.
- Maintains a complete history of the state of WM throughout the agent's existence as a set of 'snapshots' of WM, where each snapshot is called an *episode*.

© Springer International Publishing Switzerland 2015
B. Pfahringer and J. Renz (Eds.): AI 2015, LNAI 9457, pp. 343–355, 2015.
DOI: 10.1007/978-3-319-26350-2_30

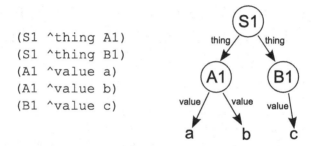

```
(S1 ^thing A1)
(S1 ^thing B1)
(A1 ^value a)
(A1 ^value b)
(B1 ^value c)
```

Fig. 1. A set of WMEs on the left and the associated graph representation on the right. In this example S1, A1 and B1 are *identifiers*, thing and value are *attributes* and a, b and c are *values*.

- Is queryable by an agent using a cue — a data structure representing the desired pattern to search for.
- Returns an episode that is the best match for the cue.

Figure 2 illustrates the relationship between Soar's WM and EpMem. A complete and detailed description of the current implementation of EpMem in Soar is given in [3]. Soar's definition of EpMem seems to be largely driven by its implementation in the Soar software and exact details of its behaviour may be open to change as the implementation changes. However its basic principles mean that it must at least be able to store and encode episodes, search episodes for a query and reconstruct a best match.

Encoding and Storage. The current EpMem implementation stores and encodes episodes by exploiting the temporal redundancy of WM — episodes close in time are likely to contain many of the same WMEs. To provide an efficient storage mechanism, EpMem only records the changes between each

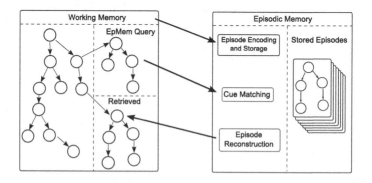

Fig. 2. The relationship between EpMem and WM in Soar. Episodes are stored in EpMem as encoded representations of WM and are retrieved by searching and matching against a cue.

episode by storing the structure of WM and the episode temporal information separately.

Retrieval. A cue is a set of WMEs that provide a pattern to search for in EpMem's stored episodes. A match for this cue is the most recent episode that contains this pattern, or the the best approximation. In the current implementation, *surface features* (see Sect. 2.2) are used as a metric to measure the quality of a match.

The current implementation performs matching in a two phase process: first by performing a temporal search for episodes that contain all the surface features of a cue (a *surface match*), and then by performing a structural match on the candidate episode by performing a graph match to determine if the cue is a subgraph of the candidate episode [2].

Reconstruction. The reconstruction algorithm takes an episode number from the cue matching algorithm and returns the full episode to the agent in WM. Further details of episode reconstruction are outside the scope of this discussion.

1.3   Contributions

Searching for an instance of a cue in EpMem is an instance of the NP-complete *subgraph isomorphism* problem. Soar reduces the search cost by implementing a two-phase matching algorithm. Soar's current retrieval algorithm terminates quickly if a surface match is likely to correspond with a complete graph match. However:

- If the stored episodes contain many structures that return perfect surface matches, but fail the structural graph match, the graph match step will be executed many times.
- If the cue contains WMEs which have the same identifier and attribute but different values (multi-valued attributes), the backtracking step could reach many dead-ends and backtrack frequently.

We address these issues by making the following contributions:

- We describe the structural match as a *constraint satisfaction problem* (CSP) (Sect. 2.3). Soar's structural match has not been extensively studied [3], so formalising the search under a known framework allows us to leverage well-known techniques to improve search time cost. We enforce *arc consistency* to tighten the constraints before backtracking, thus reducing the amount of dead-ends hit during search.
- We perform the structural match before checking that the temporal information is consistent (Sects. 2.3, 2.4). This allows us to do the most constrained search first, and ensures that the graph match is performed only once.
- We show that the proposed algorithm obtains comparable performance on most cues, and exceeds performance for cues where the original algorithm performs poorly (Sect. 3).

2   A New Episodic Memory Retrieval Algorithm

A WME w can be defined as a 3-tuple of symbols $w \in W = I \times A \times V$ where

- I, A, and V are three sets of symbols called *identifiers*, *attributes* and *values* respectively.
- The set of values $V = I \cup K$ is the disjoint union of identifier symbols and a set of *constant* symbols.
- The set of identifiers I contains a unique special element S.

There are three natural projections $\pi_I(w)$, $\pi_A(w)$, $\pi_V(w)$ that map a working memory element to its identifier, attribute and value. Note that I and K are disjoint, so π_V may be either an identifier symbol or a constant symbol.

Soar's current implementation of EpMem only considers 'snapshots', called *episodes*, of working memory taken at a fixed phase in the decision cycle. We denote the state of working memory at decision cycle t (t ranges from 0 onward, where 0 is used to reference the earliest episode) by $WM_t \subset W$, which is a set of working memory elements. An episode has the following structure:

- there is at least one $w \in WM_t$ for which $\pi_I(w) = S$
- $\forall w_n \in WM_t$ there is a finite sequence $(w_0, w_1, \ldots, w_n) \in WM_t$ s.t $\pi_I(w_0) = S$ and for all consecutive w_i and w_{i+1} in the sequence, $\pi_V(w_i) = \pi_I(w_{i+1})$

An episode can be represented as a directed multigraph where:

- N, the set of nodes, is the set of all identifiers and values from all working memory elements in the episode; $N = \{\pi_I(w) | w \in WM_t\} \cup \{\pi_V(w) | w \in WM_t\}$
- E, the set of edges, is the set of working memory elements in the episode; $E = WM_t$
- $s : E \to N$, the map from edges to their source node, is given by $s(e) = \pi_I(e)$
- $t : E \to N$, the map from edges to their target node, is given by $t(e) = \pi_V(e)$
- l_v, the map giving the label of a node is simply the identifier or constant symbol itself
- l_e, the map giving the label of an edge, is the attribute symbol $l_e(e) = \pi_A(e)$

An episode and its graph representation will always have these properties:

- A node S is always present, called the *root*.
- There is a path from S to every other node.
- There are no edges leaving nodes that are constants.
- Nodes which are identifier symbols might be inner nodes or leaf nodes whereas nodes which are constant symbols are always leaf nodes.

2.1   EpMem Encoding

EpMem's encoding step takes a sequence of episodes $WM_0, WM_1, WM_2 \ldots$ as they occur and encodes the relation $EM \subset T \times W$ such that $(t_i, w_j) \in WM$ whenever $w_j \in WM_{t_i}$. This relation then records the history of working memory and is called episodic memory.

The set of all working memory elements encoded in episodic memory is called the *working memory graph* (WMG) and stores the *structural information*.

The set $intervals_w = \{t \in T | (t, w) \in EM\}$ is a list of time points that a given working memory element w was present in working memory. This stores the *temporal information*.

2.2   Cue Matching

A *cue* is a set of working memory elements representing a pattern used to search EpMem. A cue $Q \subset W$ has the following properties:

– There is at least one $w \in Q$ for which $\pi_I(w) = S$.
– $\forall w_n \in Q$ there is a finite sequence $(w_0, w_1, \ldots, w_n) \in Q$ s.t $\pi_I(w_0) = S$ and for all consecutive w_i and w_{i+1} in the sequence, $\pi_V(w_i) = \pi_I(w_{i+1})$ and for any $i \neq j$, $w_i \neq w_j$.

The graph representation of a cue is constructed in an identical way to that of an episode. The properties above for a cue mean that its graph always contains the root node S and does not contain any cycles.

An episode WM_t is said to match a cue Q if there is a map $match : WM_t \rightarrow Q$ with the following properties:

– If $\pi_I(match(w)) = S$ for any $w \in WM_t$ that $\pi_I(w) = S$.
– For any $w \in WM_t$, $\pi_A(match(w)) = \pi_A(w)$.
– If $\pi_V(w) \in K$ then $\pi_V(match(w)) = \pi_V(w)$.
– If $\pi_V(w_i) = \pi_I(w_j)$ for $w_i, w_j \in WM_t$, then $\pi_V(match(w_i)) = \pi_I(match(w_j))$.

In other words an episode matches a cue if the cue is a subgraph of the episode, ignoring identifier node labels. Since there may not be such any episode that contains the cue, the current implementation of Soar allows *partial matches* by counting *surface features*. A surface feature is a finite sequence $(w_0, w_1, w_2 \ldots w_l)$ of elements of a a cue Q (or an episode WM_t) with the following properties:

– $\pi_I(w_0) = S$.
– $\pi_V(w_l) \neq \pi_I(w_i)$ for any $w_i \in Q$.
– For any consecutive w_i and w_{i+1} in the sequence, $\pi_V(w_i) = \pi_I(w_{i+1})$.
– For any $i \neq j$, $w_i \neq w_j$.

These properties mean that a surface feature is a path from node S to a leaf node. The most recent partial match is used when there is a tie between partial matches.

2.3   EpMem Retrieval

We can break the EpMem retrieval problem into two stages: the structural match and the temporal match. The structural match is a constraint satisfaction problem (see Algorithm 1) where:

- The set of *variables* $\mathcal{X}$ is the set of identifiers present in the cue.
- The *domain* $D_x \in \mathcal{D}$ of each variable x is the set of identifiers in the WMG.
- The domain of S is the singleton $\{S\}$.
- For each WME w_q in the cue where $\pi_V(w_q) \in K$, define a unary constraint who's scope is $\pi_I(w_q)$ and the allowed values are identifiers $i \in D_{w_q}$ such that there is a WME $w = (i, \pi_A(w_q), \pi_V(w_q)) \in WMG$.
- For every other WME w_q in the cue, define a binary constraint who's scope is $\pi_I(w_q)$ and $\pi_V(w_q)$, and the allowed values are pairs of identifiers $i_I, i_V \in D_{w_q}$ such that there is a WME $w = (i_I, \pi_A(w_q), i_V) \in WMG$.

Algorithm 1. Procedure which generates constraint network

Input: The WMG wmg and the cue Q
Output: A constraint network $\mathcal{R} = (\mathcal{X}, \mathcal{D}, \mathcal{C})$
1: **procedure** BUILDCONSTRAINTNETWORK(wmg, Q)
2:     **for all** $w \in Q$ **do**
3:         $x \leftarrow \pi_I(w)$
4:         $y \leftarrow \pi_V(w)$
5:         $\mathcal{X} \leftarrow \mathcal{X} \cup \{x\}$
6:         **if** $y \in I$ **then**
7:             $C_{xy} \leftarrow \{(i_I, i_V)|(i_I, \pi_A(w), i_V) \in wmg\}$
8:             $\mathcal{X} \leftarrow \mathcal{X} \cup \{y\}$
9:             $\mathcal{C} \leftarrow \mathcal{C} \cup \{C_{xy}\}$
10:        **else**
11:            $C_x \leftarrow \{(i|(i, \pi_A(w), \pi_V(w)) \in wmg\}$
12:            $\mathcal{C} \leftarrow \mathcal{C} \cup \{C_x\}$
13:        **for all** $x \in \mathcal{X}$ **do**
14:            $D_x \leftarrow I \cap C_x$
15:            $\mathcal{D} \leftarrow \cup\{D_x\}$

Solutions to a CSP are generated through search. Simple backtracking is a depth-first search where partial instantiations are extended by assigning values to variables and then ensuring that all constraints are still satisfied. If a constraint is violated, the search has reached a *dead-end* and must backtrack by changing the newly assigned variable's value to another in its domain. A depth first search is $O(b^d)$, where b is the branching factor (the size of the domains) and d is the depth (the number of variables). We can reduce the search space by preprocessing the problem to reduce the sizes of the domains.

A CSP is *arc-consistent* if any value in the domain of a variable can be extended consistently by any other variable. Enforcing arc consistency shrinks the domains of all variables and makes search more efficient. The *revise* operation (see Algorithm 2) enforces *directional arc-consistency* (DAC) between two variables x_i and x_j between which there is a binary constraint C_{ij} by deleting all values in the domain D_i of x_i that are not in the relation R_{ij}. The complexity of revise is $O(b^2)$, as every value in D_i must be compared with every value in D_j. However, this is reduced to $O(b)$ in practice, as domains are stored in a hashed data structure.

Algorithm 2. The revise operation [1].

Input: two variables x_i and x_j, their domains D_i and D_j, and constraint R_{ij}
Output: D_i, s.t., x_i is arc-consistent relative to x_j
1: **procedure** REVISE(x_i, x_j)
2:     $D_i \leftarrow D_i \cap \pi_i(R_{ij} \bowtie D_j)$

A *constraint graph* is a graphical representation of a CSP, where variables are nodes, and constraints are edges between nodes. If a constraint graph is a tree under some ordering $d = (x_1, \ldots, x_n)$, then we can enforce DAC along that ordering d by ensuring that every variable x_i is arc-consistent relative to every variable x_j such that $i \leq j$ (see Algorithm 3 [1]). This performs the revise operation at most c times, where c is the number of binary constraints (i.e., the number of WMEs in the cue). So enforcing DAC has time complexity $O(cb)$.

Algorithm 3. Directional arc-consistency (DAC) [1].

Input: A network $\mathcal{R} = (\mathcal{X}, \mathcal{D}, \mathcal{C})$ and an ordering $d = (x_1, ..., x_n)$.
Output: A directionally arc-consistent network
1: **procedure** DAC$(\mathcal{R}, d)$
2:     **for** $i \leftarrow n$ **to** 1 **do**
3:         **for all** $j < i$ s.t. $R_{ji} \in \mathcal{R}$ **do**
4:             REVISE(x_j, x_i) ▷ Algorithm 2

There is a natural correspondence between the graph representation of the cue and the constraint graph as each WME is an edge in the cue graph and is also a constraint edge in the constraint graph. The structure of the cue implies that the constraint graph always reduces to a tree.

The total cost of doing the structural match is the time taken to enforce DAC added to the time needed to generate all solutions through backtracking. The backtracking stage is much more efficient as dead-ends are never hit. All generated solutions are sent to the temporal match for the next phase of processing.

2.4 Temporal Match

Solutions from the structural graph match phase must be checked to ensure that each WME was present in WM at the *same* time — that is, during a single episode. We compile the solution into a data structure that tracks satisfaction of structural solutions as we walk though the intervals. This data structure is effectively a *conjuctive normal form* (CNF) statement, comprised of a conjunction of clauses (which are the solutions returned from the structural match), where each clause is a disjunction of literals (each WME assignment in the solution).

Interval endpoints are processed for every WME assignment in all solutions ordered by most recent. As these endpoints are processed, the number of literals in each clause which are 'on' are tracked by determining whether an endpoint is when a WME switches on or off. If this count reaches the maximum possible (i.e. the number of WMEs in a solution), then a solution is present in the same episode and we can return the episode number to the EpMem reconstruction algorithm. If not, we proceed until the interval walk exhausts all episodes, then return the episode with the maximum count. We track the satisfaction of each entire solution returned from the structural match. The satisfaction of a structural match is only updated when an endpoint is hit. When the satisfaction for a solution reaches the maximum value (i.e. the number of WMEs in a solution) then that solution is present in the same episode and we can return the episode number to the EpMem reconstruction algorithm. Otherwise, we proceed until the interval walk exhausts all episodes, then return the episode with the maximum count.

The definition of a *partial match* has changed from the number of surface features to the number of structural solution WMEs present. It remains to be seen how this affects agents that rely on partial match in their decision cycles. However, this temporal match algorithm can be modified to support the traditional partial match criteria by altering the compilation step. There is further scope for study and optimisation of the temporal match component.

See Algorithm 4 for a description of the temporal match and Algorithm 5 for a complete description of the retrieval procedure.

3   Experiments

We used two environments to evaluate EpMem algorithm performance:

- *TankSoar*[1] — a computer simulation that uses Soar agents to control virtual tanks that navigate a two-dimensional map; we chose TankSoar because it was used in [3] as a test environment to evaluate the original EpMem algorithm.
- A simple implementation of the popular computer game *2048*[2] (where a player, in this case a Soar agent, must move a set of tiles around on a four-by-four grid). We selected this as it produced WM structures similar to those that caused slower performance in the TankSoar environment.

The TankSoar agent *mapping-bot* receives sensor information on its input-link, and keeps track of the game's map internally. The 2048 environment models

---

[1] http://soar.eecs.umich.edu/articles/downloads/domains/176-tanksoar.
[2] Original version available: http://gabrielecirulli.github.io/2048/.

Algorithm 4. The temporal matching process

Input: all solutions generated from the structural match *solutions* and the interval table *intervals*

Output: the episode number which best matches the cue, or *null* if not found

1: **procedure** TEMPORALMATCH(*solutions, intervals*)
2:     *endpoints* ← {}   ▷ A priority queue
3:     **for all** *sol* ∈ *solutions* **do**
4:         **for all** *w* ∈ *sol* **do**
5:             *endpoints* ← *endpoints* ∪ {$\pi_{start}(intervals_w), \pi_{end}(intervals_w)$}
6:     **for all** *endpoint* ∈ *endpoints* **do**
7:         **if** *endpoint* is the *end* of an interval **then**
8:             INCREMENT *satisfaction(solution(endpoint))*
9:             **if** *satisfaction(solution(endpoint))* **is** |*sol*| **then**
10:                 **return** *endpoint*
11:         **else**
12:             DECREMENT *satisfaction(solution(endpoint))*
13:     **return** *null*

Algorithm 5. The proposed EpMem retrieval algorithm. Java implementation available in [6].

Input: the WMG *wmg*, the interval table *intervals*, the cue *Q*

Output: the episode which best matches the cue

1: **procedure** RETRIEVE(*wmg, intervals, Q*)
2:     $\mathcal{R}$ ← BUILDCONSTRAINTNETWORK(*wmg, Q*)   ▷ Algorithm 1
3:     *d* ← GENERATEORDERING($\mathcal{R}$)   ▷ Orders nodes s.t. the induced width of the graph is minimised (e.g. min-width)
4:     DAC($\mathcal{R}, d$)   ▷ Algorithm 3
5:     *solutions* ← BACKTRACKING($\mathcal{R}, d$)   ▷ Standard backtracking
6:     **return** TEMPORALMATCH(*solutions, intervals*)   ▷ Algorithm 4

---

its game state as a four-by-four grid. Each cell in the grid has a column, row and value. The 2048 agent searches for past episodes that match some number of cells (see Fig. 4).

We used the existing Soar EpMem search implementation and four different cues to search EpMem ranging in size from 1 to 320000 episodes generated from the TankSoar environment. These cues were taken from test data in [3] and can be seen in Fig. 3. We then performed the test on the same EpMem data using the proposed algorithm implemented in the Java programming language.

We also used both search implementations to search for cues containing between 1 and 16 'cells' in episodic memory of up to 900 episodes generated by the 2048 environment. The structure of these cues is shown in Fig. 4.

3.1 Results

The run times for both search implementations in the TankSoar environment is shown in Fig. 5. Both were comparable in performance, with the proposed

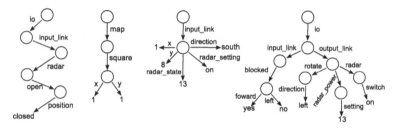

Fig. 3. From left to right, cue 1, cue 2, cue 3 and cue 4 used to search TankSoar's EpMem.

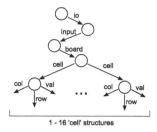

Fig. 4. The cue structure used by the 2048 agent contains 1-16 'cell' structures.

algorithm having a small advantage despite the addition of the preprocessing step. Figure 6 shows the run time for a search in the 2048 episodic memory using both algorithms with cues containing between 1 and 4 cells. The proposed algorithm was consistently able to reduce the variable domains to a single element, which resulted in backtrack-free structural match. The original Soar implementation consistently passed the surface match phase and triggered a graph match for many intervals. The unreduced domains caused near worst case time for each repeated attempt at structural graph match. We attempted to test cues with more than 4 cells, but tests required too much time to execute (greater than 30 min).

Figure 7 shows the search times for a cue consisting of two cells and an episodic memory of 1 to 900 episodes. The original Soar implementation was in some cases able to complete the search early when the graph match phase matched early in the interval walk, but in other cases ended up attempting an expensive graph match many times. The proposed algorithm consistently performed better as it performed the graph match only once in each search and was able to reduce each domain to a single element each time.

3.2   Significance

We have shown that there are cue structures which stress the current algorithm which perform well under the new algorithm. Agent designers which use Soar's

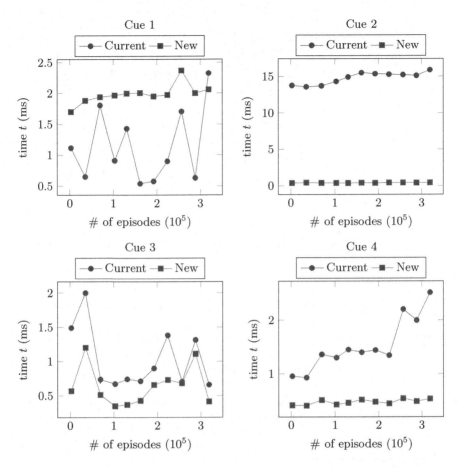

Fig. 5. The search time for the four TankSoar cues. Both algorithms are comparable except for cue 2, where the new algorithm is an order of magnitude better.

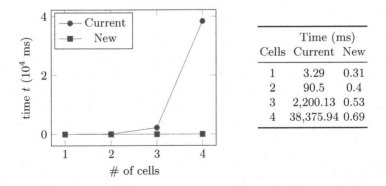

Cells	Time (ms)	
	Current	New
1	3.29	0.31
2	90.5	0.4
3	2,200.13	0.53
4	38,375.94	0.69

Fig. 6. Average search time for Soar increases exponentially as the number of cell structures in the cue are increased.

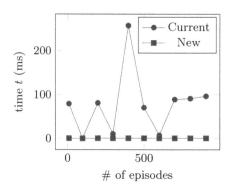

	Time (ms)	
Episodes	Current	New
10	79.32	0.4
100	0.21	$6.39 \cdot 10^{-2}$
200	81.06	0.5
300	10.62	0.52
400	257.34	0.53
500	70.48	0.53
600	7.14	0.52
700	88.89	0.52
800	91.08	0.5
900	96.34	0.52

Fig. 7. Average search time for Soar vs. current implementation with 2-cell cue. Soar's performance is highly dependent on the number of times it needs to do a graph match.

episodic memory are able to have a greater variety of WM and cue structures and have still have reactive agents. Experimentally, it has been shown that a cycle time of 50–100 ms is required to maintain reactivity in real-world situations [2]. Although the experiments we conducted in this paper are not real-world situations, they are simple and still stress the current algorithm. For agents which persist for extended periods of time or have complex memory structures, the proposed algorithm may provide better performance than the current algorithm. However, there is great diversity in both agent environments agent and the WM structures in agent memories, so further investigation is required.

4   Conclusion

Soar's EpMem system has an effective encoding and cue matching algorithm. However, certain episode and cue structures can significantly reduce its performance by causing repeated attempts at performing a full graph match, or by stressing the simple backtracking algorithm.

We have proposed an alternative cue matching system that performs an efficient graph match first before attempting the interval search. The graph match enforces arc consistency before backtracking to tighten the constraints and avoid dead-ends. An implementation of this approach in the Java programming language [6] provides comparable performance to the current algorithm in most cases, and significant improvements in situations where: (a) there are frequently perfect surface matches, yet the structural graph match fails; or (b) the structural graph match hits many dead-ends and backtracks frequently.

In addition, the proposed algorithm can extend EpMem to handle cyclic and non-rooted cues and has scope for further optimisation.

References

1. Dechter, R.: Constraint Processing. Morgan Kaufmann, San Francisco (2003)
2. Derbinsky, N., Laird, J.E.: Efficiently implementing episodic memory. In: McGinty, L., Wilson, D.C. (eds.) ICCBR 2009. LNCS, vol. 5650, pp. 403–417. Springer, Heidelberg (2009)
3. Derbinsky, N.L.: Effective and efficient memory for generally intelligent agents. Ph.D. thesis, The University of Michigan (2012)
4. Laird, J.E.: Extending the soar cognitive architecture. Front. Artif. Intell. Appl. **171**, 224 (2008)
5. Laird, J.E.: The Soar Cognitive Architecture. The MIT Press, Cambridge (2012)
6. Li, F.: epmemcsp (2015). https://github.com/fli/epmemcsp

Finding the k in K-means Clustering: A Comparative Analysis Approach

Markus Lumpe[(✉)] and Quoc Bao Vo

Faculty of Science, Engineering and Technology, Swinburne University of Technology,
P.O. Box 218, Hawthorn, VIC 3122, Australia
{mlumpe,bvo}@swin.edu.au

Abstract. This paper explores the application of inequality indices, a concept successfully applied in comparative software analysis among many application domains, to find the optimal value k for k-means when clustering road traffic data. We demonstrate that traditional methods for identifying the optimal value for k (such as gap statistic and Pham *et al.*'s method) are unable to produce meaningful values for k when applying them to a real-world dataset for road traffic. On the other hand, a method based on inequality indices shows significant promises in producing much more sensible values for the number k of clusters to be used in k-means clustering for the same road network traffic dataset.

1 Introduction

Intelligent transport systems (ITS) are innovative services that aim at utilizing information and communication technologies to sustainably deliver safer and more efficient transport. These services typically integrate traffic data and feedback from a number of sources to make timely decisions in order to improve safety and to reduce congestion. Traffic data hereby refers to time-series data collected by traffic monitoring equipments. A typical metropolitan traffic surveillance system (*e.g.*, the Melbourne network[1]) usually has tens of thousands of sensors collecting data at fixed intervals round the clock. Consequently, time-series databases for traffic data are typically very large.

To make sense of traffic data, especially datasets that have been collected over an extended period of time, unsupervised data mining techniques such as clustering have been proposed [6,12]. Figure 1 illustrates the need to identify such patterns in traffic data. Here, the data for the number of vehicles passing through the intersection of Barkers Road and Power Street in Kew (Melbourne) for the city-bound direction on Barkers Road is captured by VicRoads via induction-loop detectors. The data is aggregated in blocks of 15 min for each day in October 2010. In Fig. 1 three clear patterns can be observed that correspond to the traffic on Sundays, Saturdays, and weekdays. Thus, it is natural to ask the question: how can these patterns be effectively captured automatically?

---

[1] https://www.vicroads.vic.gov.au/business-and-industry/design-and-management/
-traffic-signals-and-systems/traffic-signals-in-victoria.

© Springer International Publishing Switzerland 2015
B. Pfahringer and J. Renz (Eds.): AI 2015, LNAI 9457, pp. 356–364, 2015.
DOI: 10.1007/978-3-319-26350-2_31

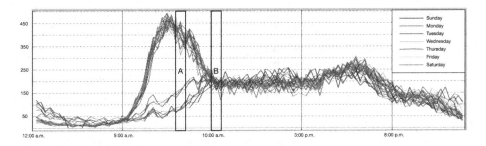

Fig. 1. Traffic pattern October 2010 (8–8:30 a.m. (A) and 10–10:30 a.m. (B)).

In many problem domains, k-means clustering offers an effective means for pattern discovery [6]. However, k-means clustering relies on the parameter k being given. Popular and classical methods for determining the optimal number of clusters include the *gap statistic* [6], the *G-means* algorithm [5], and the function $f(K)$ [13]. Unfortunately, these methods struggle to produce meaningful values of k for real-world traffic datasets such as those for the Melbourne network. In this paper, we introduce an alternative technique for determining the optimal number of clusters to be used for k-means clustering. The proposed technique is based on the notion of inequality indices, a concept that has been successfully applied to a number of application domains, including the problem of comparative software analysis [4,15,18].

Inequality indices have been developed by econometric researchers to measure the inequality in the income distribution within a society. By relating this concept to clustering, we can view a cohesive cluster as a group of individuals in the society with relatively small inequality so that merging any two such groups would lead to significantly higher inequality. Subsequently, we can identify the sensible number of cohesive clusters for a given dataset. Informally, the sensible number of clusters is comprised of (i) members within a cluster having relatively small inequality, and (ii) the inequality across each pair of clusters being relatively large. We are able to demonstrate in this paper that our proposed method based on the decomposable Theil index produces much more sensible numbers of clusters compared to methods such as the gap statistic, G-means, and the function $f(K)$ when applied to the traffic dataset for the Melbourne network in the year of 2012 provided by VicRoads.

2   Inequality-Based Data Analysis

Cluster analysis seeks to group a collection of objects into sub-sets such that those within a cluster are considered closely related [6,11]. The cluster centers (*i.e.*, centroids) capture the coordinate-wise *expected data points* for each object in a given cluster, usually obtained via a central tendency statistics (*i.e.*, *mean*) and the minimum within-cluster sum of squares, as in the case of k-means. Inequality indices, on the other hand, focus on *data entropy* [16], in econometric research usually captured as *operative differences* (*i.e.*, variation in not only value, but also frequency and location, irrespective of the shape of data distribution), in data distribution [14]. As a consequence, inequality indices are very sensitive to changes [4], and this sensitivity turns inequality indices into effective means for comparing and even categorizing datasets [2,6].

Inequality-based data analysis has a long tradition in socio-econometrics [14]. Its effectiveness has also been recently demonstrated in the area of software engineering [4,10,15,18]. For example, Vasa *et al.* [18] showed that inequality-based data analysis provides a better way to summarize highly skewed software metrics data. Furthermore, the specific sensitivity to changes in data allows for a more meaningful interpretation and understanding of evolving software systems [4,18]. Likewise, Serebrenik and Brand [15] studied Debian Linux and the effects of varying programming language use. In particular, they investigated whether the size of programs in the Debian eco-system is a function of the programming language used to implement them. Finally, inequality-based analysis also lends its hand to the study and manipulation of community structures in complex networks, where it can yield predictors for the success of partition refinement based not only on the number, but also on the density of incoming and outgoing links within a network topology [10].

Econometric research has suggested numerous inequality indices (see Sen [14] for an overview). Among them, we find the *decomposable* Theil index [17] that allows for the analysis of a population being partitioned into sub-groups with respect to some observable characteristic. Econometric researchers use this technique to answer the question *"how much inequality can be explained"* [2] in a population based on some discriminative population attribute. We adopt this method to find a k in k-means in which k yields a clustering of network traffic data that can *"explain"* the operational differences in the observed traffic volume patterns.

For a discrete population with non-negative values x_i, $1 \leq i \leq n$, and a partitioning Π, the Theil index is given by $I(\Pi) = I_B(\Pi) + I_W(\Pi)$, with

$$I_B(\Pi) = \sum_{g=1}^{|\Pi|} \frac{X_g}{X} \log \left(\frac{X_g}{X} \Big/ \frac{|\Pi_g|}{n} \right)$$

$$I_W(\Pi) = \sum_{g=1}^{|\Pi|} \frac{X_g}{X} \left(\sum_{i=1}^{n_g} \frac{x_{g_i}}{X_g} \log \left(\frac{x_{g_i}}{X_g} \Big/ \frac{1}{|\Pi_g|} \right) \right)$$

where $X = \sum_{i=1}^{n} x_i$ and $X_g = \sum_{i=1}^{|\Pi_g|} x_{g_i}$. $I(\Pi)$ ranges over the interval $[0, \log n]$ for any $n \in \mathbb{N}$, such that there exists an $x_i > 0$ with $1 \leq i \leq n$ (*i.e.*, a given dataset has to contain at least one strictly positive value). The value $I_B(\Pi)$ denotes the *"between-group"* inequality, whereas the value $I_W(\Pi)$ captures the *"within-group"* inequality of the partitioning.

How well the partitioning reflects the underlying operational differences between the sub-groups is expressed by the ratio $R_B(\Pi) = I_B(\Pi)/I(\Pi)$ [2]. The ratio $R_B(\Pi)$ indicates what part of the inequality in x_i, $1 \leq i \leq n$, is due to the partitioning Π. $R_B(\Pi)$ ranges between 0 and 1. $I_B(\Pi)$ is 0 for a trivial partitioning of a population into one group (*i.e.*, $k = 1$) – the overall differences can be completely attributed to the operational differences in the population. $I_B(\Pi) = 1$ corresponds to the case where the partitioning is *"complete"* – every element in the population is considered a group in itself (*i.e.*, $k = n$).

We can use $R_B(\Pi)$ to determine k in k-means when classifying network traffic volume patterns. According to Cowell and Jenkins [2], for any characteristics t and s, the value of $R_B(\Pi_{t\&s})$ must not be smaller than $R_B(\Pi_t)$ or $R_B(\Pi_s)$, that is, $R_B(\Pi_{t\&s}) \geq R_B(\Pi_t)$ and $R_B(\Pi_{t\&s}) \geq R_B(\Pi_s)$. Applying this principle to k-means clustering, we accept a partitioning Π_{k+1}, for some $k < k_{max}$,[2] if $R_B(\Pi_{k+1}) > R_B(\Pi_k)$. Otherwise, Π_k is taken to be the partitioning for the underlying network traffic volume data. Technically, $R_B(\Pi)$ represents the share of variance explained by the between-group differences. In $R_B(\Pi)$, the denominator $I(\Pi)$ is invariant under partitioning. Moving from any partitioning to a finer sub-partitioning, the share of the between-group variance must not decrease for the finer sub-partitioning to be valid. More precisely, we define clustering of network traffic data as a partitioning problem in which k-means yields a partitioning Π_k with minimized within-cluster distortions and an exogenous summary attribute (*i.e.*, traffic volume) provides the means to maximize the between-cluster inequality ratio $R_B(\Pi_k)$.

3   Clustering Traffic Data

Using the Melbourne traffic dataset provided by VicRoads for the year 2012, we distilled a sub-set capturing the traffic information for a route from the intersection Burwood Road and Glenferrie Road in Hawthorn to the Queen Victoria Market on Victoria Street in Melbourne's CBD. According the Google Maps, the distance between these locations is approx. 4.4 km and a trip by car would require roughly 12 min, notwithstanding the actual traffic conditions. The available data comprises 354 days (the remaining data points contained missing values) for which we extracted disjoint periods between 6:30–7:00 a.m. and 10:30–11:00 a.m., (cf. Fig. 1), to appreciate, in particular, the varying morning traffic patterns for school and non-school days due to a high school density in the area. The route itself consists of 18 monitored intersections. We encoded the data for a given day into a vector $l_i \times t_j$, where l_i is the number of relevant approaches along the route, and t_j is an interval time stamp, $j \in \{0, 15\}$.

Unfortunately, our intuition about data and its classification break down in high-dimensional settings [6,12]. Our traffic data is no exception. It is, therefore, useful to apply a "dimensionality reduction" [12] when assessing the outcome of estimating k and the corresponding classification by k-means clustering. Moreover, for the application of the Theil index, we require a *characterizing property* (or parametric variation [14]) that provides us with the means to explain the between-groups differences and that is consistent with the data being classified.

---

[2] Here k_{max} has to be a reasonably large upper bound reflecting the specific characteristics of the dataset [13].

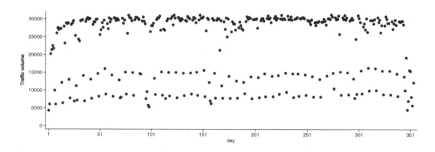

Fig. 2. Traffic volume between 8:00 and 8:30 a.m.

A natural candidate for network traffic data is *traffic volume*, which captures the *"essence"* of the data [12]. Conceptually, a path in a traffic network can be viewed as a sequence of *"acquisition queues"* that control the network throughput [8,9]. The amount of vehicles flowing through the network is usually constrained by several factors, including capacity and traffic light cycles [8]. We use the total volume of vehicles passing through the intersections of our selected route as characterizing attribute (cf. Fig. 2). For example, between 8:00 and 8:30 a.m., 30,086 cars were counted for Tuesday, March 3, whereas 7,774 cars were registered in the same time frame on Sunday, March 11, and 12,804 on Saturday, March 10. Intuitively, these values suggest that $k = 3$ yields k-means clustering assigning these days to a different cluster each. But is $k = 3$ a good fit for the data? The dispersion of the weekday traffic volume between 20,000 and 30,000 vehicles, as shown in Fig. 2, appears too broad to fit into just a single cluster, even though it might provide clustering stability [1,11].

To estimate k more systematically, we applied three popular methods, the *gap statistic* [6], the *G-means* algorithm [5], and the $f(K)$ evaluation function proposed by Pham *et al.* [13] to our traffic data. For the periods 8:00–8:30 a.m. and 10:00–10:30 a.m., $k = 2$ is the smallest k with a $Gap(k)$ not smaller than $Gap(k + 1) - s(k + 1)$, where $s(k + 1)$ is the standard error of $Gap(k + 1)$ [6]. In other words, the gap statistic suggests two clusters, which would separate weekdays from weekends. The variability in the traffic volumes for weekdays as well as the differences between Saturdays and Sundays (cf. Fig. 2) have no bearing on the gap statistic.

Using the G-means algorithm, the number k is estimated to be 22 and 18, respectively, for the periods 8:00–8:30 a.m. and 10:00–10:30 a.m. The G-means algorithm seeks to discover a k for k-means, where the clusters are Gaussian [5]. G-means relies on a statistical test (the Anderson-Darling test for normality is the default) to decide whether or not to split a given cluster further. G-means finishes when no additional refinement is possible. The number of clusters suggested by G-means for our traffic data is rather large. The recommended clustering would assign most Sundays to one clusters, would split Saturdays in 4–6 clusters, and would yields a very fine-grained classification of weekday traffic patterns, often resulting in small clusters with 5 members or less. From a practical point [6], G-means appears to significantly overestimate the number of clusters for the Melbourne network traffic data.

Finally, the function $f(k)$ computes the ratio between the real and the estimated distortion when the number of clusters is k. The value of $f(k)$ is close to 1 when the data distribution is uniform [13]. However, if objects agglomerate in discernible regions, then the value of $f(k)$ is, often significantly, smaller than 1. A k, $1 \leq k \leq k_{max}$, for which $f(k) \ll 1$, yields well-defined clusters. Pham *et al.* [13] state that any k with $f(k) < 0.85$ could be used for clustering. For the 8:00–8:30 a.m. period there are three k's with $f(k) < 0.85$: $k = 2$, $k = 5$, and $k = 7$. However, the support for $k = 2$ outweighs that for $k = 5$ and $k = 7$ by a factor of two. The 10:00–10:30 a.m. period gives rise to just $k = 2$. All other values of k fail to meet the threshold $f(k) < 0.85$.

The preceding tests for our traffic dataset were inconclusive. We know, however, that both periods, 8:00–8:30 a.m. and 10–10:30 a.m., contain at least two clusters. Otherwise, the gap statistic would have estimated the optimal number of clusters to be one. The gap statistic is one of the few competing methods that can identify the case $k = 1$ [6].

For any dataset for which a characterizing property can be defined and which can be partitioned into $k \geq 2$ groups, we can apply the decomposable Theil index and compute the ratio $R_B(\Pi_k)$. The Melbourne network traffic data satisfies theses conditions. As outlined in Sect. 2, we can use $R_B(\Pi_k)$ to find a $k < k_{max}$ with $R_B(\Pi_{k+1}) \not> R_B(\Pi_k)$. Following the gap statistic approach, we take

$$k = \min_{1 \leq k' < k_{max}} \{k' \mid R_B(\Pi_{k'+1}) \not> R_B(\Pi_{k'})\}.$$

The estimates for the periods 8:00–8:30 a.m. and 10–10:30 a.m. using the ratio $R_B(\Pi_k)$ are $k = 5$. At these values for k the between-group inequality amounts to 98 % and 87 %, respectively, of the overall inequality in the dataset. The period 10:00–10:30 a.m., in addition, yields $R_B(\Pi_2) = R_B(\Pi_3)$ at 67 %, demonstrating that a finer clustering may not reduce the between-group inequality at all. However, the variability across clusters at $k = 5$ exceeds substantially the variability within the clusters. The partitioning Π_5 "explains" the inequality [2] in both time periods, where Π_5 is a partitioning for which $R_B(\Pi_5)$ is the first maximum.

Table 1 provides an overview of the estimation performance of the gap statistic, G-means, the evaluation function $f(k)$, and the Theil index-based ratio $R_B(\Pi_k)$ for the morning traffic period between 06:30 and 11:00 a.m. The Theil index-based method works well for traffic datasets. It produces $2 \leq k \leq 5$, values of k that provide a meaningful cluster decomposition for Melbourne's network traffic data. Moreover, the total number of vehicles offers a suitable characterizing property for the estimation of k via $R_B(\Pi_k)$. The gap statistic and the function $f(k)$, on the other hand, recommend mostly $k = 2$, which would allow for the separation of weekday and weekend patterns at most. The gap statistic agrees with $R_B(\Pi_k)$ twice, whereas the function $f(k)$ does so only once.

Table 1. Performance of estimation approaches (number of clusters).

Time period	Gap statistic	G-means	Function $f(k)$	Ratio $R_B(\Pi_k)$
6:30–7:00 a.m	3	10	2	5
7:00–7:30 a.m	2	10	2	5
7:30–8:00 a.m	3	13	2	3
8:00–8:30 a.m	2	22	2	5
8:30–9:00 a.m	2	15	2	5
9:00–9:30 a.m	4	12	2	5
9:30–10:00 a.m	2	11	2	5
10:00–10:30 a.m	2	18	2	5
10:30–11:00 a.m	2	10	2	2

The value $k = 2$ for the 10:30-11:00 a.m. is sensible. Traffic patterns converge and stay similar until 10:00 p.m. The two clusters discern normal traffic conditions from low traffic in early January, public holidays, and the Christmas break period. The G-means estimates exceed our $R_B(\Pi_k)$ method by a factor 2 or more.

The actual clusterings produced by k-means for 08:00–8:30 a.m. and 10:00–10:30 a.m. are shown in Fig. 3. For 8:00–8:30 a.m., we obtained two distinct clusters for Saturdays (1) and Sundays (3). Some of the weekdays also fall into those clusters. The weekdays that fall into the Saturday cluster cover the week between December 24 and December 31, and the Christmas break period. The Sunday cluster, besides all other Sundays, comprises all weekdays that fall on a public holiday, including Melbourne Cup Day and ANZAC Day in 2012.

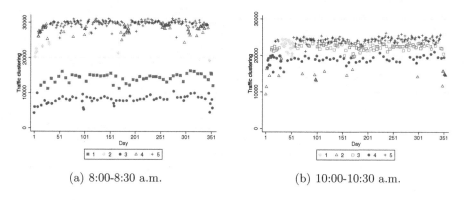

(a) 8:00-8:30 a.m.   (b) 10:00-10:30 a.m.

Fig. 3. Traffic classification at 8:00–8:30 a.m. and 10:00–10:30 a.m.

Between 8 and 8:30 a.m., there are three distinct clusters for weekdays with an average traffic volume of 22,436 (2), 28,052 (4), and 29,79 (5) vehicles using the route between Burwoord Road and Queen Victoria Market. The clustering also indicates monthly differences. Cluster 2 mostly contains days in January 2012, whereas cluster 4 captures traffic patterns when schools were on break in Victory. Figure 3(a) shows how the clusters relate to traffic volume and days between 8:00 and 8:30 a.m. in 2012.

As suggested by Fig. 1, the clusters for the period 10:00–10:30 a.m. are not as well separated as the ones between 8:00 and 8:30 a.m., even though there are discernible differences. The disparity between days of the week become less pronounced. Nevertheless, there are noticeable differences as illustrated in Fig. 3(b). There is a distinct February cluster (1). We use $R_B(\Pi_k)$ and the total traffic volume to estimate k, but rely on k-means to perform the partitioning. Traffic in February is higher than in January, but has not yet reached the same level as in the other months in 2012. The value k obtained by our method, in addition to separating weekdays from Saturdays and Sundays, also accounts for the variation in traffic data in Melbourne for the Australian summer months.

4   Conclusions

The k-means algorithm is one the most popular data clustering methods and yields meaningful groupings for datasets of quantitative type, including network traffic data. However, k-means requires a pre-selected k, the number of clusters to be populated. Several approaches to estimate k have emerged, including the gap statistic [6], the G-means algorithm [5], and the evaluation function $f(k)$ [13]. Unfortunately, for the domain of network traffic data these methods perform poorly. On the other hand, a technique based on the Theil index [17] can produce meaningful estimates of k via a "characterizing" aggregation of network traffic data. The Theil index is an inequality index originally invented to study income inequality and its effects on society [14,17]. When applied to the problem of clustering traffic data, we can iteratively compare the Theil index's "between-group" inequality of k groups to the overall inequality in the data. The optimal k is the first k that yields a partitioning for traffic data which has a greater between-group inequality than a partitioning for $k + 1$.

We have tested our approach on network traffic data for Melbourne provided by VicRoads for the year 2012. It produced promising results. However, other datasets resisted an effective estimation of k with our technique. This should not be a surprising observation given the well-known discovery by Färber et al. [3] that the groupings produced by clustering techniques for a dataset do not necessarily correspond to the known classification of that dataset. While we do not expect our technique to work optimally for all kinds of problems [1,11], we endeavor to broaden its applicability and, in particular, explore its information theoretic aspect [7,12] in more detail in the future.

References

1. Ben-David, S., von Luxburg, U., Pál, D.: A sober look at clustering stability. In: Lugosi, G., Simon, H.U. (eds.) COLT 2006. LNCS (LNAI), vol. 4005, pp. 5–19. Springer, Heidelberg (2006)
2. Cowell, F.A., Jenkins, S.P.: How much inequality can we explain? A methodology and an application to the united states. Econ. J. **105**(429), 412–430 (1995)
3. Färber, I., Günnemann, S., Kriegel, H.P., Kröger, P., Müller, E., Schubert, E., Seidl, T., Zimek, A.: On using class-labels in evaluation of clusterings. In: MultiClust: 1st International Workshop on Discovering, Summarizing and Using Multiple Clusterings held in conjunction with KDD, p. 1 (2010)
4. Goloshchapova, O., Lumpe, M.: On the application of inequality indices in comparative software analysis. In: Proceedings of 22nd Australian Software Engineering Conference (ASWEC 2013), pp. 117–126. IEEE Computer Society, Melbourne, June 2013
5. Hamerly, G., Elkan, C.: Learning the k in k-means. In: Thrun, S., Saul, L.K., Schölkopf, B. (eds.) Advances in Neural Information Processing Systems 16, pp. 281–288. The MIT Press, Cambridge (2004)
6. Hastie, T., Tibshirani, R., Friedman, J.: The Elements of Statistical Learning: Data Mining, Inference, and Prediction, 2nd edn. Springer, New York (2013)
7. Kasturi, J., Acharya, R., Ramanathan, M.: An information theoretic approach for analyzing temporal patterns of gene expression. Bioinformatics **19**, 449–458 (2003)
8. Le, T., Vu, H.L., Nazarathy, Y., Vo, Q.B.: Hoogendoorn: linear-quadratic model predicative control for urban traffic networks. J. Transp. Res. Part C: Emerg. Technol. **36**, 498–512 (2013)
9. van Leeuwaarden, J.S.H., Lefeber, E., Nazarathy, Y., Rooda, J.E.: Model predictive control for the acquisition queue and related queueing networks. In: Proceedings of 5th International Conference on Queueing Theory and Network Applications (QTNA 2010), pp. 193–200. ACM, New York, July 2010
10. Lumpe, M.: Partition refinement of Component Interaction Automata. Sci. Comput. Program. **78**, 27–45 (2012)
11. von Luxburg, U.: Clustering stability: an overview. Found. Trends Mach. Learn. **2**(3), 235–274 (2010)
12. Murphy, K.P.: Machine Learning: A Probabilistic Perspective. The MIT Press, Cambridge (2013)
13. Pham, D.T., Dimov, S.S., Nguyen, C.D.: Selection of K in K-means clustering. J. Mech. Eng. Sci. **219**(Part C), 103–119 (2005)
14. Sen, A.K.: On Economic Inequality. Oxford University Press, Oxford (1973)
15. Serebrenik, A., van den Brand, M.: Theil index for aggregation of software metrics values. In: Proceedings of 26th IEEE International Conference on Software Maintenance (ICSM 2010), pp. 1–9. IEEE Computer Society, Timişoara, September 2010
16. Shannon, C.E.: A mathematical theory of communication. SIGMOBILE Mob. Comput. Commun. Rev. **5**(1), 3–55 (2001)
17. Theil, H.: Economics and Information Theory. North-Holland Publishing Company, Amsterdam (1967)
18. Vasa, R., Lumpe, M., Branch, P., Nierstrasz, O.: Comparative analysis of evolving software systems using the gini coefficient. In: Proceedings of 25th IEEE International Conference on Software Maintenance (ICSM 2009), pp. 179–188. IEEE Computer Society, Edmonton, September 2009

Discovering Causal Structures from Time Series Data via Enhanced Granger Causality

Ling Luo[1,2(✉)], Wei Liu[2,3], Irena Koprinska[1], and Fang Chen[2]

[1] School of Information Technologies, University of Sydney, Sydney, Australia
{ling.luo,irena.koprinska}@sydney.edu.au
[2] Australian Technology Park Laboratory, NICTA, Sydney, Australia
fang.chen@nicta.com.au
[3] Faculty of Engineering and Information Technologies, University of Technology Sydney, Sydney, Australia
wei.liu@uts.edu.au

Abstract. Granger causality has been applied to explore predictive causal relations among multiple time series in various fields. However, the existence of non-stationary distributional changes among the time series variables poses significant challenges. By analyzing a real dataset, we observe that factors such as noise, distribution changes and shifts increase the complexity of the modelling, and large errors often occur when the underlying distribution shifts with time.

Motivated by this challenge, we propose a new regression model for causal structure discovery – a Linear Model with Weighted Distribution Shift (linear WDS), which improves the prediction accuracy of the Granger causality model by taking into account the weights of the distribution-shift samples and by optimizing a quadratic-mean based objective function. The linear WDS is integrated in the Granger causality model to improve the inference of the predictive causal structure. The performance of the enhanced Granger causality model is evaluated on synthetic datasets and real traffic datasets, and the proposed model is compared with three different regression-based Granger causality models (standard linear regression, robust regression and quadratic-mean-based regression). The results show that the enhanced Granger causality model outperforms the other models especially when there are distribution shifts in the data.

Keywords: Data mining algorithms · Causal inference · Time series regression

1 Introduction

When a temporal modeling method involves multiple time series, it is desirable to explore the causal relations among these time series beyond the statistical correlation, as the causal relations can reveal insights of the data generation process and increase the prediction accuracy. Therefore, learning causal structures among time series is a significant task in analyzing and predicting time series.

The time series predictive causal inference problem can be formally defined as follows. Given a set of multivariate time series $\{X_t^i\}$ ($i \in [1, N]$) measured at successive time points $t \in T$, where T is the set of time indices and $i \in [1, N]$ is the i-th time series, the task is to

© Springer International Publishing Switzerland 2015
B. Pfahringer and J. Renz (Eds.): AI 2015, LNAI 9457, pp. 365–378, 2015.
DOI: 10.1007/978-3-319-26350-2_32

infer the causal relations among the N time series, which can help predict the future values X_t^i accurately from the previous lag values $\{X_{t-k}^i\}$ ($i \in [1, N]$, $k \in [1, d]$), where d is the maximum lag value involved in predicting X_t^i.

Granger causality is one possible solution to this task. The Granger causality model aims to provide a testable and explicit definition of the causality and existence of feedback between two related temporal variables [1]. However, the original Granger causality model is based on standard linear regression, which may generate suboptimal temporal models when handling real data with undesirable features such as high skewness, noise or distribution changes. We aim to improve the performance of the causal inference by using enhanced temporal regression models, which take into consideration the distribution information in the model construction.

Motivating Example. Consider a real travel time dataset with 530 samples as shown in Fig. 1 (more details about this dataset are provided in the experimental section). We computed and sorted the Squared Errors (SEs) produced by a standard linear regression model of all data samples. We notice that most of the SEs are very small, but the SEs ranked after 480 start to increase rapidly. The sum of all 530 SEs is 72.93, while the sum of the SEs which rank after 480 is 66.44. This means that about 10 % of the points with large errors account for about 91 % of the total SEs.

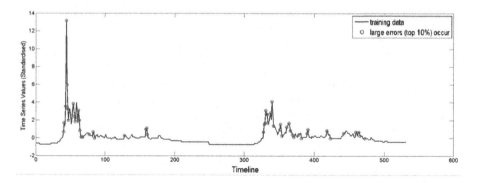

Fig. 1. Plot of a real travel time dataset with 530 samples. The samples with large prediction errors (ranked top 10 %) are highlighted by red circles.

To examine the learning procedure in details, the time points with large prediction errors (errors ranked top 10 %) are marked with red circles in Fig. 1. We can see that the large errors tend to occur when the current time point has an abrupt increase or decrease comparing with the preceding points, which forms a *local distribution shift*. This observation indicates that if we can identify the samples which may produce large errors (such as the local-distribution-shift points), and reduce their errors specifically, it would be possible to decrease the overall error on the dataset.

Motivated by the observation above, we design a novel linear regression method, which handles the points that may produce large errors due to distribution shifts separately, and we use this regression model to enhance the performance of the Granger causality models. Our goals are to reduce the overall prediction errors and to strengthen the capability of causal structure learning. Our contributions are as follows:

- We introduce a new linear regression learning process – Linear Model with Weighted Distribution Shift (linear WDS), which identifies the local-distribution-shift points and adapts the regularized empirical risk function to decrease the large errors made by the local-distribution-shift samples effectively.
- We use the linear WDS to improve the performance of the traditional Granger causality models to infer causal structure from multiple time series accurately.
- We empirically evaluate the performance of the enhanced Granger causality with linear WDS on synthetic datasets and real multivariate traffic datasets and compare the results with three different regression-based Granger causality models (standard linear regression, robust regression and quadratic-mean based regression).

2   Related Work

Granger Causality   Since its introduction in [1], the Granger causality model has been proven to be an effective temporal causal model in different areas such as econometrics [2], climate analysis [3] and neuroscience [4]. The main idea of the Granger causality is the following: assuming that there are two time series X^i and X^j, we would like to predict the future values of X^j. Two regression models should be built – the model that only uses the past values of X^j (called the *restricted model*) and the model uses both the past values of X^i and X^j (called the *full model*):

$$restricted\ model{:}X_t^j = \beta_0 + \sum_{k=1}^{p} \beta_k X_{t-k}^j + \varepsilon_t \tag{1}$$

$$full\ model{:}X_t^j = \theta_0 + \sum_{k=1}^{p} \beta_k X_{t-k}^j + \sum_{n=1}^{q} \theta_n X_{t-n}^i + \varepsilon_t \tag{2}$$

where p and q are lag values for X_t^j and X_t^i, β_k and θ_n are coefficients of the regression models, ε_t is the noise. If the full model is significantly more accurate than the restricted model, it is said that X^i Granger causes X^j, and X^i is the cause X^j is the effect. The F-test or other statistical test is used to compare whether the full model significantly reduces the Residual Sum of Squares (RSS). If multiple time series are involved, the Granger causality test is conducted pair-wisely. The full model uses all variables to make predictions, and the restricted model removes one of the variables to assess whether the absence of that variable will decrease the accuracy significantly.

It is important to notice that the Granger causality (also called predictive causality) used in this paper may not infer the real "cause" [5]. Rather, it is a useful mathematical model that can improve the prediction of future variables.

Regularized Objective Function and Quadratic Mean Learning Methods.   Many machine learning algorithms construct models by optimizing the regularized objective function: $w^* = \arg\min_w \left(R_{emp}(w) + \|\lambda w\|_p\right)$ where w is the coefficient of the model, $R_{emp}(w)$ is the empirical risk function, and $\|\lambda w\|_p$ is the L-p norm regularizer, which is used to avoid overfitting on the training data. Conventional learners minimize the Arithmetic Mean (AM) of the prediction errors of n samples, $1/n* \sum_{i=1}^n l\left(x_i, y_i, w\right)$, in the empirical risk function $R_{emp}(w)$, where $l\left(x_i, y_i, w\right)$ is the loss function for sample (x_i, y_i), and w is the coefficient of the model.

If we apply the regression method on a skewed dataset, which has an asymmetrically distributed dependent variable, the model would be "levered" towards the skewed side – the "long tail", in order to minimize the large squared errors produced by the data in the "long tail" [6]. The Quadratic Mean Learning methods (QMLearn) [6] aim to construct unbiased models on the skewed datasets by using the quadratic-mean based regularized empirical risk functions. Take regression problems as an example, the QMLearn method groups the loss functions of all training samples into two equal-sized subgroups based on the sorted dependent variable, and the empirical risk function is defined as the Quadratic Mean (QM) of the loss funcion as Eq. 3. If we define the first term $\sum_{i=1}^{n/2} l\left(x_i, y_i, w\right) / (n/2)$ in Eq. 3 as g_1, the second term $\sum_{i=n/2+1}^{n} l\left(x_i, y_i, w\right) / (n/2)$ as g_2, $AM = \left(g_1 + g_2\right)/2$, then we can derive the Eq. 4 which shows the relation between QM and AM of two groups.

$$R_{emp}^{QM}(w) = \sqrt{\left[\left(\frac{\sum_{i=1}^{n/2} l\left(x_i, y_i, w\right)}{n/2}\right)^2 + \left(\frac{\sum_{i=n/2+1}^{n} l\left(x_i, y_i, w\right)}{n/2}\right)^2\right]/2} \tag{3}$$

$$\begin{aligned} QM &= \sqrt{(g_1^2 + g_2^2)/2} = \sqrt{\left((g_1 + g_2)/2\right)^2 + \left((g_1 - g_2)/2\right)^2} \\ &= \sqrt{AM^2 + \left((g_1 - g_2)/2\right)^2} \end{aligned} \tag{4}$$

It means the QM can minimize both the sum of errors (AM) and the difference between the two groups of errors simultaneously. If there are large discrepancies between these two groups, QM can make a trade-off decision to balance the two groups.

Distribution Shifts. The distribution shifts of time series increase the complexity of data modelling and make accurate prediction difficult [7]. Distribution shifts refer to the value fluctuations that change the distribution of the series when comparing with the preceding points. In this paper, we concentrate on the *local distribution shifts*, which have temporary and local effects to the time series [8]. There are two important aspects to consider for the local distribution shifts: (1) the distribution-shift samples often carry useful dynamic information, such as the change of environment and occurrence of influential events, so the removal of these samples would lead to the loss of valuable information; (2) the impact of the local distribution shifts is not as significant as the level shift outliers. Level shift outliers are intense changes in the process and the changes are permanent, which require a separate model for the following points, while the local distribution shifts are temporary and relatively mild [8]. Therefore, we keep the shift points in the modelling and adjust the global model to keep balance between shift points and normal points. The methods to detect the changes or shift points in the time series have been discussed in [9]. The main detection methods in previous work include: (1) a statistical hypothesis test such as the Wilcoxon Rank Sum, which compares the distributions of two samples [7]; (2) kernel change-point analysis [10] and (3) detection via the local mean and standard deviation [8].

3   Methodology

This section describes each component of our proposed method in details, including detecting the local-distribution-shift points, utilizing the distribution-shift information in the objective function of the optimization problem and integrating the new regression model in the Granger causality model.

3.1   Local Distribution Shift Detection

The first step in our method is identifying the shift samples. A simple and effective method to detect the distribution-shift samples can be analyzing the mean and standard deviation of the preceding points as described in [8].

Assume the regression model uses d lag values $\{X_{t-p}\}$ ($p \in [1, d]$) to predict the value X_t, and the lag values can be considered as staying in a sliding window (X_{t_win}). Let the mean of values in the current window be $mean\,(X_{t\_win})$ and the standard deviation of values be $std\,(X_{t\_win})$, the target point X_t is tagged as a distribution shift if

$$X_t \notin \left[mean\,(X_{t\_win}) - k * std\,(X_{t\_win})\,, mean\,(X_{t\_win}) + k * std\,(X_{t\_win})\right] \qquad (5)$$

where k is the parameter to control the strength of the detection (meaning k determines the portion of distribution-shift samples in a dataset). When k is larger, less X_t will be tagged as the distribution-shift samples, as the non-shift range becomes larger which means the detection is less sensitive. The default value of k is set to 2 based on the analysis in [8], but it can be adjusted based on the dataset.

Equation 5 is a heuristic to identify the data points that will produce large prediction errors. There are two points to note about the local distribution shifts detection: (1) the detection is only based on the d preceding points, which means it is not relevant to the preceding points that are outside the current window. For example, if the subsequent points of a shift remain at the same value as the shift point without going back to the normal level immediately (e.g. forming a plateau), the subsequent points on the plateau will not be considered as the local-distribution-shift points as long as they are in the range given in Eq. 5; (2) samples which are far away from the majority may not be distribution shifts. From Fig. 1 we can observe that the high errors do not necessarily occur at the points far away from the global mean value. They can still be predicted accurately if the correlation between these points and their preceding values can be captured by the regression model.

3.2   The Proposed Regularized Loss Function

The loss functions $l\left(x_i, y_i, w\right)$ in the AM and QM empirical risk functions presented in Sect. 2 are in general forms which can be adapted for various methods for different learning requirements. For the standard linear regression, the loss function is the squared error $l\left(x_i, y_i, w\right) = \left(y_i - \hat{y}_i\right)^2 = \left(y_i - w^T * x_i\right)^2$ when predicting the sample (x_i, y_i), where y_i is the true value and $\hat{y}_i$ is the predicted value.

Our method also splits the samples into two subgroups as the QMLearn method [6] (see Eq. 3). However, instead of grouping samples evenly based on the sorted dependent variable, we put normal points in one group and the local-distribution-shift points in the other group. To adjust the weight of the local-distribution-shift group, we add a parameter α in the objective function. The empirical risk function is defined as:

$$R^{QM}_{dist\_shift}(w) = \sqrt{\left(\left(\frac{\sum_{i=1}^{m_1}(y_i - w^T * x_i)^2}{m_1}\right)^2 + \alpha\left(\frac{\sum_{i=1}^{m_2}(y_i - w^T * x_i)^2}{m_2}\right)^2\right)\bigg/(1 + \alpha)} \qquad (6)$$

where w is the coefficient vector for the regression model, m_1 is the number of normal samples, m_2 is the number of distribution-shift samples, $\alpha \in [0, \infty)$ is the weight for the distribution-shift samples. Finally, the function takes a weighted quadratic mean of the two subgroups.

Based on Eq. 6, the mean squared errors of the normal samples $\sum_{i=1}^{m_1}\left(y_i - w^T * x_i\right)^2 / m_1$ is g_1 and the mean squared errors of the distribution-shift samples $\sum_{i=1}^{m_2}\left(y_i - w^T * x_i\right)^2 / m_2$ is g_2. Normally, g_2 is larger than g_1, so minimizing the difference between these two groups will lead to a better trade-off than just minimizing the sum of them. The parameter α is a weighting parameter to adjust the balance between the two groups when g_1 and g_2 are not in the same order of magnitude, i.e. g_1 is too small compared to g_2; otherwise the larger term g_2 would dominate the optimization process and lead to suboptimal regression models over the whole dataset.

Taking the regularized term into consideration, our proposed method linear WDS finds the optimal solution to the following objective function:

$$w^* = \arg\min_w \left(R^{QM}_{dist\_shift}(w) + \lambda w^T w\right) \qquad (7)$$

The reason of choosing $\lambda w^T w$, the squared value of L-2 norm $\|\lambda w\|_2$, is to simplify the derivative of the formula. The new empirical risk function $R^{QM}_{dist\_shift}(w)$ preserves the convexity of its subgroups, which is the mean squared errors, so we are solving a convex optimization problem. The trust-region method [11] is applied to search the solution for this unconstraint minimization problem.

3.3   Enhanced Granger Causality

We propose an enhanced Granger causality that applies the weighted quadratic-mean based empirical risk function (as Eq. 7) in the Granger test to explore the causal structure among multiple time series. Both of the restricted model and the full model in the Granger test are built by placing the local-distribution-shift samples in a separate group and optimizing the weighted quadratic mean of these two groups. The workflow of the enhanced Granger causality algorithm is presented in Table 1.

Table 1. Enhanced Granger causality algorithm

Input	Multivariate time series X_t^i, $(i \in [1, N], t \in T)$, max lag
Output	Causal structure graph ; prediction errors for each time series
1	Standardize the data and check the prerequisite: covariance stationary
2	Construct model and find the lag value d based on the AIC
3	**for** i = 1 to N **do** // *go through each variable X^i*
4	Construct the full model by *linear WDS* for X^i
5	Compute the error RSS_{full}
6	**for** j = 1 to N, j ≠ i **do** // *go through each variable except X^i*
7	Construct the restricted models by *linear WDS* for X^i without using the variable X^j
8	Compute the error RSS_{res}
9	Use F-test to determine whether X^j is the Granger causes of X^i
10	**end for**
11	**end for**
12	**return** prediction errors for each time series and visualize the structure graph

First of all, the data is standardized by computing the z-score. Then, the method checks whether the data is covariance stationary by running the ADF and KPSS tests [12]. These tests are required to make sure the causal structure is reliable, as the Granger test assumes that the time series is stationary. The best time lag is determined by comparing the Akaike Information Criterion (AIC) for models with lag values from 1 to the user specified max lag. The best lag value is used in the following steps to build the regression models. The Granger test is conducted for each pair of variables $\{X^i, X^j\}$ to determine whether X^j is the Granger cause of X^i, so when there are N variables, we need $N(N-1)$ tests. If excluding the variable X^j will lead to a statistically significant increase in the RSS, then X^j is the Granger cause of X^i. The variable identified as the "cause" plays an indispensable role in predicting the consequence or effect variable. The causal relation between X^i and X^j will be represented as a link from X^j to X^i in the graph of the causal structure. Although the Granger causal structure may be inconsistent with the true causal structure, it is useful in improving the temporal modelling accuracy and understanding of the dependency among time series.

4   Experimental Evaluation

The experiment was conducted on the synthetic and real travel time datasets. We evaluated the performance of the enhanced Granger test in terms of accuracies of the prediction and causal structure inference.

4.1 Datasets and Performance Measures

Synthetic Datasets. The synthetic multivariate time series are used to evaluate the accuracy of the causal structure inference. We design two types of datasets: the data with the normal distribution and the data with local-distribution-shift samples. The datasets were generated following the methodology in [12]. Each dataset contains 5 time series, and the causal relationships among variables are shown in Fig. 2. This causal graph shows that the variable 1 is the cause for the variables 2, 3 and 4, while the variables 4 and 5 are mutual causes and effects. For the datasets with the normal distribution, we firstly generated five independent normal-distributed random time series as the base. Then, we added the causal relationships on them according to the set of equations [13] as follows

$$
\begin{cases}
X_t^1 = X_t^1 + 0.95 * \sqrt{2} * X_{t-1}^1 - 0.9025 * X_{t-2}^1 \\
X_t^2 = X_t^2 + 0.5 * X_{t-2}^1 \\
X_t^3 = X_t^3 - 0.4 * X_{t-3}^1 \\
X_t^4 = X_t^4 - 0.5 * X_{t-2}^1 + 0.25 * \sqrt{2} * X_{t-1}^4 + 0.25 * \sqrt{2} * X_{t-1}^5 \\
X_t^5 = X_t^5 - 0.25 * \sqrt{2} * X_{t-1}^4 + 0.25 * \sqrt{2} * X_{t-1}^5
\end{cases}
\tag{8}
$$

where X_t^i is the value at time t for time series i ($i \in [1, 5]$).

Fig. 2. Causal structure of synthetic datasets

For the datasets with local distribution shifts, the first step was the same as the generation of the normal-distributed dataset. Then, we randomly selected 20 % of the points in the time series and increased or decreased the values by 100 %–200 % to simulate the local distribution shifts in the series. After that, we integrated the causal relationships among the 5 series. We generated 100 synthetic datasets in total, 50 sets for each type, and the length of each time series is 2000.

Traffic Datasets. The traffic datasets contain the travel time for road segments of a major Australian city over one month from June 17th 2013 to July 14th 2013. Each time series has 7,656 records of the travel time from the origin to the destination of a certain road segment, and the average sampling interval is 5 min. As the aim is to investigate the causal relation between road segments, we concatenated on pairs of road segments which are adjacent to each other based on the spatial information. We selected pairs of segments where the origin of the second road segment is the same as the destination of the first one. For each pair of road segments, the traffic flow is directional – from the first segment to the second one. We used 18 pairs of road segments in the experiment.

Performance Measures. To evaluate the causal structure, we use the F1 score to measure the similarity between the true causal graph and the output of the method [14].

The true positive is the directional link in the output that is also in the true causal graph; the false positive is the link in the output that does not appear in the true graph; the false negative is the absent link in the output comparing to the true graph; and the true negative is the link neither in the output nor in the true graph. Higher F1 score means the inference is more accurate. For the real datasets without the true causal graph, we check whether the causal structure can improve the prediction accuracy. If a causal model can decrease the prediction errors, it is considered as an accurate causal model.

We used 10-fold cross validation as an evaluation procedure. The performance measure for the regression models is the average Root-Mean-Square Errors (RMSE) on the testing set across the 10 folds. The two-sample paired *t*-test was used to statistically compare the RMSE between methods and the baseline LR. The non-parametric Friedman test was also conducted to statistically compare the performance rank of different Granger causality tests [15].

4.2   Experimental Results and Analysis

We compared the performance of Granger causality tests with 4 different regression models: (1) the standard Linear Regression (LR), (2) the Robust Regression (RR), (3) the Linear Regression based on Quadratic Mean (linear QM) [6] and (4) our proposed method – linear WDS. The reasons for selecting those methods are that LR is the most frequently used regression model in the Granger test, so it is a simple and effective baseline; RR is a regression method with robust performance on noisy data, which is a possible solution to deal with the local distribution shifts; and linear QM is the state-of-the-art method which applies the quadratic-mean based loss function.

Our implementation is based on the Granger Causal Connectivity Analysis toolbox [12]. This toolbox consists of different components, in which the core component – regression model constructions were adapted to the proposed objective function. The parameters were set as follows: λ in the *L-2* norm regularizer (Eq. 7) was set to the reciprocal of the number of instances; RR used a bisquare weighting function; for the linear WDS, $k = 2$ and α was configured for each dataset using a validation set with 10 % of data randomly selected from the whole set. In more details, we started from the range $\alpha = [0.001, 0.1]$, which normally contains the best condition, and stopped the iterations when we found the value with the least RMSE on the validation set.

Results on the Synthetic Datasets.   Table 2 presents the average F1 scores of the causal structures generated by the Granger causality with LR, linear QM, RR and linear WDS on two types of datasets. The p-value of paired t-test is computed to check whether the F1 score of a certain model is significantly different from the F1 score of the baseline method LR.

The results from Table 2 show that the Granger causality model with the linear WDS achieves the best performance, with average F1 scores 0.9784 and 0.9379 on the normal datasets and the datasets with distribution shifts, respectively. The t-test results confirm that the linear WDS has statistically significantly better performance than LR, p = 1.59E-11 on the normal datasets, and p = 1.09E-5 on the shift datasets. As the datasets with distribution shifts are more complex to model, all the F1 scores on them are about 3 % lower than the scores on the normal data. As RR is robust to the outliers, it also outperforms the LR and linear QM on the datasets with distribution shifts, and the average scores are just slightly lower than linear WDS. The results verify that the Granger causality with linear WDS could infer the causal structure more accurately than the Granger causality tests with LR, linear QM and RR.

Table 2. Average of F1 scores and statistical test results of the causal structures produced by Granger causality tests with four different regression methods

Datasets	Statistics	LR	Linear QM	RR	Linear WDS
Normal	**F1 score**	0.9597	0.9599	0.9598	**0.9784**
	p-value	Base	>0.05	>0.05	1.59E-11
Distribution-Shift	**F1 score**	0.9277	0.9278	0.9352	**0.9379**
	p-value	Base	>0.05	0.0002	1.09E-5

Results on the Traffic Datasets. For the traffic datasets, Fig. 3 displays three representative pairs of road segments (for Set 1, 5 and 12) on the map, and the causal structures inferred by the Granger causality with four different regression methods. In the table of causal structures, $1 \rightarrow 2$ means the travel time on the first road segment (denoted by ① in the graph) is the cause of the travel time on the second road segment; if there is no arrow between 1 and 2, it means segments 1 and 2 do not have causal dependency.

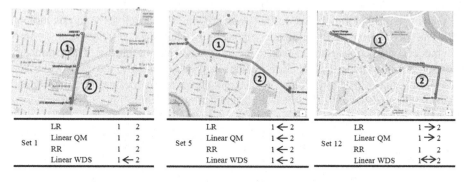

Fig. 3. Causal structures of three sets (1, 5 and 12) inferred by Granger causality with LR, Linear QM, RR and Linear WDS

From practical viewpoints, if there are no congestions on the road, we would normally get $1 \rightarrow 2$, as the traffic flow is from the first segment to the second segment; however, if the segments are busy, the congestions will propagate backwards from the second segment to the first segment, so we could get $2 \rightarrow 1$. Taking Set 5 (the figure in the middle) as an example, we observe that the connection of two road segments of this dataset, which has been highlighted in orange, is a shopping center that may cause traffic delays and congestions. The causal structures of four methods on Set 5 are all $2 \rightarrow 1$, so the Granger causality test results are consistent with our observation. For the other two datasets, the Granger test with linear WDS gets 2 → 1 on Set 1, and 1↔2 on Set 12; while the tests with LR, linear QM only get 1 → 2 on Set 12.

Table 3. Comparison of Granger causalaity tests with 4 different regression models. Each row contains the RMSE of 4 constrained models based on the Granger causal inference results on the dataset, and the rank of each method is inside the bracket. The lowest RMSE on each dataset has been highlighted in bold. The paired t-test is to statistically compare the average RMSE of a certain method with the base case LR, and the Friedman test is to statistically compare the rank of linear WDS with the other methods.

Dataset	LR	Linear QM	RR	Linear WDS
1	0.3514 (2)	0.3531 (3)	0.3533 (4)	**0.3479** (1)
2	0.3881 (3)	0.3883 (4)	0.3851 (2)	**0.3831** (1)
3	0.3942 (3)	0.3944 (4)	**0.3851** (1)	0.3875 (2)
4	0.4033 (2)	0.4255 (4)	0.4174 (3)	**0.3947** (1)
5	0.4564 (2)	0.4570 (3)	0.4710 (4)	**0.4482** (1)
6	0.4641 (3)	0.4639 (2)	0.4732 (4)	**0.4563** (1)
7	0.4831 (2)	0.4836 (3)	0.4867 (4)	**0.4570** (1)
8	0.4837 (3)	0.5136 (4)	**0.4337** (1)	0.4549 (2)
9	0.4934 (2)	0.4945 (3)	0.4966 (4)	**0.4876** (1)
10	**0.5347** (1)	0.5548 (3)	0.5551 (4)	0.5350 (2)
11	0.6233 (2)	0.6239 (3)	0.6372 (4)	**0.6171** (1)
12	0.6389 (3)	0.6384 (2)	0.6538 (4)	**0.6100** (1)
13	0.6393 (2)	0.6404 (3)	0.6644 (4)	**0.6108** (1)
14	**0.6547** (1)	0.6548 (2)	0.7128 (4)	0.6649 (3)
15	0.6809 (3)	0.6817 (4)	0.6653 (2)	**0.6226** (1)
16	0.7000 (3)	0.7380 (4)	**0.4685** (1)	0.6061 (2)
17	0.7218 (3)	0.7799 (4)	0.5944 (2)	**0.5841** (1)
18	0.7328 (2)	0.7360 (3)	0.8233 (4)	**0.7298** (1)
Average	0.5469	0.5568	0.5376	0.5221
t-test	Base	0.0241	0.5838	0.0119
Average rank	2.33	3.22	3.11	1.33
Friedman test	0.001	0.0002	0.0047	Base

Although most of the road segments involved in the experiment are busy roads in the urban area, simply taking $2 \to 1$ as the ground truth might be unreasonable as the traffic may leave the first segment without entering the second segment or the traffic could also enter the second segment from the other roads. To evaluate the performance quantitatively, we build the constrained models based on the causal structures inferred from the Granger causality with different regression methods, to verify the correctness of the predictive causal structures. In more details, to build the constrained model for the series X^i, which means X^i is the dependent variable, we only used X^i and the identified causes of X^i as the independent variables. The constrained model with the smaller prediction errors (RMSE) is preferred, so the corresponding causal structure and the regression method which produced that causal structure are considered to be more accurate than their counterparts. Table 3 presents the RMSE and the accuracy rank of the models built by different regression methods.

In Table 3, each row contains the results for one dataset, and the least RMSE for each dataset among four methods has been highlighted in bold. The ranks of the methods are also listed in the brackets beside RMSE. Our proposed method linear WDS has the smallest RMSE on 13 out of 18 datasets. The t-test result, which compares the average RMSE with the baseline method LR across all datasets, shows that the prediction error of linear WDS 0.5221 is statistically significantly smaller than the error of the LR 0.5469 ($p = 0.0119 < 0.05$). Although RR also outperforms LR on average, the decrease of the RMSE is not statistically significant.

Additionally, a nonparametric test, the Friedman test, was used to compare the ranks of four methods. Friedman test checks whether the methods have stable performance across all datasets. The average rank of linear WDS is 1.33, which is statistically significantly higher than the other three methods, with all the p values less than 0.05. Moreover, LR comes the second, and followed by RR, although RR has a lower average RMSE. The linear QM did not perform very well on the traffic datasets. The reason might be that the global skewness is not very high, although the datasets have high portion of local-distribution-shift points. The linear QM is expected to have advantages on more skewed datasets. Considering the comparison of all four methods, the Granger causality with linear WDS outperforms the other baselines in discovering causal structures.

5   Conclusions

In this paper, we address a challenging issue for discovering Granger causality among multiple time series variables where temporal distributional changes exist. We propose an enhanced Granger causality model with a new linear regression learner, linear WDS, which adjusts the weight of local-distribution-shift samples and optimizes the quadratic mean of the loss function during model construction. The linear WDS aims to achieve lower overall errors by focusing on the large errors produced by the local-distribution-shift samples.

The performance of the Granger causality with linear WDS was empirically evaluated on the synthetic and real traffic datasets, and compared to three existing regression methods (the standard linear regression, the linear QM and the robust regression). The results showed that the enhanced Granger causality could infer the causal structure more accurately than the other methods on synthetic datasets with or without local distribution shifts. On the traffic datasets, the enhanced Granger causality can achieve good performance on major road segments with busy traffic. When building constrained models based on the causal structures inferred by different Granger causality tests, the constrained model of linear WDS had significantly lower average RMSE compared with the linear regression and the linear QM model, and it was also comparable with the robust regression model. The qualitative evaluation verified the advantages of using enhanced Granger causality with linear WDS.

In future work, we plan to integrate the weighted-distribution-shift loss function in other regression methods, such as quantile regression and support vector regression, to improve the prediction accuracy and the correctness of the causal inference.

References

1. Granger, C.W.: Investigating causal relations by econometric models and cross-spectral methods. Econometrica J. Econometric Soc. **37**, 424–438 (1969)
2. Hiemstra, C., Jones, J.D.: Testing for linear and nonlinear granger causality in the stock price-volume relation. J. Finance **49**, 1639–1664 (1994)
3. Lozano, A.C., Li, H., Niculescu-Mizil, A., Liu, Y., Perlich, C., Hosking, J., Abe, N.: Spatial-temporal causal modeling for climate change attribution. In: Proceedings of the 15th ACM SIGKDD International Conference on Knowledge Discovery and Data Mining, pp. 587–596. ACM, 1557086 (2009)
4. Roebroeck, A., Formisano, E., Goebel, R.: Mapping directed influence over the brain using Granger causality and fMRI. Neuroimage **25**, 230–242 (2005)
5. Diebold, F.: Elements of Forecasting. Cengage Learning, Mason (2006)
6. Liu, W., Chawla, S.: A quadratic mean based supervised learning model for managing data skewness. In: SDM, pp. 188–198 (2011)
7. Liu, X., Wu, X., Wang, H., Zhang, R., Bailey, J., Ramamohanarao, K.: Mining distribution change in stock order streams. In: 2010 IEEE 26th International Conference on Data Engineering, pp. 105–108 (2010)
8. Ristanoski, G., Liu, W., Bailey, J.: A time-dependent enhanced support vector machine for time series regression. In: Proceedings of the 19th ACM SIGKDD International Conference on Knowledge Discovery and Data Mining, pp. 946–954. ACM (2013)
9. Gama, J., Žliobaitė, I., Bifet, A., Pechenizkiy, M., Bouchachia, A.: A survey on concept drift adaptation. ACM Comput. Surv. (CSUR) **46**, 44 (2014)
10. Harchaoui, Z., Moulines, E., Bach, F.R.: Kernel change-point analysis. In: Advances in Neural Information Processing Systems, pp. 609–616 (2009)
11. Branch, M.A., Coleman, T.F., Li, Y.: A subspace, interior, and conjugate gradient method for large-scale bound-constrained minimization problems. SIAM J. Sci. Comput. **21**, 1–23 (1999)

12. Seth, A.K.: A MATLAB toolbox for Granger causal connectivity analysis. J. Neurosci. Methods **186**, 262–273 (2010)
13. Baccalá, L.A., Sameshima, K.: Partial directed coherence: a new concept in neural structure determination. Biol. Cybern. **84**, 463–474 (2001)
14. Arnold, A., Liu, Y., Abe, N.: Temporal causal modeling with graphical granger methods. In: Proceedings of the 13th ACM SIGKDD International Conference on Knowledge Discovery and Data Mining, pp. 66–75. ACM (2007)
15. Demšar, J.: Statistical comparisons of classifiers over multiple data sets. J. Mach. Learn. Res. **7**, 1–30 (2006)

Automating Marine Mammal Detection in Aerial Images Captured During Wildlife Surveys: A Deep Learning Approach

Frederic Maire[1]([envelope]), Luis Mejias Alvarez[1], and Amanda Hodgson[2]

[1] Science and Engineering Faculty, Queensland University of Technology,
Brisbane, Australia
{f.maire,luis.mejias}@qut.edu.au
[2] Murdoch University Cetacean Research Unit, Murdoch University, Perth, Australia
a.hodgson@murdoch.edu.au

Abstract. Aerial surveys conducted using manned or unmanned aircraft with customized camera payloads can generate a large number of images. Manual review of these images to extract data is prohibitive in terms of time and financial resources, thus providing strong incentive to automate this process using computer vision systems. There are potential applications for these automated systems in areas such as surveillance and monitoring, precision agriculture, law enforcement, asset inspection, and wildlife assessment. In this paper, we present an efficient machine learning system for automating the detection of marine species in aerial imagery. The effectiveness of our approach can be credited to the combination of a well-suited region proposal method and the use of Deep Convolutional Neural Networks (DCNNs). In comparison to previous algorithms designed for the same purpose, we have been able to dramatically improve recall to more than 80 % and improve precision to 27 % by using DCNNs as the core approach.

1 Introduction

Aerial surveys conducted in light aircraft are a common technique for monitoring various species of wildlife throughout the world (e.g. [5,16,19]). As imaging technologies improve, researchers are moving towards replacing human observers with camera systems (e.g. [7,14,26]), and replacing manned aircraft with Unmanned Aerial Vehicles (UAVs) ([13,25]). These technologies have the potential to improve safety, reduce costs, increase the reliability of the data, and allow surveys to be conducted in remote or inaccessible areas.

The few published works in this area include those by Advanced Coherent Technologies who have developed techniques to detect and track whales using multispectral cameras and computer vision [17,18,20,21]. However the algorithms used in this system are not available to other researchers. Groom et al. [12] describe an object based image analysis method for surveying marine birds that reduces the number of images requiring manual review.

© Springer International Publishing Switzerland 2015
B. Pfahringer and J. Renz (Eds.): AI 2015, LNAI 9457, pp. 379–385, 2015.
DOI: 10.1007/978-3-319-26350-2_33

An alternative to handcrafting image features is to learn them from annotated image datasets. In this paper, we introduce an approach aimed at automating the review process (or parts of it) based on Deep Learning, leveraging off the enormous momentum created by the deep learning community. We have been able to dramatically improve recall (achieving more than 80 %) and improve precision (reaching 27 %) by using Deep Convolutional Neural Networks (DCNNs) as a core approach. Here we compare two different deep learning architectures and advocate a simple *region proposals* method. This method involves performing clustering in the, we can select the compactness parameter (that trades off color-similarity and proximity), and the number of centers for K-means so that the superpixels are approximately the size of a dugong. 5D space that consists of color information and image location.

This paper is structured as follows. Section 2 outlines the different components in the pipeline of our detector. Section 3 explains the *region proposals* module. Section 4 gives details about the DCNNs investigated. Section 5 describes the experiments carried out.

2   System Overview

Figure 1 illustrates the high level view of our dugong detector. We combine the efficiency of Simple Linear Iterative Clustering (SLIC), a region proposals algorithm reviewed in Sect. 3, and deep convolutional neural networks in a pipeline to detect dugongs in large images.

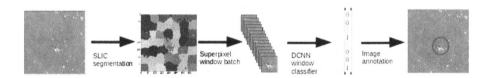

Fig. 1. The actual size of the input images is 6034 × 4012 pixels. For clarity, the input image used in the diagram is only 400 × 400. A dugong can be seen close to the center. The pipeline of the classifier starts with segmentation of the input image in superpixels using the SLIC algorithm. In the next step, a batch of windows centred at the superpixels is generated. This batch of windows is then fed to a deep convolutional neural network. Finally, a copy of the input image is annotated with the position of the detected dugongs.

To build this detector, we extracted training examples from an annotated image collection in a format suitable for a DCNN. The initial database of positive examples was enlarged by applying rotations and scaling transforms to this dataset (see Fig. 2). We then sampled windows away from the marked locations to create an initial set of negative examples. Better negative examples (more ambiguous windows) were also obtained by training a first generation DCNN, and evaluating this DCNN on the training images. The false positives produced

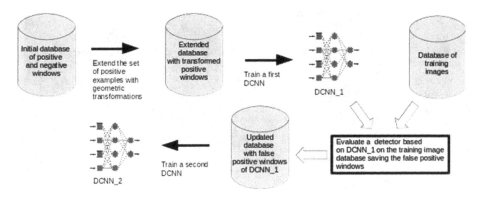

Fig. 2. Two generations of DCNNs are trained. The role of the first DCNN is to identified negative windows that are hard to classify. These difficult examples enrich the initial database, and allow to re-train DCNNs with better performance.

from this process replaced some of the initial negative window examples. A second generation DCNN was then trained on this updated database.

3   Region Proposals

From a computational complexity point of view, the size of the neural network is a predominant factor determining the running time as the evaluation of the window classifier is the major time cost [15]. Unfortunately, CNNs do not have an amortized cost like the Viola-Jones detector [24]. For a 20 Megapixel image, millions of candidate windows have to be processed.

A number of recently introduced methods generate category independent region proposals. These generic methods include objectness [2], selective search [22], category-independent object proposals [8], constrained parametric min-cuts (CPMC) [6], multi-scale combinatorial grouping [3], and low-level image segmentation methods like [1,9,23]. SLIC [1] is a simple and efficient clustering method that works in the 5D space consisting of color information and image location using K-means. Moreover, it is particularly well suited to our application as we know the expected apparent size of the objects of interest (in our case, dugongs). The (approximate) number of superpixels in the segmented output image is a parameter n of the SLIC algorithm that we set to the number of dugongs that could be packed on a regular grid. That is, we set

$$n = \left(\frac{\sqrt{w \times h}}{d} \right)^2$$

where d is the expected side length of a bounding box for a dugong, and w and h are respectively the width and height of the full image Fig. 3.

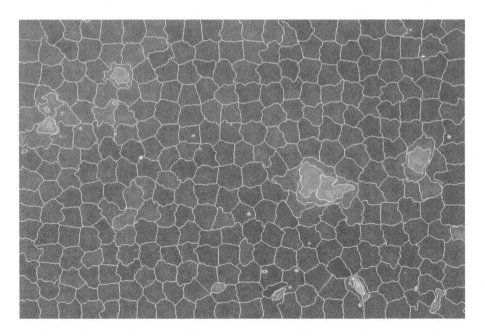

Fig. 3. Contour image of the SLIC segmentation. We can observe that the contours of the bodies of the dugongs match superpixels.

4   Convolutional Neural Network Architecture

We leveraged the Pylearn2 framework [10] to implement our deep CNNs. However, a few low-level functions had to be written using the Theano library [4] on which Pylearn2 is built. We compared two CNN architectures for our application (see Fig. 4).

5   Experimental Results

The DCNNs were trained with weight decay using Stochastic Gradient Descent with a batch size of 100. The size of the layers of Rectilinear CNNs was determined empirically. The learning rate was 5.0×10^{-4}, and the final momentum was 0.8. The Maxout CNNs were trained with a learning rate of 1.0×10^{-3}. The weight include probability was set to 0.5. The size of the layers of the Maxout CNNs was constrained by the GPU implementation in Pylearn2.

We evaluated the detectors by computing the *precision, recall* and *F1-score* on a test set of large images. We use *TP*, *FP* and *FN* to denote respectively the number of true positives, false positives and false negatives respectively. By definition, the precision is *TP / (TP + FP)*, the recall is *TP / (TP + FN)*, and the *F1-score* is the harmonic mean of the precision and the recall. That is, *2 TP / (2 TP + FP + FN)*.

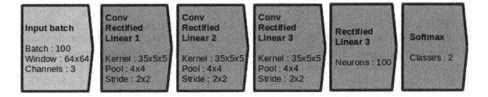

Fig. 4. Our first type of convolutional neural network starts with three rectilinear convolutional layers followed by a hidden layer, and finally a logistic regression layer. The output from the latter is a prediction of whether or not the input window contains a dugong. The second architecture we experimented with was a deep convolutional neural network with a Maxout activation function [11].

Table 1. Best detection results for the two types of DCNN

DCNN	TP	FP	FN	Precision	Recall	F1-score
First generation maxout	45	1245	6	0.0348	0.8823	0.0671
Second generation maxout	41	230	10	0.1512	0.8039	0.2546
First generation rectilinear	45	1909	6	0.0230	0.88235	0.04488
Second generation rectilinear	41	110	10	0.27152	0.8039	0.4059

Table 1 shows the benefit of training a second generation network. Although the recall performance of the second generation CNN decreased slightly compared to the first generation network, the precision significantly improved. The F1-score increased dramatically for the two types of networks considered. More details are provided in an extended version of this paper available at http://eprints.qut.edu.au/view/person/Maire,_Frederic.html.

6   Conclusion

In this paper, we have demonstrated that the combination of region proposals based on SLIC segmentation and classification based on CNNs is well suited for the processing of aerial images from marine fauna surveys. In situations where photos are taken at a known height, and the altitude of the aircraft stays relatively constant, there is little variation in the apparent size of the animals, and no pyramidal analysis of the images is required. Our experimental results showed that deep architectures (three convolutional layers) do help improve detection performance.

References

1. Achanta, R., Shaji, A., Smith, K., Lucchi, A., Fua, P., Susstrunk, S.: Slic superpixels compared to state-of-the-art superpixel methods. IEEE Trans. Pattern Anal. Mach. Intell. **34**(11), 2274–2282 (2012)

2. Alexe, B., Deselaers, T., Ferrari, V.: Measuring the objectness of image windows. IEEE Trans. Pattern Anal. Mach. Intell. **34**(11), 2189–2202 (2012)

3. Arbelaez, P., Pont-Tuset, J., Barron, J., Marques, F., Malik, J.: Multiscale combinatorial grouping. In: 2014 IEEE Conference on Computer Vision and Pattern Recognition CVPR, pp. 328–335. IEEE (2014)

4. Bergstra, J., Breuleux, O., Bastien, F., Lamblin, P., Pascanu, R., Desjardins, G., Turian, J., Warde-Farley, D., Bengio, Y.: Theano: a CPU and GPU math expression compiler. In: Proceedings of the Python for Scientific Computing Conference (SciPy) (2010)

5. Calverley, P.M., Downs, C.T.: Habitat use by nile crocodiles in ndumo game reserve, south africa: a naturally patchy environment. Herpetologica **70**(4), 426–438 (2014)

6. Carreira, J., Sminchisescu, C.: Constrained parametric min-cuts for automatic object segmentation. In: 2010 IEEE Conference on Computer Vision and Pattern Recognition CVPR, pp. 3241–3248. IEEE (2010)

7. Conn, P.B., Ver Hoef, J.M., McClintock, B.T., Moreland, E.E., London, J.M., Cameron, M.F., Dahle, S.P., Boveng, P.L.: Estimating multispecies abundance using automated detection systems: ice-associated seals in the bering sea. Methods Ecol. Evol. **5**(12(Sp. Iss. SI)), 1280–1293 (2014)

8. Endres, I., Hoiem, D.: Category independent object proposals. In: Daniilidis, K., Maragos, P., Paragios, N. (eds.) ECCV 2010, Part V. LNCS, vol. 6315, pp. 575–588. Springer, Heidelberg (2010)

9. Felzenszwalb, P.F., Huttenlocher, D.P.: Efficient graph-based image segmentation. Int. J. Comput. Vis. **59**(2), 167–181 (2004)

10. Goodfellow, I.J., Warde-Farley, D., Lamblin, P., Dumoulin, V., Mirza, M., Pascanu, R., Bergstra, J., Bastien, F., Bengio, Y.: Pylearn2: A Machine Learning Research Library. ArXiv e-prints, August 2013

11. Goodfellow, I.J., Warde-Farley, D., Mirza, M., Courville, A., Bengio, Y.: Maxout networks. arXiv preprint (2013). arXiv:1302.4389

12. Groom, G., Stjernholm, M., Nielsen, R.D., Fleetwood, A., Petersen, I.K.: Remote sensing image data and automated analysis to describe marine bird distributions and abundances. Ecological Informatics **14**, 2–8 (2013)

13. Hodgson, A., Kelly, N., Peel, D.: Unmanned aerial vehicles (uavs) for surveying marine fauna: a dugong case study. PLoS ONE **8**(11), e79556 (2013)

14. Koski, W.R., Thomas, T.A., Funk, D.W., Macrander, A.M.: Marine mammal sightings by analysts of digital imagery versus aerial surveyors: a preliminary comparison. J. Unmanned Veh. Syst. **01**(01), 25–40 (2013)

15. Maire, F., Mejias, L., Hodgson, A.: A convolutional neural network for automatic analysis of aerial imagery. In: Wang, L., Ogunbona, P., Li, W., (eds.) Digital Image Computing: Techniques and Applications (DICTA 2014), Wollongong, New South Wales, Australia (2014)

16. Michaud, J.-S., Coops, N.C., Andrew, M.E., Wulder, M.A., Brown, G.S., Rickbeil, G.J.M.: Estimating moose (alces alces) occurrence and abundance from remotely derived environmental indicators. Remote Sens. Environ. **152**, 190–201 (2014)

17. Podobna, Y., Schoonmaker, J., Boucher, C., Oakley, D.: Optical detection of marine mammals. In: Proceedings SPIE 7317, Ocean Sensing and Monitoring, vol. 7317 (2009)

18. Podobna, Y., Sofianos, J., Schoonmaker, J., Medeiros, D., Boucher, C., Oakley, D., Saggese, S.: Airborne multispectral detecting system for marine mammals survey. In: Proceedings SPIE 7678, Ocean Sensing and Monitoring II, 76780G, 20 April 2010

19. Rekdal, S.L., Hansen, R.G., Borchers, D., Bachmann, L., Laidre, K.L., Wiig, O., Nielsen, N.H., Fossette, S., Tervo, O., Heide-Jorgensen, M.P.: Trends in bowhead whales in west greenland: Aerial surveys vs. genetic capture-recapture analyses. Mar. Mammal Sci. **31**(1), 133–154 (2015)
20. Schoonmaker, J., Podobna, Y., Boucher, C., Sofianos, J., Oakley, D., Medeiros, D., Lopez, J.: The utility of automated electro-optical systems for measuring marine mammal densities. In: OCEANS 2010, pp. 1–6 (2010)
21. Schoonmaker, J., Wells, T., Gilbert, G., Podobna, Y., Petrosyuk, I., Dirbas, J.: Spectral detection and monitoring of marine mammals. In: SPIE 6946, Airborne Intelligence, Surveillance, Reconnaissance (ISR) Systems and Applications V, 694606 (2008)
22. Uijlings, J.R.R., van de Sande, K.E.A., Gevers, T., Smeulders, A.W.M.: Selective search for object recognition. Int. J. Comput. Vis. **104**(2), 154–171 (2013)
23. Vedaldi, A., Soatto, S.: Quick shift and kernel methods for mode seeking. In: Forsyth, D., Torr, P., Zisserman, A. (eds.) ECCV 2008, Part IV. LNCS, vol. 5305, pp. 705–718. Springer, Heidelberg (2008)
24. Viola, P., Jones, M.: Rapid object detection using a boosted cascade of simple features. In: Proceedings of the 2001 IEEE Computer Society Conference on Computer Vision and Pattern Recognition, CVPR 2001, IEEE, vol. 1, pp. I-511 (2001)
25. Watts, A.C., Perry, J.H., Smith, S.E., Burgess, M.A., Wilkinson, B.E., Szantoi, Z., Ifju, P.G., Percival, H.F.: Small unmanned aircraft systems for low-altitude aerial surveys. J. Wildl. Manag. **74**(7), 1614–1619 (2010)
26. Wilson, S., Bazin, R., Calvert, W., Doyle, T., Earsom, S.D., Oswald, S.A., Arnold, J.M.: Abundance and trends of colonial waterbirds on the large lakes of southern manitoba. Waterbirds **37**(3), 233–244 (2014)

Path Algebra for Mobile Robots

Frederic Maire[(✉)] and Gavin Suddrey

School of Electrical Engineering and Computer Science,
Science and Engineering Faculty, Queensland University of Technology,
Gardens Point, Brisbane, QLD 4000, Australia
{f.maire,g.suddrey}@qut.edu.au
http://www.qut.edu.au

Abstract. In this paper, we introduce a path algebra well suited for navigation in environments that can be abstracted as topological graphs. From this path algebra, we derive algorithms to reduce routes in such environments. The routes are reduced in the sense that they are shorter (contain fewer edges), but still connect the endpoints of the initial routes. Contrary to planning methods descended from Disjktra's Shortest Path Algorithm like D^*, the navigation methods derived from our path algebra do not require any graph representation. We prove that the reduced routes are optimal when the graphs are without cycles. In the case of graphs with cycles, we prove that whatever the length of the initial route, the length of the reduced route is bounded by a constant that only depends on the structure of the environment.

1 Introduction

Most current navigation systems generate global, metric representation of the environment with either obstacle-based grid maps or feature-based metric maps [4]. While suitable for small areas, global metric maps have inefficiencies of scale [7,9]. Arguably, topological mapping is more efficient as it concisely represents a partial view of the world as a graph [13]. The vertices of the graph may correspond to anchored points in the environment with physical meaning that the robot can precisely navigate to, like a door or a corridor intersection as illustrated in Fig. 1. The vertices can also correspond to the vertices of a Voronoi diagram [5].

In this paper, we focus on mobile robots equipped with a sensor suite enabling the robots to

- detect when they reach a topological node.
- determine the degree of a node (number of incident edges).
- select an edge, and drive along this edge.

We show that when the graph is without cycles, a map is superfluous for navigation purpose. More precisely, we define a path algebra that provides a very compact representation of the route followed by the robot. This path algebra

© Springer International Publishing Switzerland 2015
B. Pfahringer and J. Renz (Eds.): AI 2015, LNAI 9457, pp. 386–397, 2015.
DOI: 10.1007/978-3-319-26350-2_34

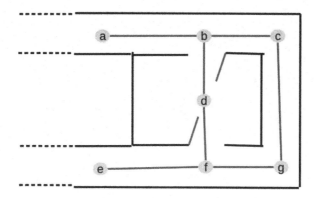

Fig. 1. A topological map derived from a floor map. The node d represents a room. The two doorways of this room correspond to the edge bd and the edge df.

enables the computation of return routes as well as the automatic simplification of long routes containing detours. We also extend this method to graph containing cycles.

In Sect. 2, we review previous work on mapping and localisation. In Sect. 3, we introduce a path algebra. In Sect. 4, we demonstrate how the path algebra can be used on different types of graphs.

2   Related Work

The major mobile robot navigation schemes can roughly be classified into two categories: navigation with position information and navigation without position information [8].

Metric maps belong to the first category. They are popular because they are suitable for path planning with a high degree of accuracy. Their drawback is that they are often expensive to calculate, and do not scale well in large, unstructured and dynamic environments like most outdoor places [1].

The alternative to metric maps are topological maps, which are graphs that in their pure form do not store metric information. Topology is mainly concerned with the connectivity properties of a space. In the context of an indoor mobile robot, the space refers to the expanse containing fixed structural components of the building like rooms, walls, doors and corridors, as well as the mobile entities like people, furniture and equipment [14]. Distinct places are represented as nodes, adjacency between different locations is represented by the graph edges (see Fig. 1). Because of their sparse representation, topological maps can be a memory efficient representation of the environment and provide good scalability [4].

Hybrid approaches build topometric maps, allowing the robot to use the topological information to plan a global path and to exploit metric information to find shortcuts [3].

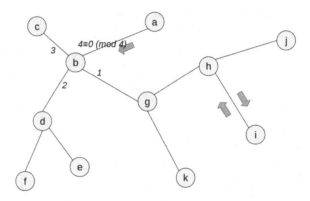

Fig. 2. The route from the arc ab to the arc hi can be coded as the tuple $(1, 1, 2)$. The labels on the edges incident to Node b are the indices with respect to arc ab (Color figure online).

Appearance-based mapping and localisation approaches include appearance-only systems like FAB-MAP [2] which can be considered as an extreme case of topological map. Topometric systems using a more primitive data association than FAB-MAP, like RatSLAM [11] or CAT-Graph [10] require a rough estimate of the distance travelled. More recently, these approaches have been combined in [6] to achieve better robustness to variation in the environment appearance and changes in illumination and structure.

Although the $D^\star$ algorithm accepts partially known environments, it has to build a graphical representation of its environment to plan paths [12]. The path algebra that we introduce does not have this requirement.

3   Route Representation

When following the instructions of a navigation GPS device, the driver of a car receives instructions of the form "at the next roundabout, take the third exit". This command format is well suited for logging the itinerary followed by a mobile robot in an environment that has a graphical topology like a road-network or the corridors of a building. Figure 2 illustrates such an environment. The agent/robot traverses the graph following its edges and using its vertices as roundabouts. In order to indicate the direction the robot is facing on an edge, we specify the location of the robot by an arc. We adhere to the standard terminology of graph theory, and reserve the term *arc* for an edge that has been given an orientation. The blue arrow on the left of arc $a \rightarrow b$ represents the position of the robot going from Node a to Node b.

The navigational route instructions from the arc $a \rightarrow b$ as the starting position, to the arc labeled $h \rightarrow i$ as the destination, would sound as follows if told by a GPS device:

1. "drive to the next the intersection"
2. "take the first left"

3. "drive to the next the intersection"
4. "take the first left"
5. "drive to the next the intersection"
6. "take the second left"
7. "drive to the next the intersection"
8. "you have arrived at your destination"

A less verbose representation would code the route into the integer sequence $(1, 1, 2)$.

3.1 Forward Route

Definition 1. *A **forward route** of length k is a sequence of integers $(c_1, c_2, \ldots, c_k)$ coding for each node the agent traverses, the relative edge, with respect to the entering edge, at which the agent should leave the node. At the jth node of its journey, the agent takes the c_j^{th} relative edge counting clockwise from the entering edge.*

The integers $(c_1, c_2, \ldots, c_k)$ are signed. The command c for a node with n edges can be reduced modulo n. In the graph shown in Fig. 2, if the agent is on the arc ab, and the driving command c is equal to 6, then the agent will continue its journey via the arc bd. The commands 6 and 2 have the same effect at this node because 6 is equivalent to 2 modulo 4. Formally, we write $6 \equiv 2 \pmod 4$. More generally,

Proposition 1. *When arriving at a node with n edges, the commands c and c' have the same effect if and only if $c \equiv c' \pmod n$.*

Observing that $-c \equiv n - c \pmod n$, we derive the following special case that becomes useful when backtracking:

Proposition 2. *When arriving at a node with n edges, the commands $-c$ and $n - c$ have the same effect.*

Considering again Fig. 2, if the agent is on the arc ab and the driving command c is equal to 0 or 4, then the agent will perform a U-turn at b, and ends up on arc ba. More generally,

Proposition 3. *When arriving at a node with n edge, the command c will trigger a U-turn if and only if $c \equiv 0 \pmod n$.*

When the agent arrives at a leaf, the value of the command c does not matter:

Proposition 4. *At a leaf node (node with exactly one incident branch), all commands c have the same effect. Whatever the value c takes, the agent performs a U-turn.*

Proof. It is enough to observe that $\forall c \in \mathbb{Z}, \quad c \equiv 0 \pmod 1$

3.2   Return Route

While driving, if we take the 2nd leftmost exit at a roundabout during the forward leg of a trip, we should take the 2nd rightmost exit at the same roundabout during the return trip. More generally, suppose that an agent on a forward journey visits the nodes $x_1, x_2, \ldots, x_k$, and that the agent traverses the node x_i with the command c_i on its way forward. If the agent wants to backtrack to its starting position, the agent should traverse the node x_i with the command $-c_i$ on its return journey. In this paper, the reverse arc of an arc α will be denoted by $\overline{\alpha}$. That is, if $\alpha = x \to y$, then $\overline{\alpha} = y \to x$.

Proposition 5. *If command c takes an agent positioned on an arc α to an arc β, then the command $-c$ will take an agent positioned on the arc $\overline{\beta}$ to the arc $\overline{\alpha}$.*

Proposition 5 can be generalized to longer paths by a simple induction on the number of nodes in the path.

Proposition 6. *If $(c_1, c_2, \ldots, c_k)$ is a command sequence that takes an agent positioned on an arc α to an arc β via the nodes $x_1, x_2, \ldots, x_k$, then the command sequence $(-c_k, -c_{k-1}, \ldots, -c_1)$ will take an agent positioned on the arc $\overline{\beta}$ to the arc $\overline{\alpha}$ via the nodes $(x_k, x_{k-1}, \ldots, x_1)$.*

3.3   Route Simplification

Consider Fig. 3 where a robot starting at the position of the blue arrow executes the command sequence $c = (1, 5, 2)$ that takes the robot through Node a, then to Node b, where a U-turn is made, then back to Node a and onto Node c. In this example, $c_2 = 5$ is equal to the number of incident edges to the second node $x_2 = b$ of the route. Therefore, we could have used the equivalent sequence $c' = (1, 0, 2)$. This sequence can be further reduced to the singleton $c'' = (1+2) = (3)$.

More generally, we always have

Proposition 7. *If (c_1, c_2, c_3) is a command triplet that takes an agent positioned on an arc α to an arc β via the nodes x_1, x_2 and x_3; and if moreover $c_2 \equiv 0 \pmod{n_2}$ where n_2 is the degree of the node x_2, then the sequence (c_1, c_2, c_3) has the same effect as the singleton sequence $(c_1 + c_3)$. In other words, if an agent starts on the arc α and executes the single command $(c_1 + c_3)$, it will ends up positioned on the arc β. These conditions also imply that the nodes x_1 and x_3 are the same.*

Proof. Without loss of generality, we assign the index 0 to the arc α entering x_1. When the U-turn at Node x_2 is performed, the robot returns to Node $x_1 = x_3$ via the edge of x_1 indexed c_1. The index of the exit edge is obtained by adding c_3. Therefore the relative index of the arc β with respect to the arc α is $c_1 + c_3$ as illustrated in Fig. 4.

Often, we are mainly concerned with the starting arc α and the destination arc β of a given route c. This leads us to the following definition;

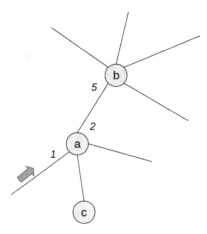

Fig. 3. The robot starts at the location of the blue arrow. The integer on the left of an arc corresponds to the command c_i for the ith node. In this, example when arriving at Node a the robot takes the first exit, then follows the arc $a \to b$ to Node b where it takes the 5th exit. This U-turn brings back the robot on the arc $b \to a$. At the second visit of Node a, the robot takes the 2nd exit with respect to arc $b \to a$. The whole sequence $c = (1, 5, 2)$ can be contracted into $c = (3)$ (Color figure online).

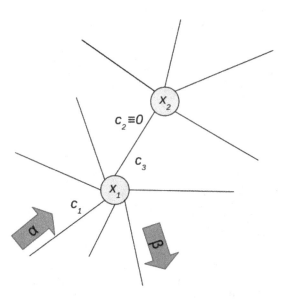

Fig. 4. Sequence reduction; whenever $c_2 \equiv 0 \pmod{n_2}$ where n_2 is the degree of the node x_2, the sequence (c_1, c_2, c_3) has the same effect as the singleton sequence $(c_1 + c_3)$.

Definition 2. *If two routes c and c' start with the same arc α and end with the same destination arc β, then the routes are said to be* equivalent. *We will write $c \sim c'$ to express this equivalence.*

In Fig. 5, the robot starts on the arc labeled 1, and executes the command sequence $c = (1, 0, 2, 2, 0, 2, 0, 2, 3, 2, 0, 2, 1, 0, 1)$ that leads the robot to the arc labeled 16. Proposition 7 allows us to simplify the route by iteratively reducing command triplets of the form $(c_{i-1}, 0, c_{i+1})$. The first occurrence of 0 is in the triplet $(1, 0, 2)$. The triplet reduces to (3). This first reduction shows that the sequence c is equivalent to the sequence $(3, 2, 0, 2, 0, 2, 3, 2, 0, 2, 1, 0, 1)$. Further reductions can be performed. The next occurence of 0 is $(2, 0, 2)$ which reduces to (4). Therefore, we now have $c \sim (3, 4, 0, 2, 3, 2, 0, 2, 1, 0, 1)$. Continuing the reduction, $(4, 0, 2)$ is replaced with (6). Therefore, $c \sim (3, 6, 3, 2, 0, 2, 1, 0, 1)$. The next reduction replaces $(2, 0, 2)$ with (4), and entails that $c \sim (3, 6, 3, 4, 1, 0, 1)$. Finally, $(1, 0, 1)$ is replaced with (2), yielding $c \sim (3, 6, 3, 4, 2)$. Although, our automatic substitutions have eliminated all occurrences of 0, we have not completely exploited Proposition 7. Indeed, the degree sequence $(4, 3, 4, 3, 3)$ of the nodes traversed in the reduced route $(3, 6, 3, 4, 2)$ requires more attention.

As $6 \equiv 0 \pmod{3}$, we have $(3, 6, 3, 4, 2) \sim (3, 0, 3, 4, 2) \sim (6, 4, 2)$. As the degree sequence of the nodes traversed for the command sequence $(6, 4, 2)$ is $(4, 3, 3)$, c can be put in the more canonical form $(2, 1, 2)$. To sum up, by iterating the reduction described in Proposition 7, we have shown that the command sequence $c = (1, 0, 2, 2, 0, 2, 0, 2, 3, 2, 0, 2, 1, 0, 1)$ is equivalent to the command sequence $(2, 1, 2)$.

From this toy example, we can generalize our approach to a generic algorithm for the minimization of any route (pseudo-code listed in Algorithm 1).

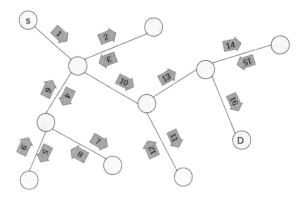

Fig. 5. A route traversing 15 nodes taking the robot from the arc labeled 1 to the arc labeled 16. The command sequence is $c = (1, 0, 2, 2, 0, 2, 0, 2, 3, 2, 0, 2, 1, 0, 1)$. Thanks to Proposition 7, we can reduce the long route $(1, 0, 2, 2, 0, 2, 0, 2, 3, 2, 0, 2, 1, 0, 1)$ to the direct route $(2, 1, 2)$.

input :
a route with the degrees of the traversed nodes
$\mathcal{C}$: a command sequence $(c_1, c_2, \ldots, c_n)$ where the c_i's are integer values
(possibly negative)
$\mathcal{D}$: a degree sequence $(d_1, d_2, \ldots, d_n)$ where d_i is the number of edges incident
to the ith traversed node
output:
the shortest route $\mathcal{C}'$ equivalent to $\mathcal{C}$

```
1  begin
2  |    C' = C                        /* initialize the reduced route */
3  |    D' = D            /* initialize the associated degree sequence */
4  |    repeat
5  |    |   k = length(C')    /* recompute the length of the reduced route */
6  |    |   for i ∈ [2, k − 1] do
7  |    |   |   if c_i ≡ 0 (mod d_i) then
   |    |   |   |   /* apply Proposition 7 to contract (c_{i−1}, c_i, c_{i+1}) into
   |    |   |   |      (c_{i−1} + c_{i+1}) in C'                            */
8  |    |   |   |   C' ⟵ (c_1, c_2, …, c_{i−2}, c_{i−1} + c_{i+1}, c_{i+2}, …, c_k)
   |    |   |   |   /* remove the degrees of the i^{th} and (i+1)^{th} nodes from
   |    |   |   |      D'                                                   */
9  |    |   |   |   D' ⟵ (d_1, d_2, …, d_{i−2}, d_{i−1}, d_{i+2}, …, d_k)
10 |    |   |   |   break
11 |    |   |   end
12 |    |   end
13 |    until no index i ∈ [2, k − 1] such that c_i ≡ 0 (mod d_i) can be found
14 end
```

Algorithm 1. Route Minimization

4 Navigation Applications

In this section, we consider in turn the navigation task on graphs without cycles, then on graphs with cycles.

4.1 Navigation on Trees

Given a route $\mathcal{R} = (c_1, c_2, \ldots, c_n)$, we write $\overline{\mathcal{R}} = (-c_n, -c_{n-1}, \ldots, -c_1)$ to denote the reverse of route $\mathcal{R}$.

Proposition 8. *If a route $\mathcal{R}_1$ takes a robot from a place α_0 to a place α_1, and the route $\mathcal{R}_2$ takes the robot from the same α_0 to a second place α_2, then the route $\overline{\mathcal{R}_1}\mathcal{R}_2$ will take the robot from the $\overline{\alpha_1}$ to α_2.*

The place α_0 could be the base of the robot (charging station), with α_1 and α_2 being places of interest for which the robot has stored the respective routes $\mathcal{R}_1$ and $\mathcal{R}_2$. By applying Algorithm 1 on $\overline{\mathcal{R}_1}\mathcal{R}_2$ we derive a direct route from $\overline{\alpha_1}$ to α_2.

4.2   Navigation on Graphs with Cycles

In Sect. 3, we introduced a route reduction algorithm that is applicable to any graph. However, it is only for trees (connected graphs without cycles) that we can guarantee that the returned contracted route corresponds to the shortest path between the starting arc α and the destination arc β. Without extra information, it is impossible to detect a loop closure.

The route $c = (1, 2, 1, 2, 1)$ executed on the graphs of Figs. 6 and 7 provides the same topological experience to the robot, in the sense that the robot visits two sequences of nodes with identical degree sequences, and selects edges with the same relative indices in the two cases.

To deal with cycles, it is sufficient to mark enough arcs with unique identifiers so that all cycles of the graph have at least one marked arc. In the

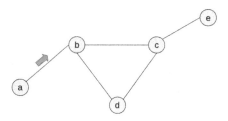

Fig. 6. Starting on the blue arrow, the route $c = (1, 2, 1, 2, 1)$ brings the robot to the arc $c \to e$. Unfortunately, Algorithm 1 cannot simplify c into the equivalent route $c' = (1, 1)$. Moreover, from a topological point of view, the robot cannot distinguish the journey on this graph from the journey experienced on the graph of Fig. 7 (Color figure online).

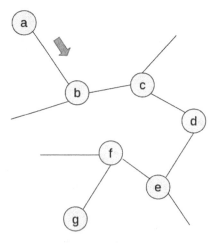

Fig. 7. Starting on the blue arrow, the route $c = (1, 2, 1, 2, 1)$ brings the robot to the arc $f \to g$. From a topological point of view, the robot cannot distinguish the journey on this graph from the journey experienced on the graph of Fig. 6 (Color figure online).

example of Fig. 6, imagine that the arc $b \rightarrow c$ is marked with γ. The robot can then annotate the sequence $c = (1, 2, 1, 2, 1)$ with the marker γ to indicate the traversing of a distinguished arc. The annotated version of the sequence c is $(1, \gamma, 2, 1, 2, \gamma, 1)$. The annotated sequence $(\gamma, 2, 1, 2, \gamma)$ can be reduced to (γ). Therefore the sequence $(1, \gamma, 2, 1, 2, \gamma, 1)$ can be replaced with the sequence $(1, \gamma, 1)$ In practice, either the path between Place b and Place c corresponding to the arc $b \rightarrow c$ is uniquely identifiable thanks to natural features, or we have to install a uniquely identifiable landmark on this path.

Algorithm 2 extends Algorithm 1 to graphs with cycles. We need first to create a set $\mathcal{M}$ of uniquely marked arcs such that $\mathcal{M}$ intersects every cycle of the graph. To reduce an annotated route $\mathcal{C}$ on the graph, we first replace any subsequence of the form $(\gamma, \mathcal{P}, \gamma)$ of $\mathcal{C}$ with (γ), where $\mathcal{P}$ is a route and $\gamma \in \mathcal{M}$. Finally, when no further substitutions are applicable, we run Algorithm 1.

---

input :
$\mathcal{M}$: a set of marked arcs such that $\mathcal{M}$ intersects every cycle of the graph
$\mathcal{C}$: an annotated route with marked arcs from $\mathcal{M}$
output:
a reduced route $\mathcal{C}'$ equivalent to $\mathcal{C}$

1 **begin**
2     $\mathcal{C}' = \mathcal{C}$ `/* initialize the reduced route */`
3   **repeat**
4   scan $\mathcal{C}'$ for multiple occurrences of any marked arc γ
5   let $(\gamma, \mathcal{P}, \gamma)$ be the longest subsequence containing the same marked arc γ
6   replace $(\gamma, \mathcal{P}, \gamma)$ in $\mathcal{C}'$ with (γ)
7   **until** *no $\gamma \in \mathcal{M}$ appears more than once in $\mathcal{C}'$*
8   run Algorithm 1 on $\mathcal{C}'$
9 **end**

Algorithm 2. Route Reduction on Graphs with Cycles

Theorem 1. *Algorithm 2 reduces any route $\mathcal{C}$ (possibly long random walks) to an equivalent route $\mathcal{C}'$ whose length is bounded by $q + m \times (q + 1)$, where m is the number of edges of the graph, and q is the number of marked arcs.*

Proof. Consider $\mathcal{C}'$ the reduced route returned by Algorithm 2. The route $\mathcal{C}'$ is the concatenation of runs of non-marked arcs and marked arcs. Each marked arc appears at most once in $\mathcal{C}'$. This accounts for the term q in the upper bound formula. Recall that the graph induced by the non-marked arcs is a tree, as it contains no cycles by construction of $\mathcal{M}$. Therefore each run of non-marked arcs is of length at most m as Algorithm 1 has eliminated all U-turns in $\mathcal{C}'$. We have at most $q + 1$ runs of non-marked arcs in $\mathcal{C}'$. This accounts for second term $m \times (q + 1)$ in the upper bound formula.

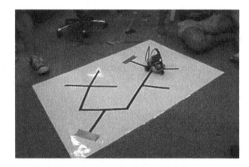

Fig. 8. Testing Algorithm 1 on a Lego-Mindstorms robot. The robot wanders on the network while logging its route. Upon detecting the second grey patch or hearing a clap, it returns to the first grey patch using the reduced route.

5   Experiments

We validated the algorithms of Sect. 4 in simulation as well as with real robots. We programmed Lego Mindstorms robots to navigate in networks drawn on laminated posters (see Fig. 8). The robot performed a random walk with route logging. Upon an external trigger (detecting a grey patch on the floor, or sound clap), the robot had to return to some specific location (either the starting position or the place where another grey patch was found).

6   Conclusion

In this paper, we have introduced a path algebra that is well suited for navigation in environments whose topology is a graph. We showed that when the graph contains no cycles, our route minimization algorithm returns the optimal route. The path algebra is not restricted to planar environments. For example, using a lift in a building can be abstracted as traversing a node with each destination floor corresponding to an edge of the node.

If cycles are present in the graph, distinguishing some places becomes necessary for localisation. We showed that in order to navigate, it is sufficient to be able to recognize a set of places on the ground that corresponds to a set of arcs that intersects every cycle of the topological graph representing the environment. In future work, we plan to test our approach in a building environment with a mobile robot capable of detecting doors and reading signs on doors.

References

1. Augustine, M., Ortmeier, F., Mair, E., Burschka, D., Stelzer, A., Suppa, M.: Landmark-tree map: a biologically inspired topological map for long-distance robot navigation. In: 2012 IEEE International Conference on Robotics and Biomimetics (ROBIO), pp. 128–135. IEEE (2012)

2. Cummins, M., Newman, P.: Appearance-only slam at large scale with fab-map 2.0. Int. J. Robot. Res. **30**(9), 1100–1123 (2011)
3. Dayoub, F., Morris, T., Upcroft, B., Corke, P.: Vision-only autonomous navigation using topometric maps. In: 2013 IEEE/RSJ International Conference on Intelligent Robots and Systems (IROS), pp. 1923–1929. IEEE (2013)
4. Filliat, D., Meyer, J.-A.: Map-based navigation in mobile robots: I. A review of localization strategies. Cognitive Syst. Res. **4**(4), 243–282 (2003)
5. Garrido, S., Moreno, L., Blanco, D., Jurewicz, P.: Path planning for mobile robot navigation using voronoi diagram and fast marching. Int. J. Robot. Autom **2**(1), 42–64 (2011)
6. Glover, A.J., Maddern, W.P., Milford, M.J., Wyeth, G.F.: Fab-map+ ratslam: appearance-based slam for multiple times of day. In: 2010 IEEE International Conference on Robotics and Automation (ICRA), pp. 3507–3512. IEEE (2010)
7. J. Hartmann, Klussendorff, J.H., Maehle, E.: A unified visual graph-based approach to navigation for wheeled mobile robots. In: 2013 IEEE/RSJ International Conference on Intelligent Robots and Systems (IROS), pp. 1915–1922, November 2013
8. Jehn-Ruey, J., Yung-Liang, L., Fu-Cheng, D.: Mobile robot coordination and navigation with directional antennas in positionless wireless sensor networks. Int. J. Ad Hoc Ubiquitous Comput. **7**(4), 272–280 (2011)
9. Konolige, K., Marder-Eppstein, E., Marthi, B.: Navigation in hybrid metric-topological maps. In: 2011 IEEE International Conference on Robotics and Automation (ICRA), pp. 3041–3047, May 2011
10. Maddern, W., Milford, M., Wyeth, G.: Towards persistent indoor appearance-based localization, mapping and navigation using cat-graph. In: 2012 IEEE/RSJ International Conference on Intelligent Robots and Systems (IROS), pp. 4224–4230, October 2012
11. Milford, M., Jacobson, A., Chen, Z., Wyeth, G.: Ratslam: using models of rodent hippocampus for robot navigation and beyond (2013)
12. Stentz, A., Mellon, I.C.: Optimal and efficient path planning for unknown and dynamic environments. Int. J. Robot. Autom. **10**, 89–100 (1993)
13. Tully, S., Kantor, G., Choset, H.: A unified Bayesian framework for global localization and SLAM in hybrid metric/topological maps. Int. J. Robot. Res. **31**(3), 271–288 (2012). doi:10.1177/0278364911433617
14. Worboys, M.: Modeling indoor space. In: Proceedings of the 3rd ACM SIGSPATIAL International Workshop on Indoor Spatial Awareness, pp. 1–6. ACM (2011)

Abduction in PDT Logic

Karsten Martiny[1]([⊠]) and Ralf Möller[2]

[1] Hamburg University of Technology, Hamburg, Germany
karsten.martiny@tuhh.de
[2] Universität zu Lübeck, Lübeck, Germany
moeller@uni-luebeck.de

Abstract. Probabilistic Doxastic Temporal (PDT) Logic is a formalism
to represent and reason about belief evolutions in multi–agent systems.
In this work we develop a theory of abduction for PDT Logic. This gives
means to novel reasoning capabilities by determining which epistemic
actions can be taken in order to induce an evolution of probabilistic
beliefs into a desired goal state. Next to providing a formal account of
abduction in PDT Logic, we identify pruning strategies for the solution
space, and give a sound and complete algorithm to find minimal solutions
to the abduction problem.

1 Introduction and Related Work

Epistemic and doxastic logics are used to reason about agents' knowledge. Formalizing the analysis of knowledge and belief through such logics has been an
active topic of research in diverse fields. Numerous extensions to modal epistemic logic have been made to reason about knowledge in multi–agent settings
(e.g., [8]), to add probabilistic knowledge (e.g., [7]), and to analyze the dynamic
evolution of knowledge (e.g., [4]).

In realistic scenarios an agent usually has only incomplete and inaccurate
information about the actual state of the world, and thus considers several different situations as actually being possible. As it receives new information (e.g.,
it observes some facts), it has to update its beliefs about these possible worlds
such that they are consistent with the new information. These updates can for
example result in regarding some worlds as impossible, or judging some worlds
to be more likely than before. Thus, in addition to analyzing the set of worlds
an agent believes to be possible, it is also expedient to quantify these beliefs in
terms of probabilities. This provides means to specify fine–grained distinctions
between the range of worlds that an agent considers possible but highly unlikely,
and worlds that seem to be almost certainly the actual world.

When multiple agents are involved in such a setting, an agent may not only
have varying beliefs regarding the facts of the actual world, but also regarding
the beliefs of other agents. In many scenarios, the actions of one agent will not
only depend on its belief in facts of the actual world, but also on its beliefs in
some other agent's beliefs.

© Springer International Publishing Switzerland 2015
B. Pfahringer and J. Renz (Eds.): AI 2015, LNAI 9457, pp. 398–410, 2015.
DOI: 10.1007/978-3-319-26350-2_35

To analyze the belief evolution of multiple agents, problem domains can be modeled using *Probabilistic Doxastic Temporal (PDT) Logic* [12]. When analyzing these problem domains, one is often interested in determining what could be done to bring about a certain belief state of some agent. To illustrate this, consider the following example:

Example 1 (Cyber security). Suppose that an adversary is trying to break into a computer system. This is usually done by using an attack graph to detect and exploit potential vulnerabilities of the system. An attack graph specifies a set of paths (i.e., sequences of actions) to carry out an attack. Several paths of the attack graph might be used in parallel, potentially by different agents (for instance, a number of infected computers controlled by a botnet). Usually, attack patterns specified by one attack graph are used multiple times. This has two important ramifications. The adversary will learn from experience which of the paths yield a high probability of successful attacks to a system. Defenders in turn will be able to gain knowledge of the attack graph through the repeated observation of certain patterns. Thus, when a system is under attack, the defender will have beliefs about both the chosen attack paths and the adversary's belief regarding the success of the respective path. Naturally, the defender's goal is to choose countermeasures such that the attacker believes that further attacks are useless.

A formal analysis of belief evolutions in such a cyber security setting using PDT Logic has been presented in [14]. However, previous work only provides means for deductive reasoning about the consequences of given events. In this paper, we show how abduction can be formalized in PDT Logic. This enables us to determine a required minimal set of actions that one has to take in order to bring about a desired goal belief state. Next to cyber security settings as in the above example, this approach may be useful in various domains. To name only a few examples, in financial markets it might be critical for a company to determine what kind of information has to be released to the public such that the shareholders' belief in a positive outlook is sufficiently high. In cooperative multi–agent scenarios, it is useful to determine minimal required communication acts among agents, such that all agents obtain all relevant information. In this work, we focus on the theoretical aspects of the abduction problem. Due to space constraints, we can only provide examples to a very limited extend. However, detailed modeling examples using PDT Logic can be found for example in [12,14].

Abduction has been a subject of extensive research (e.g., [5,9]), with extensions to temporal logic (e.g., [3]) and uncertainty (e.g., [16]). However, there is little work that studies abduction in the context of both time and uncertainty. A recent study of abduction in settings involving both time and uncertainty has been introduced in [15]. This approach considers abduction for the single–agent case and uses time-invariant probabilities. By extending this work such that probabilistic multi–agent beliefs and their dynamic evolution can be represented, we develop a novel abductive formalism that is able to determine necessary actions to induce desired beliefs in a multi–agent scenario.

The remainder of this paper is structured as follows. In the next section, a brief overview of PDT Logic as introduced in [12] is given. Section 3 shows how abduction can be formalized using PDT Logic and — after deriving some conditions to prune the search space — presents an algorithm to solve the abduction problem. Finally, Sect. 4 concludes this work.

2   PDT Logic

We now briefly summarize the syntax and semantics of PDT Logic from [12]. A function–free first order logic language $\mathcal{L}$ with finite sets of constant symbols $\mathcal{L}_{cons}$ and predicate symbols $\mathcal{L}_{pred}$, and an infinite set of variable symbols $\mathcal{L}_{var}$ is given. Every predicate symbol $p \in \mathcal{L}_{pred}$ has an *arity*. A *term* is any member of the set $\mathcal{L}_{cons} \cup \mathcal{L}_{var}$. A term is called a *ground term* if it is a member of $\mathcal{L}_{cons}$. If $t_1, .., t_k$ are (ground) terms and p is a predicate symbol in $\mathcal{L}_{pred}$ with arity n, then $p(t_1, ..., t_k)$ with $k \in \{0, ..., n\}$ is a (ground) atom. If a is a (ground) atom, then a and $\neg a$ are (ground) *literals*. The set of all ground literals is denoted by $\mathcal{L}_{lit}$. As usual, $\mathcal{B}$ denotes the Herbrand Base of $\mathcal{L}$.

Time is modeled as a set τ of discrete time points $\tau = \{1, ..., t_{max}\}$. The set of agents is denoted by $\mathcal{A}$. To describe what a group of agents $\mathcal{G} \subseteq \mathcal{A}$ observes, observation atoms are defined as follows:

Definition 1. *For a non-empty group of agents $\mathcal{G} \subseteq \mathcal{A}$ and ground literal $l \in \mathcal{L}_{lit}$, $Obs_{\mathcal{G}}(l)$ is an* observation atom. *The set of all observation atoms is denoted by $\mathcal{L}_{obs}$.*

Both atoms and observation atoms are formulae. If F and G are formulae, then $F \wedge G$, $F \vee G$, and $\neg F$ are formulae.

Note that the formal concept of observations is not limited to express passive acts of observing facts, but can instead be used to model a wide range of actions: for instance, communication between agents could be modeled as group observations for the respective agents — the ramifications of the communication act are exactly the same as they would be in a shared observation (assuming that agents do not lie). In this sense, observations in PDT Logic represent the effects of epistemic actions in the line of [2] and are used to alter the belief state of agents — we will build on this below when formalizing the abduction problem.

Ontic facts and according observations form *worlds* (or *states* in the terminology of [8]). A world ω consists of a set of ground atoms and a set of observation atoms, i.e., $\omega \in 2^{\mathcal{B}} \cup 2^{\mathcal{L}_{obs}}$ With a slight abuse of notation, $a \in \omega$ and $Obs_{\mathcal{G}}(l) \in \omega$ are used to denote that an atom a (resp. observation atom $Obs_{\mathcal{G}}(l)$) holds in world ω. Since agents can only observe facts that actually hold in the respective world, admissibility conditions of worlds w.r.t. the set of observations can be defined:

Definition 2. *A world ω is admissible, iff for every observation $Obs_{\mathcal{G}}(l) \in \omega$ the observed fact holds (i.e., $x \in w$ if l is a positive literal x, and $x \notin w$ if l is a negative literal $\neg x$) and for every subgroup $\mathcal{G}' \subset \mathcal{G}$, $Obs_{\mathcal{G}'}(l) \in \omega$.*

We use Ω to denote the set of all admissible worlds. Satisfaction of a ground formula F by a world ω (denoted by $\omega \models F$) is defined the usual way [11].

Example 2. In a (highly simplified) formalization of 1, we could have a set of two agents $\mathcal{A} = A, D$, representing an attacker and defender, respectively. We assume that we have two different computer systems C_1, C_2, and actions $att(c), def(c)$ to represent that system c is attacked resp. defended. In this scenario, possible ontic facts could for example be $att(C_1)$ and $def(C_1)$. Corresponding observations to this could for example be $Obs_{\{D\}}(att(C_1))$, representing that the defender observes an attack on system C_1, or $Obs_{\{A\}}(def(C_1))$, representing that the attacker observes a defensive action to protect system C_1. In this scenario, the following could be examples of possible worlds:

$$\omega_1 = \{att(C_1), Obs_{\{A\}}(att(C_1))\}\},$$
$$\omega_2 = \{att(C_1), Obs_{\{A\}}(att(C_1)), Obs_{\{D\}}(att(C_1))\},$$
$$\omega_3 = \{att(C_1), Obs_{\{A,D\}}(att(C_1))\},$$

Naturally, if the attacker carries out an attack, she observes this attack, represented by $Obs_{\{A\}}(att(C_1))$ in all worlds. In ω_1, her attack is undetected, as the defender does not observe it. In worlds ω_2 and ω_3, the defender detects this attack. The difference between the latter two worlds is that in ω_3, there is a *shared* observation about the attack, i.e., both agents know that the respective opponent has the same observation, while in ω_2, the defender observes the attack, but the attacker is unaware of this.

Using the concept of admissible worlds, we can represent the evolution of time as sequences of worlds:

Definition 3. *A* thread *is a mapping* $Th : \tau \to \Omega$

Thus, a thread is a sequence of worlds and $Th(i)$ identifies the actual world at time i according to thread Th. The set of all possible threads (i.e., all possible sequences constructible from τ and Ω) is denoted by $\mathcal{T}$. A method of constructing such a set of threads induced by a set of PDT Logic belief formulae $\mathfrak{B}$ is described in [13]. Due to space constraints, we do not discuss this method here and instead simply assume that a specification of possible threads $\mathcal{T}$ induced by a set of PDT Logic belief formulae $\mathfrak{B}$ is given. For notational convenience, we assume that there is an additional prior world $Th(0)$ for every thread.

Temporal relationships between events can be expressed through temporal rules:[1]

Definition 4. *Let* F, G *be two formulae, and* Δt *a time interval. Then* $r_{\Delta t}(F, G)$ *is called a temporal rule.*

The meaning of such an expression is to be understood as "F is followed by G after exactly Δt time units".

---

[1] The introduction of PDT Logic in [12] enables the expression of a variety of temporal relationships through an axiomatic definition. Due to space constraints, we present an adapted simplified version that suffices for the purpose of this work.

2.1 Kripke Structures

With the definition of threads, a slightly modified version of Kripke structures [10] can be adopted. For a set $\mathcal{A}$ of n agents, a Kripke structure is defined as a tuple $\langle \Omega, \mathcal{K}_1(t), ..., \mathcal{K}_n(t) \rangle$, with the set of admissible worlds Ω and binary relations $\mathcal{K}_i$ on Ω for every agent $i \in \mathcal{A}$ and every time point $t \in \tau$. Intuitively, $(\omega, \omega') \in \mathcal{K}_i$ specifies that at time t in world ω, agent i considers ω' as a possible world as well.

These Kripke structures are initialized for each agent such all worlds that occur at time $t = 1$ in some thread Th' are considered possible.

$$\forall Th \in \mathcal{T} : \ \mathcal{K}_i(Th(0)) = \bigcup_{Th' \in \mathcal{T}} \{Th'(1)\}, \ i = 1, ..., n \tag{1}$$

Note that the set of time points τ ranges over $1, ..., t_{max}$. We use the auxiliary time point $t = 0$ only to simplify the subsequent presentation: by initializing the Kripke structures as specified above, we can express the Kripke structures for all time points $t \in \tau$ as results of successive updates to the respective $\mathcal{K}_i$.

With the evolution of time, each agent can eliminate the worlds that do not comply with its respective observations. Through the elimination of worlds, an agent will also reduce the set of threads it considers possible (if — due to some observation — a world ω is considered impossible at a time point t, then all threads Th with $Th(t) = \omega$ are considered impossible). It is assumed that agents have perfect recall and therefore will not consider some thread possible again if it was considered impossible at one point. Thus, $\mathcal{K}_i$ is updated w.r.t. the agent's respective observations such that it considers all threads possible that both comply with its current observations and were considered possible at the previous time point:

$$\mathcal{K}_i(Th(t)) := \{Th'(t) : (Th'(t-1) \in \mathcal{K}_i(Th(t-1)) \wedge$$
$$\{Obs_{\mathcal{G}}(l) \in Th(t) : i \in \mathcal{G}\} = \{Obs_{\mathcal{G}}(l) \in Th'(t) : i \in \mathcal{G}\})\} \tag{2}$$

Note that — depending on the actual observations — different Kripke structures $\mathcal{K}_i$ may occur at a specific time point t. $K_i(t)$ is used to denote the set of all possible Kripke structures for agent i at time t.

Example 3. Consider the set of worlds $\omega_1, ..., \omega_3$ from the previous example. In the absence of any other information, the resulting Kripke structures in this case would be

$$\mathcal{K}_A(\omega_1) = \mathcal{K}_A(\omega_2) = \{\omega_1, \omega_2\}, \mathcal{K}_A(\omega_3) = \{\omega_3\},$$
$$\mathcal{K}_D(\omega_1) = \{\omega_1\}, \mathcal{K}_D(\omega_2) = \{\omega_2\}, \mathcal{K}_D(\omega_3) = \{\omega_3\},$$

i.e., the attacker cannot distinguish between the worlds where her attack went undetected and where the attack was detected without her knowing about this. The defender in turn is able to distinguish between all three worlds, as his respective observations are unique in each of these worlds.

2.2   Subjective Posterior Temporal Probabilistic Interpretations

Each agent has probabilistic beliefs about the expected evolution of the world over time. This is expressed through subjective temporal probabilistic interpretations:

Definition 5. *Given a set of possible threads $\mathcal{T}$, some thread $\overset{\circ}{Th} \in \mathcal{T}$, a time point t' and an agent i, $\mathcal{I}_{i,t'}^{\overset{\circ}{Th}} : \mathcal{T} \rightarrow [0,1]$ specifies the subjective posterior probabilistic temporal interpretation from agent i's point of view at time t' in thread $\overset{\circ}{Th}$, i.e., a probability distribution over all possible threads: $\sum_{Th \in \mathcal{T}} \mathcal{I}_{i,t'}^{\overset{\circ}{Th}}(Th) = 1$. $\overset{\circ}{Th}$ is called the point of view (pov) thread of interpretation $\mathcal{I}_{i,t'}^{\overset{\circ}{Th}}$.*

The prior probabilities of each agent for all threads are then given by $\mathcal{I}_{i,0}^{\overset{\circ}{Th}}(Th)$. Since all threads are indistinguishable a priori, there is only a *single* prior distribution for each agent. Furthermore, in order to be able to reason about nested beliefs, it is assumed that prior probability assessments of all agents are commonly known (i.e., all agents know how all other agents assess the prior probabilities of each thread). This in turn requires that all agents have exactly the same prior probability assessment over all possible threads: if two agents have different, but commonly known prior probability assessments, we essentially have an instance of Aumann's well-known problem of "agreeing to disagree" [1]. Intuitively, if differing priors are commonly known, it is common knowledge that (at least) one of the agents is at fault and should revise its probability assessments. As a result, there is only one prior probability distribution which is the same from all viewpoints, denoted by $\mathcal{I}$.

Even though there is only a single prior probability distribution over the set of possible threads, it is still necessary to distinguish the viewpoints of different agents in different threads, as the definition of interpretation updates shows:

Definition 6. *Let i be an agent, t' a time point, and $\overset{\circ}{Th}$ a pov thread. Then, if the system is actually in thread $\overset{\circ}{Th}$ at time t', agent i's probabilistic interpretation over the set of possible threads is given by the update rule:*

$$\mathcal{I}_{i,t'}^{\overset{\circ}{Th}} = \begin{cases} \frac{1}{\alpha_{i,t'}^{\overset{\circ}{Th}}} \cdot \mathcal{I}_{i,t'-1}^{\overset{\circ}{Th}}(Th) & \text{if } Th(t') \in \mathcal{K}_i(\overset{\circ}{Th}(t')) \\ 0 & \text{if } Th(t') \notin \mathcal{K}_i(\overset{\circ}{Th}(t')) \end{cases} \tag{3}$$

with $\frac{1}{\alpha_{i,t'}^{\overset{\circ}{Th}}}$ being a normalization factor to ensure that $\sum_{Th \in \mathcal{T}} \mathcal{I}_{i,t'}^{\overset{\circ}{Th}}(Th) = 1$:

$$\alpha_{i,t'}^{\overset{\circ}{Th}} = \sum_{\substack{Th \in \mathcal{T}, \\ Th(t') \in \mathcal{K}_i(\overset{\circ}{Th}(t'))}} \mathcal{I}_{i,t'-1}^{\overset{\circ}{Th}}(Th) \tag{4}$$

Essentially, the update rule assigns all impossible threads a probability of zero and scales the probabilities of the remaining threads such that they are proportional to the probabilities of the previous time point.

2.3 The Belief Operator

Now, with the definitions of subjective posterior probabilistic temporal interpretations, the belief operator $B_{i,t'}^{\ell,u}(\varphi)$ to express agents' beliefs can be defined. Intuitively, $B_{i,t'}^{\ell,u}(\varphi)$ means that at time t', agent i believes that some fact φ is true with a probability $p \in [\ell,u]$. The probability interval $[\ell,u]$ is called the *quantification* of agent i's belief. F_t is used to denote that formula F holds at time t and, accordingly, $Obs_{\mathcal{G}}(l)_t$ to denote that an observation $Obs_{\mathcal{G}}(l)$ occurs at time t. These expressions are called time–stamped formulae and time–stamped observation atoms, respectively.

Definition 7. *Let i be an agent, t' a time point, and $[\ell,u] \subseteq [0,1]$. Then,* belief formulae *are inductively defined as follows:*

1. *If F is a formula and t is a time point, then $B_{i,t'}^{\ell,u}(F_t)$ is a belief formula.*
2. *If $r_{\Delta t}(F,G)$ is a temporal rule, then $B_{i,t'}^{\ell,u}(r_{\Delta t}(F,G))$ is a belief formula.*
3. *If F and G are belief formulae, then so are $B_{i,t'}^{\ell,u}(F)$, $F \wedge G$, $F \vee G$, and $\neg F$.*

For a belief $B_{i,t'}^{\ell,u}(\varphi)$ about something, φ is called the belief object.

The semantics of this operator is defined as follows:

Definition 8. *Let i be an agent and $\mathcal{I}_{i,t'}^{\mathring{T}h}(Th)$ be agent i's interpretation at time t' in pov thread $\mathring{T}h$. Then, it follows from this interpretation that agent i believes at time t' with a probability in the range $[\ell,u]$ that*

1. *(Belief in ground formulae)*
   a formula F holds at time t (denoted by $\mathcal{I}_{i,t'}^{\mathring{T}h} \models B_{i,t'}^{\ell,u}(F_t)$) iff:

$$\ell \le \sum_{Th \in \mathcal{T},\, Th(t) \models F} \mathcal{I}_{i,t'}^{\mathring{T}h}(Th) \le u. \tag{5}$$

2. *(Belief in rules)*
   a temporal rule $r_{\Delta t}(F,G)$ holds (denoted by $\mathcal{I}_{i,t'}^{\mathring{T}h} \models B_{i,t'}^{\ell,u}(r_{\Delta t}(F,G))$) iff:

$$\ell \le \sum_{Th \in \mathcal{T}} \mathcal{I}_{i,t'}^{\mathring{T}h}(Th) \cdot \mathsf{fr}(Th,F,G,\Delta t) \le u. \tag{6}$$

   with the function fr giving the frequency of rule $r_{\Delta t}(F,G)$, i.e., fr divides the number of occurrences where F is followed by G in Δt time units by the total number of occurrences of F in thread Th.
3. *(Nested beliefs)*
   a belief $B_{j,t}^{\ell_j,u_j}(\varphi)$ of some other agent j holds at time t' (denoted by $\mathcal{I}_{i,t'}^{\mathring{T}h} \models B_{i,t'}^{\ell,u}(B_{j,t}^{\ell_j,u_j}(\varphi)))$ iff:

$$\ell \le \sum_{\substack{Th \in \mathcal{T} \\ \mathcal{I}_{j,t}^{\mathring{T}h} \models B_{j,t}^{\ell_j,u_j}(\varphi)}} \mathcal{I}_{i,t'}^{\mathring{T}h}(Th) \le u. \tag{7}$$

Note that with respect to this semantics, a belief $B_{i,t}^{1,1}(\varphi)$ with a quantification $\ell = u = 1$ represents certainty. Thus, $B_{i,t}^{1,1}(\varphi)$ represents *knowledge* regarding a fact φ and is therefore equivalent to the established knowledge operator $K_i(\varphi)$ (cf. e.g., [8]).

As the semantics of the belief operator is defined with respect to the subjective posterior interpretations of the respective agent, it is clear that beliefs change according to the interpretation updates as given in Definition 6. As the interpretations are updated with the occurrence of observations, it is clear that the beliefs of an agent can be influenced by ensuring that the respective agent makes certain observations. We will use this below to identify possible actions to induce the abduction goal. A detailed analysis on the resulting belief evolutions over time can be found in [12]. Further detailed examples that illustrate how PDT Logic can be used as a modeling language to formally specify a problem domain are discussed in [13].

A set of belief formulae $\mathfrak{B}$ *entails* a belief formula G (denoted by $\mathfrak{B} \models G$), iff every thread Th in the set of threads $\mathcal{T}$ induced by $\mathfrak{B}$ satisfies G.

3   Abduction in PDT Logic

Given a set of PDT Logic formulae $\mathfrak{B}$ describing a specific scenario, it is often useful to know what actions one could take to induce a certain belief $B_{i,t'}^{\ell,u}(\varphi)$ of some agent at a specific time t'. As the beliefs in PDT Logic change due to observations, it is natural to define possible actions as a set of observations that can be induced.

Definition 9. *Let $\mathfrak{B}$ be a set of PDT Logic formulae, H be a set of PDT Logic formulae representing observations $Obs_\mathcal{G}(l)_t$ and let $G \equiv B_{i,t_g}^{\ell,u}(\varphi)$ be an atomic belief formula. Then, the triple $\langle \mathfrak{B}, H, G \rangle$ is an instance of the PDT Abduction Problem. $S \subseteq H$ is a solution to the abduction problem iff $\mathfrak{B} \cup S$ is satisfiable and $\mathfrak{B} \cup S \models G$. A solution S is a minimal solution to the abduction problem if there exists no solution S' with $|S'| < |S|$ so that $\mathfrak{B} \cup S' \models G$.*

Intuitively, $\mathfrak{B}$ constitutes the background knowledge that models a specific environment, G describes the goal we want to achieve, and the hypotheses space H represents information that we can share with the agents in order to induce the belief described by G.

3.1   The Hypotheses Space $\mathcal{H}$

As the background knowledge $\mathfrak{B}$ induces a set of possible threads $\mathcal{T}$ (cf. [13]), we do not need to specify the hypotheses space H explicitly, but instead we can determine a set of hypothesis candidates H' from $\mathcal{T}$ as the set of all observations that can possibly occur:

$$H' = \{Obs_\mathcal{G}(l)_t : (\exists Th \in \mathcal{T} : Obs_\mathcal{G}(l) \in Th(t))\} \tag{8}$$

Before actually trying to solve the abduction problem specified in Definition 9, we can identify necessary preconditions that an observation $Obs_G(l)_t \in H'$ has to satisfy in order to be able to contribute to a solution of the abduction problem: The set H' collects all observations that can possibly occur in the situation described by $\mathfrak{B}$. However, not all of these observations have the means to alter the quantification of the goal belief G. With a slight abuse of notation, we use $i \in G$ to denote that agent i is involved in the goal belief G, i.e., G contains a belief operator $B_{i,t'}^{\ell,u}$ (possibly as part of a nested belief). Then, we can define a dependency property $dep(G, Obs_G(l)_t)$ between the goal and an observation as follows:

Definition 10 (Goal dependency). *Let G be the abduction goal and let $Obs_G(l)_t$ be an observation. G is dependent on $Obs_G(l)_t$, denoted by $dep(G, Obs_G(l)_t)$, iff*

$$i \in G \ \wedge \ i \in \mathcal{G} \tag{9}$$

Naturally, any observation $Obs_G(l)_t \in H'$ that does not satisfy this dependency property is unable to contribute to achieving the goal and can therefore be neglected when searching for a solution to the abduction problem. Thus, we can define the set of relevant atomic hypotheses as

$$H = \{Obs_G(l)_t \in H' : \ dep(G, Obs_G(l)_t)\} \tag{10}$$

Whenever an observation occurs for some agent i, the set of threads it considers possible is reduced such that only those threads remain where the respective observation holds. We use $K_i^S(t_g)$ to denote the set of possibility relations for agent i at the time t_g of the goal belief[2] induced by a potential solution $S \subseteq H$. We can then leverage the semantics of the belief operator (cf. Definition 8) to obtain another necessary precondition: for $G \equiv B_{i,t_g}^{\ell,u}(\varphi)$ with $0 < \ell$ and $u < 1$, in every distinguishable situation $\mathcal{K}_i(t_g)$ that i considers possible at time t_g, there need to be two threads Th_1, Th_2 so that the respective belief object φ is satisfied in one thread and unsatisfied in another. If the belief is quantified with $\ell = u = 1$, all threads in all distinguishable situations $\mathcal{K}_i(t_g)$ have to satisfy the belief object φ. Otherwise, if these conditions are not met, it is clear that the goal belief is not valid, independently of any specific probability assignment. These conditions can be checked syntactically prior to evaluating the semantic entailment $\mathfrak{B} \cup S \models G$. Using $sp(S)$ to denote the syntactic possibility of a solution S, we can formally express these considerations as

$$sp(S) = \begin{cases} true & \text{if } 0 < \ell, u < 1 \text{ and } \forall \mathcal{K}_i(t_g) \in K_i^S(t_g) : \\ & \exists Th_1, Th_2 \in \mathcal{K}_i(t_g) : Th_1(t_g) \models \varphi \wedge Th_2(t_g) \models \neg\varphi \\ true & \text{if } \ell = u = 1 \text{ and } \forall \mathcal{K}_i(t_g) \in K_i^S(t_g) : \\ & (\forall Th \in \mathcal{K}_i(t_g) : Th(t_g) \models \varphi) \\ false & \text{otherwise} \end{cases} \tag{11}$$

---

[2] To simplify the presentation, we assume that (even for nested beliefs) the goal formula G involves only a single time point t_g. The proposed methods are also applicable to goal formulae involving multiple time points, but this will significantly increase the complexity of presentation.

With these considerations we can define the entire search space $\mathcal{H}$ for possible solutions to the abduction problem as

$$\mathcal{H} = \{S \in 2^H : sp(S)\} \tag{12}$$

3.2   The Abduction Process

To determine whether a candidate solution $S \subseteq \mathcal{H}$ is actually a solution to the abduction problem, i.e., S together with the background knowledge $\mathfrak{B}$ entails the goal G, we can reformulate the entailment problem as a satisfiability problem in the usual way [6], provided that $\mathfrak{B} \cup \{G\}$ is consistent:

$$\mathfrak{B} \cup S \models \{G\} \;\equiv\; \neg sat(\mathfrak{B} \cup S \cup \{\neg G\}) \tag{13}$$

Checking satisfiability of a set of PDT Logic can be performed as described in [13]. The complexity of satisfiability checking is as follows.

Theorem 1 (Complexity of PDT SAT). *Reference [13] Checking satisfiability of a set of PDT Logic belief formulae $\mathfrak{B}$ is NP–complete.*

Building on this result, we obtain the following complexity result for deciding whether a solution exists for an instance of the PDT Logic abduction problem:

Theorem 2 (Complexity of PDT Abduction). *Let $A = \langle \mathfrak{B}, H, G \rangle$ be an instance of the PDT Logic abduction problem. Deciding whether a solution exists is Σ_2^P–complete.*

Proof. Due to space constraints, we only give a proof sketch here. The complete proof works analogously to the proof of Theorem 4.2 in [15].

Showing membership is straightforward: We can guess a potential solution $S \subseteq H$. Using (13) and Theorem 1, it is easy to see that this solution can be verified in polynomial time by querying an NP oracle.

A known Σ_2^P–complete problem [17] is validity checking of a quantified Boolean formula Φ of the form $\exists X \forall Y \, \psi(X, Y)$ with mutually distinct Boolean variables $X = \langle x_1, ..., x_n \rangle$ and $Y = \langle y_1, ..., y_m \rangle$, respectively and $\psi(X, Y)$ a Boolean formula over the variables x_i and y_j. Intuitively, this problem has a close connection to the PDT Logic abduction problem, as we need to find some assignment to X (i.e., an abductive solution) such that the goal Y is always satisfied. Thus, we use the respective x_i as potential observation objects of the abduction problem, and set $\psi(X, Y)$ as the abduction goal; i.e., we do not restrict the set of possible threads by leaving the background knowledge $\mathfrak{B}$ empty, pick an arbitrary agent a and define hypotheses and goal belief for this agent as follows:

$$\mathfrak{B} = \emptyset, \quad H = \bigcup_{i=1}^{n} \{Obs_a(x_i)_1, Obs_a(\neg x_i)_1\}, \quad S = B_{a,t}^{1,1}(\psi(X, Y))$$

Using this formulation, we can transform validity checks of any Boolean formula Φ of the above form to an instance of the PDT Logic abduction problem and thus show that the problem is Σ_2^P–hard.   □

Adapting the approach from [15] by substituting geometric polytope operations with according satisfiability checks, we can identify several distinct cases that will guide the abduction procedure:

Proposition 1. *Let $\langle \mathfrak{B}, H, G \rangle$ be an instance of the PDT Logic abduction problem and let $S \subseteq H$ be a potential solution to this problem. Then, the following observations hold for the abduction problem:*

1. *$\neg sat(\mathfrak{B} \cup \{G\})$, background knowledge and goal are inconsistent. Then, there is no solution to the abduction problem and no hypothesis S has to be tested. Otherwise, if background knowledge and goal are consistent, we can identify the following scenarios:*
2. *$\neg sat(\mathfrak{B} \cup \{\neg G\})$, the background knowledge always entails the goal. Then, $\emptyset$ is already a solution to the abduction problem and no hypothesis S has to be tested.*
3. *$\neg sat(\mathfrak{B} \cup S)$, the potential solution is inconsistent w.r.t. the background knowledge. Then, every potential solution S' with $S \subseteq S' \subseteq H$ is also inconsistent, and therefore cannot be a solution to the abduction problem. Then, we can remove S' from $\mathcal{H}$ to prune the hypotheses space when searching for solutions to the abduction problem.*

The first two checks determine whether it is at all required to search for a solution to the abduction problem. The third case provides a pruning condition for the hypotheses search space $\mathcal{H}$: if a solution candidate is not satisfiable together with the background knowledge, it is futile to test any superset of this solution. Using these properties, we obtain the abduction procedure depicted in Algorithm 1: after checking whether it is required to search for a solution at all (lines 2–5), the procedure iterates through all potential solutions from $\mathcal{H}$, ordered by their respective size (lines 4–15), and prunes the search space whenever some potential solution S is inconsistent w.r.t. the background knowledge (line 14). The procedure terminates if a solution is found or the search space is empty.

Remark 1. Reference [15] provides another pruning condition for abductive reasoning in APT Logic by arguing that for $\mathfrak{B} \cup S \not\models G$ (with $sat(\mathfrak{B} \cup S)$), any subset $S' \subseteq S$ cannot solve the abduction problem, either. This is not applicable in PDT Logic, because beliefs change with additional observations, and thus it is possible that S' is indeed a solution to the abduction problem, while S with additional observations is not.

Iterating through the search space in increasing order with respect to the solution size has to important ramifications: First, it is ensured that any pruning operations due to inconsistent combinations of background knowledge and solution candidates are carried out as early as possible. The smaller the respective solution, the larger is the respective pruned superset and thus, pruning operations are applied most effectively. Second, any solution S returned by Algorithm 1 is a minimal solution to the abduction problem.

Algorithm 1. Abduction Algorithm for PDT Logic

```
 1: procedure ABDUCE(𝔅,H,G)
 2:     if ¬sat(𝔅 ∪ G) then                              ▷ case 1: 𝔅 ∪ G is inconsistent
 3:         return false
 4:     if ¬sat(𝔅 ∪ ¬G) then                            ▷ case 2: 𝔅 ⊨ G
 5:         return ∅
 6:     H ← {S ∈ 2^H :  sp(S)}           ▷ init search space as set of syntactically possible solutions
 7:     i ← 1
 8:     while (H ≠ ∅  and  i ≤ |H|) do                   ▷ test solutions in order of simplicity
 9:         for S ∈ H  with  |S| = i  do
10:             if ¬sat(𝔅 ∪ S ∪ ¬G) then                ▷ S is a solution
11:                 return S
12:             else
13:                 if ¬sat(𝔅 ∪ S) then     ▷ case 3: 𝔅 ∪ S is inconsistent, prune supersets
14:                     H ← H \ {S' : S' ∈ H ∧ S' ⊇ S}
15:             i ← i + 1
16:     return false
```

Theorem 3. *Let $A = \langle 𝔅, H, G \rangle$ be an instance of the PDT Logic abduction problem. If A has a solution, then Algorithm 1 returns a minimal solution S so that $𝔅 \cup S \models G$. Otherwise, the algorithm returns $false$.*

Proof. We start with showing that any set discarded in the pruning step (line 14) cannot be a solution to the abduction problem. If $𝔅 \cup S$ is unsatisfiable, this set is already overly constrained so that no thread remains that could possibly satisfy all formulae in this set. Then, as observed in Proposition 1, adding further constraints will clearly still result in an empty set of possible threads. Thus, it is unnecessary to test any set $S' \supseteq S$ for possible solutions to the abduction problem.

If the abduction problem has a solution, it is clear that the loop in lines 4–15 will eventually find and return a solution, as all solution candidates are tested iteratively unless they are discarded as above. Since the algorithm iterates over the set of possible solutions by increasing size of the solution, any returned solution S will necessarily be minimal. If there had been a smaller solution S' with $|S'| < |S$, the algorithm would have terminated earlier by returning this solution S'. ☐

4 Conclusion

In this paper, we have presented how abduction can be formalized in the context of Probabilistic Doxastic Temporal (PDT) Logic. We have discussed how relevant hypotheses space can be determined automatically from a set of threads and have developed a sound and complete algorithm to give a minimal solution to the abduction problem. We have shown that the problem of searching for a solution to the abduction problem is Σ_2^P–complete and we have derived several criteria for effectively pruning the solution search space.

To the best of our knowledge, this is the first work that studies abduction in the context of dynamically evolving beliefs for multi–agent systems, and thus, the methods introduced in this work provide means for novel reasoning capabilities.

References

1. Aumann, R.J.: Agreeing to disagree. Ann. Stat. **4**(6), 1236–1239 (1976)
2. Baltag, A., Moss, L.: Logics for epistemic programs. Synthese **139**(2), 165–224 (2004)
3. Baral, C.: Abductive reasoning through filtering. Artif. Intell. **120**(1), 1–28 (2000)
4. van Ditmarsch, H., van der Hoek, W., Kooi, B.: Dynamic Epistemic Logic, 1st edn. Springer, New York (2007)
5. Eiter, T., Gottlob, G.: The complexity of logic-based abduction. J. ACM **42**(1), 3–42 (1995)
6. Etchemendy, J.: Logical Consequence: The Cambridge Dictionary of Philosophy. Cambridge University Press, Cambridge (1999)
7. Fagin, R., Halpern, J.Y.: Reasoning about knowledge and probability. J. ACM **41**, 340–367 (1994)
8. Fagin, R., Halpern, J.Y., Moses, Y., Vardi, M.Y.: Reasoning About Knowledge. MIT Press, Cambridge (1995)
9. Josephson, J.R., Josephson, S.G. (eds.): Abductive Inference: Computation, Philosophy, Technology. Cambridge University Press, Cambridge (1996)
10. Kripke, S.A.: Semantical considerations on modal logic. Acta Philosophica Fennica **16**(1963), 83–94 (1963)
11. Lloyd, J.W.: Foundations of Logic Programming, 2nd edn. Springer, New York (1987)
12. Martiny, K., Möller, R.: A probabilistic doxastic temporal logic for reasoning about beliefs in multi-agent systems. In: 2015 Proceedings of the 7th International Conference on Agents and Artificial Intelligence, ICAART 2015. SciTePress (2015)
13. Martiny, K., Möller, R.: PDT logic - a probabilistic doxastic temporal logic for reasoning about beliefs in multi-agent systems. Technical report (2015). http://www.ifis.uni-luebeck.de/index.php?id=publikationen
14. Martiny, K., Motzek, A., Möller, R.: Formalizing agents beliefs for cyber-security defense strategy planning. In: Proceedings of the 8th International Conference on Computational Intelligence in Security for Information Systems, CISIS 2015, 15–17 June 2015, Burgos, Spain (2015)
15. Molinaro, C., Sliva, A., Subrahmanian, V.S.: Super-solutions: succinctly representing solutions in abductive annotated probabilistic temporal logic. ACM Trans. Comput. Logic **15**(3), 18:1–18:35 (2014)
16. Poole, D.: The independent choice logic for modelling multiple agents under uncertainty. Artif. Intell. **94**(12), 7–56 (1997). Economic Principles of Multi-Agent Systems
17. Stockmeyer, L.J., Meyer, A.R.: Word problems requiring exponential time (preliminary report). In: Proceedings of the Fifth Annual ACM Symposium on Theory of Computing, STOC 1973, pp. 1–9. ACM, New York (1973)

Exploiting Innocuousness in Bayesian Networks

Alexander Motzek$^{(\boxtimes)}$ and Ralf Möller

Institute of Information Systems, Universität zu Lübeck, Lübeck, Germany
{motzek,moeller}@ifis.uni-luebeck.de

Abstract. Boolean combination functions in Bayesian networks, such as noisy-or, are often credited a property stating that inactive dependences (e.g., observed to *false*) do not "cause any harm" and an arc becomes vacuous and could have been left out. However, in classic Bayesian networks we are not able to express this property in local CPDs. By using novel ADBNs, we formalize the innocuousness property in CPDs and extend previous work on context-specific independencies. With an explicit representation of innocuousness in local CPDs, we provide a higher causal accuracy for CPD specifications and open new ways for more efficient and less-restricted reasoning in (A)DBNs.

1 Introduction

Boolean combination functions in Bayesian networks (BNs) are often credited a property stating that if a dependence is observed to be inactive (i.e., a precondition observed to be *false*) it shall not "cause any harm" and its arc becomes vacuous, i.e., could have been left out. We call such a property an *"innocuousness"* property of conditional probability distributions (CPDs). We cannot specify such an innocuousness property in CPDs, nor formalize it up to now. Notwithstanding, vacuous dependencies have shown to be of valuable interest for efficient reasoning in Bayesian networks, and an innocuousness property is widely associated with, e.g., noisy-or combination functions. Further, being able to explicitly specify an innocuousness property in CPDs would allow for more precise representations of the world, as demanded and emphasized by Pearl [8]. Formalizing an innocuousness property can almost be achieved with context-specific independencies (CSIs) introduced by Boutilier et al. [2], but consider the following example: Say, random variable X is conditionally dependent on Y and C and we specify a CPD $P(X|Y,C)$. With CSIs we can specify that X becomes independent of Y in a specific context $C = c \in \text{dom}(C)$, but X stays dependent on Y in another context $C = c' \in \text{dom}(X)$. Boutilier et al. [2] formalize: $P(X|Y,c) = P(X|c)$ holds, if $\forall y, y' \in \text{dom}(Y), \forall x \in \text{dom}(X) : P(x|y,c) = P(x|y',c)$, but there exists a $c' \in \text{dom}(C)$ s.t. $\exists y, y' \in \text{dom}(Y), \exists x \in \text{dom}(X) : P(x|y,c') \neq P(x|y',c')$. However, if we would like to formalize an innocuousness property stating that a context $C = c \in \text{dom}(C)$ "removes" *itself* (and not only *another* variable), i.e., we would like to specify "$P(X|Y,c) = P(X|Y)$" in a CPD, we run into the problem that a formal definition is neither available nor easily possible:

© Springer International Publishing Switzerland 2015
B. Pfahringer and J. Renz (Eds.): AI 2015, LNAI 9457, pp. 411–423, 2015.
DOI: 10.1007/978-3-319-26350-2_36

The allegedly irrelevant random variable C in question is in fact the one that ought to be relevant for specifying the independence.

Motzek and Möller [7] describe a novel form of Bayesian networks, called Activator Dynamic Bayesian Networks (ADBN). While they focus on the exploitation of activators in terms of graphical models, activators, as we will see in this paper, allow a formal definition of an innocuousness property. Further, they only consider activator sets when studying new possibilities in graphical models. We show that by considering innocuousness properties in CPD specifications, previously imposed restrictions of [7] can be significantly relaxed.

Independencies in Bayesian networks and graphical models in general have been extensively studied for efficient inference, notably by Zhang and Poole exploiting causal independencies [13], and have been extended with Boutilier et al.'s contextual independencies [2] in [9]. Still, a contextual independence where a context itself becomes independent was not considered in these works, and this hampers ways of more efficient reasoning and representations of causalities. Boolean combination functions have undergone notable considerations in works by Henrion [5], Srinivas [11], and Antonucci [1] introducing extensions to cope with imprecision. Cozman [3] provides formal definitions and specifies properties of combination functions leading to an axiomization of the noisy-or function, where we find an "accountability" property, which goes into the direction of defining an innocuousness property. However, a formal definition of innocuousness itself as "an inactive node does not cause any harm" is still missing. Although, the counterpart "only an active node causes harm" is mentioned as an "amechanistic" property in Heckerman and Breese [4] as well as Zagorecki and Druzdzel [12], but their work follows a different direction focusing on easier parametrization of CPDs.

The contribution of this paper can be summarized as follows. By formalizing a yet unexpressed innocuousness property in CPDs, we are able to more accurately represent causalities in CPDs, and we relax restrictions previously posed on graphical models. Based on graph enumeration techniques we quantitatively explore new relaxations of syntactic restrictions of graphical models for Bayesian networks.

We discuss preliminaries on ADBNs in Sect. 2 and introduce novel principles of graphical models using a running example. Afterwards, in Sect. 3 we introduce the innocuousness property and provide a formal definition using activator random variables. Subsequently we exploit the innocuousness property for relaxing restrictions posed on (A)DBNs in Sect. 4 and we show in Sect. 5 that the utility of (A)DBNs is significantly enhanced by exploiting innocuousness properties. We conclude in Sect. 6.

2   Activator Dynamic Bayesian Networks (ADBN)

After initial notations and definitions used throughout this paper, we demonstrate a running example for ADBNs. We consider an example from [7], which outlines restrictions of classic DBNs and motivates the use of *cyclic* ADBNs.

Notation 1 (State Variables). *Let X_i^t be the random variable for the ith state X_i at time t, where X_i^t is assignable to a value $x_i \in \mathrm{dom}(X_i^t)$. Let $\boldsymbol{X}^t$ be the vector of all n state variables at time t, s.t.,*

$$\boldsymbol{X}^t = \left(X_1^t, \ldots, X_n^t \right)^{\mathsf{T}}.$$

Let $P(X_i^t = x_i)$ (or $P(x_i^t)$ for brevity) denote the probability of state X_i having x_i as a value at time t. If $\mathrm{dom}(X) = \{true, false\}$ we write $+x^t$ for the event $X^t = true$ and $\neg x^t$ for $X^t = false$ as usual. If X_i^t is unspecified and not fixed by evidence, $P(X_i^t)$ denotes the probability distribution of X_i^t w.r.t. all possible values in $\mathrm{dom}(X_i)$.

Definition 1 (Dynamic Bayesian Network). *A DBN is a tuple $(B_0, B_\rightarrow)$ with B_0 defining an initial Bayesian network (BN) representing time $t = 0$, containing all state variables X_i^0 in $\boldsymbol{X}^0$, and a consecutively repeated Bayesian network fragment $B_\rightarrow$ defining state dependencies between X_i^s and X_j^t, with $X_i^s \in \boldsymbol{X}^s, X_j^t \in \boldsymbol{X}^t, s \leq t$. By repeating $B_\rightarrow$ for every time step $t > 0$, a DBN $(B_0, B_\rightarrow)$ is unfolded into a BN defining its semantics as a joint probability over all random variables $P(\boldsymbol{X}^{0:t^{\mathsf{T}}})$. Notwithstanding, for every random variable X_i^t a local CPD, e.g., as a CPT, is defined.*

State dependencies defined in $B_\rightarrow$ are limited, s.t. no cyclic dependencies are created during unfolding.

Example 1 (Running Example & Motivation for ADBNs). Let us assume that in a company one is concerned with regulatory compliance over time. Business documents are exchanged and might contain manipulated information. Receiving such documents might influence an employee becoming corrupt at time t, which, further, might influence other employees. As an employee might *indeliberately* become corrupt, we say he becomes *credulous*. We represent the credulousness state of an employee, say, Claire, Don, and Earl, by respective random variables C^t, D^t, E^t. An influence can only occur if a message is passed from employee X to Y at t. We represent a message exchange by a random variable M_{XY}^t. We assume messages are only passed from Claire to Don to Earl. We can model these influences correctly in a DBN with $B_\rightarrow$ consisting of state random variables C^t, D^t, E^t and message variables M_{CD}^t, M_{DE}^t. Every state X^t depends on its previous state X^{t-1}, and D^t conditionally depends on C^t and M_{CD}^t, and respectively, E^t on D^t, M_{DE}^t. For every random variable, an appropriate CPD is defined. ▲

This example shows that simple influences can be correctly modeled in a DBN. However, we would like to model that messages can potentially be passed between *every* employee, which would render every state variable dependent on every other state variable. Clearly, this would cause cycles in $B_\rightarrow$, which are syntactically forbidden in Bayesian networks. In DBNs, cycles are usually resolved over time in a *diagonal* fashion, where states of t only influence $t + 1$ (Fig. 1, gray). However, we already used "time" for our modeling perspective, and bending dependencies over time causes conflicts with causality. The diagonal structure implies that receiving a message does not ultimately render another person cred-ulous, but only in the consecutive timeslice. This immediately constrains the use of timeslices to infinitesimally small timeslices. If our observations of passed messages are temporally coarser, say, daily, or if high-frequency updates are too costly, then we clash with causality as indirect influences (e.g., C influences E *through D*) are not covered anymore, and we are bound to observations which do not require the anticipation of indirect influences (see later Proposition 1).

Fortunately, Motzek and Möller [7] show that ADBNs can actually be based on *cyclic* graphs completely sound with Bayesian network semantics as long as observations (message transfers) fulfill certain restrictions, based on the following definitions and theorems.

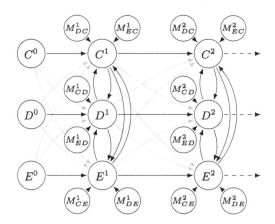

Fig. 1. A correctly represented world using an ADBN (black) for Example 1. Syntactic DAG constraints of BNs prevented desired cyclic intra-state dependencies and diagonal inter-state dependencies were enforced (hinted in light gray). In the diagonal case, M_{XY}^t represents $M_{XY}^{t-1\,t}$, i.e., M_{XY}^t affects the dependency of state Y^t on X^{t-1}.

Definition 2 (Activator Random Variables). *We use the notation A_{XY} for a so called activator random variable which activates a dependency of random variable Y on X in a given context. Let* $\mathrm{dom}(A_{XY}) = \{true, false\}$ *(extensions to non-boolean domains are straightforward). We define the* deactivation *criterion* $A_{XY} = false$ *as*

$$\forall x, x' \in \mathrm{dom}(X), \forall y \in \mathrm{dom}(Y), \forall z \in \mathrm{dom}(Z) :$$
$$P(y|x, \neg a_{XY}, z) = P(y|x', \neg a_{XY}, z) = P(y|*, \neg a_{XY}, z), \quad (1)$$

where $*$ represents a wildcard and Z further dependencies of Y.

The activation *criterion describes a situation where Y becomes dependent on X, i.e., the CPD entry for y is not uniquely identified by just $+a_{XY}$ and z, hence*

$$\exists x, x' \in \mathrm{dom}(X), \exists y \in \mathrm{dom}(Y), \exists z \in \mathrm{dom}(Z) :$$
$$P(y|x, +a_{XY}, z) \neq P(y|x', +a_{XY}, z). \quad (2)$$

Let A^{st} describe a matrix of activator random variables between time s and t,

$$A^{st} = \begin{pmatrix} A_{11}^{st} & \cdots & A_{1n}^{st} \\ \vdots & \ddots & \vdots \\ A_{n1}^{st} & \cdots & A_{nn}^{st} \end{pmatrix}.$$

Let A_i^{st} denote the ith column of A^{st} and let $\mathcal{A}^{st}$ denote the corresponding column vector of all entries of A^{st}. For brevity, we write A^t for A^{tt} (excluding A_{kk}^{tt}), and correspondingly for A_{ij}^t, A_i^t and $\mathcal{A}^t$.

In fact, in Example 1 message transfers (M_{XY}^t) take the role of activator random variables and we actually obtain an *Activator* DBN from Example 1.

Definition 3 (Activator Dynamic Bayesian Network). *A repeated ADBN fragment $B'_\rightarrow$ consists of dependencies between state variables X_i^s and X_j^t, $t-1 \le s \le t$ (Markov-1) and matrices A^{st} of activators. Let A_{ij}^{st} be the activator random variable influencing X_j^t regarding a dependency on X_i^s, such that X_j^t's local CPD follows Eqs. 1 and 2. Every activator is assigned a prior probability. An ADBN is then syntactically defined by $(B_0, B'_\rightarrow)$ defining its semantics as a well-defined joint probability over all random variables $P(\mathbf{X}^{0:t^\mathsf{T}}, \mathcal{A}^{01:tt^\mathsf{T}})$.*

Note that activators in an ADBN are classic random variables and are part of the modeled domain (message transfers in Example 1), i.e., activators are no auxiliary variables. ADBNs are restricted in order to comply with a Bayesian network:

Theorem 1 (Bayesian Network Soundness). *For every combination, i.e., an arbitrary instantiation $\mathbf{A}_*^{1:t}$ of $\mathcal{A}^{1:t}$, an ADBN $(B_0, B'_\rightarrow)$ corresponds to a Bayesian network, if for all t, $\mathbf{A}_*^t$ satisfies the acyclicity constraint:*

$$\forall x, y, z \in \mathbf{X}^t : \mathfrak{A}(x, z)^t, \mathfrak{A}(z, y)^t \rightarrow \mathfrak{A}(x, y)^t \tag{3}$$
$$\neg \exists q : \mathfrak{A}(q, q)^t,$$

with a predicate $\mathfrak{A}(i, j)^t$ defined as

$$\mathfrak{A}(i, j)^t = \begin{cases} false & if \ \neg a_{ij}^t \\ true & otherwise \end{cases}.$$

A proof for this theorem can be found in [7, Sect. 3].

Theorem 1 means that if a specific structure of an DBN is not known in advance or is changing over time, an ADBN can intrinsically adapt itself to observations. Well-defined semantics is obtained in a (cyclic) ADBN, if only certain combinations of $\mathcal{A}^{1:t}$ are instantiated, enforced by minimal sets of observations.

Example 2 (Restriction Example). Continuing Example 1, we observe a message transfer from Claire to Don $(+m_{CD}^1)$ and from Don to Earl $(+m_{DE}^1)$, and we can neglect all other transfers, i.e., $\neg m_{DC}^1, \neg m_{ED}^1, \neg m_{CE}^1, \neg m_{EC}^1$. These observations satisfy Theorem 1 and thus lead to a valid Bayesian network, even though it is based on a cyclic graph. To fully evaluate all implications of the observations, we have to anticipate an indirect influence from Claire to Earl through Don during timeslice 1. A diagonal "classic" DBN (as in Fig. 1, gray) cannot anticipate this indirection and would lead to spurious results.   ▲

While cyclic ADBNs are syntactically restricted, Examples 1 and 2 demonstrate that diagonal acyclic alternatives are significantly more restricted:

Proposition 1 (Diagonal ADBN Restrictions). *A classic, "diagonal" (as in Fig. 1, gray) (A)DBN is restricted in its usage to special observation sets. Indirect influences are spread over multiple timesteps and possible indirect influences inside one timestep cannot be considered. This restricts a DBN to observations where indirect influences strictly do not occur, i.e., no two activators A_{*i}^t and A_{i*}^t are allowed to be active, i.e. the set of probably active activators must form a bipartite digraph with uniformly directed edges (cf. [7, Proposition 1]).*

In the following, we introduce a novel property of CPDs, which significantly relaxes restrictions opposed by Theorem 1.

3   Innocuousness

We introduced innocuousness informally as "an inactive node does not cause any harm", but were unable to give a formal definition for such a property in CPDs of classic (D)BNs. Often "accountability", i.e., $P(+x|\neg*) = 0$ [3], is confused with the innocuousness property, but causally $P(+x|\neg*) = 0$ can also represent that exactly one *false*-dependence is responsible for $P(+x|\neg*)$ being 0.

As an extension to context-specific independencies (CSIs) from Boutilier et al. [2], we define a concept of *innocuousness contexts*, with fewer restrictions of CSIs. CSIs provide a formal definition for a variable X becoming independent of a variable Y in a context $\boldsymbol{C} = \boldsymbol{c} \in \mathrm{dom}(\boldsymbol{C})$, where $X, Y \notin \boldsymbol{C}$. This allows us to specify properties such as $P(X|Y, \boldsymbol{c}) = P(X|\boldsymbol{c})$ in local CPDs. But, $X, Y \notin \boldsymbol{C}$ prevents us from specifying that a context $\boldsymbol{C} = \boldsymbol{c}$ removes one of *its own* random variables $C \in \boldsymbol{C}$, e.g., "$P(X|Y, \boldsymbol{c}) = P(X|Y)$". Using activators in ADBNs we extend Boutilier's work to innocuousness contexts. We formally define that in a context $\boldsymbol{C} = \boldsymbol{c}$, a context variable $C \in \boldsymbol{C}$ can itself becomes irrelevant, which we call self-reflexive independence. Let us say a context $C = c \in \mathrm{dom}(C)$, if it shall represent that X becomes independent of C, given $C = c$, i.e. $P(X|c, A_{CX}, \boldsymbol{Z}) = P(X|A_{CX}, \boldsymbol{Z})$. Using an activator-enriched CPD we define this to hold for binary activator random variables if

$$\forall x \in \mathrm{dom}(X), \forall \boldsymbol{z} \in \mathrm{dom}(\boldsymbol{Z}) :$$
$$P(x|c, +a_{CX}, \boldsymbol{z}) = P(x|c, \neg a_{CX}, \boldsymbol{z}) = P(x|*, \neg a_{CX}, \boldsymbol{z}), \tag{4}$$

where $\boldsymbol{Z}$ represents remaining further dependencies of X. Extensions to non-boolean activator random variables are straightforward.

This means, given $C = c$, A_{CX} becomes irrelevant for X, i.e., X becomes independent of A_{CX}. As A_{CX} can be instantiated in any form now, we can also say $\neg a_{CX}$. According to the deactivation criterion of an activator, X then becomes independent of C given $\neg a_{CX}$, or rather X becomes independent of C given $\neg c$, which is exactly what we intended.

Now, assume to specify a CPD $P(X|C, A_{CX}, Q, \boldsymbol{Z})$, where the innocuousness property of a variable is only in place in a further context. For example, there exists a variable Q that activates the innocuousness property of C only given $Q = q \in \mathrm{dom}(X)$. In this case, Eq. 4 only holds for specific $\boldsymbol{z} \in \mathrm{dom}(\boldsymbol{Z})$. This means, one innocuousness context is defined by instantiations of multiple random variables. Moreover, one random variable might stand in multiple different innocuousness contexts.

Notation 2 (Innocuousness Contexts). *Activator random variables are marked with a dot, e.g., $\dot{A}_{YX}$, if they are subject to become irrelevant in specific contexts. We denote a context in which Y is innocuous for X as a so called innocuousness context as a left superscript on A_{YX}. If a context is met and Y is innocuous for X, we say that A_{YX} stands in the innocuousness context. For the first example, this would be*

$$P(X|C, {}^{C=c}\dot{A}_{CX}, \boldsymbol{Z}),$$

with which we can also denote a previously discussed toggle variable Q: Only in the context $Q = q$ and $C = c$, X becomes independent of C, as A_{CX} becomes freely instantiable. For this situation we write

$$P(X|C, {}^{Q=q,C=c}\dot{A}_{CX}, Q, \boldsymbol{Z}),$$

where $\boldsymbol{Z}$ represents further dependencies of X, but without X, Q, C and A_{CX}.

Notation 3 (Innocuousness Context Vectors). *Variables might become innocuous in multiple contexts. Multiple innocuousness contexts $\varphi_{A_{YX}}$ of one activator A_{YX} are encapsulated in a vector $\boldsymbol{\varphi}_{A_{YX}}$ and are delimited by ; . An innocuousness context vector $\boldsymbol{\varphi}_{A_{YX}}$ can also be seen as a Boolean formula, where all contexts are disjunctions and a context is a conjunction of instantiations.*

This notation allows to mark contexts, in which an activator becomes irrelevant and could have been chosen to be deactive, and thus modifies the topology. Definition 4 describes the explicit specification of innocuousness in CPDs.

Definition 4 (Activator Innocuousness). *Let $\Phi_{A_{YX}}$ be the vector of random variables used in a context $\varphi_{A_{YX}}$ associated with A_{YX}. Every innocuousness context $\varphi_{A_{YX}} \in \boldsymbol{\varphi}_{A_{YX}}$ is then defined to hold*

$$
\begin{aligned}
\forall x \in \mathrm{dom}(X), \forall \boldsymbol{z} \in \mathrm{dom}(\boldsymbol{Z}) : P(x|\varphi_{A_{YX}}, {+}a_{YX}, \boldsymbol{z}) &= P(x|\varphi_{A_{YX}}, \neg a_{YX}, \boldsymbol{z}) \\
= P(x|\{\varphi_{A_{YX}} \backslash y \in \mathrm{dom}(Y)\}, y, \neg a_{YX}, \boldsymbol{z}) &= P(x|\{\varphi_{A_{YX}} \backslash y\}, *, \neg a_{YX}, \boldsymbol{z}),
\end{aligned}
\tag{5}
$$

with remaining arbitrary dependencies of X on other random variables $\mathbf{Z}$ and z as an arbitrary instantiation of those, excluding A_{YX} and $\Phi_{A_{YX}}$.

Frankly, with Definition 4 we can formulate the same CSIs as Boutilier et al. [2], but, further, we can specify previously mentioned self-reflexive independences. We are thus able to explicitly express $P(X|\{\varphi_{A_{YX}} \backslash y\}, y, A_{YX}, \mathbf{Z}) = P(X|\{\varphi_{A_{YX}} \backslash y\}, A_{YX}, \mathbf{Z})$ as demonstrated in the following example.

Example 3 (Activator Innocuousness). Continuing Example 2, let us assume a noisy-or combination function for each CPD of a state X^t. With a noisy-or combination, every activator random variable (a message transfer) $\dot{M}_{XY}^t$ stands in the innocuousness context $\varphi_{M_{XY}^t} = \neg x^t$. We can now explicitly represent that Claire is not influenced by a non-credulous Earl, i.e., $P(C^t|C^{t-1}, D^t, \neg e^t, A_{DC}^t, A_{EC}^t) = P(C^t|C^{t-1}, D^t, A_{DC}^t, A_{EC}^t)$, by fixing

$$\forall\, C^t, C^{t-1}, D^t, A_{DC}^t, A_{EC}^t :$$
$$P(C^t|C^{t-1}, D^t, \neg e^t, A_{DC}^t, {+}a_{EC}^t) = P(C^t|C^{t-1}, D^t, \neg e^t, A_{DC}^t, \neg a_{EC}^t)$$
$$\overset{\text{(by Def. 2)}}{=} P(C^t|C^{t-1}, D^t, {+}e^t, A_{DC}^t, \neg a_{EC}^t)$$

in the respective CPD specification of C^t (likewise for non-creduluos Don). ▲

We see that an arc in $B_\rightarrow$ representing a dependency of X on Y can become vacuous in a context of the variable Y itself, which was previously impossible to formalize and impossible to define in a CPD without activators. This is beneficial for more efficient reasoning and a higher causal accuracy of independence declarations in all DBNs with activator random variables.

Further, Motzek and Möller [7] did not consider any properties of CPDs for possible acyclicity constraints in ADBNs, and only focus on defined activator sets. In the next section, we consider innocuousness properties of CPDs and relax restrictions posed on graphical models.

4   Exploiting Innocousness

By considering properties of CPDs of state variables $\mathbf{X}^t$, we relax restrictions of Theorem 1 by supporting innocuousness contexts as further acyclicity constraints. Note that in an ADBN, these checks and constraints are only sufficient conditions for achieving sound results and are not required for necessary calculations, if, e.g., observations can be trusted to fulfill these restrictions.

Theorem 2 (Bayesian Network Soundness Revised). *For every set of arbitrary instantiation of $\mathbf{A}^{1:t}$ and $\mathbf{X}^{0:t}$, written $\mathbf{A}_*^{1:t}$, $\mathbf{X}_*^{0:t}$, an ADBN $(B_0, B'_\rightarrow)$ corresponds to a Bayesian network, if for all t, $\mathbf{A}_*^t$ and $\mathbf{X}_*^t$ satisfy the new acyclicity constraint:*

$$\forall x, y, z \in \mathbf{X}^t : \mathfrak{A}(x, z)^t, \mathfrak{A}(z, y)^t \rightarrow \mathfrak{A}(x, y)^t$$
$$\neg\exists q : \mathfrak{A}(q, q)^t, \tag{6}$$

with a predicate $\mathfrak{A}(i,j)^t$ defined as

$$\mathfrak{A}(i,j)^t = \begin{cases} false & if \quad \neg a_{ij}^t \vee \varphi_{A_{ij}^t} \\ true & otherwise \end{cases},$$

with the innocuousness context vector $\varphi_{A_{ij}^t}$ seen as a disjunction of multiple contexts $\varphi_{A_{ij}^t}$ for activator A_{ij}^t, as defined in Definition 4. Given a correspondence to a Bayesian network an ADBN's semantics is well-defined and the complete joint probability over all random variables is straightforwardly specified by the product of all locally defined CPDs,

$$P(\boldsymbol{X}^{0:t^\mathsf{T}}, \boldsymbol{A}^{1:t^\mathsf{T}}) = P(\boldsymbol{X}^{0:t-1^\mathsf{T}}, \boldsymbol{A}^{1:t-1^\mathsf{T}}) \cdot$$
$$\prod_i P(X_i^t | \boldsymbol{X}^{t^\mathsf{T}} \backslash X_i^t, \dot{\boldsymbol{A}}_i^{t^\mathsf{T}}, X_i^{t-1}) \cdot P(\boldsymbol{A}^{t^\mathsf{T}}). \quad (7)$$

Theorem 2 means, every (cyclic) ADBN corresponds to a sound Bayesian network, if only certain combinations of $(\boldsymbol{X}^{0:t}, \boldsymbol{A}^{1:t})$ are instantiated. Minimal sets of observations, i.e., partial instantiations of $(\boldsymbol{X}^{0:t}, \boldsymbol{A}^{1:t})$, have to enforce that during inference only valid combinations are used.

In the following, we prove Theorem 2 by showing that any instantiation of $\boldsymbol{A}^{1:t}$, $\boldsymbol{X}^{0:t}$ holding Eq. 6 is topologically equivalent to some instantiation of $\boldsymbol{A}^{1:t}$ holding Eq. 3, and thus is a valid Bayesian network with straightforward joint probability.

Proof (Theorem 2). According to Theorem 1 every (cyclic) ADBN is a Bayesian network and its semantic joint probability is well-defined as the product of all locally defined CPDs, if an instantiation of $\boldsymbol{A}^{1:t}$ holds Eq. 3. Motzek and Möller [7, Proof 1] show, that for every of such combinations, a topological order $\prec$ must exist, s.t. by reversing Bayes' chain rule in Eq. 7 we obtain a joint probability distribution, which belongs to a valid Bayesian network.

Definition 5 (Topology Equivalence). *Given an ADBN $(B_0, B_\rightarrow)$, an instantiation $(\boldsymbol{X}^{0:t}, \boldsymbol{A}^{1:t})_1$ is topologically equivalent to an instantiation $(\emptyset, \boldsymbol{A}^{1:t})_2$, if for both the same topological order $\prec$ exists in $(B_0, B_\rightarrow)$.*

Generally, in an acyclic ADBN, for every arbitrary instantiation $(\boldsymbol{X}^{0:t}, \boldsymbol{A}^{1:t})_*$ the same topological order $\prec_*$ exists. In a cyclic ADBN, the topological order is defined (at "runtime") by a minimal set of deactive $\boldsymbol{A}^{1:t}$ holding to Theorem 1. In that case, some state variables $\boldsymbol{X}^t$ become independent of state variables $\boldsymbol{X}_E^t$, which previously created cycles and prohibited a topological order $\prec$. Note that the set of activators $\boldsymbol{A}^{1:t}$ only are necessary conditions for creating a topological order and only follow a lexicographic order. However, under Definition 4, an active activator A_+^t might stand in a context $\varphi_{A_+^t}$, which renders A_+^t innocuous or irrelevant. It is straightforward from Definition 4 that A_+^t can then be seen as deactive from a topological perspective, which we call topologically deactive. Two sets of instantiations $(\boldsymbol{X}^{0:t}, \boldsymbol{A}^{1:t})_1$, $(\emptyset, \boldsymbol{A}^{1:t})_2$ then share the same topological order $\prec$, if the set of topologically-deactive activators in $(\boldsymbol{A}^{1:t})_1$ is a superset of deactive activators in $(\boldsymbol{A}^{1:t})_2$ and a topological order exists for $(\boldsymbol{A}^{1:t})_2$ (i.e., holds Theorem 1).

With Definition 5, every instantiation $(\boldsymbol{X}^{0:t}, \boldsymbol{A}^{1:t})_1$ holding Theorem 2 is *topologically* equivalent to an instantiation $(\emptyset, \boldsymbol{A}^{1:t})_2$ holding Theorem 1 for which a joint probability function is well-defined based on the same topological order $\prec$. Under this topological order $(B_0, B_\rightarrow)$ is a Bayesian network and Proof 1 in [7] is analogous. □

Note that while the joint probability of two topologically equivalent instantiations follows the same topological order, the results/outcomes of both must not be the same, as we need to consider priors of $\boldsymbol{A}^{1:t}$.

Theorem 2 shows and it is proven that ADBNs cannot only be based on cyclic graphs and handle *acyclic* activator observations, but also *cyclic* activator observations can be made when considering specific CPD properties. The following example demonstrates the observation of a cyclic activator constellation.

Example 4 (Restriction Relaxation Example). Continuing Example 3, we now, observe $\neg e^1$ (in addition to previous observations), but $+m_{ED}^1$. The observations $+m_{ED}^1$, $+m_{DE}^1$ obviously lead to a cycle, which is not allowed according to Theorem 1 and is neither allowed in diagonal networks (Proposition 1). However, we find that the observation $\neg e^1$ meets the innocuousness context $\varphi_{M_{ED}^1}$ (Example 3, noisy-or), i.e., E^1 is innocuous for D^1 given $\neg e^1$. Therefore, this observation fixes $(\boldsymbol{X}^{0:t}, \boldsymbol{A}^{1:t})$ to instantiations that an ADBN can handle (Theorem 2). In fact, observations in this example fix all instantiations of $(\boldsymbol{X}^{0:t}, \boldsymbol{A}^{1:t})$ to be *topologically equivalent* to the one from the previous example, i.e., both observations are topologically equivalent.

Note that still this observation cannot be handled by a diagonal alternative, as we need to anticipate an indirect influence: $\neg e^1$ tells us *indirectly* something about C^1, e.g., that $\neg c^1$ is now more likely than without the new observations. ▲

This example shows that the cyclic model can handle a larger set of observation constellations in contrast to a diagonal alternative. The next section generalizes these advantages for a general model.

5   Discussion and Comparison

In this section, we investigate how cyclic ADBNs compare to classic diagonal (A)DBNs. As discussed before, only certain combinations of $(\boldsymbol{X}^t, \boldsymbol{A}^t)$ in a timestep t lead to valid Bayesian networks, which means (partial) observations of $(\boldsymbol{X}^t, \boldsymbol{A}^t)$ have to fulfill certain restrictions. Further, we explore how the exploitation of innocuousness properties can relax these restrictions. We find that this exploitation significantly allows for more observation sets to be handled, and that cyclic ADBNs heavily outperform their diagonal counterparts w.r.t. expressivity.

In cyclic ADBNs, instantiations of $(\boldsymbol{X}^t, \boldsymbol{A}^t)$ during a timestep t were restricted according to Theorem 1 and are relaxed due to Theorem 2. For diagonal ADBNs, instantiations are restricted, s.t. no indirect influences can occur (Proposition 1). Notwithstanding, innocuousness properties can also relax this restriction. For a comparison, let us consider Example 1 consisting of N employees, i.e., state variables X^t, and likewise $N(N-1)$ message exchange variables in every network fragment $B_\rightarrow$.

Without considering CPD innocuousness properties, i.e., we do not exploit contexts from X^t, we find that the number of possible $\mathcal{A}^t$ combinations in a cyclic ADBN corresponds the number of DAGs [10, Seq. A003024]. In a classic diagonal ADBN no indirect effects are anticipated, and thus, no "interlocking" (possibly active) activator combinations of $\mathcal{A}^t$ are allowed. We find this as the number of uniformly directed bipartite graphs, where isolated nodes belong to a fixed group [10, Seq. A001831]. For every of these combinations we have 2^N combinations of all X^t.

To emphasize the effect of exploiting innocuousness context, we consider that $q\%$ out of all N state variables X^t in an ADBN fragment $B'_\rightarrow$ are innocuous states X_Q, meaning that every state X_i^t "is not harmed" by any of these $X_Q^t \in X_Q^t$ if $\neg x_Q^t$. This implies that every activator $\dot{A}_{ij}^t$ has the context $\varphi_{A_{ij}^t} = \neg x_i^t$, if $X_i^t \in X_Q^t$. Thus, $\mathrm{rank}(X_Q^t) = Q = \lfloor N \cdot q \rfloor$, for which flooring operations lead to wavy lines in Fig. 2.

In an cyclic ADBN we obtain the total number $\mathcal{N}^{\mathcal{O}}{}_{N,Q}$ of allowed combinations $(X^t, \mathcal{A}^t)$ in a timestep t with Q innocuous nodes according to Theorem 2 as

$$\mathcal{N}^{\mathcal{O}}{}_{N,Q} = 2^{N-Q} \cdot \sum_{k=0}^{Q} 2^{k(N-1+N-k)} \cdot \mathrm{A003024}_{N-k} \cdot \binom{Q}{k}. \tag{8}$$

$\mathcal{N}^{\mathcal{O}}{}_{N,Q}$ origins from the consideration that we can have between $k=0$ to $k=Q$ "deactive" innocuous nodes. Thus, activators between $N-k$ nodes are still bound to DAG combinations, for which we have $\mathrm{A003024}_{N-k}$ many with 2^{N-Q} instantiations of X^t. For every of those DAG combinations we have k deactive nodes, whose $N-1$ activators are free, i.e. $2^{k(N-1)}$ combinations, and $N-k$ active nodes, whose activators with the k deactive nodes are free, i.e. $2^{(N-k)k}$ further combinations. For each combination, we have $\binom{Q}{k}$ options to choose which (labeled) innocuous states are deactive.

Notwithstanding, the restriction that only indirect-free combinations of $\mathcal{A}^t$ are allowed in diagonal ADBNs (Proposition 1) is also relaxed by considering innocuousness properties of X^t. To enumerate these, we need the number of uniformly directed bipartite graphs with groups of size n, m, $N = n + m$, which is

$$\mathrm{A001831}'_{N,n} = \binom{N}{n} \cdot (2^n - 1)^{N-n}. \tag{9}$$

With Q innocuous-nodes we then find the total number $\mathcal{N}'{}_{N,Q}$ of allowed combinations in diagonal (A)DBNs to be

$$\mathcal{N}'{}_{N,Q} = 2^{N-Q} \cdot \sum_{k=0}^{Q} \sum_{n=0}^{N-k} 2^{k(N+n-1)} \cdot \mathrm{A001831}'_{N-k,n} \cdot \binom{Q}{k} \tag{10}$$

$\mathcal{N}'_{N,Q}$ origins from the same considerations as $\mathcal{N}^{\mathcal{O}}_{N,Q}$, but the activators of the k deactive nodes are not completely free anymore. As (active) activators of the second group of size m interlock with activators of the deactive nodes.

Figure 2 shows a comparison of $\mathcal{N}^{\mathcal{O}}_{N,Q}$ and $\mathcal{N}'_{N,Q}$ for $0 < N \leq 25$ and different Q. Note that even in a logarithmic plot, a cyclic ADBN has an exponential advantage in favor of a classic acyclic (A)DBN.

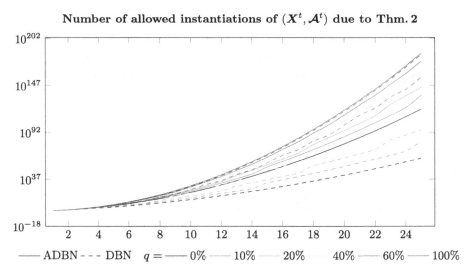

Fig. 2. Cyclic ADBNs ($\mathcal{N}^{\mathcal{O}}_{N,Q}$, solid) clearly outperform classic diagonal DBNs ($\mathcal{N}'_{N,Q}$, dashed) in the number of allowed instantiations of $(X^t, \mathcal{A}^t)$. Note that for a full noisy-or network (100 %) all possible graph structures ($\mathcal{A}^t$ combinations) are allowed in the case of $\forall i \neg x_i$, which draws $\mathcal{N}'_{N,N}$ near $\mathcal{N}^{\mathcal{O}}_{N,N}$. Still, even in this extreme case a cyclic ADBN outperforms a classic DBN by two orders of magnitude (*semi-logarithmic plot*).

6   Conclusion

In this paper we have formalized an innocuousness property of random variables, which is often associated with Boolean combination functions for general CPDs. Based on a formalization with random variables taking the role of activators, we relax restrictions on graphical models for the use in Bayesian networks and have given a quantitative evaluation of restrictions posed on such networks. This is beneficial for working with graphical models representing a process over time requiring the anticipation of indirect influences under a free choice of temporal granularity. Further, by providing a formal definition for innocuousness in CPDs, we gain the ability to formally represent that in specific contexts a dependency is causally irrelevant, opening new ways for more efficient inference and a higher causal accuracy in specifying CPDs.

Still, like in any other DBN, operations remain computationally intractable with respect to dimension complexity (number of state variables), and this demands approximate inference techniques. Considering our formalization that certain dependencies, i.e. arcs, are irrelevant in specific situations and an resulting BN might turn out to be singly connected, approximate inference techniques can heavily benefit from ADBNs and, here newly defined, innocuousness properties. Future work is dedicated to new inference techniques and extensions to relational Bayesian networks [6].

References

1. Antonucci, A.: The imprecise noisy-OR gate. In: 14th International Conference on Information Fusion, pp. 1–7. IEEE (2011)
2. Boutilier, C., Friedman, N., Goldszmidt, M., Koller, D.: Context-specific independence in Bayesian networks. In: 12th Conference on Uncertainty in Artificial Intelligence, pp. 115–123 (1996)
3. Cozman, F.G.: Axiomatizing noisy-OR. In: 16th Eureopean Conference on Artificial Intelligence, p. 979 (2004)
4. Heckerman, D., Breese, J.S.: Causal independence for probability assessment and inference using Bayesian networks. IEEE Trans. Syst. Man Cybern. Part A Syst. Hum. **26**(6), 826–831 (1996)
5. Henrion, M.: Practical issues in constructing a Bayes belief network. Int. J. Approximate Reasoning **2**(3), 337 (1988)
6. Jaeger, M.: Relational Bayesian networks. In: 13th Conference on Uncertainty in Artificial Intelligence, pp. 266–273 (1997)
7. Motzek, A., Möller, R.: Indirect causes in dynamic Bayesian networks revisited. In: 24th International Joint Conference on Artificial Intelligence, pp. 703–709. AAAI (2015)
8. Pearl, J.: Reasoning with cause and effect. AI Mag. **23**(1), 1–83 (2002)
9. Poole, D., Zhang, N.L.: Exploiting contextual independence in probabilistic inference. J. Artif. Intell. Res. **18**, 263–313 (2003)
10. Sloane, N.J.A.: The on-line encyclopedia of integer sequences. OEIS Foundation Inc., Sequences A003024 & A001831 (2015). http://oeis.org/
11. Srinivas, S.: A generalization of the noisy-OR model. In: 9th International Conference on Uncertainty in Artificial Intelligence, pp. 208–215 (1993)
12. Zagorecki, A., Druzdzel, M.J.: Probabilistic independence of causal influences. In: 3rd European Workshop on Probabilistic Graphical Models, pp. 325–332 (2006)
13. Zhang, N.L., Poole, D.: Exploiting causal independence in Bayesian network inference. J. Artif. Intell. Res. **5**, 301–328 (1996)

A Tweet Classification Model Based on Dynamic and Static Component Topic Vectors

Parma Nand, Rivindu Perera[(✉)], and Gisela Klette

School of Computer and Mathematical Science,
Auckland University of Technology, Auckland, New Zealand
{parma.nand,rivindu.perera,gisela.klette}@aut.ac.nz

Abstract. This paper presents an unsupervised architecture for retrieving and ranking conceptually related tweets which can be used in real time. We present a model for ranking tweets with respect to topic relevance in order to improve the accuracy of information extraction.

The proposed architecture uses concept enrichment from a knowledge source in order to expand the concept beyond the search keywords. The enriched concept is used to determine similarity levels between tweets and the given concept followed by a ranking of those tweets based on different similarity values. Tweets above a certain similarity threshold are considered as useful for providing relevant information (this is not part of this paper). We obtained precision values up to 0.81 and F values up to 0.61 for a tweet corpus of 2400 Tweets on the topic related to 2014 NZ general elections.

Keywords: Topic modeling · Natural language processing · Text mining · Social media

1 Introduction

Twitter, as a Social Media Platform (SMP), has been the subject of extensive number of studies for two reasons. Firstly, Twitter is used for posting short public messages rather than group or person to person postings. It is designed for users to post messages with common interests, for example information about politics, events, products, people inter alia. This makes Twitter a unique and rich repository of knowledge on both public, and private aspects of society. The other major reason is that it is easy to retrieve tweets and use them as data for research.

The *Twitter API*[1] allows real time retrieval of tweets based on keywords and other limited criteria such as posting time, tweets from specific accounts and topic popularity. The most common technique to retrieve tweets on a topic is to use a logical combination of keywords or phrases, which will retrieve all tweets containing the keywords. In most cases, this will also download a large amount of unrelated tweets, since the keywords could have been used in a different context

---

[1] Twitter API: https://dev.twitter.com/.

© Springer International Publishing Switzerland 2015
B. Pfahringer and J. Renz (Eds.): AI 2015, LNAI 9457, pp. 424–430, 2015.
DOI: 10.1007/978-3-319-26350-2_37

and/or used with a different sense. In order to be able to filter out the irrelevant tweets, there has been substantial efforts (Eg. [1–5])to re-rank the retrieved tweets from the Twitter API according their appropriateness to the topic.

Several AI researchers have exploited the social media attributes of tweets as features to train machine learning algorithms in attempts to rank them according to topic appropriateness. [6] define Twitter Building Blocks (TBBs) as structural blocks in a Twitter message and use these for higher level informational characteristics. As an example, tweets with the same structure as BBC news tweets are likely to be news tweets. The TBBs consist of non-content features such as neighbour TBB type, TBB count, TBB length and OOV (out of vocabulary words). Other works such as [7] and [8] also exploit generic social media as well as Twitter specific features to rank tweets retrieved from the Twitter API using a keyword based query. Some sample features used in these works are the presence or absence of a URL, author authorities, whether the current tweet is a repost(re-tweet), number of re-tweets, hash tag score, whether the current tweet is a reply-tweet and the ratio of OOV words to total number of words. In addition to the social media structural features Duan et al., also use a content feature. They compute the cosine similarity between each pair of tweets and then use this score in combination with Term Frequency Inverse Document Frequency (TF-IDF) to rank the individual tweets retrieved by the query.

In this paper we present a language model which uses concept enrichment to retrieve and rank tweets by capturing the dynamics of a given topic on a micro-blogging platform (MBP). The model is based on the fact that a trending topic on a MBP consists of two components; a persistent component and a dynamic component. The model uses the information content of the tweets for dynamic component and an external knowledge base, DBpedia, for the persistent component. The model was tested on data from Twitter, however the model can be translated on any MBP since it is based entirely on content, rather than MBP-specific structure related features.

2   Background

Previous works that have dealt with the task of twitter ranking can be categorized into those that make use of a machine learning algorithm and those that use some form of similarity measure. The work in [6] presents a model which uses a fifteen dimension feature vector to train a SVM (Support Vector Machine) model by using a corpus of 2000 human tagged tweets. Each tweet is split into Twitter Building Blocks (TBBs) consisting of the tokens with at least one TBB containing the query term used to search the tweets. The other TBBs contain structural features such as whether the TBB is an URL, whether there are followers of the author and whether the current tweet was retweeted. The paper ([7]) presents a very similar model using 3 sets of features; content relevance, twitter specific features and account authority. This study concludes that account authority and the length of tweets are the best conjunction as features for learning to rank tweets. [8] present another model, however this is completely based on structural features. This study affirms the conclusions from [7], however emphasize that the presence of URL is a stronger feature relevant to ranking.

Unlike above papers, we focus on ranking tweets based entirely on the content of the message rather than other unrelated structural features. The proposed model is similar to the one presented in [16]. In this work the authors use TF-IDF for ranking new tweets based on a background corpus consisting of 150,000 Twitter message corpus. The ranked messages are then merged into topic clusters using Jaccard similarity exceeding 65 %. The limitation in O'Connor et al.'s model is that the ranking is biased by a static corpus, hence is not completely realtime. [9] present a model which mitigates this limitation by capturing the dynamics of the topics using query expansion. The authors of this work build a background corpus by selecting messages posted closer to the query time using the original query terms. The rationale for this is that messages temporally closer to the query time are more relevant compared to older messages. A weighted mixture of the original query and top n terms from the generated corpus is then used to expand the query to retrieve further messages and rank them. Our model is an extension of the notion of query expansion from [9].

The rationale for query expansion is that tweets about an entity can be expressed by a wide a set of keywords rather than a single or a couple of keywords. When a user wishes to search for tweets pertaining to a topic, he would normally enter either a single or a very small logical combination of keywords resulting in selecting tweets which directly contain the keywords. This would leave out a large proportion of texts which use other keywords relevant to the topic.

A topic on a MBP can be broken down into a persistent and a dynamic component. The dynamic component accounts for the current conversation about an entity and this can rapidly change over time. The persistent component consists of conversation about the more static attributes of the topic. In order to be able to identify a balanced set of tweets one would need to use some combination of the dynamic and the persistent components. We propose a model which uses knowledge infusion from DBpedia to account for the persistent component and the MBP itself for the dynamic component. The information from these sources is combined to form word vectors followed by using a selection of similarity calculators in order to rank the messages. The architecture and the experiments are described in detail in the next section.

3   Model Description and Experiment

The proposed query expansion model uses *DBpedia*[2] as the knowledge source, however any other knowledge source such as the *Google* search may be effective as well. DBpedia provides persistent knowledge of about 4.0 million entities, categorized under 529 classes (person, organization, places, etc.). The knowledge is organized as predicates called triples, approximately consisting of a subject, predicate and object. For example the triple "⟨*New Zealand National Party,*

---

[2] DBPedia: http://www.dbpedia.org.

leader, John Key⟩" contains the information about the party leader and similarly "⟨*New Zealand National Party, type, Liberal-conservativeParties*⟩" contains information about the type of party. DBpedia knowledge base is organized into pages corresponding to the pages in Wikipedia, however Wikipedia may contain slightly more information as free text which might not have been structured in the DBpedia knowledge base.

Information was extracted from DBpedia by extracting the predicates Resource Description Framework (RDF) files as described in [10,11]. The predicates objects and subjects were then extracted from the predicates and sent through a pre-processing module. This module cleaned the noun phrases by removing non English characters, numbers, URLS, punctuations, duplicates and noun phrases which were longer than 50 tokens. The resulting noun phrases was tokenized and the tokens were used to construct the persistent component of the topic vector.

The dynamic component of the word vector was constructed using a set of *seed* tweets. The set of seed tweets is constructed by retrieving the first 100 tweets using only the noun phrase corresponding the topic entity using the Twitter API. The tweets retrieved from the Twitter API was first filtered for locality compatibility. In the case of a location mention in the content of a tweet, we used the location miner from [12] to eliminate tweets which did not belong to the locality of the topic, which was New Zealand for this project, extracted from the query terms. This gives us locality specific tweets that are directly related to the topic entity, however, will also contain other related entities that are typical at the time of the retrieval. The seed tweets were POS tagged using a HMM POS tagger from [13,14] which is able to identify the syntactical components as well as tweeter specific components such as hash tags URLs, and user mentions.

We downloaded a set of 2400 tweets before the New Zealand general elections at the end of 2014 for a larger research project on the influence of social media on party popularity.

The tweets were download using the keywords pertaining to New Zealand elections such as "*John Key*", "*National Party*" and "*NZ elections*". This resulted in a wide variety of tweets belonging to the wider topic of elections in NZ. The objective for the experiment was to rank the tweets that are relevant to the "National Party" which could be used later for downstream tasks such as sentiment detection. The tweets were manually annotated by a group of 15 post-graduate NLP students as being relevant to National Party or not. The annotators were instructed to take into account the topics relevant directly to National Party as well as the evolving topic temporally relevant to National Party. A selection of 100 tweets were annotated by 4 different annotators with a Cohens Kappa coefficient of 0.87. The annotators classified a total of 591 out of 2400 tweets belonging to the topic of "*National Party in New Zealand*"

The word vector consisting of the persistent and dynamic components were tested with various weights for similarity calculations. We used the following similarity measures: Cosine Similarity, Euclidean Distance, MongeElkan Similarity, Levenshtein Similarity, JaroWinkler Distance, Jaccard Distance and TFIDF

Distance. Similarity algorithms were implemented as described in [15]. Various components of POS components were tested in the similarity computations, the best performance was achieved using combinations of nouns, proper nouns and hash tags, hence all the results reported in this paper are based on tokens corresponding to these three tags. The tagging used in the ranking computations were directly from the POS tagger, hence the ranking results incorporate the propagated errors to all upstream tasks such as tokenization, tagging and chunking.

4   Ranking Results and Discussion

Table 1. Performance comparison of similarity algorithms for top 250 ranked tweets

Algorithm	Accuracy	Precision	Recall	F-value
Cosine	0.85	0.66	0.57	0.61
Jaccard	0.83	0.61	0.53	0.57
Euclid	0.77	0.43	0.37	0.40
Mongee	0.73	0.33	0.28	0.30
Levins	0.71	0.28	0.24	0.26
Jarowr	0.71	0.28	0.24	0.26
TFIDF	0.69	0.21	0.18	0.20

Initial experiments were done to determine the best similarity algorithm. We applied 50 % weight for the persistent and the same for the dynamic components using a word vector size of 100. The static component words were chosen based on the first 50 tokens from DBpedia triples. The 50 words for the persistent component were chosen from the top, most frequent set comprising of tokens from the noun phrases and hash tags. This topic word vector was then compared with the tweet word vector consisting only of noun phrases and hash tags. Experiments including other components such as @mentions and verb phrases did not yield good results. Table 1 summarizes the results of the similarity computations.

The results show that Cosine similarity was a clear winner with an F-value of 0.61 for selecting the top 250 ranked tweets from the corpus of 2400. Jaccard distance also had a relatively high F-value of 0.57 compared to the rest of the algorithms which had F-values less then 0.4. The rest of the experiments were done using the best performing similarity calculator, Cosine Similarity.

Tweeter ranking was done by using various combinations of topic vectors and various tweet components. The best results were obtained using about half of the topic vector from the persistent topics from DBpedia and the other half from the noun phrases and hash tags from the seed tweets. The comparison vector for the topic specific Tweets was constructed using the most frequent terms from Tweets downloaded using the keywords "John Key", "National Party" and "NZ elections".

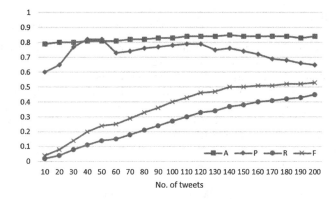

Fig. 1. Results for retrieving relevant tweets with equal proportions of persistent and dynamic topics in the word vector

The tweets were ranked using amplified cosine similarity values above an arbitrary threshold value of 100. The accuracy, precision, recall and F-values were computed for top tweets ranging from 10, in steps of 10 up a total of 200. The first experiment was done by using a topic vector of only persistent topics consisting of 100 words from the DBpedia page for the New Zealand National Party. The graph in Fig. 1 shows results obtained with equal proportion of topics from both persistent and dynamic components. The highest F-value obtained was 0.64, a precision value of 0.81 and a recall value of 0.53 for 190 tweets. The next best result was for persistent only topics with values 0.58, 0.73 and 0.48 respectively.

The results show that external infusion of knowledge for downloading and ranking tweets significantly increases the accuracy. Our proposed language model uses the fact that tweets relevant to a topic would revolve partly around the persistent topic and partly around evolving temporal topics current at the time of retrieval. Model tests show that the best results are achieved when a combination of both persistent topics derived from an external knowledge source and dynamic topic derived from a set of seed tweets are combined.

5   Conclusion and Future Work

We presented a non-learning language model which can be used to retrieve ranked tweets relevant to a topic of interest. The model divides a topic into knowledge around more persistent aspects and those that are transient and temporally relevant. We used DBpedia for the persistent component and a small set of seed tweets for the dynamic component. The dynamic and the persistent combined word vector used with cosine similarity calculator gave an F-value of 0.64 with a precision of 0.81 with a sample size of 2400 tweets. In future we are going to verify the model with more extensive range of topics and with multiple sources of knowledge such as Google Search results.

References

1. Luo, Z., Osborne, M., Petrovic, S., Wang, T.: Improving twitter retrieval by exploiting structural information. In: AAAI (2012)
2. Dong, A., Zhang, R., Kolari, P., Bai, J.: Time is of the essence: improving recency ranking using twitter data. In: Proceedings of the 19th International Conference on World Wide Web (2010)
3. Han, Z., Li, X., Yang, M., Qi, H., Li, S., Zhao, T.: Hit at trec 2012 microblog track. In: Proceedings of Text Retrieval Conference (2012)
4. Efron, M., Golovchinsky, G.: Estimation methods for ranking recent information. In: SIGIR Conference on Research and Development in Information Retrieval (2011)
5. Luo, Z., Osborne, M., Tang, J., Wang, T.: Who will retweet me? finding retweeters in Twitter. In: Proceedings of the 36th International ACM SIGIR Conference on Research and Development in Information Retrieval (2013)
6. Luo, Z., Osborne, M., Saša, P.: Improving twitter retrieval by exploiting structural information. In: AAAI Proceeding, pp. 22–26, T.W (2012)
7. Duan, Y., Jiang, L., Qin, T., Zhou, M., Shum, H.Y.: An empirical study on learning to rank of tweets. In: Proceedings of the 23rd International Conference on Computational Linguistics, pp. 295–303 (2010)
8. Nagmoti, R., Teredesai, A., De Cock, M.: Ranking approaches for microblog search. In: 2010 IEEE/WIC/ACM International Conference on Web Intelligence and Intelligent Agent Technology, vol. 1, pp. 153–157. IEEE (2010)
9. Massoudi, K., Tsagkias, M., de Rijke, M., Weerkamp, W.: Incorporating query expansion and quality indicators in searching microblog posts. In: Clough, P., Foley, C., Gurrin, C., Jones, G.J.F., Kraaij, W., Lee, H., Mudoch, V. (eds.) ECIR 2011. LNCS, vol. 6611, pp. 362–367. Springer, Heidelberg (2011)
10. Perera, R., Nand, P.: The role of linked data in content selection. In: Pham, D.-N., Park, S.-B. (eds.) PRICAI 2014. LNCS, vol. 8862, pp. 573–586. Springer, Heidelberg (2014)
11. Perera, R., Nand, P.: Real text-cs- corpus based domain independent content selection model. In: IEEE 26th International Conference on Tools with Artificial Intelligence (ICTAI), pp. 599–606 (2014)
12. Nand, P., Perera, R., Sreekumar, A., Lingmin, H.: A multi-strategy approach for location mining in tweets: AUT NLP group entry for ALTA-2014 shared task. In: Proceedings of the Australasian Language Technology Association Workshop 2014, pp. 163–170, Brisbane, Australia (2014)
13. Nand, P., Lal, R., Perera, R.: A HMM POS tagger for micro-blogging type texts. In: Proceedings of the 13th Pacific Rim International Conference on Artificial Intelligence (PRICAI 2014) (2014)
14. Nand, P., Perera, R.: An evaluation of POS tagging for tweets using HMM modeling. In: 38th Australasian Computer Science Conference (2015)
15. Chapman, S.: Simmetrics. Simmetrics is a similarity metric library, eg from edit distances (Levenshtein, Gotoh, Jaro etc) to other metrics,(eg Soundex, Chapman). Work provided by UK Sheffield University funded by (AKT) an IRC sponsored by EPSRC, grant number GR N 15764 (2009). URL http://sourceforge.net/projects/simmetrics/
16. O'Connor, B., Krieger, M., Ahn, D.: TweetMotif: exploratory search and topic summarization for Twitter. In: ICWSM (2010)

Understanding Toxicities and Complications of Cancer Treatment: A Data Mining Approach

Dang Nguyen$^{(\boxtimes)}$, Wei Luo, Dinh Phung, and Svetha Venkatesh

Centre for Pattern Recognition and Data Analytics, School of Information Technology, Deakin University, Geelong, Australia
{ngdang,wei.luo,dinh.phung,svetha.venkatesh}@deakin.edu.au

Abstract. Cancer remains a major challenge in modern medicine. Increasing prevalence of cancer, particularly in developing countries, demands better understanding of the effectiveness and adverse consequences of different cancer treatment regimes in real patient population. Current understanding of cancer treatment toxicities is often derived from either "clean" patient cohorts or coarse population statistics. It is difficult to get up-to-date and local assessment of treatment toxicities for specific cancer centres. In this paper, we applied an Apriori-based method for discovering toxicity progression patterns in the form of temporal association rules. Our experiments show the effectiveness of the proposed method in discovering major toxicity patterns in comparison with the pairwise association analysis. Our method is applicable for most cancer centres with even rudimentary electronic medical records and has the potential to provide real-time surveillance and quality assurance in cancer care.

1 Introduction

A cancer diagnosis is often the most unfortunate news in one's life. Sadly, more and more people have to face such unfortunate news [1]. Acknowledging the relatively low cure rate of cancer, whether to receive radical treatments is a valid question for a patient to ask. For many, the life quality after the cancer diagnosis is the most important. Unfortunately, treatments, in particular radical treatments, cause painful toxicities, from mild hair loss, ulcers in mouth, to more permanent damage to the body [2] and even life-threatening conditions [3].

Although the types of toxic effects of cancer treatments are well-known [4], their exact realisation at different patient groups is often not well understood. With more and more cancer patients facing the treatment decision, more and more variability in their conditions, and more and more treatment options, we need understand the toxicity better among the real patient population. The good news is that many of the treatment toxicities and complications have already been documented, mainly for billing purposes, in the hospital medical coding. Can we use these codes to better understand treatment toxicity in the *true* patient population? Or even better, can we use the codes to measure toxicities, to predict

© Springer International Publishing Switzerland 2015
B. Pfahringer and J. Renz (Eds.): AI 2015, LNAI 9457, pp. 431–443, 2015.
DOI: 10.1007/978-3-319-26350-2_38

how they will unfold as the treatment progress, and in the end to gauge an *optimal* treatment plan for each individual patient with different cancer diagnoses in different stages. This paper presents a step toward such a quest. We ask: With the medical coding, can we discover the most common progressive patterns of toxicities/complications among patients who receive radical radiation treatment?

To answer the question, one solution is to apply a recently proposed method for temporal comorbidity analysis [5]. In this approach, we assess pairwise associations among diagnosis codes to determine dependence/independence and then run binomial test to determine the temporal direction of the association between two codes. This approach, as we show in this paper, is relatively insensitive to temporal gap constraints among cancer toxicities, and hence is ineffective in revealing temporal progression of toxicity and complications of cancer treatments. Therefore, our goal is to overcome this weakness by a method based on temporal association rule discovery.

The remaining of paper is organized as follows. In Sect. 2, related work on applications of data mining techniques to disease diagnosis and treatment is briefly given. The main contributions are presented in Sect. 3, in which some definitions to extend Apriori [6] to discover temporal association rules are described. Experimental results are discussed in Sect. 4 while conclusions and future work are represented in Sect. 5.

2   Related Work

Data mining for disease diagnosis and treatment has become increasingly important recently. In this section, we review data mining applications in the area of health-care.

In [7], the authors developed a method based on Apriori [8] to extract temporal dependencies between gene expressions. Their method was able to detect various sizes of time delay between associated genes and sets of co-regulators for the target genes. With same topic, Alves et al. reviewed relevant methods of frequent itemset mining in gene association analysis [9]. Yang and Chen [10] combined decision tree and association rule mining to find the correlation between the clinical information and the pathology report to support lung cancer pathologic staging diagnosis. Data mining were also applied to patient medical records to improve the quality of clinic care [11], to data of Voluntary Counseling and Testing for HIV/AIDS to identify high-risk populations [12], and analysis of circadian rhythms from online communities [13].

In comorbidity study, Kim et al. applied association rule mining to analyze comorbidity in patients with type 2 diabetes mellitus [14]. Three data mining techniques, namely artificial neural network, logistic regression, and random forest were used to predict overall survivability in comorbidity of cancers [15]. Hanauer and Ramakrishnan described an approach for modeling temporal relationships in a large scale association analysis of electronic health record data [5]. The work of Hanauer has been applied to other clinical settings. For example, Munson et al. [16] used this method to discover associations and temporal relationships with charcot foot.

3   Methods

A high-level overview of the analytic approach used in this study is illustrated in Fig. 1. In the following sections, we describe the data source and each step involved in the process.

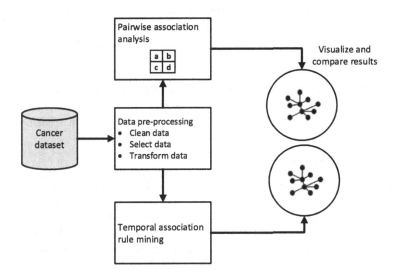

Fig. 1. A high-level overview of our analytic method

3.1   Clinical Dataset

The study is set in a regional cancer centre in an Australia tertiary hospital. The cancer centre covers a population over 200,000 and provides cancer treatment/ palliative care to over 6,000 patients each year. The study cohort consists of 717 patients who received radical radiation treatments from January 2010 to April 2015. The data includes medical coding (in ICD-10 [17]) for each admitted hospital episodes from the start of radiation treatment to the end of study period. Episodes for dialysis and same day chemo treatment were excluded as these episodes contain mostly repeated treatment information, reflect little information on toxicity and complications. The procedure codes (in Australian Classification of Health Interventions codes) for radiation treatments were also obtained.

Ethics approval was obtained from the Hospital and Research Ethics Committee at Barwon Health (number 12/83). Deakin University has reciprocal ethics authorisation with Barwon Health.

3.2   Data Pre-processing

We aim to find temporal association rules of the form $LH \rightarrow RH$, where LH (left hand side) and RH (right hand side) are sets of diagnosis codes. RH and LH

are expected to follow an temporal order, where the occurrence of RH follows the occurrence of LH for most patients. To distinguish associations that co-occur of LH and RH from associations that have genuine temporal order, we follow the approach in [5] to impose temporal gaps between LH and RH. We say rule $LH \rightarrow RH$ is valid within time frame t if the first instance of RH and the first instance of LH is at least t apart. When LH is a set, the first instance of LH is defined to be the $\max_p \{t_p, \forall p \subseteq LH\}$; when RH is a set, the first instance of RH is defined to be $\min_p \{t_p, \forall p \subseteq RH\}$ in which p is a patient whose codes are in LH or RH.

In this study, we are particularly interested in two time frames for the temporal associations: 1 week and 1 month. Additionally, we pruned the dataset with three conditions: (1) we included only codes which appeared in at least 30 patients; (2) we considered all diagnoses within the 3 days to be co-occurring, to accommodate the delay in medical coding; and (3) we selected only diagnosis codes in ICD-10 regarding ACHI procedure codes for radiation treatment.

3.3   Pairwise Association Analysis

The method finds frequently temporal relations among comorbidities in longitudinal hospital discharge records. The comorbidity pairs take the form of pairs of diagnosis codes in which one code follows the other temporally in different episodes of a patient. To find such pairs, the method takes two steps. In the first step, all frequent pairs were discovered, ignoring the temporal order of the first incident; this is achieved through Chi-square test between each pair of codes. The pairs with a correlation stronger enough were passed to the second step. In the second step, the correlated pairs were further filtered to retain only those showing significant sequential order. This is achieved through Binomial test on the temporal precedence/antecedence of codes. Further details can be found in [5].

Such a method for detecting code pairs, however, often leads to a large number of code pairs with redundant information. This is partly due to information redundancy among medical codes themselves. For example, a cancer diagnosis (Cxx) is often accompanied by a cancer morphology code (Mxxxxx). In the discovered pairs, a code pair Cxx→Y will often accompanies another code pair Mxxxxx → Y. In other words, the method itself does not attempt to summarize repeated information among the medical codes. And hence the number of pairs can be too large for human interpretation. Although the method has several weaknesses, it is easily implemented to draw an overview picture of temporal associations among diagnosis codes. In this section, we briefly introduce how we apply the method. All statistical analyses were conducted using R version 3.2 and results were visualized by network graph using Gephi version 0.8.2 beta [18] with the Fruchterman Reingold layout.

We first conducted a pairwise association analysis among the codes in the dataset. The number of possible pairs among codes is $(n^2 - n)/2$, where n is the number of distinct codes, resulting in 7,381 combinations. For each pair of codes (i.e., diagnosis), we calculated chi-square χ^2 statistic and its associated p-value on 2×2 contingency tables to determine if two codes were significantly associated with each other. Each 2×2 table includes (a) the number of patients who have both code A and code B; (b) the number of patients who have code A but not code B; (c) the number of patients who have code B but not code A; and (d) the number of patients who have neither code A nor code B.

We further used the binomial test to assess whether there was a significant temporal relationship between each pair of codes. That is, we tested if one code preceded another code. Given two codes A and B, we counted the number of times code A occurred before code B and vice versa in all patient records which contain both codes A and B, using only the initial instance of each code. The counts for each code pair were then compared using an exact binomial test with a hypothesized probability of success $= 0.5$ and confidence $= 0.95$. The direction between two codes was identified by the code which occurred first more often and the magnitude was represented by the p-value obtained from the binomial test.

3.4   Association Rule Analysis

As we see the previous section, the pairwise comorbidity analysis does not consolidate codes with similar information and hence can lead to too many rules for easy interpretation. One way to address this problem is to first summarize co-occurred codes. In data mining, a common approach is to use itemsets to capture frequently co-occurred events. Therefore, we adapt association rules (ARs) to provide a more succinct way to represent sequentially dependent comorbidities. Another advantage of association rule mining (ARM), in comparison to the pairwise approach, is that ARs are by design asymmetric (so is time) and therefore more suitable to represent temporal dependency.

In this section, we first explain some preliminary concepts of conventional ARM [6]. We then define some new terminologies used in our proposed method for mining temporal association rules (TARs).

Conventional Association Rule Mining (CARM)

Definition 1. *Let $\mathcal{I} = \{x_1, x_2, ..., x_m\}$ be a set of elements called items. A set $X \subseteq \mathcal{I}$ is called an itemset. Let $\mathcal{T} = \{t_1, t_2, ..., t_n\}$ be another set of elements called transaction identifiers or tids. A set $T \subseteq \mathcal{T}$ is called a tidset.*

Definition 2. *The support of an itemset X in a dataset D, denoted $\sup(X)$, is the number of transactions in D which contain X.*

Definition 3. *An association rule has a form of $X \to Y$ where X and Y are two itemsets, $X, Y \subseteq \mathcal{I}$ and $X \cap Y = \emptyset$. The support of the rule is the number of transactions in which both X and Y co-occur as subsets:*

$$\sup(X \to Y) = \sup(XY)$$

Definition 4. *The confidence of a rule is the conditional probability that a transaction contains Y given that it contains X:*

$$\text{conf}(X \to Y) = P(Y \mid X) = \frac{\sup(XY)}{\sup(X)}$$

In the context of this study, an item denotes a diagnosis code and an itemset represents a set of codes. The support of a code A ($\sup(code A)$) is the number of patients who have code A. An association rule R has a form of $A \to B$, where A and B are two disjoint set of codes. Although an association rule can show how likely are two set of codes to co-occur, it does not consider temporal patterns during the mining process. In other words, CARM has limit on extracting temporal dependencies between two set of codes.

Example 1. An example transaction dataset is shown in Table 1. It contains three patients and 5 codes (A, B, C, D, and E). For example, consider rule $R : A \to B$. We have $\sup(R) = \sup(AB) = 2$ and $\mathrm{conf}(R) = \frac{\sup(AB)}{\sup(A)} = \frac{2}{3}$.

Table 1. An example transaction dataset

Patient_ID	Code
1	A, B, C, D, E
2	A, B, D
3	A, D, E

Temporal Association Rule Mining (TARM). The CARM (as introduced in the previous section) assumes that data were from a single episode. For example, in market basket analysis, the benchmark application of CARM, each purchase order is considered independent. This is in contrast with our purpose, where we are not interested in comorbidities in one hospitalization, but comorbidities that are temporally dependent while appearing at different stages of the disease process.

To enable the discovery of such rules, we propose the following constructs that extend the traditional association rule discovery.

Definition 5. *A temporal item is an item associated with a list of patients and time stamps $\{p_i, t_j\}$ where p_i is a patient ID and t_j is a time stamp. A temporal itemset is a set of temporal items.*

Definition 6. *A temporal association rule R has a form of $LH \to \triangle RH$ where LH and RH are two temporal itemsets, $LH \cap RH = \emptyset$ and $\triangle$ is a time frame for difference between two temporal itemsets. $\sup(R) = \sup(LH \cup RH)$ (i.e., the number of patients who have both LH and RH) and $\mathrm{conf}(R) = \frac{\sup(LH \cup RH)}{\sup(LH)}$.*

Definition 7. *The direction support of a temporal rule $R : LH \to \triangle RH$ is defined as follows:*

$$\mathrm{dirsup}(R) = |\{p_i | \min(RH) - \max(LH) \geq \triangle\}|,$$

where p_i is a patient who has both LH and RH, $\min(RH) = \min\{X.t_j, \forall X \subseteq RH\}$ and $\max(LH) = \max\{Y.t_k, \forall Y \subseteq LH\}$.

Definition 8. *The direction confidence of a temporal rule* $R : LH \rightarrow \triangle RH$ *is defined as follows:*

$$\text{dirconf}(R) = \frac{\text{dirsup}(R)}{\text{sup}(R)}$$

Example 2. An example temporal dataset is shown in Table 2. It also contains three patients (1, 2, and 3) and 5 codes (A, B, C, D, and E). However, each temporal item is associated with a list of patients and time stamps. For example, item A is associated with $\{1, t_1\}$, $\{2, t_4\}$, and $\{3, t_8\}$, in which $\{1, 2, 3\}$ are Patient IDs and $\{t_1, t_4, t_8\}$ are time stamps. Consider rule $R : A \rightarrow D$. We have sup(R) = sup(AD) = 3 and conf(R) = $\frac{\text{sup}(AD)}{\text{sup}(A)} = \frac{3}{3} = 1$. Assume that $\triangle = 30$ days. Given patient 1, min(D) $-$ max(A) = $t_2 - t_1 \geq \triangle$. Given patient 2, min(D) $-$ max(A) = $t_5 - t_4 \geq \triangle$. Given patient 3, min(D) $-$ max(A) = $t_7 - t_8 < \triangle$. As a result, dirsup(R) = 2 and dirconf(R) = $\frac{2}{3}$.

Table 2. An example temporal dataset

(a) Patients with diagnosis codes

Patient_ID	Code
1	$\{A,t_1\}, \{B,t_1\}, \{C,t_1\}, \{D,t_2\}, \{E,t_3\}$
2	$\{A,t_4\}, \{B,t_4\}, \{D,t_5\}$
3	$\{A,t_8\}, \{D,t_7\}, \{E,t_6\}$

(b) Time stamps

Time stamp	Date
t_1	2010/08/20
t_2	2015/03/12
t_3	2013/07/17
t_4	2013/06/21
t_5	2014/04/07
t_6	2015/02/27
t_7	2012/01/12
t_8	2012/11/14

TARM usually consists of two steps: (1) discover frequent temporal itemsets from the dataset; and (2) generate TARs from these frequent temporal itemsets. In this paper, we extended the Apriori algorithm [8] to discover TARs.

3.5   Visualization of Comorbidities

Visualize Pairwise Associations. We visualize pairwise associations using directed graphs where each directed edge represents a discovered code pairs. This is the same as the visualization method mentioned in [5]. The interpretation of graph is quite simple because each edge represents only one direction from one code to another code. However, we cannot identify combined temporal relations which are regarded an important characteristic of radiation treatment and other information in such kind of graphs.

Visualize Temporal Association Rules. For TARs, we use the graph visualization introduced in [19]. A graph is constructed from the rules discovered from the dataset. For each TAR, a meta node M is introduced so that each item in the left hand side (LHS) of the rule represented by a node on graph has an edge pointing towards M and each item in the right hand side (RHS) of the rule which is also denoted by a node has an edge from M. For example, consider an example graph shown in Fig. 2. This graph represents rule "1526900: Radiation treatment megavoltage $>= 2$ fields dual modality linear accelerator→K590: Constipation and Z515: Palliative care". The meta M is denoted by a large blue node in the middle of graph. Its size and color are set based on *conviction* [6] and *lift* [6] of the rule.

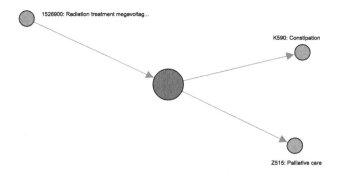

Fig. 2. An example of TAR visualization (Color figure online)

4   Results and Discussion

From January 2010 to April 2015, there were 717 patients who started radical radiation treatment in the cancer centre. There are 3,364 hospitalizations following the first radiation treatment (on average 4.7 episodes per patient). These episodes contain 28,667 diagnosis codes (on average 8.34 codes per episode). Among these codes, 122 appeared more than 30 different patients; they were used for analyses.

In this section, we represent the results obtained from pairwise association analysis and our approach (temporal association rule mining) with two time gaps: one week and one month.

4.1   One Week

Figure 3 displays network graphs constructed from the temporal relationships with **Radiation treatment** (1526900) at a time frame of one week where Fig. 3(a) shows associations obtained in pairwise association analysis while Fig. 3(b) represents the temporal associations generated by temporal association rule mining.

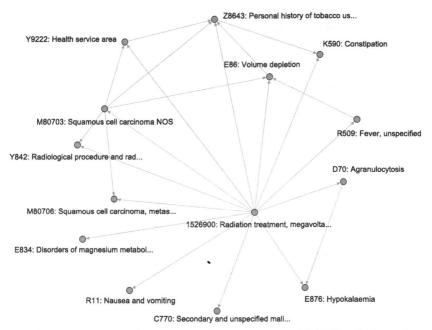

(a) Temporal associations between **Radiation treatment** (1526900) and other codes

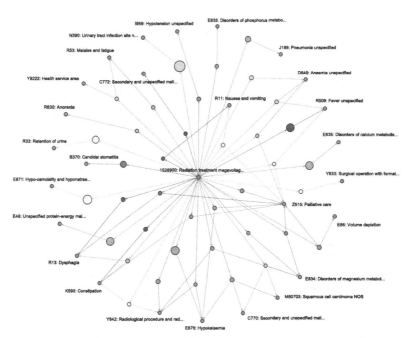

(b) Temporal association rules with **Radiation treatment** (1526900)

Fig. 3. Network graphs related to **Radiation treatment** (1526900) with a time interval of one week

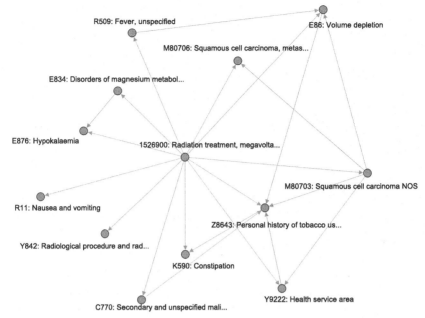

(a) Temporal associations between **Radiation treatment** (1526900) and other codes

(b) Temporal association rules with **Radiation treatment** (1526900)

Fig. 4. Network graphs related to **Radiation treatment** (1526900) with a time interval of one month

Figure 3(a) confirms common complications following a radiation treatment, including: (1) agranulocytosis and hypokalaemia, (2) nausea and vomitting (3) magnesium decificiency, (4) fever, (5) constipation, and (6) dehydration.

Figure 3(b) contains most of these and some others. One advantage of the TAR results is that it shows complications that co-occur. In particular, we can see two types of complications: the ones occurring with palliative care and the ones without palliative care. The ones with palliative care include most of the complications discovered using the pairwise method, such as dehydration, magnesium deficiency, anaemia, fever, and constipation. These are mostly well-known complications. The complications without palliative care include urinary tract infection, pneumonia (without mentioning radiation treatment as a cause). These complications tend to be more patient specific, and are less likely to be associated with radiation treatment.

4.2   One Month

Figure 4(a-b) represent the temporal relationships between **Radiation treatment** (1526900) and other diagnoses generated based on pairwise association analysis and temporal association rule analysis respectively.

We expect the one-month results to be a subset of the one-week results showing delayed complications. Among the pairwise associations, only D70 (agranulocytosis) disappeared from the one month results. This shows the gap constraint did not work effectively for the pairwise method. In contrast, the temporal association rules for one month contained significantly fewer rules. In particular, urinary tract infection and disorder of phosphorus metabolism were removed from the one-month result.

The results returned as pairwise associations and temporal association rules confirmed many known complications of radiation treatments. They also provide new insights into less understood complication patterns, such as pneumonia without being attributed to radiation treatment. This might be due to inconsistent coding, but it warrants further investigation in any case.

5   Conclusion and Future Work

We showed that data mining tools can be used to understand the intricacies of adverse effects of radiation treatments and their progressions in real patient populations. Our extension of conventional association rules to capture the temporal order of clinical events is effective and comprehensive compared to pairwise association analysis. It opens up a number of future research directions that may interest the artificial intelligence (AI) community.

Our approach is driven by real clinical questions, applied to real-life and up-to-date data, and revealed clinical meaningful patterns and evidence for care. It exemplifies that potential of AI methods in emerging clinical domains. The itemset concepts in the association rules provide a simple method to group co-occurring diagnosis codes. An alternative way is to apply topic model concept [20]; we are in the process of exploring that direction. Another further topic is to apply our method to mine TARs from different data sources [21, 22] which should provide more interesting information than those come from a single data source.

References

1. Stewart, B.W., Wild, C.P. (eds.) World Cancer Report 2014. World Health Organization (2014)
2. Plenderleith, I.H.: Treating the treatment: toxicity of cancer chemotherapy. Can. Fam. Physician **36**, 1827–1830 (1990)
3. Shanholtz, C.: Acute life-threatening toxicity of cancer treatment. Crit. Care Clin. **17**(3), 483–502 (2001)
4. Yoshida, K., Yamazaki, H., Nakamara, S., Masui, K., Kotsuma, T., Akiyama, H., Tanaka, E., Yoshioka, Y.: Comparison of common terminology criteria for adverse events v3.0 and radiation therapy oncology group toxicity score system after high-dose-rate interstitial brachytherapy as monotherapy for prostate cancer. Anticancer Res. **34**(4), 2015–2018 (2014)
5. Hanauer, D.A., Ramakrishnan, N.: Modeling temporal relationships in large scale clinical associations. J. Am. Med. Inform. Assoc. **20**(2), 332–341 (2013)
6. Zaki, M.J., Meira Jr., W.: Data Mining and Analysis: Fundamental Concepts and Algorithms. Cambridge University Press (2014)
7. Nam, H., Lee, K., Lee, D.: Identification of temporal association rules from time-series microarray data sets. BMC Bioinf. **10**(Suppl. 3), S6 (2009)
8. Agrawal, R., Srikant, R.: Fast algorithms for mining association rules in large databases. In: Proceedings of the 20th International Conference on Very Large Data Bases, VLDB 1994. Morgan Kaufmann Publishers Inc., San Francisco, pp. 487–499 (1994)
9. Alves, R., Rodriguez-Baena, D.S., Aguilar-Ruiz, J.S.: Gene association analysis: a survey of frequent pattern mining from gene expression data. Briefings Bioinf. **11**(2), 210–224 (2010)
10. Yang, H., Chen, Y.-P.P.: Data mining in lung cancer pathologic staging diagnosis: correlation between clinical and pathology information. Expert Syst. Appl. **42**(15), 6168–6176 (2015)
11. Jensen, P., Jensen, L., Brunak, S.: Mining electronic health records: towards better research applications and clinical care. Nat. Rev. Genet. **13**(6), 395–405 (2012)
12. Nguyen, D., Vo, B., Le, B.: CCAR: an efficient method for mining class association rules with itemset constraints. Eng. Appl. Artif. Intell. **37**, 115–124 (2015)
13. Dao, B., Nguyen, T., Venkatesh, S., Phung, D.: Analysis of circadian rhythms from online communities of individuals with affective disorders. In: International Conference on Data Science and Advanced Analytics (DSAA 2014), pp. 463–469. IEEE (2014)
14. Kim, H.S., Shin, A.M., Kim, M.K., Kim, Y.N.: Comorbidity study on type 2 diabetes mellitus using data mining. Korean J. Intern. Med. **27**(2), 197–202 (2012)
15. Zolbanin, H.M., Delen, D., Zadeh, A.H.: Predicting overall survivability in comorbidity of cancers: a data mining approach. Decis. Support Syst. **74**, 150–161 (2015)
16. Munson, M.E., Wrobel, J.S., Holmes, C.M., Hanauer, D.A.: Data mining for identifying novel associations and temporal relationships with charcot foot. J. Diabetes Res. 2014 (2014)

17. World Health Organization, International Classification of Diseases (ICD) (2013). http://www.who.int/classifications/icd/en/
18. Bastian, M., Heymann, S., Jacomy, M.: Gephi: an open source software for exploring and manipulating networks. ICWSM **8**, 361–362 (2009)
19. Hahsler, M., Chelluboina, S.: Visualizing association rules: introduction to the R-extension package arulesViz. In: R project module, pp. 223–238 (2011)
20. Luo, W., Phung, D., Nguyen, V., Tran, T., Venkatesh, S.: Speed up health research through topic modeling of coded clinical data. In: The 2nd International Workshop on Pattern Recognition for Healthcare Analytics (2014)
21. Zhao, Y., Zhang, H., Figueiredo, F., Cao, L., Zhang, C.: Mining for combined association rules on multiple datasets. In: Proceedings of the 2007 International Workshop on Domain Driven Data Mining, pp. 18–23. ACM (2007)
22. Huynh, V., Phung, D., Nguyen, L., Venkatesh, S., Bui, H.H.: Learning conditional latent structures from multiple data sources. In: Cao, T., Lim, E.-P., Zhou, Z.-H., Ho, T.-B., Cheung, D., Motoda, H. (eds.) PAKDD 2015. LNCS, vol. 9077, pp. 343–354. Springer, Heidelberg (2015)

A Representation Theorem for Spatial Relations

Özgür Lütfü Özçep$^{(\boxtimes)}$

Institute of Information Systems (IFIS), University of Lübeck, Lübeck, Germany
oezcep@ifis.uni-luebeck.de

Abstract. Spatial relations have been investigated in various inter-related areas such as qualitative spatial reasoning (for agents moving in an environment), geographic information science, general topology, and others. Most of the results are specific constructions of spatial relations that fulfill some required properties. Results on setting up axioms that capture the desired properties of the relations are rare. And results that characterize spatial relations in the sense that they give a complete set of axioms for the intended spatial relations still have to be presented. This paper aims at filling the gap by providing a representation theorem: It shows that there is a finite set of axioms that are fulfilled by a binary relation if and only if it can be constructed as a binary spatial relation based on a nested partition chain.

Keywords: Spatial relation · Axiomatization · Representation

1 Introduction

Spatial relations have been investigated in various inter-related areas such as qualitative spatial reasoning [16], geographic information science [17], general topology [9], and others. Most of the results achieved are specific constructions of spatial relations that fulfill some desired properties—which may vary according to the application/modeling context. Although the axiomatic method is a well-proven approach for the description of entities, results on setting up axioms that capture the desired properties of spatial relations are rare. And results that characterize spatial relations in the sense that they give a complete set of axioms for the intended spatial relations still have to be achieved.

The present paper aims at filling this gap with a semantic analysis of a specific class of spatial relations: Not only does it set up an axiom set that the intended spatial relation should fulfill but it takes a deeper look into the structure of the models for the axioms. The general idea is to systematically characterize the models by grouping them into disjoint, mathematically well-defined classes. A particularly interesting case is the one in which the set of models is described by exactly one class of models built according to some construction principle. In this case, the axioms really characterize the intended concepts, providing a canonical representation according to the construction principle of the class. A well-known

© Springer International Publishing Switzerland 2015
B. Pfahringer and J. Renz (Eds.): AI 2015, LNAI 9457, pp. 444–456, 2015.
DOI: 10.1007/978-3-319-26350-2_39

application of this methodology is Stones representation theorem [15]. A nice side-product of a representation theorem is that unintended models, which could result from an incomplete axiomatization, are excluded.

The spatial relations for which this paper gives a representation theorem are defined on the basis of a special structure, a hierarchal structure of nested partitions [5, 10, 11]. Typical examples of such total orders of nested partitions are made up of administrative units where the administrative units in a rougher granularity (e.g., districts) are the unions of administrative units of the lower level (e.g., municipalities). For example, think of two partitions of Switzerland, where the first partition consists of municipalities and where the second consists of districts. All districts are municipalities or are unions of two or more municipalities.

The interest in such types of spatial relations stems from observations regarding the context-dependency of spatially relatedness: The criterion for deciding whether a is considered spatially related to b depends on the type of the object that has b as its spatial extension. If b is a natural object such as a mountain, then the spatial criteria (be it geometric, topological, or metric) for regarding objects spatially related may depend on scaling contexts of (big) natural borders such as those of forests or rivers etc. But if the object with spatial extension b is a non-natural artifact such as a house, then different criteria have to be taken account: say borders made up by cadastral data.

The situation, as the house example demonstrates, may even be more complicated: it may be the case that the same spatial area b (house area) is the spatial extension of two different objects: the house considered as the pure geometrical object or the house considered as a legal object which has to adhere to planning laws. Depending on which objects are relevant for the use case different criteria are relevant in order to decide whether an object is spatially related to a house: In the first case, a purely metric criterion is in order, in the second case the district or even the country in which the house is situated is in order.

In this paper, only objects of the same type are considered. (With respect to the example above this means: we either consider only houses as pure geometrical objects or consider only houses as legal objects.) Hence, for spatially relatedness there is one context criterion fixed according to which two objects are considered to be related or not. This criterion is formalized by nested partitions of a spatial domain X. A partition provides a granularity or scale w.r.t. which the spatial relatedness of two regions is fixed; the main idea is to consider one of the arguments (here the second one) as the one determining the scaling context, i.e., the level on the ground of which two regions are defined to be spatially related or not. The results of this paper can be easily generalized to the case of regions with different types by considering collections of nested partition chains.

With this model in mind, the representation theorem now reads as follows: There is a finite set of axioms such that any binary relation fulfilling these can be represented as a spatial relation based on a nested partition chain.

The rest of the paper is structured as follows. Section 2 recapitulates the definitions of partition chains and spatially relatedness. Section 3 gives a comparison of spatially relatedness with proximity. Section 4 defines the upshift operator used in the axioms. Section 5 contains the main axioms for the representation theorem, which is proved in Sect. 6. The last two sections deal with related work and give a conclusion.[1]

2   Partitions and Spatially Relatedness

This work builds on the nearness framework developed by [11], which in turn is an abstraction of the framework by [10]—the abstraction being the transition from regions [12] to arbitrary sets. Following [11], it is assumed in this paper that X (the domain of objects which are the candidates for extensions of regions) is just a finite set. Hence, the results of this paper are relevant for discrete/digital topology [4,13].

We recapitulate the main technical concepts of a partition and of a normal partition chain. The usual partition concept of set theory will be called *set partition*. That is, given X and a family of sets $\{a_i\}_{i \in I}$ where the a_is are pairwise disjoint is a set partition iff X is the union of all the a_is, formally $X = \biguplus_{i \in I} a_i$.

Definition 1 (Partition). *A partition of a set X on level $i \in \mathbb{N}$ is a family of pairs $(i, a_j)_{j \in J}$ s.t. $(a_j)_{j \in J}$ is a set partition of X. A pair $c = (i, a_j)$ is called a cell of level i. Its underlying set a_j (the second argument) is denoted $us(c)$.*

Partition chains are partitions of X that are nested.

Definition 2 (Partition Chain). *Consider a collection of $n+1$ different partitions of X where all partitions have only finitely many cells. This set of partitions is called a partition chain pc iff*

1. *all cells $(i + 1, a_j)$ of level $i + 1$ (for $i \in \{0, \dots, n - 1\}$) are unions of i-level cells, i.e., there exist (i, b_k), $k \in K$, such that $a_j = \biguplus_{k \in K} b_k$;*
2. *and the last partition (level n) is made up by (X).*

Every cell has a unique upper cell. For a cell (i, a_j) (with $1 \le i \le n - 1$) let $(i, a_j)^{\uparrow, pc} = (i + 1, a_k)$ be the unique cell of the upper level in this partition chain pc such that $a_j \subseteq a_k$. For the cell of level n set $(n, X)^{\uparrow, pc} = (n, X)$. The cell $(i, a_j)^{\uparrow, pc}$ is called the upper cell of (i, a_j). If the partition chain is clear from the context, let $(i, a_j)^{\uparrow}$ stand for $(i, a_j)^{\uparrow, pc}$.

A partition chain is normal iff all set partitions underlying the partitions are pairwise distinct. A partition chain is strict iff for every level i, $i > 0$ and every cell (i, a_j) there is no cell on the level below with the same underlying set $(i - 1, a_j)$.

Example 1. An example of a (strict) partition chain with three levels is illustrated in Fig. 1, where we give a region oriented presentation (left) and the tree structure (right) with the associated levels. In order to make the example fully concrete we assume that $us(X) = \{1, 2, 3, 4, 5, 6\}$ and $c_i = (0, \{i\})$, for $i \in us(X)$.

---

[1] An extended version of this paper with all proofs can be found at the following URL: https://dl.dropboxusercontent.com/u/65078815/AI15representation.pdf.

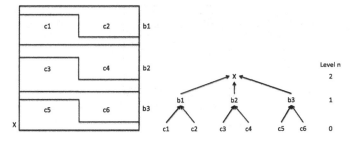

Fig. 1. A strict partition chain $(c_i)_{i \in \{1,2,3,4,5,6\}} \leq (b_i)_{i \in \{1,2,3\}} \leq (X)$

For a subset $b \neq \emptyset$ of X let $\tilde{b}^{pc}$ denote the cell (i, a_j) such that $b \subseteq a_j$ and i is minimal. The integer $i = l_{pc}(b)$ is called the level of b in pc. If the partition chain pc is unique in the used context, it is not mentioned in the subscripts. As a shorthand for $(\tilde{b}^{pc})^{\Uparrow,pc}$ one may write $b^{\Uparrow,pc}$.

Example 2. Consider again Fig. 1. For the set $\{5,6\} = us(c_5) \cup us(c_6)$ we have $\widetilde{\{5,6\}} = b_3$ and so $\{5,6\}^{\Uparrow,pc} = X$. For the set $\{3,6\} = us(c_3) \cup us(c_6)$ we already have $\widetilde{\{3,6\}} = X$, and so again $\{3,6\}^{\Uparrow,pc} = X$.

Definition 3 (Spatially Relatedness sr). *For a normal partition chain pc over X spatially relatedness sr_{pc} is defined by:*

$$\mathsf{sr}_{pc}(a,b) \;\; iff \;\; a \cap us(b^{\Uparrow,pc}) \neq \emptyset \tag{1}$$

So the main idea of the spatial relation is that the second argument (here b) determines the partition level w.r.t. which the first argument (here a) is considered to be related; if b is a cell, then one checks whether the intersection of a with the upper cell of b is non-empty. If it is non-empty, then a is spatially related to b. Otherwise a is not spatially related to b. If b is not the underlying set of a cell, then one looks for the smallest upper cell whose underlying set contains b and then proceeds as before.

Example 3. We consider the partition chain in Fig. 2. It is similar to that of Fig. 1, but here let $X = \{1, 2, \ldots, 7, 8\}$, $c_i = (0, i)$ for $i \in \{1, 2, 3, 4\}$ and $c_5 = (0, \{5, 7\})$, $c_6 = (0, \{6, 8\})$. Moreover, there is a set (region) $z = \{7, 8\}$ which overlaps with the cells c_5 and c_6 and a set $w = \{6\}$ contained in the cell c_6. We have $\tilde{z} = b_3 = (1, \{5, 6, 7, 8\})$ and hence $z^{\Uparrow,pc} = X$. So every set $b \subseteq us(X)$ is spatially related to z, i.e., $\mathsf{sr}_{pc}(b, z)$. In contrast, consider the set w. Here one has $\tilde{w} = c_6 = (0, \{6, 8\})$ and hence $w^{\Uparrow,pc} = b_3 = (1, \{5, 6, 7, 8\})$. So only sets not disjoint from $\{5, 6, 7, 8\}$ are near w.

Example 4. This example illustrates the difference between (metrical) nearness and spatial relatedness. Assume that we consider cadastral data covering two different nations and we consider houses as legal objects adhering to planning laws. Two houses a and b are sited on different sides of the border line of two

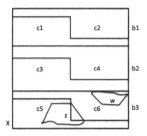

Fig. 2. Partition Chain with Non-Cells for Illustrating Spatially Relatedness

nations that have two completely different planning laws. Then, the legal object a would not stand in sr relation to the legal object b (w.r.t. the partition made of the cadastral data) though the areas a and b are clearly metrically near. Because of this we use the more neutral term *spatially relatedness* instead of *nearness* as used by [10].

3   Spatially Relatedness vs. Proximity

The partition based spatial relations have some connections to but nonetheless are different from minimal proximity relations δ. Structures (X, δ) with domain X and a binary relation δ over X are *minimal proximity structures* [4] iff the following axioms are fulfilled (where $a, b, c \subseteq X$):

(P1) If $\delta(a, b)$, then a and b are nonempty.
(P2*) $\delta(a, b)$ or $\delta(a, c)$ iff $\delta(a, (b \cup c))$.
(P3) $\delta(a, c)$ or $\delta(b, c)$ iff $\delta((a \cup b), c)$.

Obviously, sr_{pc} fulfills (P1) and (P3), but only the following weakening of (P2*):

(P2) If $\delta(a, b)$ or $\delta(a, c)$, then $\delta(a, (b \cup c))$

Moreover one can show that sr_{pc} fulfills the following two properties:

(P4) If $a \cap b \neq \emptyset$, then $\delta(a, b)$ and $\delta(b, a)$.
(P5) For all $a \subsetneq X$ with $a \neq \emptyset$: $\delta(a, (X \setminus a))$ or $\delta((X \setminus a), a)$.

The following proposition [10] summarizes these results.

Proposition 1. *All* sr_{pc} *for normal partition chains pc fulfill the axioms (P1), (P2), (P3), (P4), and (P5).*

As the following example shows, this set of axioms is incomplete in the following sense: There are still models where δ is interpreted by a binary relation that cannot be represented as sr_{pc} for an appropriate partition chain pc; in other words, these axioms do not completely characterize/represent the relations of the type sr_{pc}.

Example 5. Assume $X = \{1, 2, 3, 4\}$ and the following δ-relations are given:

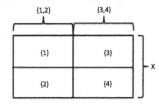

- None of the following holds: $\delta(\{4\}, \{2\})$, $\delta(\{4\}, \{1\})$, $\delta(\{1\}, \{3\})$, $\delta(\{1\}, \{4\})$
- for all other $a, b \subseteq X$ with $a, b \neq \emptyset$ it holds that $\delta(a, b)$.

It can be easily checked that δ fulfills (P1)–(P5), but that it is not representable as sr_{pc} for a normal partition chain.

The last assertion is proved as follows: Take the assertion $\delta(3, 2)$. Assume that there is a normal pc such that $\delta = \mathsf{sr}_{pc}$. Consider the following cases:

1. $c := \widetilde{\{2\}} = (0, \{2\})$. As $\delta(\{1\}, \{2\})$ and $\delta(\{3\}, \{2\})$ we must have $\{1, 2, 3\} \subseteq us(c^\uparrow)$. As not $\delta(\{4\}, \{2\})$, $us(c^\uparrow) = \{1, 2, 3\}$. That means that on level 1 one can have only the sets $\{1, 2, 3\}$ and $\{4\}$ as underlying cells. But this means that $\widetilde{\{4\}} = (0, \{4\})$ and $4^\Uparrow = (1, \{4\})$. But this contradicts the fact that $\delta(\{2\}, \{4\})$ holds while one would have to have not $\mathsf{sr}_{pc}(2, 4)$.
2. In the other cases $c := \widetilde{\{2\}} = (0, a)$ for a set a with $\{2\} \subsetneq a$. But then $c^\uparrow = (1, b)$ for a set b which must again be $b = \{1, 2, 3\}$ for the same reasons as in the former case. But then one gets a contradiction again.

4   The Upshift Operator

The main idea for the representation theorem is to reconstruct the levels by referring only to δ. A first step towards this end is to define the *upshift operator* $\cdot^{\Uparrow\delta}$, an abstract analogue of the level shifting operator $\cdot^{\Uparrow\uparrow, pc}$. The upshift operator is going to be defined below as a unique function based on δ. In all axioms where $\cdot^{\Uparrow\delta}$ occurs it can be unfolded to its defining formula to get rid of the new symbol.

Given δ, the equivalence relation $^\bullet\!\sim$ is defined as follows:

$$a \overset{\bullet}{\sim} b \text{ iff } \{c \subseteq X \mid \delta(c, a)\} = \{c \subseteq X \mid \delta(c, b)\} \tag{2}$$

This equivalence relation can be formulated for any relation δ, independently of the specific properties of δ. As usual, for any equivalence relation $\sim$, $[a]_\sim$ denotes the equivalence class of a w.r.t. $\sim$. A simple observation is the following:

Proposition 2. *For partition chains pc and $a, b \subseteq X$ s.t. $\tilde{a} = (i, a)$, $\tilde{b} = (i, b)$, and $a^{\Uparrow, pc} = b^{\Uparrow, pc}$ it holds that $a \overset{\bullet}{\sim} b$.*

Definition 4. *Given a binary relation δ, the upshift operator $\cdot^{\Uparrow\delta}$ for δ is defined for any nonempty set $b \subseteq X$ as follows:*

$$b^{\Uparrow\delta} = \bigcup [b] \overset{\bullet}{\sim} \tag{3}$$

So the set $b^{\Uparrow\delta}$ is just the union of all sets a that have the same set of sets δ-near it as b. If a partition chain pc over X is given, then one has two different shift operators, the operator $\cdot^{\Uparrow,pc}$, which calculates the upper cell w.r.t. pc, and the sr_{pc} level shift operator $\cdot^{\Uparrow\mathsf{sr}_{pc}}$. As the following proposition shows, the δ-shift operator is nothing else than the level shifting operator in case of $\delta = \mathsf{sr}_{pc}$.

Proposition 3. *Let pc be a partition chain over X. Then for any nonempty $b \subseteq X$ the following equality holds:* $b^{\Uparrow,pc} = b^{\Uparrow\mathsf{sr}_{pc}}$.

5   The Main Axioms

In the following subsections the main axioms are introduced that make up the representation theorem.

5.1   Spatially Relatedness is Grounded

The following axiom states a necessary and sufficient condition for the spatially relatedness of two sets with reference to the upshift operator $\cdot^{\Uparrow\delta}$. It says that a δ-related to b if and only if a has a non-empty intersection with the upshift of b.

(Pgrel) For all $a, b \subseteq X$: $\delta(a, b)$ iff $a \cap b^{\Uparrow\delta} \neq \emptyset$.

The axiom expresses a principle on the connection between the abstract δ relation and the set-theoretic element-of relation $\in$: Namely that δ is **grounded** in the element relation (hence the acronym grel).

Unfolding (Pgrel) w.r.t. the definition of $\cdot^{\Uparrow\delta}$ results in the axiom (Pgel'):

(Pgrel') For all $a, b \subseteq X$: $\delta(a, b)$ iff there is some c such that $c \,^{\bullet}\!\!\sim b$ and $a \cap c \neq \emptyset$.

Further, the relation symbol $^{\bullet}\!\!\sim$ can be eliminated—leading to (Pgrel") which refers only to δ (and some set operations).

(Pgrel") For all $a, b \subseteq X$: $\delta(a, b)$ iff there is some c such that for all z: $\delta(z, c)$ iff $\delta(z, b)$, and $a \cap c \neq \emptyset$.

Intuitively speaking, (Pgrel") says that a is δ-related to b iff it has a non-empty intersection with a set c that is similar ($^{\bullet}\!\!\sim$ equivalent to) to b. Looking at the unfolding, it is no surprise that (Pgrel) on its own is not expressive enough to characterize sr_{pc} and hence is far away from being a definition of sr_{pc}. This is demonstrated by Example 6.

Example 6. Let $X = \{1, 2, 3\}$ and δ be as follows:

- for all $a \subseteq X$: $\delta(a, \{1\})$
- $\neg\delta(1, 2)$
- $\neg\delta(1, 3)$

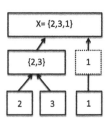

One calculates $2^{\Uparrow\delta} = 3^{\Uparrow\delta} = \{2, 3\}$ and $\{2, 3\}^{\Uparrow\delta} = X$ and shows that (Pgrel) is fulfilled. Nonetheless this δ is not representable as sr_{pc} for some normal pc.

The reason for non-representability in the above example is that there is no appropriate level notion. All the sets $\{1\}, \{2\}$, and $\{3\}$ would have to be of level 0. Applying $\cdot^{\Uparrow\delta}$ to $\{2\}$ and $\{3\}$ gives $\{2,3\}$, but the application to $\{1\}$ already gives X. Hence $\{1\}$ would have to appear on two levels (serving also as a cell on the level of $\{2,3\}$), but this would mean that the only set δ-related to $\{1\}$ is X—which is not the case.

Nonetheless, (Pgrel) has some consequences for the other axioms. In order to give a more detailed view, (Pgrel) is divided into two sub-axioms.

(Pgreln) For all $a, b \subseteq X$: If $\delta(a, b)$, then $a \cap b^{\Uparrow\delta} \neq \emptyset$.

The added "n" stands for "necessary condition" as a necessary condition is specified for δ.

(Pgrels) For all $a, b \subseteq X$: $\delta(a, b)$ if $a \cap b^{\Uparrow\delta} \neq \emptyset$.

The "s" stands for sufficient condition.

Proposition 4. *The following entailment relations hold:*
(1.) (Pgreln), (Pgrels) ⊨ (P3) and (2.) (Pgrels) ⊨ (P4)

So, with (Pgreln) and (Pgrels) the axiom (P3) becomes redundant, and (Pgrels) already entails (P4). (Pgrels) is already entailed by (P4). And hence (Pdefs) and (P4) are equivalent.

Proposition 5. *(P4) ⊨ (Pgrels)*

A simple consequence of axioms (Pgreln), (P2), (P4) is the monotonicity of the upshift operator.

Proposition 6. *If the axioms (Pgreln), (P2), (P4) hold, then monotonicity holds: For all $a \subseteq b$ one has $a^{\Uparrow\delta} \subseteq b^{\Uparrow\delta}$.*

So, the upshift operator $\cdot^{\Uparrow\delta}$ fulfills one of the conditions of a closure operator in a topological sense. But $\cdot^{\Uparrow\delta}$ is not a closure operator, not even a pre-closure/Cech-operator, i.e., it does not fulfill the following conditions for an operator $f : \text{Pot}(X) \longrightarrow \text{Pot}(X)$: (i) $f(\emptyset) = \emptyset$; (ii) $a \subseteq f(a)$; (iii) $f(a \cup b) = f(a) \cup f(b)$. (Here, $\text{Pot}(X) =$ the power set of X). Condition (i) is not fulfilled as it is not defined for empty sets—but this could be remedied. Condition (ii) is fulfilled (under (P4)), but Condition (iii) states distributivity w.r.t. the union of sets.

5.2 Alignment of Upshift Close-Ups

The upshift operator is intended to produce cells only. One aspect of this property is captured by the following nestedness condition.

(Pnested) For $a, b \subseteq X$: Either $a^{\Uparrow\delta} \subseteq b^{\Uparrow\delta}$ or $b^{\Uparrow\delta} \subseteq a^{\Uparrow\delta}$ or $a^{\Uparrow\delta} \cap b^{\Uparrow\delta} = \emptyset$.

As mentioned above, $\cdot^{\Uparrow\delta}$ is not a closure operator. Nonetheless, one can state the following axioms characterizing the behavior of the double application of the operator—replacing idempotence—and characterizing the outcome of applying it to a union of sets—replacing distributivity over unions of sets.

(Pdoubleshift) If $a^{\Uparrow\delta} \subsetneq b^{\Uparrow\delta}$, then $a^{\Uparrow\delta\,\Uparrow\delta} \subseteq b^{\Uparrow\delta}$.

The axiom (Pdoubleshift) states that if the upshift of a is properly contained in a cell (the upshift of b), then another upshift application will keep it in this cell.

(Punionshift) If $a^{\Uparrow\delta\,\Uparrow\delta} = b^{\Uparrow\delta\,\Uparrow\delta}$ and $a^{\Uparrow\delta} \not\subseteq b^{\Uparrow\delta}$ and $b^{\Uparrow\delta} \not\subseteq a^{\Uparrow\delta}$, then

$$a \cup b^{\Uparrow\delta} = a^{\Uparrow\delta\,\Uparrow\delta\,\Uparrow\delta} = b^{\Uparrow\delta\,\Uparrow\delta\,\Uparrow\delta}.$$

Proposition 7. *All* sr_{pc} *over a normal partition chain pc fulfill (Pdoubleshift) and (Punionshift).*

The axioms above do not capture the effect of the $\tilde{}$ operator, which makes spatially relatedness being determined by its underlying cells. The main observation here is given by the following axiom. Intuitively, it says that subsets of two sets which are not upshift comparable lead to the same upshift.

(Pcelldet) If $a^{\Uparrow\delta} \not\subseteq b^{\Uparrow\delta}$ and $b^{\Uparrow\delta} \not\subseteq a^{\Uparrow\delta}$, then for all $a' \subseteq a$ and $b' \subseteq b$ (with $a', b' \neq \emptyset$) it follows that $a' \cup b'^{\Uparrow\delta} = a \cup b^{\Uparrow\delta}$.

Proposition 8. *All* sr_{pc} *over a normal partition chain pc fulfill (Pcelldet).*

5.3 Isolated Points

An interesting point regarding $\cdot^{\Uparrow\delta}$ is that it may contain fixed points or *isolated points*—as they are denoted in the following. In fact, for normal partition chains in which you may have sets a that occur on more than one level, lets call them pc-fixpoints, it holds that $a^{\Uparrow\mathsf{sr}_{pc}} = a$: $a = a^{\Uparrow,pc} \stackrel{(Prop.3)}{=} a^{\Uparrow\mathsf{sr}_{pc}}$.

Definition 5 (Upshift-Isolated). *A set* $a \subseteq X$ *is upshift isolated,* $\mathsf{uiso}(a)$, *iff* $a^{\Uparrow\delta} = a$.

Now let us look again at points a in a normal partition pc that are pc-fixpoints. Another property these sets have is the following: If $\mathsf{sr}_{pc}(x, a)$, then $a \cap x \neq \emptyset$. Hence one may define the following equivalent notion of isolation:

Definition 6 (Set-Isolated). *A set* $a \subseteq X$ *is set-isolated,* $\mathsf{siso}(a)$, *iff: For all* $x \subseteq X$: *if* $\delta(x, a)$, *then* $x \cap a \neq \emptyset$.

A simple observation is that these notions are the same if (Pgreln) and (P4) are fulfilled.

Proposition 9. $(Pgreln), (P4) \models \forall a.\mathsf{uiso}(a) \leftrightarrow \mathsf{siso}(a)$.

5.4   Splittings

In general, sr_{pc} relations do not fulfill the other direction in axiom (P2*) which states that if a is δ-related to $b \cup c$, then a is δ-related b or c. Following [11], call the pair (b, c) with $b \cap c = \emptyset$ an *irregular split* of $b \cup c$ w.r.t. a. The main observation is that for any a there can be at most one irregular split.

Proposition 10. *For* sr_{pc}*, every set* a *has at most one irregular split.*

This property will now be formulated as an axiom over δ:

(PirrSplit) For δ, every a has at most one irregular split.

Any relation δ fulfilling (P2) and (PirrSplit) has a partition of X into cells which can serve as the cells of level 0. The crucial concept is the following.

Definition 7 (Cell-Equivalence). *For all* $x, y \in X$ *let the relation of cell-equivalence,* $\sim_0$ *for short, be defined by*

$$x \sim_0 y \; \text{iff} \; \{x\} \bullet\!\sim \{x,y\} \; \text{and} \; \{y\} \bullet\!\sim \{x,y\} \tag{4}$$

The cell-equivalence relation is indeed an equivalence relation:

Proposition 11. *Assume* δ *fulfills (P2) and (PirrSplit). Then the relation* $\sim_0$ *is an equivalence relation, i.e., it is symmetric, transitive, and reflexive.*

Actually, using the same proof idea, it is possible to prove the following theorem, which generalizes the result of the proposition.

Theorem 1. *For all subsets* $b_1, b_2 \subseteq [x]_{\sim_0}$*:* $b_1 \bullet\!\sim b_2$*.*

So, this result gives the base on which to build the partition chain, namely the partition consisting of cells $[x]_{\sim_0}$.

6   Representation Theorem

This section gives the proof for the representation theorem for those spatial relations that are based on strict partition chains. A problem on building further cells upon cells $[x]_{\sim_0}$ are isolated sets. Hence, we explicitly exclude isolated sets.

(Pnoiso) For every $a \subsetneq X$ one has: $a \neq a^{\Uparrow\delta}$.

The main problem in representing spatially relatedness is to capture the fact that all paths from the root to the leaves in the pc have the same length. So consider the following notion of rank for any binary relation δ on X.

Definition 8 (Rank). *For any* $a \in \mathsf{Pot}(X) \setminus \{\emptyset\}$ *we define by induction on* $n \in \mathbb{N}$*:* $a^0 = a$ *and* $a^{n+1} = a^{n\Uparrow\delta}$*. Then the* rank *of* a *is:*

$$r(a) = \begin{cases} m & s.t. \; there \; is \; m' \; with \; a^{m'} = X \; and \\ & m \; is \; the \; minimal \; one \; from \; the \; m' \\ \infty & else \end{cases}$$

The second case comes into play when there are isolated sets. Now one can formulate the following axiom which says that every pair of singleton sets $\{x\}$, $\{y\}$ over the domain X have the same rank.

(Psamerank) For all $x, y \in X$: $r(\{x\}) = r(\{y\})$.

Proposition 12. sr_{pc} *over a strict partition chain pc fulfills (Psamerank).*

With these additional axioms, the representation theorem for spatial relations generated by strict partition chains can be proved.

Theorem 2. *If δ fulfills (P1), (P2), (Pgreln), (Pgrels), (Pnoiso), (PirrSplit), (Psamerank), (Pcelldet), (Pdoubleshift), and (Punionshift), then there is a strict pc, such that $\delta = \mathsf{sr}_{pc}$.*

Together with the propositions proved before one gets the following corollary.

Corollary 1. *A binary relation δ fulfills (P1), (P2), (Pgreln), (Pgrels), (Pnoiso), (PirrSplit), (Psamerank), (Pcelldet), (Pdoubleshift), and (Punionshift) if and only if there is a strict pc, such that $\delta = \mathsf{sr}_{pc}$.*

7 Related Work

The idea of a scaling context for spatial relations (more concretely: nearness relations) goes back to the work of [16]. This work and following work [3,8,17] do not deal with axiomatic characterizations as done in this paper.

The definition of spatially relatedness in this paper follows a general "information processing" strategy that can be found in different areas of computer science. Belief revision [1] is concerned with the general task of integrating a new piece of information a into a knowledge base b. If a is not compatible (associate: spatially related) to b, then one weakens b to a set b' with $b \models b'$ and $b' \cup \{a\} \not\models \bot$ by throwing out elements from b. The KB b' is less strict and thus less informative than b (associate: b' is becoming more similar to X). A similar situation appears in the area of abduction, where one has to find explanations for observations [7]. In most cases, observations cannot be deduced from the theory or facts at hand, but have to be found within a space of possible explanations. The idea is to keep the creativity effort needed as low as possible—going only a minimal step upwards in the explanation space.

The underlying structure of sr_{pc} are partition chains, which are special trees. The work of [6,14] focuses on the dynamics of such tree structures, called adjacency trees. In contrast, this paper uses these tree structures as a basis for a spatially relatedness definition and gives a full axiomatic characterization.

8   Conclusion and Outlook

The presented work gave a fine-grained semantical analysis of spatially relatedness based on partition chains—resulting in a representation theorem for the special case where the partition chain is strict. So, in fact, the work presented in this paper completed the characterization of the spatially relatedness relations in [10,11]. Users of information systems such as agents that move, act and plan in an environment according to some internal qualitative spatial map or semantic web systems [2] relying on spatio-temporal ontologies may benefit from this representation result because it completely characterizes the spatial relations at hand.

More concretely, if an agent or a query answering system would rely only axiomatic characterizations of a spatially relatedness relation, then it would have to incorporate a deduction engine in order to do planning or query answering: Because only by considering all entailments of the axioms for spatially relatedness relations, the agent will guarantee that he will reach all possible plan configurations or all possible answers, respectively. On the other hand, knowing that the axioms for spatially relatedness have exactly one model (modulo renaming of the domain elements), the agent/the system may directly work with the model and apply, e.g., model checking—which is usually more feasible regarding complexity than calculating the deductive closure of a set of axioms.

Another benefit of the representation theorem for an AI planning or query answering agent is the possibility to focus on nested partition relations. In many situations these kinds of structures are not dynamic but may change due to some natural operations such as: two regions merge to an upper region, or an additional level of regions is established etc. (see [11], where these kinds of change are called global change). The change of the underlying partition structure leads to changes of the induced spatial relation. But one still knows that also the new spatial relation induced by the new nested partition chain still fulfills all properties as described by the axioms.

Regarding future work we note first that within the representation theorem a construction for strict partition chains on top of the upshift operator was used—relying on the exclusion of isolated sets. Spatial relations based on normal partition chains may have isolated sets, hence a different construction is called for. Moreover, this paper presumed finite domains X. This causes no problem as long as the upper structure of partitions is finite.

Further future work concerns the adaptations to overlapping hierarchies of regions. This is needed to model spatially relatedness induced by micro-functional areas with natural borderlines. Such areas (such as a forest, a valley, etc.) clearly have an influence on spatially relatedness but are not necessarily mutually disjoint.

References

1. Alchourrón, C.E., Gärdenfors, P., Makinson, D.: On the logic of theory change: partial meet contraction and revision functions. J. Symbolic Logic **50**, 510–530 (1985)
2. Berners-Lee, T., Hendler, J., Lassila, O.: The semantic web. Sci. Am. **284**(5), 34–43 (2001)
3. Brennan, J., Martin, E.: Spatial proximity is more than just a distance measure. Int. J. Hum. Comput. Stud. **70**, 88–106 (2012)
4. Düntsch, I., Vakarelov, D.: Region-based theory of discrete spaces: a proximity approach. Ann. Math. Artif. Intell. **49**, 5–14 (2007)
5. Grütter, R., Scharrenbach, T., Waldvogel, B.: Vague spatio-thematic query processing: a qualitative approach to spatial closeness. Trans. GIS **14**(2), 97–109 (2010)
6. Jiang, J., Worboys, M.: Event-based topology for dynamic planar areal objects. Int. J. Geogr. Inf. Sci. **23**(1), 33–60 (2009)
7. Kowalski, R.: Computational Logic and Human Thinking: How to be Artificially Intelligent. Cambridge University Press (2011)
8. Mata, F.: Geographic information retrieval by topological, geographical, and conceptual matching. In: Fonseca, F., Rodríguez, M.A., Levashkin, S. (eds.) GeoS 2007. LNCS, vol. 4853, pp. 98–113. Springer, Heidelberg (2007)
9. Naimpally, S., Warrack, B.D.: Proximity Spaces. Cambridge Tracts in Mathematics and Mathematical Physics, vol. 59. Cambridge University Press (1970)
10. Özçep, Ö.L., Grütter, R., Möller, R.: Nearness rules and scaled proximity. In: Raedt, L.D., Bessiere, C., Dubois, D. (eds.) Proceedings of ECAI 2012, pp. 636–641 (2012)
11. Özçep, Ö.L., Grütter, R., Möller, R.: Dynamics of a nearness relation-first results. In: Proceedings of SteDy 2012 (2012)
12. Randell, D.A., Cui, Z., Cohn, A.G.: A spatial logic based on regions and connection. In: Proceedings of the 3rd International Conferecence on Knowledge Representation and Reasoning, pp. 165–176 (1992)
13. Smyth, M.B., Webster, J.: Discrete spatial models. In: Aiello, M., Pratt-Hartmann, I., Benthem, J. (eds.) Handbook of Spatial Logics, pp. 713–798. Springer, Netherlands (2007)
14. Stell, J.G., Worboys, M.F.: Relations between adjacency trees. Theor. Comput. Sci. **412**(34), 4452–4468 (2011)
15. Stone, M.H.: The thoery of representations for boolean algebras. Trans. Am. Math. Soc. **40**(1), 37–111 (1936)
16. Worboys, M.F.: Nearness relations in environmental space. Int. J. Geogr. Inf. Sci. **15**(7), 633–651 (2001)
17. Yao, X., Thill, J.C.: How far is too far? - a statistical approach to context-contingent proximity modeling. Trans. GIS **9**(2), 157–178 (2005)

Stream-Query Compilation with Ontologies

Özgür Lütfü Özçep[⊠], Ralf Möller, and Christian Neuenstadt

Institute of Information Systems (IFIS),
University of Lübeck, Lübeck, Germany
{oezcep,moeller,neuenstadt}@ifis.uni-luebeck.de

Abstract. Rational agents perceiving data from a dynamic environment and acting in it have to be equipped with capabilities such as decision making, planning etc. We assume that these capabilities are based on *query answering* with respect to (high-level) *streams of symbolic descriptions*, which are grounded in (low-level) data streams. Queries need to be answered w.r.t. an ontology. The central idea is to compile ontology-based stream queries (continuous or historical) to relational data processing technology, for which efficient implementations are available. We motivate our query language STARQL (Streaming and Temporal ontology Access with a Reasoning-Based Query Language) with a sensor data processing scenario, and compare the approach realized in the STARQL framework with related approaches regarding expressivity.

Keywords: Stream · Rewriting · Ontology · Description logic

1 Introduction

Many of the main problems, topics, and activities of AI research are related to the design and analysis of rational agents. Rational agents perceive sensor data from an environment and act therein so "as to achieve one's goals, given one's beliefs" [13, p. 7]. This hints to the challenging problem of transforming sensor data into high-level conceptualizations—required for representing and reasoning with beliefs and goals—and transforming operations on the higher level to operations on the lower level. We assume that the agent has to transform lower-level data streams to higher-level streams of symbolic descriptions.

The challenging transformation problems mentioned above are not new and have been approached in different projects such as the DyKnow project [10] or projects using logic-based situation awareness techniques [4,5]. In contrast to these, the approach of this paper tackles the problem of performant transformations by generating only "views" of high-level streams where the views are defined by declarative mappings from low-level data streams to high-level streams of symbolic descriptions. Hence, the high-level stream is actually not materialized by the mappings but is considered as a virtual stream on the abstraction level

---

This work has been supported by the European Commission as part of the FP7 project Optique.

© Springer International Publishing Switzerland 2015
B. Pfahringer and J. Renz (Eds.): AI 2015, LNAI 9457, pp. 457–463, 2015.
DOI: 10.1007/978-3-319-26350-2_40

needed to accomplish (goal-based) reasoning. In order to realize inference services, such as ontology-based query answering on high-level streams, one has to reduce them to operations on low-level streams. This kind of reduction is an essential element of the paradigm of *ontology-based data access* (OBDA) [8] and involves *rewriting a query to SQL (or stream SQL such as CQL [1]) w.r.t. the ontology* as well as *unfolding a query with respect to mapping rules*.

Recent work on temporalizing and streamifying OBDA [2,6,7,9,11,12] provides solutions for adapting methods, concepts, and ideas from temporal logics and relational stream languages to the ontological layer. Nonetheless, the picture of relations between different approaches, in particular w.r.t. the semantics of specific logical operators such as the window operator, is still incomplete.

The contributions of this paper are the following: STARQL's semantics is tailored towards a plug-in architecture. In order to extend the expressivity picture, we show that an OBDA instantiation of STARQL has a different semantics that leads to the same set of answers as w.r.t. the original semantics (Theorem 1). As a corollary we get first-order logic (FOL) rewritability for answering STARQL queries (Theorem 2). We show that (a safe fragment of) TCQs [6] is embeddable into STARQL (Proposition 2).[1]

2   Logical Preliminaries

We shortly recapitulate necessary terminology. Details can be found in the longer version. An ontology is a triple $\mathcal{O} = \langle Sig, \mathcal{A}, \mathcal{T} \rangle$ with a signature Sig, an ABox $\mathcal{A}$ (set of assertional axioms), and a TBox $\mathcal{T}$ (set of terminological axioms). The STARQL framework allows for plugging in different DLs, but here we focus on the DL-Lite family [8], which is tailored towards the OBDA paradigm. In OBDA, query answering over an ontology is reduced to model checking on the data by rewriting the query w.r.t. the TBox and unfolding it w.r.t. mappings. In the whole paper let $[n] = \{0, 1, \ldots, n\}$ be the natural numbers up to $n \in \mathbb{N}$. We assume familiarity with (unions of) conjunctive queries (UCQs). The set of answers for a query $\phi(\boldsymbol{x})$ in interpretation $\mathcal{I}$ is $ans(\phi(\boldsymbol{x}), \mathcal{I}) = \{\boldsymbol{a} = (a_1, \ldots, a_n) \mid a_i$ a constant in Sig and $\mathcal{I} \models \phi(\boldsymbol{a})\}$. The *certain answers* w.r.t. an ontology $\langle \mathcal{A}, \mathcal{T} \rangle$ are: $cert(\phi(\boldsymbol{x}), \langle \mathcal{A}, \mathcal{T} \rangle) = \bigcap_{\mathcal{I} \models \langle \mathcal{A}, \mathcal{T} \rangle} ans(\phi, \mathcal{I})$. For inconsistent ontologies $\boldsymbol{x}$ is bound to all possible bindings $\boldsymbol{a}$ over all constants in the ontology.

For every ABox $\mathcal{A}$ let $DB(\mathcal{A})$ be its minimal Herbrand model. Let QL_1 and QL_2 be query languages over the same signature and OL be an ontology language. QL_1 allows for QL_2-rewriting of query answering w.r.t. OL iff for all queries ϕ in QL_1 and TBoxes $\mathcal{T}$ in OL there exists a query $\phi_{\mathcal{T}}$ in QL_2 such that for all ABoxes $\mathcal{A}$ it holds that: $cert(\phi, \langle \mathcal{A}, \mathcal{T} \rangle) = ans(\phi_{\mathcal{T}}, DB(\mathcal{A}))$. A particularly interesting case is QL_2 = first-order logic (FOL) queries. A well-known fact [8] is: UCQs are FOL rewritable w.r.t. DL-Lite ontologies.

A query ϕ is *domain independent* iff for all interpretations $\mathcal{I}, \mathcal{J}$ such that $\mathcal{I}$ is a substructure of $\mathcal{J}$: $ans(\phi, \mathcal{I}) = ans(\phi, \mathcal{J})$ [3].

---

[1] For a longer version of this paper see: https://dl.dropboxusercontent.com/u/65078815/AI15Stream.pdf.

A flow of time $(T, \leqslant)$ consists of a set T of time points and a linear, transitive, reflexive, and antisymmetric relation $\leqslant$ on T. A *temporal ABox* is a finite set of timestamped ABox axioms $ax\langle t \rangle$, with $t \in T$. We call structures of the form $\langle (\mathcal{A}_i)_{i \in [n]}, \mathcal{T} \rangle$ consisting of a finite sequence of ABoxes and a pure TBox a *sequenced ontology (SO)*. The index i of the ABox $\mathcal{A}_i$ is called its *state*.

A *stream* is a sequence of elements of the form $d\langle t \rangle$ where d is an element from the *stream domain* D and t is a timestamp from a *flow of time*. We assume that the time ordering respects the arrival ordering in the sequence.

3   The STARQL Framework

The following example illustrates the main constructors of STARQL. A rational agent has different sensors, in particular different temperatures attached to different components. The agent receives high-level messages within an input stream Sin. The agent has some background knowledge on the sensors. The agent has to recognize critical temperatures: Mark every 10 s all temperature sensors as critical such that the following holds: In the last 5 min there was a monotonic increase on some interval followed by an alert message of category A. A STARQL formalization is given in the following listing.

```
1   CREATE STREAM Sout AS
2   CONSTRUCT  GRAPH NOW { ?s a inCriticalState }
3   FROM Sin[NOW-5min, NOW]->10s, <http://ABox>, <http://TBox>
4   USING PULSE AS START = 0s, FREQUENCY = 10s
5   WHERE { ?s a TempSens }
6   SEQUENCE BY StdSeq AS SEQ
7   HAVING
8   EXISTS i1, i2, i3 IN SEQ
9    0 < i1 AND i2 < MAX AND plus(i2,1,i3) AND
10   GRAPH i3 { ?s :message ?m  . ?m a A-Message } AND
11   FORALL i, j in SEQ FORALL ?x,?y:
12     IF   i1 <= i  AND i<= j  AND j <= i2  AND
13         GRAPH i { ?s :val ?x }  AND GRAPH j { ?s :val ?y }
14     THEN ?x <= ?y
```

The CONSTRUCT operator fixes the format of the output stream. The window operator [NOW-5min, NOW]->10 s gives snapshots of the stream with the frequency of 10 s and range of 5 min. The WHERE clause (l. 5) determines the sensors ?s. For every binding of ?s, the query evaluates conditions that are specified in the HAVING clause (l. 7–14). A sequencing method (here StdSeq) maps an input stream to a sequence of ABoxes (annotated by states i, j) according to a grouping criterion. Testing for conditions at a state is done with the SPARQL sub-graph mechanism. So, e.g., GRAPH i3 {?s :message ?m . ?m a A-Message} (l. 10) asks whether ?s showed a message of Type A at state i3. Here, i3 is the successor of the end state i2 in the interval [i1, i2] for which one tests monotonicity (FORALL condition, l. 12–14).

The grammar of STARQL(OL, ECL) (Fig. 1) contains parameters that have to be specified in its instantiations, the ontology language OL and the embedded condition language ECL. ECL is a query language referring to the signature of the ontology language. STARQL uses ECL conditions in its WHERE and HAVING clauses. In Sect. 4 the instantiation STARQL(DL-Lite, UCQ) is considered.

$$starql \longrightarrow [prefix]\ createExp$$
$$createExp \longrightarrow \texttt{CREATE STREAM } sName \texttt{ AS}$$
$$constrExp$$
$$pulseExp \longrightarrow \texttt{PULSE WITH}$$
$$\texttt{START = } start,$$
$$\texttt{FREQUENCY = } freq$$
$$constrExp \longrightarrow \texttt{CONSTRUCT } cHead(\boldsymbol{x}, \boldsymbol{y})$$
$$\texttt{FROM } listWStrExp$$
$$[,\ \underline{URIsToA/TBoxes}]$$
$$[\texttt{USING } pulseExp]$$
$$\texttt{WHERE } whereCl(\boldsymbol{x})$$
$$\texttt{SEQUENCE BY } seqMeth$$
$$\texttt{HAVING } safeHCl(\boldsymbol{x}, \boldsymbol{y})$$
$$cHead(\boldsymbol{x}, \boldsymbol{y}) \longrightarrow \texttt{GRAPH } timeExp\ triple(\boldsymbol{x}, \boldsymbol{y})$$
$$\{\ .\ cHead(\boldsymbol{x}, \boldsymbol{y})\}$$
$$listWStrExp \longrightarrow (sName \mid constrExp)\ winExp$$
$$[,\ listWStrExp]$$
$$winExp \longrightarrow [timeExp_1, timeExp_2]\texttt{->}sl$$
$$timeExp \longrightarrow \texttt{NOW} \mid \texttt{NOW - } constant$$

$$whereCl(\boldsymbol{x}) \longrightarrow \boxed{ECL(\boldsymbol{x})}$$
$$seqMeth \longrightarrow \texttt{StdSeq} \mid seqMeth(\sim)$$
$$term(i) \longrightarrow i$$
$$term() \longrightarrow \texttt{MAX} \mid \texttt{0} \mid \texttt{1}$$
$$arAt(i_1, i_2) \longrightarrow term_1(i_1)\ op\ term_2(i_2)$$
$$(op \in \{<, <=, \ =, \ >, \ >=\})$$
$$arAt(i_1, i_2, i_3) \longrightarrow \texttt{plus}(term_1(i_1),$$
$$term_2(i_2),$$
$$term_3(i_3))$$
$$stateAt(\boldsymbol{x}, i) \longrightarrow \texttt{GRAPH i } \boxed{ECL}\ (\boldsymbol{x})$$
$$atom(\boldsymbol{x}) \longrightarrow arAt(\boldsymbol{x}) \mid stateAt(\boldsymbol{x})$$
$$hCl(\boldsymbol{x}) \longrightarrow atom(x) \mid hCl(\boldsymbol{x}) \texttt{ OR } hCl(\boldsymbol{x})$$
$$hCl(\boldsymbol{x}, \boldsymbol{y}) \longrightarrow hCl(\boldsymbol{x}) \texttt{ AND } hCl(\boldsymbol{y})$$
$$hCl(\boldsymbol{x}) \longrightarrow hCl(\boldsymbol{x}) \texttt{ AND FORALL } \boldsymbol{y}$$
$$\texttt{IF } hCl(\boldsymbol{x}, \boldsymbol{y}) \texttt{ THEN } hCl(\boldsymbol{x}, \boldsymbol{y})$$
$$hCl(\boldsymbol{x}, \boldsymbol{z}) \longrightarrow \texttt{EXISTS } \boldsymbol{y}\ hCl(\boldsymbol{x}, \boldsymbol{y}) \texttt{ AND}$$
$$hCl(\boldsymbol{z}, \boldsymbol{y})$$
$$safeHCl(\boldsymbol{x}) \longrightarrow hCl(\boldsymbol{x})$$
$$(\boldsymbol{x} \text{ contains no i variable})$$

Fig. 1. Syntax for STARQL(OL, ECL) template.

Regarding semantics, we consider only the HAVING clause. Details for the other constructs can be found in the longer version of the paper.

HAVING Clause. We have to define the semantics of $\phi(\boldsymbol{a}_{wh}, \boldsymbol{y})$ for every binding $\boldsymbol{a}_{wh}$ from the evaluation of the WHERE clause. We declare for every t how to get bindings for $\boldsymbol{y}$. Assume that the sequence of ABoxes at t is $seq = (\mathcal{A}_0, \ldots, \mathcal{A}_k)$. We define the *separation-based* certain answers: $cert_{sep}(\phi(\boldsymbol{a}_{wh}, \boldsymbol{y}), \langle \mathcal{A}_i \cup \mathcal{A}_{st}, \mathcal{T} \rangle)$. For t one constructs a sorted first order logic structure $\mathcal{I}_t$: The domain of $\mathcal{I}_t$ consists of the index set $\{0, \ldots, k\}$ as well as the set of all individual constants of the signature. For every state atom $stateAt$ GRAPH i $ECL(\boldsymbol{z})$ in $\phi(\boldsymbol{a}_{wh}, \boldsymbol{y})$ with free variables $\boldsymbol{z}$ having length l, say, introduce an $(l + 1)$-ary symbol R and replace GRAPH i $ECL(\boldsymbol{z})$ by $R(\boldsymbol{z}, i)$. The denotation of R in $\mathcal{I}_t$ is then defined as the set of certain answers of the embedded condition $ECL(\boldsymbol{z})$ w.r.t. the ith ABox $\mathcal{A}_i$: $R^{\mathcal{I}_t} = \{(\boldsymbol{b}, i) \mid \boldsymbol{b} \in cert(ECL(\boldsymbol{z}), \langle \mathcal{A}_i \cup \mathcal{A}_{st}, \mathcal{T} \rangle)\}$. Constants denote themselves in $\mathcal{I}_t$. This fixes a structure $\mathcal{I}_t$ with finite denotations of its relation symbols. The evaluation of the HAVING clause is then nothing more than evaluating the FOL formula (after substitutions) on the structure $\mathcal{I}_t$.

Properties. The separation-based semantics has an immediate consequence for perfect rewritability, which is adapted to the sequenced setting as follows. Let $\mathcal{O} = \langle (\mathcal{A}_i)_{i \in [n]}, \mathcal{T} \rangle$ be a SO. Let the canonical model $DB((\mathcal{A}_i)_{i \in [n]})$ of a sequence of ABoxes be defined as the sequence of minimal Herbrand models $DB(\mathcal{A}_i)$ of the ABoxes $\mathcal{A}_i$. Let QL_1 and QL_2 be two query languages over the same signature of a SO and OL be a language for the sequenced ontologies SO.

Definition 1. QL_1 *allows for* QL_2-*rewriting of query answering w.r.t. the ontology language* OL *iff for all queries* ϕ *in* QL_1 *and TBoxes* $\mathcal{T}$ *in* OL *there exists a query* $\phi_\mathcal{T}$ *in* QL_2 *such that for all* $n \in \mathbb{N}$ *and all sequences of ABoxes* $(\mathcal{A}_i)_{i \in [n]}$ *it holds that:* $cert(\phi, \langle(\mathcal{A}_i)_{i \in [n]}, \mathcal{T}\rangle) = ans(\phi_\mathcal{T}, DB((\mathcal{A}_i)_{i \in [n]}))$.

We call such an ECL language *rewritability closed* w.r.t. OL iff ECL allows for perfect rewriting s.t. the rewritten formula is again an ECL condition.

Proposition 1. *Let ECL be a rewritability-closed condition language and consider the instantiation of* HAVING *called* QL_1. *Then* QL_1 *allows for* QL_1 *rewriting for separation-based certain query answering w.r.t. OL.*

4   Separation-Based versus Holistic Semantics

We define a new, *holistic* semantics STARQL HAVING clauses and show that for a fragment of the HAVING clauses we get the same answers as w.r.t. the separation-based semantics. This allows to compare STARQL's expressivity with temporal-logic oriented query languages. In particular, we show that the HAVING fragment captures (a safe fragment) of the LTL inspired query language of temporal conjunctive queries (TCQs) [6]. With every time point t an SO $\langle(\mathcal{A}_i)_{i \in [n_t]}, \mathcal{T}\rangle$ is associated. The length of the sequence n_t may depend on the time point t. As we now fix it, we write just n for n_t.

Definition 2. *Let* $\mathcal{O} = \langle(\mathcal{A}_i)_{i \in [n_t]}, \mathcal{T}\rangle$ *be a SO. Let* $\hat{\mathcal{I}} = (\mathcal{I}_i)_{0 \leqslant i \leqslant n}$ *be a sequence of interpretations* $\mathcal{I}_i = (\Delta, \cdot^{\mathcal{I}_i})$ *over a fixed non-empty domain* Δ *where the constants' interpretations do not change for different* $\mathcal{I}_i, \mathcal{I}_j$. *Then* $\hat{\mathcal{I}}$ *is a model of* $\mathcal{O}$ *(written* $\hat{\mathcal{I}} \models \mathcal{O}$*) if* $\mathcal{I}_i \models \langle\mathcal{A}_i, \mathcal{T}\rangle$ *for all* $i \in [n]$.

All interpretations $\mathcal{I}_i$ have the same domain Δ. The constants' denotations does not change from state to state, hence they are considered rigid.

The satisfaction relation between sequences of interpretations Boolean HAVING clauses is defined as follows.

Definition 3. *Let* $\hat{\mathcal{I}} = (\mathcal{I}_i)_{i \in [n]}$ *be a sequence of interpretations and* σ *be an assignment of individuals* $d \in \Delta$ *to individual variables and numbers* $i \in [n]$ *to state variables* i, j. *Let* $\sigma[j \mapsto \underline{j}]$ *be the variant of* σ *where* j *is assigned* $\underline{j}$. *The semantics are defined as follows:* $\hat{\mathcal{I}}, \sigma \models$ GRAPH $i\,\alpha$ *iff* $\mathcal{I}_{\sigma(i)} \models \alpha$; $\hat{\mathcal{I}}, \sigma \models$ EXISTS $i\,\phi$ *iff* $\exists \underline{i} \in [n].\hat{\mathcal{I}}, \sigma[i \mapsto \underline{i}] \models \phi$; $\hat{\mathcal{I}}, \sigma \models$ FORALL $i\,\phi$ *iff* $\forall \underline{i} \in [n].\hat{\mathcal{I}}, \sigma[i \mapsto \underline{i}] \models \phi$; $\hat{\mathcal{I}}, \sigma \models$ EXISTS $x\,\phi$ *iff* $\exists d \in \Delta.\hat{\mathcal{I}}, \sigma[x \mapsto d] \models \phi$; $\hat{\mathcal{I}}, \sigma \models$ FORALL $x\,\phi$ *iff* $\forall d \in \Delta\hat{\mathcal{I}}, \sigma[x \mapsto d] \models \phi$; $\hat{\mathcal{I}}, \sigma \models \phi_1$ AND ϕ_2 *iff* $\hat{\mathcal{I}}, \sigma \models \phi_1$ *and* $\hat{\mathcal{I}}, \sigma \models \phi_2$; $\hat{\mathcal{I}}, \sigma \models \phi_1$ OR ϕ_2 *iff* $\hat{\mathcal{I}}, \sigma \models \phi_1$ *or* $\hat{\mathcal{I}}, \sigma \models \phi_2$; $\hat{\mathcal{I}}, \sigma \models \phi_{arAt}$ *iff* $\mathcal{I}_0, \sigma \models \phi_{arAt}$; $\hat{\mathcal{I}} \models \phi$ *iff* $\hat{\mathcal{I}}, \phi \models \phi$. *For a* HAVING *clause* $\phi(x)$ *and a sequence of interpretations* $\hat{\mathcal{I}}$, *the set of answers is* $ans(\phi(x), \hat{\mathcal{I}}) = \{a \mid \hat{\mathcal{I}} \models \phi(a)\}$. *The set of certain answers w.r.t. a SO* $\mathcal{O} = \langle(\mathcal{A}_i)_{i \in [n]}, \mathcal{T}\rangle$ *is:* $cert_h(\phi, \mathcal{O}) = \bigcap_{\hat{\mathcal{I}} \models \mathcal{O}} ans(\phi, \hat{\mathcal{I}})$.

We consider a fragment of the HAVING clauses for OL = DL-Lite and ECL = UCQ and denote it by $\mathcal{L}^\exists_{HCL}$: We disallow the operators FORALL x and EXISTS x. The implicit existential quantifiers in the UCQs are kept.

Theorem 1. *For any SO $\mathcal{O} = \langle (A_i)_{i \in [n]}, \mathcal{T} \rangle$ and any $\phi \in \mathcal{L}_{HCL}^{\exists}$ the following equality holds: $cert_h(\phi, \mathcal{O}) = cert_{sep}(\phi, \mathcal{O})$.*

As a corollary we get rewritability for STARQL queries in the holistic semantics.

Theorem 2. *Let QL_1 be the instantiation of the HAVING clause language with $ECL = UCQ$ and $OL = DL\text{-}Lite$. Then QL_1 allows for QL_1 rewriting for holistic certain query answering w.r.t. OL on the ABox sequence.*

5 Comparison with TCQs

The fragment of the HAVING clause language, for which we showed the equivalence of the holistic and the separation-based semantics, is still expressive enough to simulate query languages that combine temporal logic operators with lightweight DL languages such as the language of temporal conjunctive queries (TCQs) [6].

TCQs are defined by following a weak integration of conjunctive queries (CQs) and a linear temporal logic (LTL) template. The syntax is as follows:

$$\phi \longrightarrow CQ \mid \phi \wedge \phi \mid \phi \vee \phi \mid \bigcirc \phi \mid \bullet \phi \mid \bigcirc^- \phi \mid \bullet^- \phi \mid \phi \, \mathsf{U} \, \phi \mid \phi \, \mathsf{S} \, \phi$$

For the LTL like semantics confer [6] or the longer version of the paper. As the TCQs are not domain independent, we consider the following safe fragment TCQ^s where the rule for $\vee$ is replaced by: If ϕ_1, ϕ_2 are in TCQ^s and have the same free variables, then $\phi_1 \vee \phi_2$ is in TCQ^s; $\bullet$, $\bullet^-$ are disallowed; the rules for S, U are replaced by: If ϕ_1, ϕ_2 are in TCQ^s and have the same free variables, then $\phi_1 \, \mathsf{U} \, \phi_2$ and $\phi_1 \, \mathsf{S} \, \phi_2$ are in TCQ^s. TCQ^ss are embeddable into STARQL.

Proposition 2. *For all SOs $\mathcal{O}$ and TCQ^s ϕ: $cert(\phi, \mathcal{O}) = cert_h(\theta(\phi), \mathcal{O})$*

Due to some unsafe operators in TCQs, STARQL embeds not all TCQs. This is an intended feature of STARQL as it was intended to be applicable for strict OBDA scenarios with safe query languages s.a. SQL. On the other hand, STARQL is more expressive than the safe fragment of TCQs: it offers (different) means to generate ABox sequences, whereas for TCQs it is assumed that these are given in advance. Moreover, TCQs allow for quantifiers only within embedded CQs, but cannot handle outer quantification, which is needed in order to, e.g., express the monotonicity condition as in the example from the beginning.

6 Conclusion

The STARQL query language framework is part of the recent venture of adapting OBDA for stream-temporal reasoning and as such is a candidate for performant high-level stream processing within rational agents. In this paper we extended the picture on the different approaches by positioning STARQL w.r.t. the language of TCQs, which was possible by defining a temporal-logic oriented semantics for STARQL in addition to the original semantics.

References

1. Arasu, A., Babu, S., Widom, J.: The CQL continuous query language: semantic foundations and query execution. VLDB J. **15**, 121–142 (2006)
2. Artale, A., Kontchakov, R., Wolter, F., Zakharyaschev, M.: Temporal description logic for ontology-based data access. In: IJCAI 2013, pp. 711–717 (2013)
3. Avron, A.: Constructibility and decidability versus domain independence and absoluteness. Theor. Comput. Sci. **394**(3), 144–158 (2008)
4. Baader, F., Bauer, A., Baumgartner, P., Cregan, A., Gabaldon, A., Ji, K., Lee, K., Rajaratnam, D., Schwitter, R.: A novel architecture for situation awareness systems. In: Giese, M., Waaler, A. (eds.) TABLEAUX 2009. LNCS, vol. 5607, pp. 77–92. Springer, Heidelberg (2009)
5. Baader, F., Borgwardt, S., Lippmann, M.: Temporalizing ontology-based data access. In: Bonacina, M.P. (ed.) CADE 2013. LNCS, vol. 7898, pp. 330–344. Springer, Heidelberg (2013)
6. Borgwardt, S., Lippmann, M., Thost, V.: Temporal query answering in the description logic DL-lite. In: Fontaine, P., Ringeissen, C., Schmidt, R.A. (eds.) FroCoS 2013. LNCS, vol. 8152, pp. 165–180. Springer, Heidelberg (2013)
7. Calbimonte, J.P., Jeung, H., Corcho, O., Aberer, K.: Enabling query technologies for the semantic sensor web. Int. J. Semant. Web Inf. Syst. **8**(1), 43–63 (2012)
8. Calvanese, D., De Giacomo, G., Lembo, D., Lenzerini, M., Poggi, A., Rodriguez-Muro, M., Rosati, R.: Ontologies and databases: the DL-lite approach. In: Tessaris, S., Franconi, E., Eiter, T., Gutierrez, C., Handschuh, S., Rousset, M.-C., Schmidt, R.A. (eds.) Reasoning Web. LNCS, vol. 5689, pp. 255–356. Springer, Heidelberg (2009)
9. Della Valle, E., Ceri, S., Barbieri, D.F., Braga, D., Campi, A.: A first step towards stream reasoning. In: Domingue, J., Fensel, D., Traverso, P. (eds.) FIS 2008. LNCS, vol. 5468, pp. 72–81. Springer, Heidelberg (2009)
10. Heintz, F., Kvarnström, J., Doherty, P.: Bridging the sense-reasoning gap: Dyknow - stream-based middleware for knowledge processing. Adv. Eng. Inform. **24**(1), 14–26 (2010)
11. Özçep, Ö.L., Möller, R., Neuenstadt, C.: A stream-temporal query language for ontology based data access. In: Lutz, C., Thielscher, M. (eds.) KI 2014. LNCS, vol. 8736, pp. 183–194. Springer, Heidelberg (2014)
12. Le-Phuoc, D., Dao-Tran, M., Xavier Parreira, J., Hauswirth, M.: A Native and adaptive approach for unified processing of linked streams and linked data. In: Aroyo, L., Welty, C., Alani, H., Taylor, J., Bernstein, A., Kagal, L., Noy, N., Blomqvist, E. (eds.) ISWC 2011, Part I. LNCS, vol. 7031, pp. 370–388. Springer, Heidelberg (2011)
13. Russell, S.J., Norvig, P.: Artificial Intelligence - A Modern Approach. Prentice Hall, Egnlewood Cliffs (1995)

Vote Counting as Mathematical Proof

Dirk Pattinson[1] and Carsten Schürmann[2]($\boxtimes$)

[1] The Australian National University, Canberra, Australia
[2] IT University of Copenhagen, Copenhagen, Denmark
carsten@itu.dk

Abstract. Trust in the correctness of an election outcome requires proof of the correctness of vote counting. By formalising particular voting protocols as rules, correctness of vote counting amounts to verifying that all rules have been applied correctly. A proof of the outcome of any particular election then consists of a sequence (or tree) of rule applications and provides an independently checkable certificate of the validity of the result. This reduces the need to trust, or otherwise verify, the correctness of the vote counting software once the certificate has been validated. Using a rule-based formalisation of voting protocols inside a theorem prover, we synthesise vote counting programs that are not only provably correct, but also produce independently verifiable certificates. These programs are generated from a (formal) proof that every initial set of ballots allows to decide the election winner according to a set of given rules.

1 Introduction

In many countries scrutineers observe traditional, paper-based elections. Their role is to verify that the voting scheme is applied correctly, thus establishing trust in the election outcome.

For electronic vote counting, the situation is much more difficult. The components of the vote counting software are usually written in a high-level programming language such as C++, PL/SQL, or Java, and hinges not only on the correct translation of a voting protocol into a chosen programming language, but also on the correct implementation of the protocol and usually on the correctness of library functions used in the implementation of the software. This truly herculean task (e.g. [2]) hinges on

1. the formalisation of the voting protocol in formal logic in a theorem prover
2. the formalisation of the semantics of the chosen programming language
3. a formal correctness proof.

Thus, the trust in the correct counting of votes lies with an extremely small number of highly-trained experts that have the necessary technical skills to affirm that the verification task has been carried out with due diligence. This is in sharp contrast to paper-based elections where trust is established by the presence of election observers.

© Springer International Publishing Switzerland 2015
B. Pfahringer and J. Renz (Eds.): AI 2015, LNAI 9457, pp. 464–475, 2015.
DOI: 10.1007/978-3-319-26350-2_41

When using technology in elections, it becomes much more difficult to observe an election. Computer programs and systems are so complex that additional trust assumptions are required that are difficult to discard: is the software version correct? Was the software properly audited and tested? Was the computer that runs the program compromised? The difficulty with such software systems is the fact that they just delivers results, but no means to independently ascertain their correctness, so that trust in election results requires a careful analysis of the whole computation stack and the tools involved in their determination, in sharp contrast to verifiable voting [4] where no such trust is needed.

This difficulty has been recognised in [7], where *linear logic* [8] is used to specify and analyse for voting schemes and also as a certificate language that allows to validate election results retroactively. Unlike classical logic, linear logic is sensitive to the notion of resource in the form of assumptions that may only be used once. E.g. linear implication $a \Rightarrow b$ expresses that (the resource, e.g. a ballot) is *consumed* to yield b (e.g. an updated tally). This allows us to formulate vote counting algorithms abstractly and concisely, because votes are guaranteed to be counted only once. A proof search algorithm uses the program as input, and while constructing a proof for correct counting, it determines the election result in passing. Checking the correctness of a count corresponds to (machine-) checking the (linear logic) proof generated by the proof search algorithm. This task can be performed by a small proof checker for linear logic (anyone can implement such a checker) that is independent of the particular counting algorithm. In other words, the linear logic proof plays the role of an independently verifiable certificate that attests to the correctness of an individual election count. This de-couples the task of certificate generation and verification.

This paper goes one step further and analyses the process of vote counting from the perspective of mathematical proof. Rather than translating voting protocols into a linear logical formalism, we interpret the voting protocol as a set of (counting) rules that have the same status as proof rules in mathematical logic.

The similarity between vote counting rules and rules used in (mathematical) proof theory is indeed striking. Consider for example the following rule of disjunction (or) introduction and counting a single vote:

$$\frac{\Gamma \vdash A}{\Gamma \vdash A \vee B} \qquad \frac{b \vdash (u \, \S \, [c] \, \S \, u', t)}{b \vdash (u \, \S \, u', t[c \mapsto t(c) + 1)]}$$

In the logical rule on the left, Γ is a set of *assumptions* and the premiss reads as 'A is provable from the assumptions Γ'. The conclusion states that in this case, also the formula $A \vee B$ (A or B) is provable (from the same assumptions). For vote counting (on the right), we read b as the collection of ballots cast (each ballot being a vote for a single candidate), $u \, \S \, [c] \, \S \, u'$ as the list of currently uncounted votes, composed of two lists u and u' and a singleton list $[c]$, and $t : C \to \mathbb{N}$ as the tally, a function from the set C of candidates to the natural numbers, with $t[c \mapsto t(c) + 1]$ denoting the function that maps $c \mapsto t(c) + 1$ and $d \mapsto t(d)$ for $d \neq c$. In words: if we have reached a situation where the uncounted votes contain a (single) vote for candidate c and we have a running tally t, then removing c from this set of uncounted votes, together with the updated tally $t[c \mapsto t(c) + 1]$ is also a correct state of FPTP vote counting.

Formal mathematical proofs are based on axioms, i.e. statements for which no further justification is necessary. These axioms correspond to the initial states of vote counting. Again, for simple FPTP elections, the similarity is striking. Consider e.g. the rules

$$\overline{\Gamma \cup \{A\} \vdash A} \qquad \overline{b \vdash (b, \mathrm{nty})}$$

where $\mathrm{nty} : C \to \mathbb{N}$ is the null tally, i.e. the function $c \mapsto 0$ for all $c \in C$. The logical axiom on the left says that every formula A is a consequence of a set of assumptions containing A. The vote counting axiom on the right stipulates that taking all of b as uncounted, together with a tally that records 0 cast votes for all candidates, is a correct state of vote counting.

In mathematical logic, we accept a statement as provable if we can produce a proof, a tree or sequence of correct rule applications with axioms at the leaves. By analogy, a tree or sequence of correctly applied vote counting rules that determine the outcome of an election is a proof of the correctness of the count. In the same way in which a formal mathematical proof is both machine checkable and independent from the means to produce it, a sequence of vote counting rules can be checked independently from the software by which it was generated.

Compared to the encoding of voting protocols in existing logical formalisms, the use of protocol-specific (proof) rules minimises the gap between the natural language specification of a voting protocol and its formalisation, while retaining machine-checkable certificates: a certificate is nothing more but a sequence (or tree) of rule applications that can be verified for correctness by simply checking that all rules have been applied correctly. This has two advantages. First, such verifiers can be implemented in low level languages with only a few lines of code reducing the need to trust compilers and libraries. Second, the technical skills required to independently produce code that verifies the correctness of certificates do not extend beyond the skills acquired in a first programming course which dramatically increases the pool of (electronic) scrutineers. While the complexity of both correctness checking and counting itself is both polynomial, certificates can be used to precisely pinpoint *where* a count went wrong, and the rule-based formulation of protocols themselves serves as mathematical specification.

This paper exemplifies this approach using two voting protocols, first-past-the-post (FPTP) elections, and single transferable vote (STV). To obtain a provably correct vote counting program, we formalise the rules that describe both formalisms in the theorem prover Coq [3] and define a type of proofs, where a proof is a sequence of correctly applied vote counting rules. We use FPTP and STV as examples to enable comparison with [7]. In contrast to *op. cit.*, the higher expressive power of a general-purpose theorem prover such as Coq (in particular the availability of general arithmetic) allows us to extend our approach to more expressive schemes such as range voting.

We then establish, again in Coq, that there exists both a candidate, or set of candidates that win the election, and a proof (in the above sense) of this fact. This allows us to use Coq's program extraction facility to automatically construct a (provably correct) program that determines winners, and constructs a formal proof of this fact using the counting rules. This proof is represented by a standard data structure in a mainstream programming language (we choose Haskell) and so open to (electronic) scrutiny by means of proof checking programs. Crucially, these proof checking programs are short and easy to implement.

Related Work. We have already mentioned the linear logic formalisation of [7]. While this enables us to use off-the-shelf proof checkers to verify the correctness of an election count, it requires a substantial degree of familiarity with linear logic to ascertain that linear logic specifications indeed reflect the protocol. Trust in the correctness of election outcomes in [7] rests in the trust in the correctness of proof checkers. In our approach, proof checkers can be implemented in main stream programming languages, and thus many more than just a few experts can convince themselves about the correctness of the election result.

The idea of not only computing a result, but also provide an independently checkable trace of all intermediate steps was also advocated in [13] and is orthogonal to Necula's proof carrying code [12] where a program executable is accompanied with guarantees. We also advocate programs that not only provide a result, but also an independently verifiable certificate of its correctness.

The relationship between the representation of voting protocols in linear logic [7] and our approach is as follows. In linear logic, linear contexts are multisets of facts; rules when read in forward-chaining way correspond to multiset rewriting. The Coq formalisation in this paper makes this reasoning explicit: rules are state transformers, and they are implemented as such directly in Coq.

The task of verifying vote counting software has been tackled, with various levels of success, for instance in [2,5]. Both approaches focus on verifying the process used to compute election results, but no certificates are produced. Our trust in the correctness of election outcomes therefore relies on a whole tool chain whose integrity can only be verified by very few experts with in-depth technical knowledge as already outlined above. Here, no such trust is needed. We would accept the proof of the correctness of an election outcome as long as it passes (independent) verification, irrespective of the method used to obtain it. Software that is currently used to compute election results, such as the EVACS system used in the Australian Capital Territory [9], represent the voting protocol in a main-stream programming language. This results in a non-trivial gap between the textual and the formal specification of the voting protocol in use, and does not produce independently verifiable certificates. The correctness of the outcome not only relies on the correctness of the program *per se* but also on the vast amount of third-party libraries used in the implementation.

Our work addresses the auditable correctness of vote counting rather than the overall security of electronic voting protocols and is therefore independent of, but complementary to the verification of security properties of voting schemes, as e.g. carried out in [6].

Coq Proofs and Haskell Code. The Coq source code from which the vote counting code is generated, proof checkers, and examples are available from http://users. cecs.anu.edu.au/~dpattinson/Software/.

Notation. We write $\mathsf{List}(X)$ for the set of lists where elements are drawn from the set X, $[a_0, a_1, \ldots, a_n]$ for the list with elements $a_0, a_1, \ldots, a_n$, and denote a list with first element f and remainder fs by $f\mathord{:}fs$. The length of a list l is written as $\mathsf{len}(l)$ and the concatenation of two lists l and m by $l \, \S \, m$. Otherwise, we use set notation and say, e.g. $c \in l$ to express that c occurs in the list l, i.e. sometimes identify a list with the set of its elements.

2   First Past the Post Vote Counting

We use the FPTP elections as a simple, introductory example. A textual specification of FPTP is, for example, the following.

1. Mark all ballots as being uncounted votes.
2. To count a single vote, pick an uncounted vote, mark it as counted, and increase the candidate's tally by one.
3. If no uncounted votes remain, the candidate with the highest tally will be declared the winner.

This informal specification relies on contextual knowledge (e.g. that the initial tally of each candidate is 0) which needs to be made precise in the mathematical formalisation.

Mathematical Formalisation. For the mathematical formalisation of this protocol in terms of (proof) rules, we fix a set C of candidates and the two judgements

$$b \vdash \mathsf{winner}(c) \quad \text{and} \quad b \vdash \mathsf{state}(\mathsf{u}, \mathsf{t})$$

where $b \in \mathsf{List}(C)$ is the list of votes cast (we identify a vote for a candidate with the actual candidate). The judgement on the left asserts that $c \in C$ wins the election where the list b of ballots have been cast, and the judgement on the right represents a state of counting the ballots b where $u \in \mathsf{List}(C)$ are uncounted votes and $t : C \to \mathbb{N}$ is the running tally. The rules themselves, given in Fig. 1, are direct transcriptions of the protocol above.

The first (premiss-free) rule plays the role of an axiom in formal logical reasoning and bootstraps the process of vote counting. Here nty is the null tally $\mathsf{nty}(c) = 0$. The rule (C1) formalises the counting of a single ballot, where

$$\mathsf{inc}(c, t, t') \equiv t'(c) = t(c) + 1 \land \forall d \in C.(d \neq c \to t'(d) = t(d))$$

expresses that t' is the tally resulting from t by incrementing c's votes by one, and (Dw) declares a candidate with a maximal number of votes as the winner. We ignore ties, i.e. any candidate with a maximal number of votes can be declared winner, but we could break ties by adding one more rule.

$$(Ax)\frac{}{b \vdash \mathsf{state}(b, \mathsf{nty})} \qquad (C1)\frac{b \vdash \mathsf{state}(u_0 \, \text{?}\, [c] \, \text{?}\, u_1, t)}{b \vdash \mathsf{state}(u_0 \, \text{?}\, u_1, t')} \quad (\mathsf{inc}(c, t, t'))$$

$$(Dw)\frac{b \vdash \mathsf{state}([], t)}{b \vdash \mathsf{winner}(c)} \quad (\forall d \in C.\, t(d) \leq t(c))$$

Fig. 1. Proof rules for FPTP counting

Logical Formalisation. To generate a provably correct counting program, we have formalised the rules and judgements in the Coq theorem prover. The formalisation is a simple textual translation from the mathematical representation. For example, the rule (C1) becomes

```
| c1 : forall u0 c u1 nu t nt,        (** count one vote **)
    Pf b (state  (u0 ++[c]++u1, t)) -> (* if we have an uncounted vote for c *)
    inc c t nt ->                      (* and the new tally increments c's votes by one *)
    nu = u0++u1 ->                     (* and the vote has been removed from the uncounted votes *)
    Pf b (state (nu, nt))              (* we continue with new tally and consume the vote for c *)
```

and the auxiliary predicate used in the side condition is formalised below.

```
Definition inc (c:cand) (t: cand -> nat) (nt: cand -> nat) : Prop :=
    (nt c = t c + 1)%nat /\  forall d, d <> c -> nt d = t d.
```

This defines a (dependent inductive) type that represents derivations using the rules described above. The formalisation in Coq guarantees that the rules can only be applied if the side conditions are met.

Certifiably Correct Counting. Based on this definition, we prove a simple sanity theorem (in Coq): all elections have a winner. That is, for all sets b of ballots cast (assuming all ballots are formal), there is a candidate w that wins the election *provably*, and a proof of this fact, a derivation using the rules where all side conditions have been obeyed. This theorem is proved for an arbitrary type cand under the assumption that cand is finite, inhabited, and has decidable equality. These assumptions hold for any type that represents a finite set.

Crucially, we give a *constructive* proof of the theorem above. As consequence, the proof contains enough information to in fact *find* an election winner together with a proof of this fact. We then use Coq's extraction mechanism [10] to generate a vote counting program that (a) delivers the election result together with a proof of this fact, and (b) does this in a provably correct way, i.e. the generated program *provably* generates a correct outcome/proof pair. This program not only determines the winner, but also a derivation of this fact, using the rules above. This derivation can then be independently checked for correctness. We have used the programming language Haskell [11], other possible choices are OCaml and Scheme. This generates a data type for derivations

```
data Pf cand =
    Ax (List cand) (cand -> Nat)
  | C1 (List cand) (cand -> Nat) (Pf cand)
  | Dw cand (Pf cand)
```

```
(ax)-------------------------------------------------------------------------------
     state([Alice, Bob, Bob, Claire, Darren], Alice[0] Bob[0] Claire [0] Darren [0])
(c1)------------------------------------------------------------------------------
     state([Bob, Bob, Claire, Darren], Alice[1] Bob[0] Claire [0] Darren [0])
                         . . .
     state([Darren], Alice[1] Bob[2] Claire [1] Darren [0])
(c1)--------------------------------------------------
     state([], Alice[1] Bob[2] Claire [1] Darren [1])
(dw)-----------
     winner(Bob)
```

Fig. 2. An example proof of the outcome of an FPTP election

and visualising the above data type, we obtain the run displayed in Fig. 2 where ... indicates the omission of (C1)-steps, and we have elided the $b \vdash$ prefix.

While the use of the extraction mechanism guarantees provable correctness of the generated program (we can guarantee both that election winners are computed correctly and correctness of the generated proof), the proof serves as a machine-checkable certificate that can be verified independently.

Proof Checking. It is routine to implement a proof-checker that confirms whether or not a proof of an election outcome, i.e. an element of the data type Pf Cand, represents a correctly formed proof tree whose last judgement declares the claimed winner. Under the proof-as-certificate interpretation, this is a certificate verifier. The certificate verifier is written in a general-purpose programming language (we choose Haskell as proofs are already Haskell data types) and is of the same length as the specification (about ten lines of Haskell code), a programming task that requires modest skill. We argue that the simplicity of certificate checkers entails a substantial gain in the trust in the election outcome once the certificate is checked by a (n ideally large) number of checkers, constructed by different individuals. One such checker is contained in the source code that accompanies this paper.

3   Single Transferable Vote

Several variations of STV voting are in use in various countries around the world (e.g. Malta, Ireland, India and Australia). Every member of the electorate does not vote for a single candidate, but instead ranks the candidates in order of her personal preference. We use a vanilla version of STV to demonstrate our approach. STV is parametrised by the number of available seats and a quota that determines the number of votes a candidate must achieve to be elected. The most commonly used quota is the Droop quota, given by

$$q = \frac{\sharp\text{ballots}}{\sharp\text{seats} + 1} + 1$$

but our development is independent of the particular quota used. STV counting proceeds as follows.

1. if candidate has enough first preference to meet the quota, (s)he is declared elected. Any surplus votes for this candidate are transferred.
2. if all first preference votes are counted, and the number of seats is (strictly) smaller than the number of candidates that are either continuing (have not been eliminated) or elected, a candidate with the least number of first preference votes is eliminated, and her votes are transferred.
3. if a vote is transferred, it is assigned to the next candidate (in preference order) on the ballot.
4. vote counting finishes if either the number of elected candidates is equal to the number of available seats, or if the number of remaining continuing candidates plus the number of elected candidates is less than or equal to the number of available seats.

In this (simple) formulation of STV, we ignore transfer values and ties.

Mathematical Formalisation. As for FPTP, we express the election protocol in terms of (proof) rules that we then formalise in the Coq theorem prover. As before, we have a set C of electable candidates, and represent ballots by (rank-ordered) lists of candidates so that a single ballot is of type $B = \mathsf{List}(C)$. The formalisation uses two judgements:

$$(b, q, s) \vdash \mathsf{state}(u, a, t, h, d) \quad \text{and} \quad (b, q, s) \vdash \mathsf{winners}(w).$$

Both judgements are parameterised by a triple (b, q, s) where $b \in \mathsf{List}(B)$ is the list of ballots cast, $q \in \mathbb{N}$ is the quota used and $s \in \mathbb{N}$ is the total number of seats available. The judgement on the left represents an intermediate state of vote counting, where $u \in \mathsf{List}(B)$ is the list of uncounted ballots, $a : C \to \mathsf{List}(B)$ is an assignment that records, for each candidate $c \in C$, the votes $a(c)$ with first preference c that have been counted in c's favour (the first preference of each ballot $b \in a(c)$ is c). The remaining components are the tally $t : C \to \mathbb{N}$ that records, for each candidate, the number of first preferences already counted in favour of c, the list h of continuing (hopeful) candidates that are still in the running, and e is the list of elected candidates. The judgement on the right above asserts that w is the list of election winners. We describe the protocol above by the rules given in Fig. 3. Our formalisation deliberately does not enforce any particular order of rule applications (this could be encoded using side conditions) so that our results pertain to *any* count that is correct according to the given rules.

The first rule describes the initial state of vote counting where nty is the null tally and nas is the null assignment, that is, $\mathsf{nty}(c) = 0$ and $\mathsf{nas}(c) = []$ for all $c \in C$. The rule (C1) describes the counting of one first preference, where inc is defined as in Sect. 2, and we use

$$\mathsf{eqe}(c, l, l') \equiv \exists l_1, l_2.\, (l = l_1 \mathbin{\S} l_2 \wedge l' = l_1 \mathbin{\S} [c] \mathbin{\S} l_2)$$

to expresses that l and l' are equal, except that l' additionally contains c (at an arbitrary position), equivalently l' contains c and l arises by removing one

$$(\text{Ax})\dfrac{}{(b,q,s) \vdash \mathsf{state}(u,a,t,h,e)} \quad \begin{array}{l} - \ u = b, \ a = \mathsf{nas}, \ t = \mathsf{nty}, \ e = [] \\ - \ h \ \text{pairwise distinct}, \ h = C \end{array}$$

$$(\text{C1})\dfrac{(b,q,s) \vdash \mathsf{state}(u,a,t,h,e)}{(b,q,s) \vdash \mathsf{state}(u',a',t',h,e)} \quad \begin{array}{l} - \ \mathsf{eqe}((f{:}fs),u',u)), \ f \in h, t(f) < q, \\ - \ \mathsf{add}(f,f{:}fs,a,a'), \ \mathsf{inc}(f,t,t') \end{array}$$

$$(\text{EI})\dfrac{(b,q,s) \vdash \mathsf{state}(u,a,t,h,e)}{(b,q,s) \vdash \mathsf{state}(u,a,t,h',e')} \quad \begin{array}{l} - \ c \in h, t(c) = q, \mathsf{len}(e) < s \\ - \ \mathsf{eqe}(c,h',h), \ \mathsf{eqe}(c,e,e') \end{array}$$

$$(\text{Tv})\dfrac{(b,q,s) \vdash \mathsf{state}(u,a,t,h,e)}{(b,q,s) \vdash \mathsf{state}(u',a,t,h,e)} \quad \begin{array}{l} - \ f \notin h \\ - \ \mathsf{repl}((f{:}fs),fs,u,u') \end{array}$$

$$(\text{Ey})\dfrac{(b,q,s) \vdash \mathsf{state}(u,a,t,h,e)}{(b,q,s) \vdash \mathsf{state}(u',a,t,h,e)} \quad - \ \mathsf{eqe}([],u',u)$$

$$(\text{TI})\dfrac{(b,q,s) \vdash \mathsf{state}([],a,t,h,e)}{(b,q,s) \vdash \mathsf{state}(u,a,t,h',e)} \quad \begin{array}{l} - \ \mathsf{len}(e) + \mathsf{len}(h) > s, \ c \in h, \ u = a(c) \\ - \ \forall d \in h.(t(c) \le t(d)), \ \mathsf{eqe}(c,h,h') \end{array}$$

$$(\text{Hw})\dfrac{(b,q,s) \vdash \mathsf{state}(u,a,t,h,e)}{(b,q,s) \vdash \mathsf{winners}(w)} \quad \begin{array}{l} - \ \mathsf{len}(e) + \mathsf{len}(h) \le s \\ - \ w = e \,\mathbin{\S}\, h \end{array}$$

$$(\text{Ew})\dfrac{(b,q,s) \vdash \mathsf{state}(u,a,t,h,e)}{(b,q,s) \vdash \mathsf{winners}(w)} \quad \begin{array}{l} - \ \mathsf{len}(e) = s \\ - \ w = e \end{array}$$

Fig. 3. Proof rules for STV counting

occurrence of c from l. Adding a vote v as counted in favour of c to assignment a is represented by

$$\mathsf{add}(c,v,a,a') \equiv \exists l_1, l_2.(a(c) = l_1 \,\mathbin{\S}\, l_2 \wedge a'(c) = l_1 \,\mathbin{\S}\, [v] \,\mathbin{\S}\, l_2) \wedge$$
$$\forall d \in C.(d \neq c \rightarrow a'(d) = a(d))$$

where a' is the assignment resulting from this operation. The rule (EI) applies once a candidate has reached the quota, (Tv) formalises the transfer of votes (where the first preference has either been eliminated or is already elected). Here, we use

$$\mathsf{repl}(c,d,l,l') \equiv \exists l_1, l_2.\,(l = l_1 \,\mathbin{\S}\, [c] \,\mathbin{\S}\, l_2 \wedge l' = l_1 \,\mathbin{\S}\, [d] \,\mathbin{\S}\, l_2)$$

to say that l' is the list l with one occurrence of c replaced by d. The rule (Ey) discards empty votes (with no remaining preferences), (TI) eliminates a candidate with a minimal number of first preferences in a situation where all first preferences are counted and there are still seats to fill, by removing a candidate with least number of first preferences from the list of continuing candidates, and marking all ballots cast in her favour as uncounted. When counting these votes, preferences will be transferred using (Tv). The termination conditions are (Hw) that asserts that once the number of hopefuls and elected candidates falls below the number of available seats, both hopefuls and elected candidates shall be declared winners, and (Ew) that asserts that the elected candidates win, once their number is equal to the number of available seats. We now express the rules above in the internal logic of Coq.

Logical Formalisation. As before, we formalise the rules and the associated notion of proofs, in a (parameterised, dependent) inductive type which (again) is a matter of changing the syntax, as for FPTP counting. E.g. the rule (Tl) becomes:

```
| tl : forall u a t h nh e c,       (** transfer least **)
   Pf b q s (state ([], a, t, h, e)) -> (* if we have no uncounted votes *)
   length e + length h > s ->       (* and there are still too many candidates *)
   In c h  ->                       (* and candidate c is still hopeful *)
   (forall d, In d h-> t c <= t d) ->  (* but all others have more votes *)
   eqe c nh h ->                    (* and c has been removed from the new list of hopefuls *)
   u = a(c) ->                      (* we transfer c's votes by marking them as uncounted *)
   Pf b q s (state (u, a, t,nh, e))  (* and continue in this new state *)
```

Certifiably Correct Counting. As for FPTP, we do not implement a program, but instead extract a provably correct program from a constructive proof that winners always exist. From this, we generate a program that produces both election winners and a sequence of rule applications that justify this. The output of running this program on a sample election is depicted in Fig. 4 where some intermediate steps, as well as the initial $(b, q, s) \vdash$, have been elided.

For presentation purposes, we only show the first uncounted vote, the remaining uncounted votes are indicated with ellipses, and we elide the assignment of

```
(ax)-------------------------------------------------------------------------
    state([[Ana,Bob], ...], Ana[0] Bob[0] Chris[0] Deb[0], [Ana,Bob,Chris,Deb], [])
(c1)-------------------------------------------------------------------------
    state([[Ana,Chris], ...], Ana[1] Bob[0] Chris[0] Deb[0], [Ana,Bob,Chris,Deb], [])
                     . . .
    state (*, Ana[2] Bob[1] Chris[1] Deb[1], [Bob,Chris,Deb], [Ana])
(tl)-----------------------------------------------------------------
    state([[Deb,Chris], ...], Ana[2] Bob[1] Chris[1] Deb[1], [Bob,Chris], [Ana])
(tv)----
    state([[Chris], ...], Ana[2] Bob[1] Chris[1] Deb[1], [Bob,Chris], [Ana])
(c1)-----------------------------------------------------------------
    state(*, Ana[2] Bob[1] Chris[2] Deb[1], [Bob,Chris], [Ana])
(el)-----------------------------------------------------------
    state(*, Ana[2] Bob[1] Chris[2] Deb[1], [Bob], [Chris,Ana])
(ew)-----------------------------------------------------------
    winners ([Chris,Ana])
```

Fig. 4. An example STV proof

candidates to votes (counted in their favour). More examples are contained in the source code that accompanies this paper.

Proof Checking. In contrast to FPTP, proof checking involves the verification of eight (rather than three) proof rules. Each rule can be verified using at most six lines of Haskell code, leading to a proof checker that is of about the same size as the specification. As proof rules are independent, this only adds quantity (not complexity) to the task of implementing a proof checker. As for FPTP, we claim that the proof checker can be implemented with basic programming skills.

4   Discussion

We have presented an approach for certifiably correct vote counting by formalising election protocols as rules that resemble logical deduction rules in spirit. Our implementations of vote counting are provably correct (as they are obtained from program extraction in a theorem prover) and deliver a machine-checkable certificate of the correctness of the count that can be validated with basic programming skills. For any system to be used in real elections, it needs to be demonstrated that the computed election outcome is in fact consistent with the vote counting method formalised in legislation. This amounts to showing that the execution of the computer code to generate the election result in fact conforms to the textual specification, enshrined in law. Various methods provide different types of evidence.

Open-Source Software such as e.g. open-source EVACS System [9] used e.g. in the Australian Capital Territory, can be scrutinised by any member of the general public. This means that any member of the public can scrutinize the computer *code* that has generated the election outcome, but no evidence of its correct *execution* is available. Moreover, even to ascertain the correctness of computer code is an extremely laborious and skill-intensive task.

Verified Vote Counting such as described e.g. in [2] provides a mathematical proof of program correctness with respect to its mathematical specification. To attest soundness of the overall system, we need to be sure that (a) the logical specification is correct with respect to the legislation, and (b) that the verification is carried out with due diligence, both of which require considerable skill. Even after both are carried out, no evidence of the correct *execution* is available.

Certifiably Correct Vote Counting. The approach presented in [7] goes one step further and eliminates trust in the verification process, as every count comes with a certificate (the linear logic proof) that can be checked independently. Moreover, this can be done with a proof checker that is verifiably correct down to the level of machine instructions [1]. On the other hand, the rather special formalism used (linear logic) requires substantial training and limits the pool of possible scrutineers.

Domain-Specific Certifiable Vote Counting. In the approach presented in this paper, only basic familiarity with mathematical concepts (and no knowledge of logic) is required to ascertain the correctness of the specification of the protocol. The certificates generated by our method can be verified by relatively simple computer programs that can be written and verified by anyone with basic programming skills. Running certificate checkers written in different programming languages running on different hardware back up the computed election result.

Conclusion. Clearly more research into the nature of trust in electronic elections is needed. The approach outlined here generates (independently verifiable) evidence for the correctness of election outcomes. Our formalisation only requires little mathematical knowledge and is very close to a textual specification of the voting protocol. As a consequence, correctness of the specification can be asserted by a significantly larger group of people compared to other approaches (that heavily rely on logical notions). The same applies to the certificates generated: their grounding in mathematically simple structures makes it possible to independently implement certificate checkers with moderate programming skills.

References

1. Appel, A.W., Michael, N.G., Stump, A., Virga, R.: A trustworthy proof checker. J. Autom. Reasoning **31**(3–4), 231–260 (2003)
2. Beckert, B., Goré, R., Schürmann, C., Bormer, T., Wang, J.: Verifying voting schemes. J. Inf. Sec. Appl. **19**(2), 115–129 (2014)
3. Bertot, Y., Castran, P., Huet, G., Paulin-Mohring, C.: Interactive Theorem Proving and Program Development: Coq'Art: The Calculus of Inductive Constructions. Texts in Theoretical Computer Science. Springer, Heidelberg (2004)
4. Chaum, D.: Secret-ballot receipts: true voter-verifiable elections. IEEE Secur. Priv. **2**(1), 38–47 (2004)
5. Cochran, D., Kiniry, J.: Votail: a formally specified and verified ballot counting system for irish PR-STV elections. In: Pre-proceedings of the 1st International Conference on Formal Verification of Object-Oriented Software (FoVeOOS) (2010)
6. Delaune, S., Kremer, S., Ryan, M.: Verifying privacy-type properties of electronic voting protocols: a taster. In: Chaum, D., Jakobsson, M., Rivest, R.L., Ryan, P.Y.A., Benaloh, J., Kutylowski, M., Adida, B. (eds.) Towards Trustworthy Elections. LNCS, vol. 6000, pp. 289–309. Springer, Heidelberg (2010)
7. DeYoung, H., Schürmann, C.: Linear logical voting protocols. In: Kiayias, A., Lipmaa, H. (eds.) VoteID 2011. LNCS, vol. 7187, pp. 53–70. Springer, Heidelberg (2012)
8. Girard, J.: Linear logic. Theor. Comput. Sci. **50**, 1–102 (1987)
9. Software Improvements. Electronic and voting and counting sytems. http://www.softimp.com.au/evacs/index.html (2015). Accessed 12 May 2015
10. Letouzey, P.: A new extraction for Coq. In: Geuvers, H., Wiedijk, F. (eds.) TYPES 2002. LNCS, vol. 2646, pp. 200–219. Springer, Heidelberg (2003)
11. Marlow, S., Peyton Jones, S.: The glasgow haskell compiler. In: The Architecture of Open Source Applications, vol. 2. Lulu (2012)
12. Necula, G.C.: Proof-carrying code. In: Lee, P., Henglein, F., Jones, N.D. (eds.) Proceedings of the POPL 1997, pp. 106–119. ACM Press (1997)
13. Schürmann, C.: Electronic elections: trust through engineering. In: Proceedings of the RE-VOTE 2009, pp. 38–46. IEEE Computer Society (2009)

Answer Presentation with Contextual Information: A Case Study Using Syntactic and Semantic Models

Rivindu Perera[✉] and Parma Nand

School of Computer and Mathematical Sciences, Auckland University of Technology,
Auckland 1010, New Zealand
{rivindu.perera,parma.nand}@aut.ac.nz

Abstract. Answer presentation is a subtask in Question Answering that investigates the ways of presenting an acquired answer to the user in a format that is close to a human generated answer. In this research we explore models to retrieve additional, relevant, contextual information corresponding to a question and present an enriched answer by integrating the additional information as natural language. We investigate the role of Bag of Words (BoW) and Bag of Concepts (BoC) models to retrieve the relevant contextual information. The information source utilized to retrieve the information is a Linked Data resource, DBpedia, which encodes large amounts of knowledge corresponding to Wikipedia in a structured form as triples. The experiments utilizes the QALD question sets consisted of training and testing sets each containing 100 questions. The results from these experiments shows that pragmatic aspects, which are often neglected by BoW (syntactic models) and BoC (semantic models), form a critical part of contextual information selection.

Keywords: Contextual information · Semantic models · Syntactic models · DBpedia

1   Introduction

Question Answering (QA) systems mostly comprise of four steps; question processing, answer search, answer extraction, and answer presentation. These four steps collectively contribute for the overall performance of QA systems. Several studies have examined the first three steps, focusing on delivering correct answers as short statement facts [1–3]. This paper focuses on the last stage of QA systems aiming to present answers as if it was delivered by human being. This involves three aspects; searching for extra relevant contextual information, ranking and selecting it, and presenting it as text similar to human composed text.

Following the recent roadmap proposed by Mendes and Coheur [4], this study presents peripheral contextual information for factoid questions which require factual answers. We study two main factoid question types (single entity and multiple entity) apply various models to retrieve the contextual information.

© Springer International Publishing Switzerland 2015
B. Pfahringer and J. Renz (Eds.): AI 2015, LNAI 9457, pp. 476–483, 2015.
DOI: 10.1007/978-3-319-26350-2_42

The paper examines the performance of syntactic and semantic models to retrieve peripheral contextual information for both aforementioned question types implemented on a generic framework targeting QA systems.

The paper is structured as follows. Section 2 introduces the triple weighting methods that are considered in the research. The section explores the process of adopting various BoW and BoC models to retrieve contextual information to enrich answers. In Sect. 3, we present the experimental framework with results and discuss the findings from the experiment. Section 4 describes the relevant related work. Section 5 concludes the paper with an outlook on future work.

2 Content Selection Using Weighted Triples

This section presents models to rank triples focusing on open domain questions as communicative goals. Our objective is to select a set of triples from a linked data resource (i.e. DBpedia) which can be used to generate a more informative answer for a given question. We investigate the problem from two perspectives; as a Bag of Words (BoW) and as a Bag of Concepts (BoC).

2.1 Problem as a Bag of Words

Token Similarity. In this approach triples are ranked by calculating the cosine similarity between the question/answer and the triple. Both question/answer and triples are tokenized and the cosine similarity was computed using (1).

$$sim_{cosine}(\overrightarrow{Q}, \overrightarrow{T}) = \frac{\overrightarrow{Q} \cdot \overrightarrow{T}}{|\overrightarrow{Q}||\overrightarrow{T}|} = \frac{\Sigma_{i=1}^n Q_i T_i}{\sqrt{\Sigma_{i=1}^n Q_i^2}\sqrt{\Sigma_{i=1}^n T_i^2}} \tag{1}$$

Here, Q and T represent the question and the triple respectively.

Term Frequency – Inverse Document Frequency (TF-IDF). In our problem we considered the triple collection as a document collection and the query was provided as an augmented domain corpus. The TF-IDF is then able to provide a rank to each term (t) present in the triple (T) compared to the rest of the triples. The weight of a triple is the sum of weights assigned to all the terms present in the triple. The TF-IDF takes a document collection and rank each document based on the presence of query terms. The TF-IDF can be explained as follows:

$$TF - IDF(Q, T) = \sum_{i \in Q,T} tf_i.idf_i = \sum_{i \in Q,T} tf_i.log_2\frac{N}{df_i} \tag{2}$$

Where *tf* represents the term frequency, N stands for number of documents in the collection and df is the number of documents with the corresponding term. Q represents the question, however in our experiment we tested the possibility of utilizing a domain corpus instead of the original question or the question with the answer.

Okapi BM25. The Okapi ranking function can be defined as follows:

$$Okapi(Q,T) = \sum_{i \in Q,T} \left[log \frac{N}{df_i} \right] \cdot \frac{(k_1 + 1) \, tf_{i,T}}{k_1 \left((1 - b) + b \left(\frac{L_T}{L_{ave}} \right) \right) + tf_{i,T}} \cdot \frac{(k_3 + 1) \, tf_{i,Q}}{k_3 + tf_{i,Q}}$$

$$(3)$$

Where, L_T and L_{ave} represent the length of the triple and average of length of a triple. The Okapi also uses set of parameters where b is usually set to 0.75 and k_1 and k_3 are ranging between 1.2 and 2.0. The and k_3 can be determined through optimization or can be set to range within 1.2 and 2.0 in the absence of development data.

Residual Inverse Document Frequency (RIDF). The idea behind the RIDF is to find content words based on actual IDF and predicted IDF. The widely used methods to IDF prediction is Poisson and K mixture. Since K mixture fits with term distribution very well, we modelled that lower the residual (between actual IDF and IDF predicted by K mixture), the term tends to be a content term. Given term frequencies in triple collection, predicted IDF can be used to measure the RIDF for a triple as follows:

$$RIDF = \sum_{i \in T} \left(idf_i - \widehat{idf_i} \right) = \sum_{i \in T} \left(idf_i - log \frac{1}{1 - P(0; \lambda_i)} \right) \quad (4)$$

Where λ_i represents the average number of occurrences of term and $P(0; \lambda_i)$ represents the Poisson prediction of df where term will not be found in a document. Therefore, $1 - P(0; \lambda_i)$ can be interpreted as finding at least one term and can be measured using:

$$P(k; \lambda_i) = e^{-\lambda_i} \frac{\lambda_i^k}{k!} \quad (5)$$

Based on the same RIDF concept, we can moderate this to work with term distribution models that fits well with actual df such as K mixture. The definition of the K-mixture is given below.

$$P(k; \lambda_i) = (1 - \alpha)\delta_{k,0} + \frac{\alpha}{\beta + 1} \left(\frac{\beta}{\beta + 1} \right)^k \quad (6)$$

In K-mixture based RIDF we interpreted the deviation from predicated df to make the term as a non-content term.

2.2 Problem as a Bag of Concepts

Latent Semantic Analysis. This method analysed how triples in the collection can be concept wise ranked and retrieved related to the question and answer where triples are represented in a semantic space. To retrieve the triples based on this new representation the question and answer must also be transformed to the latent semantic space. Our initial experiment identified that the transformation of question and answer to latent semantic space cannot perform well for contextual information selection. Due to this fact in the experiment we used the augmented domain corpus as the query.

Corpus Based Log Likelihood Distance. The idea behind the implementation of this method is to identify domain specific concepts compared to the general concepts and rank triples which contain such concepts. For this we employed the domain corpus (see Sect. 2.3) and a general reference corpus (see Sect. 2.4). We utilized the log likelihood distance [5] to measure the importance as mentioned below:

$$w_t = 2 \times \left(\left(f_t^{dom} \times log \left(\frac{f_t^{dom}}{f\_exp_t^{dom}} \right) \right) + \left(f_t^{ref} \times log \left(\frac{f_t^{ref}}{f\_exp_t^{ref}} \right) \right) \right) \quad (7)$$

where, f_t^{dom} and f_t^{ref} represent frequency of term (t) in domain corpus and reference corpus respectively. Expected frequency of a term (t) in domain $(f\_exp_t^{dom})$ and reference corpora $(f\_exp_t^{ref})$ were calculated as follows:

$$f\_exp_t^{dom} = s_{dom} \times \left(\frac{f_t^{dom} + f_t^{ref}}{s_{dom} + s_{ref}} \right) \quad (8)$$

$$f\_exp_t^{ref} = s_{ref} \times \left(\frac{f_t^{dom} + f_t^{ref}}{s_{dom} + s_{ref}} \right) \quad (9)$$

where, s_{dom} and s_{ref} represent total number of tokens in domain corpus and reference corpus respectively. Next, we can calculate the weight of a triple ($\langle$subject, predicate, object$\rangle$) by summing up the weight assigned to each term of the triple.

2.3 Domain Corpus

The domain corpus is a collection of text related to the domain of the question being considered. However, finding a corpus which belongs to the same domain as the question is challenge in its own. To overcome this, we have utilized a unsupervised domain corpus creation based on a web snippet extraction with the input as extracted key phrases from questions and answers.

2.4 Reference Corpus

The reference corpus is an additional resource utilized for the LLD based contextual information selection. In essence, to facilitate the LLD calculation to determine whether a term is important for particular domain, a balanced corpus is needed. The reference corpus represents a balanced corpus which contains text from different genres. We have used the British National Corpus (BNC) as the reference corpus.

2.5   Triple Retrieval

The model employs the Jena RDF framework for the triple retrieval. We have implemented the Java library to query and automatically download necessary RDF files from DBpedia.

2.6   Threshold Based Selection

After associating each triple with calculated weight, we then have to limit the selection based on a particular cut-off point as the threshold (θ). Due to absence of knowledge to measure the θ at this stage, it is considered as a factor that needs to be tuned based on experiments. Further discussion on selecting the θ can be found in Sect. 4.

3   Experimental Framework

3.1   Dataset

We used the QALD-2 training and test datasets and removed the invalid questions. The invalid questions include the questions marked as "out of scope" by dataset providers and questions where DBpedia triples do not exist. Table 1 provides the statistics of the dataset, including the distribution of questions in two different question categories, single entity and multiple entity questions.

We have also built a gold triple collection for each question. These gold triples were selected by analysing community provided answers for the questions in our dataset. Using this gold triples in the evaluation and statistics will be discussed in Sect. 3.2.

3.2   Results and Discussion

The gold standard evaluation method is utilized for the task [6,7]. The training question set is used to measure the threshold (θ) which need to be used as the cut-off point for the ranked triples. The idea behind using this threshold value is that the accurate model should rank all relevant triples higher compared to

Table 1. Statistics related question dataset. Invalid questions are which are marked by dataset providers or questions where for which triples cannot be retrieved from DBpedia

	Training set	Test set
All questions	100	100
Invalid questions	5	10
Single entity questions	47	42
Multiple entity questions	48	48

the irrelevant triples. Therefore, with the increase in θ for an accurate model (a model that rank all relevant triples higher than irrelevant), the precision will remain constant until it starts selecting the irrelevant triples and then it will gradually decrease. The recall will increase with θ and will be constant after it starts selecting irrelevant triples. Therefore, the θ which gives the highest F-score will be the turning point for both precision and recall.

Using the θ identified from training set, we can then test the model using testing dataset. When measuring the θ based on training dataset it is also important to measure the percentage of gold triples from the total triples. This is because if the percentage deviates from mean significantly, then it is hard to find a threshold value that can satisfy the entire question set. A set of statistics related to this calculation is shown in Table 2.

Table 2. Statistics related to the gold triple percentage in total triple collection in training dataset

	μ	σ	Max%	Min%
Single entity type	68.89	4.28	78.79	63.58
Multiple entity type	30.43	3.88	37.06	22.93

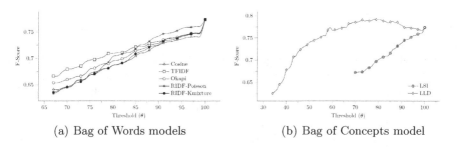

(a) Bag of Words models   (b) Bag of Concepts model

Fig. 1. F-score gained for single entity type questions using Bag of Words models (left) and Bag of Concepts models (right)

According to statistics shown in Table 2 it is clear that there is a possibility to find threshold values for both question sets. Figure 1(a) and (b) depicts the evaluation performed on the single entity question category from training dataset, for both Bag of Words and Bag of Concepts models. Figure 2(a) and (b) depicts the evaluation performed on the multiple entity question category from training dataset, for both Bag of Words and Bag of Concepts models.

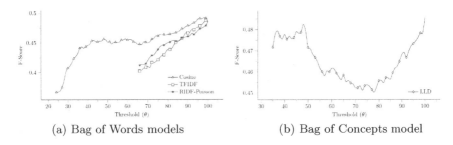

(a) Bag of Words models   (b) Bag of Concepts model

Fig. 2. F-score gained for multiple entity type questions using Bag of Words models (left) and Bag of Concepts models (right)

4   Related Work

Benamara and Dizier [8] present the cooperative question answering approach which generates natural language responses for given questions. In essence, a cooperative QA system moves a few steps further from ordinary question answering systems by providing an explanation of the answer.

Bosma [9] incorporates the summarization as a method of presenting additional information in QA systems. He coins the term, an intensive answer to refer to the answer generated from the system. The process of generating intensive answer is based on summarization using rhetorical structures. Several other summarization based methods for QA such as Demner-Fushman and Lin [10], Yu et al. [11], and Cao et al. [12] also exist with different methods. However, the common drawback that they all shares is the inability to select cohesive information units (e.g., triples).

Vargas-Vera and Motta [13] present an ontology based QA system, AQUA. Although AQUA is primarily aimed at extracting answers from a given ontology, it also contributes to answer presentation by providing an enriched answer. The AQUA system extracts ontology concepts from the entities mentioned in the question and present those concepts in aggregated natural language.

5   Conclusion

This study has examined the role of syntactic and semantic models in contextual information selection for answer presentation. The results of this investigation show that although some semantic models (e.g., LLD) performs well for single entity based questions, in general, pragmatic aspects become more important for this task. However, as of our knowledge this is the first study that investigated the syntactic and semantic models in the contextual information selection to enrich answers as a method of presentation. In future, we expect to extend the work by integrating other possible methods to select contextual information. In addition to these extensions, the contextual information selection will be integrated to our Natural Language Generation (NLG) project [14–16] as the content selection module.

References

1. Perera, R.: IPedagogy: question answering system based on web information clustering. In: T4E 2012 (2012)
2. Perera, R.: Scholar: cognitive computing approach for question answering. Honours thesis, University of Westminster (2012)
3. Perera, R., Nand, P.: Interaction history based answer formulation for question answering. In: Klinov, P., Mouromtsev, D. (eds.) KESW 2014. CCIS, vol. 468, pp. 128–139. Springer, Heidelberg (2014)
4. Mendes, A.C., Coheur, L.: When the answer comes into question in question-answering: survey and open issues. Nat. Lang. Eng. **19**(01), 1–32 (2013)
5. Gelbukh, A., Sidorov, G., Lavin-Villa, E., Chanona-Hernandez, L.: Automatic term extraction using log-likelihood based comparison with general reference corpus. In: Hopfe, C.J., Rezgui, Y., Métais, E., Preece, A., Li, H. (eds.) NLDB 2010. LNCS, vol. 6177, pp. 248–255. Springer, Heidelberg (2010)
6. Perera, R., Nand, P.: Real text-CS - corpus based domain independent content selection model. In: ICTAI 2014, pp. 599–606 (2014)
7. Perera, R., Nand, P.: The role of linked data in content selection. In: Pham, D.-N., Park, S.-B. (eds.) PRICAI 2014. LNCS, vol. 8862, pp. 573–586. Springer, Heidelberg (2014)
8. Benamara, F., Dizier, P.S.: Dynamic generation of cooperative natural language responses in webcoop. In: 9th European Workshop on Natural Language Generation, Budapest, Hungary, ACL (2003)
9. Bosma, W.: Extending answers using discourse structure. In: Recent Advances in Natural Language Processing, Borovets, Bulgaria, Association for Computational Linguistics (2005)
10. Demner-Fushman, D., Lin, J.: Answer extraction, semantic clustering, and extractive summarization for clinical question answering. In: Proceedings of the 21st International Conference on Computational Linguistics and the 44th Annual Meeting of the ACL - ACL 2006, Morristown, NJ, USA, pp. 841–848. Association for Computational Linguistics, July 2006
11. Yu, H., Lee, M., Kaufman, D., Ely, J., Osheroff, J.A., Hripcsak, G., Cimino, J.: Development, implementation, and a cognitive evaluation of a definitional question answering system for physicians. J. Biomed. Inform. **40**, 236–251 (2007)
12. Cao, Y., Liu, F., Simpson, P., Antieau, L., Bennett, A., Cimino, J.J., Ely, J., Yu, H.: AskHERMES: an online question answering system for complex clinical questions. J. Biomed. Inform. **44**, 277–288 (2011)
13. Vargas-Vera, M., Motta, E.: AQUA – ontology-based question answering system. In: Monroy, R., Arroyo-Figueroa, G., Sucar, L.E., Sossa, H. (eds.) MICAI 2004. LNCS (LNAI), vol. 2972, pp. 468–477. Springer, Heidelberg (2004)
14. Perera, R., Nand, P.: A multi-strategy approach for lexicalizing linked open data. In: Gelbukh, A. (ed.) CICLing 2015. LNCS, vol. 9042, pp. 348–363. Springer, Heidelberg (2015)
15. Perera, R., Nand, P.: Realtext-lex: a lexicalization framework for linked open data. In: ISWC 2015 Demonstration (2015)
16. Perera, R., Nand, P.: Generating lexicalization patterns for linked open data. In: Second Workshop on Natural Language Processing and Linked Open Data Collocated with 10th Recent Advances in Natural Language Processing (RANLP) (2015)

A Multi-Agent Approach for Decentralized Voltage Regulation by Considering Distributed Generators

Fenghui Ren[1]([⊠]), Minjie Zhang[1], and Chao Yu[2]

[1] School of Computing and Information Technology, University of Wollongong,
Wollongong, Australia
{fren,minjie}@uow.edu.au
[2] School of Computer Science and Technology, Dalian University of Technology,
Dalian, China
cy496@dlut.edu.cn

Abstract. Distributed generators (DGs) are considered as significant components to modern micro grids in recent years because they can provide instant and renewable electric power to consumers without using transmission networks. However, the use of DGs may affect the use of voltage regulators in a micro grid because the DGs are usually privately owned and cannot be centrally managed. In this paper, an innovative multi-agent approach is proposed to perform automatic and decentralized controls of distributed electric components in micro grids. Autonomous software agents are employed to make local optimal decisions on voltage regulation by considering multiple objectives and local information; and agent-based communication and collaboration are employed toward a global voltage regulation through dynamic task allocation. The proposed approach contains three layers for representing the micro grid, the multi-agent system and the human-computer interface, and is implemented by using three Java-based packages, i.e. InterPSS, JADE and JUNG respectively.

1 Introduction

Distributed Generators (DGs), which emerge as alternative power resources in recent years, are considered as one of the most significant technologies in power grid systems [1]. In general, by comparison with conventional bulk generations, DGs are smaller scale and located closer to loads. There is debate amongst engineers as to the benefits of DGs. On one hand, DGs can supply power to consumers in a micro grid (MG) without needing a transmission network, so as to significantly decrease power loss, voltage drop and cost, and share the loads with bulk generations [1]. On the other hand, most DGs can only provide intermittent power to a MG due to the intermittent nature of energy resources such as wind and sun [11]. Also, an utility usually does not own DGs, and has difficulties in controlling DGs. Therefore, with an increasing level of DGs penetrations, a MG may behave quite differently from conventional operations.

© Springer International Publishing Switzerland 2015
B. Pfahringer and J. Renz (Eds.): AI 2015, LNAI 9457, pp. 484–497, 2015.
DOI: 10.1007/978-3-319-26350-2_43

Maintaining consistent and stable voltage levels in a MG is very important because under-voltage can cause overheating of induction motors, and over-voltage can cause equipments damage [7]. Voltage regulation is a procedure to keep voltages within normal limits, which is usually ±5 % of the rated voltage [2]. Usually, through collecting sensor readings from predefined measurement points, a Load Tap Changer (LTC) or a Voltage Regulator (VR) can estimate the status of a MG, and perform corresponding operations to regulate voltages [3,9]. However, such a regulation mechanism may no longer be suitable due to the connection of DGs because (i) most DGs are privately owned, and will mislead a LTC or VR to perform incorrect operations [1,7]; and (ii) the LTC or VR cannot provide fast enough regulation, which DGs need to ride through voltage drops during emergency conditions [8], and will result in some disconnections of other DGs and more serious voltage drops.

To solve such a problem, extra methods must be considered in order to get a fast regulation. Theoretically, voltage levels are impacted by power delivered through it. If power injected to a MG can be quickly modified, then voltages will be adjusted in a short period accordingly. Conventional bulk generations are impractical due to their large scales, but such a problem does not exist for DGs. Therefore, adjusting DGs power outputs is considered as a matter of course for a fast voltage regulation. However, because of increasing penetration levels, DGs need to collaborate with other devices in order to provide an effective voltage regulation. Therefore, how to efficiently manage DGs to coordinate with other electrical components in voltage regulation by considering the dynamics of a MG and different regulation objectives is a big challenge in power engineering. Conventional approaches fail to solve such a challenge due to their limitations of flexibility, communication, cooperation, and decision making [4].

Several approaches were proposed to address the above challenge in recent years. In [3], voltage regulation problem was formulated as an optimization problem on reactive power dispatching by considering DGs, and was solved through a large amount of calculation. Although technologies, such as distributed computing [5,14], adaptive computing [9] and fuzzy control [13] were employed to increase the efficiency of voltage regulation, the lack of interactions between electrical components still limits dispatching efficiency by considering the dynamics of a MG and the uncertainties of DGs. Agent techniques were employed to search for a solution of efficient voltage regulation through agent-based communication, reasoning and decision making. In [7,10], several Multi-Agent Systems (MASs) for voltage regulation and reactive power dispatching are introduced. However, all the MASs employed a central controller to manage the regulation by using global information. Therefore, such centralized mechanisms will face difficulties in some applications when global information is not available [10,12]. Even through some decentralized MASs were also proposed to overcome such a limitation [6,10], practical issues such as regulation cost and time, communication protocol, and system development were not properly addressed.

In this paper, a novel multi-agent approach is proposed to solve the challenge of voltage regulation. The proposed MAS employs a decentralized organization to dynamically monitor voltage levels of a MG and handles voltage fluctuations by considering the connections of DGs. Multiple objectives and constraints are considered by the proposed agents during distributed voltage regulations, and agents make individual regulation plans through autonomous decision making based on local information and/or collaborations with neighboring agents through communications.

2   Voltage Regulation Considering Distributed Generators

2.1   Principle

In a MG, the measurement of the controlled sensors allows a LTC or VR to estimate voltage fluctuations, and so as to perform adjustments to regulate voltages accordingly. However, because DGs are usually not an utility owned, a LTC or VR's estimations on voltages at problem nodes will be disturbed by the operations from DGs, and the LTC or VR's operations might be incorrect. Also, LTC and VR are slow acting devices and will take about 10 s to adjust about $5\% - 16\%$ of the rated voltage. On the other hand, all DGs are sensitive to voltage fluctuations, and can be automatically disconnected from a MG by the protection circuit within 2 s when a voltage fluctuation is more than $\pm 10\%$ of the nominal voltage [8]. Furthermore, the disconnection of a DG will usually cause a further voltage fluctuation, which may trigger a chain reaction of other DGs disconnections and cascading voltage fluctuations. Therefore, a conventional LTC and VR cannot provide a fast enough voltage regulation to protect DGs in emergency situations. To solve such a problem, extra methods must be considered in order to get fast regulations. Theoretically, voltage levels are impacted by active and reactive power. If power injected to a MG can be quickly modified, then voltage levels will be adjusted in a short period accordingly. Conventionally, a bulk generation is the only power resource, and it is impractical to modify the bulk generation power output in a short period. However, with the connection of DGs, adjusting DGs power outputs is considered as a matter of course for a fast voltage regulation.

Traditionally, all DGs are required to work in a power factor control model [8], where the power factor $(PF = P/Q)$ indicates the ratio between active power output (P) and reactive power output (Q). When DGs work in a power factor control model, a constant PF is maintained. However, if a DG's voltage approaches statutory limits, i.e. V_{min} or V_{max}, the DG can deactivate the power factor control model and regulate its voltage through adjusting its power output. Basically, in order to keep P at a requested level, a DG will increase Q when its voltage drops to the lower threshold V_{min}^{PFC}, so as to increase its voltage. On the other hand, if its voltage reaches its upper threshold V_{max}^{PFC}, the DG will decrease Q, which leads to a decrement of its voltage. Therefore, based on the Jacobian matrix of the Newton power flow [14], the linear relationship between a DG's changes on its power output and voltage is displayed in Formula (1):

$$\Delta V = \Lambda_{VQ} \cdot \Delta Q + \Lambda_{VP} \cdot \Delta P. \tag{1}$$

where ΔP and ΔQ are a DG's changes on active and reactive power, ΔV is DG's corresponding voltage change, and Λ_{VP} and Λ_{VQ} are the correlations between changes of voltage, active and reactive power, respectively.

Usually, in order to minimize impacts to a MG, active power output will not be changed, i.e. $\Delta P = 0$, and a DG will only adjust its reactive power output during a voltage regulation.

2.2 Objectives and Constraints

In this paper, three objectives for a voltage regulation are set by considering DGs, which are:

(i) Time Objective: In order to get a fast regulation on voltage to protect DGs in emergency situations, total time spent on the regulation should be minimized, i.e.

$$\min \sum_i t(\Delta v_i), \tag{2}$$

where $t(\Delta v_i)$ is the time spent on regulating i's voltage.

(ii) Cost Objective: A MG may connect multiple DGs, and costs of the DGs on voltage regulations will also be different by considering their motor types, resources and locations. We also want to minimize the total cost, i.e.

$$\min \sum_i \Delta Q_i \cdot c_i, \tag{3}$$

where c_i is DG i's cost of adjusting a unit reactive power.

(iii) Population Objective: In case multiple voltage fluctuations occur, voltage regulations should recover problem nodes as much as possible to their normal limits, i.e.

$$\max\{v_i 0.85 \ (p.u.) \le v_i \le 1.05 \ (p.u.)\}, \tag{4}$$

The fulfillment of the objectives should not lead to violation of operating other components; hence, several constraints are reinforced.

(i) Current Limit: For each electrical component i, current through it should be not greater than its limit, i.e.

$$\forall i, |I_i| \le |I_i^{max}|. \tag{5}$$

(ii) Voltage Limit: The voltage regulation should not cause any new voltage fluctuation to other components, i.e.

$$\forall i, 0.95 \ (p.u.) \le v_i \le 1.05 \ (p.u.). \tag{6}$$

(iii) Reactive Power Output Limit: An DG's reactive power output should not exceed its surplus capability, i.e.

$$\forall i, |Q_i| \le |Q_i^{max}|. \tag{7}$$

3   A Multi-Agent Approach

3.1   Principle of the Proposed Approach

In order to fulfill the above objectives by considering all requested constraints, a multi-agent approach is introduced in this section. As shown in Fig. 1, the proposed approach contains three layers. First, the power system layer locates in the bottom and presents a MG. In this paper, we consider five key electrical components for voltage regulation purposes, i.e. *substation* (controlling LTC), *feeder* (controlling VR), *busbar*, *load* and *DG*. Second, the multi-agent layer locates in the middle and presents a MAS to dominate the electrical components. Five types of agents are proposed in this layer to control the five identified electrical components correspondingly. Third, the interface layer locates on the top and visualizes the whole system.

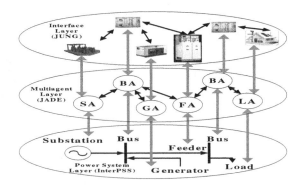

Fig. 1. A three-layer view of the proposed approach.

By comparison with conventional centralized voltage regulations, the proposed approach has the following advantages. (i) A *decentralized management* is employed by the proposed MAS, which means that there is no central controller, and agents work automatically based on information they receive from corresponding electrical components and neighboring agents. (ii) Agents are represented as nodes in a peer-to-peer network, and can communicate with their neighboring agents. Non-adjacent agents can communicate and share information through in-between agents. And (iii) there is no dependency relationship between agents, and the system architecture is extendable. Agents act as a "plug and operate" component. In the following subsections, characteristics of proposed agents will be introduced firstly, then three mechanisms will be introduced to dynamically control the agents in distributed voltage regulation. Finally, implementation of the proposed MAS will be also briefly introduced.

3.2 Agent Design

We propose five agents as follows.

Substation Agent (*SA*): A *SA* represents a secondary substation, and monitors current, voltage and power output of the substation. During a normal operation, the *SA* continuously exchanges information with neighboring agents, and operates a LTC under requests to perform a conventional voltage regulation. The response time and cost of a *SA* are two crucial factors for its neighboring agents to decide whether the *SA* should be requested to involve in a regulation process.

Feeder Agent (*FA*): A *FA* represents a physical feeder which delivers power to downstream components, and monitors current and voltage drop on the feeder through communicating with upstream and downstream agents. A *FA* checks cables transmission abilities to decide whether required power can be delivered. In case a *FA* is requested to join in a voltage regulation process, it will operate corresponding VRs to fulfill the request. Usually, a *FA* can provide a faster regulation than a *SA*, but a slower regulation than a GA. A *FA*'s regulation cost is impacted by the distance between its VRs and problem nodes.

Bus Agent (*BA*): A *BA* represents a physical busbar that conducts power between electrical components. A *BA* records information on connected electrical components, such as current and voltage. During a voltage regulation, a *BA* can make its local decisions on a local regulation plan in order to reach its local objectives. Usually, once a *BA* receives a regulation request from a neighboring agent, the *BA* will firstly search for a local solution by using only local resources. and then will request help from its upstream agents.

Generator Agent (*GA*): A *GA* represents a DG. During normal operations, a *GA* monitors current, voltage and power output of a DG, and maintains the DG's power factor. During a voltage regulation process, a *GA* deactivates the DG's power factor control model and provides voltage supports to a MG through adjusting the DG's reactive power output. Also, a *GA* should ensure that the DG's reactive power output does not exceed its limit. Usually, a DG is ranked by considering its response time, cost and effect on a voltage regulation, and a *GA* also makes individual decisions on how to respond to neighboring agents regulation requests by considering the DG's capacity.

Load Agent (LA): A *LA* represents a load in a MG. A *LA* monitors current and voltage level of the load, and reports to its upstream *BA* once a voltage fluctuation is detected. Each *LA* is assigned a priority to indicate the significance of the load. Usually, a *LA* with a high priority is handled earlier than a *LA* with a low priority during voltage regulation. Once a regulation plan is determined, a *LA* will confirm with its upstream agent for execution.

3.3 Mechanism Design

In order to efficiently manage electrical components to perform distributed voltage regulations by considering the existence of DGs, three novel mechanisms are proposed to control agents and to regulate voltage during three typical operations on electrical components, i.e. the connection, the disconnection, and the voltage fluctuation.

Connection Mechanism. When a new electrical component i needs to be connected to a MG, a corresponding agent a_i will be firstly generated to represent the new component. Let a_i be represented by a seven-tuple $a_i =< AID_i, I_i^{max}, T_i^{max}, Q_i^{max}, V_i^t, C_i^t, P_i^t >$ (where AID_i is a_i's ID, $I_i^{max}, T_i^{max}, Q_i^{max}, V_i^t, C_i^t, P_i^t$ indicates a_i's max current, max regulation time, max reactive power, voltage, regulation cost and priority, respectively), and the nine-tuple $n_{i,j} =< AID_j, I_{i,j}^{max}, Q_{i,j}^{max}, T_{i,j}^{max}, I_{i,j}^t, Q_{i,j}^t, C_{i,j}^t, \Lambda_{i,j}^t, P_i^t >$ be a_i's record on its neighboring agent a_j. Then the connection process is as follows:

Step (i): a_i is created to represent the electrical component i, and is initialized according to component i's features.

Step (ii): a_i sends a connection request with information $< AID_i, I_i^{max}, Q_i^{max}, T_i^{max}, C_i^t, P_i^t >$ to a_j, and waits for a_j's response. If component i cannot provide reactive power, then $Q_i^{max} = 0$, $T_i^{max} = +\inf$, and $C_i^t = +\inf$.

Step (iii): a_j receives a_i's connection request. If the connection is not allowed, a_j denies the request, and the procedure goes to **Step (v)**. Otherwise, the procedure goes to **Step (iv)**.

Step (iv): Firstly, a_j creates a new neighboring agent record according to information sent by a_i, i.e. $n_{j,i} =< AID_i, \min(I_i^{max}, I_j^{max}), Q_i^{max}, T_i^{max}, 0, 0, (C_i^t + L_{j,i}), 0, P_i^t >$ (where $L_{j,i}$ indicates a cost of power loss on a cable between components i and j), and adds $n_{j,i}$ to its neighboring agents set, i.e., $\mathbf{N_j} \leftarrow \{n_{j,i}\} \cap \mathbf{N_j}$. Secondly, a_j informs other existing neighboring agents about its update on reactive power supply, cost and priority by sending $(Q_i^{max}, T_i^{max}, (C_i^t + L_{j,i}), P_i^t)$. Thirdly, a_j's neighboring agents update their records on a_j, i.e., $Q_{k,j}^{max} \leftarrow (Q_{k,j}^{max} + Q_i^{max})$, $T_{k,j}^{max} \leftarrow \min(T_{k,j}^{max}, T_i^{max})$, $C_{k,j}^t \leftarrow \min(C_{k,j}^t, (C_i^t + L_{j,i} + L_{k,j}))$, and $P_{k,j}^t \leftarrow \max(P_{k,j}^t, P_i^t)$. Lastly, a_j's neighboring agents inform their updates to their neighboring agents, and concurrently, a_j replies a_i with an agreement.

Step (v): If a_i receives an agreement from a_j, a_i creates a new neighboring agent record according to information sent by a_j, i.e. $n_{i,j} =< AID_j, \min(I_i^{max}, I_j^{max}), \sum_k Q_{j,k}^{max}, \min(\min_k\{T_{j,k}^{max}\}, T_j^{max}), 0, 0, (\min(\min_k\{C_{j,k}^t\}, C_j^t) + L_{i,j}), 0, \max(\max_k\{P_{j,k}^t\}, P_j^t) >$, and adds $n_{i,j}$ to its neighboring agents set, i.e. $\mathbf{N_i} \leftarrow \{n_{i,j}\} \cap \mathbf{N_j}$. After that, a_i connects to the MG. Otherwise, the procedure is terminated.

Disconnection Mechanism. An existing electrical component may also need to be disconnected from a MG. Suppose that agent a_i wants to disconnect from a MG, and agent a_j is its upstream component, then the disconnection process is given as follows:

Step (i): a_i sends a disconnection request to a_j, and waits for a_j's response.

Step (ii): a_j receives the request, and then activates the *voltage regulation mechanism* to re-dispatch reactive power without considering a_i. If a_j fails to re-allocate reactive power, then the disconnection is not allowed and the procedure goes to **Step (iv)**. Otherwise, the procedure goes to **Step (iii)**.

Step (iii): Firstly, a_j deletes the record of a_i from its neighboring agents set, i.e. $\mathbf{N_j} \leftarrow \mathbf{N_j}/n_{j,i}$. Secondly, a_j informs other existing neighboring agents about its update on reactive power supply, cost and priority by sending (Q_i^{max}, $\min(\min_k\{T_{j,k}^{max}\}, T_j^{max})$, $\min(\min_k\{C_{j,k}^t\}, C_j^t)$, $\max(\max_k\{P_{j,k}^t\}, P_j^t)$) (where $k \in \mathbf{N_j}, k \neq i$). Thirdly, a_j's neighboring agents update their records on a_j, i.e., $Q_{k,j}^{max} \leftarrow (Q_{k,j}^{max} - Q_i^{max})$, $T_{k,j}^{max} \leftarrow \min(\min_k\{T_{j,k}^{max}\}, T_j^{max})$, $C_{k,j}^t \leftarrow \min(\min_k\{C_{j,k}^t\}, C_j^t)$, and $P_{k,j}^t \leftarrow \max(\max_k\{P_{j,k}^t\}, P_j^t)$. Lastly, a_j's neighboring agents inform their updates to their neighboring agents, and concurrently, a_j replies a_i with an agreement on disconnection.

Step (iv): If a_i receives an agreement from a_j, a_i will delete the record of a_j from its neighboring agents set, i.e. $\mathbf{N_i} \leftarrow \mathbf{N_i}/n_{i,j}$, and then a_i disconnects from components j. Otherwise, a_i should keep the connection with a_j, and seeks for another disconnection from the MG in future.

Distributed Voltage Regulation Mechanism. If any voltage fluctuation happens on any electrical component, this mechanism will be activated to regulate voltages by considering all the objectives and constraints mentioned in Subsect. 3.1. Basically, a decentralized design is employed in this mechanism. Agents make local reasoning and decision making on their regulation plans based on their local information, which includes the calculation of regulation solutions, reactive power resource selections, and reactive power dispatching. A recursive strategy is employed during the regulation when multiple agents are involved. The regulation process is introduced as follows.

Step (i): Let a_k be the agent which firstly notices a voltage fluctuation, i.e. its voltage is beyond its limit $\pm 5\%$ (p.u.), and V_k^t be the voltage value. Then a_k firstly calculates the difference between its existing voltage and its target voltage using Formula (8). In this paper, the target voltage is set to 0.85 (p.u.) for any existing voltage lower than 0.85 (p.u.), and is set to 1.05 (p.u.) for any existing voltage higher than 1.05 (p.u.).

$$\Delta V_k^t = \begin{cases} 0.85 - V_k^t, & \text{if } V_k^t < 0.85, \\ 1.05 - V_k^t, & \text{if } V_k^t > 1.05. \end{cases} \tag{8}$$

Step (ii): In order to choose a right adjustment for a voltage regulation, a_k makes a combined consideration on different factors, i.e. regulation speed, cost and effectiveness. Let a_i be a_k's *ith* neighboring agent, and a_k firstly evaluates a_i by using Formula (9).

$$E(a_k, a_i) = \frac{1/T_{k,i}^{max}}{\sum_j 1/T_{k,j}^{max}} \cdot W_k^s + \frac{1/C_{k,i}^t}{\sum_j 1/C_{k,j}^t} \cdot W_k^c + \frac{\Lambda_{k,i}^t}{\sum_j \Lambda_{k,j}^t} \cdot W_k^e, \tag{9}$$

where W_k^s, W_k^c, and W_k^e are a_k's preferences on the speed, cost and effectiveness of the regulation respectively, and $W_k^s + W_k^c + W_k^e = 1$.

Then, a_k ranks all neighboring agents as $\mathbf{N_k^r}$, i.e. $\forall a_i, a_j \in \mathbf{N_k^r}, a_i \geq a_j \Rightarrow E(a_k, a_i) \geq E(a_k, a_j)$. Let a_i be a next agent in $\mathbf{N_k^r}$, then a_k calculates a voltage change that a_i should provide by considering a line's loss as $\Delta V_{k,i}^t = \Delta V_k^t + L_{k,i}$. Also, a_k calculates a possible change on a_i's reactive power output in order to cover $\Delta V_{k,i}^t$ according to Formula (1) under an assumption that $\Delta P = 0$, i.e. $\Delta Q_{k,i}^t = \Delta V_{k,i}^t / \Lambda_{k,i}^t$. If a_k believes that a_i can afford such a modification, i.e. $\Delta Q_{k,i}^t + Q_{k,i}^t \leq Q_{k,i}^{max}$, a_k will send the voltage change request $req_{k,i}^t = \Delta V_{k,i}^t$ to a_i. Otherwise, the voltage change request will be updated by considering a_i's maximum reactive power output as $req_{k,i}^t = \Delta V_{k,i}^{u,t} = \Lambda_{k,i}^t \cdot (Q_{k,i}^{max} - Q_{k,i}^t)$, and leave the remaining voltage change, i.e. $\Delta V_{k,i}^{r,t} = \Delta V_{k,i}^t - \Delta V_{k,i}^{u,t}$, to a next neighboring agent in $\mathbf{N_k^r}$.

Step (iii): Once a_i receives a_k's regulation request, the request will be inserted into a_i's request queue, i.e. $\mathbf{req_i}$, by considering a_k's priority and time when the request was received. Let $req_{k,i}^t$ and $req_{j,i}^t$ be two requests in $\mathbf{req_i}$, then $req_{k,i}^t$ is in front of $req_{j,i}^t$ iff $R(i, req_{k,i}) > R(i, req_{j,i})$, where $R(i, req_{k,i})$ is defined in Formula (10).

$$R(i, req_{k,i}) = \frac{1/(t_k - t_1)}{\sum_k 1/(t_k - t_1)} \cdot W_i^t + \frac{P_{i,k}^t}{\sum_k P_{i,k}^t} \cdot W_i^p, \tag{10}$$

where t_k is time when the request $req_{k,i}^t$ was received, and $P_{i,k}^t$ is a_i's record on a_k's priority. W^t and W^p are a_i's weighting on time and priority, respectively. Each time when a_k receives a new request, queue $\mathbf{req_i}$ will be updated.

Let us assume that a_i already completes all requests in front of $req_{k,i}^t$, and starts to process request $req_{k,i}^t$. If a_i represents an electrical component which can adjust reactive power directly (i.e. a DG, a feeder or a substation), then a_i can make a decision on the request $req_{k,i}^t$ without contacting other agents. In order to do that, a_i firstly calculates its remaining supply ability to a_k as $Q_{i,k}^{r,t} = Q_i^{max} - \sum_k Q_{i,k}^t$, and replies to a_k to indicate the actual amount that a_i can supply, i.e. $rsp_{i,k} = \min(Q_{i,k}^{r,t}, |req_{k,i}^t|)$. However, if a_i cannot adjust reactive power directly, a_i needs to contact its neighboring agents for a_k's request. To do that, a_i needs to employ *voltage regulation mechanism* again by seeking $req_{k,i}^t$ change on its voltage. Obviously, such an recursive procedure will be repeated until an electrical component, which can adjust reactive power directly, is reached.

Step (iv): Suppose that a_i receives a response from a neighboring agent a_j, i.e. $rsp_{j,i}^t$. If a_i's request can be fully satisfied by a_j, i.e. $rsp_{j,i}^t = req_{i,j}^t$, then a_i will respond $rsp_{i,k}^t \leftarrow rep_{j,i}^t$ to a_k directly. Otherwise, a_i will seek for the remaining voltage $\Delta V_{i,m}^{r,t} \leftarrow (\Delta V_{k,i}^t - rsp_{j,i}^t \cdot \Lambda_{i,j}^t)$ from its next neighboring agent by sending an request $req_{i,m}^t = \Delta V_{i,m}^{r,t} / \Lambda_{i,m}^t$. Such a procedure will be repeated until a_i's request is fully satisfied by its neighboring agents or no more neighboring agent can be contacted. Finally, a_i responds to a_k by combing all the responses from neighboring agents, i.e. $rsp_{i,k}^t = \sum_j rsp_{j,i}^t$. Then a_i is ready for executing operations and waits for a_k's confirmation. However, if a_i receives a cancellation request from a_k before operations can be executed, a_i will cancel the regulation and forward the cancellation to related neighboring agents.

Step (v): Once a_k receives a_i's response, a_k will reply to a_i with a confirmation for executing. If a_k's request can be fully satisfied by a_i, i.e. $rsp_{i,k}^t = req_{k,i}^t$, then the regulation is complete. Otherwise, a_k will seek for the remaining voltage change $\Delta V_{k,m}^{r,t} \leftarrow (\Delta V_{k,i}^t - rsp_{i,k}^t \cdot \Lambda_{k,i}^t)$ from its next neighboring agent by sending an request $req_{k,m}^t = \Delta V_{k,m}^{r,t}/\Lambda_{k,m}^t$. Then the steps (ii)–(iv) will be repeated until a_k's original request is fully satisfied by its neighboring agents cumulatively. Because conventional LTC and VR are involved in the procedure and represented by *SAs* or *FAs*, we assume that a_i's original request on voltage change can be satisfied eventually.

Step (vi): a_i receives a_k's confirmation, and forwards the confirmation to related neighboring agents. The agents, which receive the confirmation, start to adjust their reactive power as promised.

3.4 System Development

As shown in Fig. 1, our MAS solution contains three layers and we employ three well-known Java-based packages, i.e. InterPSS (Internet technology based Power System Simulator), JADE (Java Agent Development Framework), and JUNG (Java Universal Network/Graph Framework), for the development of each layer, respectively. InterPSS is an open-source Java-based development project to enhance power system design, analysis, diagnosis and operation. We employ InterPSS for the development of the power system layer. JADE is a free agent development framework, and the communication among agents in JADE is carried out according to FIPA-specified Agent Communication Language (ACL). We employ JADE on top of InterPSS to develop the middle layer to monitor and control electrical components. JUNG is a free software library that provides a common and extendable language for modeling, analysis, and visualization of data that can be represented as a graph or network. We employ JUNE on the top of InterPSS and JADE to visualize the whole system.

4 Simulation

In this section, we demonstrate the performance of the proposed MAS through a case study. In Fig. 2, a MG is firstly output by using InterPSS. The MG contains one substation, two feeders, five buses, six loads, and one generator. The limits of reactive power flow for the substation, buses and feeders are set to 500 MVar. The maximum reactive power supply for the substation is set to 300 MVar, and the MG is also connected to a 100 MVar DG. It is also assumed that the DG's response time on a voltage regulation is much shorter than a LTC or VR, and we set those two response times to 0.1 p.u./sec and 0.02 p.u./sec,

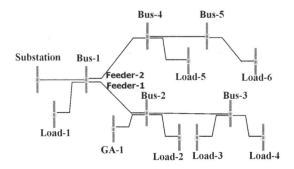

Fig. 2. An InterPSS output showing a power micro grid.

respectively. We set the cost for adjusting 1 MVar as \$20 through a LTC and VR, and as \$10 through a DG. The delivery of 1 MVar through 1 km is assumed to be \$1, and the distance between any two electrical components is assumed to be 1 km. In Fig. 3, the multi-agent simulation of the MG using JADE and JUNE is illustrated. The graph illustrates reactive power dispatching in the MG at a certain moment. Information about reactive power such as direction, amount and price are displayed.

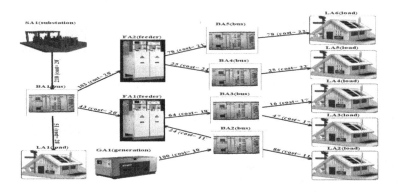

Fig. 3. A multi-agent simulation of a micro grid.

In order to test the proposed mechanisms, another generator, i.e. DG2 (rated at 50 MVar), is proposed to connect the MG through *BA5*. In Fig. 4, communications between agents during DG2's connection, and a voltage regulation through *GA2* are displayed. Explanations are given below.

(**Messages 1-2**): *GA2* sends a request to *BA5* for connection, and *BA5* agrees with the connection. (**Messages 3-16**): *BA5* informs its updates (i.e., limit, cost and sensitivity) to its neighboring agents, i.e. *FA2* and *LA6*. Then *FA2* further informs its neighboring agents, i.e. *BA1* and *BA4*, about its update. Such a procedure is executed by other agents recursively, and eventually all agents

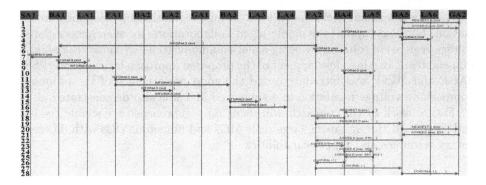

Fig. 4. Communications between agents during component connection and voltage regulation

receive update notices from their neighboring agents. (**Messages 17-20**): *LA5* sends a voltage regulation request to *BA4*, and *BA4* forwards such a request to *FA2*. Because *BA5* already informed *FA2* that a faster, cheaper, and more efficient voltage regulation service can be provide after *GA2*'s connection, through comparison with the voltage regulation service provided by *BA1* (i.e. provided by *SA1* through adjusting LTC actually), *FA2* decides to contact *BA5* firstly, and then *BA5* forwards the request to *GA2*. (**Messages 21-24**): *GA2* agrees with *BA5*'s request to provide a voltage regulation through adjusting its reactive power output. *GA2* replies an agreement to *BA5*'s request, and waits for *BA5*'s confirmation for executing. Then *BA5* forwards the agreement to *FA2*. Eventually, the agreement is received by the original requestor, i.e. *LA5*. All involved agents, i.e. *GA2*, *BA5*, *FA2*, and *BA4*, are waiting for *LA5*'s confirmation for executing. (**Messages 25-28**): *LA5* confirms with *BA4* that it is ready for the execution, and such a confirmation is eventually forwarded to *GA2* through *BA4*, *FA2* and *BA5*. Then *GA2* adjusts its reactive power output, and *LA5*'s voltage is regulated.

The above case study demonstrated that the proposed MAS solution can effectively manage the dynamics of a MG, and perform distributed voltage regulations by using of local information and agent communication. The proposed agents can make decentralized decisions to control corresponding electrical components and perform self-adaptive voltage regulation services. The procedures, i.e. selecting reactive power resources by considering their limits, costs and sensitivities, planing reactive power dispatching by considering the dynamics of neighboring agents, and executing of voltage regulation plans, have demonstrated the good performance of the proposed agents.

5   Conclusion and Future Work

In this paper, a decentralized multi-agent approach for dynamic and distributed voltage regulation by considering the uncertainty of DGs was proposed. The proposed approach not only provides sufficient autonomy for an individual agent to make local optimal decisions on local voltage regulation by using local information, but also supports dynamic agent collaborations for searching a global voltage regulation solution by using agent communication, dynamic task allocation and team forming. development of the proposed approach by using InterPSS, JADE and JUNE was introduced, and the good performance of the proposed approach on voltage regulation in a simulated MG was also demonstrated.

Future work of this research will focus on comprehensive systemic testing and evaluation through using large scale MGs and numerous DGs with different energy resources and supply capabilities.

References

1. Basak, P., Chowdhury, S., Halder nee Dey, S., Chowdhury, S.: A literature review on integration of distributed energy resources in the perspective of control, protection and stability of microgrid. Renew. Sustain. Energy Rev. **16**(8), 5545–5556 (2012)
2. Bolognani, S., Carli, R., Cavraro, G., Zampieri, S.: Distributed reactive power feedback control for voltage regulation and loss minimization. IEEE Trans. Autom. control **60**(4), 966–981 (2015)
3. Deshmukh, S., Natarajan, B., Pahwa, A.: Voltage/VAR control in distribution networks via reactive power injection through distributed generators. IEEE Trans. Smart Grid **3**(3), 1226–1234 (2012)
4. Dou, C.X., Liu, B.: Multi-agent based hierarchical hybrid control for smart microgrid. IEEE Trans. Smart Grid **4**(2), 771–778 (2013)
5. Elmitwally, A., Elsaid, M., Elgamal, M.: Multi-agent-based voltage stabilization scheme considering load model effect. Int. J. Electr. Power Energy Syst. **55**, 225–237 (2014)
6. Fakham, H., Colas, F., Guillaud, X.: Real-time simulation of multi-agent system for decentralized voltage regulation in distribution network. In: IEEE Power and Energy Society General Meeting, pp. 1–7 (2011)
7. Farag, H., El-Saadany, E., Seethapathy, R.: A two ways communication-based distributed control for voltage regulation in smart distribution feeders. IEEE Trans. Smart Grid **3**(1), 271–281 (2012)
8. Hamad, A., Farag, H.E., El-Saadany, E.F., et al.: A novel multiagent control scheme for voltage regulation in DC distribution systems. IEEE Trans. Sustain. Energy **6**(2), 534–545 (2015)
9. Li, H., Li, F., Xu, Y., Rizy, D., Kueck, J.: Adaptive voltage control with distributed energy resources: algorithm, theoretical analysis, simulation, and field test verification. IEEE Trans. on Power Syst. **25**(3), 1638–1647 (2010)
10. Rahman, M., Mahmud, M., Pota, H., Hossain, M.: Distributed multi-agent scheme for reactive power management with renewable energy. Energy Convers. Manag. **88**, 573–581 (2014)
11. Ramchurn, S., Vytelingum, P., Rogers, A., Jennings, N.: Agent-based homeostatic control for green energy in the smart grid. ACM Trans. Intell. Syst. Technol. **2**(4), 35 (2011)
12. Rogers, A., Ramchurn, S., Jennings, N.: Delivering the smart grid: challenges for autonomous agents and multi-agent systems research. In: Proceedings of the 26th AAAI Conference on Artificial Intelligence, pp. 2166–2172 (2012)

13. Spatti, D., da Silva, I., Usida, W., Flauzino, R.: Real-time voltage regulation in power distribution system using fuzzy control. IEEE Trans. Power Deliv. **25**(2), 1112–1123 (2010)
14. Yu, L., Czarkowski, D., de León, F.: Optimal distributed voltage regulation for secondary networks with DGs. IEEE Trans. Smart Grid **3**(2), 959–967 (2012)

Turning Gaming EEG Peripherals into Trainable Brain Computer Interfaces

Manisha Senadeera[1], Frederic Maire[1(✉)], and Andry Rakotonirainy[2]

[1] Science and Engineering Faculty, Queensland University of Technology,
2 George Street, Brisbane 4078, Australia
manisha.senadeera@connect.qut.edu.au, f.maire@qut.edu.au
[2] Faculty of Health, Queensland University of Technology,
2 George Street,
Brisbane 4078, Australia
r.andry@qut.edu.au

Abstract. Companies such as NeuroSky and Emotiv Systems are selling non-medical EEG devices for human computer interaction. These devices are significantly more affordable than their medical counterparts, and are mainly used to measure levels of engagement, focus, relaxation and stress. This information is sought after for marketing research and games. However, these EEG devices have the potential to enable users to interact with their surrounding environment using thoughts only, without activating any muscles. In this paper, we present preliminary results that demonstrate that despite reduced voltage and time sensitivity compared to medical-grade EEG systems, the quality of the signals of the Emotiv EPOC neuroheadset is sufficiently good in allowing discrimination between imaging events. We collected streams of EEG raw data and trained different types of classifiers to discriminate between three states (rest and two imaging events). We achieved a generalisation error of less than 2 % for two types of non-linear classifiers.

Keywords: EEG · Machine learning · Device control · BCI · K-nearest Neighbors · SVM

1 Introduction

Attempts to decipher the brain and interface it with hardware can be traced back to the 1970's when Brain Computer Interface (BCI) research started [18]. BCI devices and methods enable their users to have their brain activity monitored. These devices were in the past confined to the medical arena however in recent years, technological advances have significantly lowered the price of these devices and allowed for the development of non-medical applications [10,11]. An important factor in the usability of EEG devices is their setup and calibration time. New EEG biosensors are easier to operate to the point that they are even used in games to offer novel experience to players [11].

© Springer International Publishing Switzerland 2015
B. Pfahringer and J. Renz (Eds.): AI 2015, LNAI 9457, pp. 498–504, 2015.
DOI: 10.1007/978-3-319-26350-2_44

In this paper, we investigate the capabilities of one such device, the Emotiv EPOC neuroheadset [1]. This headset comes with the ability to extract raw EEG data from 14 sensors, positioned according to the 10/20 International System, as pictured in Fig. 1.

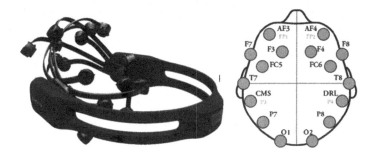

Fig. 1. Left: the Emotiv EPOC headset. Right: the location of its electrodes. The CMS and DRL electrodes act as reference nodes [1].

We tried a variety of classifiers using standard machine learning techniques to determine whether it was possible to discriminate between three different mental states. Section 2 retraces briefly the evolution of BCI technology. Section 3 describes the experimental methodology. Section 4 presents experimental results.

2 Related Work

Although EEG has been primarily used for medical applications such as diagnosing sleep disorders or epilepsy, they have also enabled BCI applications [13,17]. EEG waves are obtained using electrodes attached to the scalp. These sensitive electrodes pick up postsynaptic potentials, created by inhibitory and excitatory potentials in the dendrites of neurons in the cerebral cortex [16]. Berger suggested that the complex EEG was composed of two fundamental waveforms: the larger α waves correlated with mental activity, and the smaller β waves associated with the metabolic activities of cortical tissue [6]. This observation was refined and led to the identification of five frequency ranges [12]. Each range corresponds to a particular state of mind.

The simplest form of BCI is switching. This can be achieved through blinking, or the acts of having the eyes open or closed [14]. EEG-based BCI can also be controlled by means of steady state visual evoked potentials, P300 evoked potentials and motor imagery. Many BCI systems share the architecture shown in Fig. 2 where an EEG data stream is fed to a classifier whose output is then used to generate control commands.

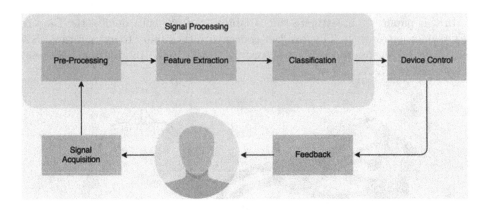

Fig. 2. Architecture of a BCI with feedback loop. The generated output command is displayed to the user as feedback.

P300 Evoked Potentials. P300 brainwaves were first discovered by Sutton in 1965 and are evoked via a visual stimulus that a user concentrates upon while different non-target stimuli are also presented [15]. The P300 signal is evoked 250 to 500 ms after the subject detects the target stimulus among the several non-target stimuli [2,9]. The P300 speller is a device that operates on the P300 evoked potentials. The device presents 36 letters in a 6 x 6 matrix. Each row or column flashes up in turn and the subject concentrates on a particular letter that they wish to write. When the selected letter is lit up, a signal is evoked in the brain and appears in the EEG data [2].

Steady State Visual Evoked Potentials. Steady State Visual Evoked Potentials (SSVEP) are the responses evoked from visual stimulus at different frequencies. The SSVEP characteristic increases activity in the EEG signal at the stimulus frequency. An example of SSVEP applications in BCI are demonstrated by Guneysu and Akin who used the SSVEP induced by observing LED lights flashing at different frequencies to control a humanoid robot to draw a square [3].

Motor Imagery. Motor imagery is the act of mentally imagining a particular action. Imagining the movement of limbs induces significant changes in the cortical area [5]. This in turn results in changes in potentials allowing for measurements to be made with electrodes. A key component of utilising motor imagery for BCI control is training. Within a few days a tetraplegic patient can learn to control a hand orthosis with mental imaging of left and right hand movements as was demonstrated by Pfurtscheller and Neuper [8].

3   Stimulus Classification with the Emotiv EPOC

In order to test the suitability of the Emotiv EPOC headset as a BCI device with multiple control outputs, an experiment was set up to record EEG signals from three subjects (the authors of this paper). Subjects were shown three different types of visual stimuli and data were collected with an Emotiv EPOC headset. Several classifiers were investigated as will be explained in the following sections.

3.1 Experimental Setup

The three tasks the subjects performed were watching a white screen for 10 s, watching a blue cube being pushed away from the subject for 5 s, and watching a blue cube being pulled towards the subject for 5 s. Screen shots of the cube being pushed and pulled are displayed in Fig. 3. The cube animation starts at the same size and then becomes either larger (Pull) or smaller (Push).

Fig. 3. Animation of the Cube Pull

The processing pipeline for exploiting the trained classifiers is illustrated in Fig. 2. Upon receiving the data packets from the Emotiv headset, all raw EEG data for each subject was extracted and loaded into the Python workspace as a Numpy matrix. From these concatenated task data from each subject, information which was not relevant to the study, like the inertial measurement unit (IMU) data, was removed. This created a data matrix with 14 columns with each row corresponding to a time sample (at a rate of 128 samples per second).

4 Experimental Results

All classifiers tested benefited from centering and scaling the data to unit variance. All results reported are on the dataset centered and scaled to unit variance. Scaling was essential for the success of non-linear SVM classifiers. PCA preprocessing did not help.

4.1 Support Vector Machine

There is a parameter C, common to all SVM kernels, that controls how smooth the decision surface between classes is. This parameter trades off misclassification of training examples against the simplicity of the decision surface. The larger C is, the more emphasis is put on classifying all training examples correctly, whereas a smaller C will make the decision surface smoother.

The accuracy results for a SVM with a linear kernel trained on the dataset of Subject B with a 3 fold cross-validation for different values of the regularisation parameter C did not exceed 54%. A random classifier would score around 33% because we have three classes. The same procedure was repeated for the other subjects and led to very similar results. The performance of other classifiers based on linear frontiers like *linear discriminant analysis* and *logistic regression* was slightly worse (around 50%).

Table 1 displays the accuracy for a SVM with a Gaussian kernel trained on the data of Subject B for different values of the regularisation parameter C. The best choice of C is 1000, giving an accuracy rate of 97%.

Table 1. Subject B - SVM (Gaussian Kernel)

	$C = 0.1$	$C = 1.0$	$C = 10$	$C = 100$	$C = 1000$	$C = 10000$
Score 1	0.754766	0.865313	0.926875	0.959609	0.971354	0.967839
Score 2	0.756406	0.865234	0.930547	0.959297	0.972786	0.969531
Score 3	0.752578	0.863047	0.931016	0.960703	0.969661	0.969141

The experiment was repeated for the other subjects and led to the same choice for C. Gaussian SVM's were then trained for all three subjects with a 10 fold cross validation. The average accuracy values were 99.76%, 97.30% and 98.13% for Subject's A, B and C respectively.

4.2   Decision Tree

The decision trees performed better than the linear classifiers, but not as well as the Gaussian SVM. Pruning the trees, and limiting the size of their leaves did not help significantly. Using a 10 fold cross validation, the accuracy values were on average 97% for Subject A, 84.8% for Subject B and 88.9% for Subject C.

4.3   K-Nearest Neighbors

The best performance was obtained with the conceptually simplest classifier. The idea behind the nearest neighbor method is to find a predefined number k of training samples closest in distance to the unlabeled query point, and predict the most common label of these k neighbors as the class label of the query point.

Table 2 shows the accuracy values of different k-Nearest Neighbors classifiers using 10 fold cross validation. We observed that $k = 1$ provides the best results.

5   Discussion

Collecting data and training classifiers can be time consuming. A natural question to ask is whether a classifier trained on one person generalizes well to other

Table 2. k Nearest Neighbors

	k = 1	k = 3	k = 5	k = 7
Subject A Avg	0.99996875	0.9995625	0.99946875	0.9993125
Subject A Std	9.375e-05	0.0004677	0.0004204	0.0004146
Subject B Avg	0.991432292	0.987526042	0.984375	0.98145833
Subject B Std	0.0011076	0.00131478	0.00159259	0.00145833
Subject C Avg	0.99199218	0.98858817	0.9863560	0.98334263
Subject C Std	0.00088277	0.00143063	0.00265419	0.002260045

people. The answer is unfortunately negative. For example, the SVM classifier built from Subject B data achieved an accuracy of 97 % on this person's test data. However when applied to Subject A and Subject C, the accuracy of the same SVM dropped to 26 % and 36 % respectively. That is, the classifier performed as badly as random guessing. This observation leads to the conclusion that trained classifiers are not transferable between subjects.

In the future, we would like also to assess the suitability of the Emotiv EPOC neuroheadset for classifying mental workloads. Our interest in assessing mental workloads stems from the road transport field where it is consistently shown that high mental workloads (e.g. conversation on a mobile phone) distracts drivers and causes crashes. Identifying the most representative objective measures of mental workload whilst driving is a complex and significant area of research in transport as cognitive workload is a function of situation complexity and driving experience [4,7]. Mental workload influences the driving performance and its measure can be used to tailor BCI based interventions which will reduce crashes related to driver distractions.

6   Conclusion

In this paper, we demonstrated that the Emotiv EPOC neuroheadset which was primarily designed for use in entertainment, market research and neurotherapy, is also suitable for BCI control systems. We showed that using the sensor values of its 14 electrodes, we can train classifiers to discriminate between three mental states with high accuracy. The best results were obtained with a 1-nearest neighbor classifier achieving a generalisation error of less than 1 % across all the subjects. Our experiments on the collected EEG data show that non-linear classifiers perform substantially better than linear classifiers with the highest accuracy of a linear classifier being only 54 %. More details are provided in an extended version of this paper available at http://eprints.qut.edu.au/view/person/Maire,_Frederic.html.

In the future, we plan to explore further the capabilities of the device to determine what is the maximum number of classes that can be discriminated at a given level of accuracy. Our preliminary results are encouraging because they show that the difficult classification task of distinguishing between an object moving closer or moving away can be performed with high accuracy.

References

1. Emotiv Systems: Emotiv EPOC / EPOC+ (2014). https://emotiv.com/epoc.php
2. Farwell, L.A., Donchin, E.: Talking off the top of your head: toward a mental prosthesis utilizing event-related brain potentials. Electroen. Clin. Neuro. **70**(6), 510–523 (1988)
3. Guneysu, A., Akin, H.L.: An SSVEP based BCI to control a humanoid robot by using portable EEG device, vol. 2013, pp. 6905–6908. IEEE, United States (2013)
4. Hood, D., Joseph, D., Rakotonirainy, A., Sridharan, S., Fookes, C.: Use of brain computer interface to drive: preliminary results. In: Proceedings of the 4th International Conference on Automotive User Interfaces and Interactive Vehicular Applications, AutomotiveUI 2012, pp. 103–106. ACM (2012). http://doi.acm.org/10.1145/2390256.2390272
5. Jeannerod, M.: The representing brain: neural correlates of motor intention and imagery. Behav. Brain Sci. **17**(2), 187–202 (1994)
6. Millett, D.: Hans berger: from psychic energy to the EEG. Perspect. Biol. Med. **44**(4), 522–542 (2001)
7. Paxion, J., Galy, E., Berthelon, C.: Mental workload and driving. Front. Psychol. **5**, 1344 (2014)
8. Pfurtscheller, G., Neuper, C.: Motor imagery and direct brain-computer communication. P. IEEE **89**(7), 1123–1134 (2001)
9. Polich, J.: Updating P300: an integrative theory of P3a and P3b. Clin. Neurophysiol. **118**(10), 2128–2148 (2007)
10. Prindle, D.: Thoughts into motion: Amazing brain-controlled devices that are already here (2012). http://www.digitaltrends.com/cool-tech/brain-control-the-user-interface-of-the-future/
11. Raajan, N.R., Jayabhavani, G.N.: A smart way to play using brain machine interface (BMI), pp. 1130–1135. IEEE (2013)
12. Sanei, S., Chambers, J.: EEG Signal Processing: Fundamentals of EEG Signal Processing. Wiley, New York (2007)
13. Sheth, D., Benbadis, R.: EEG in common epilepsy syndromes (2014). http://emedicine.medscape.com/article/1138154-overview
14. Singla, R., Chambayil, B., Khosla, A., Santosh, J.: Comparison of SVM and ANN for classification of eye events in EEG. J. Biomed. Sci. Eng. **4**(1), 62–69 (2011)
15. Sutton, S., Braren, M., Zubin, J., John, E.R.: Evoked-potential correlates of stimulus uncertainty. Science **150**(3700), 1187–1188 (1965)
16. Szafir, D., Signorile, R.: An exploration of the utilization of electroencephalography and neural nets to control robots. In: Campos, P., Graham, N., Jorge, J., Nunes, N., Palanque, P., Winckler, M. (eds.) INTERACT 2011, Part IV. LNCS, vol. 6949, pp. 186–194. Springer, Heidelberg (2011)
17. Vidal, J.J.: Toward direct brain-computer communication. Annu. Rev. Biophys. Bio. **2**(1), 157–180 (1973)
18. Wolpaw, J.R., Birbaumer, N., Heetderks, W.J., McFarland, D.J., Peckham, P.H., Schalk, G., Donchin, E., Quatrano, L.A., Robinson, C.J., Vaughan, T.M.: Brain-computer interface technology: a review of the first international meeting. IEEE T. Rehabil. Eng. **8**(2), 164–173 (2000)

Event Classification Using Adaptive Cluster-Based Ensemble Learning of Streaming Sensor Data

Ahmad Shahi, Brendon J. Woodford, and Jeremiah D. Deng$^{(\boxtimes)}$

Department of Information Science, University of Otago, PO Box 56,
9054 Dunedin, New Zealand
jeremiah.deng@otago.ac.nz

Abstract. Sensor data stream mining methods have recently brought significant attention to smart homes research. Through the use of sliding windows on the streaming sensor data, activities can be recognized through the sensor events. However, it remains a challenge to attain real-time activity recognition from the online streaming sensor data. This paper proposes a new event classification method called Adaptive Cluster-Based Ensemble Learning of Streaming sensor data (ACBE-streaming). It contains desirable features such as adaptively windowing sensor events, detecting relevant sensor events using a time decay function, preserving past sensor information in its current window, and forming online clusters of streaming sensor data. The proposed approach improves the representation of streaming sensor-events, learns and recognizes activities in an on-line fashion. Experiments conducted using a real-world smart home dataset for activity recognition have achieved better results than the current approaches.

Keywords: On-line learning · Streaming · Activity recognition · Smart home

1 Introduction

With fast advances in sensor networking and sensor technologies in recent years, smart-home environments have become a research direction that attracts increasing research interests. A smart home that is equipped with different types of sensors assists inhabitants in living comfortably and safely [1,2]. For smart homes to be able to respond to residents' requirements in a context-aware way, activity recognition is a key technique that continues to challenge ongoing research [3–6].

An activity can be regarded as a sequence of activated sensors performed by a resident to achieve a certain goal [2,4]. The sensor data arrive continuously at high speed and sometimes with various sampling rates [7]. It poses challenges for these stream data to be stored and processed in batch mode, and they also need to be processed in an online, responsive manner so that prompt system or human actions are enabled. There is a need of developing algorithms to manage

© Springer International Publishing Switzerland 2015
B. Pfahringer and J. Renz (Eds.): AI 2015, LNAI 9457, pp. 505–516, 2015.
DOI: 10.1007/978-3-319-26350-2_45

and process the data streams with these time and space constraints, and many resort to the approach of employing sliding windows. Generally, there are two schemes to design a window-based model in data stream mining: time-based windows and count-based windows.

In the time-based window approach, several studies have been done in human activity recognition [8–10]. Although this approach is relatively simple and has a low computational complexity during the training phase, it is often difficult to select an ideal window length for the time interval. For example, if the time interval is too small, the window may not be able to cover the relevant activities and allow sensible decision-makings; on the other hand, if the time interval is too extensive, multiple activities may be embedded into it, also raising difficulty in correct classification. To make it worse, this problem manifests itself when dealing with sensors that do not have a constant sampling rate [6].

The count-based approach divides the sensor data sequence into windows each containing an equal number of sensor events, but the window size can vary during the operation [6]. Although this approach offers computational advantages over the explicit segmentation process and does not require future sensor events for classifying past sensor events, it also has inherent shortcomings. For instance, it is hard to cope with the time lags among sensors. If a resident leaves the home and comes back after an hour, the current sensor has a time lag with the preceding sensor event of an hour. On the other hand, a sensor-based approach should wait for a future activated sensor to fulfil the windowing and classify the past sensor data, which turns it into a non-streaming approach [6]. In addition, with the increasing use of sensor technologies in smart homes, there seems to be a growing need for decision making on these systems to be promptly adaptive to newly arriving sensor data [1].

To address these issues, we propose a new approach to process sensor events and conduct event classification using an adaptive ensemble-based method, termed "adaptive cluster-based ensemble learning of streaming sensor data" (ACBEstreaming), which trains the classifier ensemble in real time. The ACBE-streaming model is adaptive and it solves the time lag problem among activated sensors, also considering previous sensor data that might be meaningful in a current window. For this purpose we modify the cluster-based ensemble learning algorithm [2] that would be applicable in on-line learning to recognize different activities.

The remainder of this paper is organized as follows. Section 2 introduces the proposed approach (ACBEstreaming) in detail. Section 3 presents some results and discussion of the proposed approach in comparison with other methods. Lastly, we conclude the paper in Sect. 4, outlining some future work.

2   ACBEstreaming: Our Approach

It has been demonstrated that a trained classifier ensemble can often outperform a single classifier [2, 11, 12]. In this study, we focus on the Cluster-Based Classifier Ensemble (CBCE) method which was proposed by [2], a technique that combines

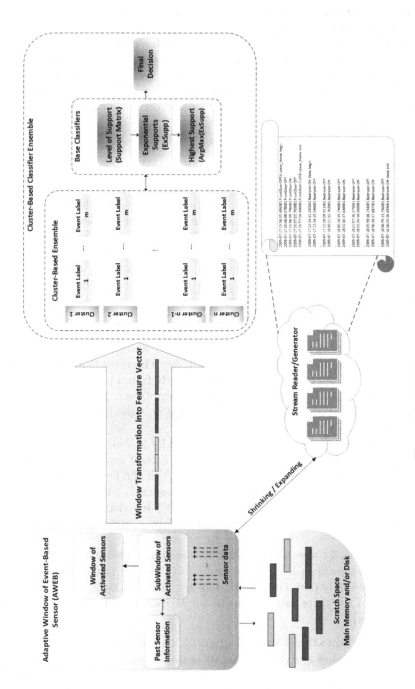

Fig. 1. The ACBEstreaming framework.

clustering and a k-NN based ensemble to reach higher classification accuracy over a number of benchmark datasets. It is however not readily usable for unbounded, streaming sensor data.

This paper adapts the CBCE approach for streaming data use, and introduces two major modifications, namely an adaptive windowing model and an on-line clustering process that can deal with an unbounded sensor streams, and create the adaptive cluster-based ensemble model through on-line learning. Figure 1 shows our ACBEstreaming approach. The details of the proposed method are elaborated in the following sections.

2.1   Adaptive Windowing Model

As indicated earlier, sliding window models have been widely used in existing research of data stream mining. However, there is no universal solution to tackle a data stream in any application domain, and it is often necessary to employ specific criteria for various data streams [4]. This paper proposes an adaptive windowing method for streaming sensor data using three elements: window length adaptation (shrinking and/or expanding the window), a time decay function, and a scheme for accommodating past sensor information. The details of these elements are elaborated in the following text.

Algorithm 1 lists the pseudocode of our adaptive windowing method. Firstly, we initialize the window size ($initWS$) to the average of the minimum and the maximum duration of activities, denoted as $MinSW$ and $MaxSW$ respectively, with SW standing for the activity as a sub-window. The extension size (ext) for window expansion is calculated by deducting $MaxSW$ from $initWS$, and the Past Sensor Information (PSI) pool is assigned to an empty set. From then on, the sensor data are streamed from the dataset and are processed along with PSI if it is not empty.

These streamed sensor data are also evaluated by a *time decay function* for checking distance time of sensors to assign to the correct sensor window (SW). After reading the sensor data ($SW \leftarrow SW \cup \{S_t\}$), if the length of SW is less than or equal to $initWS$ and the activity is recognized, SW is added to the window data matrix (W). Otherwise, if $Size(SW)$ is exhausted, $initWS$ will be expanded by ext, but this is expanded once only, controlled by the $extTag$ boolean variable.

Finally, if an activity is not recognized during reading the sensors and $Size(SW)$ is exhausted after the expansion, the sensor data will be added as to the PSI pool, which stores potentially useful information for the next window.

The details of the three mechanisms of our adaptive windowing method are outlined as follows.

Window-Size Adaptation (Shrinking and Expanding): The shrinking and expanding of the window size from streaming sensor events are based on a sensor-based window scheme rather than the time-based window which was developed in [10]. Due to occurrences of time lags in activities in a smart home, employing

---

Algorithm 1. Adaptive Windowing

Input: Receiving sensor data S_t

Output: A matrix of Window of Sensor vectors (W) with Activity labels (y_i)

1 Set the initial Window Size $(initWS) \leftarrow \frac{MinSW+MaxSW}{2}$;
2 Set $ext \leftarrow MaxSW - initWS$ // extension size;
3 Set $extTag \leftarrow true$ // A control variable to extend window size once;
4 Set Past Sensor Information $(PSI) \leftarrow \varnothing$// an empty set;
5 **while** *active* **do**
6   Streaming Sensor from dataset (S_t);
7   **if** $PSI \neq \varnothing$ **then**
8   Consider PSI with streamed sensor (S_t);
9        $(PSI) \leftarrow \varnothing$;
10   **end**
11   Checking distance time of sensors using time decay function $(P(t_i, t_k))$;
12    $SW \leftarrow SW \cup \{S_t\}$ // Add S_t to SW. SW stands for Sub-Window;
13   **if** $size(SW) \leq initWS$ **then**
14   **if** *activity is recognized* **then**
15   **if** $size(SW) < initWS$ **then**
16   Shrink the SW size;
17   **end**
18          $SW =< attrVec(S_t), ClassLabel(y_t) >$;
19          $W \leftarrow W \cup \{SW\}$;
20   **end**
21   **else if** *(extTag==true)* **then**
22       $initWS \leftarrow initWS + ext$ // Expand the initial window size;
23       $extTag = false$;
24   **else**
25   Set PSI←SW;
26       $(initWS) \leftarrow \frac{MinSW+MaxSW}{2}$;
27   **end**
28 **end**
29 Return W;

---

sensor-based windowing is more reasonable than using a fixed time interval or a fixed count.

Time Decay Function: In a sensor network, there is a possibility that two sensors that are triggered between a long interval but irrelevant to each other are included in the same window. Thus, a time decay function is used to reduce this effect and allow relevant sensor events to be included in a window. The time decay function is given as:

$$P(t_i, t_k) = P_0 \exp(-\lambda(t_i - t_k)) \tag{1}$$

where $P(t_i, t_k)$ is the sensor values at time t_i as decayed from t_k, P_0 is an initial amount at time $t_i = t_k$, λ is a decay rate, and t is the number of sensor periods.

If $\lambda > 1$, it is only the sensor events that are temporally rather close to the last event that contribute to the feature vector. With $0 < \lambda < 1$, the feature vector is under the influence of a temporally wider range of sensor events. When $\lambda = 0$, the temporal distance has no influence on the feature vector, making it a simple aggregation of different sensor events.

Past Sensor Information: In stream data mining, most methods usually take into account the sensor events of the current window, which contains no information about the past sensor events. We argue that past sensor information is sometimes an important factor that may affect the activity in the current window. For instance, '*Enter Home*' is an activity that may occur after a '*Leave Home*' activity. This motivates us to add the past sensor information to the current window in order to enhance the sensor event capturing and allow better activity classification.

2.2   Feature Vector Formation

Feature vectors are created using a fixed dimensionality to represent readings from different sensors. In our experiment using 14 sensors, each sensor reading vector $S_i = <s_1, s_2, \ldots, s_{14}>$ is tagged with a label y_i as the activity class. A collection of S_i and the corresponding y_i tags are fed into the ACBEstreaming algorithm to learn the activity cluster-based ensemble.

2.3   On-line Clustering

The standard k-means clustering was used in some previous work [2]. This is unfortunately not suitable for streaming sensor data which need to be handled in an on-line manner. Here we present a modified online k-means clustering algorithm to cluster streaming sensor data, as described in Algorithm 2.

Algorithm 2. On-line K-means Clustering of Streaming Sensors

Input: Receives the streaming of sensor in W which $W = W \cup SW$

Output: Cluster centres $\{C_i\}$, $i = 1, \cdots, k$, with their relevant membership counts n_i

1 Make initial guesses for the means $C_1, \ldots, C_k$ // k : number of clusters;
2 Set the counts $n_1, \ldots, n_k$ to zero **while** *not interrupted* **do**
3 $\quad$ Acquire the next window W, $W = W \cup SW$;
4 $\quad$ // W stands for the Window and SW stands for the Sub-Window;
5 $\quad$ **if** C_i *is closest to* SW **then**
6 $\quad\quad$ Update centre $C_i = \frac{n_i C_i + m C}{n_i + m}$;
7 $\quad\quad$ Update membership counts: $n_i = n_i + m$;
8 $\quad$ **end**
9 **end**

In the streaming setting, sensor data arrive in windows, with potentially many sensor data vectors per window. The simplest extension of the standard k-means algorithm would begin with initializing the cluster centres. They are usually random locations, because any arriving data have not been seen yet. For each new window of data vectors, the same step operations are performed. The centre that matches closest to the current sub-window will be updated. The centres thus keep the formed clusters in track and adapt to the changes over time as introduced by the arriving new sensor data. We briefly outline the online k-means centre updating rule as follows, annexed by a proof.

Lemma: Let C_i be the winning cluster centre to be updated, n_i be its membership counts thus far; Let C be the new cluster centre from the current window W, and m the number of points added to the cluster in the current window. Then, the new cluster centre can be calculated as

$$C_i' = \frac{n_i C_i + mC}{n_i + m}. \tag{2}$$

Proof: Since C_i is the average of n_i feature vectors (which are denoted as $\{x_s\}$, $s = 1, \cdots, n_i$) that belong to the i-th cluster, we have

$$\sum_{s=1}^{n_i} x_s = n_i C_i \tag{3}$$

For the new batch stream data, there are m vectors, from x_{n_i+1} to x_{n_i+m}, added to cluster i with a temporary mean C, hence

$$\sum_{s=1}^{m} x_{n_i+s} = mC \tag{4}$$

Combining the previous two sums, we obtain the new cluster centre as the mean of *all* data points assigned to the cluster:

$$C' = \frac{\sum\limits_{s=1}^{n_i+m} x_s}{n_i + m}$$

$$= \frac{\sum\limits_{s=1}^{n_i} x_s + \sum\limits_{s=1}^{m} x_{n_i+s}}{n_i + m} \tag{5}$$

$$= \frac{n_i C_i + mC}{n_i + m}$$

∎

After the centre updating, the number of vectors counted to cluster i is updated as $n_i \leftarrow n_i + m$.

2.4   Creating the Base Classifier

Hereafter we present the classification scheme of the ACBEstreaming method, which is largely based on that of [2]. With the ACBEstreaming method, one base classifier is a collection of clusters built on the training sensor dataset with different subsets of features. When a new instance arrives, it is assigned to its closest cluster centre from each collection. Later on, a final prediction is calculated based on the class labels of the instances that belong to the selected clusters. Details follow.

The same as in [2], it is assumed that each cluster supports one or more classes. A cluster provides a degree of support to class c if it contains at least one instance that belongs to this class. The level of support allocated for class c is dependent on the number of instances from class c and the total size of the cluster. In the next step, a support matrix A_k is constructed. This matrix stores the level of the support for each class by each of the clusters in the collection. Each row in the matrix represents a cluster and each column represents a class. The level of support of class j from cluster i is given as:

$$A_k[i,j] = \begin{cases} \dfrac{MN_{ij} - N_i}{MN_i - N_i} & \text{if } N_{ij} - \dfrac{N_i}{M} \geq 0 \\[2ex] \dfrac{MN_{ij} - N_i}{N_i} & \text{otherwise} \end{cases} \qquad (6)$$

where N_i refers to the total number of instances in cluster i, N_{ij} represents the number of instances from class j that belong to cluster i, and M is a number of classes in the classification problem.

For detailed discussion regarding Eq. (6) one can refer to [2]. From the clustering process, it should not be expected that the majority of the instances belong to the same class. Rather, it is however assumed that one cluster can consist of several classes. Based on this assumption, an average number of instances from each class within cluster i is given by N_i/M. This step guarantees that the level of support provided for a class from one cluster is dependent on all instances within this cluster, not only the instances from the class being considered.

Subsequently, the value of level of support ranges between $[-1, 1]$. If the maximum level of support is equal to 1, it means all instances within the cluster belong to the same class. If there are no instances in the cluster for some classes, the level of support will be equal to -1.

2.5   Classifier Outputs

In the classification process, when a new instance has arrived, the cluster with the closest centre based on the Euclidean distance is selected. The level of support is computed for each class based on the value of the selected row and the distance between the new instances and the cluster centres. We sum up the support generated by the new instance on the class c_j from all clusters:

$$\Omega(c_j) = \sum_{k=1}^{K} \begin{cases} \exp\left(\dfrac{A_k[i_k, j]}{1 + d(x, C_k)}\right) & \text{if } A_k[i_k, j] > -1 \\ 0 & \text{otherwise} \end{cases} \tag{7}$$

where $d(x, C_k)$ is the Euclidean distance between the cluster centre C_k and the new instance x. Eventually x will be classified to the class that attains the maximum support as the final decision.

3   Results and Discussion

We now present the experimental results on sensor stream mining using the proposed ACBEstreaming algorithm. As the baseline method [2] can only model sensor data in an off-line mode and is applicable in a static environment, we also compare our proposed approach with other existing online methods.

For performance evaluation, we have conducted experiments on a real-life dataset. The Van Kasteren dataset is collected in the house of a 26-year-old man [9]. The resident lived alone in a three-room apartment where 14 state-change sensors were installed. Similar to the baseline [2], ACBEstreaming is evaluated using the K-fold cross validation method.

The accuracy (which is also called observed accuracy) of total window of streams is calculated as follows:

$$Accuracy = \frac{1}{MK} \sum_{w=1}^{M} \sum_{k=1}^{K} \frac{P_k^w}{N_k^w}, \tag{8}$$

where M is a total number of received window of activities (W), K is the number of folds which is set to 10, P is a number of instances that were classified correctly by the classifier methods, and N is a number of instances in each of k folds. Therefore, for each stream w, the classifiers are evaluated and their accuracy are computed. The total average accuracy will be divided by K fold and finally it is divided again by M, the number of the received windows W. In ACBEstreaming, the sensor data with class labels are streamed in by windows (the default $Size(W)$ is 10).

In addition to accuracy, another performance metrics, the Cohen Kappa [13], is also evaluated and compared with the Naïve Bayes (NB) and K-Nearest Neighbor (KNN) methods implemented in the MOA package [14]. The Kappa statistic is a metric that compares an observed accuracy with an expected accuracy. The Kappa value is used not only to evaluate a single classifier, but also to evaluate classifiers among themselves. In addition, it takes into account random chance (agreement with a random classifier), which generally means it is less misleading than simply using accuracy. The Kappa statistic is computed as follows:

$$Kappa = \frac{Obs\_Acc - Exp\_Acc}{1 - Exp\_Acc}, \tag{9}$$

where the observed accuracy, Obs_Acc, is the proportion of the instances correctly classified to the total number of instances, while the expected accuracy, Exp_Acc, is directly related to the number of instances of each class along with the number of instances that the classifiers agreed with the ground truth labels. In a mathematical form, we have:

$$Exp\_Acc = \frac{1}{MK} \sum_{w=1}^{M} \sum_{k=1}^{K} \sum_{c=1}^{C} \frac{O_{k,c}^{w} P_{k,c}^{w}}{N_k^w N_k^w}, \tag{10}$$

where O is a number of instances labeled as class c according to ground truth, P is a number of instances classified as the same class c by the classifiers. After calculating the average of each class c, the last step is to add all these values together, and divide the sum again by the total number of instances in each fold of received window (N_k^w). Further, to get an average of a K fold, it will be divided by K and finally by the total number of widows, M.

Table 1 shows the accuracy and Kappa evaluation of ACBEstreaming compared with CBCE, Naïve Bayes and K-Nearest Neighbour.

Table 1. Results obtained for ACBEstreaming compared with the state-of-the-art.

Methods	Accuracy (%)	Kappa
CBCE (Baseline)	**97.5**	**0.97**
ACBEstreaming	89.73	0.74
StreamingNaïveBayes	73.68	0.45
StreamingKNN	78.40	0.56

The effectiveness of ACBEstreaming is shown in Table 1 in terms of accuracy and Kappa values. To analyze the accuracy and Kappa, experiments were conducted with the CBCE and ACBEstreaming approaches from sensor data in an off-line and an on-line fashion respectively. Although it is shown from the Table 1 that CBCE method somehow outperforms the proposed method in terms of accuracy and Kappa, the latter has the best performance among online methods when compared with other work [6]. It should be noted that CBCE needs to be used with all training data are available at one scan which is only possible in an off-line mode, while the ensemble in ACBEstreaming is incrementally learnt and adaptively updated from streaming sensor data.

In addition, Table 1 also illustrates the comparison of the ACBEstreaming approach with other two single classifiers, Naïve Bayes (`moa.classifiers.bayes.NaiveBayes`) and K-Nearest Neighbour (`moa.classifiers.lazy.KNN - k 3`) with 'class.moa.tasks.EvaluateModel' (Evaluate a model on a stream) in MOA. We initialized the '$numPasses$' variable in '$LearnModel$' to 10 (i.e. window size W is 10). These settings were chosen empirically to obtain the presented results.

Table 1 shows that our proposed method outperforms the other two methods. Generally, in contrast to single leaning approaches that try to build one learner from training data, ensemble methods train multiple learners to solve the same problem. Therefore, the generalization capability of an ensemble is often much stronger than a single base learner. Indeed, ensemble methods are appealing mainly because they are able to boost weak learners (that may be just slightly better than random guessers) to strong learners capable of making very accurate predictions [12].

4 Conclusion

The key contribution of this paper is to provide a new approach for mining and recognizing activities over streaming sensor data, called ACBEstreaming, which employs adaptive windowing and trains the classifier ensemble through online learning of sensor data streams. The proposed method modifies earlier work on ensemble-based classification and applies it to on-line learning in the context of activity recognition. The ACBEstreaming approach is incrementally learnt and constructed by sensor data streams, therefore promising for finding applications in real-time scenarios. As demonstrated by the experimental results using a real-world smart home dataset, the new method produces comparable or better performance while operating in the online streaming mode.

In the future, we plan to develop a stronger cluster ensemble based model that operates in a autonomous manner for clustering binary sensor data, and in particular is enabled to choose the number of clusters adaptively in an on-line fashion. Another possible direction is to evaluate the performance of the proposed method for identifying possible concept drifts in a data stream.

References

1. Shahi, A., Sulaiman, M.N., Mustapha, N., Perumal, T.: Naive bayesian decision model for the interoperability of heterogeneous systems in an intelligent building environment. Automat. Construct. **54**, 83–92 (2015)
2. Jurek, A., Bi, Y., Shengli, W., Nugent, C.D.: Clustering-based ensembles as an alternative to stacking. IEEE Trans. Knowl. Data Eng. **26**(9), 2120–2137 (2014)
3. Gopalratnam, K., Cook, D.J.: Online sequential prediction via incremental parsing: the active LeZi algorithm. IEEE Intell. Syst. **22**(1), 52–58 (2007)
4. Rashidi, P.: Stream sequence mining for human activity discovery. In: Sukthankar, G., Geib, C., Bui, H., Pynadath, D., Goldman, R.P. (eds.) Plan, Activity, and Intent Recognition. Elsevier (2014)
5. Rashidi, T., Cook, D.J.: Adapting to resident preferences in smart environments. In: Proceedings of the AAAI Workshop on Advances in Preference Handling (2008)
6. Krishnan, N.C., Cook, D.J.: Activity recognition on streaming sensor data. Pervasive Mob. Comput. **10**, 138–154 (2014)
7. Rashid, M.M., Gondal, I., Kamruzzaman, J.: Regularly frequent patterns mining from sensor data stream. In: Lee, M., Hirose, A., Hou, Z.-G., Kil, R.M. (eds.) ICONIP 2013, Part II. LNCS, vol. 8227, pp. 417–424. Springer, Heidelberg (2013)
8. Huỳnh, T., Blanke, U., Schiele, B.: Scalable recognition of daily activities with wearable sensors. In: Hightower, J., Schiele, B., Strang, T. (eds.) LoCA 2007. LNCS, vol. 4718, pp. 50–67. Springer, Heidelberg (2007)

9. van Kasteren, T., Noulas, A., Englebienne, G., Kröse, B.: Accurate activity recognition in a home setting. In: Proceedings of the 10th International Conference on Ubiquitous Computing, UbiComp 2008, pp. 1–9. ACM, New York (2008)

10. Okeyo, G., Chen, L., Wang, H., Sterritt, R.: Dynamic sensor data segmentation for real-time knowledge-driven activity recognition. Pervasive Mob. Comput. **10**, 155–172 (2014)

11. Melville, P., Mooney, R.J.: Creating diversity in ensembles using artificial data. Inf. Fusion **6**(1), 99–111 (2005)

12. Zhou, Z.-H.: Ensemble Methods: Foundations and Algorithms, 1st edn. Chapman & Hall/CRC, Boca Raton (2012)

13. Cohen, J.: A coefficient of agreement for nominal scales. Educ. Psychol. Measur. **20**(1), 37–46 (1960)

14. Bifet, A., Holmes, G., Kirkby, R., Pfahringer, B.: MOA: massive online analysis. J. Mach. Learning Res. **11**, 1601–1604 (2010)

Standoff-Balancing: A Novel Class Imbalance Treatment Method Inspired by Military Strategy

Michael J. Siers[(✉)] and Md Zahidul Islam

School of Computing and Mathematics,
Charles Sturt University, Bathurst, Australia
{msiers,zislam}@csu.edu.au

Abstract. A class imbalanced dataset contains a disproportionate number of a certain class' records compared to other classes. Classifiers which are built from class imbalanced datasets are biased and thus underperform for the minority class. Treatment methods such as sampling and cost-sensitivity can be used to negate the bias induced by class imbalance. In this study, we present an analogy between class imbalance and war. By creating this analogy, we make it possible for military strategies to be applied to class imbalanced datasets. We propose a novel class imbalance treatment method *Standoff-Balancing* which uses a well-known mathematical law from military strategy literature. We compare the proposed technique with four existing techniques on five real world data sets. Our experiments show that the proposed technique may provide a higher AUC to existing techniques.

Keywords: Class imbalance · Classification · Cost-sensitive

1 Introduction

Classification is a task in data mining and machine learning. In classification, a dataset is typically a set of records R. Each record in R contains a set of attributes $A = \{A_1, A_2, \ldots, A_{|A|}\}$. Each attribute A_i in a record R_j has a value. For example, in R_1, A_1 may be 7.3 and A_2 may be *"true"*. In addition to a set of attributes, each record in R also contains a classification c. The classification c is often referred to as the class attribute. Thus, we can express a record as: $R_j = \{A, c\}$, and a dataset as $D = \{R_1, R_2, \ldots, R_{|R|}\}$. Learning algorithms can be applied to a training dataset D_T to train a classifier. During the learning process, a classifier observes the patterns in D_T. Based on these patterns, the classifier may then classify records that do not have class values. The performance of a classifier is dependent on multiple factors. One factor is the quality of D_T. Various issues may negatively affect the quality of D_T. This study addresses the class imbalance issue [8].

Consider the task of training a classifier which classifies patients as either cancerous or non-cancerous. That is, the class attribute's value is either *cancerous* or *non-cancerous*. Assume that D_T contains records of 90 non-cancerous patients but only 10 records of cancerous patients. D_T is considered

© Springer International Publishing Switzerland 2015
B. Pfahringer and J. Renz (Eds.): AI 2015, LNAI 9457, pp. 517–525, 2015.
DOI: 10.1007/978-3-319-26350-2_46

imbalanced because there are many more records of one class than the other class. The classifier trained on D_T will likely be biased towards classifying patients as non-cancerous. This is unacceptable since one or more cancerous patients might forgo life-saving treatment. Class imbalance treatment methods can be used to balance D_T. For example, each record of a cancerous patient could be duplicated, and two thirds of the non-cancerous patients are removed from D_T. This would result in a dataset of 30 cancerous patients and 30 non-cancerous patients. Thus, the resulting classifier will not be biased. Another approach could involve attaching weights to each patient. The weight of a cancerous patient would be higher (9 times) than the weight of a non-cancerous patient. In this study, we create an analogy between the class imbalance problem and war. By creating this analogy, we make it possible for military strategies to be used as class imbalance treatment methods.

Imbalance may also occur in military warfare. That is, consider two forces that are at war with one another. If one is considerably larger than the other, then it is an imbalanced battle. In this study we present an analogy between class imbalance and war. By creating this analogy, we make it possible to apply military strategies to the class imbalance problem. We demonstrate the usefulness of this analogy by proposing a novel class imbalance treatment method *Standoff-Balancing* which is based on a well known military strategy. By doing so, we are applying a technique from a highly developed literature to a comparatively immature problem area. Our experiments demonstrate that the proposed treatment method results in higher performing classifiers than existing treatment methods. We expect that the demonstrated superiority of the proposed method will encourage more research into the incorporation of military strategies in data mining and machine learning.

The rest of the paper is structured as follows. Section 2 introduces the related works on class imbalance treatment methods and the related military strategy. Section 3 first introduces our analogy between class imbalance and war. We present our analogy between class imbalance and war in Sect. 3.1. We describe the steps in the proposed algorithm Standoff-Balancing in Sect. 3.2. Then our proposed method *Standoff-Balancing* is presented. We compare the proposed technique against four existing methods over five datasets in Sect. 4. We provide the concluding remarks in Sect. 5. Following the conclusion, we provide information regarding the availability of the source code for the proposed method.

1.1   Main Contributions

– By presenting an analogy between class imbalance and war, we allow the use of military strategy in class imbalance.
– We propose a class imbalance treatment method based on our extension of a well known military strategy.
– The proposed method's AUC is compared against 4 existing methods on 5 datasets.

2   Previous Work

In Sect. 2.1, we introduce the existing methods for treating class imbalance. The military strategy that we will use in our proposed method is introduced in Sect. 2.2.

2.1   Class Imbalance Treatment Methods

When a training dataset has only two possible class values it is called a binary dataset. When there are more than two possible class values, it is called a multi-class dataset. Typically, treatment methods are designed for two class values. These are usually *negative* and *positive*. To apply these treatment methods to a multi-class dataset, it is common-practice to use a one-vs-all approach [5]. That is, to consider all but one class value to be negative. The remaining class value is considered positive. This converts the multi-class dataset to a binary dataset. Later, in Sect. 4, two of the five datasets (glass0 and ecoli1) in our experiments use the one-vs-all approach. These two datasets are available with the one-vs-all approach already applied.

The class value which is the most frequent in a training dataset D_T is considered the majority class. Similarly, the class value which is the least frequent is considered the minority class. By applying treatment methods to a class imbalanced D_T, the class imbalance problem can be negated. Most treatment methods can be categorized into one or more of the following categories: *Undersampling*, *Oversampling*, *Cost-Sensitive*, and *Ensembling*.

Undersampling methods reduce the size of the majority class. This is accomplished by removing majority records from the training dataset D_T. The simplest way of undersampling is to remove records at random. By doing so, there is no bias during undersampling. However, important records can potentially be lost. A strong focus in undersampling research is how to avoid removing important records from D_T. Training classifiers on undersampled datasets is faster since the dataset is smaller. However, the speed of building a classifier is not always a concern.

The opposite of undersampling is oversampling. Oversampling methods increase the size of the minority class. The simplest way of oversampling is to duplicate minority records at random. By training a classifier on a dataset with duplicated records, the classifier may become too specific to the training data. This leads to poor generalization on future records. SMOTE [2] or Synthetic Minority Oversampling TEchnique, solves this issue by creating new minority records which are between two existing records. For each minority record N_i, one of its nearest neighbours n is randomly chosen from D_T. A new record is then created with values which are randomly placed at some point between N_i and n's values. This process can be repeated multiple times. If the user specifies that $x00\%$ of the original minority size is desired, SMOTE is repeated x times for each N_i. Adasyn [7] improves upon this by calculating how many iterations of SMOTE should be used for each N_i. This is done by considering how many majority records are near. The more majority records surrounding a minority record, the more iterations of SMOTE are used for that minority record. One weakness of Adasyn is that only integer amounts of records may be designated for creation around a record. Class errors may also cause Adasyn to perform poorly.

Classifiers can be trained cost-sensitively. This means that certain records are given more importance over others during the training process. By associating a high cost with a record in D_T, the record is given more importance. In binary classification, a classification is either positive or negative. The classification can then be either true or false. For example, classifying a positive record as negative would be a false positive classification. Cost-sensitive methods often require the use of a cost-matrix. The cost-matrix defines the costs associated with true positive (TP), true negative (TN), false positive (FP), and false negative (FN) classifications. For readability we refer to these costs as C_{TP}, C_{TN}, C_{FP}, and C_{FN} respectively.

Cost-sensitive classification can be used to treat class imbalance. This is done by adjusting the cost-matrix to reflect the class imbalance. C_{TP} and C_{TN} are set to 0 since there should be no penalty for correct classifications. If C_{FP} is set to 1 then C_{FN} is set to $|M| / |N|$. This way, $|N| \times C_{FP} = |M| \times C_{FN}$. MetaCost [4] is a cost-sensitive method which first builds an ensemble on random subsets of D_T. Then for each record x in D_T, the probability of x belonging to M or N is calculated by averaging the outputs of the classifiers in the ensemble. Consider $P(j|x)$ to be the averaged probability of x belonging to class j, and $C(i|j)$ to be the cost of classifying a class j record as class i. MetaCost then edits the class value $v \in x$ according to Eq. 1. The edited D_T is then used to build a final classifier which is used for cost-sensitive classification. Since MetaCost edits the class values, it is referred to as a relabeling method. MetaCost is highly popular within the cost-sensitive literature.

$$v = argmin_i \sum_j P(j|x)C(i|j) \qquad (1)$$

CSTree [14] is a cost-sensitive classifier which builds a decision tree used for classification. If every record in D_T is classified as positive, it is possible to count the resulting number of TP, TN, FP, and FN classifications. These counts can then be multiplied by their corresponding costs then added together. The resulting number is called C_P in CSTree, or the cost of classifying every record in a dataset as positive. Similarly C_N is the cost of labeling every record in a dataset as negative. These four costs are used to build a C4.5 like decision tree. A weakness of using CSTree or MetaCost for class imbalance treatment is that only a fixed cost-matrix may be used.

2.2 Lanchester's Laws

Lanchester's laws [10] help military strategists to understand the outcome of a battle between two forces. For example, we can consider two armies. Army B has 6 soldiers and army A has 23 soldiers. By using Lanchester's linear law we assume that each soldier can only fight one other soldier at a time. This simply means that the size of both armies decreases one at a time until one army is completely wiped out. Thus, Lanchester's linear law would suggest that for army B to have an equal force to army A there are three solutions. The first is increasing its size to 23. The second is increasing its firepower to be 23/6 times stronger than army A's firepower. The third is a combination of both. Lanchester's linear law is used for ancient combat, that is, using simple weapons such as swords and spears.

However, in modern combat soldiers do not always fight one-on-one. Thus, Lanchester also proposed a variant for modern combat called Lanchester's squared law. In the squared variant, army B will need to increase its firepower to be the square of the division of the army A's size by army B's size. Thus, army B will need to increase its firepower by a factor of $(23/6)^2$. In practice, rather than use the linear or squared law, 1.5 is used as the exponent [16]. Therefore, in order to fully emulate what is done in military practice, we also use 1.5 as the exponent in this study.

3 Our Method

3.1 Analogy Between Class Imbalance and War

War and class imbalance share many similarities. In war, two factions engage in an ongoing fight. Within a training dataset D_T, the majority class records M and minority class records N are "fighting" for adequate representation. In this fight, each record represents a soldier. In war, each faction may have multiple armies stationed at different locations. Within M and N, groups of records which have similar values may exist. These groups are commonly referred to as clusters. In our analogy, we consider clusters to be armies. In an army, each soldier is armed with weaponry. Some soldiers may have superior weaponry to others. This is similar to how some records may have higher weights than other records in cost-sensitive classification. These four key similarities are the core of our analogy.

Consider a situation in which two opposing armies are of equal strength. If these armies fight each other, almost all soldiers from both parties are likely to die. Therefore, neither party wishes to provoke the other party since it will result in mutually assured destruction. This situation is known as a standoff. If M and N are considered to be factions at war with each other, then creating a standoff situation will prevent M from fighting N for adequate representation.

3.2 Standoff-Balancing

The proposed method *Standoff-Balancing* aims to balance the impact of each cluster within the dataset. To do this, Lanchester's laws are used to calculate the required increase in firepowers for minority class records. The steps for *Standoff-Balancing* are outlined in Fig. 1. First, for every attribute in the dataset, the values are normalized to be between 0.0 and 1.0. The reason for this is to make sure that each attribute has equal impact when calculating distance. The dataset is then split into two datasets, one comprising of minority records, and the other of majority records. These two datasets are then clustered independently. The discovered clusters can now be considered as armies within the minority faction and majority faction. By considering the clusters in this way, we can now apply Lanchester's laws to calculate the required increase in firepower for every record within that cluster. Clusters that are far apart from each other are easier to differentiate by a classifier. Therefore, when calculating the strength of the majority class compared to a minority cluster, rather than just take the count of records in the opposing majority clusters, the count is multiplied by the distance. This way, clusters which are farther away have less impact on each other.

Parameters:
D_T - the training dataset
L - a cost-sensitive classification method
G - a clustering method.
Output:
A balanced classifier.

Steps:
1: Normalize all attribute values in D_T to be between 0.0 and 1.0.
2: Split D_T into two datasets: One comprising of only majority records M, and the other of minority records N.
3: Let M^G and N^G be the clusters found in M and N by using G.
4: Let M^F and N^F be the set of required firepowers for each cluster in M and N respectively such that M_i^F is the required firepower of the i'th majority cluster and N_j^F is the required firepower of the j'th minority cluster. $\forall\ M_i^F$ and N_j^F: calculate the needed firepower using Equation 2.
5: $\forall M_i^F$ and N_j^F, generate a corresponding cost-matrix using the minority and majority cluster cost-matrixes in Table 1.
6: Return the classifier that is the result of applying L to D_T using the corresponding cost-matrixes for each record.

Fig. 1. The proposed algorithm: Standoff-Balancing

Equation 2 is used to calculate the required firepower F_i for each minority cluster N_i^G in the set of minority clusters N^G. Each minority cluster is balanced against the sum of the number of records in the majority clusters. However, the euclidean distance from the minority cluster to the majority cluster is used as a weight for each majority cluster's size. The distances from the minority cluster's centroid[1] S_i^N to each majority cluster's centroid S_j^M is normalized such that the

---

[1] A cluster's centroid is a record whose values are the average of the records within that cluster.

farthest centroid is 1.0 and the closest is 0.0. Thus, the farther a majority cluster is to a minority cluster, the less firepower is required.

$$F_i = (\frac{\sum_{j=1}^{|M^G|}(|M_j^G| \times (1 - dist(S_j^M, S_i^N))}{|N_i^G|})^{1.5} \qquad (2)$$

These firepowers can then be used to create cost-matrices for the minority clusters by using the minority cost-matrix in Table 1. Each majority record uses the standard majority cluster cost-matrix shown in Table 1. Finally when training the classifier, the corresponding cost-matrix is used for each record.

Table 1. Our minority and majority cluster cost-matrixes

Minority cluster		Majority cluster	
$C_{TP} = 0$	$C_{FN} = F_i$	$C_{TP} = 0$	$C_{FN} = 1$
$C_{FP} = 1$	$C_{TN} = 0$	$C_{FP} = 1$	$C_{TN} = 0$

Every majority record uses the same costs as shown in Table 1. Finally when training the classifier, the corresponding cost-matrix is used for each record. The core component of standoff-balancing is to create cluster-specific weights to use in cost-sensitive classification. Thus, standoff-balancing may be used with most cost-sensitive methods given a small adjustment from a fixed cost-matrix to cluster-specific cost-matrixes as input. Threshold methods handle costs a posteriori [9], and therefore could handle Standoff-Balancing easily.

4   Experimental Results

We compare the proposed method against four existing methods over five real world datasets. Overall accuracy or error rate are not useful metrics for evaluating classifiers built from class imbalanced datasets. Therefore we use the area under the receiver operating characteristic curve (AUC) due to its popularity in the class imbalance literature. We use C4.5 [13], MetaCost [4] and k-means [12] as the classifier, cost-sensitive method, and clustering method respectively. These methods were chosen for their popularity within their respective area of literature. Since k-means is highly dependent on correct parameter settings, we use EM [3] with k-means.

We now provide information on the four existing methods that we compare against. Namely, CSTree, SMOTE, Adasyn and MetaCost. We use 10 fold stratified cross validation and the folds used for each method are the same. For SMOTE and Adasyn we use $k = 5$ since it is highly popular within the literature. For SMOTE we set the percentage to 200 % as suggested by the experiments by Chawla and Bowyer [2]. Due to the randomness in SMOTE and Adasyn, we present the average of 3 independent runs. We use the implementations of

SMOTE, MetaCost and EM with k-means from the WEKA [6] machine learning library. MetaCost uses the default parameters specified in its WEKA implementation. We turn off pruning in CSTree to build deeper trees which achieve better performance as suggested in the literature [15].

The proposed technique Standoff-Balancing is referred to as Standoff in Table 2 in order to conserve space. We use five datasets available from KEEL [1] and the UCI [11] machine learning repository. The first four are from KEEL whereas the BreastCancer dataset is from the UCI machine learning repository.

Table 2. AUC comparison of the methods (higher the better)

Dataset	Original	CSTree	SMOTE	Adasyn	MetaCost	Standoff
Pima	0.735	0.72	0.740	0.700	0.725	**0.757**
glass0	0.77	0.82	0.817	0.764	0.784	**0.836**
haberman	0.604	0.572	0.627	0.641	0.646	**0.647**
ecoli1	0.924	0.899	0.925	0.877	0.866	**0.929**
BreastCancer	0.939	0.943	0.935	0.901	0.945	**0.952**

The results of the comparison are presented in Table 2. In each dataset Standoff-Balancing achieves a higher AUC than both MetaCost and CSTree. This suggests that the costs generated by the proposed method are superior than the costs traditionally used to treat class imbalance. The proposed technique achieves the highest AUC in all five datasets.

5   Conclusion

In this study we present an analogy between war and class imbalance. By doing so we make it possible to apply military strategies to the class imbalance problem. We proposed a novel class imbalance treatment method *Standoff-Balancing* which uses Lanchester's laws to create a balance in strength between the clusters within a dataset. An experimental comparison on five publicly available real world datasets suggests the efficacy of the proposed technique. The proposed method's efficacy was demonstrated by two different factors. First, it achieved higher AUC than the existing techniques on all five datasets. Secondly, it was the only method to consistently achieve a higher AUC than the original dataset. Our experiments not only demonstrate the efficacy of the proposed method, but also the usefulness of the presented analogy. The usefulness of this analogy suggests that further research into the application of military strategy to the class imbalance problem has strong potential.

Our future work includes extending Standoff-Balancing to deal with datasets which contain many class values. We also plan to run our experiments on a large number of datasets. This will allow us to make insights as to which characteristics of a dataset are best for Standoff-Balancing.

Code Availability

The code used to run our experiments is available at the following websites. http://www.mikesiers.com/software/ and http://csusap.csu.edu.au/~zislam/.

References

1. Alcalá, J., Fernández, A., Luengo, J., Derrac, J., García, S., Sánchez, L., Herrera, F.: Keel data-mining software tool: data set repository, integration of algorithms and experimental analysis framework. J. Multiple-Valued Logic Soft Comput. **17**(2–3), 255–287 (2010)
2. Chawla, N.V., Bowyer, K.W., Hall, L.O., Kegelmeyer, W.P.: SMOTE: synthetic minority over-sampling technique. J. Artif. Intell. Res. **16**, 321–357 (2002)
3. Dempster, A.P., Laird, N.M., Rubin, D.B.: Maximum likelihood from incomplete data via the EM algorithm. J. Roy. Stat. Soc. Series B (Methodological) **39**, 1–38 (1977)
4. Domingos, P.: Metacost: a general method for making classifiers cost-sensitive. In: Proceedings of the Fifth ACM SIGKDD International Conference on Knowledge Discovery and Data Mining, pp. 155–164. ACM (1999)
5. Fernández, A., García, S., del Jesus, M.J., Herrera, F.: A study of the behaviour of linguistic fuzzy rule based classification systems in the framework of imbalanced data-sets. Fuzzy Sets Syst. **159**(18), 2378–2398 (2008). http://dx.doi.org/10.1016/j.fss.2007.12.023
6. Hall, M., Frank, E., Holmes, G., Pfahringer, B., Reutemann, P., Witten, I.H.: The WEKA data mining software: an update. ACM SIGKDD Explor. Newsl. **11**(1), 10–18 (2009)
7. He, H., Bai, Y., Garcia, E.A., Li, S.: ADASYN: adaptive synthetic sampling approach for imbalanced learning. In: 2008 IEEE International Joint Conference on Neural Networks (IEEE World Congress on Computational Intelligence). IEEE, June 2008. http://dx.doi.org/10.1109/IJCNN.2008.4633969
8. He, H., Garcia, E.: Learning from imbalanced data. IEEE Trans. Knowl. Data Eng. **21**(9), 1263–1284 (2009). http://dx.doi.org/10.1109/TKDE.2008.239
9. Hernández-Orallo, J., Flach, P., Ferri, C.: A unified view of performance metrics: translating threshold choice into expected classification loss. J. Mach. Learn. Res. **13**(1), 2813–2869 (2012). http://dl.acm.org/citation.cfm?id=2503308.2503332
10. Lanchester, F.W.: Mathematics in warfare. World Math. **4**, 2138–2157 (1956)
11. Lichman, M.: UCI Machine Learning Repository (2013). http://archive.ics.uci.edu/ml
12. Lloyd, S.P.: Least squares quantization in PCM. IEEE Trans. Inf. Theory **28**(2), 129–137 (1982)
13. Quinlan, J.R.: C 4.5: Programs for Machine Learning. The Morgan Kaufmann Series in Machine Learning. Morgan Kaufmann, San Mateo (1993)
14. Sheng, V.S., Gu, B., Fang, W., Wu, J.: Cost-sensitive learning for defect escalation. Knowl. Based Syst. **66**, 146–155 (2014). http://dx.doi.org/10.1016/j.knosys.2014.04.033
15. Siers, M.J., Islam, M.Z.: Software defect prediction using a cost sensitive decision forest and voting and a potential solution to the class imbalance problem. Inf. Syst. **51**, 62–71 (2015). http://dx.doi.org/10.1016/j.is.2015.02.006
16. Simpkin, R.E.: Race to the Swift: Thoughts on Twenty-First Century Warfare, vol. 1. Potomac Books, Herndon (1985)

Wisdom of Crowds: An Empirical Study of Ensemble-Based Feature Selection Strategies

Teo Susnjak[1]([⊠]), David Kerry[1], Andre Barczak[1], Napoleon Reyes[1], and Yaniv Gal[2]

[1] Massey University, Auckland, New Zealand
t.susnjak@massey.ac.nz
[2] Compac Ltd., Auckland, New Zealand

Abstract. The accuracy of feature selection methods is affected by both the nature of the underlying datasets and the actual machine learning algorithms they are combined with. The role these factors have in the final accuracy of the classifiers is generally unknown in advance. This paper presents an ensemble-based feature selection approach that addresses this uncertainty and mitigates against the variability in the generalisation of the classifiers. The study conducts extensive experiments with combinations of three feature selection methods on nine datasets, which are trained on eight different types of machine learning algorithms. The results confirm that the ensemble based approaches to feature selection tend to produce classifiers with higher accuracies, are more reliable due to decreased variances and are thus more generalisable.

Keywords: Ensemble feature selection · Dimensionality reduction · Machine learning · Classification · Data mining · Ensemble classifiers

1 Introduction

The main purpose of machine learning is to produce classifiers that generalise in their predictive accuracy beyond the datasets used to train them. To a large degree, their final accuracy is dependent on the descriptive strength and quality of the features that constitute the training dataset. It is often tempting to simply provide a machine learning algorithm with as many features as are available for a given dataset. However, doing so has been consistently shown to be associated with negative outcomes [15–17].

The inclusion of large feature numbers in a training dataset presents computational challenges that mostly arise during the training phase and can be prohibitive for some algorithms [28], but can also be a strain during the detection time for real-time systems processing high-volume data streams. Unnecessary and redundant features increase the search space for a machine learning algorithm. This in turn dilutes the signal strength of a true pattern and makes it more likely that due to the presence of noisy and irrelevant features, a spurious pattern will be discovered instead.

© Springer International Publishing Switzerland 2015
B. Pfahringer and J. Renz (Eds.): AI 2015, LNAI 9457, pp. 526–538, 2015.
DOI: 10.1007/978-3-319-26350-2_47

In general it is not known *a priori* which features are meaningful, and finding the optimal feature subset has been proven to be a NP-complete problem [2]. Nonetheless, it is still imperative that feature selection algorithms be applied to a dataset as a pre-processing step before training classifiers, in order to reduce feature dimensionality [14]. Not only are both the computational complexity and the generalisability improved by selecting the most concise subset, but the resulting model is more interpretable due to the fact that it is generated with the fewest possible number of parameters [10].

Research into feature selection has produced a wide array of techniques and algorithms. Each technique provides a different perspective on the data and thus a different assessment of how meaningful individual features are. Some techniques perform considerably better than others on different datasets, sample sizes, feature numbers and problem domains [12] and it is generally uncertain which technique will be most suitable for a problem at hand. Prior to performing machine learning, it is not uncommon in some domains where stability in feature subsets is important [1], to initially investigate the results from several feature selection algorithms before choosing the best feature selection technique which satisfies a required criterion [6].

Feature selection techniques can generally be divided into two broad categories. Filter methods are univariate techniques which consider the relevance of a particular feature in isolation to the other features and rank the features according to a metric. These algorithms are computationally efficient since they do not integrate the machine learning algorithm in its evaluation. However, they can be susceptible to selecting subsets of features that may not produce favourable results when combined with a chosen machine learning algorithm [30]. These methods lack the ability to detect interactions among features as well as feature redundancy. On the other hand, wrapper methods overcome some of these shortcomings. They explicitly use the chosen machine learning algorithm to select the feature subsets and tend to outperform filter methods in predictive accuracy [30]. However, these techniques exhibit bias in favour of a specific machine learning algorithm, and since they are computationally more intensive, they are also frequently impractical on large datasets.

Hybrid filter-wrapper methods have been a subject of recent research due to their ability to exploit the strengths of both strategies [17,18]. Hybrid approaches essentially allow any combination of filter and wrapper methods to be combined. Due to this, some novel and interesting hybrid approaches have recently been proposed such as: using the union of feature-subset outputs from Information Gain, Gain Ratio, Gini Index and correlation filter methods as inputs to the wrapper Genetic Algorithm [20], hybridization of the Gravitational Search Algorithm with Support Vector Machine [23] and using Particle Swarm Optimisation-based multi-objective feature selection approach in combination with k-Nearest-Neighbour [27]. Given their flexibility, hybrid approaches thus offer some degree of tuning the trade-offs between accuracy and performance. Nonetheless, devising a feature selection algorithm that is both highly accurate and computationally efficient is still an open question [10].

The ubiquity of data acquisition technologies and the affordability of ever increasing data storage capacities means that datasets are now larger in both sample numbers as well as feature vectors. In the age of Big Data, it is not uncommon to encounter datasets having many thousands of features [4, 12] in a variety of problem domains. This presents considerable challenges and for these reasons, feature selection is an active and important part of ongoing research. The challenge is to some degree amplified since machine learning has entered into mainstream use and is becoming more frequently utilised in numerous industrial [24] as well as business sectors, where in-depth expertise in the intricacies of this field are not always readily available.

Motivation. Our motivation is to devise a strategy for performing feature selection which increases the likelihood of generating good and robust feature subsets that can effectively be combined with a wide range of machine learning algorithms and datasets spanning numerous domains. The aim is to address the need to formulate a strategy that generates a reliable feature subset in a timely manner, and can particularly be useful in industrial and business settings where machine learning is employed by practitioners who have not necessarily had expert training. The purpose is to automate the process of feature selection and to eliminate the possibilities of generating poor subsets, while foregoing the goals of finding the optimal solution due to its impracticality.

This research investigates combining outputs of multiple feature selection algorithms in order to produce an effective feature subset for machine learning. The inspiration is drawn from the theory of ensemble-based classifiers. Its foundational principle states that while any one classifier may perform more accurately than a combined classification of all available classifiers on a given dataset, across the space of all possible problems, the aggregate decisions of multiple classifiers will however outperform any one available individual classifier. Ensemble-based classifiers have demonstrated superior results compared to individual classifiers in a wide range of applications and scenarios [22]. In empirical studies, it has been shown that ensembles yield better results provided that there is significant diversity among the classifiers.

Our research builds and extends on previous work by Tsai and Hsiao [25] who experimented with combining the feature subset selections of Principal Component Analysis, Information Gain and the Genetic Algorithm, as inputs for the Neural Network classifier on the domain of stock price prediction. This research goes further and the key contributions lie in demonstrating how the ensemble-based feature selection strategy can be generalised to a much broader set of domains, and can be combined with a wider number of machine learning algorithms. We empirically show how this strategy performs in conjunction with eight machine learning algorithms using different combinations of the Information Gain (IG) [13], Linear Discriminant Analysis (LDA) and Chi^2 [19] techniques for generating feature subsets, using nine datasets for testing.

The extensive experiments in this research show that the ensemble-based strategy for feature selection does indeed generalise to multiple machine learning

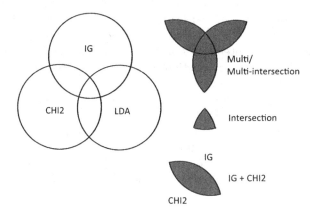

Fig. 1. Three different strategies for combining the outputs of the feature selection algorithms.

algorithms and different problem domains. The results confirm that relying on multiple sources for input on feature selection does outperform any one single feature selection algorithm in the long run. In addition, the results also provide some insights and rough rules-of-thumb for machine learning practitioners as to which individual feature selection strategies have a tendency to work well in combination with machine learning algorithms.

2   Experimental Design

IG, LDA and Chi^2 were selected for the experiments since they are readily available in most data mining software packages and, individually, are widely used for the purposes of feature selection [10, 28, 29]. These methods were also chosen due to the slightly different perspectives each one has on what constitutes 'meaningfulness' of a given feature. The success of ensemble-based decision making lies with the existence of disagreement amongst individual methods. LDA's strength lies in projecting the data cloud onto new axes which maximise the variance and in the process identify redundancy, while taking class membership into account. Meanwhile, Chi^2 and IG do not consider feature redundancy but rank the feature according to different criteria. Chi^2 tests the independence of each feature in respect to its class label, while IG similarly evaluates how commonly occurring a feature value is for a given class, compared to its frequency amongst all other classes.

Given that a sufficient degree of diversity exists within an ensemble system, it is then important to devise an appropriate aggregation strategy. Ensemble-based classifiers usually assign a weight to each of its constituent classifiers which reflects their discriminative strength, calculated during the training process. Since feature selection algorithms cannot easily be assigned a confidence weight, a different strategy must be applied. This research applies set theory in order to aggregate the outputs of the feature selection algorithms, which has been shown to be effective by [25]. The strategy is depicted in Fig. 1, showing how features that are an (1) intersection of all, (2) multi-intersection of all, or (3) an intersection of two feature selection algorithms, can be combined. Different permutations of the intersections, together with the individual feature selection algorithms provided eight strategies for the experiment, plus the control which did not apply any feature selection.

Experiments were conducted using nine datasets whose properties are outlined in Table 1. The datasets originated from two sources, the first three fruit datasets were obtained from an industrial source[1]. The datasets represent fruit surface features from three different varieties. Though the datasets originated from entirely different fruit, they were all generated by their proprietary software and there is thus an element of risk that the datasets capture some artefacts of the feature extraction process, which may bias them towards certain feature selection algorithms. The remaining datasets were sourced online from the UCI Machine Learning Repository [3]. Procurement of datasets with a variety of feature and sample numbers as well as domains of origin was the goal.

Table 1. Dataset properties.

Dataset name	Classes	Instances	Features
Nectarines	4	587	13
Peaches	3	240	10
Plums	3	141	13
Waveform	3	5000	21
Fac profile	10	2000	216
Fourier	10	2000	76
Karhunen-Love	10	2000	64
Pixel avg	10	2000	240
Zernike Moments	10	2000	47

The various feature subsets were trained on eight different machine learning algorithms, listed together with their tunable parameters in Table 2. Given that a total of 6840 classifiers were trained across the entire experimental process (9 datasets × 5 folds × 19 thresholds × 8 machine learning algorithms), tuning the machine learning algorithms for optimal training parameters was not feasible. Therefore, most classifiers were trained with default parameters for each of their respective algorithms.

The experimental workflow is depicted in Fig. 2. Each dataset was passed into the feature selection stage where eight subsets were created. The first subset is

---

[1] These datasets were provided by Compac Sorting Ltd., a company that specialises in automated fruit sorting via image processing.

Table 2. Machine learning training parameters.

Classifier	Training settings	Implementation source
kNN	k = 3	scikit-learn [21]
SVM [26]	linear kernel, regularisation parameter C = 0.025	scikit-learn [21]
Decision Tree	maximum depth = 10	scikit-learn [21]
Random Forest [5]	maximum depth = 10, estimators = 20	scikit-learn [21]
AdaBoost [9]	number of estimators = 100	scikit-learn [21]
Naive Bayes	Gaussian default setting	scikit-learn [21]
AdaBoost.ECC [11]	100 boosting iterations	authors' C++ implementation
RIPPER [7]	2 rounds of optimisations with pruning enabled	authors' C++ implementation

the control containing all features. The next three subsets were created from applying Chi^2, IG and LDA to rank the features in order of how informative they are according to their respective feature evaluation criteria.

Subsequently, the classification stage trains and tests the classifiers on each feature subset using five fold cross-validation. The process continues with the thresholding stage. The feature selection algorithms rank the features in the order of their apparent usefulness but are not able to determine if a given feature is 'informative' or 'poor'; a threshold therefore needs to be selected as the cut-off for the percentage of features to keep in the training/testing subset. The testing was repeated exhaustively with a threshold range of 5 % to 95 % of the features accepted, with 5 % intervals. This enabled every combination of classifier, approach and dataset the opportunity to achieve its optimal feature subset size as a proportion of the ranked features.

The last stage in the process gathered the performance data for every combination of dataset, feature selection subset and classifier. The accuracy and the geometric-mean scores with the corresponding standard deviation were collected from the thresholding stage. Geometric-mean was calculated in addition to accuracy due to the greater ability of the geometric-mean to convey classifier generalisation on datasets with significant class imbalances. Some degree of class imbalance was present on the Nectarines, Peaches and Plums datasets; however, the negligible differences between the geometric-mean and accuracy scores showed that this was not at a significant level. For this reason, the geometric-mean scores are not reported in the results.

3 Results

The performance results from the experiments presented here involve several thousand classifiers. Space limitations preclude us listing all accuracy results in

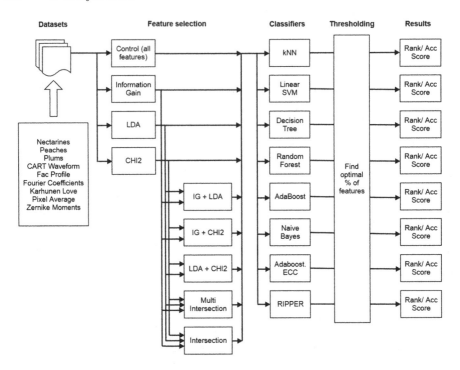

Fig. 2. The schematic representation of the stages in the experimental process together with the workflow.

their raw form. Instead, we provide small snapshots of the underlying accuracy results from each feature selection algorithm, while relying mostly on their summaries in the form of mean ranks as is acceptable practice [8]. In addition the non-parametric Friedman statistical test will also be employed in order to verify statistical significances in the findings.

Table 3 lists the accuracies of each feature selection algorithm across all datasets for the Random Forest classifier. The table is summarised in the form of mean ranks which aggregate the performances of all the feature selection algorithms for this particular classifier. The table shows that across all the datasets, the ensemble-based multi-intersection method outperformed all remaining methods in this study. Performing machine learning without first conducting feature subset selection has predictably generated the least generalisable classifiers.

A further seven tables in the same format as Table 3 were generated for the remaining classifiers used in these experiments. The mean rank summaries were extracted from the tables and all combined together in order to render a graphical depiction in Fig. 3. The mean ranks in Table 3 for the Random Forest classifier can be traced in this figure.

Figure 3 is a box-and-whisker plot, in which each of the feature selection methods are ordered based on the average of their mean ranks from each of the classifiers on all datasets. The median, inter-quartile and occasional outliers are

Table 3. Example of Random Forest classifiers' raw accuracy results on all datasets using each of the feature selection methods. The results in the table represent one of eight tables generated from which a set of mean ranks were calculated.

	Multi	Chi^2	LDA	LDA+ Chi^2	IG+ Chi^2	IG	LDA+ IG	Inters	Control
Nectarines	**0.625**	0.612	0.576	0.574	**0.625**	0.617	0.589	0.576	0.588
Peaches	**0.804**	0.796	0.78	0.788	0.775	0.78	0.776	0.792	0.759
Plums	0.623	0.579	0.623	0.579	0.595	0.621	**0.636**	0.566	0.589
Waveform	**0.844**	0.841	0.838	0.839	0.836	0.832	0.837	0.835	0.83
Fac profile	0.947	**0.949**	0.944	0.943	0.947	0.948	0.94	0.943	0.942
Fourier	**0.825**	0.817	0.821	0.817	0.774	0.723	0.784	0.779	0.714
Karhunen Love	0.922	**0.928**	0.919	0.923	0.843	0.808	0.838	0.857	0.852
Pixel Average	0.953	0.951	0.949	0.947	0.949	0.945	0.952	**0.956**	0.943
Zernike	**0.73**	0.727	0.729	0.714	0.727	**0.73**	0.729	0.722	0.724
Mean Ranks	**1.9**	3.6	4.4	5.7	5.3	5.4	5.2	5.9	7.7

displayed. As expected, the worst performing method is the control whereby no feature selection was performed. The best performing strategy is the ensemble-based multi-intersection method, followed closely by Chi^2 and LDA methods. The two clear outliers in the graph indicate the positive responsiveness of the

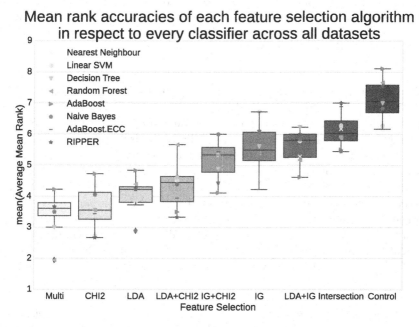

Fig. 3. Box-and-whisker plot showing the accuracies of each feature selection techniques in terms of their mean rank score for each machine learning algorithm, when combined across all datasets. The feature selection algorithms are listed in the order from best to worst, based on to the average of their mean rank scores.

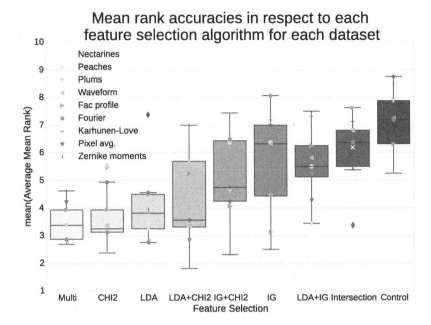

Fig. 4. Box-and-whisker plot showing the accuracies of each feature selection technique in terms of their mean rank score for each dataset, when combined across all machine learning algorithms. The feature selection algorithms are listed in the order from best to worst according the average of their mean rank scores.

Random Forest and Naive Bayes classifiers to the multi-intersection and LDA feature selection methods respectively.

Importantly, the figure also conveys the degree of variance for each of the feature selection algorithms. Smaller variability is desirable as it indicates more consistent and predictable performances. The multi-intersection method exhibits the lowest variance of all the strategies examined. Ensemble-based methods are known to reduce the variance and thus the results are not altogether surprising. However, it is noteworthy that variance is reduced, while comparatively the best accuracies are achieved, thus indicating that there has been no increase in bias.

In addition, the differences in the mean ranks in Fig. 3 are confirmed to be statistically significant. The critical value for the Friedman Rank Sum Test at $\alpha = 0.05$ is $\chi^2 = 15.507$. The test produces a test statistic $\chi^2_F(8) = 47.3$ and p-value $= 1.365e - 07$.

The effectiveness of feature selection algorithms is not only determined by their suitable combination with specific machine learning algorithms, but also by the actual underlying datasets. Presented in Fig. 4 is an alternative perspective on the result from the previous figure which illustrates the effectiveness of the feature selection algorithms on each of the datasets, using the combined accuracies of the classifiers. The overarching message from the results data has not changed from Fig. 3. The ensemble-based multi-intersection method is still the

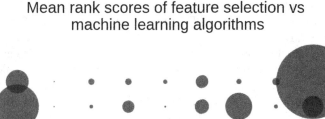

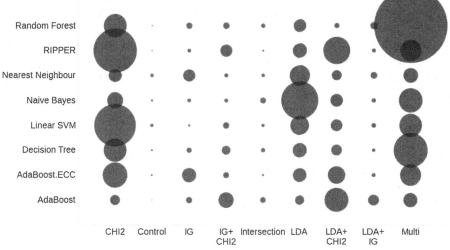

Fig. 5. Visual matrix of the effectiveness of each feature selection method in combination with every machine learning algorithm, across the aggregate of all datasets. Effectiveness is expressed in terms of mean ranks from accuracy scores and projected on to the exponential scale in order to emphasise strengths in the patterns.

best performing, while the ordering of the remaining methods is intact. Importantly, the findings indicate that the ensemble-based method is also invariant to the variety of datasets and domains, and not only to the type of machine learning methods used. Overall, Fig. 3 demonstrates the variances have increased across all feature selection algorithms. Most notably the combination of LDA+Chi^2 and IG+Chi^2, as well as for the IG, the variances have increased markedly. Of note is the negative effect of one dataset in particular (Nectarines) on the performance of LDA which once again demonstrates the point that while each feature selection algorithm will perform acceptably on some datasets, there also exist datasets on which a given method will perform very poorly.

Even though the number of datasets used in this research is limited and thus precludes us from making definitive claims, it is nonetheless useful to be able to draw out some insights and very general rules-of-thumb from these empirical findings as to which feature selection methods and machine learning algorithms have a tendency to work well in combination. Figure 5 attempts to graphically convey this information and draw out insights from the experimental data conducted here. The figure demonstrates the responsiveness of a classifier to the various feature selection methods using the mean ranks, to which the exponential function has been applied in order to emphasise patterns. Clearly the combination of multi-intersection and the Random Forest classifier stands out as a suitable combination. A strong signal can be observed between the Naive Bayes classifier with LDA, as with the combination Chi^2 with Ripper and Linear SVM.

It should be noted that there are also very poor pairings of classifier and feature selection combinations for a given data set. An example of this is shown as Chi^2 generally performs very well, but performs poorly in combination with AdaBoost. A good starting point to find the optimum and avoid potential poor performance, therefore, would be to use the performance of an ensemble approach as a benchmark and use only combinations that outperform the benchmark for further performance refinement.

4   Conclusion

Feature selection is an indispensable component of machine learning. While there are numerous feature selection methods in existence, their robustness is affected by both the chosen machine learning algorithms they are intended to be used with, as well as the characteristics of the underlying datasets and the problem domains themselves. While some feature selection methods will generally tend to work well with certain machine learning algorithms, there are times when they will perform poorly in combination with given datasets which have specific properties. *A priori* knowledge of which combinations will work well together is usually inaccessible, and computationally deriving an optimal combination is a NP-complete problem.

This research considers the ensemble-based approach to selecting feature subsets. Instead of relying on a single feature-selection algorithm that might work well on some occasions but poorly on others, the ensemble-based approach advocates thoughtfully combining the outputs of several different methods. The ensemble-based approach does not guarantee that an optimal solution will be produced, but it does ensure that a very good solution will be found instead, and this is sufficient for many machine learning domains. Ensemble-based approaches do however guarantee that the provided solution will always be better than the weakest performing feature selection algorithms within its ensemble.

The study considered three frequently used filter methods: Chi^2, LDA and Information Gain. It applied different combinations of these methods on nine datasets from a wide range of domains, and tested the feature subsets against eight different classifiers from a broad range of machine learning algorithms.

Extensive experiments were conducted. The data shows that across a number of problem domains and different machine learning algorithms, the ensemble-based approach to feature selection tend to outperform the usage of single feature selection methods explored here. Ensemble-based approaches are more resistant to the variability that different machine learning algorithms and datasets bring to the classification accuracies, and are therefore generally more robust. Given the uncertainty as to which feature selection and machine learning algorithms will combine effectively, this research confirms the suitability of presented methods for domains which process broad varieties of datasets and require timely and consistently reliable solutions.

References

1. Abeel, T., Helleputte, T., Van de Peer, Y., Dupont, P., Saeys, Y.: Robust biomarker identification for cancer diagnosis with ensemble feature selection methods. Bioinformatics **26**(3), 392–398 (2010)
2. Albrecht, A.A.: Stochastic local search for the feature set problem, with applications to microarray data. Appl. Math. Comput. **183**(2), 1148–1164 (2006)
3. Asuncion, A., Newman, D.: UCI machine learning repository (2007). http://www.ics.uci.edu/~mlearn/MLRepository.html
4. Bermejo, P., de la Ossa, L., Gámez, J.A., Puerta, J.M.: Fast wrapper feature subset selection in high-dimensional datasets by means of filter re-ranking. Knowl.-Based Syst. **25**(1), 35–44 (2012)
5. Breiman, L.: Random forests. Mach. Learn. **45**(1), 5–32 (2001)
6. Chandrashekar, G., Sahin, F.: A survey on feature selection methods. Comput. Electr. Eng. **40**(1), 16–28 (2014)
7. Cohen, W.: Fast effective rule induction. In: Proceedings of the Twelfth International Conference on Machine Learning, pp. 115–123 (1995)
8. Demšar, J.: Statistical comparisons of classifiers over multiple data sets. J. Mach. Learn. Res. **7**, 1–30 (2006)
9. Freund, Y., Schapire, R.: A decision-theoretic generalization of on-line learning and an application to boosting. J. Comput. Syst. Sci. **55**(1), 119–139 (1997)
10. Gheyas, I.A., Smith, L.S.: Feature subset selection in large dimensionality domains. Pattern Recogn. **43**(1), 5–13 (2010)
11. Guruswami, V., Sahai, A.: Multiclass learning, boosting, and error-correcting codes. In: Proceedings of the Twelfth Annual Conference on Computational Learning Theory, COLT 1999, pp. 145–155. ACM, New York (1999)
12. Hua, J., Tembe, W.D., Dougherty, E.R.: Performance of feature-selection methods in the classification of high-dimension data. Pattern Recogn. **42**(3), 409–424 (2009)
13. Hunt, E.B., Marin, J., Stone, P.J.: Experiments in induction. Academic Press, New York (1966)
14. Inbarani, H.H., Azar, A.T., Jothi, G.: Supervised hybrid feature selection based on pso and rough sets for medical diagnosis. Comput. Methods Programs Biomed. **113**(1), 175–185 (2014)
15. Inza, I., Larrañaga, P., Blanco, R., Cerrolaza, A.J.: Filter versus wrapper gene selection approaches in dna microarray domains. Artif. Intell. Med. **31**(2), 91–103 (2004)
16. Kohavi, R., John, G.H.: Wrappers for feature subset selection. Artif. Intell. **97**(1), 273–324 (1997)
17. Kotsiantis, S.: Feature selection for machine learning classification problems: a recent overview. Artif. Intell. Rev. **42**, 1–20 (2011)
18. Leung, Y., Hung, Y.: A multiple-filter-multiple-wrapper approach to gene selection and microarray data classification. IEEE/ACM Trans. Comput. Biol. Bioinform. (TCBB) **7**(1), 108–117 (2010)
19. Liu, H., Setiono, R.: Chi2: Feature selection and discretization of numeric attributes. In: TAI, p. 388. IEEE (1995)
20. Oreski, S., Oreski, G.: Genetic algorithm-based heuristic for feature selection in credit risk assessment. Expert Syst. Appl. **41**(4), 2052–2064 (2014)
21. Pedregosa, F., Varoquaux, G., Gramfort, A., Michel, V., Thirion, B., Grisel, O., Blondel, M., Prettenhofer, P., Weiss, R., Dubourg, V., Vanderplas, J., Passos, A., Cournapeau, D., Brucher, M., Perrot, M., Duchesnay, E.: Scikit-learn: machine learning in Python. J. Mach. Learn. Res. **12**, 2825–2830 (2011)

22. Polikar, R.: Essemble based systems in decision making. IEEE Circuits Syst. Mag. **6**(3), 21–45 (2006)
23. Sarafrazi, S., Nezamabadi-pour, H.: Facing the classification of binary problems with a gsa-svm hybrid system. Math. Comput. Model. **57**(1), 270–278 (2013)
24. Susnjak, T., Barczak, A., Reyes, N.: On combining boosting with rule-induction for automated fruit grading. In: Kim, H.K., Ao, S.-L., Amouzegar, M.A. (eds.) Transactions on Engineering Technologies, pp. 275–290. Springer, Netherlands (2014)
25. Tsai, C.F., Hsiao, Y.C.: Combining multiple feature selection methods for stock prediction: union, intersection, and multi-intersection approaches. Decis. Support Syst. **50**(1), 258–269 (2010)
26. Vapnik, V.N., Vapnik, V.: Statistical Learning Theory, vol. 1. Wiley, New York (1998)
27. Xue, B., Zhang, M., Browne, W.N.: Particle swarm optimization for feature selection in classification: a multi-objective approach. IEEE Trans. Cybern. **43**(6), 1656–1671 (2013)
28. Yang, Y., Pedersen, J.O.: A comparative study on feature selection in text categorization. In: ICML, vol. 97, pp. 412–420 (1997)
29. Ye, J., Li, Q.: A two-stage linear discriminant analysis via QR-decomposition. IEEE Trans. Pattern Anal. Mach. Intell. **27**(6), 929–941 (2005)
30. Zhu, Z., Ong, Y.S., Dash, M.: Wrapper-filter feature selection algorithm using a memetic framework. IEEE Trans. Syst. Man Cybern. Part B: Cybern. **37**(1), 70–76 (2007)

Modeling Ice Storm Climatology

Ranjini Swaminathan[1,3](✉), Mohan Sridharan[2], Gillian Dobbie[1],
and Katharine Hayhoe[3]

[1] Department of Computer Science, The University of Auckland,
Auckland, New Zealand
ranjinis@gmail.com, gill@cs.auckland.ac.nz
[2] Department of Electrical and Computer Engineering, The University of Auckland,
Auckland, New Zealand
m.sridharan@auckland.ac.nz
[3] Climate Science Center, Texas Tech University, Lubbock, USA
katharine.hayhoe@ttu.edu

Abstract. Extreme weather events such as ice storms cause significant damage to life and property. Accurately forecasting ice storms sufficiently in advance to offset their impacts is very challenging because they are driven by atmospheric processes that are complex and not completely defined. Furthermore, such forecasting has to consider the influence of a changing climate on relevant atmospheric variables, but it is difficult to generalise existing expertise in the absence of observed data, making the underlying computational challenge all the more formidable. This paper describes a novel computational framework to model ice storm climatology. The framework is based on an objective identification of ice storm events by key variables derived from vertical profiles of temperature, humidity, and geopotential height (a measure of pressure). Historical ice storm records are used to identify days with synoptic-scale upper air and surface conditions consistent with an ice storm. Sophisticated classification algorithms and feature selection algorithms provide a computational representation of the behavior of the relevant physical climate variables during ice storms. We evaluate the proposed framework using reanalysis data of climate variables and historical ice storm records corresponding to the north eastern USA, demonstrating the effectiveness of the climatology models and providing insights into the relationships between the relevant climate variables.

1 Introduction

Extreme winter weather events such as snow, sleet and ice storms can cause significant damage to property and life. Perhaps the most dangerous of these are ice storms that result from freezing rain. Ice storms are a globally occurring phenomena that can account for billions of dollars of damage to urban and natural systems.

Ice storms account for roughly 60 % of winter storm losses within the United States, and they have caused billions of dollars of loss in the continental United

© Springer International Publishing Switzerland 2015
B. Pfahringer and J. Renz (Eds.): AI 2015, LNAI 9457, pp. 539–553, 2015.
DOI: 10.1007/978-3-319-26350-2_48

States and Canada [11,16]. The catastrophic effects of ice storms are being seen all over the world, with the 2008 ice storm in China resulting in 129 human fatalities, \$22.3 billion in direct economic losses and structural damage, and the displacement of $\approx$1.7 million people [31]. The increasing concentration of human populations in areas vulnerable to ice storms, such as the north eastern United States, will only exacerbate the impact of such ice storms in the future.

It is very challenging to accurately forecast ice storms sufficiently in advance to offset the extent of the associated losses, primarily because the complex atmospheric processes driving ice storms are difficult to model. Traditional weather models forecast ice storms using very high resolution observed data for climate variables, and extensive interpretation by humans with domain expertise. Besides overcoming the difficulty of fully understanding and modeling the complex physical processes in the atmosphere that cause ice storms, the influence of a changing climate is a crucial factor to be considered when detecting ice storms in the future. Global Climate Models (GCMs) provide projections for climate variables decades and even a century in advance under changing climate scenarios, thereby providing the scientific community with a framework to study and model future climate events.

Existing expertise, however, is difficult to generalise and apply to future projections in order to quantify the impact of human-induced climate change on the occurrence and severity of ice storms. Such analysis is challenged by the lack of "observed" data and by the necessity to consider the influence of a changing climate on the atmospheric variables that drive ice storms. The work described in this paper is thus motivated by the need to represent complex ice storm climatologies in order to project ice storms in the future under a changing climate. We present a computational framework to model the complex phenomena of ice storms, and make the following key contributions:

- we build on classification algorithms to model ice storm climatology, and to identify fundamental climate patterns that are indicative of this climatology;
- we exploit feature selection algorithms to improve both the computational efficiency and the effectiveness of the climatology models; and
- we provide a computational representation of the relevant physical climate variables and their behaviour during ice storms, affording atmospheric scientists better insights into the relationships between these variables.

The proposed computational framework is illustrated and experimentally evaluated using reanalysis data, comprising relevant physical climate variables, and historical ice storm records, corresponding to the north eastern USA.

2 Related Work

There are no published studies in the literature that apply sophisticated AI or machine learning algorithms to model ice storm climatology. This section motivates our work by describing how ice storm climatology has been modeled historically, and by summarizing the broader applicability of computational models for climate science.

Ice Storm Forecasting: Ice storms develop due to many complex atmospheric processes that have traditionally been studied at different spatial (micro, meso and synoptic or cyclone scale) and temporal (from hourly to a few hours) resolutions. Researchers have studied the effect of topology on freezing rain, and investigaed the influence of cold-air damming due to the Appalachian mountains in the north eastern USA and the Rocky Mountain areas [2,26]. Researchers have also attributed the spatial variation of freezing rain events around the Great Lakes region in the USA, to the frequency of surface cyclone tracks, availability of moisture in the Atlantic ocean, and regional topography [8]. Others have used composites of relevant climate variables over a period of five days to describe how large scale circulation patterns of these variables fostered freezing rain events in the Great Lakes region [16]. The sensitivity of the Atlantic ocean's sea surface temperatures to ice events in the north eastern USA has been investigated [9], and the atmospheric conditions that typically coincide with north eastern ice storms, including synoptic scale movement of moisture and temperature, have been identified [5].

The literature on studying and modeling ice storms shows that the complex factors driving ice storms are yet to be completely defined. While it is established that synoptic scale events play a key role in their occurrence, existing methods continue to use composites of high resolution observational data for climate variables. The data is further processed by computationally intensive numerical weather models, which limits these studies to single or selected features of freezing events in geographically localized areas. Furthermore, to forecast ice storms in the present or near future by observing salient weather patterns, meteorologists still rely on human expertise to make the final decision regarding intensity and occurrence. The requirements of high resolution data and extensive human input highlight the need for alternative models for ice storm climatology. This need is amplified by the question of how human-induced climate change may be affecting the atmospheric circulation patterns that lead to ice storms. Global Climate Models (GCMs) use the laws of physics to simulate many processes and interactions in the climate system. GCMs are the main tool used to study future climate; they can generate three-dimensional projections for different atmospheric variables under future climate change scenarios. GCM simulations provide data at coarser spatial granularities (250 km vs 32 km) and lower temporal resolutions (daily averages instead of a few hours) than weather stations. Researchers have generated future ice storm projections for the north eastern USA and eastern Canada by using statistical downscaling techniques [6,7]. These techniques establish statistical relationships between large-scale GCM projections and high-resolution observations, and use these relationships to obtain "observation-like" local data with future GCM projections [29]. However, statistical downscaling makes the assumption that the relationships between large-scale and local climate variables is stationary over time, and does not exploit the complex inter-variable relationships within and across scales. These limitations make it challenging to use statistically downscaled data for modeling complex interactions between climate variables, such as those driving ice storms.

Computational Models for Climate Science: Research in climate and atmospheric sciences poses a plethora of scientific questions that could greatly from advances in AI. A recent tutorial at a premier machine learning conference highlighted the complex nature of many problems of interest, and briefly summarized some of the computational solutions that have been developed for a subset of these problems [1]. A recent article describes that we are yet to explore advanced computational techniques in the context of climate and large-scale meteorological processes such as ice storms [14]. Our research study is motivated by our strong belief that computational solutions for problems in climate science (and other research domains) should go beyond just a superficial application of techniques—they should actively improve our understanding of the problem and the domain [28]. We describe a framework that addresses key computational challenges to address a hard problem in climate science. This framework also serves as an integral first step towards improving the climate science community's understanding and ability to detect the complex climate phenomenon of ice storms.

3   Materials and Methods

In this section, we first summarize what is currently known about the climatology of ice storms, identifying the variables that we consider in the proposed computational framework. Next, we identify and describe the sources of data corresponding to the study area in which we illustrate and evaluate our framework. Finally, we describe the computational models for classification and feature selection that are included in the proposed framework.

3.1   Ice Storm Climatology

Ice storms are caused by freezing rain and start out in a cold layer of moist air up in the atmosphere quite the same way as snow and sleet, i.e., as a snowflake. In the case of snowfall, the snowflake continues to drop down through cold air and settles as snow when it hits the surface of the earth. Sleet forms when the snowflake first melts through a layer of warm air and freezes while moving through a cold layer of air again before it hits the ground. In the case of freezing rain, the snowflake moves down through a layer of warm air and melts into rain first, then into another layer of cold air where it turns into a supercooled droplet. If the surface temperature of the earth is below freezing, then this supercooled droplet freezes as soon as it touches any surface. We see that *certain specific and relatively rare atmospheric conditions must be in place for freezing rain to occur.* This freezing rain then creates a film of ice on the ground and the surface of every object it falls on. Figure 1 shows a schematic representation of how freezing rain forms and how it differs from other forms of precipitation.

To consider the atmospheric circulation patterns that are consistent with ice storm events, we analyze the vertical profiles of three climate variables: temperature, humidity, and geopotential height. While humidity and temperature at

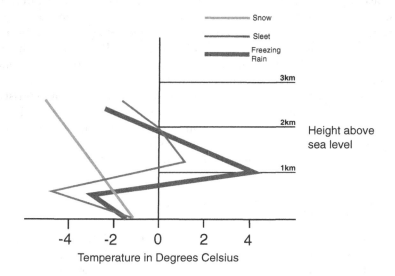

Fig. 1. Freezing rain forms when a snowflake falls through alternating cold and warm layers before it gets supercooled and freezes on contact with the ground. The figure also shows how sleet and snow form in comparison to freezing rain. Figure adapted and modified from [27].

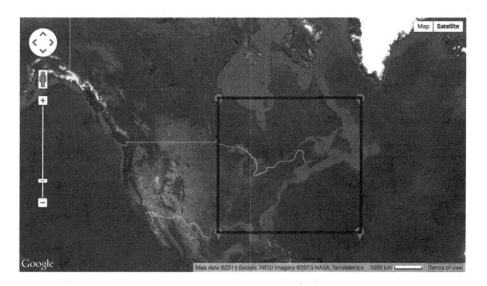

Fig. 2. Synoptic scale plot showing the geographic range of the climate variables considered for modeling ice storm climatology in north eastern USA. The range extends from 24° N to 55° N and from 50° W to 94° W.

different atmospheric pressure levels account for the moisture and alternating warm and cold layers, the geopotential height approximates the actual height of a pressure surface above mean sea level. In other words, it is the height of a constant pressure surface (e.g., 750 millibars) above the surface of the earth measured in meters. We also consider these atmospheric patterns at *synoptic scales*, as seen in Fig. 2. For the chosen study area of north eastern USA, this means that we capture upper atmosphere dynamics that are characteristic of ice storms in the large area that comprises the north eastern USA, and do not attempt to localize our projections to county-level geographic locations within the area. Considering the patterns at synoptic scale helps develop more reliable models of the ice storm climatology without human feedback and downscaling errors.

3.2 Data Selection

Although we rely on climate variable projections from GCMs for estimating ice storm occurrences in the future, GCM simulations are themselves prone to uncertainties that are difficult to quantify. Uncertainty in GCM output can broadly be attributed to: (1) natural variability in the climate system; (2) scientific limitations in understanding and modeling climate; and (3) human choices that will determine the extent of future emissions [19,20]. Our initial model development is therefore done with data generated by a process called "reanalysis", which is a systematic approach to generate consistent data for climate monitoring and research. Reanalysis is a consistent reprocessing of archived weather observations with modern forecasting systems producing gridded datasets that estimate a large variety of atmospheric, sea-state and land surface parameters [10]. Reanalysis data is also well-suited for our task because there is no physical way to obtain comprehensive weather observations for climate variables at different heights above the earth's surface. Furthermore, reanalysis data is available at different spatial and temporal resolutions. Reanalysis data thus enables us to build more robust models than with just the GCM output. Since GCM simulations are independent of observations, the models developed also provide a framework to evaluate different GCMs and identify those that would be best suited for future ice storm projections. For our work, we used reanalysis data from the National Centers for Environmental Prediction (NCEP) provided by the National Oceanic and Atmospheric Administration, USA [24].

We model ice storm climatology in the north eastern USA by studying the atmospheric patterns that characterized historical ice storm events in the region. We chose the same 175 significant ice storm events selected in [5], all occurring in winter periods (October 1st to April 30th) between the years 1993 and 2010. These events were selected from the National Climatic Data Center's (NCDC, USA) Storm Database and a given event qualified as an ice storm if it met one of three established criteria [5]. Any ice storm events occurring in the northeastern USA is also influenced by atmospheric processes at much larger geographic scales (synoptic), which could potentially include larger masses of land and even the ocean. The synoptic scale considered sufficiently large (by climate scientists) for studying these ice storms is shown in Fig. 2, spanning the area between 24° N and 55° N from 50° W to 94° W. This translates to a 145 × 149 grid of climate variable estimates from the reanalysis data. Data for temperature (2m, 850hPa, 925hPa, 1000hPa), specific humidity (2m, 750hPa, 850hPa, 925hPa, 1000hPa) and geopotential height(200hPa, 500hPa, 700hPa, 850hPa) was used for each cell in this grid since these values were considered relevant to ice storm formation by domain experts. The temporal component of the data consists of estimates of climate variables at three-hour intervals each day.

3.3   Experimental Framework

Ice storms are characterized by signature patterns defined by climate variables at various atmospheric pressure levels. The task of learning computational models that capture this signature from historical data, and use the models to identify similar patterns in the future, can be posed as a supervised learning problem. The *Learning Task* is to learn models for climate variable patterns during ice storm and non-ice storm winter periods in the past. The *Performance Task* is to apply the learned models to identify ice storms in future GCM projections.

Classification Models for Ice Storm Climatology: We considered some popular algorithms such as decision trees, rule-based system, neural networks, and Support Vector Machines (SVMs) to classify ice storm and non-ice storm patterns defined by climate variables. The SVM-based classifier was (experimentally) observed to provide the best classification accuracy, and was hence used in all subsequent experiments. Specifically, we adapted an SVM with a linear kernel to use the climate variables as input features, providing class labels ("positive" for ice storms and "negative" for non-ice storms) as the corresponding outputs. SVMs transform the task of determining boundaries around patterns in feature spaces into the dual convex quadratic problem. Features and patterns are projected to high(er) dimensions and loss functions are used to penalize errors, resulting in sparse representations and robust decision boundaries [18,22,25]. SVMs have a simple geometric interpretation, provide robustness to overfitting, and use structural risk minimization methods [3]. SVMs are well-suited to model the complex relationships that characterize the behavior of ice storms. Since the reanalysis data was available at very high resolutions, the larger 145 × 149 grid was subsampled by averaging values (of each variable) across 5 × 5 grid cells— this size of the subsampling mask provides a granularity consistent with GCM projections. Furthermore, the three-hourly readings for the different climate variables were averaged to provide the equivalent of daily average readings, which is (again) the resolution at which GCM projections will be available. To evaluate the sensitivity of classification to different spatio-temporal scales, we considered other grid sizes and temporal resolutions—see Sect. 4.

We considered the relevant climate variables, i.e., temperature, humidity and geopotential height, at pressure levels ranging from the surface (2 m) to 1000 millibars. Traditional ice storm forecasting methods consider five-day composite maps for climate variables. Instead, we build a classification model for each day in a five-day window (three days before, the day of the storm, and one day after). A total of 175 vectors of the historical record of ice storms, each with 10092 features, were used to build the classification model for each day in the window. We can now apply a sliding window on GCM projections for winter periods during future decades (or centuries) to identify five-day sequences that signal the potential for an ice storm. Section 4 describes the outcome of 10-fold cross validation on this dataset for the baseline (SVM) classifier.

Feature Selection for Ice Storms: Given the complexity of the underlying atmospheric processes, and the relatively small number of ice storm events under consideration, learning the classification model for each day can be challenging and computationally expensive with input vectors of ≈ 10000 features. Also, the learned models will be used to identify ice storms in time periods spread over several decades. One objective of this work is to identify features that contribute most significantly towards the ice storm signature so that climate scientists can gain better insights into the physical characteristics of the atmospheric processes driving ice storms. We would also like to see how the features selected change over the five-day window to better understand the change in relevance of the climate variables as the ice storm develops. Dimensionality reduction methods such as Principal Component Analysis [21] are not applicable since they transform the original features into a different dimensional space, and we lose the correspondence information needed to identify the contribution of individual climate variables. Guidelines laid out by [15] recommend first applying domain knowledge to perform feature selection and reduce the dimensionality of features for classification tasks. However, we are unable to use domain knowledge to reduce the set of features—the complexity of the relationships between climate variable driving ice storms makes it difficult for even a domain expert to identify the most relevant features.

We investigated two different measures for ranking features: (1) Pearson's correlation coefficient; and (2) a mutual information-based information-theoretic measure. The Pearson's correlation coefficient is computed between individual features and the class labels:

$$\rho_{Class, Attribute} = \frac{cov(Class, Attribute)}{\sigma_{Class} \sigma_{Attribute}} \tag{1}$$

where $\rho_{Class, Attribute}$ is the Pearson's coefficient, $cov(Class, Attribute)$ is the covariance between random variables $Class$ and $Attribute$, and σ is the standard deviation. The information gain measure, on the other hand, evaluates the worth of each feature as the information gained about each class given the individual feature:

$$InfoGain(Class, Attribute) = H(Class) - H(Class|Attribute) \tag{2}$$

where $H(Class)$ is the marginal entropy of the class and $H(Class|Attribute)$ is the conditional entropy of the class given the feature. We experimentally evaluated how classification performance changed with different feature set sizes, to deduce the optimal set of features for the problem of modeling ice storm climatology. As described in Sect. 4, the feature selection algorithms reduced the number of features considered, and improved the (overall) accuracy of the classifier built using the reduced number of features. As before, we used 10-fold cross validation to evaluate the classification performance of the combination of feature selection algorithms and the SVM classifier.

4   Experimental Results

We present results from various experiments conducted in relation to ice storm climatology modeling. All algorithms were implemented and evaluated on the datasets using the WEKA data mining software [17]. As stated in Sect. 3, the results reported below are obtained by performing 10-fold cross validation. We primarily used the F-measure to reflect classification performance in all the experimental trials—a higher value of this measure represents better classification accuracy. We also used some other measures (see below) to evaluate the significance of the performance of different algorithms.

Table 1. Classification performance of ice storm vs. non-ice storm dates in winter periods. Values represent the F-measure for a linear kernel support vector machine classifier. Every 5×5 grid in the input data becomes a single supercell for the classification.

Readings per day	Day 1	Day 2	Day 3	Day 4	Day 5
Four	0.594	0.640	0.717	0.780	0.780
Two	0.606	0.623	0.708	0.754	0.814
One	0.528	0.614	0.634	0.740	0.814
Daily avg.	0.571	0.606	0.683	0.769	0.751

Table 2. Classification performance of ice storm vs. non-ice storm dates in winter periods. Values represent the F-measure for a linear kernel support vector machine classifier. Every 3×3 grid in the input data becomes a single supercell for the classification.

Readings per day	Day 1	Day 2	Day 3	Day 4	Day 5
Four	0.594	0.640	0.711	0.797	0.786
Two	0.608	0.629	0.717	0.760	0.806
One	0.531	0.609	0.648	0.749	0.820
Daily avg.	0.557	0.600	0.668	0.774	0.754

Classification Performance at Varying Spatio-Temporal Scales: As stated in Sect. 3, we considered classification performances with two different spatial resolutions, one where the gridded data is averaged over a mask of 3×3 cells and another with 5×5 cells. Reanalysis data is also available at different temporal resolutions and the values of the climate variables can be considered multiple times through the day. The daily average reading is obtained by averaging all available readings for an entire day. GCM simulation data for estimating ice storms in the future will be available to us in the form of daily average readings and hence we consider the daily average f-measure to be our baseline performance. In order to provide a more comprehensive overview, we also conducted experiments with different combinations of spatial scales and temporal scales, with the corresponding classification performance (i.e., F-measure scores) summarized in Tables 1 and 2. We observe that classification performance peaks on Day 4, which is the day of the storm. In some cases, the classifier for the day after the storm does better and can be attributed to the fact that many of the storm events typically occur over a period of two days. In general, this tells us that the ice storm signal is strongest on the day of the storm, and that our model is able to capture this signature.

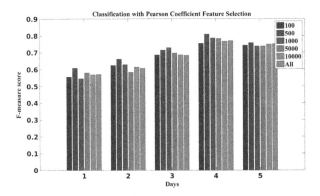

Fig. 3. Classification performance for different sized feature sets, using the Pearson's coefficient for feature selection on the days before, during, and after the ice storm.

As expected, we also observe that the classification performance improves for both spatial coverages as the temporal resolution increases (i.e., if we consider multiple readings of the climate variables in a day). However, we notice that the classification performance does not drop significantly when the spatial and temporal resolutions are set to match those of the GCM projections. In other words, we are able to demonstrate that our classification framework's performance is reasonably robust to changes in spatial and temporal resolutions. For the remaining experiments, we thus used only the GCM-scale resolution data (5×5 smoothing and daily averages).

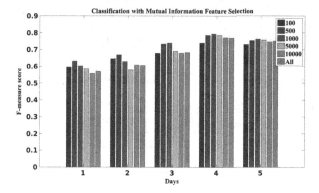

Fig. 4. Classification performance for different sized feature sets, using the mutual information gain for feature selection on the days before, during, and after the ice storm

Classification Performance with Feature Selection: Next, we describe the classification performance when feature selection is included with the baseline classifier evaluated above. As described in Sect. 3.3, we used the correlation coefficient measure and the mutual information measure to rank features. Figure 3 shows the F-measure scores when the features were ranked using the correlation between individual features and the class label. Figure 4 shows the corresponding results for when the features were ranked based on a mutual information theoretic gain measure. In both cases, the number of features was varied to find an optimal set smaller than the entire set of features (10092) that best captured the signatures of ice storms. From Figs. 3 and 4, we observe that for both methods of feature selection, we get the best classification performance for Day 4, which is the day of the ice storm. We also observe that using the top 500 features, as ranked by the feature ranking measures, results (on average) in the best classification performance—this *much smaller* set of features can be considered as being representative of ice storm patterns.

Table 3. Classification performance and statistics for the day of the storm (Day 4). The Kappa measure increases with feature selection and the RMSE decreases. The best score for each column is highlighted in bold—feature selection using the Pearson coefficient provides the best results.

Feature subsets	F-measure	Kappa statistic	RMSE
All	0.769	0.537	0.481
Top 500 (Pearson)	**0.809**	**0.617**	**0.438**
Top 500 (Info-gain)	0.786	0.571	0.463

Many methods have been developed to evaluate the significance of classification performance [4,12,13]. We used three measures to evaluate whether the

Table 4. The z scores for the McNemar's test comparing classification performance with and without feature selection. The arrow points in the direction of the better performing algorithm—using the Pearson coefficient for feature selectionprovides the best results.

Feature Subsets	All	Top 500 (Pearson)	Top 500 (Info-gain)
All		↑ 1.625	↑ 0.5893
Top 500 (Pearson)			←1.4289
Top 500 (Info-gain)			

ability to detect ice storms improves significantly with feature selection. First, the *Kappa statistic* [30] compares the observed accuracy with the expected accuracy or random chance. It is a measure of the agreement between the predicted and the actual classifications. A higher Kappa statistic value indicates better classification accuracy. Second, the *Root Mean Square Error* (RMSE) [30] is a measure of the difference between the values predicted by the classifier and the ground truth (i.e., known) values. A lower RMSE indicates better classification accuracy. Finally, we used the *McNemar test* [23], a non-parametric test used to analyze matched pairs of data. It compares the classification accuracy of two algorithms on a per-instance basis and computes a z score that can be translated into confidence levels. A *One-tailed Prediction* is used to determine when one algorithm is better than the other. Tables 3 and 4 summarize the result of using these measures. Table 3 shows that selecting and using the top 500 features raises the Kappa value and lowers the RMSE. Table 4 shows that the classification performance improvement obtained using the Pearson's coefficient for feature selection corresponds to a 94.8 % confidence level. Due to space limitations, we only show the results for the day of the storm.

Computational Representation of Atmospheric Processes: In addition to modeling the ice storm climatology, the results summarized above can also be used to computationally represent the climate variables that are most relevant to characterize ice storms. For instance, Fig. 5 shows the distribution of the climate variables at different spatial locations (specific latitude and longitude cells after 5×5 subsampling) and at different pressure levels on the day of the storm (i.e., Day 4), using the Pearson correlation coefficient for feature selection—both methods of feature selection show similarity in the atmospheric patterns (i.e., the spatial distribution of climate variables) on the day of the storm. The geopotential height variables selected indicate that a confluence of temperature and moisture variables occurs at ≈ 700 mb. This observation is consistent with the formation of snowflakes at this height. We also observe three different layers of temperature where the snowflake alternatively cools, melts and then freezes just as it hits the surface of the earth. This is clearly indicated in the representations by both methods of feature selection. Similar figures that depict the computational representations for the days before and after the storm (not shown in this paper) show the ice storm conditions slowly start to form during

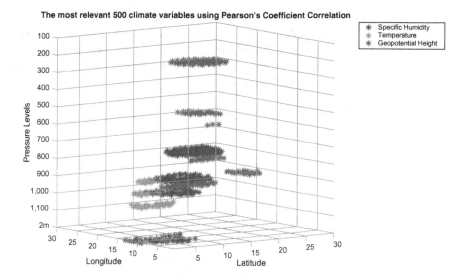

Fig. 5. The computational representation of the most relevant climate variables selected using Pearson's Coefficient for ice storm modeling is shown for the day of the storm.

Day 2, peak on Day 4 and disappear on Day 5—these observations are (once again) completely consistent with the observed behavior of the ice storms. *The key finding is that we are able to computationally confirm these patterns using a much smaller subset of features typically considered while modeling ice storms.*

5 Conclusions and Future Work

Ice storms are potentially catastrophic weather events that can cause significant damage to life and property. The ability to project ice storm prevalence and intensity in the future can help us minimize their negative impact. This paper presents a novel computational framework to model ice storm climatology, a complex problem in climate science. This framework learns atmospheric patterns that characterize historical ice storm occurrences using reanalysis data for the relevant climate variables. We show that the classification models in our framework are robust (according to climatolgists) across spatial and temporal scale changes in climate data. In addition, the framework includes feature selection algorithms, which identify a smaller subset of features that increase the classification accuracy and significantly reduce the computational cost of building (and testing) the classification models on decadal data. Finally, our computational models help confirm and improve the existing understanding of what causes ice storms, in ways that will allow climate scientists to develop new insights into this complex atmospheric phenomenon.

This study opens up many directions for inter-disciplinary research. For instance, we would like to develop and use more sophisticated feature selection algorithms that can better ground the relationships between the climate variables driving ice storms. We would also like to use our models constructed with reanalysis data to evaluate Global Climate Model output for historical time periods, which will allow us to choose the Global Climate Models that are best suited for generating future ice storm projections.

Acknowledgments. This work was supported in part by the US National Science Foundation under Grant No. DEB-1457875.

References

1. Banerjee, A., Monteleoni, C.: Tutorial: climate change: challenges for machine learning. In: Neural Information Processing Systems (2014)
2. Bernstein, B.C.: Regional and local influences on freezing drizzle, freezing rain, and ice pellet events. Weather Forecast. **15**(5), 485–508 (2000)
3. Bishop, C.: Pattern Recognition and Machine Learning. Springer, New York (2006)
4. Bostanci, B., Bostanci, E.: An evaluation of classification algorithms using Mc Nemar's test. In: Bansal, J.C., Singh, P.K., Deep, K., Pant, M., Nagar, A.K. (eds.) (BIC-TA 2012). AISC, pp. 15–26. Springer, Heidelberg (2013)
5. Castellano, C.M.: Synoptic and mesoscale aspects of ice storms in the Northeastern US. Master's thesis in atmospheric science, University at Albany, SUNY, USA (2012)
6. Cheng, C.S., Auld, H., Li, G., Klaassen, J., Li, Q.: Possible impacts of climate change on freezing rain in south-central Canada using downscaled future climate scenarios. Nat. Hazards Earth Syst. Sci. **7**(1), 71–87 (2007)
7. Cheng, C.S., Li, G., Auld, H.: Possible impacts of climate change on freezing rain using downscaled future climate scenarios: updated for Eastern Canada. Atmos.-Ocean **49**(1), 8–21 (2011)
8. Cortinas, Jr., J.: A climatology of freezing rain in the great lakes region of North America. Mon. Weather Rev. **128**(10), 3574–3588 (2000)
9. da Silva, R.R., Bohrer, G., Werth, D., Otte, M.J., Avissar, R.: Sensitivity of ice storms in the Southeastern United States to Atlantic SST-Insights from a case study of the december 2002 storm. Mon. Weather Rev. **134**(5), 1454–1464 (2006)
10. Dee, D.P., Balmaseda, M., Balsamo, G., Engelen, R., Simmons, A.J., Thépaut, J.-N.: Toward a consistent reanalysis of the climate system. Bull. Am. Meteorol. Soc. **95**(8), 1235–1248 (2014)
11. DeGaetano, A.T.: Climatic perspective and impacts of the 1998 Northern New York and New England ice storm. Bull. Am. Meteorol. Soc. **81**(2), 237–254 (2000)
12. Demšar, J.: Statistical comparisons of classifiers over multiple data sets. J. Mach. Learn. Res. **7**, 1–30 (2006)
13. Dietterich, T.G.: Approximate statistical tests for comparing supervised classification learning algorithms. Neural Comput. **10**(7), 1895–1923 (1998)
14. Grotjahn, R., Black, R., Leung, R., Wehner, M.F., Barlow, M., Bosilovich, M., Gershunov, A., Gutowski, Jr., W.J., Gyakum, J.R., Katz, R.W., et al. North American Extreme Temperature Events and Related Large Scale Meteorological Patterns: A Review of Statistical Methods, Dynamics, Modeling, and Trends. Climate Dynamics, pp. 1–34 (2015)
15. Guyon, I., Elisseeff, A.: An introduction to variable and feature selection. J. Mach. Learn. Res. **3**, 1157–1182 (2003)

16. Gyakum, J.R., Roebber, P.J.: The 1998 ice storm-analysis of a planetary-scale event. Mon. Weather Rev. **129**(12), 2983–2997 (2001)
17. Hall, M., Frank, E., Holmes, G., Pfahringer, B., Reutemann, P., Witten, I.H.: The WEKA data mining software: an update. ACM SIGKDD Explor. Newsl. **11**(1), 10–18 (2009)
18. Hastie, T., Tibshirani, R.: Classification by pairwise coupling. In: Advances in Neural Information Processing Systems, vol. 10. MIT Press (1998)
19. Hawkins, E., Sutton, R.: The potential to narrow uncertainty in regional climate predictions. Bull. Am. Meteorol. Soc. **90**(8), 1095–1107 (2009)
20. Hawkins, E., Sutton, R.: The potential to narrow uncertainty in projections of regional precipitation change. Clim. Dyn. **37**(1–2), 407–418 (2011)
21. Jolliffe, I.: Principal Component Analysis. Wiley Online Library (2002)
22. Keerthi, S.S., Shevade, S.K., Bhattacharyya, C., Murthy, K.R.K.: Improvements to Platt's SMO algorithm for SVM classifier design. Neural Comput. **13**(3), 637–649 (2001)
23. McNemar, Q.: Note on the sampling error of the difference between correlated proportions or percentages. Psychometrika **12**(2), 153–157 (1947)
24. Mesinger, F., DiMego, G., Kalnay, E., Mitchell, K., Shafran, P.C., Ebisuzaki, W., Jovic, D., Woollen, J., Rogers, E., Berbery, E.H., et al.: North American regional reanalysis. Bull. Am. Meteorol. Soc. **87**(3), 343–360 (2006)
25. Platt, J.: Fast training of support vector machines using sequential minimal optimization. In: Schoelkopf, B., Burges, B., Smola, A. (eds.) Advances in Kernel Methods - Support Vector Learning. MIT Press, Cambridge (1998)
26. Rauber, R.M., Olthoff, L.S., Ramamurthy, M.K., Miller, D., Kunkel, K.E.: A synoptic weather pattern and sounding-based climatology of freezing precipitation in the united states east of the rocky mountains. J. Appl. Meteorol. **40**(10), 1724–1747 (2001)
27. Spector, D.: Why Icestorms are so Dangerous, Business Insider Australia, 12 February 2014
28. Wagstaff, K.L.: Machine learning that matters. In: Proceedings of the 29th International Conference on Machine Learning (ICML-12), pp. 529–536 (2012)
29. Wilby, R.L., Wigley, T.M.L., Conway, D., Jones, P.D., Hewitson, B.C., Main, J., Wilks, D.S.: Statistical downscaling of general circulation model output: a comparison of methods. Water Resour. Res. **34**(11), 2995–3008 (1998)
30. Witten, I.H., Frank, E.: Data Mining: Practical Machine Learning Tools and Techniques. Morgan Kaufmann, San Francisco (2005)
31. Zhou, B., Gu, L., Ding, Y., Shao, L., Wu, Z., Yang, X., Li, C., Li, Z., Wang, X., Cao, Y., et al.: The great 2008 Chinese ice storm: its socioeconomic-ecological impact and sustainability lessons learned. Bull. Am. Meteorol. Soc. **92**(1), 47–60 (2011)

Investigating Japanese Ijime (Bullying) Behavior Using Agent-Based and System Dynamics Models

Chaiwat Thawiworadilok[1(✉)], Mohsen Jafari Songhori[1,2], and Takao Terano[1]

[1] Interdisciplinary Graduate School of Science and Engineering,
Computational Intelligence and System Science, Tokyo Institute of Technology, J2 Bldg., Room 1704, 4259 Nagatsuta-cho, Midori-ku, Yokohama, Kanagawa, 226-8502, Japan
teerachait@trn.dis.titech.ac.jp
[2] JSPS Research Fellow, Tokyo, 102-0083, Japan

Abstract. In the recent years, observing *Ijime* (i.e. Japanese bully) among Japanese students has been a growing concern. To understand the effect of this behavior in a detail level, we build an Agent-Based Model (ABM) and conduct a set of experiments. In the model, interactions occur between victim and bully as Prisoner Dilemma game. A System Dynamics model is also built and simulated to verify robustness and examine effects of the assumptions in ABM model. The results indicate that students can attend higher social standing by being not cooperative. As such, if one is victim, it is recommended not to comply with the bully. In the case of being bully, one needs to take the role seriously. Thus, by encouraging victim to be aggressive toward bully, the effect of *Ijime* can be alleviated.

Keywords: Iterated prisoner's dilemma · Ijime · Bully · Agent-based modeling · System Dynamics

1 Introduction

In the recent years, *Ijime* or bullying has been one of the concerns among Japanese students at school. It is one of the major factors that leads to suicide in many cases [7] especially in younger ages. The root of *Ijime* comes from the Japanese emphasize on conformity as it focuses on harmony of group over individuality. According to [8], mental aggression toward victim is a main method of action such as peer ignorance. Furthermore, the effect of *Ijime* is also amplified by its structure. Comparing to traditional bully, *Ijime* structure consists of four actors: (I) bully, (II) victim, (III) supporter and (IV) bystander. Bystander, special to *Ijime*, refers to a student who avoids being involved with *Ijime*. Such avoidance can be represented as to change his/her relationship with the victim or being neutral in regard to bully. The bystander's role become important in *Ijime* because he/she passively allows its happening.

Agent-based modeling (ABM) is one of the computational models that uses interactions among agents to understand a system as a whole. Hence, ABM, as an alternative model, is used to simulate the classroom situation because of difficulties in real classroom experiment.

© Springer International Publishing Switzerland 2015
B. Pfahringer and J. Renz (Eds.): AI 2015, LNAI 9457, pp. 554–558, 2015.
DOI: 10.1007/978-3-319-26350-2_49

The main parts of the model are the bully and victim agents. In addition, Prisoner's Dilemma (here after PD) game is used to represent the interaction between the two agents. The similarity to PD game comes from the fact that either party can choose to defect or cooperate [4, 5]. Other researchers, [9] also uses PD game to simulate the *Ijime* situation on bystander interactions. However, this paper focuses more on the victim and bully interactions.

To compare and validate the results of the ABM model, a System Dynamics (SD) model is also developed and simulated. Development of SD model for the same phenomena, *Ijime*, has two advantages. First, it provides one independent validation model [12]. Secondly, lack of heterogeneity in SD model helps us to understand if this factor plays an important role in the macro results seen in ABM model [11].

2   Agent-Based Model

The ABM model only includes one entity which is students in a class. At each time step, each agent is assigned one of the four roles (i.e. bully, victim, bystander, and supporter). The state variables and description as well as controlled parameters are explained in detailed in the former paper by author [13].

There have been published papers in the literature that suggest the role of bully is rotated among friends in the same group [1–3]. Thus, p_i^{bully} is assumed to be uniformly distributed. In addition, the probability of becoming victim is calculated relatively within a group. There exists dynamics not only within groups but also between groups. Hence, students may change their groups at different times. When selecting group, agents use the characteristic and probability of becoming victim as criteria.

From the perspective of *Ijime*, the four cases of the game can be described as follow. (I) the case where bully cooperate means bully make little joke on victim. (II) The case where bully defects refers to when bully does *Ijime* seriously (III) The case where victim cooperates means the victim complies with bully. (IV) Finally, the case where victim defects refers to when to victim fights back.

Once assigned the roles, the bully and victim agents of each group play PD game. Afterwards, the supporter's score in each group is calculated based on the number of supporter in a group together with bully' score. This assumption appears to be consistent with intuition and the fact that they join and encourage the bully.

For calculating the bystanders' score, three cases are considered. In the case of avoiding Ijime, the bystanders' score will depend on the number of supporters as the larger number of bystanders, the more permissive and safer environment they will have. In the case of becoming helper, if the bystander is not noticed, the victim's score is increased proportional to number of helper. However, if the bystander becomes visible, the bystander will get punishment. This is to represent the severity of the inconformity. At the end of time step, the score will be normalize. Detailed explanation and formulas of the model can be found in [13]

As agents learn over time and may revise their strategies toward *Ijime* cases, they are assumed to revise their strategies at some fixed time intervals by using survival of the fittest concept [6].

3   System Dynamics Model

System dynamic model is an alternative to model social problem. Different from ABM, SD emphasizes macro approaches to model a problem [10]. In particular, SD model assumes homogeneity in the population. The advantage of SD model is a wide range of feedback effects. In this research, SD is used to verify the robustness of the model. Homogeneity in our SD model refers to conformed classroom. To build the SD model, the causal loop and stock and flow diagrams should be developed [11]. However, only casual loop is presented in this paper due to the limited space[1]. Figure 1 shows how variables affect the other variables. The flow starts from the entering rate. For model simplicity, the focus of model is on only one group. Also, the entering rate is kept constant. The entering rate affects the population of the current observed group. Then, the population will affect the score of strategies. The mean of score changes the class standard deviation of scores. Afterward, the leaving rate is modified.

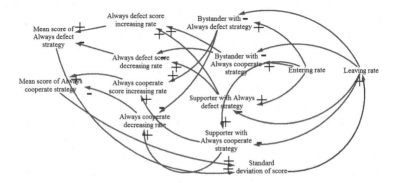

Fig. 1. Causal loop diagram (SD model)

Similar to ABM, a strategy with a high mean score shows that the strategy dominates the other strategy. In the same manner, strategies also affect standard deviation of score on all students in the group. To calculate the standard deviation of scores, the scores of students in the current group is used.

In succession, the standard deviation also affects the dynamics within the group via the leaving rate of the group. This leaving rate controls the number of students in the group. The rate becomes high as the variance becomes high as the diversity in the group would encourage students to search for more similar peers.

4   Experiments and Results

In the ABM, the simulation results are generated using the parameters defined in the ABM section of our previous paper [13]. To control randomness, one thousand simulation runs are generated. As most of the results have similar trend, the following setting

---

[1]   Stock and flow diagram is available with author upon request.

is selected: (I) five members per group (II) supporter to bystander ratio = 0.25 (III) probability of bystander becoming helper = 0.075 and (IV) $T_{sim} = 1000$ ticks

Figure 2 shows the average number of agents for each strategy using ABM and the average score of each strategy respectively. According Fig. 2, "Always defect" strategy is doing better than "Always cooperate" strategy.

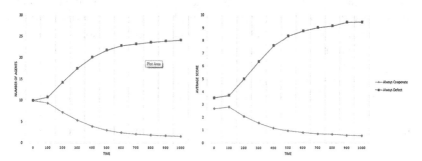

Fig. 2. Left-Average agent per strategy, Right- Average score per strategy (ABM)

Results from the SD model also yield similar phenomena. Figure 3 shows the average number of agents per strategy and the average score of each strategy respectively. Similarly, the "Always defect" strategy dominates the "Always cooperate" strategy.

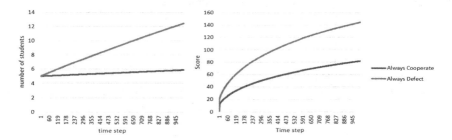

Fig. 3. Left- Agent for each strategy, Right-Score mean per strategy (SD model)

Although SD model yields similar result with ABM model, SD graph has tendency to become diverged whereas ABM graph becomes converged.

5   Discussion and Conclusion

We have developed both ABM and SD models for *Ijime* behavior with the use of PD game. The emergence can be observed through the presented graphs. The results from the experiments are also robust throughout parameters range. The generated simulation results imply the following: (I) Cooperation does not work well in this environment as defective decision becomes overwhelmed in all scenarios. This reflects current *Ijime* situation in Japan where victims are forced to cooperate with the group. This continuously put stress on the victim. Repeatedly, it can lead to serious conditions. (II) Alternatively, the victim can disrupt the class harmony by expressing his/her feeling. The victim can gain higher social position if he/she raises up the dissatisfaction. Hence, the effects of *Ijime* can be reduced through fighting back.

The results of both models differ in both number of student and also the score of each strategy. As the ABM environment becomes converged, the SD model becomes diverged. This may stem from the fact that the SD model lacks group dynamics that exists in the ABM. Group dynamics helps pushing the student to stay in the most suitable place while the SD model only focuses at one group at a time. Despite difference, both ABM and SD model produce similar result through macro perspective modeling. The results help confirming the robustness of ABM. In a class with homogenous student, they are also encouraged to try and fight back the bully.

In summary, using an ABM, we have gained a better understanding on how to deal with *Ijime*. By encouraging the victim to be brave, we can reduce the effect of *Ijime*. Similar to any other study, we have made some assumptions/limitations in this paper. For future work, we plan to compare our model with empirical observation.

References

1. Mitsuru, T.: Japanese school bullying: ijime. Ponencia presentada en la jornada "Comprendiendo y preveyendo el acoso escolar: una perspectiva internacional", *celebrada el*, 19 (2001)
2. Erica, P.: Adolescent Suicide in Japan: The Fatal Effects of Ijime (2011)
3. Shoko, Y.: The era of bullying: Japan under Neoliberalism (2008)
4. Robert, A., William, D.H.: The evolution of cooperation. Science **211**(4489), 1390–1396 (1981)
5. Wu, J., Axelrod, R.: How to cope with noise in the iterated prisoner's dilemma. J. Conflict Resolut. **39**(1), 183–189 (1995)
6. Morita, Y.: Bullying as a contemporary behaviour problem in the context of increasing 'societal privatization' in Japan. Prospects **26**(2), 311–329 (1996)
7. Naito, T., Gielen, U.P.: Bullying and Ijime in Japanese School: a sociocultural perspective. Violence in Schools, pp. 169–190. Springer, Berlin (2005)
8. Taki M: 'Ijime bullying': characteristic, causality and intervention, Oxford-Kobe Seminars (21–25 May 2003)
9. 矢野翔太; 近匡; 小柳文子. 囚人のジレンマを用いたいじめ発生メカニズムの解析と対策 (2013)
10. Hazhir, R., John, S.: Heterogeneity and network structure in the dynamics of diffusion: comparing agent-based and differential equation models. Manag. Sci. **54**(5), 998–1014 (2008)
11. Sterman, J.D.: Business Dynamics: Systems Thinking and Modeling for a Complex World. Irwin/McGraw-Hill, Boston (2000)
12. Chen, L.-C., Kaminsky, B., Tummino, T., Carley, K.M., Casman, E., Fridsma, D., Yahja, A.: Aligning simulation models of smallpox outbreaks. In: Chen, H., Moore, R., Zeng, D.D., Leavitt, J. (eds.) ISI 2004. LNCS, vol. 3073, pp. 1–16. Springer, Heidelberg (2004)
13. Thawiworadilok, C., Jafari Songhori, M., Takao, T.: Coping with bullying in the classroom through agent based modeling. In: The 9th Conference Workshop on Agent-Based Approach in Economic and Social Complex System, pp. 168–179 (2015)

On the Krom Extension of $\mathcal{CFDI}_{nc}^{\forall-}$

David Toman$^{(\boxtimes)}$ and Grant Weddell

Cheriton School of Computer Science, University of Waterloo, Waterloo, Canada
{david,gweddell}@cs.uwaterloo.ca

Abstract. We consider the consequences on basic reasoning problems of the Krom extension to the description logic dialect $\mathcal{CFDI}_{nc}^{\forall-}$, that is, of allowing negated primitive concepts on left-hand-sides of inclusion dependencies. Specifically, we show that TBox consistency and concept satisfiability remain in PTIME, but that this extension leads to intractability for both knowledge base consistency and instance retrieval. We then trace the roots of intractability by presenting tight conditions that recover PTIME complexity for both of these problems. The conditions relate to the structure of functional constraints in $\mathcal{CFDI}_{nc}^{\forall-}$ and to the unique name assumption.

1 Introduction

The $\mathcal{CFD}$ family of attribute-based *description logic* (DL) dialects are designed primarily to support efficient PTIME reasoning services about object relational data sources. The dialects share a common ability to support terminological cycles with universal (value) restrictions over functional roles together with a rich variety of functional constraints such as keys and functional dependencies over functional role paths.

A recent addition to this family is called $\mathcal{CFDI}_{nc}^{\forall-}$ [12]. In terms of expressive power, $\mathcal{CFDI}_{nc}^{\forall-}$ is a strict generalization of DL-Lite$_{core}^{\mathcal{F}}$, a DL dialect closely related to the W3C OWL 2 QL profile. In this paper, we consider the consequences on a number of basic reasoning problems of the Krom extension to $\mathcal{CFDI}_{nc}^{\forall-}$, that is, in which negated primitive concepts are permitted on left-hand-sides of inclusion dependencies occurring in a $\mathcal{CFDI}_{nc}^{\forall-}$ terminology or TBox. The reasoning problems we consider are as follows: (a) to determine if a given TBox is consistent, (b) to determine if a given concept is satisfiable in the context of a given TBox, and (c) to determine if a given knowledge base is consistent.

We proceed by first defining a new member of the $\mathcal{CFD}$ family called $\mathcal{CFDI}_{nc}^{\neg}$, a generalization of DL-Lite$_{Krom}^{\mathcal{F}}$. This dialect, however, imposes three restrictions on the Krom extension of $\mathcal{CFDI}_{nc}^{\forall-}$. The restrictions relate to functional constraints and to the so-called ABox part of a knowledge base, the part that captures facts about individuals in the underlying domain. We show that all the above-mentioned reasoning problems remain tractable for $\mathcal{CFDI}_{nc}^{\neg}$. We then show that the restrictions are tight for knowledge base consistency: determining

© Springer International Publishing Switzerland 2015
B. Pfahringer and J. Renz (Eds.): AI 2015, LNAI 9457, pp. 559–571, 2015.
DOI: 10.1007/978-3-319-26350-2_50

if a given knowledge base is consistent becomes NP-complete whenever *any one* of the three restrictions is removed. In contrast, we show that TBox consistency and concept satisfiability are tractable in the full Krom extension of $\mathcal{CFDI}_{nc}^{\forall-}$.

We begin in the next section by introducing the syntax and semantics of $\mathcal{CFDI}_{nc}^-$, and discuss its key features and limitations. In Sect. 3, we consider the problem of TBox consistency and concept satisfiability, and then more general knowledge base consistency in Sect. 4. A review of related work and summary comments follows in Sect. 5.

2   The Description Logic $\mathcal{CFDI}_{nc}^-$

All members of the $\mathcal{CFD}$ family of DLs are fragments of FOL with underlying signatures based on disjoint sets of unary predicate symbols called *primitive concepts*, constant symbols called *individuals* and unary function symbols called *attributes*. Note that incorporating attributes deviates from the normal practice of using binary predicate symbols called *roles*. However, among other things, attributes make it easier to incorporate concept constructors suited to the capture of relational data sources that include various dependencies by a straightforward reification of arbitrary n-ary predicates (including binary roles). An illustration of this will be given after first introducing a new member of the $\mathcal{CFD}$ family.

Definition 1 ($\mathcal{CFDI}_{nc}^-$ Concepts). Let F, PC and IN be disjoint sets of (names of) *attributes*, *primitive concepts* and *individuals*, respectively. A *path function* Pf is a word in F^* with the usual convention that the empty word is denoted by *id* and concatenation by ".". The set of $\mathcal{CFDI}_{nc}^-$ concepts augments PC with additional expressions given by the following forms:

$$\neg A, \quad \forall f.A, \quad \forall f.\neg A, \quad \exists f^-, \text{ and } A : \mathsf{Pf}_1, \mathsf{Pf}_2, \dots, \mathsf{Pf}_k \to \mathsf{Pf}, \qquad (1)$$

where $A \in$ PC, $f \in$ F, $k > 0$, and where Pf and Pf_i are path functions such that Pf is a prefix of Pf_1. We write L, possibly subscripted, to denote a *literal concept*, that is, either a primitive concept or the negation of a primitive concept.

Semantics is defined in the standard way with respect to an interpretation $\mathcal{I} = (\triangle, (\cdot)^{\mathcal{I}})$, where $\triangle$ is a domain of "objects" and $(\cdot)^{\mathcal{I}}$ an interpretation function that fixes the interpretation of primitive concepts A to be subsets of $\triangle$, attributes f to be total functions on $\triangle$, and individuals a, b to be elements of $\triangle$. The interpretation function is extended to path expressions by interpreting *id* as the identity function and concatenation as function composition. The semantics of the remaining $\mathcal{CFDI}_{nc}^-$ concepts are then defined as follows:

$$(\neg A)^{\mathcal{I}} = \triangle \setminus A^{\mathcal{I}}$$
$$(\forall f.L)^{\mathcal{I}} = \{x \mid f^{\mathcal{I}}(x) \in L^{\mathcal{I}}\}$$
$$(\exists f^-)^{\mathcal{I}} = \{x \mid \exists y \in \triangle \mid x = f^{\mathcal{I}}(y)\}$$
$$(A : \mathsf{Pf}_1, \dots, \mathsf{Pf}_k \to \mathsf{Pf})^{\mathcal{I}} = \{x \mid \forall y \in A^{\mathcal{I}} :$$
$$\bigwedge_i \mathsf{Pf}_i^{\mathcal{I}}(x) = \mathsf{Pf}_i^{\mathcal{I}}(y) \to \mathsf{Pf}^{\mathcal{I}}(x) = \mathsf{Pf}^{\mathcal{I}}(y)\} \qquad \square$$

A concept resembling the last of the forms in (1) is called a *key path functional dependency* (key PFD). Informally, such a concept denotes a set of objects, each of which, whenever agreeing with any A-object on all left-hand-side path functions, also agrees with that object on Pf.

Definition 2 ($\mathcal{CFDI}_{nc}^{-}$ Knowledge Bases). Metadata and data in a $\mathcal{CFDI}_{nc}^{-}$ knowledge base $\mathcal{K}$ are respectively defined by a *TBox $\mathcal{T}$* and an *ABox $\mathcal{A}$*. A TBox $\mathcal{T}$ consists of a finite set of *inclusion dependencies*, each of which adheres to one of the following forms:

$$L_1 \sqsubseteq L_2, \quad L_1 \sqsubseteq \forall f.L_2, \quad A \sqsubseteq \exists f^-, \text{ and } A_1 \sqsubseteq A_2 : \mathsf{Pf}_1, \ldots, \mathsf{Pf}_k \to \mathsf{Pf},$$

where $\{A, A_1, A_2\} \subseteq \mathsf{PC}$, $f \in \mathsf{F}$, L_1 and L_2 are literal concepts, and Pf and Pf_i are path functions. An ABox $\mathcal{A}$ consists of a finite set of facts, each of which is either a *concept assertion $A(a)$* or a *basic function assertion $f(a) = b$*, where $A \in \mathsf{PC}$, $f \in \mathsf{F}$, and $\{a, b\} \subseteq \mathsf{IN}$.

An interpretation $\mathcal{I}$ satisfies an inclusion dependency $C_1 \sqsubseteq C_2$ if $(C_1)^{\mathcal{I}} \subseteq (C_2)^{\mathcal{I}}$, a concept assertion $A(a)$ if $a^{\mathcal{I}} \in A^{\mathcal{I}}$ and a basic function assertion $f(a) = b$ if $f^{\mathcal{I}}(a^{\mathcal{I}}) = b^{\mathcal{I}}$. $\mathcal{I}$ satisfies a knowledge base $\mathcal{K}$ if it satisfies each inclusion dependency and fact in $\mathcal{K}$. We also assume $\mathcal{I}$ satisfies UNA (the *unique name assumption*), i.e., $a^{\mathcal{I}} \neq b^{\mathcal{I}}$ for any distinct pair of individuals a and b occurring in $\mathcal{K}$. □

Like $\mathcal{CFDI}_{nc}^{\forall-}$, an additional restriction on the interactions between inverse attributes and value restrictions must be imposed on a $\mathcal{CFDI}_{nc}^{-}$ TBox, namely every TBox $\mathcal{T}$ is also assumed to satisfy the following syntactic condition:

If $\{A \sqsubseteq \exists f^-, \ L_1 \sqsubseteq \forall f.L_2\} \subseteq \mathcal{T}$ then either
$$A \sqsubseteq \neg L_2 \in \mathcal{T}, \neg L_2 \sqsubseteq A \in \mathcal{T}, or A \sqsubseteq L_2 \in \mathcal{T}. \quad (2)$$

It has been shown that, without these restrictions, TBox reasoning alone becomes intractable (EXPTIME-complete) [12]. Also, to simplify further development (and w.l.o.g.), we assume an ABox will also satisfy a syntactic attribute functionality condition, in particular, that, for any individual a and attribute f, there is at most one basic function assertion of the form $f(a) = b$. Indeed, multiple distinct assertions for a given individual and attribute pair immediately lead to inconsistency under UNA.

Note that $\mathcal{CFDI}_{nc}^{-}$ falls short of the Krom extension of $\mathcal{CFDI}_{nc}^{\forall-}$. In particular, the latter does not assume UNA and allows more flexibility: an TBox and ABox may contain a variety of non-key PFDs and more general function assertions of the form $\mathsf{Pf}_1(a) = \mathsf{Pf}_2(b)$, respectively. One feature of $\mathcal{CFDI}_{nc}^{\forall-}$ that remains possible, however, is a capacity for value restrictions on left-hand-sides of inclusion dependencies, e.g., the inclusion dependency $\forall f.A_1 \sqsubseteq A_2$ is equivalent to $\neg A_2 \sqsubseteq \forall f.\neg A_1$.

To illustrate using $\mathcal{CFDI}_{nc}^{-}$ to capture relational data sources, consider a ternary ENROLL relation for a particular data source that records information about students, courses and grades. The relation can be reified as the *Takes* primitive concept by including the following in a TBox:

$$Takes \sqsubseteq \forall student.Student$$
$$Takes \sqsubseteq \forall course.Course$$
$$Takes \sqsubseteq \forall grade.Integer$$
$$Takes \sqsubseteq Takes : student, course, grade \rightarrow id$$

The first three inclusion dependencies ensure that component attributes are appropriately "typed" (observe how unary foreign keys are naturally captured); the fourth ensures that reification captures the underlying set semantics for relation ENROLL (no pair of distinct tuples can have the same values for all attributes) with the help of a key PFD. It is also straightforward to further restrict who may enroll in graduate courses. For example, by adding the following inclusion dependencies to the TBox:

$$\neg GradTakes \sqsubseteq \forall course.\neg GradCourse$$
$$GradTakes \sqsubseteq \forall student.GradStudent$$
$$GradTakes \sqsubseteq Takes$$

one ensures that only graduate students can be enrolled in graduate courses.

Negation on left-hand-sides can also be used to introduce a "top" concept, say *Thing*, by including the following in a TBox:

$$Student \sqsubseteq Thing$$
$$\neg Student \sqsubseteq Thing$$

The primitive *Thing* concept can then be used enforce general range restrictions for attributes:

$$Thing \sqsubseteq \forall student.Student.$$

Note that this ability to reify relationships and thus roles might seem to enable simulating role hierarchies ($\mathcal{H}$) via subsumptions between primitive concepts corresponding to reified roles. However, condition (2) above will interfere and lead to spurious inferences not implied by the role hierarchy alone if one is not careful [13]. This, however, should be expected since the unrestricted combination of role hierarchies and functionality constraints leads to intractability already in DL-Lite [1].

3 TBox Consistency and Concept Satisfiability

In this section, we introduce an NLOGSPACE procedure for deciding TBox consistency for $\mathcal{CFDI}_{nc}^{\neg}$.

Definition 3. Let PC and F be finite sets of primitive concepts and attributes, respectively. We define an *implication graph* over PC and F to be a node and edge labeled graph $G = (V, E)$ whose nodes are

$$V = \mathsf{PC} \cup \{\neg A \mid A \in \mathsf{PC}\} \cup \{\forall f.A \mid A \in \mathsf{PC}, f \in \mathsf{F}\}$$
$$\cup \{\forall f.\neg A \mid A \in \mathsf{PC}, f \in \mathsf{F}\}$$

and whose edges are labeled by $\mathsf{F} \cup \{\epsilon\}$ and such that $\forall f.A \xrightarrow{f} A$ and $\forall f.\neg A \xrightarrow{f} \neg A$ are among the edges occurring in E.   □

In the following, the nodes of implication graphs are identified with their labels. We call the pairs $(A, \neg A)$ and $(\forall f.A, \forall f.\neg A)$ complementary, and, if N is a node in an implication graph G, write $\neg N$ to denote the node of G for the concept label that is complementary to N. We call a node N in the implication graph $G = (V, E)$ an *empty node* if $N \xrightarrow{\epsilon} \neg N \in E$. We consider *paths* in G to be synonymous with words in a finite automaton based on G in which ϵ denotes the empty transition. Thus we identify, e.g., $\epsilon f \epsilon$ with ϵf, with f, etc.

Definition 4 (Implication Graph for $\mathcal{T}$). Let $\mathcal{T}$ be a $\mathcal{CFDI}_{nc}^{-}$ TBox and $G = (V, E)$ be an implication graph. We say that G is *closed under consequences* if for all nodes N, N_1, and N_2

1. $N \xrightarrow{\epsilon} N \in E$,
2. if $N_1 \xrightarrow{\epsilon} N_2 \in E$ then $\neg(N_2) \xrightarrow{\epsilon} \neg(N_1) \in E$,
3. if $N_1 \xrightarrow{\epsilon} N \in E$ and $N \xrightarrow{\epsilon} N_2 \in E$ then $N_1 \xrightarrow{\epsilon} N_2 \in E$,
4. if $N_1 \xrightarrow{\epsilon} N_2 \in E$ then $\forall f.N_1 \xrightarrow{\epsilon} \forall f.N_2 \in E$ for N_1 and N_2 primitive or negations of primitive concepts, and
5. if $N \sqsubseteq \exists f^- \in \mathcal{T}$, $N \xrightarrow{\epsilon} N_1 \in E$ and $\forall f.N_1 \xrightarrow{\epsilon} \forall f.N_2 \in E$ then $N \xrightarrow{\epsilon} N_2 \in E$.

We define an *implication graph for $\mathcal{T}$* to be the least implication graph $G = (V, E)$ over the primitive concepts and attributes occurring in $\mathcal{T}$ that is closed under consequences and such that if $C \sqsubseteq D \in \mathcal{T}$ then $C \xrightarrow{\epsilon} D \in E$ for D not a PFD or $\exists f^-$.   □

The implication graph for $\mathcal{T}$ makes implications between literals implied by $\mathcal{T}$ *explicit*: $N_1 \xrightarrow{\epsilon} N_2 \notin E$ implies $N_1 \sqcap \neg N_2$ is satisfiable w.r.t. $\mathcal{T}$ since it is not possible for two paths to exist that (a) traverse the same attributes and inverses (for literals subsumed by $\exists f^-$), (b) originate in N_1 and $\neg N_2$, respectively, and (c) end in a complementary pair of concepts. This is enforced by Definition 4 and yields the following theorem:

Theorem 5. Let $\mathcal{T}$ be a $\mathcal{CFDI}_{nc}^{-}$ TBox in normal form. Then $\mathcal{T}$ is consistent if and only if no two complementary nodes are empty in the implication graph G for $\mathcal{T}$.

Proof (sketch): If two complementary nodes, e.g., A and $\neg A$, are empty then $\mathcal{T} \models A \sqsubseteq \neg A$ and $\mathcal{T} \models \neg A \sqsubseteq A$; and thus $\mathcal{T}$ is not consistent (as the above implies $\triangle = \emptyset$).

If no such pair exists then we define a tree interpretation $\mathcal{I}$ by repeating the following construction starting from an initial root object o with an empty label (we write $N \in n$ for N is an element of the label of the node n):

1. Let n be a node that has *not* been labeled by neither N nor $\neg N$ for some literal N. Then if $N' \xrightarrow{\epsilon} N \in G$ and $N' \in n$ we add N to the label of n. Otherwise, either N or $\neg N$ must be non-empty (by assumption) and neither

of the two literals is implied by any of the existing labels of n. We add N if N is non-empty and $\neg N$ otherwise. Definition 4 (1–3) guarantees that this cannot lead to inconsistency.

2. When (1) does not apply to n and an f successor of n does not exist we create a new node m and add to its label all concepts A such that $\forall f.A \in n$ and $\neg A$ for $\neg \forall f.A \in n$. Definition 4 (1–2, 4) guarantees that this cannot lead to inconsistency.

3. When (1) does not apply to n, $A \in n$, $A \sqsubseteq \exists f^- \in \mathcal{T}$, n is not an f successor of another node, and an f^- successor n (which, in the final interpretation, will correspond to an f predecessor of n) does not exist, we create a new node m and add to its label all concepts $\forall f.N \in m$ whenever $N \in n$. Definition 4 (1–2, 5) guarantees that this cannot lead to inconsistency.

The result of the construction yields a tree model $\mathcal{I}$ for $\mathcal{T}$ whose domain consists of all nodes created above, and whose class membership is defined by primitive concepts present in their labels. The interpretation of attributes then corresponds to the edges created during the construction.   □

Note that, in step (3), one f^- successor is sufficient since the logic is not able to enforce the existence of multiple distinct f^- successors. Also note that the above interpretation is tree-like. As a consequence, key PFDs in $\mathcal{T}$ are satisfied vacuously and therefore need not be taken into account in the above construction nor in the definition of an implication graph. This would also be the case for non-key PFDs were they permitted in $\mathcal{CFDI}_{nc}^-$. Since the remaining restrictions manifest in $\mathcal{CFDI}_{nc}^-$ are also not a factor in the above construction (they only concern an ABox), our results extend to the full Krom extension of $\mathcal{CFDI}_{nc}^{\forall-}$.

It is easy to see that a simple saturation algorithm that applies the rules from Definition 4 runs in $\mathcal{O}(|\mathcal{T}|^2)$; a more refined NLOGSPACE algorithm utilizes graph (path) search for inconsistency for each pair of complementary nodes in G. A lower bound for TBox consistency is inherited from the Krom fragment of propositional logic.

Corollary 6. $\mathcal{CFDI}_{nc}^-$ TBox consistency is NLOGSPACE-complete.

Whenever $\mathcal{T}$ is consistent, the implication graph for $\mathcal{T}$ also records which primitive concepts (and/or their negations) are empty in every model, those for which $A \xrightarrow{\epsilon} \neg A \in E$.

Corollary 7. Concept satisfiability with respect to a $\mathcal{CFDI}_{nc}^-$ TBox is NLOGSPACE-complete.

4   ABox Reasoning and $\mathcal{K}$ Satisfiability

The complexity landscape for ABox reasoning, in particular for knowledge base consistency, depends crucially on the restrictions imposed on the Krom extension of $\mathcal{CFDI}_{nc}^{\forall-}$ manifest in $\mathcal{CFDI}_{nc}^-$: allowing key PFDs and basic function assertions only, and on the adoption of UNA. Later on in this section, we show that relaxing any of these restrictions (in an attempt to extend $\mathcal{CFDI}_{nc}^-$ to be the Krom extension of $\mathcal{CFDI}_{nc}^{\forall-}$) leads to intractability.

The Tractable Cases. We first focus on the tractable cases starting by considering KB consistency for $\mathcal{CFDI}_{nc}^-$ knowledge bases without PFDs:

Definition 8 (2-CNF for an ABox). Let $\mathcal{T}$ be a consistent TBox without any mention of the PFD concept constructor and $\mathcal{A}$ an ABox. We define 2-CNF$(\mathcal{T}, \mathcal{A})$ to be the set of propositional 2-clauses over (ground) literals $L_i(a)$ of the form $A(a)$ and $\neg A(a)$ defined as follows:

– $A(a)$ for $A(a) \in \mathcal{A}$,
– $L_1(a) \to L_2(a)$ for $\mathcal{T} \models L_1 \sqsubseteq L_2$,
– $L_1(a) \to L_2(b)$ for $f(a) = b \in \mathcal{A}, \mathcal{T} \models L_1 \sqsubseteq \forall f.L_2$,
– $L_1(b) \to L_2(a)$ for $f(a) = b \in \mathcal{A}, \mathcal{T} \models \forall f.L_1 \sqsubseteq L_2$,

where a is an ABox individual, A a primitive concept, and f an attribute.

Recall that we have shown in Sect. 3 that any implication questions of the form $\mathcal{T} \models L_1 \sqsubseteq L_2$ can be *read* directly from the implication graph G for $\mathcal{T}$ since all implications of this form are explicit in G. Thus, we have the following:

Theorem 9. Let $\mathcal{T}$ be a $\mathcal{CFDI}_{nc}^-$ TBox without any mention of the PFD concept constructor and $\mathcal{A}$ an ABox. Then the knowledge base $\mathcal{K} = (\mathcal{T}, \mathcal{A})$ is consistent if and only if $\mathcal{T}$ is consistent and 2-CNF$(\mathcal{T}, \mathcal{A})$ is satisfiable.

Proof (sketch): A satisfying assignment for 2-CNF$(\mathcal{T}, \mathcal{A})$ yields an interpretation for the ABox of $\mathcal{K}$: $a^{\mathcal{I}} \in A^{\mathcal{I}}$ if the proposition $A(a)$ is true in that assignment. To complete this interpretation, since ABox individuals may be missing some attributes, we follow the construction used in Theorem 5.

Conversely, if $\mathcal{K}$ is consistent, the class membership of ABox objects yields a satisfying assignment for 2-CNF$(\mathcal{T}, \mathcal{A})$.   □

Note that the theorem can be trivially strengthened by allowing an ABox to contain general assertions of the form $\mathsf{Pf}_1(a) = \mathsf{Pf}_2(b)$ since, in the absence of PFDs, there is never a need to equate ABox objects. For the same reason, UNA need not be assumed since, in this case, reasoning under UNA coincides with reasoning without UNA.

Corollary 10. $\mathcal{CFDI}_{nc}^-$ knowledge base consistency is NLOGSPACE-complete.

Adding Keys Under UNA. Testing for consistency in the presence of key constraints relies on the observation that, under UNA, the *structure* of an ABox (i.e., how individuals are connected to others via attributes) is *fixed*. Hence, key violations reduce to determining whether, for a pair of individuals, their membership in concept descriptions triggers a *violation* of a particular key constraint: we *prohibit* such interpretations by adding additional 2-clauses to the ABox propositionalization. A prototypical example of such a violation (for a pair of individuals a and b in $\mathcal{A}$) and a key PFD

$$A \sqsubseteq B : \mathsf{Pf}.\mathsf{Pf}_1, \ldots, \mathsf{Pf}_k \to \mathsf{Pf} \tag{3}$$

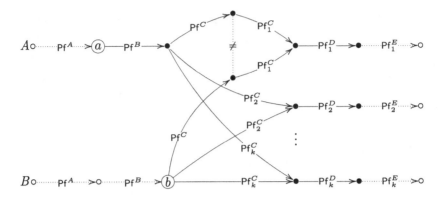

Fig. 1. Potential Key PFD violation by a and b in $\mathcal{A}$.

where, in general, Pf is the composition $\mathsf{Pf}^A\,\mathsf{Pf}^B\,\mathsf{Pf}^C$, Pf_1 is $\mathsf{Pf}_1^C\,\mathsf{Pf}_1^D\,\mathsf{Pf}_1^E$, and Pf_i are $\mathsf{Pf}^A\,\mathsf{Pf}^B\,\mathsf{Pf}_i^C\,\mathsf{Pf}_i^D\,\mathsf{Pf}_i^E$ for $1 < i \le k$ as depicted in Fig. 1: the solid nodes and edges depict (part of) the explicit ABox assertions, empty nodes and dotted edges are objects and attribute paths *implied* by attribute totality and *inverse* constraints. Note that while no path agreements can be realized outside of the ABox, the path functions in the key constraint may start and finish *outside* of the explicit ABox. We formalize this idea using the following definitions:

Definition 11. Let $\mathcal{T}$ be a TBox, $\mathsf{Pf} = f_k.\cdots.f_1$ be a path function, and A_0 and A primitive concepts. We say that A_0 has a Pf-*predecessor* A in $\mathcal{T}$ if (i) $\mathcal{T} \models A_i \sqsubseteq \exists f_{i+1}^-$ and $\mathcal{T} \models \forall f_{i+1}.A_i \sqsubseteq A_{i+1}$ for $0 \le i < k$, and (ii) $\mathcal{T} \models A_k \sqsubseteq A$ for some concepts $A_1,\ldots,A_k$. □

Definition 12. Let $\mathcal{K} = (\mathcal{T},\mathcal{A})$ be a KB and a and b two distinct individuals in $\mathcal{A}$ and A_0 and B_0 two primitive concepts. We say that a PFD key constraint of the form (3) is *potentially violated in* $\mathcal{K}$ *by a pair of assertions* $a : A_0$ *and* $b : B_0$ if

1. A_0 has an Pf^A-predecessor A and B_0 has an $\mathsf{Pf}^A\,\mathsf{Pf}^B$-predecessor B in $\mathcal{T}$,
2. $\mathsf{Pf}^B\,\mathsf{Pf}^C\,\mathsf{Pf}_i^C$ and $\mathsf{Pf}^C\,\mathsf{Pf}_i^C$ reach the same $\mathcal{A}$ object starting from a and b, respectively,
3. $\mathsf{Pf}^B\,\mathsf{Pf}_i^C$ and Pf_i^C reach the same $\mathcal{A}$ object starting from a and b, respectively, for all $1 < i \le k$, and
4. $\mathsf{Pf}^B\,\mathsf{Pf}^C$ and Pf^C do not reach the same $\mathcal{A}$ object starting from a and b. □

Under UNA, such potentially violated key PFDs can only be satisfied if the objects a and b do *not* belong simultaneously to the descriptions A_0 and B_0, respectively, since UNA prevents us from *repairing* the violation by equating the offending objects. Consequently, one can simply add the following 2-CNF clauses,

$A_0(a) \to \neg B_0(b)$ and $B_0(b) \to \neg A_0(a)$, for all pairs of individuals a, b in $\mathcal{A}$ and key PFDs $L_1 \sqsubseteq L_2 : \mathsf{Pf}_1,\ldots,\mathsf{Pf}_k \to \mathsf{Pf}$ in $\mathcal{T}$ for which the latter is potentially violated by the former w.r.t. a and b and A_0 and B_0, called KEY-2-CNF$(\mathcal{T}, \mathcal{A})$, to the set of clauses 2-CNF$(\mathcal{T}, \mathcal{A})$ to account for this situation (recall Definition 8 above for the latter). Hence:

Theorem 13. Let $\mathcal{T}$ be a $\mathcal{CFDI}_{nc}^{\neg}$ TBox with arbitrary occurrences of key PFDs and let $\mathcal{A}$ be an ABox. Then, assuming UNA, the knowledge base $\mathcal{K} = (\mathcal{T}, \mathcal{A})$ is consistent if and only if $\mathcal{T}$ is consistent and 2-CNF$(\mathcal{T}, \mathcal{A})$ $\cup$ KEY-2-CNF$(\mathcal{T}, \mathcal{A})$ is satisfiable.

Proof (sketch): The proof is essentially the same as for Theorem 9 since the interpretation of the ABox makes all PFDs in $\mathcal{T}$ satisfied (the KEY-2-CNF$(\mathcal{T}, \mathcal{A})$ clauses guarantee that) and since the additional anonymous objects cannot violate PFDs since corresponding parts of the interpretation are tree shaped.   □

Note that KEY-2-CNF$(\mathcal{T}, \mathcal{A})$ contains at most $\mathcal{O}(|\mathcal{T}||\mathcal{A}|^2)$ clauses.

Corollary 14. $\mathcal{CFDI}_{nc}^{\neg}$ knowledge base consistency is NLOGSPACE-complete.

The results above assume that the ABox contains concept and basic function assertions only. However, this condition can be relaxed without harm to allow path assertions of the form "$f(a) = g(b)$" for which the same construction and proof argument would apply.

Intractable Extensions. Unfortunately, relaxing any of the conditions imposed on $\mathcal{CFDI}_{nc}^{\neg}$ KBs, namely, UNA, allowing key PFDs only, and requiring primitive ABoxes, leads to intractability (hence, $\mathcal{CFDI}_{nc}^{\neg}$ is incomparable to $\mathcal{CFDI}_{nc}^{\forall-}$ and there is no tractable extension that would encompass $\mathcal{CFDI}_{nc}^{\forall-}$). For each of the cases, we show a reduction of 3-SAT to knowledge base consistency. Figure 2 illustrates the ABox *widgets* used to capture the behavior of 3-clauses in the three respective cases: an assignment that makes a clause false will correspond to an interpretation that makes the corresponding widget unsatisfied.[1]

The Keys Without UNA Case. We reduce 3-CNF satisfiability to KB consistency. Let φ be a propositional formula in 3-CNF with clauses $C_1, \ldots, C_k$ and propositional variables $x_1, \ldots, x_n$. Define a $\mathcal{CFDI}_{nc}^{\neg}$ knowledge base $\mathcal{K}_\varphi = (\mathcal{T}, \mathcal{A})$ as follows:

1. For each propositional variable x_i in φ, introduce an $\mathcal{A}$ individual x_i.
2. For each clause C_j introduce individuals a_j, b_j, and c_j.
3. For the variables x_{i_1}, x_{i_2}, and x_{i_3} that appear in a clause C_j, include the following basic function assertions in $\mathcal{A}$: $l_{j,1}(x_{i_1}) = a_j, l_{j,2}(x_{i_2}) = a_j$, and $l_{j,3}(x_{i_3}) = b_j$.
4. Add assertions $f(a_j) = c_j$ and $f(b_j) = c_j$ to $\mathcal{A}$.
5. Add concept assertion $B(b_j)$ to $\mathcal{A}$.

---

[1] The widgets, in principle, can support 4-clauses. For 3-CNF we fixed concept membership of one of the objects to B.

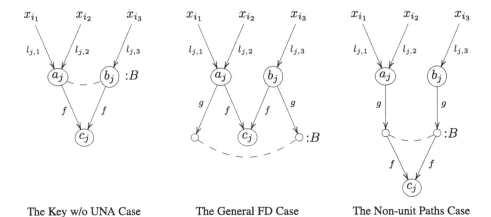

The Key w/o UNA Case   The General FD Case   The Non-unit Paths Case

Fig. 2. 3-SAT reduction widgets for clause C_j.

6. Add the following inclusion dependencies to $\mathcal{T}$:
   - $T \sqsubseteq \forall l_{j,1}.\neg A$ and $\neg T \sqsubseteq \forall l_{j,1}.A$ when x_{i_1} appears positively in C_j, and $T \sqsubseteq \forall l_{j,1}.A$ and $\neg T \sqsubseteq \forall l_{j,1}.\neg A$ when x_{i_1} appears negatively in C_j;
   - $T \sqsubseteq \forall l_{j,2}.B$ and $\neg T \sqsubseteq \forall l_{j,2}.\neg B$ when x_{i_2} appears positively in C_j, and $T \sqsubseteq \forall l_{j,2}.\neg B$ and $\neg T \sqsubseteq \forall l_{j,2}.B$ when x_{i_2} appears negatively in C_j; and
   - $T \sqsubseteq \forall l_{j,3}.\neg A$ and $\neg T \sqsubseteq \forall l_{j,3}.A$ when x_{i_3} appears positively in C_j, and $T \sqsubseteq \forall l_{j,3}.A$ and $\neg T \sqsubseteq \forall l_{j,3}.\neg A$ when x_{i_3} appears negatively in C_j.
7. Finally, add the key PFD $A \sqsubseteq A : f \to id$ to $\mathcal{T}$.

A truth assignment to the propositions x_i occurring in φ is then encoded by an interpretation $\mathcal{I}$ of $\mathcal{K}_\varphi$: $x_i^{\mathcal{I}} \in T^{\mathcal{I}}$ is where x_i is true, and $x_i^{\mathcal{I}} \in \neg T^{\mathcal{I}}$ is where x_i is false. With this in mind, it is straightforward to confirm that any interpretation of $\mathcal{K}_\varphi$ will only encode a *satisfying* truth assignment and that, in turn, an interpretation of $\mathcal{K}_\varphi$ can always be constructed by a satisfying truth assignment (see the proof sketch to Lemma 15 below).

The General Functional Dependencies Case. Allowing *functional dependencies* of the form $A : f_1, \ldots, f_k \to f$ where f is not one of the f_i (i.e., non-key PFD) even under UNA also leads to intractability. The reduction shares the first four steps with the previous case:

5. Add concept assertion $\forall g.B(b_j)$ to $\mathcal{A}$.
6. Add the following to $\mathcal{T}$:
   - $T \sqsubseteq \forall l_{j,1}.\neg A$ and $\neg T \sqsubseteq \forall l_{j,1}.A$ when x_{i_1} appears positively in C_j, and $T \sqsubseteq \forall l_{j,1}.A$ and $\neg T \sqsubseteq \forall l_{j,1}.\neg A$ when x_{i_1} appears negatively in C_j.
   - $T \sqsubseteq \forall l_{j,2}.\forall g.B$ and $\neg T \sqsubseteq \forall l_{j,2}.\forall g.\neg B$ when x_{i_2} appears positively in C_j, and $T \sqsubseteq \forall l_{j,2}.\forall g.\neg B$ and $\neg T \sqsubseteq \forall l_{j,2}.\forall g.B$ when x_{i_2} appears negatively in C_j; and
   - $T \sqsubseteq \forall l_{j,3}.\neg A$ and $\neg T \sqsubseteq \forall l_{j,3}.A$ when x_{i_3} appears positively in C_j, and $T \sqsubseteq \forall l_{j,3}.A$ and $\neg T \sqsubseteq \forall l_{j,3}.\neg A$ when x_{i_3} appears negatively in C_j.
7. $A \sqsubseteq A : f \to g \in \mathcal{T}$.

The Non-unit Path Agreements Case. Similarly, allowing non-primitive ABoxes with path function assertions of the form $\mathsf{Pf}_1(a) = \mathsf{Pf}_2(b)$ even under UNA leads to intractability. The reduction again shares the first three steps with the previous cases:

4. Add $g.f(a_j) = c_j$, $g.f(b_j) = c_j$.
5. Add $\forall g.B(b_j)$ to $\mathcal{A}$.
6. Add the following to $\mathcal{T}$:
   - $T \sqsubseteq \forall l_{j,1}.\forall g.\neg A$ and $\neg T \sqsubseteq \forall l_{j,1}.\forall g.A$ when x_{i_1} appears positively in C_j, and $T \sqsubseteq \forall l_{j,1}.\forall g.A$ and $\neg T \sqsubseteq \forall l_{j,1}.\forall g.\neg A$ when x_{i_1} appears negatively in C_j.
   - $T \sqsubseteq \forall l_{j,2}.\forall g.B$ and $\neg T \sqsubseteq \forall l_{j,2}.\forall g.\neg B$ when x_{i_2} appears positively in C_j, and $T \sqsubseteq \forall l_{j,2}.\forall g.\neg B$ and $\neg T \sqsubseteq \forall l_{j,2}.\forall g.B$ when x_{i_2} appears negatively in C_j; and
   - $T \sqsubseteq \forall l_{j,3}.\forall g.\neg A$ and $\neg T \sqsubseteq \forall l_{j,3}.\forall g.A$ when x_{i_3} appears positively in C_j, and $T \sqsubseteq \forall l_{j,3}.\forall g.A$ and $\neg T \sqsubseteq \forall l_{j,3}.\forall g.\neg A$ when x_{i_3} appears negatively in C_j.
7. Finally, add the key PFD $A \sqsubseteq A : f \rightarrow id$ to $\mathcal{T}$.

Note the use of paths of length 2 in step 4 to construct anonymous objects that evade UNA. In all three cases it is easy to confirm that:

Lemma 15. Let φ be a propositional formula in 3-CNF. Then φ is satisfiable if and only if $\mathcal{K}_\varphi$ is consistent.

Proof (sketch): It is easy to verify that a satisfying assignment for φ can be used to create a model for $\mathcal{K}$, essentially by assigning the class membership of the individuals x_{i_j} to the primitive concepts T and $\neg T$ and then extending this assignment to the a_j, b_j and c_j individuals as dictated by the constraints in $\mathcal{T}$.

Conversely, a model $\mathcal{I}$ for $\mathcal{K}$ induces an assignment of x_i to true if and only if $x_i^{\mathcal{I}} \in T^{\mathcal{I}}$. This in turn guarantees that at least one of the literals in every clause is true due to step (6) in the reductions since the assignment together with the concept B would otherwise make the widget corresponding to the particular 3-CNF clause inconsistent: it would force the two individuals connected by the dashed lines in Fig. 2 to be equal as a consequence of the PFD and to belong to B and $\neg B$ at the same time, thus disqualifying such an interpretation as a model of $\mathcal{K}$. □

Hence, knowledge base consistency without assuming UNA or when allowing functional dependencies in $\mathcal{T}$ or non-primitive path equalities in $\mathcal{A}$ is NP-complete: membership in NP is established by guessing primitive classes and equalities for ABox individuals.

Instance Retrieval. Since $\mathcal{CFDI}_{nc}^{\neg}$ is closed under negation, *instance retrieval*, the problem of determining if $\mathcal{K} \models C(a)$, reduces naturally to determining knowledge base (in)consistency: $\mathcal{K} \models C(a)$ if and only if $(\mathcal{T}, \mathcal{A} \cup \{\neg C(a)\})$ is inconsistent for $\mathcal{K} = (\mathcal{T}, \mathcal{A})$. Thus, instance retrieval inherits the computational properties of testing for consistency:

– $\mathcal{CFDI}_{nc}^{-}$ instance retrieval is NLOGSPACE-complete.
– Instance retrieval is coNP-complete for all extensions of $\mathcal{CFDI}_{nc}^{-}$ introduced in Sect. 4.

The instance retrieval result cannot, however, be extended to conjunctive queries: it is a straightforward consequence of results of Schaerf that allowing negated primitive concepts on left-hand-sides of inclusion dependencies alone makes CQ answering coNP-complete [7].

5   Related Work and Summary Comments

PFD-based constraints were first introduced and studied in the context of graph-oriented data models (similar to RDF) and its refinements [5,14]. Subsequently, an FD concept constructor was proposed and incorporated in Classic [2], an early DL with PTIME reasoning capabilities, without changing the complexity of its implication problem. However, this paper shows that tractability is lost for the Krom extensions of such logics whenever non-key PFDs are allowed.

Logics in the $\mathcal{CFD}$ family that allow conjunctions on the left-hand-side of subsumptions [11] (i.e., Horn fragments) cannot support tractability in the presence of, e.g., disjointness. The most notable cases are as follows.

– Allowing conjunction "⊓" yields the logic $\mathcal{CFD}^{\perp}$ and therefore makes reasoning PSPACE-complete [11].
– Allowing conjunction and value restriction "∀" makes reasoning EXPTIME-complete, even in the absence of negation (or the empty concept $\perp$) [6].
– Allowing unrestricted PFDs (alone) also leads to EXPTIME-completeness of logical implication [6], and makes general reasoning (in particular, knowledge base consistency) undecidable [8,10].

Adding various forms of functional dependencies and keys to other DLs—usually as additional separate varieties of constraints, often called a *key box*—have also been considered, e.g., by [4]. They show that their dialect is undecidable for description logics with inverse roles, but becomes decidable when unary functional dependencies are disallowed. This line of investigation is continued in the context of PFDs and inverse attributes, with analogous results [9]. Subsequently, Calvanese et al. have shown how DL-Lite can be extended with a path-based variety of identification constraints analogous to PFDs without affecting the complexity of reasoning problems [3].

In this paper, we have explored the consequences of the Krom extension of $\mathcal{CFDI}_{nc}^{\forall-}$. We have shown that TBox consistency is no longer trivial but still tractable (NLOGSPACE-complete). We have also seen that tractability of knowledge base consistency is preserved when additional syntactic conditions are imposed on the extension, yielding $\mathcal{CFDI}_{nc}^{-}$. On the other hand, we have also shown that relaxing any of these restrictions leads to intractability.

References

1. Artale, A., Calvanese, D., Kontchakov, R., Zakharyaschev, M.: The DL-Lite family and relations. J. AI Res. **36**, 1–69 (2009)
2. Borgida, A., Weddell, G.: Adding uniqueness constraints to description logics. In: Bry, F. (ed.) DOOD 1997. LNCS, vol. 1341, pp. 85–102. Springer, Heidelberg (1997)
3. Calvanese, D., De Giacomo, G., Lembo, D., Lenzerini, M., Rosati, R.: Path-based identification constraints in description logics. In: Principles of Knowledge Representation and Reasoning, pp. 231–241 (2008)
4. Calvanese, D., De Giacomo, G., Lenzerini, M.: Identification constraints and functional dependencies in description logics. In: Proceedings of International Joint Conference on Artificial Intelligence (IJCAI), pp. 155–160 (2001)
5. Ito, M., Weddell, G.: Implication problems for functional constraints on databases supporting complex objects. J. Comput. Syst. Sci. **49**(3), 726–768 (1994)
6. Khizder, V.L., Toman, D., Weddell, G.: On decidability and complexity of description logics with uniqueness constraints. In: Van den Bussche, J., Vianu, V. (eds.) ICDT 2001. LNCS, vol. 1973, pp. 54–67. Springer, Heidelberg (2000)
7. Schaerf, A.: On the complexity of the instance checking problem in concept languages with existential quantification. J. Intell. Inf. Syst. **2**(3), 265–278 (1993)
8. Toman, D., Weddell, G.: On keys and functional dependencies as first-class citizens in description logics. In: Furbach, U., Shankar, N. (eds.) IJCAR 2006. LNCS (LNAI), vol. 4130, pp. 647–661. Springer, Heidelberg (2006)
9. Toman, D., Weddell, G.E.: On the interaction between inverse features and path-functional dependencies in description logics. In: Proceedings of International Joint Conference on Artificial Intelligence (IJCAI), pp. 603–608 (2005)
10. Toman, D., Weddell, G.E.: On keys and functional dependencies as first-class citizens in description logics. J. Aut. Reasoning **40**(2–3), 117–132 (2008)
11. Toman, D., Weddell, G.E.: Applications and extensions of PTIME description logics with functional constraints. In: Proceedings of International Joint Conference on Artificial Intelligence (IJCAI), pp. 948–954 (2009)
12. Toman, D., Weddell, G.: On adding inverse features to the description logic $\mathcal{CFD}_{nc}^{\forall}$. In: Pham, D.-N., Park, S.-B. (eds.) PRICAI 2014. LNCS, vol. 8862, pp. 587–599. Springer, Heidelberg (2014)
13. Toman, D., Weddell, G.E.: On the utility of $\mathcal{CFDI}$. In: Proceedings of the 28th International Workshop on Description Logics (2015)
14. Weddell, G.: A theory of functional dependencies for object oriented data models. In: International Conference on Deductive and Object-Oriented Databases, pp. 165–184 (1989)

Understanding People Relationship: Analysis of Digitised Historical Newspaper Articles

Sharon Torao-Pingi[1(✉)] and Richi Nayak[2]

[1] University of Papua New Guinea, Port Moresby,
National Capital District, Papua New Guinea
spingi@upng.ac.pg
[2] Queensland University of Technology, Brisbane
Queensland, Australia
r.nayak@qut.edu.au

Abstract. The study of historical persons and their relationships gives an insight into the lives of people and the way society functioned in early times. Such information concerning Australian history can be gleaned from Trove's digitized collection of historical newspapers (1803–1954). This research aims to mine Trove's articles using closed and maximal association rules mining along with visualization tools to discover, conceptualize and understand the type, size and complexity of the notable relationships that existed between persons in historical Australia. Before the articles could be mined, they needed vigorous cleaning. Given the data's source, type and extraction methods, estimated word-error rates were at 50–75 %. Pre-processing efforts were aimed at reducing errors originating from optical character recognition (OCR), natural language processing and some co-referencing both within and between articles. Only after cleaning were the datasets able to return interesting associations at higher support thresholds.

Keywords: Association rule mining · Natural language processing · OCR errors

1 Introduction

For various reasons, people have always been interested in the lives and interactions of other people. This is evident from user searches on the web; 11–17 % of web queries include a person's name and 4 % of web queries contain just a person's name [1]. The study of historical persons helps us learn about the lives of people and the way society functioned in early times. Trove digitised newspapers of 1803–1954, hosted by the National Library of Australia, in essence would contain interesting news and events taking place in Australia at the time, these may involve vital information about the lives of ordinary as well as extraordinary persons who lived in historical Australia and the interesting relationships that they shared with others of their time.

In this project, closed and maximal frequent itemsets and association rules mining were used to find popular individuals and interesting associations between those mentioned in the articles of Trove; especially with respect to persons related to

© Springer International Publishing Switzerland 2015
B. Pfahringer and J. Renz (Eds.): AI 2015, LNAI 9457, pp. 572–588, 2015.
DOI: 10.1007/978-3-319-26350-2_51

particular "concepts" or themes associated with Australian history. These concepts came in the form of 20 queries; due to space constraints, only 10 are discussed here.

The text mined had been extracted from poor quality historical newspapers that had been digitised using OCR software. Person names were extracted from this corpus using natural language processing (NLP). These extraction processes, including the papers' quality gave rise to high rate of word-errors, making pre-processing necessary for interesting associations to be obtained from the datasets at higher support thresholds.

The R package arulesViz [5] and the network visualisation software Gephi[1] were used not only to visualise the associations between people names, but also to analyse and understand the type, size and complexity of the relationships that existed between persons in the different query-related datasets or querysets.

This research is significant in two ways; from the historical point of view, where interesting relationships between commonly mentioned individuals in Australia's history could be learned; and from the experimental point of view, where methods were tested and refined to best find and analysis these associations given the data's type and source.

The outline of this paper is as follows. A discussion of pre-processing methods focusing on OCR removal is covered in Sect. 2; Sect. 3 discusses the data methods used and the results obtained; and Sect. 4 offers some conclusions.

2   Methods and Experiments

The data's extraction methods, its source and type gave rise to 3 sources of noise to which pre-processing efforts were directed. First, the process of transcribing the old newspaper manuscripts into editable text by the OCR system, ABBYY FineReader[2], resulted in high amounts of OCR errors. Then, there were errors which occurred from Named Entity (NE) omissions and misclassifications during NE Recognition (NER) using the Natural Language Processing (NLP) software, SENNA[3] on the OCRed text, which resulted in word error rates estimated up to 50–75 %. The last source of errors came from person name ambiguities that exist naturally both within and across articles.

This section details the data pre-processing methods employed for producing a quality data for mining non-ambiguous association rules between persons. We were interested in finding associations of persons connected under a particular concept or theme associated with Australian history, and then analysing the type of relationships shared. These concepts, which were identified by historian Paul Turnbull who was expert advisor to the project, were submitted as queries to retrieve data of interest for mining purpose. A total of 20 queries (Table 1) were used, however only the first 10 are discussed in this paper, *omitted queries (q11 to q20) possess characteristics similar to the first 10.

---

[1] http://gephi.github.io.

[2] http://www.abbyy.com/finereader/.

[3] http://ronan.collobert.com/senna/.

Table 1. Queries used in data acquisition

q#	Query string	q#	Query string	q#	Query string	q#	Query string
1	aborigines	6	gold	11	*camels	16	*locusts
2	celestials	7	inquest	12	*cyclone	17	*manslaughter
3	corona	8	myall creek	13	*drought	18	*natives
4	emancipists	9	native police	14	*flooding	19	*plaque
5	fossils	10	snake	15	*influenza	20	*sheep scab

2.1   Acquisition of Datasets

About 77 million news articles were parsed using SENNA's Tokenizer and the NLP-tagged documents tarred. Trove APIs were then utilised to extract articleIds from top-2000 articles most relevant to each query formulated as the search. The articleIds were then used to extract corresponding articles from the tagged collection. Only people names tagged were extracted onto text files. Resulting in 20 query-related text files or *querysets* (*q* for individual querysets, e.g. *q2* for *queryset 2*), with an average of 1400 articles (lines) per queryset; since not all articles contained person names.

Figure 1 shows part of *q2* with person names (separated by hash tags) from articles on separate lines as picked up by SENNA. The level of noise in the datasets as well as empty files (articles with no person names) can be observed here.

```
64 #murray #thisdeno #chinamnera #launaundrymen #egglestone #mongoongolians #dotor #kavanagh #murray
57 #h.#
05 #chinamen #mr. a. #s. #roe #ewing #sin ling #kavanagh #mr. ewing #sdetective ka #sin #ling #roe #(
38 #hocking #ohinaman #sevenal chinamen #io how#
17
35 #f. #neyork tesin #thfire jeinsuitance #stigat0,000,000 #berkele #fourlooters #tho #cliff horse# )
62 #s. roe #t. wood #mr. roe #mr. roe #mr. wood #wood #mr. roe #ah tuck #b. mills #mr. m3ills #inspec
66 #ssdisgraeefdl #chihinesehi#
93 #su sing #sang #su sing #sang lhock #sang ihock #sang hock #sang #hock #sang hook #sang hock# su :
87 #su3nday# ←
```

Fig. 1. A section of *q2* showing person names per article/line

Noise Levels in Querysets. Due to the presence of word errors and the absence of complete formatting information, NER from historical documents is more difficult to study than natively digital data [8]. Since the Trove text is historical in nature and the data type mined was person named entities, the querysets were extremely noisy.

A count of unique names for the entire dataset returned 120 million names. Names appearing just once constituted ≈75 % of the total names. >95 % of names occurred ≤10 times, from direct observation, these low frequency "names" were almost all nonsensical, a clue to noise levels for the Trove dataset, which could be as high as 75 %. Physical examination showed that high word (name) error rates caused large quantities of partial names to be retrieved. ≈65 % of names were single-tokened; a little more than 10 % were 2-tokened (mostly with an initial); and names with just initials and/or titles made up ≈25 % of the total names. This is reflected in all the querysets.

2.2   Data Pre-processing

Historical data digitally transcribed using even state-of-the-art OCR systems can produce word error rates of over 50 % [3]. Berg-Kirkpatrick et al., point out that this high rate of word errors makes it difficult for many research to be conducted using historical data. Factors like unknown fonts, uneven baseline, fragmented text, over-inking, touching/overlapping letters etc. caused word and character errors that contributed to high OCR errors in Trove's translated text. Figure 2 shows part of a translated article featured on 02/04/1903 from the Trove website. The OCRed text is shown on the left and its unedited digital translation on the right.

As seen, the amount of OCR errors in such low quality image translations was fundamental and cannot be understood without the help of humans. OCR errors decline with later documents being better preserved with presentation and layout standardised. Pre-processing of the querysets involved 4 stages as discussed below.

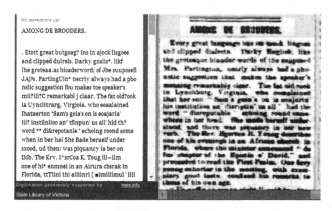

Fig. 2. A digital image of a news article (right) in Trove and its translation (left).

Removal of Garbage Names and Query-Related Terms. After initial removal of non-alphabetical characters (except the space), garbage strings resulting from OCR were identified using regex and removed using a general rule-based approach [11]. According to Taghva et al., garbage names are >40 characters long, have more symbols than characters, have >4 occurrences of identical characters in a row and are made up of the same character. The last rule was modified to exclude name tokens *ii* and *iii* to allow for names of monarchs such as *"king george iii"*.

Lists of query-related terms, like *"chinese"* and *"chinaman"* for *q2* were compiled for each query. Such terms were removed from articles only if they appeared in an article as a single-tokened term.

Detection and Correction of OCR Errors. With slight modification to Tong and Evans' [13] definition of real-word and non-word errors; a non-*name* error appears in a source article and is interpreted (under OCR) as a string that did not correspond to any valid name in a given namelist; e.g. *"willliaiiii"*. A real-*name* error occurs when a source-text name is interpreted as a name that actually does occur in a given namelist, but is not identical with the source-text name; e.g. *"louise"* for *"louisa"*, where both names exist in the given namelist.

Person Namelists or Person Name Dictionaries. In order to detect and correct non-names, namelists of "correct" names were needed. These were constructed from high frequency names from source text based on the assumption that frequent names were more likely to be correct [9]. Namelists for each queryset, as it is important for any dictionary or lexicon to be domain-specific [12]. There was the problem though of how often a name should occur in the dataset for it to be considered correct. The namelists were subjected to some pre-processing, such as the removal of names containing more than 2 repeating characters. As real-word errors need context to correct [13]; the query-specific namesets sufficed to that means also.

Similarity Reduction Technique. The right combination of in-text name frequency for the namelists with n-gram similarity thresholds were empirically determined to minimise the number of false negatives being corrected and false positives being left in text. As Taghva and Stofsky [12] also observed, low frequency thresholds created larger namelists with the possibility of too many false positives; while high frequency thresholds meant smaller namelists with too many false negatives being corrected, candidate lists for correction of misspelling being also limited; which coincided with our conclusions.

Candidate names from namelists chosen for substitution with non-names had highest in-text name frequency at or above the similarity thresholds set. Each token in names were checked at decreasing similarity thresholds. With an average of 7 names per article, about 2 were detected as non-names and corrected through this process.

Similarity thresholds were set at 0.85, 0.80 and 0.75 for 3 passes through the similarity reduction program with name frequency of namelists set at 75, 100 and 100 respectively. It was seen that similarity thresholds <0.75 caused correct names to be changed, e.g. *"francisco"* to *"francis"* and *"stephenson"* to *"stephen"*. Longer names with repeating character sequences such as *"williamilliamil"* were rightly changed to *"william"* and shorter names such as *"mury"* to *"mary"*. However, due to the risk of "correcting" already correct names, minimum similarity was set at 0.75; leaving names <5 characters long and longer names from repeating character sequences uncorrected in text. Similarity values were seen to be affected by the type of errors, whether from character(s) deletion, addition or substitution.

Figure 3 shows formatted results from pass 3 of a queryset through the similarity program. It highlights 2 of the difficulties encountered.

The left case shows *"hayler"* being picked up by the similarity program as a non-name. Candidate *"hayley"* with the highest frequency is chosen to replace it. Of course, any of the 3 candidates could be correct, or *"hayler"* may be a legitimate name itself. The right case shows a non-name being "corrected" with a high-frequency non-name.

Within-Document Name Disambiguation. Multiple references to individuals were removed on a rudimental level to avoid rules that related individuals to themselves; see subset of such rules in Fig. 4A for *q2*. The persons *"s roe"*, *"mr roe"* and *"roe"* are

```
hayler//hayles   (SIM: 0.75, FREQ: 233)        lilli//ililli    (SIM: 0.75, FREQ: 255)
hayler//hayley   (SIM: 0.75, FREQ: 282)        lilli//illili    (SIM: 0.75, FREQ: 114)
hayler//tayler   (SIM: 0.75, FREQ: 256)        ILILLI choosen
HAYLEY choosen
```

Fig. 3. Examples of non-names corrected at Pass 3

```
['mr wood'=1, 'b mills'=1]: 2 ==> ['roe'=1, 'mr roe'=1, 't wood'=1]: 1
['mr roe'=1, 'mr wood'=1, 'b mills'=1]: 2 ==> ['roe'=1, 't wood'=1]: 1
['t wood'=1, 'mr wood'=1, 'b mills'=1]: 2 ==> ['roe'=1, 'mr roe'=1]: 1
['mr roe'=1, 't wood'=1, 'mr wood'=1, 'b mills'=1]: 2 ==> ['roe'=1]: 1
['mr roe'=1]: 10 ==> ['mr wood'=1]: 2    <conf:(0.2)> lift:(30.2
['mr roe'=1]: 10 ==> ['mr wood'=1, 'ah tuck'=1]: 2    <conf:(0.2
['mr roe'=1]: 10 ==> ['mr wood'=1, 'mr mills'=1]: 2    <conf:(0.
['mr roe'=1]: 10 ==> ['mr wood'=1, 'inspector fleming'=1]: 2
```

Fig. 4. A. (top) and B. (bottom) showing subset of rules from ambiguous and disambiguated names in *q2* respectively.

probably references to a "*mr s roe*", however participate in rules as different individuals. Figure 4B shows rules after disambiguation. After disambiguation, number of rules decreased by up to 50 %.

If a name's title-less component is a subsequence of another name in an article, it was considered an alias. E.g, a "*mr john smith*" could be referred to as "*john smith*", "*mr smith*" or "*smith*"; these multiple references would be removed to leave the proper name [7] with the longest-string, "*mr john smith*". Also, the assumption of "*one person per document*" [2], was used to disambiguate persons as ambiguous personal names in news article rarely refer to multiple entities due to the nature of news articles being short in length [4].

Summary of Pre-processing. Other "noise" also removed were names composed of titles and/or initials only and single letters. Table 2 shows the results of pre-processing on the different noise types.

Table 2. Average % of noise in querysets

Source	Av. % reduced per queryset	Range (min %– max %)
Garbage, single-lettered names and other noise)	2.15	0.8–5.2
Titles and/or Initials only	3.1	1.7–4.15
Duplicate names	23	15–37.95
Name aliases	0.01	0.0–0.01
Query-related terms	0.1	0.0–0.67
Total	**28.4**	**21.7–44.3**

After pre-processing the querysets, the number of rules reduced by up to 90 % and support measures improved by 99 %; that is from *minSupp* 0.001 % for unprocessed querysets (e.g. in Fig. 5A) to 1 % for pre-processed querysets (e.g. in Fig. 5B).

```
 1.  ['kong'=1]: 1 ==> ['chinamen'=1]: 1    <conf:(1)> lift:(11.35) lev:(0) conv:(0.91)
 2.  ['wong tew'=1]: 2 ==> ['chinamen'=1]: 2    <conf:(1)> lift:(11.35) lev:(0) conv:(1.82)
 3.  ['kow'=1]: 1 ==> ['chinamen'=1]: 1    <conf:(1)> lift:(11.35) lev:(0) conv:(0.91)
 4.  ['ihe'=1]: 1 ==> ['chinamen'=1]: 1    <conf:(1)> lift:(11.35) lev:(0) conv:(0.91)
 5.  ['gan'=1]: 2 ==> ['chinamen'=1]: 2    <conf:(1)> lift:(11.35) lev:(0) conv:(1.82)
 6.  ['do'=1]: 1 ==> ['chinamen'=1]: 1    <conf:(1)> lift:(11.35) lev:(0) conv:(0.91)
 7.  ['c richardson'=1]: 2 ==> ['chinamen'=1]: 2    <conf:(1)> lift:(11.35) lev:(0) conv:(1.82)
 8.  ['zep'=1]: 1 ==> ['chinamen'=1]: 1    <conf:(1)> lift:(11.35) lev:(0) conv:(0.91)
 9.  ['y j'=1]: 1 ==> ['chinamen'=1]: 1    <conf:(1)> lift:(11.35) lev:(0) conv:(0.91)
10.  ['xow'=1]: 1 ==> ['chinamen'=1]: 1    <conf:(1)> lift:(11.35) lev:(0) conv:(0.91)
```

```
 1.  ['chapel'=1]: 10 ==> ['macaulay'=1]: 10    <conf:(1)> lift:(105.18) lev:(0.01) conv:(9.9)
 2.  ['louis xvi'=1]: 9 ==> ['macaulay'=1]: 9    <conf:(1)> lift:(105.18) lev:(0.01) conv:(8.91)
 3.  ['louis xvi'=1]: 9 ==> ['chapel'=1]: 9    <conf:(1)> lift:(115.7) lev:(0.01) conv:(8.92)
 4.  ['sing yum'=1]: 7 ==> ['ah gee'=1]: 7    <conf:(1)> lift:(128.56) lev:(0.01) conv:(6.95)
 5.  ['sing yum'=1]: 7 ==> ['b thurston'=1]: 7    <conf:(1)> lift:(144.63) lev:(0.01) conv:(6.95)
 6.  ['william iv'=1, 'macaulay'=1]: 9 ==> ['chapel'=1]: 9    <conf:(1)> lift:(115.7) lev:(0.01) conv:
 7.  ['william iv'=1, 'chapel'=1]: 9 ==> ['macaulay'=1]: 9    <conf:(1)> lift:(105.18) lev:(0.01) conv
 8.  ['william iv'=1, 'louis xvi'=1]: 8 ==> ['macaulay'=1]: 8    <conf:(1)> lift:(105.18) lev:(0.01) c
 9.  ['william iv'=1, 'louis xvi'=1]: 8 ==> ['chapel'=1]: 8    <conf:(1)> lift:(115.7) lev:(0.01) conv
10.  ['louis xvi'=1]: 9 ==> ['macaulay'=1, 'chapel'=1]: 9    <conf:(1)> lift:(115.7) lev:(0.01) conv:(
```

Fig. 5. A. Top-10 rules, ordered by *conf* from *q2* (*celestials*) BEFORE pre-processing. B. Top-10 rules, ordered by *conf* from *q2* (*celestials*) AFTER pre-processing.

3   Association Rules of Person Names

With cleaned data, interesting relationships could now be mined using association rules. In *market-basket analysis* terms; articles containing names are *transactions*, a person's name an *item*, and a set of names (i.e. a nameset), an *itemset*. The *size* of a nameset is the number of person names in the set. An *association rule*, is represented as {A} =>{B}, where A and B are namesets, the rule shows the relationship between A and B in a queryset, where A, the *antecedent*, implies the presence of B (the *consequent*).

3.1   Measuring Interestingness of Associations

Support (σ) and *confidence* (*c*) are interest measures used to measure the significance or frequency of an association, and its strength or predictive power respectively [7]. An association is considered *objectively interesting* if persons participating in the rule appear in the queryset together frequently enough to meet some minimum support level set (i.e. *minSupp*); and the chance of the consequent appearing in an article given that the antecedent is mentioned, i.e. the confidence *c*, is greater than or equal to some minimum confidence level set (i.e. *minConf*) [13].

Lift measures correlation between namesets involved in the rule. High lift indicates consequent has high probability of appearing in an article given that the consequent is mentioned. Low lift means otherwise.

Frequent itemsets (denoted *FI*) are sets of person(s) or namesets that appear in the queryset frequently enough to be considered interesting, that is, where its $\sigma \geq minSupp$. *Closed* frequent itemsets (denoted *CFI*) are considered *most* interesting as these are closed groups of individual(s) mentioned together most frequently in the querysets. *Maximal* frequent namesets (denoted *MFI*) are namesets with the most number of persons who appeared together in the queryset enough times to meet *minSupp*. The relationship is such that $MFI \subseteq CFI \subseteq FI$. Using R's *arules, Apriori* and *Eclat* [6] were used to discover association rules and the *FI* sets respectively.

3.2   Closed and Maximal Namesets

Association rules are derived from *FI* with namesets of size 2 or more, "singleton" *FI* are frequently mentioned individuals in the querysets. The *Total No. of FI* column in Table 3 gives each queryset's *FIs* at *minSupp* 0.0025 %. The % composition of *CFI* and *MFI* over *FI* are shown. k denotes size of the nameset and t the number of tokens in a name.

Table 3. Frequent namesets returned

Query	No. of articles	Av. # of names per article	Total no. of FI	% of FI		% k = 1		% k > 1		% 1t-1k	
				CFI	MFI	CFI	MFI	CFI	MFI	CFI	MFI
1	1359	9.0	232	86.6 %	80.2 %	81.9 %	75.4 %	4.7 %	4.7 %	59.5 %	57.8 %
2	1609	6.1	792	28.0 %	24.1 %	24.6 %	21.8 %	3.4 %	2.3 %	18.4 %	16.2 %
3	1209	11.8	502	72.7 %	42.6 %	34.5 %	26.7 %	38.2 %	15.9 %	29.7 %	22.3 %
4	1856	44.2	2389	94.2 %	72.9 %	45.0 %	32.0 %	49.2 %	40.9 %	30.1 %	22.2 %
5	1457	9.3	276	96.0 %	87.0 %	88.0 %	79.7 %	8.0 %	7.2 %	56.9 %	51.4 %
6	1604	6.3	417	97.4 %	77.5 %	70.7 %	54.7 %	26.6 %	22.8 %	46.3 %	35.5 %
7	1524	11.7	335	96.7 %	81.8 %	77.0 %	65.1 %	19.7 %	16.7 %	61.2 %	52.5 %
8	1327	13.6	1423	31.9 %	23.6 %	23.2 %	17.4 %	8.7 %	6.2 %	17.3 %	13.5 %
9	1609	10.7	574	70.7 %	50.7 %	50.3 %	37.8 %	20.4 %	12.9 %	38.5 %	28.7 %
10	1495	5.3	149	55.0 %	54.4 %	52.3 %	51.7 %	2.7 %	2.7 %	45.6 %	45.0 %

It was observed that *CFIs* made up $\approx \frac{3}{4}$ of the querysets' *FIs*; however, >70 % were singletons; leaving <20 % of *FIs* making up the set of interesting rules. Of the singleton sets, more than half were single-tokened; which may well be word or name errors that only partially identify individuals.

Significant Individuals from CFI. Closed frequent namesets with highest support or frequency can be seen as most popular or significant individuals in the news with relation to the queries. E.g. see 10 most frequent names in *q1* (*aborigines*) and their related frequencies in Table 4.

As observed for other querysets also, most of CFI namesets were singletons; these individuals appeared most often on their own then with others. Unfortunately, a large portion of the names returned at high support levels were single-tokened, leaving a lot of ambiguity in the names. It is difficult to tell if such *1t-1k* namesets are partial references to notable persons or frequently occurring word errors.

Table 4. Top-10 CFI rules showing most frequently mentioned persons in $q1$

Top-10 CFI	1	2	3	4	5	6	7	8	9	10
Nameset	{murray}	{esq}	{dr basedow}	{philip}	{macleay}	{john}	{mr taplin}	{mr hamilton}	{michael sawtell}	{yorke}
Support	3.1 %	1.8 %	1.6 %	1.3 %	1.3 %	1.1 %	1.1 %	1.1 %	1.1 %	0.9 %

3.3 Rules from Closed and Maximal Namesets

Of the frequent namesets returned by the FI categories, it was important to observe the strength of their relationships. Rows in Table 5 show % of rules returned by *FI*, *CFI* and *MFI* at various σ and c levels. The 3rd row gives the number of *FI* rules at the lowest σ and c set (*minSupp* 0.25 % and *minConf* 50 % respectively); the rest of the rows show % of these produced by the *FI* categories at each σ and c level.

None of the querysets returned rules with $\sigma > 0.01$. As observed, higher σ and c thresholds saw less rules produced with higher % of closed and maximal sets. These indicate that only a very small fraction of associations discovered from the querysets are objectively interesting to note.

Table 5. Fraction of rules produced at various support and confidence thresholds

σ	0.0025									0.005									0.01								
c	0.5			0.7			0.9			0.5			0.7			0.9			0.5			0.7			0.9		
q#	FI	CFI	MFI	FI	CFI	MFI	FI	CFI	MFI	FI	CFI	MFI	FI	CFI	MFI	FI	CFI	MFI	FI	CFI	MFI	FI	CFI	MFI	FI	CFI	MFI
1	84	.155	.155	.095	.095	-	.786	.060	.060	-	-	-	-	-	-	-	-	-	-	-	-	-	-	-	-	-	-
2	2307	.036	.024	.832	.031	.020	.801	.027	.017	.016	.007	-	.016	.007	-	.009	.006	-	-	-	-	-	-	-	-	-	-
3	634	.587	.281	.628	.355	.188	.303	.148	.103	.289	.254	.014	.147	.128	.008	.030	.027	.002	.074	.074	-	.027	.027	-	-	-	-
4	580	.603	.514	.672	.324	.286	.376	.112	.102	.043	.043	.014	.010	.010	.002	.003	.003	-	.002	.003	-	-	-	-	.002	-	-
5	28	.893	.821	.607	.500	.500	.429	.321	.321	.071	.071	.071	-	-	-	-	-	-	-	-	-	-	-	-	-	-	-
6	45	.911	.800	.489	.422	.378	.267	.200	.200	.067	.067	.044	.022	.022	.022	-	-	-	-	-	-	-	-	-	-	-	-
7	37	.973	.811	.622	.595	.514	.297	.270	.243	.108	.108	.054	.081	.081	.054	.027	.027	-	.027	.027	-	.027	.027	-	.027	.027	-
8	3623	.071	.047	.922	.059	.039	.822	.046	.031	.535	.017	.002	.493	.015	.002	.425	.012	.002	.179	.007	-	.164	.007	-	.161	.007	-
9	481	.410	.281	.796	.335	.227	.522	.216	.162	.019	.019	.002	.017	.017	.002	.004	.004	-	-	-	-	-	-	-	-	-	-
10	191	.052	.052	1.00	.052	.052	1.00	.031	.031	-	-	-	-	-	-	-	-	-	-	-	-	-	-	-	-	-	-

3.4 Interestingness of CFI Rules

CFI rules show relationships between closed groups of individuals who appear most frequently together in the querysets. These associations are here observed at "high" and "low" interest levels. Rules with $\sigma \geq 0.5$ % and $c \geq 70$ % were considered "high interest" associations, those below these thresholds were considered "low interest". In Table 6, *Total Rules* represents the number of *FI* rules returned at $\sigma = 0.25$ % and $c = 50$ %. The % columns show proportion of *CFI* rules out of *Total Rules* at the various interest levels (H-*high*, L-*low*, S-*support*, C- *confidence*).

Table 6. (Left) High/Low support and confidence rules

q#	Total Rules	HS-HC	HS-LC	LS-HC	LS-LC
1	84	-	-	.095	.060
2	2307	.007	-	.024	.004
3	634	.117	.137	.227	.106
4	580	.010	.033	.309	.252
5	28	.000	.071	.500	.321
6	45	.022	.044	.378	.467
7	37	.081	.027	.486	.378
8	3623	.015	.002	.044	.010
9	481	.017	.002	.318	.073
10	191	-	-	.052	-

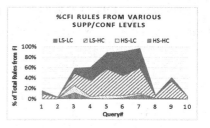

Fig. 6. (Right) Graph of High/Low support and confidence rules

Table 6 and Fig. 6 showed less CFI rules in querysets with high number of *FI* rules; compare *q2* & *q8* to *q5* & *q6*. This indicates high presence of redundant rules (having similar associations but with less support) in querysets with large number of rules.

A large fraction of CFI rules ($\approx$60 %) were LSHC, meaning stronger relationships were infrequent in the querysets. Also, <4 % of *CFI* rules were HS-HC, a further indication of the scarcity of noteworthy relationships to be found in the querysets.

Table 7 shows 7 out of the 580 rules or associations returned from *q4*'s CFI. Associated interest measures are also given. σ gives the fraction of articles containing both antecedent and consequent. c gives the probability of the consequent appearing in an article given that the antecedent has been mentioned.

Table 7. Example of association rules derived from *q4*

Association rule {antecedent} => {consequent}	σ	c	Interest
{lord earl grey} => {van diemen}	1.00 %	94.7 %	HS-HC
{mr berry,mr jones} => {john jamison}	0.67 %	80.0 %	HS-HC
{angus} => {robertson}	0.62 %	78.0 %	HS-HC
{bourke, charles wentworth, ralph darling} => {george gipps}	0.34 %	80.0 %	LS-HC
{william bligh} => {major johnston}	0.39 %	58.0 %	LS-LC
{mary hope} => {thomas brisbane}	0.28 %	55.0 %	LS-LC
{d gilbert} => {mr warburton}	0.28 %	55.0 %	LS-LC

3.5   Modularity and Average Degree Scores

FI rules at *minSupp* 0.0025 and *minConf* 80 % were passed through *Gephi* and the rules *modularised* into "communities" of nodes or namesets to assess the size and complexity of the relationships shared among individuals in the querysets.

In Table 8, *Nodes* are unique namesets that participate in the rules (*edges*) for each queryset. *Communities* (of namesets) are formed out of association rules that share common namesets. The number of rules a consequent participates (i.e. the *in-degree*) is counted and averaged as the *Average* (In-)*Degree* (av°). Modularity scores (m) within range [−0.5 to 1], measure strength of participation of namesets in relationships in and out of their communities. See Table 8 and Fig. 7A, B for results from the querysets.

Table 8. Modularity and average weights of querysets

Query	No. of nodes	No. edges	No. of communities	Modularity (m)	Average degree (av°)
1	38	76	8	0.262	2.000
2	590	1919	13	0.238	3.253
3	244	398	13	0.521	1.631
4	289	390	28	0.746	1.349
5	23	17	9	0.771	0.739
6	31	22	10	0.757	0.710
7	28	23	5	0.488	0.821
8	1052	3342	28	0.362	3.177
9	249	383	23	0.673	1.538
10	68	191	7	0.207	2.809

Table 8 and Fig. 7A, B show lower m and higher $av°$ for querysets with more rules. This supports findings from Sect. 3.4 that a bulk of the large number of rules in querysets like $q2$ & $q8$ are redundant, having same or similar namesets in most rules but with less support; while smaller number of associations discovered involved individuals more confined to their sphere of relationships.

3.6 Visualising Association Rules

Various plots were constructed in *R* and *Gephi*, using the querysets' FI rules produced at *minSupp* 0.0025 % and *minConf* 50 %. These visualisations helped to further understand the nature and interestingness of the relationships between persons in querysets. To save space, plots for $q3$ only are shown here and discussed with references to other querysets.

Scatter and 2-Key plots like in Fig. 8A, B (for $q3$) respectively were used to test correlations between the interest measures plotted.

Most of the scatter plots showed confidence and lift negatively correlated to support. Thus, strong relationships were less frequent; an established fact for association rules [10]. The probability of consequents appearing in articles given that the antecedents were present decreased with the frequency the namesets appearing together. This could be attributed to querysets' sparseness, aggravated by presence of OCR errors.

2-Key plots showed size of the namesets. Unlike Fig. 8B, 40 % of the querysets' rules had highest order of 3, indicating a maximum of 3 people in a relationship. The confidence of rules did not seem affected by the number of persons in the rule' antecedent. This is supported by Unwin et al. [14] who state that given at least one "good" or strong 2-level/order rule; adding further "assumptions" (persons) to will lower support but do little for the confidence. This coincided with results.

Most querysets displayed similar behaviours to Fig. 8A, B, unique patterns were attributed to either small number of rules or uniform interest measures in rules.

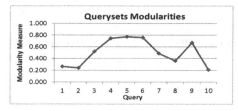
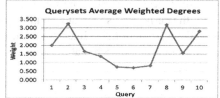

Fig. 7. A. (left) and B. (right) show modularity and average degrees respectively

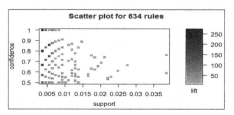
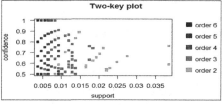

Fig. 8. A. (left) and B. (right) showing scatter plot and 2-Key plot respectively for $q3$

Popular Persons and Associates. Grouped matrix-based plots like Fig. 9 (for $q3$) show 20 groups resulting from the clustering of the queryset's rules. LHS (of rules) show antecedents and RHS the consequents. Most frequent names in antecedents for each cluster are displayed as column labels. Numbers to the left of these names (outside

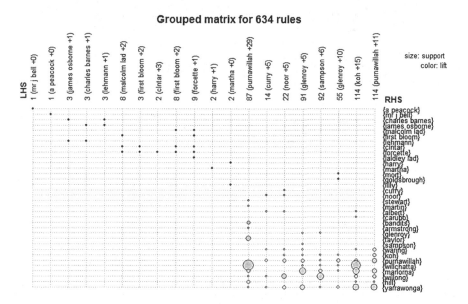

Fig. 9. Grouped matrix-based Plot for $q3$

the curly braces) represent the number of rules these persons participate in, numbers within the curly braces indicate the number of other persons or names that make up the antecedents in these rules. Interest measures are aggregated in the groups, with columns and rows re-ordered such that the most interesting rules, according to *lift*, are on the top left corners.

The shading of the bubbles indicate strong correlations between namesets on the LHS and those on RHS, which starts to decrease with increasing number of participants in the rules. This means that more persons involved in a relationship (antecedents on LHS), decreases the chance of another person (consequent on RHS) being added to the relationship.

As with most of the querysets, low support meant high lift; that is, less significant associations have stronger correlation or dependency of the consequent nameset on the antecedent nameset. Grouped matrix plots for other querysets showed similar behaviours but with varying number of namesets on both the LHS and RHS.

Network of Communities. The graph in Fig. 10 (*for* q3) produced by *Gephi* uses *Force Atlas* layout with vertex and label size proportional to the in-degree. The node size indicates its in-degree, i.e. how many rules or relationships the nameset participates in.

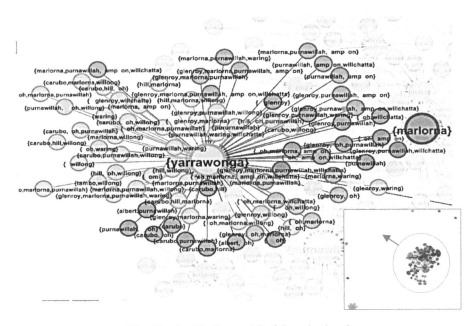

Fig. 10. Gephi's "network" of the rules in *q3*

A distant view of the relationship clusters or communities is shown on the bottom right insert, with a magnified view of the largest cluster shown.

Smaller communities with no shared nodes (namesets) are detached from each other. Algorithm *Label Adjust* was used to adjust the position of the nodes to prevent labels overlapping. Zeroing in on the main cluster (circled), and highlighting prominent node {*yarrawonga*} shows how this nameset connects to the many other namesets.

The relationship representations for querysets in *Gephi* differed in size, number of communities and modularity. Gephi graphs gave a visual of how prominent some of the namesets were in the querysets and the size of their associations. Querysets with large number of rules produced complex network of associations with very few prominent namesets. Some relationship clusters were isolated with individuals in these clusters having no connection to others in other clusters or communities.

3.7   Categorisation of Querysets

Careful observation of results showed similar behaviours shared by the querysets. Four general categories, according to querysets' type, size and complexity of relationships shared by historical figures who were in some way related to the queries, were identified. Most of *q11* to *q20* (not discussed in this paper) could be identified with these groups.

Type 1 included *q2* (*celestials*) and *q8* (*myall creek*) which showed *large and complex relationships*. The querysets returned highest FI with only 30 % closed. Of the CFI, ≈80 % were single-tokened singletons, indicating word/name errors. With large number of rules returned, only 4 and 7 % were CFI rules, with only 1–2 % having high support (interesting). Higher order rules, involving up to 8 persons with low m and higher $av°$ indicated complex relationships with strong links between persons. The large communities of mostly similar relationships contained small groups of key individuals with notable relationships to each other and diverse but less significant relationships with many others. This could mean less diversity in the news stories surrounding the historical topic (query) in question, with many persons involved, but only few significant individuals connected.

Type 2 included *q5* (*fossils*), *q6* (*gold*) and *q7* (*inquest*) which showed *few small, isolated relationships*. Each queryset returned the lowest number of *FI* and rules with highest % *CFI* (>96 %) and *CFI* rules (>90 %). Only 6 % of *CFI* rules were high interest. Highest order of rules was 3, with ≈80 % of them being of order 2, meaning most of the relationships here involved only 2 persons. Less number of communities and unique namesets with moderate m and low $av°$ indicate sparse relationships between these small groups of individuals. These could mean isolated news stories regarding these queries with small number of persons involved who share nominal relationships with others.

Type 3 included *q3* (*corona*), *q4* (*emancipists*) and *q9* (*native police*) which showed *disjointed relationships of diverse sizes*. Each queryset returned relatively large number of *FI* with larger portions (≈80 %) of *CFI*, from which about half were singletons. Of the rules returned, ≈50 % were from closed sets with only 2–4 % being high support. Modularity levels m were middling with $av°$ moderately low and high order rules of up to 6 persons. Their grouped matrix and *Gephi* plots helped to show distinct communities of varied sizes; majority of the communities involved only a few related individuals; these meaningfully-related individuals may be from unconnected news stories, with varying number of persons who share little or no connections with each other.

Type 4 included *q1* (*aborigines*) and *q10* (*snake*) which showed *small, highly knotted relationships*. These sets returned less *FI*s with least % of *CFI* (≈3.7 %). All their *CFI* were *MFI*, which indicated lowest support values for these closed sets. Almost all their association were low support and high confidence; ≈10 % were *CFI* rules, but none of high support. There were less number of communities and namesets, lowest *m*, higher *av°* and higher order of rules. These showed that persons involved in news stories surrounding these topics were almost equally associated with every other person in the querysets, meaning from a single news interest that was reported on frequently enough to be interesting.

4   Discussion and Conclusions

This paper presents a case study of mining interesting associations between persons related to important topics in Australia history. Professor Paul Turnbull, the historian associated with the project, was extremely satisfied with and excited by the outcomes of the project. According to his words, it marked a significant advance in applying advanced analytic techniques to the trove data to gain new insights into Australia's history. Searches on the indexed entities (queries) in connection with historical persons revealed were found to confirm existing assumptions about phenomena, such as anti-Chinese agitation, for example. However, the means of exploring the Trove data offered by this project disclosed interesting new avenues of inquiry. Newspaper reportage of fossils, for example, highlighted the existence of hitherto neglected localised scientific interests and activities in rural and remote Australia during the second half of the nineteenth century.

In the project, we utilised frequent, closed and maximal itemsets to extract objectively interesting associations between person names extracted from pre-processed querysets.

The querysets from the Trove archive were processed similarly. Pre-processing reduced ambiguous rules by up to 50 %; increased support thresholds for association rules by almost 100 % and reduced number of rules by ≈90 % with increased support and confidence. The large amount (≈75 %) of single-tokened, singleton sets produced at higher support were an indication of high word error rates during OCR, causing token omissions or misclassifications during NLP. Single-tokened names were found in ≈60 % of namesets in rules, which still left high degrees of ambiguity in the associations.

Low support in more than 95 % of rules can be expected from news article data given the 150 years timespan and nature of reporting where the lifespan of a news "story" is quite short, and this added to the high presence of OCR errors.

Combinations of namelist frequency and similarity thresholds may not have been ideal for all the querysets as some were sparser than others, resulting in uneven sizes of the namelists and the increased likelihood of false positives/negatives in some. It is also likely that name-tokens of popular persons may have influenced corrections with their high frequencies, resulting in multi-token names that did not actually exist.

Not all the 20 querysets could be categorised as some had unique behaviours. Further and more refined groupings could be made with more querysets. The number of persons associated and their frequency may not necessarily reflect on stories' popularity, or even the theme (the query) itself, but on the nature of the news stories. Popular news items would have been covered by multiple publishers over a certain length of time; which would give persons associated high support; results show that such were mostly individuals and small groups of 2 or 3 persons; although this may not be always the case. The number and sizes of communities involved reflected on person-name density of the querysets, the type of stories associated with the historical theme (query), and the number of stories with associated individuals appearing frequently enough to meet the support threshold.

As a future work, we would like to build more querysets and perform association. Additionally, it is hoped that a dataset prepared with all NEs, assuming the place, organisation, location and other miscellaneous entities surrounding a person, would help in disambiguating an ambiguous person during the process of clustering.

Acknowledgements. We would like to acknowledge CRC Smart Services and Prof Kerry Raymond for facilitating data acquisition from National Library of Australia. Our appreciation goes to Prof Paul Turnbull (paul.turnbull@utas.edu.au), Professor of Digital Humanities at the University of Tasmania, who was also expert advisor to the project. We would like to thank Sangeetha Kutty for initial parsing of Trove articles using SENNA, and insights into tasks; thanks also to Taufik Edy Sutanto for attempting to cluster the dataset. Also, thanks to Yeshey Peden for helpful discussion on aspects of the project and to Ashley Wright at QUT HPC for helping with data scalability issues.

References

1. Artiles, J., Borthwick, A., Gonzalo, J., Sekine, S., Amigó, E.: WePS-3 evaluation campaign: overview of the web people search clustering and attribute extraction tasks. In: CLEF (Notebook Papers/LABs/Workshops) (September 2010)
2. Bagga, A., Baldwin, B.: Entity-based cross-document coreferencing using the vector space model. In: Proceedings of the 36th Annual Meeting of the Association for Computational Linguistics and 17th International Conference on Computational Linguistics-Volume 1, pp. 79–85. Association for Computational Linguisitcs (August 1998)
3. Berg-Kirkpatrick,T., Durrett, G., Klein, D.: Unsupervised Transcription of Historical Documents Computer Science Division University of California at Berkeley. In: Proceedings of the 51st Annual Meeting of the Association for Computational Linguistics, pp. 207–217, Sofia, Bulgaria. c2013 Association for Computational Linguistics (4–9 August 2013)
4. Chen, Y., Martin, J.: Towards robust unsupervised personal name disambiguation. In: EMNLP-CoNLL, pp. 190–198 (June 2007)
5. Hahsler, M., Chelluboina, S.: Visualising association rules: introduction to the R-extension package arulesViz. R project module, pp. 223–238 (2011)
6. Hahsler, M., Grün, B., Hornik, K.: Introduction to arules – mining association rules and frequent itemsets. SIGKDD Explor. (2007)

7. Han, J., Kamber, M., Pei, J.: Data Mining: Concepts and Techniques: Concepts and Techniques. Elsevier, New York (2011)
8. Packer, T., Lutes, J., Stewart, A., Embley, D., Ringger, E., Seppi, K.: Extracting person names from diverse and noisy OCR text. Department of Computer Science Brigham Young University Provo, Utah, USA (2010). doi:10.1145/1871840.1871845
9. Reynaert, M.: Non-interactive OCR post-correction for giga-scale digitization projects. In: Gelbukh, A. (ed.) CICLing 2008. LNCS, vol. 4919, pp. 617–630. Springer, Heidelberg (2008)
10. Seno, M., Karypis, G.: Finding frequent patterns using length-decreasing support constraints. Data Min. Knowl. Discov. **10**(3), 197–228 (2005)
11. Taghva, K., Nartker, T., Condit, A., Borsack, J.: Automatic removal of "garbage strings" in OCR text: an implementation. In: The 5th World Multi-Conference on Systemics, Cybernetics and Informatics (July 2001)
12. Taghva, K., Stofsky, E.: OCRSpell: an interactive spelling correction system for OCR errors in text. Int. J. Doc. Anal. Recogn. **3**(3), 125–137 (2001)
13. Tong, X., Evans, D.A. A statistical approach to automatic OCR error correction in context. In: Proceedings of the Fourth Workshop on Very Large Corpora, pp. 88–100 (August 1996)
14. Unwin, A., Hofmann, H., Bernt, K.: The TwoKey plot for multiple association rules control. In: Siebes, A., De Raedt, L. (eds.) PKDD 2001. LNCS (LNAI), vol. 2168, pp. 472–483. Springer, Heidelberg (2001)

Towards the Elimination of the Miscommunication Between Users in Twitter

Tweet Classification Based on Expected Responses by User

Tomoaki Ueda[✉], Ryohei Orihara, Yuichi Sei, Yasuyuki Tahara,
and Akihiko Ohsuga

Graduate School of Information Systems, University of Electro-Communications,
Tokyo, Japan
ueda.tomoaki@ohsuga.is.uec.ac.jp, ryohei.orihara@toshiba.co.jp,
{sei,tahara}@is.uec.ac.jp, ohsuga@uec.ac.jp

Abstract. In recent years, a Twitter response from another user who does not share the intentions and expectations of the original poster may cause discomfort and stress, which is a social phenomenon known as SNS fatigue. For example, a user may receive answers that are different from her/his expectation after the user posts a question on the timeline. In the background of such responses there is a miscommunication between users. In order to resolve the problem, it is important to know what the original poster expected as responses to her/his tweet. In this paper, we propose a classification method of tweets according to the response that users expect, and experimentally evaluate it. As a result, we have shown that tweets which the poster does not expect any replies can be classified with 76.2 % of the average precision.

Keywords: Twitter · Support vector machine · SNS fatigue · Data mining

1 Introduction

In recent years, *SNS fatigue* [1–5] has become a social issue among SNS (Social Networking Services) users, such as micro-blogs (e.g. Twitter) and Facebook. In order to resolve SNS fatigue, it is necessary to prevent the negative emotions. However, in SNS, there are many opportunities for users to interact with people without acquaintance. Moreover, there are factors such as omission of the description, the presence or absence of the background knowledge on particular user or topic and lack of context created by Retweeting (RT), which further promote miscommunication between users. It is supposed that cause is mutual misunderstanding by the gap between the original poster and the recipient. In online communication, because there are barriers for users to make fruitful conversation, users often opt for easier methods such as blocking or simply disregarding replies. These behaviors are regarded as socially negative and often make situations worse. It is necessary to prevent a quarrel by detecting the gap between the original poster of a tweet and other users because the negative behaviors may lead to social disbenefits.

© Springer International Publishing Switzerland 2015
B. Pfahringer and J. Renz (Eds.): AI 2015, LNAI 9457, pp. 589–595, 2015.
DOI: 10.1007/978-3-319-26350-2_52

The purpose of this study is to prevent the mutual misunderstanding between users and to reduce SNS fatigue. The outline of our proposed method is the following. Using machine learning techniques, the expected response by the original poster of a tweet is recognized. Then, if she/he receives replies that conflict with the expectation, the replies are temporarily muted (hidden from the timeline) and the arrival of the unexpected reply is notified without revealing its contents. Upon the notification, the original poster has autonomy to read the reply or not. In this paper, as the initial stage of this study, we define the categories of expected responses by the original poster. In addition, we propose classification method and evaluate classifiers.

The remainder of the paper is organized as follows. We show the related work in Sect. 2. We explain our proposed method in Sect. 3. In Sect. 4, we show the results of the experimental evaluation. Finally, Sect. 5 presents the conclusion and future work.

2 Related Work

There are various studies dealing with Twitter and SNS. We show researches on SNS fatigue reduction, emotion prediction by the message, and extraction and classification of useful information from tweets. Raiko et al. [6] claimed senses of obligation for response and the expectation for other's response is a cause of *mental fatigue*. As an attempt to reduce the sense of obligation of the recipient, they implemented a prototype application with a function that make messages invisible after a certain amount of time passes. Hasegawa et al. [7] performed emotion prediction on recipients of online messages. As an application of their study, they proposed a system to filter out unpleasant messages from conversation. Zhao and Mei [8] extracted Information Needs tweets (tweets of the question format), and analyzed them as time-series data. They used Support Vector Machine (SVM) as classifiers to extract Information Needs. Furthermore they enhanced accuracy by a feature selection technique called Bi-Normal Separation (BNS) and a boosting technique called Ada-boostDIV. We referred to their method because these techniques of SVM and BNS are powerful. Hara et al. [9] proposed the classification method of informal writing in the micro-blog, such as Twitter. The informal writings are unofficial messages such as poster's own feelings or state. The classifier is the combination of SVM and Parametric Mixture Models.

3 Proposed Method

3.1 Classification Based on Expected Responses by User

There are possible responses, such as a reply, Fav (favorite), RT and the combination of them, to a tweet. In order to detect the gap between the original poster of the tweet and recipients, it is necessary to recognize expected responses by the original poster when the tweet is posted. We define categories of the expected responses and classify tweets into them. When recipient's response is not matching the classification result of the original poster's message, it can be said that there is the gap between users. One advantage of this approach is the fact that it does not always require the analysis of the contents of replies: sometimes any replies are against the expectation. It can save execution time and goes well with the real-time nature of Twitter.

We assumed that a twitter user posts a tweet to accomplish a purpose. For example, a twitter user posts a message of a question format when the user wants to know information asked in the message. Sometimes posting itself is the purpose to post. For example, a monologue tweet (e.g. *I'm hungry.*) does not affect other users. In this case the user does not expect any responses from other users. The tweet is unlikely to be retweeted except in the case the user is a celebrity. Classification of tweets based on expected responses could be also useful for information extraction from SNS, such as extraction of opinions and interests of users and identification of spam messages, entertaining images and texts.

3.2   Definition of Categories

In this study, we define categories of tweets based on expected responses by user, as show in Table 1. We describe example tweets of each category.

3.3   Training of Classifier

In this study, we use the Support Vector Machine (SVM), which is one of supervised learning methods, to create classifiers for each category.

Supervised Tagging. We have collected Japanese tweets through the Streaming API[1]. The tweets covers a period of 105 days, from November 1st, 2014 to February 13th, 2015. For our experiment we include 22,010 tweets from the official client and relatively popular third-party clients only, throwing away other tweets such as ones made by bots. We have tagged the tweets manually with categories introduced in 3.2. The manual tagging has been done taking account of the clue expressions and the other semantic features of tweets. Although the tagging have been done by one of the authors, we are considering to collect tagged data from multiple annotators through a client in the future.

Preprocessing and Feature for Learning. In order to keep the number of features manageable, we have normalized strings representing URLs (http://~) and user names (@xxx). The URL strings and the user names have been converted into "URL" and "@USERNAME" respectively.

As features for classification by machine learning, we use N-gram (N = 1, 2 and 3), because in preliminary experiments the N-gram has yielded better results than Japanese morphological analyzer: kuromoji [10].

---

[1] Twitter Streaming API: https://dev.twitter.com/streaming/overview.

Table 1. Categories of tweets based on expected responses by user.

Categories	Responses	Definitions and Examples
Monologue (ML)	No Response	ML is a tweet on personal events, appointments, current location, short notes of summarizing own thinking, physical condition and complaints. It also includes indirect tweets. This tweet involves personal contexts that cannot be known by others. Therefore, ML tweet does not expect any responses (e.g. reply, Fav and RT) from other users. **Example:** hungry & tired
Advertizing and News (AN)	RT	AN contains intention of the original poster that desires to reach to many people by RT. There is examples of advertising of stores or services and news of social events. AN tweet does not expect responses other than RT, because orig-inal poster and the subject of the information are different. **Example:** Saint Louis #weather on July 23, 2015 (URL)
Impressions, Comments and Claims (ICC)	Fav	ICC is intended to describe impressions, opinions and positions of the user on global topics (e.g. topics on Timeline, social situations and TV programs). Examples of the ICC tweets include RTs quoting AN tweets, tweets of impressions after retweeting a TP (see below), and tweets of comments on television or radio programs specified by hashtags. ICC tweets expect Favs but no replies or RTs, because criticizing replies and uncontrollable spread of the tweet to strangers through RTs often lead to flaming. **Example:** I hate when people take their personal problems to social media
Topic-providing (TP)	Fav and RT	TP is a tweet that users provide topic to timeline. This tweet expects Fav and RT by posting images and rare experiences of extraordinary scene. The original poster is exposed to irrelevant message from foreign users by RT. This tweet is not expected reply by other users, since it is often dramatized. **Example:** breaking news: look at dis (https://twitter.com/xxx/status/yyy/photo/1)
Information Needs (IN)	Reply	Information Needs (IN) is a tweet asking correctness confirmation of knowledge, details of things, recommendations, and specific personal information by the question to followers or a specific user. Example of IN tweet is to tell the song title or official name from image or lyrics. The original poster of tweet expects reply messages include answers. Example: I've had the hiccups for like 3 hrs. How do I make them go away??
Recruit (RC)	Reply and RT	RC calls for people or properties with particular quality. As with the AN tweet, RC is intended to be seen by many users. However, unlike the AN tweet, RC expects replies from other users who are interested in what is called for. Therefore, the original poster of the tweet expects reply and RT. **Example:** #recruitment #news EDUSTAFF: Psychology Graduate/Primary LSA/Learning Support Assistant (http://xxx.yyy.com)
Greeting (GT)	Reply and Fav	GT is a tweet to post a real-world greeting to the timeline. GT tweet is sup-posed to expect replies and Favs by other users, because in the real world when we greet someone we expect responses and replies from her/him. A reply to a GT tweet is often a GT tweet. **Example:** Good morning! Thank you God for another beautiful day :)
Chaintag (CT)	All	CT is a tweet that demands particular responses (reply, Fav, RT or follow) to readers. The desired response is typically described in hashtags. **Example:** Bought #USDJPY 123.728 SL 121.522 TP 128.022	Auto-copy #trade FREE. Earn upto 300% via (http://xxx.yyy.com) #fx #follow #news #RT #FF

Multiple Labels. It is desirable that all the tweets are uniquely classified into one of the eight categories for simplicity. Actually a tweet may have multiple tags because it may have the features for multiple categories, or its superficial meaning differs from the poster's intention. For the first cases, we have given them multiple tags and have excluded them from the experiment. Fortunately the number of such tweets is small. For the second cases, we have tagged them based on the superficial meaning. If the tweets have multiple labels on a client, we assume to accept all the reactions of each label.

Feature Selection and Coping with the Imbalanced Data. The number of features tends to be large when the N-gram feature is used, because there are redundant features. In such situations a feature selection method is often useful because it can not only reduce the dimensions of the data but also improve the accuracy of classification. In this study, we use the Bi-Normal Separation (BNS) [11]. BNS is employed because it is reported to be able to more accurately select features than the conventional methods.

In preliminary experiments, the accuracy of classification has been extremely low for IN. The number of the IN tweets is small comparing to other categories (only 259 tweets). We assume that the imbalance of the data is the culprit. We have removed approximately half of the negative examples by sampling. We also have sampled the positive examples for the ML tweets (more than 65 % of all).

4   Experimental Evaluation

4.1   Experimental Method

We build a binary SVM classifier for each of the eight tweet categories using the 21,991 tweets with a single label out of 22,010 tweets. We evaluate the classifiers through 10-fold cross validation. We use precision, recall and F-measure as evaluation measures. The SVM library which we use is JNI Kernel Extension for SVM^{light} [2] of Java interface of SVM^{light} [12]. The kernel which we use is the linear kernel, with parameter C being set to the default value.

4.2   Experimental Result

Experimental results are shown in Table 2. For ML and IN, averages from 10 samplings are shown, because we have sampled the imbalanced data as described above. Also the results are shown the case using keyword matching as the baseline. Set keywords are words that are the clue expressions, the other semantic features and the terms with high BNS value.

Table 2. Evaluation results of classifier.

Categories	Proposed method				Baseline		
	Precision	Recall	F-measure		Presicion	Recall	F-measure
			without BNS	with BNS			
ML	0.87464	0.88185	0.87830	0.87307	0.60964	0.62584	0.61763
AN	0.78708	0.66020	0.71808	0.74901	0.50592	0.69601	0.58593
ICC	0.75764	0.09646	0.17114	0.16470	0.18198	0.20669	0.19355
TP	0.63049	0.25428	0.36240	0.58789	0.32751	0.425	0.36994
IN	0.36016	0.32714	0.34318	0.31512	0.83077	0.0817	0.14871
RC	0.93618	0.58950	0.72345	0.77661	0.79441	0.4398	0.56615
GT	0.91749	0.81802	0.86491	0.87169	0.92875	0.7462	0.82753
CT	0.88872	0.70113	0.78386	0.83937	0.88387	0.4331	0.58133

4.3   Discussion

AN, RC, GT and CT tweets have achieved a high accuracy that exceeds 70 % in F-measure. It can be seen from the results of the baseline, this can be explained by the fact that these tweets tend to include a *keyword* specific to each category. In categories such as AN, TP, RC, GT and CT, the F-measures have been further improved by the BNS.

---

[2] JNI Kernel Extension for SVM-light: http://people.aifb.kit.edu/sbl/software/jnikernel/.

This can be explained as the BNS has filtered out some of redundant features. ICC has yielded extremely poor recall. Through the error analysis, we have found that the most true positives are ones including quotes (e.g. >RT), hashtags or simple impressions like *I agree*, and there are no comments and claims on social issues have been detected. In comments and claims a user tends to use her/his own particular expressions, unlike impressions where common phrases are used over users. We might be able to cope with this by building separate classifiers for each of subcategories. As for IN tweets, we are dealing with an issue more difficult than one in [8] because our definition of the IN does not require to include '?' and finding appropriate features is not trivial. Although the poor accuracy is understandable, the fact that there are not many IN tweets in our data seems to make things worse. It might be possible to improve the accuracy by including IN tweets in a reply format, which are excluded in the experiment above, because it means the increase of positive examples.

Finally, we discuss the performance of the classifiers taking real use cases into account. As we have seen in Sect. 3, there are tweet categories that expect no replies at all. The categories, ML, AN, ICC and TP are shown in gray background in Table 2. According to the questionnaire, we have carried out in our preliminary study, we considered that evaluations of the mute function should focus on the precision rather than recall. For the four categories, the classifiers have scored 76.2 % as the average precision. It means that if we use the classifiers to mute unwanted responses from other users, almost 80 % of the muted tweets are indeed ones should be muted, when the original tweet belongs to one of the four categories. Therefore we can say that we have detected an aspect of the gap between the original poster and the recipients in twitter.

5   Conclusion

In this paper, we have proposed categorization of tweets based on expected responses by a user, built classifiers for the categories by SVM and evaluated the classifiers. We have hypothesized that pointless conversations, one of the major causes of stressful twitter experiences, are generated by the miscommunication between users, which is resulted from the gap between the intension of the original poster and the interpretation of the recipient. In order to detect the gap, it is necessary to know what the original poster expects as responses. Therefore we have proposed the categorization of tweets based on the expected responses by them. We have tagged Japanese twitter data based on the categorization, and learned SVM classifiers for the categories. Through the experiment, we have obtained 64.7 % of the average F-measure over all the categories. Furthermore, for the four categories where the poster does not expect any replies, the classifiers have scored 76.2 % of the average precision, suggesting that it is possible to use the classifiers to mute the unwanted replies to reduce the SNS fatigue. It is also possible to use the proposed method as spam filters. Finally future works are classification of reply Tweets, tagging by multiple supervisors, Platform development of crowdsourcing for general corpus building and Evaluation through twitter client applications.

Acknowledgements. This work was supported by JSPS KAKENHI Grant Numbers 24300005, 26330081, 26870201.

References

1. Yamakami, T.: Towards understanding SNS fatigue: exploration of social experience in the virtual world. In: 7th International Conference on Computing and Convergence Technology (ICCCT), pp. 203–207 (2012)
2. Ravindran, T., Chua, A.Y.K., Hoe-Lian, G.D.: Characteristics of social network fatigue. In: 10th International Conference on Information Technology: New Generations, pp. 431–438 (2013)
3. Maier, C., Laumer, S., Eckhardt, A., Weitzel, T.: When social networking turns to social overload: explaining the stress, emotional exhaustion, and quitting behavior from social network sites' users. In: The 20th European Conference on Information Systems (ECIS), pp. 10–13 (2012)
4. Yao, X., Phang, C.W., Ling H.: Understanding the influences of trend and fatigue in individuals' SNS switching intention. In: 48th Hawaii International Conference on System Sciences (HICSS) pp. 324–334 (2015)
5. Kato, C.: Reality of negative experiences connected with "SNS Fatigue": on interviews to fifteen high school students (refereed studies). Socio-Informatics **2**(1), 31–43 (2013). (In Japanese)
6. Raiko, N., Ogasawara, N., Sato, K., Nunokawa, H.: The communication tool which reduce the feeling of duty by the disappearing message, the special interest group technical reports of IPSJ. EC Entertain. Comput. 2014-EC **31**(1), 1–6 (2014) (In Japanese)
7. Hasegawa, T., Kaji, N., Yoshinaga, N., Toyoda, M.: Predicting and evoking listener's emotion in online dialogue. J. Jpn. Soc. Artif. Intell. **29**(1), 90–99 (2014). (In Japanese)
8. Zhao, Z., Mei, Q.: Questions about questions: an empirical analysis of information needs on Twitter. In: The 22nd International Conference on World Wide Web (WWW), pp. 1545–1555 (2013)
9. Hara, M., Asai, T., Takahashi, H., Tajima, Y., Kikui, G.: Automatic categorization of informal messages in microblogs, the special interest group technical reports of IPSJ. MP 2014-MPS **97**(25), 1–2 (2014). (In Japanese)
10. Japanese morphological analyzer: kuromoji (2011). http://www.atilika.org/
11. Forman, G.: BNS feature scaling: an improved representation over tf-idf for svm text classification. In: the 17th ACM Conference on Information and Knowledge Management (CIKM), pp. 263–270 (2008)
12. SVM-Light support vector machine (2008). http://svmlight.joachims.org/

Reinforcement Learning of Pareto-Optimal Multiobjective Policies Using Steering

Peter Vamplew$^{(\boxtimes)}$, Rustam Issabekov, Richard Dazeley,
and Cameron Foale

Federation Learning Agents Group, Federation University Australia,
Ballarat, VIC, Australia
{p.vamplew,r.issabekov,r.dazeley,c.foale}@federation.edu.au

Abstract. There has been little research into multiobjective reinforcement learning (MORL) algorithms using stochastic or non-stationary policies, even though such policies may Pareto-dominate deterministic stationary policies. One approach is steering which forms a non-stationary combination of deterministic stationary base policies. This paper presents two new steering algorithms designed for the task of learning Pareto-optimal policies. The first algorithm (w-steering) is a direct adaptation of previous approaches to steering, and therefore requires prior knowledge of recurrent states which are guaranteed to be revisited. The second algorithm (Q-steering) eliminates this requirement. Empirical results show that both algorithms perform well when given knowledge of recurrent states, but that Q-steering provides substantial performance improvements over w-steering when this knowledge is not available.

Keywords: Multiobjective reinforcement learning · Non-stationary policies

1 Introduction

Most reinforcement learning (RL) algorithms consider only a single objective, encoded in a scalar reward. However many decision-making tasks naturally have multiple conflicting objectives. Examples include managing water reservoirs [2] and the wet clutch control task examined by [1]. Therefore multiobjective reinforcement learning (MORL) has emerged as a growing area of research [11].

The difference between single-objective RL and MORL is straightforward – rather than a scalar reward, the agent receives a vector reward with an element for each objective. However this has significant repercussions. In particular in the multiobjective case it may be beneficial to learn non-stationary or stochastic policies, as these policies may Pareto dominate the best possible deterministic stationary policies. Despite the potential benefits there has been relatively little work so far on MORL algorithms able to learn non-stationary or stochastic policies. This paper proposes and evaluates two methods based on the MDRL steering approach of [7,8]. The first algorithm (w-steering) adapts MDRL to the

© Springer International Publishing Switzerland 2015
B. Pfahringer and J. Renz (Eds.): AI 2015, LNAI 9457, pp. 596–608, 2015.
DOI: 10.1007/978-3-319-26350-2_53

task of finding Pareto-optimal policies. Like MDRL it assumes that the agent has prior knowledge about a set of one or more recurrent states which are guaranteed to be revisited. The second variant proposed in this paper (Q-steering) eliminates the need for pre-existing knowledge of recurrent states, thereby increasing the applicability of steering to problems where such information is not available.

2   Background

Single-objective RL generally assumes that the decision-making task can be modelled as a *Markov decision process* (MDP) defined as follows:

- S is a finite set of *states* and A is a finite set of *actions*,
- $T : S \times A \times S \rightarrow [0, 1]$ is a *state transition function* specifying for each state, action, and next state, the probability of that next state occurring,
- $R : S \times A \times S \rightarrow \Re$ is a *reward function*, which specifies the expected immediate reward for each combination of state, action and subsequent state,
- $\mu : S \rightarrow [0, 1]$ is a probability distribution over initial states, and
- $\gamma \in [0, 1)$ is a *discount factor* specifying the relative importance of short-term and long-term rewards.

The goal of the agent is to maximize the expected *return* R_t. In a *finite horizon (or episodic)* MDP, the return is typically an undiscounted finite sum of the rewards received over all time-steps until a terminating state is reached, while in an *infinite horizon* MDP, the return is usually an infinite sum, with each term discounted according to γ, where r_t is the reward obtained at time t:

$$R_t = \sum_{k=0}^{\infty} \gamma^k r_{t+k+1},$$

Multi-objective MDPs (MOMDPs) differ from the single-objective MDP model in that the reward function $\mathbf{R} : S \times A \times S \rightarrow \Re^n$ defines a vector of n rewards rather than a scalar, with each element in the vector corresponding to the reward relative to one objective. Similarly, the return is also a vector value. For example, in the infinite horizon case the multiobjective return $\mathbf{R}_t$ would be defined as:

$$\mathbf{R}_t = \sum_{k=0}^{\infty} \gamma^k \mathbf{r}_{t+k+1},$$

While the extension from MDP to MOMDP is straightforward, there are significant ramifications. For MDPs there must exist either a single optimal policy, or a set of optimal policies with identical returns. In contrast for MOMDPs a policy can outperform another with respect to one objective, but be inferior with regards to the other objective(s). A policy is said to Pareto dominate another if it is superior on at least one objective, and at least equal on all other objectives. Any dominated policy is of little value, as clearly the dominating policy is preferable. However there may be many policies which are non-dominated with regards to any other policy. All such policies can be regarded as being optimal, and so the goal of a MORL agent is to discover one or more of these Pareto-optimal policies. The number of policies required will depend on the context in which the agent is being applied – see [11] for a detailed discussion of this issue.

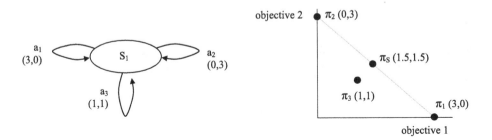

Fig. 1. An environment with 3 actions, labeled with their rewards, and the average per-action reward for each deterministic stationary policy π_1, π_2 and π_3 and for a stochastic policy π_s which selects between actions a_1 and a_2 with equal probability.

A MORL agent may also need to consider forms of policies which are not required in single-objective RL. For fully-observable single-objective MDPs a deterministic stationary optimal policy always exists [12], but for MOMDPs, stochastic or non-stationary policies may dominate the best deterministic stationary policies [3,15]. Consider the simple environment of Fig. 1. Three deterministic stationary policies π_1, π_2 and π_3 are available. If a balanced trade-off between objectives is required then policy π_3 which always selects a_3 will be preferred. However consider a stochastic policy π_s which selects between actions a_1 and a_2 with probabilities p_1 and $(1-p_1)$ respectively. If $1/3 \leq p_1 \leq 2/3$ then this policy will dominate π_3. This simple example illustrates the general point that for multiobjective problems it may be possible to generate superior results using policies which are either non-stationary or stochastic in nature.

This section has outlined the general manner in which MORL differs from single-objective RL. For the remainder of this paper it is assumed that the context is such that the MORL algorithm should exhibit the following properties:

- it will identify a single Pareto-optimal policy which will satisfy the user's preferred trade-offs between objectives;
- it will exploit the potential benefits of non-stationary policies;
- it will be well-suited to online learning, and
- it will not require any prior knowledge of the state transition or reward dynamics of the environment

3   Stochastic or Non-stationary MORL Approaches

Two main approaches for stochastic or non-stationary MORL have been explored previously. The first uses policy search methods such as policy gradient [9,12] or multiobjective evolutionary algorithms [4,13]. However empirical results on single-objective tasks show that policy search may converge more slowly and so be less suitable for on-line learning tasks [5,14,20]. As such there may be situations in which a value-based approach to MORL is preferable.

The main value-based approach used so far is to find deterministic stationary policies, and then form stochastic or non-stationary combinations of these 'base policies' [7,8,10,15]. The base policies can be found using relatively simple methods. For example, linear scalarisation takes a weighted sum of the rewards, essentially converting the problem to a single-objective MDP to which standard TD-based methods can be applied [11]. Repeated runs with differing weights will discover a range of Pareto-optimal base policies [2][1] which the agent then switches between, on either a stochastic [15] or non-stationary basis [7,8].

Multiple Directions Reinforcement Learning (MDRL) is a non-stationary approach based on geometric steering [7,8]. While MDRL was designed for stochastic games, MOMDPs can be treated as a special case of this scenario [7], and we will restrict our discussion to the case of MOMDPs. MDRL forms a non-stationary policy which aims to keep the long-term average reward vector within a defined target region in objective space. The agent learns in parallel multiple base policies based on linear scalarisation using different weight vectors, and 'steers' the average reward towards the target region by switching between base policies. The choice of active policy is made by calculating a direction vector from the average reward vector to the closest point in the target region. The policy whose weights best match the direction vector is selected as the active policy. MDRL is summarized in Algorithm 1.

MDRL has a number of desirable properties. A decision-maker can specify the desired trade-off between objectives in a straight-forward fashion via the definition of the target zone. MDRL is also a flexible 'meta-algorithm' as the underlying learning of the base policies can be achieved using any linear scalarisation method – while this paper uses tabular Q-learning [19], any off-policy algorithm and form of function approximation could be used.

However MDRL relies on two key assumptions. It is assumed that one or more states are recurrent – that is, they are guaranteed to be revisited within a finite period of time regardless of the policy being followed. Formally, if the hitting time of a state s^* is defined as $\tau_{s^*} = min\{n > 0 : s_n = s^*\}$ and E_π^s is the expectation operator when starting from state $s_0 = s$ and following policy π, then there must exist at least one state s^* and a finite constant N such that $E_\pi^s(\tau_{s^*}{}^2) < N$ for all $\pi \in \Pi$ (the set of all possible policies) and $s \in S$. In addition it is assumed that the agent will have knowledge of these recurrent states as indicated by the set S_R as an input to Algorithm 1. The decision to switch active policies is made only when the agent is in a recurrent state. For some problems identifying this set of states may be easy – for example in episodic problems the initial state

---

[1] MORL methods based on linear scalarisation are limited to discovering policies which lie on the convex hull of the Pareto front [17]. However in the context of learning base policies this is actually advantageous as policies lying inside the convex hull will not form part of any Pareto-optimal combination of base policies [15].

Algorithm 1. The Multiple Directions Reinforcement Learning algorithm

---

input: target region T, set of one or more recurrent states S_R, number of base policies j

1: initialize weight-vectors $w_1..w_j$ to be evenly distributed across the surface of a unit hypersphere with dimensionality equal to the number of objectives
2: initialise state-action values Q for j different deterministic policies $\pi_1..\pi_j$
3: initialise the average reward vector R to the null vector
4: select at random an active policy π_A
5: observe initial state s
6: **while** true **do**
7:     **if** s in S_R and R is not in T **then**
8:         let D = the vector from R to the closest point in T
9:         select active policy π_A where $A = \arg\min_{1 \leq A \leq j} |D - w_A|$
10:    **end if**
11:    select and execute action based on π_A, observe reward r and next state s'
12:    update record of average reward R
13:    update values for Q_{π_A}
14:    update values for all other $Q\pi_i$ using off-policy learning
15:    $s \leftarrow s'$
16: **end while**

---

of each episode can be treated as a recurrent state. However more generally there may no prior knowledge of the recurrent states, in which case the only viable option is for MDRL to allow switching in any state[2]. This could lead to the agent switching base policies in inappropriate states resulting in suboptimal performance.

This limitation was addressed in an alternative steering algorithm known as Current Direction Reinforcement Learning (CDRL) [8]. CDRL maintains a single policy, with scalarisation weights set to match the current direction from the average reward vector to the target set. Whenever these weights change, the agent follows this policy for an extended period of time before switching to an alternative set of weights. This switching can occur in any state, and so CDRL does not require S_R as an input. CDRL reduces the frequency of policy switching which will minimise, but not eliminate, the potential performance impact of switching between base policies in inappropriate states. However while CDRL will maintain the long-term average reward within the target region, in the short-term the average reward may lie well outside the target. For many applications this increase in the variance of the short-term reward is unacceptable. For this reason we will not consider CDRL further in this work.

4   Applying Steering to Pareto-Optimal MORL

While MDRL allows learning of non-stationary policies for MOMDPs, there are differences between the scenario for which it was designed, and the context which

---

[2] The other alternative is to treat no states as members of S_R, but this would mean the agent would never switch base policies, thereby losing the benefits of non-stationarity.

this work addresses, as identified in Sect. 2. Most significantly, MDRL treats the vector rewards as criteria to be satisfied rather than as objectives to be maximised. MDRL aims to maintain the long-term reward within the specified target region which may include both upper and lower bounds for each reward. Therefore the agent may need to steer the reward in both positive and negative directions in reward space, and so the vectors $w_1..w_j$ contain both positive and negative weights. The use of negative weights is not compatible with our goal to produce Pareto-optimal behavior, so when using steering to learn a Pareto-optimal policy all weights are constrained to be positive values. Similarly in the context of achieving Pareto-optimal performance, there may be no pre-existing knowledge on which to base the size or shape of the target region. Therefore we propose to use a target point in objective space instead. By changing the choice of target point, the agent can be directed towards different Pareto-optimal policies, and specifying a target point should be relatively intuitive for a human decision-maker. Algorithm 2 details how these amendments can be incorporated to adapt MDRL to the goal of finding a Pareto-optimal policy.

Algorithm 2. The w-steering algorithm for Pareto-optimal MORL. Differences from MDRL are indicated by highlighted background.

---

    input: target point T, number of base policies j, set of switching states S_S which will be S_R if knowledge exists of recurrent state(s) or S otherwise,

1: initialize weight-vectors $w_1..w_j$ to be evenly distributed within the positive quadrant of a unit hypersphere with dimensionality equal to the number of objectives
2: initialise state-action values Q for j different deterministic policies $\pi_1..\pi_j$
3: initialise the average reward vector R to the null vector
4: select at random an active policy π_A
5: observe initial state s
6: **while** true **do**
7:    **if** s in S_S **then**
8:        let D = T - R
9:        select active policy π_A where $A = \arg\min_{1 \leq A \leq j} |D - w_A|$
10:   **end if**
11:   select and execute action based on π_A, observe reward r and next state s'
12:   update record of average reward R
13:   update values for Q_{π_A}
14:   update values for all other $Q\pi_i$ using off-policy learning
15:   $s \leftarrow s'$
16: **end while**

While the changes introduced in Algorithm 2 adapt steering to the task of finding a Pareto-optimal policy, this algorithm still retains the fundamental assumption of MDRL – that the agent is provided with a set of recurrent states. Our contention is that both MDRL and Algorithm 2 may perform poorly in the absence of known recurrent states due to the manner in which they select the active base policy. The policy selection mechanism (lines 8 and 9 of Algorithm 2) is independent of the current state of the environment. This fails to take into account that some states may be far more desirable for some base policies than for others – for example in the Linked Rings problem shown in Fig. 2 (right) states 2, 3 and 4 are desirable states for a policy aiming to maximize the first objective, but undesirable for a policy focused on maximizing the second objective.

Algorithm 3. The Q-steering algorithm. Differences from w-steering are indicated by highlighted background.

---

    input: target point T, number of base policies j, set of switching states S_S which will be S_R if knowledge exists of recurrent state(s) or S otherwise,

1: initialize weight-vectors $w_1..w_j$ to be evenly distributed within the positive quadrant of a unit hypersphere with dimensionality equal to the number of objectives
2: initialise state-action values Q for j different deterministic policies $\pi_1..\pi_j$
3: initialise the average reward vector R to the null vector
4: select at random an active policy π_A
5: observe initial state s
6: **while** true **do**
7:     **if** s in S_S **then**
8:         let D = T - R
9:         **for** each policy π_i **do**
10:           $Q^*_i = Q_{\pi_i}(s, a^*)$ where $a^* = \arg\max_a Q_{\pi_i}(s, a)$
11:         **end for**
12:         select active policy π_A where $A = \arg\max_{1 \leq A \leq j} |D.Q^*_A|$
13:     **end if**
14:     select and execute action based on π_A, observe reward r and next state s'
15:     update record of average reward R
16:     update values for Q_{π_A}
17:     update values for all other Q_{π_i} using off-policy learning
18:     $s \leftarrow s'$
19: **end while**

We propose a modified Pareto-optimal steering algorithm to account for this, by basing policy selection not on the weights associated with each policy but on the expected value of each policy at the current state. As shown in lines 8–11 of Algorithm 3 the value of the greedy action for the current state is calculated for each base policy, and the active policy is set to be the policy whose value best matches the direction vector from the current average reward R to the target point T. This change in the policy selection method represents the sole change from Algorithm 2. For ease of reference, from hereon we will refer to Algorithm 2 as w-steering, as its policy selection is based on each policy's weight vector w, and Algorithm 3 as Q-steering, as it is based on the $Q_\pi(s, a^*)$ value of each policy.

5 Empirical Evaluation of w-Steering and Q-Steering

This section empirically compares the two Pareto-optimal steering algorithms on two benchmark problems – one where the set of switching states S_R is known, and one where this information is not available to the agent. The first benchmark is the Deep Sea Treasure (DST) task [17] as shown in Fig. 2 (left). The agent must retrieve treasure from the sea bed. More valuable treasures are located

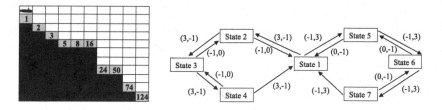

Fig. 2. (left) The Deep Sea Treasure task. (right) The Linked Rings task.

Table 1. The four Pareto-optimal deterministic stationary policies for the Linked Rings.

Policy ID	States visited when starting in State 1	Average reward per action
A	1, 2, 3, 4, 1,	3, −1
B	1, 2, 1,	1, −0.5
C	1, 5, 1,	−0.5, 1
D	1, 5, 6, 7, 1,	−1, 3

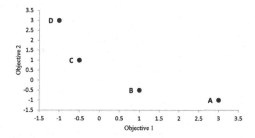

Fig. 3. The average reward per time-step for the four Pareto-optimal stationary deterministic policies for the Linked Rings problem.

further out and deeper, so the agent must compromise between the value of the treasure and the time taken to recover it. This is an episodic task with a fixed starting state, so that starting state can be used as the switching state $(S_S = \{s_{start}\})$. For episodic tasks the average reward vector R and target point T are specified on a per-episode basis.

The second benchmark is the non-episodic Linked Rings (Fig. 2 (right)). For this task, four Pareto-optimal deterministic policies exist, as shown in Table 1 and Fig. 3. Policies B and C are concave with respect to A and D, and therefore the optimal steering behaviour is to switch between policies A and D. Table 1 shows that State 1 is the only recurrent state, but it is assumed this information is not available and so both steering algorithms treat all states as potential switching states $(S_S = S)$. As the problem is non-episodic both the average reward R and the target point T are defined on a per-timestep basis.

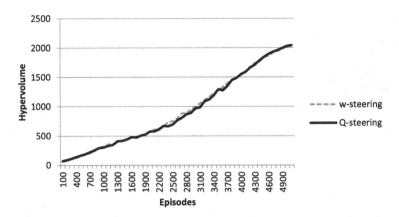

Fig. 4. Mean results achieved by w-steering and Q-steering over ten trials in terms of the online hypervolume metric for the DST environment.

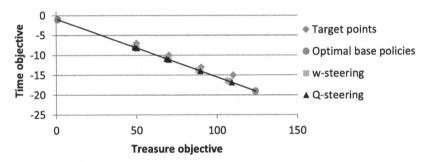

Fig. 5. The greedy return achieved by each trial of w-steering and Q-steering on the DST environment, relative to the specified target points. The line indicates the continuous Pareto front formed by linear combination of the optimal base policies.

To measure the performance of each algorithm we use the online hypervolume metric [16]. Both steering algorithms were executed multiple times on each benchmark using different target points for each run. The vector of average rewards received during learning was calculated, and these results combined across runs to produce an approximate Pareto front which was then evaluated using the hypervolume metric [21]. This entire process was repeated ten times, and the results reported are the mean across the ten repetitions. Tabular Q-learning was used as the underlying algorithm for both steering algorithms. The Q-learning parameters were set as $\alpha = 0.9$, $\lambda = 0.95$ and $\epsilon = 0.1$ for both benchmarks. No discounting was used for the episodic DST, while $\gamma = 0.9$ was used for the non-episodic Linked Rings. The reference point used for hypervolume calculations was $(0, -30)$ for Deep Sea Treasure and $(-1, -1)$ for Linked Rings.

Figure 4 shows the mean online hypervolume results for both algorithms on the DST task. Q-steering and w-steering perform very similarly on this task where knowledge of the recurrent states is available to the agent. Figure 5 illustrates the

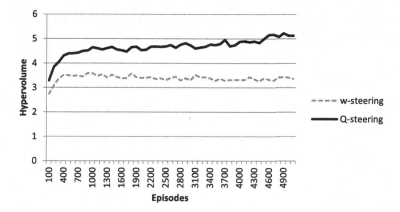

Fig. 6. Mean results achieved by w-steering and Q-steering over ten trials in terms of the online hypervolume metric for the Linked Rings environment.

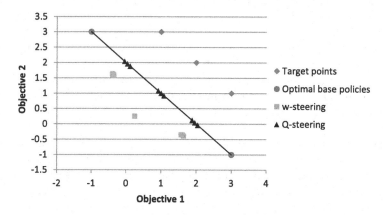

Fig. 7. The greedy return achieved by each trial of w-steering and Q-steering on the Linked Rings environment, relative to the specified target points. The line indicates the continuous Pareto front formed by linear combination of the optimal base policies.

final greedy performance of each trial over the last 100 episodes. It can be seen that both w-steering and Q-steering have correctly identified policies which lie on the Pareto front formed by linear combinations of the optimal base policies, and which are as close as possible to the optimistically-specified targets.

Figure 6 illustrates the performance of the steering algorithms on the Linked Rings where knowledge of the correct switching state is not provided. Q-steering produces a dramatic improvement over w-steering under these circumstances. Figure 7 shows that while Q-steering has correctly found the optimal combination of base policies to match the target points, w-steering's results fall well below the Pareto front. Closer examination of the actual behavior of the agent indicates that Q-steering has correctly identified state 1 as the optimal state in which to switch policies whereas the w-steering agent regularly switches between policies in states 2 and 5, giving rise to sub-optimal returns.

6   Conclusion and Future Work

The ability to identify Pareto-optimal non-stationary policies is important for MORL, as these policies may be superior to stationary deterministic policies. The MDRL algorithm [7,8] uses geometric steering to form a non-stationary mixture of deterministic base policies conditioned on the vector of rewards received so far. This paper has proposed and demonstrated a variant of MDRL (w-steering) designed for learning a Pareto-optimal policy with an objective trade-off determined by a target point specified by the user.

However, as demonstrated empirically, w-steering may perform poorly when the agent has no prior knowledge about the recurrent states in which it should switch between base policies. This paper has proposed a further steering variant (Q-steering), in which policy selection is based on the Q-values of each base policy for the current state. By taking into account the current state of the environment Q-steering substantially outperforms w-steering where prior knowledge of the correct switching states is not available. It should be noted that the q-steering policy selection mechanism is heuristic in nature and therefore not guaranteed to achieve a Pareto optimal result. Nevertheless this represents a significant improvement over the original formulation of steering, offering scope for application to a much broader class of problems.

However a number of issues still require further research. While Q-steering does not need prior knowledge of an appropriate recurrent state, such a state must exist in order for it to function correctly. Therefore Q-steering may fail on problems which are non-ergodic if certain states are reached. Automatic identification and avoidance of such states is an open research question. In addition, as noted in [15] the optimal steering policy for a problem with n objectives will utilise a maximum of n base policies. However Q-steering currently requires that base policies be learnt for many different weight combinations in order to ensure that the necessary sub-set of base policies are discovered. Methods for minimising the number of base policies learnt would be beneficial in reducing the memory overhead of Q-steering.

It should also be noted that while the benefits of non-stationary policies are clear, such policies may still be inappropriate in certain applications, such as the patient treatment scenario of [6]. A non-stationary policy which switches between one base policy which minimises symptoms but maximises side effects and another which does the reverse may appear to produce an excellent trade-off between objectives when rewards are averaged over multiple patients. However the experience of any individual patient is unlikely to be desirable. Therefore the need still exists to develop better algorithms for finding stationary deterministic policies for such tasks. The recent Pareto-set algorithm [18] represents a promising new approach for identifying such policies.

References

1. Brys, T., Van Moffaert, K., Van Vaerenbergh, K., Nowé, A.: On the behaviour of scalarization methods for the engagement of a wet clutch. In: The 12th International Conference on Machine Learning and Applications. IEEE (2013)
2. Castelletti, A., Corani, G., Rizzolli, A., Soncini-Sessa, R., Weber, E.: Reinforcement learning in the operational management of a water system. In: IFAC Workshop on Modeling and Control in Environmental Issues, pp. 325–330 (2002)
3. Chatterjee, K., Majumdar, R., Henzinger, T.A.: Markov decision processes with multiple objectives. In: Durand, B., Thomas, W. (eds.) STACS 2006. LNCS, vol. 3884, pp. 325–336. Springer, Heidelberg (2006)
4. Handa, H.: Solving multi-objective reinforcement learning problems by EDA-RL - acquisition of various strategies. In: Proceedings of the Ninth Internatonal Conference on Intelligent Sysems Design and Applications, pp. 426–431 (2009)
5. Kalyanakrishnan, S., Stone, P.: An empirical analysis of value function-based and policy search reinforcement learning. In: Proceedings of The 8th International Conference on Autonomous Agents and Multiagent Systems, vol. 2, pp. 749–756. International Foundation for Autonomous Agents and Multiagent Systems (2009)
6. Lizotte, D.J., Bowling, M., Murphy, S.A.: Efficient reinforcement learning with multiple reward functions for randomized clinical trial analysis. In: 27th International Conference on Machine Learning, pp. 695–702 (2010)
7. Mannor, S., Shimkin, N.: The steering approach for multi-criteria reinforcement learning. In: Neural Information Processing Systems, pp. 1563–1570 (2001)
8. Mannor, S., Shimkin, N.: A geometric approach to multi-criterion reinforcement learning. J. Mach. Learn. Res. **5**, 325–360 (2004)
9. Parisi, S., Pirotta, M., Smacchia, N., Bascetta, L., Restelli, M.: Policy gradient approaches for multi-objective sequential decision making. In: 2014 International Joint Conference on Neural Networks (IJCNN), pp. 2323–2330. IEEE (2014)
10. Roijers, D.M., Whiteson, S., Oliehoek, F.A.: Computing convex coverage sets for multi-objective coordination graphs. In: Perny, P., Pirlot, M., Tsoukiàs, A. (eds.) ADT 2013. LNCS, vol. 8176, pp. 309–323. Springer, Heidelberg (2013)
11. Roijers, D., Vamplew, P., Whiteson, S., Dazeley, R.: A survey of multi-objective sequential decision-making. J. Artif. Intell. Res. **48**, 67–113 (2013)
12. Shelton, C.: Importance sampling for reinforcement learning with multiple objectives. AI Technical report 2001–003, MIT, August 2001
13. Soh, H., Demiris, Y.: Evolving policies for multi-reward partially observable Markov decision processes (MR-POMDPs). In: Proceedings of the 13th Annual Conference on Genetic and Evolutionary Computation, GECCO 2011, pp. 713–720 (2011)
14. Taylor, M.E., Whiteson, S., Stone, P.: Temporal difference and policy search methods for reinforcement learning: an empirical comparison. In: Proceedings of the National Conference on Artificial Intelligence, vol. 22, p. 1675 (2007)
15. Vamplew, P., Dazeley, R., Barker, E., Kelarev, A.: Constructing stochastic mixture policies for episodic multiobjective reinforcement learning tasks. In: Nicholson, A., Li, X. (eds.) AI 2009. LNCS, vol. 5866, pp. 340–349. Springer, Heidelberg (2009)
16. Vamplew, P., Dazeley, R., Berry, A., Dekker, E., Issabekov, R.: Empirical evaluation methods for multiobjective reinforcement learning algorithms. Mach. Learn. **84**(1–2), 51–80 (2011)

17. Vamplew, P., Yearwood, J., Dazeley, R., Berry, A.: On the limitations of scalarisation for multi-objective reinforcement learning of Pareto fronts. In: Wobcke, W., Zhang, M. (eds.) AI 2008. LNCS (LNAI), vol. 5360, pp. 372–378. Springer, Heidelberg (2008)
18. Van Moffaert, K., Nowé, A.: Multi-objective reinforcement learning using sets of pareto dominating policies. J. Mach. Learn. Res. **15**, 3483–3512 (2014)
19. Watkins, C.J.C.H.: Learning from delayed rewards (1989)
20. Whiteson, S., Taylor, M.E., Stone, P.: Critical factors in the empirical performance of temporal difference and evolutionary methods for reinforcement learning. Auton. Agent. Multi-Agent Syst. **21**(1), 1–35 (2010)
21. Zitzler, E., Thiele, L., Laumanns, M., Fonseca, C.M., da Fonseca, V.G.: Performance assessment of multiobjective optimisers: an analysis and review. IEEE Trans. Evol. Comput. **7**(2), 117–132 (2003)

Absorption for ABoxes and TBoxes with General Value Restrictions

Jiewen Wu, Taras Kinash, David Toman$^{(\boxtimes)}$, and Grant Weddell

Cheriton School of Computer Science, University of Waterloo, Waterloo, Canada
{j55wu,tkinash,david,gweddell}@uwaterloo.ca

Abstract. We consider the instance checking problem for $\mathcal{SHIQ}(\mathbf{D})$ knowledge bases. In particular, we present a procedure that significantly reduces the number of ABox individuals that need to be examined for a given instance checking problem over a consistent $\mathcal{SHIQ}(\mathbf{D})$ knowledge base that contains arbitrary occurrences of value restrictions. The procedure extends earlier work that assumed value restrictions were predominantly used to establish global domain and range restrictions, and, consequently, in which other applications of value restrictions had a significant risk of requiring an infeasible number of individuals to be examined for a given problem. Finally, experimental results are given that validate the effectiveness of the procedure.

1 Introduction

We consider the instance checking problem for knowledge bases based on the *description logic* (DL) dialect $\mathcal{SHIQ}(\mathbf{D})$. A particular problem is written $\mathcal{K} \models C(a)$, where $\mathcal{K}$ and C are a $\mathcal{SHIQ}(\mathbf{D})$ knowledge base and concept, respectively, and where a is an individual occurring in $\mathcal{K}$. The problem is to determine if membership of a in C must follow from $\mathcal{K}$.

An effective procedure for solving the instance checking problem has become indispensable in SPARQL query evaluation over RDF data sources under the OWL 2 *direct semantics* entailment regime. To see why, consider a simple SPARQL query of the form

```
select ?x
from {?x in C},
```

where C is a $\mathcal{SHIQ}(\mathbf{D})$ concept, i.e., an OWL 2 complex class, and where the data source is a knowledge base $\mathcal{K}$ with a $\mathcal{SHIQ}(\mathbf{D})$ TBox $\mathcal{T}$, i.e., an OWL 2 ontology. An evaluation of the query corresponds to a potentially large number of instance checking problems; the evaluation must produce *all* individuals a occurring as URIs in $\mathcal{K}$ for which $\mathcal{K} \models C(a)$. And since $\mathcal{T}$ can be non-Horn and C not necessarily conjunctive (e.g., mentions the OWL 2 class constructor `ObjectUnionOf`), an appeal to a procedure to decide if $\mathcal{K} \models C(a)$, for *some* individuals a, will almost certainly be necessary.

© Springer International Publishing Switzerland 2015
B. Pfahringer and J. Renz (Eds.): AI 2015, LNAI 9457, pp. 609–622, 2015.
DOI: 10.1007/978-3-319-26350-2_54

In this paper, we present a procedure for the instance checking problem over a *consistent* $\mathcal{SHIQ}(\mathbf{D})$ knowledge base $\mathcal{K}$, that is, the procedure operates with the assumption that the consistency of $\mathcal{K}$ has already been established and at no time requires any additional check to ensure this.[1] Our procedure is a significant enhancement of the procedure reported in [12] that assumes value restrictions are only used to establish global domain and range restrictions, e.g., that correspond to RDFS domain and range restrictions. In this earlier procedure, other applications of value restrictions are likely to lead to what we call the *large ripple problem*, in particular, a need to examine an infeasible number of individuals for a *particular* instance checking problem. Notably, this includes cases in which value restrictions are also used to capture so-called *unary foreign keys* in the relational model, a very common variety of constraint for RDF data sources that derive from legacy relational databases. Such constraints are given in an ontology by OWL 2 *inclusion dependencies* of the form $C_1 \sqsubseteq \forall R.C_2$, where the C_i are concepts and R is a role corresponding to some RDF property. Such "foreign key" constraints ensure that, in the particular case of C_1-objects, R-values must be C_2-objects.[2]

As with the procedure reported in [12], our procedure is designed to use off-the-shelf tableau-based *subsumption checking* technology for DL dialects that include the fragment $\mathcal{SHOIQ}(\mathbf{D})$. A subsumption checking problem is given by a knowledge base $\mathcal{K}$ and an *inclusion dependency* of the form $C_1 \sqsubseteq C_2$ in which C_1 and C_2 are arbitrary concepts. The problem is to determine if $\mathcal{K}$ implies that C_1-objects are also C_2-objects, written $\mathcal{K} \models C_1 \sqsubseteq C_2$.

Our procedure also shares an assumption that such technology incorporates *enhanced binary absorption* optimization, a generalization of binary absorption optimization [7] also presented in [12]. Note that absorption is an indispensable optimization in tableau methods for subsumption checking. Generally, absorption enables *lazy unfolding* of inclusion dependencies that comprise the so-called TBox of $\mathcal{K}$ [6]. Enhanced binary absorption generalizes absorption by also allowing absorbed inclusion dependencies to have the form $(A \sqcap B) \sqsubseteq C$, where $A \sqcap B$ is a *conjunction* of an atomic concept A and *either an atomic concept or a nominal B*. As with any absorbed inclusion dependencies, such constraints are "unfolded" for an individual in a tableau chase only when it becomes known that it must be both an A-object and a B-object. This considerably enhances the overall performance of lazy unfolding and of reasoning.

Both earlier work and our procedure operate by reducing an instance checking problem over a $\mathcal{SHIQ}(\mathbf{D})$ knowledge base $\mathcal{K}$ to a subsumption checking problem over a $\mathcal{SHOIQ}(\mathbf{D})$ TBox $\mathcal{T}_\mathcal{K}$ derived from $\mathcal{K}$. Roughly, this is achieved

---

[1] This can be an important feature in cases for which a consistency check when "loading $\mathcal{K}$" would constitute a significant overhead due to the size and complexity of an included ontology.

[2] Note that unary foreign keys occur in the LUBM benchmark [2] that we appeal to in our experimental evaluation. And indeed, as confirmed by our experimental results, such keys lead to large ripple problems.

by rewriting the TBox of $\mathcal{K}$ and by introducing additional inclusion dependencies with nominals to encode the ABox assertions in $\mathcal{K}$. Note that, by relying on enhanced binary absorption to ensure the additional dependencies encoding the ABox are absorbed, the overhead of reasoning about $\mathcal{SHOIQ}(\mathbf{D})$ ontologies with *arbitrary* use of nominals that would otherwise be required is entirely avoided.

Our procedure is comprised of four steps that are largely inherited from [12]. An input $\mathcal{SHIQ}(\mathbf{D})$ knowledge base $\mathcal{K}$ is first separated into its component TBox $\mathcal{T}$ and ABox $\mathcal{A}$. A *normalized* TBox $\mathcal{T}^{\mathrm{NR}}$ is then obtained from $\mathcal{T}$ in order to *isolate* inclusion dependencies that mention value restrictions. Two subsequent $\mathcal{SHOIQ}(\mathbf{D})$ TBoxes, $\mathcal{T}_{\mathcal{K}}^1$ and $\mathcal{T}_{\mathcal{K}}^2$ are then derived from $\mathcal{T}^{\mathrm{NR}}$ and $\mathcal{A}$. In particular, the above-mentioned subsumption checking technology uses TBox $\mathcal{T}_{\mathcal{K}}^1$ to compute $\mathcal{T}_{\mathcal{K}}^2$. The same technology is then also used to solve subsequent instance checking problems over $\mathcal{K}$ by encoding them as subsumption checking problems over $\mathcal{T}_{\mathcal{K}}^2$. Our main result relates to this encoding. In particular, an instance check of the form $\mathcal{K} \models C(a)$ is mapped to a subsumption check of the form

$$\mathcal{T}_{\mathcal{K}}^2 \models \{a\} \sqcap \mathfrak{D}_C \sqsubseteq C, \tag{1}$$

where $\mathfrak{D}_C$ is a concept that initializes an appropriate "firing" of binary absorptions in $\mathcal{T}_{\mathcal{K}}^2$. This result refines the so-called *donut theorem* in [12]. The theorem is so-named since its proof relies on a way of combining the results of tableau expansion in reasoning about (1), called a *tinbit*[3], with a particular partial interpretation that must exist for any consistent $\mathcal{K}$, the *donut*, to obtain an interpretation of $\mathcal{K}$ for which a is not a C whenever $\mathcal{K} \models C(a)$ does not hold, that is, which interprets concept $\{a\} \sqcap \mathfrak{D}_C \sqcap \neg C$ as non-empty.

The remainder of the paper is organized as follows. Our primary contributions are in Sect. 3 in which we define the computation of $\mathcal{T}_{\mathcal{K}}^1$ and $\mathcal{T}_{\mathcal{K}}^2$ (see [12] for the computation of $\mathcal{T}^{\mathrm{NR}}$) and in which we present our main result. The results of an experimental evaluation of our procedure are given in Sect. 4. The results confirm that our procedure is effective in addressing performance issues that surface with TBoxes that rely on the general use of value restrictions. A review of related work and a summary discussion then follow in Sect. 5. Part of this discussion outlines refinements of our procedure that can improve its performance or increase the scope of $\mathcal{SHIQ}(\mathbf{D})$ knowledge bases for which the method can be used.

Finally, note that we do *not* address all problems relating to *instance queries* in this paper, an example of which is given as a SPARQL query above. An instance query is given by a knowledge base $\mathcal{K}$ and a concept C, and asks for the set of *all* individuals a mentioned in $\mathcal{K}$ for which $\mathcal{K} \models a : C$. For example, there is a growing body of work on exploiting precomputed or cached results by DL reasoning engines to address this problem [8,9], a topic that we briefly return to in our summary comments.[4] In this paper, we focus on the problem

---

[3] Also inspired by the Canadian word for a "donut hole" pastry called a *Timbit*.

[4] At a minimum, an interface to a cache of all individual names occurring in $\mathcal{K}$ would be required, that is, a cache of the result of evaluating the instance query given by $\mathcal{K}$ and the "top" concept $\top$.

of instance checking for cases where general reasoning is necessary, in particular where precomputed or cached results are unavailable, indeed, an unavoidable circumstance with (arbitrary) non-atomic concepts and non-Horn ontologies.

2 Preliminaries

We consider instance checking problems over knowledge bases expressed in terms of the DL dialect $\mathcal{SHIQ}(\mathbf{D})$, where $\mathbf{D}$ is the simple concrete domain of finite length strings. However, such problems will be mapped to subsumption checking problems in the more general logic $\mathcal{SHOIQ}(\mathbf{D})$ in which nominals can occur in inclusion dependencies. Although not really necessary, our definition of $\mathcal{SHOIQ}(\mathbf{D})$ introduces a number of non-terminals in a concept grammar that will help with clarity in the remainder of the paper.

Definition 1 (Description Logic $\mathcal{SHOIQ}(\mathbf{D})$). *$\mathcal{SHOIQ}(\mathbf{D})$ is a DL dialect based on disjoint infinite sets of atomic concepts* NC, *atomic roles* NR, *concrete features* NF *and nominals* NI. *Let $S \in$ NR $\cup \{R^- \mid R \in$ NR$\}$ S^{--}, we define $S^- = R$ if $S = R^-$ and $S^- = R^-$ otherwise. A role inclusion has the form $S_1 \sqsubseteq S_2$. Let $\circledast$ be the transitive-reflexive closure of $\sqsubseteq$ over the set $\{S_1 \sqsubseteq S_2\} \cup \{S_1^- \sqsubseteq S_2^- \mid S_1 \sqsubseteq S_2\}$. A role S is transitive, denoted $\mathbf{Trans}(S)$, iff $\mathbf{Trans}(R)$ or $\mathbf{Trans}(R^-)$ for some R where $R \circledast S$ and $S \circledast R$. A role S is called* complex *if $\mathbf{Trans}(S')$ for some $S' \circledast S$.*

Let $A \in$ NC, $a \in$ NI, $f, g \in$ NF, S a general role, and n be a non-negative integer. A $\mathcal{SHOIQ}(\mathbf{D})$ concept C is defined as follows:

$$
\begin{aligned}
C \quad &::= \quad C_d \mid C_1 \sqcap C_2 \mid C_1 \sqcup C_2 \mid \{a\} \mid \neg\{a\} \mid \exists^{\le n} S.C_1 \mid \exists^{\ge n} S.C_1 \\
C_d \quad &::= \quad C_b \mid f < g \mid f = k \\
C_b \quad &::= \quad L \mid \top \\
L \quad &::= \quad A \mid \neg A
\end{aligned}
$$

where k is a finite string. To avoid undecidability [5], a complex role S may occur only in concept descriptions of the form $\exists^{\le 0} S.C_1$ or of the form $\exists^{\ge 1} S.C_1$.

An interpretation $\mathcal{I}$ is a pair $\mathcal{I} = (\triangle^{\mathcal{I}} \uplus \mathbf{D}^{\mathcal{I}}, (\cdot)^{\mathcal{I}})$, where $\triangle^{\mathcal{I}}$ is a non-empty set, $\mathbf{D}^{\mathcal{I}}$ is a disjoint concrete domain of finite strings, and $(\cdot)^{\mathcal{I}}$ is a function that maps each feature f to a total function $(f)^{\mathcal{I}} : \triangle \to \mathbf{D}$, the "$=$" symbol to the equality relation over $\mathbf{D}$, the "$<$" symbol to the binary relation for an alphabetic ordering of $\mathbf{D}$, a finite string k to itself, NC to subsets of $\triangle^{\mathcal{I}}$, NR to subsets of $\triangle^{\mathcal{I}} \times \triangle^{\mathcal{I}}$, and NI to singleton subsets of $\triangle^{\mathcal{I}}$, with the interpretation of inverse roles being $(R^-)^{\mathcal{I}} = \{(o_2, o_1) \mid (o_1, o_2) \in R^{\mathcal{I}}\}$. The interpretation is extended to compound concepts in the standard way.

A TBox $\mathcal{T}$ is a finite set of constraints $\mathcal{C}$ of the form $C_1 \sqsubseteq C_2$, $S_1 \sqsubseteq S_2$ or $\mathbf{Trans}(S)$. An ABox $\mathcal{A}$ is a finite set of assertions of the form $A(a)$, $(f = k)(a)$ and $S(a, b)$. Let $\mathcal{K} = (\mathcal{T}, \mathcal{A})$ be an $\mathcal{SHOIQ}(\mathbf{D})$ knowledge base. An interpretation $\mathcal{I}$ is a model of $\mathcal{K}$, written $\mathcal{I} \models \mathcal{K}$, iff $(C_1)^{\mathcal{I}} \subseteq (C_2)^{\mathcal{I}}$ holds for each $C_1 \sqsubseteq C_2 \in \mathcal{T}$, $(S_1)^{\mathcal{I}} \subseteq (S_2)^{\mathcal{I}}$ holds for each $S_1 \sqsubseteq S_2 \in \mathcal{T}$, $\{(o_1, o_2), (o_2, o_3)\} \subseteq (S)^{\mathcal{I}}$ implying $(o_1, o_3) \in (S)^{\mathcal{I}}$ holds for $\mathbf{Trans}(S) \in \mathcal{T}$, $(a)^{\mathcal{I}} \in (A)^{\mathcal{I}}$ for $A(a) \in \mathcal{A}$,

$((a)^{\mathcal{I}}, (b)^{\mathcal{I}}) \in (S)^{\mathcal{I}}$ for $S(a, b) \in \mathcal{A}$, $(f)^{\mathcal{I}}((a)^{\mathcal{I}}) < (g)^{\mathcal{I}}((a)^{\mathcal{I}})$ for $(f < g)(a) \in \mathcal{A}$, and $(f)^{\mathcal{I}}((a)^{\mathcal{I}}) = k$ for $(f = k)(a) \in \mathcal{A}$. A concept C is satisfiable with respect to a knowledge base $\mathcal{K}$ iff there is an $\mathcal{I}$ such that $\mathcal{I} \models \mathcal{K}$ and such that $(C)^{\mathcal{I}} \neq \emptyset$. $\square$

Note that assertions of the form $S(a, b)$ and $(f = k)(a)$ correspond, respectively, to so-called *object property assertions* and *data property assertions* in RDF. By a slight abuse of grammar in the following, we allow simpler shorthand for more general concrete domain concepts C_d of the form $(t_1 \text{ op } t_2)$, where t_1 and t_2 refer to either a concrete feature or a finite string, and $\text{op} \in \{<, =\}$. Also note that we write $\forall S.C$ and $\exists S.C$ as shorthand for the respective concepts $\exists^{\leq 0} S.\neg C$ and $\exists^{\geq 1} S.C$.

Regarding value restrictions, we focus on their use in inclusion dependencies of the form

$$C_b \sqsubseteq \exists^{\leq n} S.C_b.$$

In particular, we shall be concerned with cases where $n = 0$ and where each C_b is a literal L, and refer to such dependencies as *local universal restrictions*. Indeed, the cases where C_b is not a literal correspond to domain and range restrictions of the form $\top \sqsubseteq \forall S.C_b$, a variety of dependency that has already been addressed in earlier work [12].

Recall that the first step of our procedure for instance checking separates a given $\mathcal{SHIQ}(\mathbf{D})$ knowledge base $\mathcal{K}$ into its component TBox $\mathcal{T}$ and ABox $\mathcal{A}$, and that the second step requires $\mathcal{T}$ to be mapped to a *normal form* in which local value restrictions have been extracted. The normal form is defined as follows:

Definition 2 (Normalized $\mathcal{SHIQ}(\mathbf{D})$ Terminologies). *A $\mathcal{SHIQ}(\mathbf{D})$ constraint $\mathcal{C}$ is* normalized *if it has one of the forms* $C_b \sqsubseteq \exists^{\leq n} S.C_b$, $C_L \sqsubseteq C_R$, $S_1 \sqsubseteq S_2$, *or* $\textit{Trans}(S)$, *where* C_L *and* C_R *are defined by the following grammar:*

$$C_L ::= C_d \mid C_L \sqcap C_L \mid C_L \sqcup C_L \mid \exists^{\leq n} S.C_L$$
$$C_R ::= C_d \mid C_R \sqcap C_R \mid C_R \sqcup C_R \mid \exists^{\geq n} S.C_R$$

A $\mathcal{SHIQ}(\mathbf{D})$ terminology $\mathcal{T}$ is normalized *if each constraint $\mathcal{C}$ occurring in $\mathcal{T}$ is normalized.* $\square$

It is a straightforward process to obtain an equisatisfiable normalized terminology from any $\mathcal{SHIQ}(\mathbf{D})$ terminology $\mathcal{T}$ [12].

3   Absorption for ABoxes and TBoxes with Local Universal Restrictions

We now show how local universal restrictions of the form $L_1 \sqsubseteq \forall S.L_2$ occurring in a $\mathcal{SHIQ}(\mathbf{D})$ knowledge base $\mathcal{K}$ with ABox $\mathcal{A}$ and normalized TBox $\mathcal{T}^{\text{NR}}$ can be leveraged to further optimize ABox absorption.

As prescribed in [12], the following pair of axioms are introduced for each role assertion $S(a, b)$ occurring in $\mathcal{A}$, which then qualify for binary absorption:

$$\{a\} \sqcap G_S \sqsubseteq \exists S.(\{b\} \sqcap G), \text{ and} \tag{2}$$

$$\{b\} \sqcap G_{S-} \sqsubseteq \exists S^-.(\{a\} \sqcap G). \tag{3}$$

Note that G_S, G_{S-} and G are atomic concepts functioning as *guards* that can control lazy unfolding. Also note that both (2) and (3) are needed to account for the bi-directional significance of $S(a, b)$.

A tableau algorithm might "fire" either (2) and (3) during lazy unfolding, and therefore create a new individual, either b or a, and labeled with the guard G. Unconditionally adding this label is problematic since this is likely to cause the firing of other absorptions derived from $\mathcal{A}$, thus leading to the large ripple problem mentioned in our introductory comments in which an infeasible number of individuals are "examined" for a particular instance checking problem. We now show how, under some circumstances, one can exploit local universal restrictions to eliminate guards for nominals on the right-hand side of such axioms, possibly replacing the above with the respective pair

$$\{a\} \sqcap G_S \sqsubseteq \exists S.\{b\}, \text{ and} \tag{4}$$

$$\{b\} \sqcap G_{S-} \sqsubseteq \exists S^-.\{a\}, \tag{5}$$

and thereby avoiding subsequent unfolding that might otherwise ensue if the right-hand-side G guards were not removed.

With the assumption of knowledge base consistency, the conditions that enable guard elimination for a role assertion $S(a, b)$, e.g., enable replacing (2) with (4), can be informally stated as the following pair of conditions:

(C1) S is not used in at-most restrictions other than local universal restrictions, and

(C2) for each local universal restriction $L_1 \sqsubseteq \forall S.L_2$ mentioning S, it must hold that $\mathcal{K} \models L_2(b)$.

Recall that our instance checking procedure has four steps. The first two steps are outlined in the preceding section and result in mapping a given $\mathcal{SHIQ}(\mathbf{D})$ knowledge base $\mathcal{K}$ to an ABox $\mathcal{A}$ and a normalized TBox $\mathcal{T}^{NR}$. The remaining two steps of this procedure require computing $\mathcal{T}_{\mathcal{K}}^1$ and then using off-the-shelf subsumption checking technology (with enhanced binary absorption optimization) to compute $\mathcal{T}_{\mathcal{K}}^2$. Details now follow in which the definition of $\mathcal{T}_{\mathcal{K}}^1$ is derived from the corresponding computation in [12] from which we have extracted and refined the definitions of $\mathcal{T}^{\mathcal{T}}$ and $\mathcal{T}^{\mathcal{A}}$ to properly account for the syntactic guard elimination as outlined above.

Definition 3 (Computing $\mathcal{T}_{\mathcal{K}}^1$). $\mathcal{T}_{\mathcal{K}}^1$ *is given by*

$$\mathcal{T}^{NR} \cup \mathcal{T}^{\mathcal{T}} \cup \mathcal{T}^{\mathcal{A}} \cup \mathcal{T}_{\rightarrow}^{\forall} \cup \mathcal{T}_{\rightarrow d}^{\forall} \cup \mathcal{T}_{\leftarrow}^{\forall} \cup \mathcal{T}_{\leftarrow d}^{\forall},$$

with component terminologies other than $\mathcal{T}^{NR}$ defined as follows:

$$\mathcal{T}^{\mathcal{T}} = \{L_1 \sqsubseteq G_S, L_2 \sqsubseteq G_{S^-} \mid L_1 \sqsubseteq \exists^{\leq n}S.L_2 \in \mathcal{T}^{NR}\}$$
$$\cup \{(t_1 \ op \ t_2) \sqsubseteq G_f \mid f \ in \ t_1 \ or \ in \ t_2, (t_1 \ op \ t_2) \ in \ \mathcal{T}^{NR}\}$$
$$\cup \{G_{S_2} \sqsubseteq G_{S_1}, G_{S_2^-} \sqsubseteq G_{S_1^-} \mid S_1 \sqsubseteq S_2\}$$
$$\cup \{\top \sqsubseteq G_S \sqcap G_{S^-} \mid S \ occurs \ in \ \mathcal{T}^{NR} \ and \ S \ is \ complex\}$$

$$\mathcal{T}^{\mathcal{A}} = \{\{a\} \sqcap G \sqsubseteq A \mid A(a) \in \mathcal{A}\}$$
$$\cup \{\{a\} \sqcap G_f \sqsubseteq (f \ op \ k) \mid (f \ op \ k)(a) \in \mathcal{A}\}$$
$$\cup \{\{a\} \sqcap G \sqsubseteq \exists S.\top, \{b\} \sqcap G \sqsubseteq \exists S^-.\top \mid S(a,b) \in \mathcal{A}\}$$

$$\mathcal{T}_{\rightarrow}^{\forall} = \{\{a\} \sqcap G_S \sqsubseteq \exists S.\{b\} \mid S(a,b) \in \mathcal{A}, Trans(S) \notin \mathcal{T},$$
$$and \ for \ each \ L_1 \sqsubseteq \exists^{\leq n}S.L_2 \in \mathcal{T}^{NR},$$
$$n = 0 \ and \ \{L_1(a), \text{NNF}(\neg L_2)(b)\} \cap \mathcal{A} \neq \emptyset\}$$

$$\mathcal{T}_{\rightarrow d}^{\forall} = \{\{a\} \sqcap G_S \sqsubseteq \exists S.(\{b\} \sqcap G) \mid S(a,b) \in \mathcal{A},$$
$$and \ for \ some \ L_1 \sqsubseteq \exists^{\leq n}S.L_2 \in \mathcal{T}^{NR},$$
$$n > 0 \ or \ \{L_1(a), NNF(\neg L_2)(b)\} \cap \mathcal{A} = \emptyset\}$$

$$\mathcal{T}_{\leftarrow}^{\forall} = \{\{b\} \sqcap G_{S^-} \sqsubseteq \exists S^-.\{a\} \mid S(a,b) \in \mathcal{A}, Trans(S) \notin \mathcal{T},$$
$$and \ for \ each \ L_1 \sqsubseteq \exists^{\leq n}S.L_2 \in \mathcal{T}^{NR},$$
$$n = 0 \ and \ \{NNF(\neg L_1)(a), L_2(b)\} \cap \mathcal{A} \neq \emptyset\}$$

$$\mathcal{T}_{\leftarrow d}^{\forall} = \{\{b\} \sqcap G_{S^-} \sqsubseteq \exists S^-.(\{a\} \sqcap G) \mid S(a,b) \in \mathcal{A},$$
$$and \ for \ some \ L_1 \sqsubseteq \exists^{\leq n}S.L_2 \in \mathcal{T}^{NR},$$
$$n > 0 \ or \ \{NNF(\neg L_1)(a), L_2(b)\} \cap \mathcal{A} = \emptyset\} \qquad \square$$

Intuitively, for any role assertion $S(a,b)$, the corresponding axioms in the form of (2) are checked *syntactically* for guard elimination, which are placed into $\mathcal{T}_{\rightarrow}^{\forall}$ if guards cannot be eliminated and into $\mathcal{T}_{\rightarrow d}^{\forall}$ otherwise. Similarly, "backward" axioms in the form of (3) are either in $\mathcal{T}_{\leftarrow}^{\forall}$ or $\mathcal{T}_{\leftarrow d}^{\forall}$, depending on if the right-hand side guards can be eliminated.

The syntactic check conditions presented above for ensuring (C1) and (C2) are subtle. For example, $\mathcal{T}_{\rightarrow}^{\forall}$ requires the condition $\{L_1(a), \text{NNF}(\neg L_2)(b)\} \cap \mathcal{A} \neq \emptyset$, where the presence of $L_1(a)$ in $\mathcal{A}$ guarantees $\mathcal{K} \models \neg L_2(b)$ because of the axiom $L_1 \sqsubseteq \exists^{\leq 0}S.L_2$ and the consistency assumption of $\mathcal{K}$. Hence, adding $L_1(a)$ to the conditions of $\mathcal{T}_{\rightarrow}^{\forall}$ renders syntactic checking over $\mathcal{A}$ more efficient.

Observe that computing $\mathcal{T}_{\mathcal{K}}^1$ entails simple syntactic lookups in $\mathcal{A}$ for concept assertions of the form $L_1(a)$ or $L_2(b)$. If the lookups succeed, $S(a,b)$ is consistent with all local universal restrictions of the form $L_1 \sqsubseteq \forall S.L_2$. Although such lookups are not complete, an absorbed $\mathcal{T}_{\mathcal{K}}^1$ can be useful in conducting further subsumption checks to find additional cases where role assertions are consistent with the local universal restrictions, indeed, to decide conditions (C1) and (C2) above.

Definition 4 (Computing $\mathcal{T}_{\mathcal{K}}^2$). $\mathcal{T}_{\mathcal{K}}^2$ *is given by* $(\mathcal{T}_{\mathcal{K}}^1 \backslash \mathcal{T}^{sub}) \cup \mathcal{T}^{add}$, *where* $\mathcal{T}^{sub}$
and $\mathcal{T}^{add}$ *are defined as follows:*

$$\mathcal{T}^{sub} = \{\{a\} \sqcap G_S \sqsubseteq \exists S.(\{b\} \sqcap G) \mid \mathit{Trans}(S) \notin \mathcal{T}^{NR},$$
$$\{a\} \sqcap G_S \sqsubseteq \exists S.(\{b\} \sqcap G) \in \mathcal{T}_{\rightarrow d}^\forall, \text{ and}$$
$$\text{for each } L_1 \sqsubseteq \exists^{\leq n} S.L_2 \in \mathcal{T}^{NR} : n = 0 \text{ and}$$
$$(\mathcal{T}_{\mathcal{K}}^1 \models \{a\} \sqcap G \sqsubseteq L_1 \text{ or } \mathcal{T}_{\mathcal{K}}^1 \models \{b\} \sqcap G \sqsubseteq \neg L_2) \}$$
$$\cup \{\{b\} \sqcap G_{S^-} \sqsubseteq \exists S^-.(\{a\} \sqcap G) \mid \mathit{Trans}(S^-) \notin \mathcal{T}^{NR},$$
$$\{b\} \sqcap G_{S^-} \sqsubseteq \exists S^-.(\{a\} \sqcap G) \in \mathcal{T}_{\leftarrow d}^\forall, \text{ and}$$
$$\text{for each } L_1 \sqsubseteq \exists^{\leq n} S.L_2 \in \mathcal{T}^{NR} : n = 0 \text{ and}$$
$$(\mathcal{T}_{\mathcal{K}}^1 \models \{a\} \sqcap G \sqcap \sqsubseteq \neg L_1 \text{ or } \mathcal{T}_{\mathcal{K}}^1 \models \{b\} \sqcap G \sqsubseteq L_2) \}$$
$$\mathcal{T}^{add} = \{\{a\} \sqcap G_S \sqsubseteq \exists S.\{b\} \mid \{a\} \sqcap G_S \sqsubseteq \exists S.(\{b\} \sqcap G) \in \mathcal{T}^{sub}\} \qquad \square$$

To illustrate the overall process of computing $\mathcal{T}_{\mathcal{K}}^2$, consider a knowledge base
U with component TBox Ut and ABox Ua respectively defined as follows:

$$\{Dept \sqsubseteq \forall headOf^-.Prof, Chair \sqsubseteq Prof\}, \text{ and } \{p : Chair, headOf^-(d, p)\}.$$

From Definition 2, $(Ut)^{NR} = Ut$, and Sect. 3 then defines the following.

$$\mathcal{T}^{Ut} = \{Dept \sqsubseteq G_{headOf^-}, \neg Prof \sqsubseteq G_{headOf}\}$$
$$\mathcal{T}^{Ua} = \{\{p\} \sqcap G \sqsubseteq Chair, \{d\} \sqcap G \sqsubseteq \exists headOf^-.\top, \{p\} \sqcap G \sqsubseteq \exists headOf.\top\}$$
$$\mathcal{T}_{\rightarrow}^\forall = \mathcal{T}_{\leftarrow}^\forall = \emptyset$$
$$\mathcal{T}_{\rightarrow d}^\forall = \{\{d\} \sqcap G_{headOf^-} \sqsubseteq \exists headOf^-.(\{p\} \sqcap G)\}$$
$$\mathcal{T}_{\leftarrow d}^\forall = \{\{p\} \sqcap G_{headOf} \sqsubseteq \exists headOf.(\{d\} \sqcap G)\}$$
$$\mathcal{T}_U^1 = Ut \cup \mathcal{T}^{Ut} \cup \mathcal{T}^{Ua} \cup \mathcal{T}_{\rightarrow d}^\forall \cup \mathcal{T}_{\leftarrow d}^\forall$$

Computing $\mathcal{T}_U^2$ requires a subsumption check over $\mathcal{T}_U^1$ to determine if $Dept \sqsubseteq$
$\forall headOf^-.Prof$ is consistent with the role assertion $headOf^-(d, p)$. In particular,
since

$$\mathcal{T}_U^1 \models \{p\} \sqcap G \sqsubseteq Prof,$$

we obtain the following:

$$\mathcal{T}^{sub} = \{\{d\} \sqcap G_{headOf^-} \sqsubseteq \exists headOf^-.(\{p\} \sqcap G)\}$$
$$\mathcal{T}^{add} = \{\{d\} \sqcap G_{headOf^-} \sqsubseteq \exists headOf^-.\{p\}\}$$
$$\mathcal{T}_U^2 = (\mathcal{T}_U^2 \backslash \mathcal{T}^{sub}) \cup \mathcal{T}^{add} = Ut \cup \mathcal{T}^{Ut} \cup \mathcal{T}^{Ua} \cup \mathcal{T}_{\leftarrow d}^\forall \cup \mathcal{T}^{add}$$

Thus, the final terminology $\mathcal{T}_U^2$ consists of the following axioms, all of which
will be absorbed with extended binary absorption:

(unary absorptions)   (binary absorptions)

$Dept \sqsubseteq \forall headOf^-.Prof$ $\{p\} \sqcap G \sqsubseteq Chair$

$Chair \sqsubseteq Prof$ $\{d\} \sqcap G \sqsubseteq \exists headOf^-.\top$

$Dept \sqsubseteq G_{headOf^-}$ $\{p\} \sqcap G \sqsubseteq \exists headOf.\top$

$\neg Prof \sqsubseteq G_{headOf}{}^\dagger$ $\{p\} \sqcap G_{headOf} \sqsubseteq \exists headOf.(\{d\} \sqcap G)$

                        $\{d\} \sqcap G_{headOf^-} \sqsubseteq \exists headOf^-.\{p\}$

Our main result now follows in which we show that an instance checking problem $\mathcal{K} \models C(a)$, where $\mathcal{K}$ is a $\mathcal{SHIQ}(\mathbf{D})$ knowledge base, reduces to a subsumption checking problem over either the $\mathcal{SHOIQ}(\mathbf{D})$ TBox $\mathcal{T}_\mathcal{K}^1$ or $\mathcal{T}_\mathcal{K}^2$. In preparation, we define a *derivation concept* for C:

Definition 5 (Derivative Concept). *The* derivative concept $\mathfrak{D}_C$ *for a* $\mathcal{SHIQ}(\mathbf{D})$ *concept* C *is defined as follows:*

$$\mathfrak{D}_C = \begin{cases} \top & \text{if } C = C_b; \\ \sqcap G_{f_i} & \text{if } C = (t_1 \text{ op } t_2), f_i \text{ in } t_1 \text{ or } t_2; \\ \mathfrak{D}_{C_1} \sqcap \mathfrak{D}_{C_2} & \text{if } C = C_1 \sqcap C_2 \text{ or } C = C_1 \sqcup C_2; \\ G_S \sqcap \forall S.(\mathfrak{D}_{C_1} \sqcap G) & \text{if } C = \exists^{\geq n} S.C_1 \text{ or } C = \exists^{\leq n} S.C_1. \end{cases}$$

Definition 6 and Lemma 1 that follow are reproduced from [12].

Definition 6 *Let* $\mathcal{K} = (\mathcal{T}, \mathcal{A})$ *be a* $\mathcal{SHIQ}(\mathbf{D})$ *knowledge base and* $\mathcal{T}_\mathcal{K} = \mathcal{T}_\mathcal{K}^i$ *for any* $i \in \{2,3\}$. *Let* $a : C$ *be an instance check over* $\mathcal{K}$, *and* $\{a\} \sqcap D \sqsubseteq C$, *be a subsumption check over* $\mathcal{T}_\mathcal{K}$, *where* $D = G \sqcap \mathfrak{D}_C$. *Let* $\mathcal{I}_0$ *be an interpretation that satisfies* $\mathcal{T}_\mathcal{K}$ *such that* $(\{a\})^{\mathcal{I}_0} \subseteq (D)^{\mathcal{I}_0}$ *but* $(\{a\})^{\mathcal{I}_0} \cap (C)^{\mathcal{I}_0} = \emptyset$; *also, let* $\mathcal{I}_1$ *be an interpretation that satisfies* $\mathcal{K}$ *in which all at-least restrictions are fulfilled by ABox individuals and, if necessary, anonymous objects. Without loss of generality, we assume both* $\mathcal{I}_0$ *and* $\mathcal{I}_1$ *are tree-shaped outside of the converted ABox. Define an interpretation* $\mathcal{J}$ *as follows: let* a_0 *be any ABox individual and* $\Gamma^{\mathcal{I}_0}$ *be the set of objects* $o \in \Delta^{\mathcal{I}_0}$ *such that either* $o \in (\{a_0\})^{\mathcal{I}_0}$ *and* $(\{a_0\})^{\mathcal{I}_0} \subseteq (G)^{\mathcal{I}_0}$ *or* o *is an anonymous object in* $\Delta^{\mathcal{I}_0}$ *rooted by such an object. Similarly let* $\Gamma^{\mathcal{I}_1}$ *be the set of objects* $o \in \Delta^{\mathcal{I}_1}$ *such that either* $o \in (\{a_0\})^{\mathcal{I}_1}$ *and* $(\{a_0\})^{\mathcal{I}_0} \cap (G)^{\mathcal{I}_0} = \emptyset$ *or* o *is an anonymous object in* $\Delta^{\mathcal{I}_1}$ *rooted by such an object. We set*

1. $\Delta^\mathcal{J} = \Gamma^{\mathcal{I}_0} \cup \Gamma^{\mathcal{I}_1}$;
2. $(a_0)^\mathcal{J} \in (\{a_0\})^{\mathcal{I}_0}$ *for* $(a_0)^\mathcal{J} \in \Gamma^{\mathcal{I}_0}$ *and* $(a_0)^\mathcal{J} = (a_0)^{\mathcal{I}_1}$ *for* $(a_0)^\mathcal{J} \in \Gamma^{\mathcal{I}_1}$;
3. $o \in A^\mathcal{J}$ *if* $o \in A^{\mathcal{I}_0}$ *and* $o \in \Gamma^{\mathcal{I}_0}$ *or if* $o \in A^{\mathcal{I}_1}$ *and* $o \in \Gamma^{\mathcal{I}_1}$ *for an atomic concept* A
4. $(f)^\mathcal{J}(o) \text{ op } (g)^\mathcal{J}(o)$ *if* $(f)^{\mathcal{I}_0}(o) \text{ op } (g)^{\mathcal{I}_0}(o)$ *and* $o \in \Gamma^{\mathcal{I}_0}$ *or if* $(f)^{\mathcal{I}_1}(o) \text{ op } (g)^{\mathcal{I}_1}(o)$ *and* $o \in \Gamma^{\mathcal{I}_1}$ *(and similarly for* $f \text{ op } k$*);*
5. $(o_1, o_2) \in (S)^\mathcal{J}$ *if*
    (a) $(o_1, o_2) \in S^{\mathcal{I}_0}$ *and* $o_1, o_2 \in \Gamma^{\mathcal{I}_0}$, *or* $(o_1, o_2) \in S^{\mathcal{I}_1}$ *and* $o_1, o_2 \in \Gamma^{\mathcal{I}_1}$; *or*
    (b) $o_1 \in (\{a_0\})^{\mathcal{I}_0} \cap (G)^{\mathcal{I}_0}$, $o_2 \in (\{b_0\})^{\mathcal{I}_1}$ *and* $S(a_0, b_0) \in \mathcal{A}$ *(or vice versa);* *or*
    (c) $(o_1, o_2) \in (S_1)^\mathcal{J}$ *and* $S_1 \stackrel{*}{\sqsubseteq} S$; *or*

(d) $(o_1, o') \in (S)^{\mathcal{J}}$, $(o', o_2) \in (S)^{\mathcal{J}}$ and $\textbf{Trans}(S)$.   □

Lemma 1. For $\{o_1, o_2\} \subseteq \Delta^{\mathcal{J}}$, if $(o_1, o_2) \in (S)^{\mathcal{J}}$ and $\textbf{Trans}(S) \in \mathcal{T}$, then either $\{o_1, o_2\} \subseteq \Gamma^{\mathcal{I}_0}$ or $\{o_1, o_2\} \subseteq \Gamma^{\mathcal{I}_1}$, where $\mathcal{I}_0$, $\Gamma^{\mathcal{I}_0}$, $\mathcal{I}_1$, $\Gamma^{\mathcal{I}_1}$ and $\mathcal{J}$ are given in Definition 6.   □

Theorem 1. For any consistent $\mathcal{SHIQ}(\mathbf{D})$ knowledge base $\mathcal{K}$, concept C, individual a, and $1 \leq i \leq 2$:

$$\mathcal{K} \models a : C \quad \text{iff} \quad T_{\mathcal{K}}^i \models \{a\} \sqcap G \sqcap \mathfrak{D}_C \sqsubseteq C.$$

Proof (sketch). Case $i = 1$ is implicitly included in case $i = 2$, so, it is sufficient to prove the latter. For $i = 2$, consider Definition 6. We claim that $(\{a\})^{\mathcal{J}} \cap (C)^{\mathcal{J}} = \emptyset$ (trivially) and $J \models \mathcal{K}$. To show $J \models \mathcal{K}$, note that the edges from case (4a) satisfy all dependencies in $\mathcal{K}$ as the remainder of the interpretation $\mathcal{J}$ is copied from $\mathcal{I}_0$ or $\mathcal{I}_1$. Thus, we only need to consider those S edges of the form covered by case (4b) (and the extended cases (4c) and (4d)): the edges that *cross* between the two interpretations, i.e., when $o_1 \in (\{a_0\})^{\mathcal{I}_0}$, $o_2 \in (\{b_0\})^{\mathcal{I}_1}$ and $S(a_0, b_0) \in \mathcal{A}$. Now consider an inclusion dependency expressing an *atmost* restriction $L_1 \sqsubseteq \exists^{\leq n} S.L_2 \in \mathcal{T}$. There are two possibilities: in one case, we can conclude $o_1 \notin (L_1)^{\mathcal{I}_0}$ as otherwise $o_1 \in (G_S)^{\mathcal{I}_0}$ by the definition of $\mathcal{T}_{\mathcal{T}}$ and thus $o_2 \in (G)^{\mathcal{I}_0}$ by the rules for construction of $\mathcal{T}_{\mathcal{K}}^2$, which contradicts our assumption that $(\{b_0\})^{\mathcal{I}_0} \cap (G)^{\mathcal{I}_0} = \emptyset$, hence the inclusion dependency is consistent with the role assertion vacuously; in the other case, we cannot derive a contradiction because G was removed by our optimization shown in Sect. 3, then it must be the case that the role assertion $S(a, b)$ is consistent with the axiom $L_1 \sqsubseteq \exists^{\leq 0} S.L_2 \in \mathcal{T}$, i.e., $L_1 \sqsubseteq \forall S.\neg L_2$. Lemma 1 stipulates that in case (4d) either $\{o_1, o_2, o'\} \subseteq \Gamma^{\mathcal{I}_0}$ or $\{o_1, o_2, o'\} \subseteq \Gamma^{\mathcal{I}_1}$ hold; hence any universal restriction of the form $L_1 \sqsubseteq \forall S.L_2$ (recall that concepts of the form $\exists^{\leq n} S.L_2$ are disallowed for complex S) must be satisfied by (o_1, o_2) because it is already satisfied by (o_1, o_2) in $\mathcal{I}_0$ ($\mathcal{I}_1$, respectively). Edges from case (4c) are trivial extension to all of the above. Hence all axioms in $\mathcal{K}$ are satisfied by $\mathcal{J}$.

The other direction follows by observing that if $\mathcal{K} \cup \{a : \neg C\}$ is satisfiable then the satisfying interpretation $\mathcal{I}$ can be extended to $(G)^{\mathcal{I}} = (G_f)^{\mathcal{I}} = (G_S)^{\mathcal{I}} = \Delta^{\mathcal{I}}$ for all individuals a_0, concrete features f, and roles S, and $(\{a_0\})^{\mathcal{I}} = \{a_0^{\mathcal{I}}\}$. This extended interpretation then satisfies $T_{\mathcal{K}}^2$ and $(\{a\})^{\mathcal{I}} \subseteq (D)^{\mathcal{I}} \cap (\neg C)^{\mathcal{I}}$.   □

Returning to our example above to illustrate, consider the instance checking problem

$$U \models p : Prof.$$

By Theorem 1, this can be decided by the subsumption checking problem

$$T_U^2 \models \{p\} \sqcap G \sqcap \mathfrak{D}_{Prof} \sqsubseteq Prof$$

with the assumption that U is consistent.

4   Experimental Evaluation

Earlier work has already established the efficacy of ABox absorption for instance checking problems, particularly so in cases where there are a large number of concept assertions of the form $(f = k)(a)$[5]. Indeed, [12] presents an extensive comparison of the utility of ABox absorption implemented in a tableau based DL reasoner called *CARE Assertion Retrieval Engine* (https://code.google.com/p/care-engine/) with other state-of-the-art reasoners. CARE has an underlying $\mathcal{SHI}(\mathbf{D})$ DL reasoner that implements ABox absorption, optimized double blocking [5], and dependency-directed backtracking [4]. Note that the set of finite strings is the only concrete domain supported by the reasoner.

The comparison was based on a benchmark consisting of a selection of *instance queries* over a digital camera knowledge base. Indeed, ABox absorption enabled CARE to outperform other reasoners on a number of these queries, despite provisioning ensuring that CARE itself had no access to any precomputed or cached results.

Our objectives with the experiments presented below are focused on evaluating the performance of the our new procedure with the performance of the earlier procedure in [12] used as the baseline. Thus, the empirical evaluation also uses the above-mentioned CARE system. Note that all reported times are the average of five independent runs on the 2.6 GHz AMD Opteron 6282 SE processor of a Ubuntu 12.04 Linux server, with up to 4 GB of heap.

We conducted several experiments on the LUBM benchmark [2] using one university (LUBM0) which has about 17 k individuals and 49 k role assertions.[6] Twelve queries out of the LUBM test queries[7] were used (Q_2 and Q_9 were excluded since they are not expressible as instance queries). Since the experiments focus on instance checking, the selection conditions for each of the twelve queries were reified, e.g., Q_4 was rewritten as the instance query *Professor* $\sqcap \exists worksFor.A'$, for some fresh atomic concept A', and a new concept assertion, http://www.Department0.University0.edu: A', was added to the original ABox.

The run-time results are listed in Fig. 1 in which the execution time in seconds of CARE with the new method, called OPT (the right striped bars), is compared with the execution time of CARE with the earlier version lacking the optimizations outlined in previous sections, called BASE (the left blue bars). The figure also reports the relative improvement of OPT with respect to BASE as a percentage. For LUBM0, preprocessing costs 5 and 16 s for the BASE and OPT methods, respectively.

---

[5] Recall that these are called *data property assertions* in RDF.

[6] We chose to report on experiments using the LUBM benchmark for this study because of its wider appeal, e.g., its adoption of a set of predefined queries, and because its TBox includes foreign key constraints that are not global domain and range restrictions, a property missing with the digital camera case studied in [12]. Additional experiments can be found in [11].

[7] http://swat.cse.lehigh.edu/projects/lubm/.

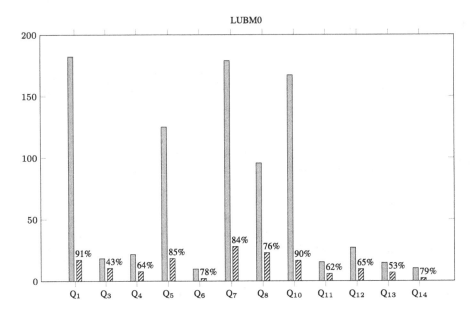

Fig. 1. Performance gains.

With OPT, about 23 % of the role assertions were optimized for LUBM's six typing constraints. Observe that, for LUBM0 instance queries, there is a dramatic improvement ranging from 43 % to 91 % by OPT. The results suggest that LUBM0 is quite sensitive to typing constraints in comparison to knowledge bases with global domain and range restrictions only.

We have also observed that, for these knowledge bases, $T_\mathcal{K}^1$ constructed using *syntactic* lookups in the ABox to approximate concept membership of individuals yields empty $T_\to^\forall$ and $T_\leftarrow^\forall$, resulting in inferior performance. Only $T_\mathcal{K}^2$, constructed using *semantic* concept membership tests, yields the desired improvements. Note, however, that $T_\mathcal{K}^1$ is essential *to compute* $T_\mathcal{K}^2$, i.e., allows more efficient class membership checks. Indeed, additional (auxiliary) experiments have shown that side-stepping either of the steps dramatically decreases the overall performance of the system.

5   Related Work and Summary

Instance queries are an important reasoning service over DL knowledge bases, and have been the subject of substantial work in the DL community. Although it is always possible to evaluate an instance query $C(x)$ by performing a sequence of instance checks $\mathcal{K} \models a : C$ for each individual a occurring in $\mathcal{K}$, reasoning engines usually try to reduce the number of such checks by using precomputed results or by "bulk processing" of a range of instance checks. An example of the latter is so-called *binary retrieval* [3], which is used to determine non-answers via a single (possibly large) satisfiability check.

Another approach to avoid checking individuals sequentially is through summarization and refinement [1], in which query evaluation is performed over a summary of the original ABox that is significantly smaller in size, and iterative refinement based on inconsistency justification is used to purge spurious answers.

There have been several approaches to exploiting precomputed results obtained at an earlier time: when a knowledge base is "loaded", or as a consequence of an explicit request [9]. Examples include the *pseudo-model merging* technique [3], presented earlier in [4] as a way to quickly falsify a subsumption check. In particular, a pseudo-model captures the deterministic consequences of concept membership for individuals. Note that model merging techniques are generally sound but incomplete. Methods on how precomputed information can be used to improve the efficiency of evaluating instance queries have also been developed [8,9]. Observe that the aforementioned optimizations concern how to reduce the number of instance checking tasks, not how to improve the instance checking problem itself.

An approach to instance checking that has much in common with our own method was introduced in [10]. In this case, an ABox is partitioned into small islands such that an instance checking problem is routed to the island "owned" by an individual. In contrast, our method simply reduces the problem of efficient instance checking to absorption.

Earlier work shows how instance checking can be improved by introducing guards that in turn prune any unnecessary consideration of individuals and the (possibly large) number of facts about individuals [13,14]. To recap, the method introduced in this earlier work assumes that knowledge bases are consistent and relies on a refinement of binary absorption to achieve efficiency. Our main result shows how the method can be refined by an additional process that effectively disables the introduction of "trigger" guards in binary absorptions, which in turn reduces the need for further lazy unfolding.

References

1. Dolby, J., Fokoue, A., Kalyanpur, A., Kershenbaum, A., Schonberg, E., Srinivas, K., Ma, L.: Scalable semantic retrieval through summarization and refinement. In: AAAI 2007, pp. 299–304 (2007)
2. Guo, Y., Pan, Z., Heflin, J.: LUBM: a benchmark for OWL knowledge base systems. Web Semant. **3**(2–3), 158–182 (2005)
3. Haarslev, V., Möller, R.: On the scalability of description logic instance retrieval. J. Autom. Reasoning (JAR) **41**(2), 99–142 (2008)
4. Horrocks, I.: Optimising tableaux decision procedures for description logics. Ph.D. thesis, the University of Manchester (1997)
5. Horrocks, I., Sattler, U.: Optimised reasoning for $\mathcal{SHIQ}$. In: ECAI 2002, pp. 277–281 (2002)
6. Horrocks, I., Sattler, U., Tobies, S.: Practical reasoning for expressive description logics. In: Ganzinger, H., McAllester, D., Voronkov, A. (eds.) LPAR 1999. LNCS, vol. 1705, pp. 161–180. Springer, Heidelberg (1999)
7. Hudek, A.K., Weddell, G.E.: Binary absorption in tableaux-based reasoning for description logics. In: Description Logics 2006 (2006)

8. Kollia, I., Glimm, B.: Cost based query ordering over OWL ontologies. In: Cudré-Mauroux, P., Heflin, J., Sirin, E., Tudorache, T., Euzenat, J., Hauswirth, M., Parreira, J.X., Hendler, J., Schreiber, G., Bernstein, A., Blomqvist, E. (eds.) ISWC 2012, Part I. LNCS, vol. 7649, pp. 231–246. Springer, Heidelberg (2012)

9. Pound, J., Toman, D., Weddell, G.E., Wu, J.: An assertion retrieval algebra for object queries over knowledge bases. In: Walsh, T. (ed.) IJCAI 2011, pp. 1051–1056 (2011)

10. Wandelt, S., Möller, R.: Towards ABox modularization of semi-expressive description logics. Appl. Ontology **7**(2), 133–167 (2012)

11. Wu, J.: Answering object queries over knowledge bases with expressive underlying description logics. Ph.D. thesis, University of Waterloo (2013)

12. Jiewen, W., Hudek, A., Toman, D., Weddell, G.: Absorption for ABoxes. J. Autom. Reasoning **53**, 215–243 (2014)

13. Wu, J., Hudek, A.K., Toman, D., Weddell, G.E.: Absorption for ABoxes. In: DL 2012 (2012)

14. Wu, J., Hudek, A.K., Toman, D., Weddell, G.E.: Assertion absorption in object queries over knowledge bases. In: KR 2012 (2012)

Optimal Hyper-Parameter Search in Support Vector Machines Using Bézier Surfaces

Shinichi Yamada[1]([✉]), Kourosh Neshatian[1],
and Raazesh Sainudiin[2]

[1] Department of Computer Science and Software Engineering,
University of Canterbury, Christchurch, New Zealand
shinichi.yamada@pg.canterbury.ac.nz
[2] School of Mathematics and Statistics, University of Canterbury,
Christchurch, New Zealand

Abstract. We consider the problem of finding the optimal specification of hyper-parameters in Support Vector Machines (SVMs). We sample the hyper-parameter space and then use Bézier curves to approximate the performance surface. This geometrical approach allows us to use the information provided by the surface and find optimal specification of hyper-parameters. Our results show that in most cases the specification found by the proposed algorithm is very close to actual optimal point(s). The results suggest that our algorithm can serve as a framework for hyper-parameter search, which is precise and automatic.

1 Introduction

Support Vector Machines (SVMs) have a special type of parameter called "hyper-parameters", which must be specified before solving the optimization problems. Learning SVMs involves solving two optimization problems:

1. Solving a convex optimization problem for each specification of hyper-parameters.
2. Determination of the best specification of hyper-parameters.

An important development for the former task is the "path analysis" whichem started with the seminal paper by Efron et al. [3]. They discovered that the optimal coefficients of LASSO (Least Absolute Shrinkage and Selection Operator) can be expressed by a piecewise linear function of a hyper-parameter. The same result was also extended to SVMs (Hastie et al. [8]). Lately it turned out that the worst case complexity of the solution path of SVMs is exponential both in the number of records and dimensions (Gärtner et al. [1]). Therefore the main interest is shifting towards approximation methods of the path analysis rather than seeking for the exact solutions (Giesen et al. [4–6]).

In this article we focus on the latter task of the optimal specification of hyper-parameters. In order to find the optimal hyper-parameters, the prediction accuracies are compared using unseen test data. It is a simple mechanism

© Springer International Publishing Switzerland 2015
B. Pfahringer and J. Renz (Eds.): AI 2015, LNAI 9457, pp. 623–629, 2015.
DOI: 10.1007/978-3-319-26350-2_55

for SVMs to prevent over-fitting and under-fitting data. Although we can not assume convexity for the optimal hyper-parameter search, the development of the path analysis helps us to quickly obtain approximate optimal solutions for each specification of hyper-parameters. Therefore we look upon the task of the optimal hyper-parameter search as a computational problem rather than an analytical one.

Bézier curves are essential tools to represent curves in CAD/CAM and computer graphics. We introduce them to estimate the surface in hyper-parameter spaces. We propose the use of the "volume" under the surface to detect the most relevant regions in the spaces.

2   Hyper-Parameters in Support Vector Machines

The maximal margin hyperplane is obtained by solving the following quadratic optimization problem:

$$\text{Maximize: } \sum_{i=1}^{n} \alpha_i - \frac{1}{4} \sum_{i,j=1}^{n} \alpha_i \alpha_j y_i y_j \langle \Phi(\mathbf{x}_i), \Phi(\mathbf{x}_j) \rangle$$

$$\text{Subject to: } \sum_{i=1}^{n} \alpha_i y_i = 0,$$

$$0 \le \alpha_i \le C, \quad \text{for } i = 1, \ldots, n$$

where C is a regularization hyper-parameter which is incorporated into SVMs to allow some misclassification in the data with some noise. $\Phi(\mathbf{x}_i)$ are nonlinear feature maps which map the input data $\mathbf{x} \in X$ into a Hilbert space H (Vapnik [12]). $\mathbf{x}_i$ are input vectors, y_i are output labels and α_i are Lagrange multipliers which are solved in this optimization problem.

The inner products of the functions $k(\mathbf{x}, \mathbf{x}') = \langle \Phi(\mathbf{x}), \Phi(\mathbf{x}') \rangle$, are called "kernels". Gaussian kernel, $k(\mathbf{x}, \mathbf{x}') = \exp(-\frac{\|\mathbf{x}-\mathbf{x}'\|^2}{\sigma^2})$, is a default choice in many cases because it is a "universal kernel" whose corresponding feature spaces are dense in the set of all continuous functions (Steinwart and Christmann [11]).

In (Keerthi and Lin [9]), the authors examined the relations between the Gaussian kernel parameter σ^2 and the regularization parameter C. They proved that when one parameter approaches to 0 or ∞ while fixing the other parameter, the SVMs under-fit or over-fit the data. Therefore there exist sector-shape boundary curves which separate the "good" region from the under-fitting and over-fitting regions. (Yan et al. [13]) observed that usually there exist "optimal" regions within the "good" regions and they claimed that the optimal pair of hyper-parameters should be taken from the "optimal" regions as in Fig. 1.

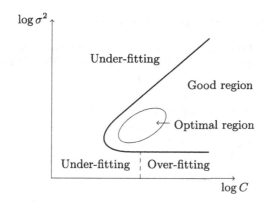

Fig. 1. Good and optimal regions

3   Hyper-Parameter Search with Bézier Surfaces

3.1   MSE as the Estimation of Variance

We denote hyper-parameter spaces as $\mathbf{P}$. For instance, if we use Gaussian kernels, a pair of the hyper-parameters is written as $\mathbf{p} = (\log C, \log \sigma^2) \in \mathbf{P} \subset \mathbb{R}^2$. Each specification of hyper-parameters is evaluated by the prediction accuracy measured by a test dataset. Therefore we consider a mapping $z = f(\mathbf{p})$ where z represents the accuracy for the hyper-parameters $\mathbf{p} \in \mathbf{P}$.

One of the most common methods of hyper-parameter search is the grid search method, which divides each search dimension into equal intervals and evaluates the points at the intersections of the divisions. The best point will be taken as the estimation of the parameters. A two-stage grid search is a reasonable strategy to optimally allocate the computational resources. It consists of a rough search in the broad region and a fine search in the restricted region. Usually the best point in the first search is used as the center of the second search. We adopt this two-stage sampling scheme but our method provides a better strategy to determine the region for the second search.

To train the SVM we use the cross-validation method in which a training dataset is divided into several partitions and each partition is repeatedly used as validation data and the rest of partitions as training data. If we use the same regular grid pattern for each partition of the cross-validation method we obtain several results z on each grid point. To combine the information of those results we compute the mean $\bar{z}$ of z for each point, which corresponds to the least square approximation of the results. Since this configuration is the same as the one-way ANOVA experiment, we also compute the mean sum of square error (MSE) as a rough estimate of the variance.

3.2 Bézier Surface and α-percent region

We estimate the mapping $z = f(\mathbf{p})$ by interpolating the mean $\bar{z}$ on grid points with the cubic Hermite tensor product Bézier surface (Prautzsch et al. [10]). We call it as the "Bézier surface" for simplicity.

α-percent region is defined to quantify the volume under the surface. Suppose that we cut the z axis by a perpendicular plane at $z = a$. When the point a is high enough, there is no intersection between the surface and the plane. As the point a goes down, at some point the surface and the plane meet together. If the point a goes down further, the intersection between the surface and the plane will form a union of closed bounded regions.

Definition 1. α-percent region (*Harlow et al.* [7]).
If the perpendicular plane at $z = a$ intersects the surface, then corresponding to the intersection, we have the bounded domain $\mathbf{D} = \{\mathbf{p} \mid f(\mathbf{p}) \geq a\}$. If the volume under the surface on the domain $\mathbf{D}$, $\int_{\mathbf{p} \in \mathbf{D}} f(\mathbf{p}) d\mathbf{p}$, occupies the α percent of the whole volume under the surface, we call the domain $\mathbf{D}$ as the α-percent region.

3.3 Search Strategy

Our search for the optimal values of hyper-parameters involves the following steps:

1. Determination of the search region
2. The first estimation — the first sampling, the construction of the Bézier surface and the computation of the α-percent region
3. The second estimation — the finer second sampling, the construction of the Bézier surface and the computation of the optimal point on the α-percent region.

4 Experiment

The purpose of the experiments in this section is to: (i) explain the application of the proposed method in a two-dimensional hyper-parameter space; (ii) verify the prediction accuracy of the proposed method. We are also interested in how the shape of the performance surface might change in re-sampling (the random assignment of training and test parts in each dataset).

We use LIBSVM (version 3.17) [2] for training and testing SVMs. Benchmark datasets are also chosen from the same website of the LIBSVM. For each dataset we randomly select 70 % of records as the training dataset and 30 % as the test dataset. We conduct experiments in the two dimensional hyper-parameter spaces using Gaussian kernels. For the α-percent region we set $\alpha = 30$. For the computation of the volume under the surface we compute the sum of the volume corresponding to the grids inside the α-percent region. By making the sides of the grids smaller and smaller the volume approaches to the true volume.

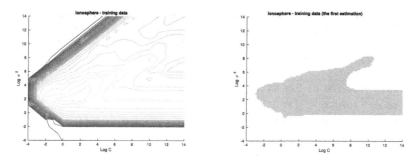

Fig. 2. The contour map (left) and the 30-percent region in the first estimation (right)

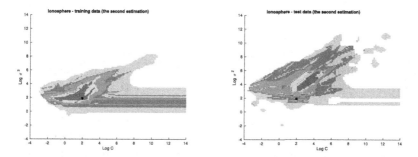

Fig. 3. The second estimation for the training data (left) and for the test data (right): The regions of red, yellow, green and magenta represent the 5-percent, 10-percent, 25-percent and 50-percent regions, respectively. The black point is the estimated value for the pair of hyper-parameters that yield maximum accuracy on the training set (Color figure online).

We allow a tolerance of 0.1 % above and below the α for the computation of the volume. We illustrate the procedure using the Ionosphere dataset.

In the first step we choose the search region so as to include the "good" region of the sector shape. In the second step we conduct a first sampling on a rough grid in the entire region, construct a Bézier surface and compute the 30-percent region, which corresponds to the region of cyan color in Fig. 2. In the third step on the 30-percent region we conduct the second finer sampling, construct the Bézier surface and compute the highest point on the surface. In the test dataset we repeat the same procedure to evaluate the prediction accuracy of the estimation. In Fig. 3 which shows the results of the second estimation for the training dataset and the test dataset.

Table 1 summarizes the results of the experiments which are repeated 100 times for each dataset. We set the number of partitions of cross-validation to be 5 for the datasets with less than 800 records (datasets above Car Evaluation in Table 1) and to be 10 for the other larger datasets. The "p-value" column is the average of the p-values of one-side t-test between the "Max" which is the maximum accuracy in the test dataset and the "Estimation" which is the accuracy of

Table 1. Results of the experiments with Gaussian kernels

Dataset	Feature size/record size	Max	Estimation	Difference	MSE	p-value	Hit percentage
Australian	14/690	88.55	85.38	3.17	3.51	0.26(5)	7,11,23,39
Breast cancer	10/683	97.85	96.66	1.19	1.86	0.32(1)	12,13,21,35
Diabetes	8/768	79.32	76.44	2.88	3.88	0.29(3)	4,9,13,28
Glass	9/214	76.56	67.02	9.55	8.27	0.22(5)	4,10,25,47
Heart	13/270	87.01	82.42	4.59	5.69	0.28(1)	7,8,14,26
Ionosphere	34/351	97.05	94.05	3.00	3.71	0.28(0)	8,12,33,62
Iris	4/150	98.33	95.72	2.61	3.81	0.32(2)	32,34,38,48
Liver disorders	6/345	76.99	70.90	6.08	6.41	0.26(3)	4,6,28,52
Sonar	60/208	90.59	86.66	3.93	6.67	0.34(1)	16,21,40,67
Svmguide2	20/391	87.46	83.28	4.18	5.56	0.30(0)	55,66,80,94
WDBC	30/569	98.77	97.51	1.26	1.87	0.32(2)	15,23,33,53
Wine	13/178	99.54	97.61	1.94	2.91	0.32(7)	34,36,41,45
Car evaluation	9/1728	99.42	98.92	0.49	1.39	0.40(0)	24,39,75,95
DNA	180/3186	96.22	95.73	0.48	1.46	0.41(0)	24,46,83,97
Fourclass	2/862	100	99.88	0.12	0.51	0.43(3)	72,72,72,72
German	24/1000	78.48	75.67	2.81	4.99	0.34(0)	5,9,18,43
Segment	19/2310	97.81	96.73	1.09	1.48	0.30(0)	16,23,40,68
Splice	60/3175	92.01	91.42	0.59	2.04	0.42(0)	34,49,78,97
Vehicle	18/846	87.12	84.02	3.10	4.97	0.33(0)	15,22,47,82
Vowel	10/990	99.18	98.47	0.71	1.99	0.40(0)	17,20,39,74

the estimation point evaluated in the test dataset. The number in the parenthesis in the "p-value" column counts the number of experiments in which p-values are less than 0.05. The "Hit percentage" column is the number of experiments (out of 100 runs) in which the estimation point falls into the 5-percent, 10-percent, 25-percent and 50-percent region in the test dataset, respectively.

5   Discussion and Conclusion

The p-values in Table 1 show that the optimal points detected by the proposed algorithm are not significantly different from the true optimal points obtained by exhaustively searching the space. That is, most often the algorithm can find optimal or near-optimal values for the hyper-parameters.

The values of MSE and *hit percentage* depict how the estimated surface can vary due to repeated sampling. The estimation with the larger datasets tends to exhibit better prediction accuracy than that with the smaller datasets.

Since the proposed method is based on volume, it is theoretically applicable to higher dimensional spaces. The usefulness of the method as a framework for hyper-parameter search will become clearer in those spaces. The time complexity of the proposed algorithm scales exponentially with respect to the number of hyper-parameters. We hope that combined with some other techniques such as *path analysis*, adaptive partitioning of the domain (Harlow et al. [7]) and parallel processing, the proposed method can carry out automatic hyper-parameter search up to several dimensional spaces.

References

1. Gärtner, B., Jaggi, M., Maria, C.: An exponential lower bound on the complexity of regularization paths. J. Comput. Geom. **3**(1), 168–195 (2012)
2. Chang, C.C., Lin, C.J.: LIBSVM: a library for support vector machines. ACM Trans. Intell. Syst. Technol. **2**, 27 (2011). http://www.csie.ntu.edu.tw/cjlin/libsvm
3. Efron, B., Hastie, T., Johnstone, I., Tibshirani, R.: Least angle regression. Ann. Stat. **32**, 407–499 (2004)
4. Giesen, J., Jaggi, M., Laue, S.: Approximating parameterized convex optimization problems. ACM Trans. Algorithms **9**(1), 10:1–10:17 (2012). http://doi.acm.org/10.1145/2390176.2390186
5. Giesen, J., Laue, S., Wieschollek, P.: Robust and efficient kernel hyperparameter paths with guarantees. In: Jebara, T., Xing, E.P. (eds.) Proceedings of the 31st International Conference on Machine Learning (ICML 2014) and JMLR Workshop and Conference Proceedings, pp. 1296–1304 (2014). http://jmlr.org/proceedings/papers/v32/giesen14.pdf
6. Giesen, J., Mueller, J., Laue, S., Swiercy, S.: Approximating concavely parameterized optimization problems. In: Pereira, F., Burges, C., Bottou, L., Weinberger, K. (eds.) Advances in Neural Information Processing Systems 25, pp. 2105–2113. Curran Associates, Inc. (2012). http://papers.nips.cc/paper/4578-approximating-concavely-parameterized-optimization-problems.pdf
7. Harlow, J., Sainudiin, R., Tucker, W.: Mapped regular pavings. Reliable Comput. **16**, 252–282 (2012)
8. Hastie, T., Rosset, S., Tibshirani, R., Zhu, J.: The entire regularization path for the support vector machine. J. Mach. Learn. Res. 5, 1391–1415 (2004). http://dl.acm.org/citation.cfm?id=1005332.1044706
9. Keerthi, S.S., Lin, C.J.: Asymptotic behaviors of support vector machines with gaussian kernel. Neural Comput. **15**(7), 1667–1689 (2003). http://dx.doi.org/10.1162/089976603321891855
10. Prautzsch, H., Boehm, W., Paluszny, M.: Bezier and B-Spline Techniques. Springer, Heidelberg (2002)
11. Steinwart, I., Christmann, A.: Support Vector Machines, 1st edn. Springer, New York (2008)
12. Vapnik, V.N.: Statistical Learning Theory, 1st edn. Wiley, New York (1998)
13. Yan, Z., Yang, Y., Yunjing, D.: An experimental study of the hyper-parameters distribution region and its optimization method for support vector machine with gaussian kernel. Int. J. Signal Process. Image Process. Pattern Recogn. **6**(5), 437–446 (2013)

Hierarchical Learning for Emergence of Social Norms in Networked Multiagent Systems

Chao Yu[1]([✉]), Hongtao Lv[2], Fenghui Ren[3], Honglin Bao[1], and Jianye Hao[4]

[1] School of Computer Science and Technology,
Dalian University of Technology, Dalian 116024, China
cy496@dlut.edu.cn,BHL19931025@mail.dlut.edu.cn
[2] School of Innovation Experiment,
Dalian University of Technology, Dalian 116024, China
lvhongtao@mail.dlut.edu.cn
[3] School of Computer Science and Software Engineering,
University of Wollongong, Wollongong 2500, Australia
fren@uow.edu.au
[4] School of Software, Tianjin University, Tianjin 300072, China
jianye.hao@tju.edu.cn

Abstract. In this paper, a hierarchical learning framework is proposed for emergence of social norms in networked multiagent systems. This framework features a bottom level of agents and several levels of supervisors. Agents in the bottom level interact with each other using reinforcement learning methods, and report their information to their supervisors after each interaction. Supervisors then aggregate the reported information and produce guide policies by exchanging information with other supervisors. The guide policies are then passed down to the subordinate agents in order to adjust their learning behaviors heuristically. Experiments are carried out to explore the efficiency of norm emergence under the proposed framework, and results verify that learning from local interactions integrating hierarchical supervision can be an effective mechanism for emergence of social norms.

Keywords: Norm emergence · Learning · Multiagent systems

1 Introduction

Social norms have been considered to be an effective mechanism to facilitate coordination and cooperation in Multi-Agent Systems (MASs). How social norms can establish themselves automatically as an emerging process is a key problem in the research of social norms. It has been well recognized that learning from individual experience is a robust mechanism to facilitate emergence of social norms [1]. A great deal of previous work has thus studied emergence of social norms either through agent learning from random interactions in a well-mixed population of agents (e.g., [1,2]), or through agent learning from local interactions with neighbors in networked agent societies (e.g., [3–7]).

© Springer International Publishing Switzerland 2015
B. Pfahringer and J. Renz (Eds.): AI 2015, LNAI 9457, pp. 630–643, 2015.
DOI: 10.1007/978-3-319-26350-2_56

Although these studies provide us with some valuable insights into general mechanisms behind emergence of norms achieved through agent learning behaviors, there exist several limitations in these studies inevitably. **First**, rich organizational characteristics such as hierarchical structure of the agent systems are not considered in these studies. For self-regulated MASs, however, these organizational characteristics can have significant impacts on opinion evolution and exchange of information among agents, thereby substantially influence the emergence of norms in the entire system. Thus, it is necessary to reflect these organizational characteristics in the local learning process of agents during the process of norm emergence. **Second**, most current studies investigate norm emergence in a small population and small action space. When the action space and population become large, subnorms may exist in the system and traditional models cannot form norms efficiently. In real-life human societies, however, social norms can always emerge efficiently in complicated organizations, ranging from enterprises to communities and nations. This is because in these organizations, there is some kind of centralized control governing the distributed interactions among agents. Therefore, it is necessary to integrate this kind of centralized control into distributed interactions among agents in order to facilitate the process of norm emergence, especially when the system is becoming more complex.

Based on the above consideration, this paper studies the influence of organizational structure on norm emergence, by proposing a hierarchical learning framework in networked MASs. In the framework, agents are separated into different clusters. Each agent in a cluster interacts with others using reinforcement learning methods, and reports its learning information to an upper layer supervisor in the cluster. After synthesizing all the interaction information in the cluster, the supervisor then generates a guide policy through learning from neighboring supervisors, and passes down this policy to subordinate agents in order to adapt their learning strategies heuristically. The main feature of the proposed framework is that through hierarchical supervision between subordinate agents and supervisors, a compromising solution can be made to elegantly balance distributed interactions and centralized control for norm emergence. Experiments show that learning from local interactions integrating hierarchical supervision can be an effective mechanism for emergence of social norms.

The remainder of the paper is organized as follows. Section 2 introduces social norms and reinforcement learning. Section 3 introduces the proposed learning framework. Section 4 presents experimental studies. Section 5 discusses related work. Finally, Sect. 6 concludes the paper with directions for future research.

2   Social Norms and Reinforcement Learning

In this study, we use the scenario of "rules of the road" [8,9] as a metaphor to study norm emergence. This rule means that when we drive on the road, we would choose driving on the left side (L) or on the right side (R), and we aim to establish a norm that one side (L or R) will be chosen by all agents in the society. This rule can be viewed as a coordination game [9] in Table 1. There are

Table 1. Payoff matrix of the coordination game

	Left (L)	Right (R)
Left (L)	1, 1	−1, −1
Right (R)	−1, −1	1, 1

two pure Nash-equilibria: both agents choose left or both agents choose right, and there is no preference on L or R in the coordination game.

In the real applications of MASs, the number of actions available to the agents is an important factor in the emergence of social norms. Most current studies on norm emergence, however, only deal with a relatively simple norm space in which a norm is chosen from two possible action alternatives. In order to study norm emergence in a larger action space, we extend the coordination game into a more general form by considering N_a actions in the action space. In every time step, agent i and agent j choose action a_i and action a_j, respectively. If $a_i = a_j$, they can get immediate payoff of 1, and −1 otherwise.

Reinforcement learning algorithms [12] have been widely used for agent interactions in previous research of norm emergence. We focus on Q-learning algorithm in this paper, in which an agent makes a decision through the estimation of a set of Q-values. Its one-step updating rule is given by Eq. 1,

$$Q(s,a) = Q(s,a) + \alpha_i[r(s,a) + \lambda \max_{a'} Q(s',a') - Q(s,a)] \tag{1}$$

where $\alpha_i \in (0,1]$ is a learning rate of agent i, $\lambda \in [0,1)$ is a discount factor, $r(s,a)$ and $Q(s,a)$ are the immediate and expected reward of choosing action a in state s at time step t, respectively, and $Q(s',a')$ is the expected discounted reward of choosing action a' in state s' at time step $t+1$.

Q-values of each state-action pair are stored in a table for a discrete state-action space. At each time step, agent i chooses the best-response action with the highest Q-value based on the corresponding Q-value table with a probability of $1-\epsilon$ (i.e., exploitation), or chooses other actions randomly with a trial-and-error probability of ϵ (i.e., exploration).

3   The Proposed Learning Framework

3.1   Overview of the Hierarchical Learning Framework

Consider the society structure in a country. There is a government to manage citizens, and a senior government to manage lower level governments, and so on. In this scenario, the government is playing a role of supervisor. It supervises the state of subordinate citizens and educates them to act properly. Inspired from the structure of society, we propose a hierarchical network as shown in Fig. 1. The agent network is separated into a series of clusters $C_x, x \in (1, 2, \cdots, X)$, where X is the number of clusters, and x denotes a supervisor for each cluster. Supervisors can be one of the subordinate agents or a dedicated agent. They are

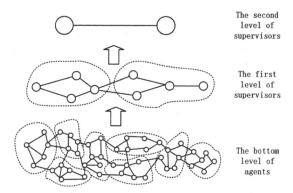

Fig. 1. An example of a hierarchical network with three levels

---

Algorithm 1. The interaction protocol of the proposed learning framework

---

1 **for** *each step $t(t = 1, \cdots, T)$* **do**
2   **for** *each agent $i(i = 1, \cdots, n)$* **do**
3   Chooses an action a_i based on its Q-values;
4   Interacts with j, $\forall j \in N(i)$, and gets a reward r_i;
5   Reports a_i and r_i to its supervisor;
6   **end**
7   **for** *each supervisor $x(x = 1, \cdots, m)$* **do**
8   Combines actions of subordinate agents into a_x;
9   Communicates with y, $\forall y \in N(x)$, and converts a_x to a_y with a probability p;
10   Passes down a_x to its subordinate agents;
11   **end**
12   **for** *each agent $i(i = 1, \cdots, n)$* **do**
13   Updates α_i or ϵ_i based on the a_x of its supervisor;
14   Updates Q-values;
15   **end**
16 **end**

---

also connected to each other based on the lower network structure. Supervisors can collect learning information of their subordinate agents, and they can also interact with their neighbors. Similarly, a higher level of supervisors can be established based on the network of the first level of supervisors. In this way, we can establish several levels of supervisors until the population of supervisors in the top level is small enough to make a final decision efficiently. In this paper, we use two-level hierarchy networks for simplification.

The interaction protocol of our hierarchical learning framework is given in Algorithm 1. In each time step, each agent i chooses an action a_i with highest Q-value or randomly chooses an action with an exploration probability of ϵ (Line 3). Then, agent i interacts with a neighbor j randomly selected from the set of neighbors $N(i)$ and gets a payoff of r_i (Line 4). The action a_i and payoff r_i are reported to agent i's supervisor x (Line 5), and supervisor x combines all actions of its subordinate agents into an action a_x (Line 8). Supervisor x then interacts with one of its neighbors and updates a_x based on the performance difference between the actions of supervision x and its neighbor (Line 9). This process will be presented in detail in the following part. When supervisor x has taken a final decision a_x, it issues a_x to its subordinate agents (Line 10). Based on this information from supervisor x, agent i adjusts its learning rate α_i or the exploration probability ϵ_i (Line 13). Finally, each agent updates its Q-values using new learning rate α_i (Line 14). This process is iterated for T time steps.

The main feature of the hierarchical learning framework is that there is hierarchical supervision governing over the distributed interactions among agents. As supervisors have a wider view than subordinate agents, they can direct their subordinate agents by learning from other supervisors. Through the feedback from supervisors, subordinate agents can understand whether their actions comply with the norms in the society, and thereby adjust their learning behaviors accordingly. By integrating supervisors' central control into agents' local learning processes, norms can be established more effectively and efficiently in the hierarchical learning framework, compared with the traditional social learning framework based on pure distributed interactions.

3.2   The Hierarchical Learning Mechanism

The Report Mechanism and Assemble Method. Each agent reports its action a_i and payoff r_i to its supervisor x. The supervisor aggregates the information into two tables QS_x and RS_x. $QS_x(a)$ means the overall acceptance of the action a in the cluster and $RS_x(a)$ means the overall reward of action a in the cluster. $QS_x(a)$ can be calculated by Eq. 2,

$$QS_x(a) = \sum_{i \in C_x} s_i \qquad (2)$$

where s_i is an indicator function given as follows:

$$s_i = \begin{cases} 1 & if \ a_i = a, \\ 0 & otherwise. \end{cases} \qquad (3)$$

$RS_x(a)$ can be calculated by Eq. 4,

$$RS_x(a) = \frac{1}{QS_x(a)} \sum_{i \in C_x, a_i = a} r_i \qquad (4)$$

especially, if $QS_x(a) = 0$, $RS_x(a)$ is set to 0.

Each supervisor then gets its aggregated action a_x. We use democratic voting mechanism in this paper, which means that a_x is simply the action most accepted by the cluster (randomly choose an action if there exists a tie). The aggregated action a_x can be given by Eq. 5,

$$a_x = arg \ max \ QS_x(a) \qquad (5)$$

The Interaction Mechanism of Supervisors. After combining actions of the cluster, each supervisor generates an assemble action a_x, which indicates the norm (i.e., action) accepted by the society. In order to generate a better supervision policy, each supervisor then interacts with a randomly chosen neighboring supervisor and learns from the neighbor by comparing the performance of their assemble actions. The motivation of this comparison comes from evolutionary game theory [13], which provides a powerful methodology to model how strategies evolve over time based on their performance. In this theory, strategies compete with each other and evolve while agents learn (mainly through imitation) from others with higher fitness. One of the widely used imitation rules is the proportional imitation, which can be given by Eq. 6 [14],

$$p = \frac{1}{1 + e^{-\beta(u_y - u_x)}} \tag{6}$$

where p is a probability for supervisor x to imitate the neighbor y, $u_x = RS_x(a_x)$ means the fitness (average payoff of the assemble action in this paper) of supervisor x, $u_y = RS_y(a_y)$ means the fitness of y, and β is a parameter to control the tensity of selection.

In Eq. 6, probability p can be large when u_x is smaller than u_y. Every supervisor x then converts a_x into a_y as the final assemble decision with probability p, or remains a_x with probability $1 - p$. In this way, supervisors imitate better actions taken by their neighbors to improve the behaviors of their clusters.

Strategy Adaption of Subordinate Agents. How to adjust learning strategies of subordinate agents using the guide information from supervisors is the key problem in the hierarchical learning framework. Based on the guide information from supervisors, subordinate agents can realize whether they have complied with the norm in the society and adjust their learning behaviors accordingly. To explicitly evaluate the agents' performance in terms of complying with the norm, we borrow the concept of WoLF (Win-or-Learn-Fast) principle [15], which is a well-known mechanism in multiagent learning, to the hierarchical learning framework. Each subordinate agent i compares a_x with its action a_i. If $a_i = a_x$, we say that agent i is "winning", and "losing" otherwise. Based on the "winning" or "losing" results, two fundamental learning parameters (i.e., learning rate α and exploration probability ϵ) can be adapted as follows.

Algorithm 2 shows the mechanism of adapting α, which is denoted as *hierarchical learning-α*. Notation s is the changing direction of learning rate. If $a_i = a_x$, the action chosen by agent i is approved by its supervisor. Agent i then decreases its learning rate α_i to maintain the "winning" state. However, if $a_i \neq a_x$, agent i increases the learning rate α_i to learn more from experience. Being more adaptive may bring about a better choice for the next round play.

Similarly, Algorithm 3 shows the mechanism of adapting ϵ, which is denoted as *hierarchical learning-ϵ*. When $a_i = a_x$, agent i decreases its exploration probability ϵ_i to maintain the "winning" state and reduces the mistake probability because of exploration. However, if $a_i \neq a_x$, agent i increases the exploration probability ϵ_i to explore more actions to get rid of the "losing" state.

The third approach is to combine the above two approaches (i.e., adapting α and ϵ at the same time), which is denoted as *hierarchical learning-both*.

Algorithm 2. hierarchical learning-α
Input: a_x, a_i, α_i;
Output: α_i;
1 **for** *each agent i* **do**
2 $\quad$ **if** $a_i = a_x$ **then**
3 $\quad\quad$ s=-1;
4 $\quad\quad$ $\Delta\alpha_i = \gamma\alpha_i$;
5 $\quad$ **end**
6 $\quad$ **else**
7 $\quad\quad$ s=1;
8 $\quad\quad$ $\Delta\alpha_i = \gamma(1 - \alpha_i)$;
9 $\quad$ **end**
10 $\quad$ $\alpha_i = \alpha_i + s\Delta\alpha_i$
11 **end**

Algorithm 3. hierarchical learning-ϵ
Input: a_x, a_i, ϵ_i;
Output: ϵ_i;
1 **for** *each agent i* **do**
2 $\quad$ **if** $a_i = a_x$ **then**
3 $\quad\quad$ s=-1;
4 $\quad\quad$ $\Delta\epsilon_i = \gamma\epsilon_i$;
5 $\quad$ **end**
6 $\quad$ **else**
7 $\quad\quad$ s=1;
8 $\quad\quad$ $\Delta\epsilon_i = \gamma(1 - \epsilon_i)$;
9 $\quad$ **end**
10 $\quad$ $\epsilon_i = \epsilon_i + s\Delta\epsilon_i$
11 **end**

4   Experiment and Results

In this section, experiments are carried out to demonstrate the performance of our framework. We compare the hierarchical learning method with the conventional social learning method in [10], and explore the influence of some parameters on norm emergence under the proposed framework.

4.1   Experimental Setting

In this paper, we test the hierarchical learning framework on a $10 * 10$ grid network. This framework consists of two levels: subordinate agents and supervisors. The network is separated into some $4 * 4$ clusters unless otherwise specified. In Q-learning algorithm, we consider each state as the same, which means that there is no state transition. Each agent can choose from 4 actions as default. Parameters α and ϵ of each agent are initially set as 0.1 and 0.01, respectively. Moreover, parameter β to control the tensity of selection in Eq. 6 and parameter γ in Algorithms 2 and 3 are both 0.1. Unless specified otherwise, the final results are averaged over 100 Monte-Carlo runs.

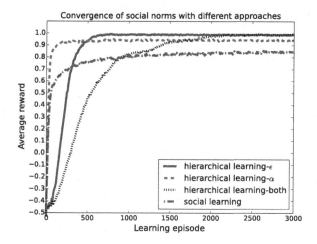

Fig. 2. The comparison of different learning approaches.

4.2   Results and Analysis

Convergence of Social Norms. Previous work has shown that norms can always emerge if agents interact with a randomly chosen neighbor. We compared the performance between the approach based on traditional social learning framework [10] and the three approaches based on our hierarchical learning framework. Figure 2 gives the learning dynamics of the four different learning approaches. From the results, we can see that, with both cases of social learning and hierarchical learning approaches, the average reward of the system increases and then keeps stable, which means the social norms emerge with all these four approaches. The convergence level and efficiency, however, are distinguishable among these approaches. The hierarchical learning approaches (especially hierarchical learning-ϵ and hierarchical learning-both) are slower at first but finally reach a higher convergence level than the social learning approach. Moreover, hierarchical learning-α emerges norms much faster than the other two hierarchical learning approaches. These results indicate that with the guide information of supervisors, agents can know more about the society, and the hierarchical mechanism is an effective way to remove subnorms in MASs to reach a higher level of norm emergence. Hierarchical learning-α is more efficient than the other two approaches because when agents adapt learning strategy with hierarchical learning-ϵ or hierarchical learning-both, the agents have to spend more time on exploring actions at first. But finally, when ϵ reduces to nearly 0, the agents can choose a correct action more firmly than other approaches.

Table 2 gives the detailed performance of the different approaches in 1000 runs. In order to better demonstrate the different performance of these approaches, we also include the results when 100 % agents have chosen the same action as the norm. Achieving 100 % norm emergence is an extremely challenging issue due to the widely recognized existence of subnorms. To calculate the

Table 2. Performance of different learning approaches (averaged over 1000 runs)

	Reward	90 % convergence		100 % convergence	
		Success rate (%)	Speed (steps)	Success rate (%)	Speed (steps)
Hierarchical learning-α	0.941	82.3	1838	81.8	1926
Hierarchical learning-ϵ	0.988	100	329	100	425
Hierarchical learning-both	0.989	99.9	590	99.9	710
Social learning	0.848	47.4	5660	45.4	5898

average steps needed for norm emergence, we take 10000 episodes as the upper limit in this run when a norm fails to emerge in 10000 episodes. The average reward, success rate (i.e., how many runs a norm can emerge successfully in the 1000 runs) and the speed (i.e., time steps needed to emerge a norm) are compared. We can see that hierarchical learning approaches are obviously more efficient and effective than traditional social learning in all these three criteria. Although hierarchical learning-α can emerge a norm more quickly than the other two approaches (refer to Fig. 2), the average steps to emerge a norm in hierarchical learning-α are larger because of a small probability of failing to emerge a norm (i.e., 10000 steps are used for calculation when a norm fails to emerge).

Dynamics of Learning Rate α and Exploration Probability ϵ. We studied the dynamics of the exploration probability ϵ and the learning rate α with different action sizes in Figs. 3 and 4 (dynamics of α and ϵ with hierarchical learning-both overlap with each other because of the identical update method). We can see that in both cases, exploration probability ϵ and learning rate α rise at first but then decline to 0 afterwards. When there are only 2 actions, both α and ϵ rise to a small extent, and then decline to nearly 0 quickly. When the action number increases to 4, it still takes a short period of time to reduce α to 0 using hierarchical learning-α. However, it takes longer time for the values to decline to 0 using hierarchical learning-ϵ and hierarchical learning-both approaches.

The dynamics of α and ϵ indicate that the agents do not know what they ought to do and therefore choose actions randomly at first. As the norm is emerging, the agents can realize which action should be the norm and thus reduce

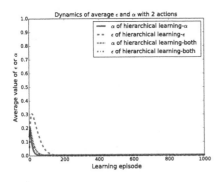

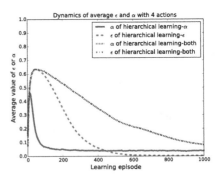

Fig. 3. Dynamics of ϵ and α (2 actions).   **Fig. 4.** Dynamics of ϵ and α (4 actions).

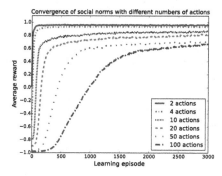
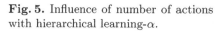
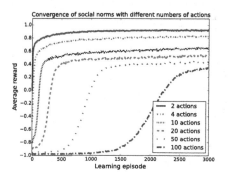

Fig. 5. Influence of number of actions with hierarchical learning-α.

Fig. 6. Influence of number of actions with traditional social learning.

the exploration probability ϵ and the learning rate α accordingly. Hierarchical learning-α is more adaptive in the scenario of a larger action space. This is because agents can find the correct action easily when ϵ is increased (agents are more likely to choose the other action with lower Q-value which is correct) in a two-action scenario. However, in a multi-action scenario, the probability to find the correct action is reduced with the increasing of exploration probability ϵ, and failing to find a correct action can lead to a further increase of ϵ. As a result, agents may be trapped in a vicious circle. But if we choose the mechanism of adapting learning rate α, the agents can learn more experience from unsuccessful interactions, which accordingly speeds up the learning process.

Influence of Number of Actions. The emergence of norms with a large action space is a difficult problem in the field of norm emergence. Figures 5 and 6 show the comparison of hierarchical learning approach (i.e., hierarchical learning-α) and conventional social learning approach. We can see that the hierarchical learning approach performs better than the social learning approach in terms of a faster convergence speed and a higher convergence level. This distinction is more apparent when the action space becomes larger. This is because the agents can receive support from the wide view of their supervisors and get access to the state of the whole society. With this information they can understand whether they have complied with the norm in the society and decide to learn/explore more or to keep their state. It is the process of imitating others with higher rewards that reduces the time to explore unnecessary actions. As a result, the norm can emerge efficiently and subnorms can be removed to a large extent.

Influence of Population Size and Cluster Size. The influence of population size on norm emergence is presented in Fig. 7. We can see that efficiency is nearly unaffected by different population sizes, but effectiveness (i.e., the level of emergence) is obviously reduced in a larger population. This is because in the hierarchical learning framework, with the same size of clusters, each supervisor

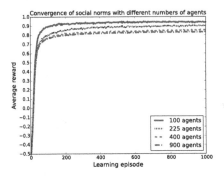
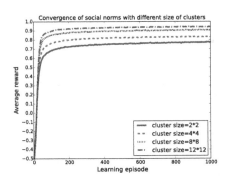

Fig. 7. Influence of population size when separated into 4 ∗ 4 clusters.

Fig. 8. Influence of cluster size in a 30∗ 30 grid network.

can have a wider view when the population size gets smaller. As a result, each supervisor can have a more powerful control force over the cluster comparatively, and the system tends to be more close to centralized control, which leads to higher efficiency. Similarly, Fig. 8 shows how the cluster size influences the level of emergence with the same size of population. As we can see, a larger cluster size can result in a higher level of emergence. This is because when the cluster size is larger, each supervisor can have a more powerful control force over the cluster comparatively, and therefore the norm can emerge more efficiently.

5   Related Work

Many studies have been done to investigate the mechanisms of norm emergence in the literature. Shoham and Tennenholtz [8] proposed a rule based on the highest cumulative reward to study the emergence of rule. Sen *et al.* [10] presented an emergence model, in which individuals learn from another agents from the group randomly. Then Sen *et al.* [2] extended this model by assessing the influence of heterogeneous agent systems and space-constrained interactions on norm emergence. Savarimuthu [1,16] recapped the research on the multiagent-based emergence mechanisms, and presented the role of three proactive learning methods in accelerating norm emergence. Airiau and Sen [11,17] extended group-based random learning model [10] to complicated networks to study the influence of topologies on norm emergence. Villatoro *et al.* [18] proposed a reward learning mechanism based on interaction history. Later, they established two rules (i.e., re-building links with neighborhood and observing neighbors' behaviors) to overcome the subnorm problems [7,19]. Mihaylo [20] came up with a learning mechanism of Win-Stay Lose-Probabilistic-Shift to make 100 % emergence of norms in complicated networks. Mahmoud *et al.* [21] further extended Axelrod's seminal model by considering topological structures among agents. Unlike these studies, which focused on norm emergence either in a fully mixed agent population or in a network agent society, our work captures the organizational characteristic of hierarchical supervision among learning agents in networked MASs. This feature sets our work apart from all these existing studies.

From another perspective, the problem that has been addressed in this paper is to employ heuristics to guide the policy search to speed up the multi-agent learning process. There is plenty of work in the research of multi-agent learning to properly address this problem. For example, Zhang et al. [22] defined a multi-level organizational structure for automated supervision and a communication protocol for exchanging information between lower-level agents and higher-level supervising agents, and Bianchi et al. [23] raised Heuristically Accelerated Minimax-Q (HAMMQ) and incorporated heuristics into the Minimax-Q algorithm to speed up convergence rate. Our work is different from the above studies in terms of targeting a different problem (i.e., norm emergence in networked MASs). Nevertheless, the main principle embodied in the proposed framework can shed some light on understanding the learning dynamics in MAL during an efficient convergence to a predefined (i.e., Nash) equilibrium.

6   Conclusions and Future Work

Norm emergence is a crucial issue in understanding and controlling complex systems. Assuming a fully centralized controller to govern the process of norm emergence is not only infeasible for large systems, but also expensive in terms of manageable cost. Therefore, it is more applicable for a norm to emerge on its own through agents' local interactions. This kind of pure distributed way of norm emergence, however, has another limitation in terms of low efficiency especially when the system becomes complex. The hierarchical learning framework proposed in this paper makes a balance between centralized control and distributed interactions during the emergence of social norms by integrating hierarchical supervision into distributed learning of agents. Experiments have indicated that this compromising solution is indeed robust and efficient for evolving stable norms in networked systems, especially when the norm space is large.

Much work remains to be done in the future. For example, the hierarchical learning framework can be conducted in some complex networks, like the small-world networks and the scale-free networks, and the influence of network structure on norm emergence can be explored in depth. Moreover, in this work, we use two-level network for experiments, and it is promising that multilevel networks can lead to a more efficient emergence of social norms.

Acknowledgments. This work is supported by the National Natural Science Foundation of China under Grant 61502072, Fundamental Research Funds for the Central Universities of China under Grant DUT14RC(3)064, and Post-Doctoral Science Foundation of China under Grants 2014M561229 and 2015T80251.

References

1. Savarimuthu, B.T.R., Arulanandam, R., Purvis, M.: Aspects of active norm learning and the effect of lying on norm emergence in agent societies. In: Kinny, D., Hsu, J.Y., Governatori, G., Ghose, A.K. (eds.) PRIMA 2011. LNCS, vol. 7047, pp. 36–50. Springer, Heidelberg (2011)

2. Mukherjee, P., Sen, S., Airiau, S.: Norm emergence under constrained interactions in diverse societies. In: Proceedings of 7th AAMAS, pp. 779–786 (2008)

3. Hao, J., Sun, J., Huang, D., Cai, Y., Yu, C.: Heuristic collective learning for efficient and robust emergence of social norms. In: Proceedings of 14th AAMAS, pp. 1647–1648 (2015)

4. Yu, C., Zhang, M., Ren, F.: Collective learning for the emergence of social norms in networked multiagent systems. IEEE Trans. Cybernet. **44**(12), 2342–2355 (2014)

5. Yu, C., Zhang, M., Ren, F., Luo, X.: Emergence of social norms through collective learning in networked agent societies. In: Proceedings of the 12th AAMAS, pp. 475–482 (2013)

6. Yu, C., Zhang, M., Ren, F., Hao, J.: Emergence of social norms through collective learning in networked agent societies. In: Proceedings of MFSC@AAMAS 2014 (2014)

7. Villatoro, D., Sabater-Mir, J., Sen, S.: Social instruments for robust convention emergence. In: Proceedings of 22nd IJCAI, pp. 420–425 (2011)

8. Shoham, Y., Tennenholtz, M.: On the emergence of social conventions: modeling, analysis, and simulations. Artif. Intell. **94**(1), 139–166 (1997)

9. Young, H.P.: The economics of convention. J. Econ. Perspect. **10**(2), 105–122 (1996)

10. Sen, S., Airiau, S.: Emergence of norms through social learning. In: Proceedings of the 20nd IJCAI, pp. 1507–1512 (2007)

11. Sen, O., Sen, S.: Effects of social network topology and options on norm emergence. In: Padget, J., Artikis, A., Vasconcelos, W., Stathis, K., da Silva, V.T., Matson, E., Polleres, A. (eds.) COIN@AAMAS 2009. LNCS, vol. 6069, pp. 211–222. Springer, Heidelberg (2010)

12. Sutton, R.S., Barto, A.G.: Reinforcement Learning: An introduction. MIT Press, Cambridge (1998)

13. Weibull, J.W.: Evolutionary Game Theory. MIT Press, Cambridge (1997)

14. Pacheco, J.M., Traulsen, A., Nowak, M.A.: Coevolution of strategy and structure in complex networks with dynamical linking. Phys. Rev. Lett. **97**(25), 1–9 (2006)

15. Bowling, M., Veloso, M.: Multiagent learning using a variable learning rate. Artif. Intell. **136**(2), 215–250 (2002)

16. Savarimuthu, B.: Norm Learning in Multi-agent Societies. University of Otago, Dunedin (2011)

17. Airiau, S., Sen, S., Villatoro, D.: Emergence of conventions through social learning. Auton. Agent. Multi-Agent Syst. **28**(5), 779–804 (2014)

18. Villatoro, D., Sen, S., Sabater-Mir. J.: Topology and memory effect on convention emergence. In: Proceedings of the 2009 WI and IAT, pp. 233–240 (2009)

19. Villatoro, D., Sabater-Mir, J., Sen, S.: Robust convention emergence in social networks through self-reinforcing structures dissolution. ACM Trans. Auton. Adapt. Syst. **8**(1), 2–21 (2013)

20. Mihaylov, M., Tuyls, K., Now, A.: A decentralized approach for convention emergence in multi-agent systems. Auton. Agent. Multi-Agent Syst. **15**(2), 1–30 (2013)

21. Mahmoud, S., Griffiths, N., Keppens, J.: Norm emergence: overcoming hub effects in scale free networks. In: Proceedings of COIN, pp. 136–150 (2012)
22. Zhang, C., Sherief, A., Lesser, V.: Integrating organizational control into multi-agent learning. In: Proceedings of the 7th AAMAS, pp. 757–764 (2009)
23. Bianchi, R.A.C., Ribeiro, C.H.C., Costa, A.H.R.: Heuristic selection of actions in multiagent reinforcement learning. In: Proceedings of 2007 IJCAI, pp. 690–695 (2007)

Information Extraction to Improve Standard Compliance

The Case of Clinical Handover

Liyuan Zhou[1] and Hanna Suominen[1,2,3,4(✉)]

[1] Machine Learning Research Group, NICTA, Canberra, ACT, Australia
{liyuan.zhou,hanna.suominen}@nicta.com.au
[2] College of Engineering and Computer Science,
The Australian National University, Canberra, Australia
[3] Faculty of Health, University of Canberra, Canberra, Australia
[4] Department of Information Technology, University of Turku, Turku, Finland

Abstract. Clinical handover refers to healthcare workers transferring responsibility and accountability for patient care, e.g., between shifts or wards. Safety and quality health standards call for this process to be systematically structured across the organisation and synchronous with its documentation. This paper evaluates information extraction as a way to help comply with these standards. It implements the handover process of first specifying a structured handover form, whose hierarchy of headings guides the handover narrative, followed by the technology filling it out objectively and almost instantly for proofing and sign-off. We trained a conditional random field with 8 feature types on 101 expert-annotated documents to 36-class classify. This resulted in good generalisation to an independent set of 50 validation and 50 test documents that we now release: 77.9 % F1 in filtering out irrelevant information, up to 98.4 % F1 for the 35 classes for relevant information, and 52.9 % F1 after macro-averaging over these 35 classes, whilst these percentages were 86.2, 100.0, and 70.2 for the leave-one-document-out cross-validation across the first set of 101 documents. Also as a result of this study, the validation and test data were released to support further research.

Keywords: Artificial intelligence applications · Clinical handover · Computer systems evaluation · Information extraction · Test-set generation

1 Introduction

During *shift-change handover*, clinicians transfer professional responsibility and accountability for patient care but document only a small part of this verbal communication. Already after a couple of shift changes, this leads to 65–100 % information

NICTA is funded by the Australian Government through the Department of Communications and the Australian Research Council through the Information and Communications Technology (ICT) Centre of Excellence Program. We thank Maricel Angel, RN at NICTA, for helping HS to create the dataset. LZ conducted all experiments under HS's supervision.

© Springer International Publishing Switzerland 2015
B. Pfahringer and J. Renz (Eds.): AI 2015, LNAI 9457, pp. 644–649, 2015.
DOI: 10.1007/978-3-319-26350-2_57

loss that is a major contributor in over 65 % of sentinel events in hospitals and associated with over 10 % of preventable adverse events [1–4]. Hence, standards for safety and quality in healthcare call for a handover process that is systematically structured across the care-giving organisation and synchronous with its documentation [5, 6].

To support the standard compliance, cascaded *speech recognition* (SR) with *information extraction* (IE) has been studied as a way of filling out a structured handover form for the nurse who is handing over to proof and sign off [7–10]. With minimal training, SR recognises up to 73.6 % of 14,095 test words correctly, even for a female nurse speaking Australian English as her second language. When considering IE as a 36-class classification task, where each word is assigned to precisely one class (i.e., the most relevant heading/subheading of the form or the class of *Irrelevant* if not relevant to any heading), the system correctly classifies 74.8 % of 8,487 test words after training on 100 documents. The cascade generates a document draft from 10–75 % of the time it takes to transcribe this by hand, whilst the proofing time is about the same. This holds the potential for reducing the loss to 0–13 % while releasing nurses from documentation – that currently takes up to 65 % of their shift – to direct caring and patient education [11–13].

In this paper, we study the generalisation capability of the IE system [10], whose data and source-code are publicly released to anyone for the purposes of testing SR and language processing algorithms. We release a new, unseen dataset of 100 synthetic but realistic handover narratives, their expert annotation with respect to the aforementioned form, and the related performance numbers of the IE application.

2 Materials and Methods

The dataset called *NICTA Synthetic Nursing Handover Data* was used in this study. This set of 201 synthetic patient cases from an imaginary Australian medical ward was developed for SR and IE related to nursing shift-change handover in 2012–15 [8, 10]; we supplemented the previous release of 101 cases for *training* with another 100 cases that were divided randomly to 50 documents for *validation* and remaining 50 documents *testing*. Case data relevant to IE consisted of a written, free-form text paragraph (i.e., the *handover narrative*) and its highlighted counterpart (i.e., the *structured handover document*). They were created by a *registered nurse* (RN) under the first author's supervision with over twelve years' experience in clinical nursing and English as her second language. The form had 5 main headings (i.e., *Introduction*, *My Shift*, *Medication*, *Appointments* and *Future Care*) that were further divided to 18, 8, 12, 3, and 3 sub or subsubheadings, respectively. The data release was organisationally approved and the RN gave consent in writing. The data license for the documents related to IE is *Creative Commons – Attribution Alone*[1] (CC-BY) for the purposes of testing language processing algorithms with the requirement to cite [10] for the first 101 cases and [8] and this paper for the remaining 100 cases.

We considered the IE task as a machine learning problem, where each word in text is considered as an entity – represented as features – and the goal is to learn to classify

---

[1] http://creativecommons.org/licenses/by/4.0/.

it automatically to one or none of the sub/subsubheadings (or main headings) present in the training documents [10]. The features were characterised as *syntactic* (the word itself; its lemma, named-entity-recognition tag, part-of-speech tag, basic dependents, basic governors, and parse tree from the sentence root; phrase that contains the word), *semantic* (top-5 candidate senses of the word in the Unified Medical Language System (UMLS), and top mapping of the word in the UMLS hierarchy to generalise the sense, and medication score of the word, based on a search on a subset of 6,373 concepts (described using nearly 12,600 words) from the WHO Anatomical Therapeutic Chemical (ATC) classification system, using the value set of 1, 0.5, and 0 for the cases of the word being a full term in ATC, otherwise included in ATC, and not in ATC, respectively], or *statistical* (location of the word on a 10-point scale from the beginning of the document to its end).

The *CRF++ 0.58* implementation of the *conditional random field* (CRF) [14] was used for classification. Its *unigram template* defined that all features of the previous, current, and next word were used first alone and then the pairwise correlations of the previous and current word as well as those of the current and next word over all features were computed. Its *bigram template* combined the features of the current word and the class of the previous word to form a new feature type. The CRF++ default settings for the *regularisation algorithm* (i.e., L_2) and the *cut-off threshold* for using only the features that occur at least once in the training documents were chosen. The CRF hyperparameter, called c in CRF++, that controlled the model fitting to the 100 training documents [i.e., larger (smaller) c tends to lead to overfitting (under-fitting)] was optimised as 50 through a *grid search* [15] on a validation set of 50 documents, which were randomly chosen from the test set of 100 documents.

3   Results

The training (test and validation) documents had 8,487 (7,730) words with 1,283 (1,240) unique lemmas (incl. punctuation) [8]. These sets were typical but fairly independent language samples, sharing only about 700 unique words. The 10 most common shared words were *and, is, he, in, for, with, she, on, the*, and *to*.

After training for the sub/subsubheading classification on 8,487 words and validation on 3,937 words, our system performance was excellent (>90.0 % F1) in extracting information about patient's age, current bed, current room, and given name for patient introduction (Table 1). It performed well (>70.0 % F1) for irrelevant information and another 2 subheadings (Admission reason/diagnosis and Last name) of Patient Introduction. In other words, it classified 2,375 words out of 3,793 test words correctly and had 77.9 % F1 in filtering out irrelevant information, up to 98.4 % for the 35 classes for relevant information, and 52.9 % after macro-averaging over these 35 classes. In comparison, the performance numbers for leave-one-document-out cross-validation on the training set were F1 of 86.2 % for the class of Irrelevant, up to 100 % for the 35 relevant classes, and 70.2 %, on macro-average over these 35 classes. In comparison, the model trained for the simpler 6-class classification (i.e., 1 of the 5 main headings or *irrelevant*), performed substantially better in filtering out

Table 1. Descritive statistics on the 101 training (Tr), 50 validation (V), and 50 test (Te) documents together with the respective performance

Class Heading *Subheading*	No. of words			Top lemmas			F1 %		
	Tr	V	Te	Tr	V	Te	Tr	V	Te
Irrelevant	3771	1588	1564	Be and in	Be and in	Be and in	84.5	79.4	80.7
Irrelevant	3771	1588	1564	Be and he	Be and in	Be and in	83.7	73.9	75.0
Patient introduction	2064	1192	1032	He she dr	He she old	He she old	90.2	81.1	82.9
Admission reason/diagnosis	414	288	256	And pain of	He of pain	Of pain severe	68.3	55.0	66.0
Age in years	246	144	137	Old yr year	Old yr year	Ole yr year	96.5	100	97.0
Allergy	14	2	1	To allergic allergy	Penicillin pollen	Nut	0.0	4.0	0.0
Care plan	36	78	78	For investigation under	Investigation monitoring under	For investigation a	0.0	5.0	0.0
Chronic condition	70	8	3	Smoker type history	Copd long blood	Asthma puffer hypercholesterolemia	26.3	0.0	33.3
Current bed	180	80	76	Bed 3 2	Bed 1 6	Bed 9 5	99.1	95.1	97.4
Current room	54	28	26	Room 1 2	Room 4 11	Room 9 5	100	84.6	96.0
Disease/problem history	147	176	90	Of and history	Of ago yr	To due a	39.4	17.9	24.1
Gender	489	206	172	Lady she he	She he	She he	94.7	33.8	29.7
Given names/initials	119	54	50	Michelle Yvonne Jeff	Mick Tanya Laurence	Judas Mirand Melanie	95.8	96.1	94.7
Last name	99	49	51	Cavedon reed clemens	Heart curtin martinez	Hughes cavalier nualart	96.4	96.9	92.6
Title	0	1	2		Mr	Mr	0.0	0.0	0.0
Under Dr_Given Names/initials	15	28	32	Dylan harry Dercy Jaime Jorge	Angelika vincent Eden Thalia Santi	Carr gabby ben Jean Saldy	56.0	0.0	0.0
Under Dr_Lastname	181	50	58	Smith Jaime chanson	Wen Yale guy	Grace Vasquez go	93.9	66.6	70.7
My shift	1353	609	544	Be all stable	Stable ob be	Ob stable caring	69.0	49.8	56.8
Activities of daily living	245	91	91	Self caring ambulant	Self caring ambulant	Self caring ambulant	72.9	56.6	61.0
Contraption	44	51	37	At iv oxygen	At 83 iv	ivt 83 at	0.0	0.0	0.0
Input/diet	101	54	25	Diet general tolerate	Diet normal to	Diet normal and	67.0	78.0	62.7
Other observation	361	154	175	Be of and	Of gc ob	ob stable neuro	26.3	29.7	26.0
Output/diuresis/bowel movement	52	28	36	fbc chart strict	Strict incontinent chart	Strict chart fluid	51.9	57.1	74.5
Risk management	12	47	47	Pressure care precaution	Assist risk contact	Assist risk fall	0.0	15.6	0.0
Status	483	169	124	All stable be	Stable new other	Stable new admission	74.3	53.2	52.3
Wounds/skin	55	15	9	And on he	To dressing dress	Dressing care for	0.0	0.0	0.0
Medication	262	170	159	Regular effect endone	Oxygen prong regular	Regular pain relief	69.6	41.8	58.4
Dosage	37	63	54	prn regular slide	Prong regular per	Regular prong prn	26.0	0.0	7.0
Medicine	157	89	88	Regular endone pain	Oxygen pain relief	Pain relief oxygen	69.6	53.1	57.8
Status	68	18	17	Effect good with	Relieve be day	The in day	73.8	20.0	0.0
Appointment/procedure	393	174	199	For be to	ct blood pend	ct blood await	36.5	27.6	34.9
Clinician last name	2	2	2	Diani dr	Cross dr	Harvey dr	0.0	0.0	0.0
Clinician title	0	7	3		Wound social nurse	Wound nurse gp	0.0	0.0	0.0
Day	40	6	20	Today tomorrow to	This today tom	Today daily yesterday	4.5	28.5	16.6
Description	157	103	119	For us to	ct blood for	ct blood with	22.6	16.8	23.7
Hospital	0	0	1			Angio	0.0	0.0	0.0
Status	159	42	49	Be result the	Pend result do	Await do result	20.1	16.1	28.2
Time	28	14	5	pm this afternoon	This hour random	Lthis afternoon later	40.0	38.1	66.6

(Continued)

Table 1. (*Continued*)

Class	No. of words			Top lemmas			F1 %		
Heading *Subheading*	Tr	V	Te	Tr	V	Te	Tr	V	Te
Future	644	204	295	For to the	Review to investigation	Doctor to for	53.7	13.7	8.4
Alert/warning/abnormal result	59	24	0	Very be to	Positive at might		12.3	0.0	0.0
Discharge/transfer plan	89	20	69	Discharge for be	To discharge tom	He be discharge	26.3	0.0	0.0
Goal/task to be completed/expected outcome	496	160	226	For to review	Review investigation to	Doctor to the	52.4	26.8	13.1

irrelevant information (80.7 % F1) and its macro-averaged F1 over the 5 main headings was also better (i.e., 67.9 %).

This system performance compared to the state-of-the-art of 90 % F1 in clinical IE in general [16]. When considering the case of handover but with only 6 classes [9], a system trained on 149 de-identified nursing handover documents in Australian English had F1 of 87.0 % for *Patient Identification*, 70.5 % for *Clinical History/Presentation*, 52.3 % for *Clinical Status*, 60.7 % for *Care Plan*, 35.2 % for *Outcomes of Care and Reminders*, and 84.7 % for *Irrelevant*.

Although there were some sub/subsubheadings for which learning the classification task with a very limited number of training data was feasible [e.g., *Patient Introduction: Current Bed* (97.4 % F1), *MyShift: Output/Diuresis/Bowel Movement* (74.5 % F1), and *Appointment: Procedure Time* (66.6 % F1)], in most cases larger amount of training data and clearer similarity between training and test cases contributed to the performance numbers. To illustrate this need for similarity, let us consider the class of *Future Goal/Task To Be Completed/Expected Outcome* with a balanced number of data in the training, validation, and test sets. Based on analyzing the most frequent lemmas in each set, we observed a clear overlap between the validation and test vocabularies. However, the vocabulary of the training set was different, and as a result, F1 was only 13.1 % on the test set.

References

1. Pothier, D., Monteiro, P., Mooktiar, M., Shaw, A.: Pilot study to show the loss of important data in nursing handover. Br. J. Nurs. **14**(20), 1090–1093 (2005)
2. Tran, D.T., Johnson, M.: Classifying nursing errors in clinical management within an Australian hospital. Int. Nurs. Rev. **57**(4), 454–462 (2010)
3. Matic, J., Davidson, P., Salamonson, Y.: Review: bringing patient safety to the forefront through structured computerisation during clinical handover. J. Clin. Nurs. **20**(1–2), 184–189 (2011)
4. Finlayson, S.G., LePendu, P., Shah, N.H.: Building the graph of medicine from millions of clinical narratives. Sci. Data **1**, 140032 (2014)
5. Australian Commission on Safety and Quality in Healthcare (ACSQHC): Standard 6: clinical handover. In: National Safety and Quality Health Standards, pp. 44–47. ACSQHC, Sydney (2012)

6. The World Health Organisation (WHO): Patient safety. Organizational tools. Communication. handover (handoff) tools. http://goo.gl/VjAikK

7. Johnson, M., Lapkin, S., Long, V., Sanchez, P., Suominen, H., Basilakis, J., Dawson, L.: A systematic review of speech recognition technology in health care. BMC Med. Inform. Decis. Mak. **14**, 94 (2014)

8. Suominen, H., Hanlen, L., Goeuriot, L., Kelly, L., Jones, G.J.F.: Task 1a of the CLEF eHealth evaluation lab 2015: clinical speech recognition. In: Cappellato, L., Ferro, N., Jones, G., San Juan, E. (eds.) CLEF 2015 Labs and Workshops, Notebook Papers. CEUR Workshop Proceedings (CEUR-WS.org) (2015)

9. Suominen, H., Johnson, M., Zhou, L., Sanchez, P., Sirel, R., Basilakis, J., Hanlen, L., Estival, D., Dawson, L., Kelly, B.: Capturing patient information at nursing shift changes: methodological evaluation of speech recognition and information extraction. J. Am. Med. Inform. Assoc. **22**(e1), e48–e66 (2015)

10. Suominen, H., Zhou, L., Hanlen, L., Ferraro, G.: Benchmarking clinical speech recognition and information extraction: new data, methods, and evaluations. JMIR Med. Inform. **3**(2), e19 (2015)

11. Poissant, L., Pereira, J., Tamblyn, R., Kawasumi, Y.: The impact of electronic health records on time efficiency on physicians and nurses: a systematic review. JAMIA **12**(5), 505–516 (2005)

12. Hakes, B., Whittington, J.: Assessing the impact of an electronic medical record on nurse documentation time. J. Crit. Care **26**(4), 234–241 (2008)

13. Banner, L., Olney, C.: Automated clinical documentation: does it allow nurses more time for patient care? Comput. Inform. Nurs. (CIN) **27**(2), 75–81 (2009)

14. Lafferty, J.D., McCallum, A., Pereira, F.C.N.: Conditional random fields: probabilistic models for segmenting and labelling sequence data. In: Proceedings of the 18th International Conference on Machine Learning, ICML 2001, pp. 282–289. Morgan Kaufmann Williamstown, MA, USA (2001)

15. Smith, A., Osborne, M.: Regularisation techniques for conditional random fields: parameterised versus parameter-free. In: Dale, R., Wong, K.-F., Su, J., Kwong, O.Y. (eds.) IJCNLP 2005. LNCS (LNAI), vol. 3651, pp. 896–907. Springer, Heidelberg (2005)

16. Meystre, S.M., Savova, G.K., Kipper-Schuler, K.C., Hurdle, J.F.: Extracting information from textual documents in the electronic health record: a review of recent research. Yearb. Med. Inform. **35**, 128–144 (2008)

Author Index

Printed in the United States
By Bookmasters